Canal Design and Construction

NIPA® GENX ELECTRONIC RESOURCES & SOLUTIONS P. LTD.
New Delhi-110 034

Canal Design and Construction

V. K. BAIRATHI
B.E. (Civil), M.E. (WRD), FIE (India)
Technical Director, Sincere Technocrates
Former Visiting Professor
Water Resources Development Training Centre
Indian Institute of Technology, Roorkee
Former Director, Irrigation Management Training Institute, Kota
&
Former Chief Engineer, Irrigation, Rajasthan, India.

NIPA® GENX ELECTRONIC RESOURCES & SOLUTIONS P. LTD.
New Delhi-110 034

NIPA® GENX ELECTRONIC
RESOURCES & SOLUTIONS P. LTD.

101,103, Vikas Surya Plaza, CU Block
L.S.C. Market, Pitam Pura, New Delhi-110 034
Ph : +91 11 27341616, 27341717, 27341718
E-mail:newindiapublishingagency@gmail.com
www: www.nipabooks.com

For customer assistance, please contact
Phone: +91-11-27 34 17 17
Fax: +91-11-27 34 16 16
E-Mail: feedbacks@nipabooks.com

Print ISBN: 978-93-80235-63-9

ebook ISBN:

The views expressed in this book are that of the author and not of any Government / Association/ Institution.

Composed and Designed by NIPA.

Preface

Canal Design is perhaps the simplest and most common amongst Irrigation Engineers. With a large number of irrigation and hydropower projects constructed in the world, the design of canals is well known. Yet, the author felt that the design is not that simple as is being practised. There are a number of theories such as Kennedy's, Lacey's and Tractive force theory and practices in vogue in the world. Even in the same country these vary considerably and give different solutions. All these are compared in chapter 6 and 7. Also many built canals are discussed and compared. Finally a unified solution has been proposed.

Generally the design of lined canals, particularly trapezoidal canals is largely done by trial and error by arbitrarily choosing the bed width depth ratio (B: D). This resulted many times in uneconomical sections. Also their operational performances have not been as was expected in the designs. **Lined canals have also shown considerable signs of damage to lining, seepage, silting weed growth, and low discharging capacity.** This book is an out come of a large experience of many engineers on various different site conditions. Many actual cases have been sited. It deals with all the practical aspects of an economic section for various discharges, topographic and soil conditions.

The canal design involves deep knowledge of following disciplines.

1. Hydraulic of flow
2. Mechanics of bed erosion and sediment transport
3. Geotechnical engineering: Soil and Rock mechanics, stability of inner and outer slopes, including that of foundation
4. Mechanics of ground water flow, seepage and drainage characteristics of soil and ground
5. Tunnelling
6. Structural Engineering and construction techniques

There are following 7 Parameters of a Channel Design

i. Discharge, Q
ii. Rugosity coefficient, N
iii. Longitudinal slope, S
iv. Side slope Z (horizontal : 1 vertical)
v. Bed width, B
vi. Depth, D or B/D ratio, X and
vii. Velocity, V

In the above no parameter is extra and none is missing. Once Q and V are uniquely decided, area A gets fixed. B/D ratio X is uniquely known from the **Manning's Master equation.**

The author found that no trial and error or adoption of any arbitrarily B/D ratio is required for best or hydraulically most efficient canal sections of any size and shape.

For optimum design, all dimensions are geometrically so fixed that hydraulic radius is always equal to D/2 (half of depth only). Anything more or less then D/2 make the section in-efficient or uneconomical. Thus normally no one should play with any dimension, unless and until required by specific site conditions. Even in specific site condition there is a specific optimum solution.

Velocity of all best sections is directly computable (see table 2.4) by one single table irrespective of any size and shape.

The effect of change of side slope, bed width and Manning's N on the hydraulic efficiency are discussed in detail in the chapter 1 to 5. This makes the decision process easy.

A suitable and economical canal cross section must satisfy following 6 important criteria's

1. Minimum cost and practical economical section, see para 9.2 and 9.6
2. Minimum seepage loss section see para 9.3
3. (i) Minimum silting and minimum abrasion on lining (for lined canals only) i.e. transport the sediment entering into the canal without damaging lining, see chapter-7 (ii) Above (i) is also applicable to unlined canals to the extent of non silting and non scouring. When clear water is entering, it should be non scouring-see chapter 7
4. Minimum weed growth section, see para 9.5
5. Maximum permissible velocity section, see para 2.3 and chapter 7
6. Minimum and Maximum permissible bed slope section, see para 2.3 and 9.13

The embankment section should also be structurally stable, see chapter 8 and para 9.9 and have maximum hydraulic efficiency, see chapter-1 in most of the normal working conditions.

Mehboob sections and trapezoidal canals with curved ends are also discussed.

To further facilitate quick design, work out various alternatives and their cost comparison, ready reckoning tables are given.

Even in the computer era these tables are very useful and serve two-fold purpose, - (i) Provide complete picture at a glance and (ii) make computer programmes and calculations easy for working out various alternatives and their cost comparison. In the end, practical examples of lined canals from a number of projects are discussed.

Lastly this volume is very friendly to field engineers and helps to avoid mistakes. It is written in such a way that the field engineers can use the relevant part directly.

The book is also very useful to engineering students to enable them to clearly understand the hydraulics and economics of canal sections, and project works.

It incorporates the work of many researchers / scholars on the subject. It also highlights future studies required.

The author gratefully acknowledges the help received from many standard works on Irrigation Engineering and allied subject, including textbooks, by Indian and foreign authors, technical papers and project reports published by various professional bodies, Governments through the unique libraries of I.I.T., Roorkee, CWC Delhi and Rajasthan Irrigation and other State Departments, without which this work would have not been possible. A list of references is given at the end of each chapter and in the end of this book also.

Lastly the author is highly grateful to all the senior and young engineers (too long list to include) of Indira Gandhi Nahar Project, Irrigation zone, Udaipur, Jaipur and Mahi Project Banswara, who patiently explained their views, problems and showed the difficult sites, without which this volume would have not been possible. The author is also grateful to the Professors of WRDTC, I. I.T. Roorkee, particularly Dr. Bharat Singh and Dr. B.N. Asthana for their useful discussion and suggestions. We do not become wiser in one day. By and large people become wiser with experience. **To gain wisdom from the experiences of many others is the motive of this book.**

All possible efforts have been made to avoid errors/omissions, yet some mistakes might have occurred. Readers may kindly inform the same for improvement in next edition or issue of corrigendum.

Any suggestion to increase the utility of this book will be gratefully received and will be carefully considered for incorporating in the subsequent editions of the book.

August, 2011

V. K. BAIRATHI
D-23, Jagan Path, Chomu house,
Sardar Patel Marg, Jaipur-302001, INDIA
e-mail : bairathi@hotmail.com
Phone : 09413101422

□□□

Contents

Units of Measurement

Values of prefixes of sub-multiples and multiples of metric units

Prefix	Numerical	Value	Indian Nomenclature
Micromicro or Pico	0.000000000001	(10^{-12})	
Millimicro or Nono	0.000000001	(10^{9})	
Micro	0.000001	(10^{6})	
Milli	0.00	(10^{3})	
Centi	0.01	(10^{2})	
Deci	0.1	(10^{-1})	
Deka	10	(10)	Ten
Hecto	100	(10^{2})	Hundred
Kilo (K)	1000	(10^{3})	Thousand
Mega or Million (M)	10,00,000	(10^{6})	Ten Lacs
Kilomega or Giga (billion)	1,00,00,00,000	(10^{9})	One Arab (Hundred Crores)
Megamega or Tira	1,00,00,00,00,000	(10^{12})	Ten Kharab

Metric Abbreviations				Length		
Kilometer	Km	Tone	t	10 millimeters	=	1 centimeter
Meter	m	Quintal	q	10 centimeters	=	1 decimeter
Centimeter	cm	Kilogram	Kg	10 decimeters	=	1 meter
Millimeter	mm	Gram	g	10 meters	=	1 decameter
Kiloleter	Kl	Acre	a	10 decameter	=	1 hectameter
Liter	l	Hectare	ha	10 hectameter	=	1 Kilometer
Milliliter	ml	Centiare	ca			

Length	
1 meter	= 1.0936 yards
	= 3.2808 feet
	= 39.3701 inches
1 Kilometer	= 0.6214 mile
1 yard	= 0.9144 meter
1 foot	= 0.3048 meter
1 inch	= 25.40 millimeters
1 mile	= 1.6093 Kilometers
1 Nautical mile	= 1.852
(International)	= 1.852 Kilometer
1 nautical mile	= 6080 feet
1 cable length	= 120 fathoms
1 fathoms	= 6 feet
1 chain (66 feet)	= 100 links
1 mile	= 8 furlongs
1 furlong	= 10 chains
1 chain	= 4 poles
1 rod, pole or perch	= 5.5 yards
1 yards	= 3 feet
1 feet	= 12 inches

Capacity	
10 milliliters	= 1 centiliter
10 centiliters	= 1 deciliter
10 deciliters	= 1 liter
10 liters	= 1 deciliter
10 deciliters	= 1 hectoliter
10 hectoliters	= 1 Kiloliter
1000 cubic centimeters	= 1 cubic decimeter
1000 cubic decimeters	= 1 Cubic Meter
1000 cu. millimeters	= 1 cu. Centimeter
1 cubic yard	= 27 cubic feet
1 cubic foot	= 1728cubic inch

Area	
1 square meter (m^2)	= 10.764 square feet
	= 100 sq. decimeters
1 square centimeter	= 0.155 square inch
	= 100 sq. millimeters
1 square Kilometer	= 100 hectares
	= 0.3861 square mile
1 hectare	= 10000 square meters
	= 107640 square feet
	= 2.471 acres
	= 100 acres
1 square foot	= 0.0929 square meter
1 square inch	= 6.452 sq. centimeters
1 acre	= 43,560 square feet
	= 0.4047 hectare
1 are	= 100 sq. meters
1 square mile	= 640 acres
	= 258.99 hectares
	= 2.59 sq. Kilometers
1 square	= 100 square meters
decimeter	= 100 sq. centimeters
1 acre	= 10 sq. chains
1 sq. chain	= 484 sq. yards
1 acre 4 rods	= 4840sq. yards
1 rood	= 40 sq. rods
1 sq. rod, pole or perch	= 30.25 sq. yards
1 sq. yard	= 9 sq feet
1 sq. foot	= 144 sq. inches

Weight	
10 milligrams	= 1 centigram
10 centigrams	= 1 decigram
10 decigrams	= 1 gram
10 grams	= 1 decagram
10 decagrams	= 1 hectogram
10 hectograms	= 1 KILOGRAM
100 kilograms	= 1 quintal
10 quintals	= 1 metric ton (tone)

SI UNITS

The use of the different systems of units by the scientists and the engineers and also by the dififferent countries of the world often leads to a lot of confusion. Therefore, it was decided at the Eleventh General Conference of Weights and Measures held in Paris in 1960 to adopt a unifield, systematically constituted, coherent system of units for international use. This system of units is called the *International System of Units* and designated by the abbreviation *SI Units.*

Most of the countries in the world have subsequently decided to adopt SI units. In India, a statutory decision was taken in 1956 by an Act of Parliament to change over to the metric system which is conformity with the SI concept.

The salient features of the SI units are :

1. *Comprehensive*: Only 6 base units and 2 supplementary units to cover the entire spectrum of science and engineering.
2. *Coherent* : Product or quotient of the units of two or more quantities comprises the unit of the resultant quantity.
3. *Rational* : It absorbs the rationalized MKS system and also the multiples and submultiples concept.
4. *Brief* : The symbol and punctuation in writing the abbreviation of units are standardised for the sake of complete uniformity.

There are six base units in SI system of units which are given below:

S. No.	Quantity	Unit	Symbol	Dimensions
1.	Length	Meter	m	L
2.	Mass	Kilogram	Kg	M
3.	Time	Second	s	T
4.	Electric Current	Amper	A	I
5.	Thermodynamic temperature	Kelvin	K	t
6.	Luminous intensity	Candela	cd	

In addition to the above noted base units there are two supplementary units which are as given below:

S. No.	Quantity	Unit	Symbol
1.	Plane angle	radian	rad
2.	Solid angle	steradian	sr

The unit of a derived quantity is obtained by taking the physical law connecting it with the basic (or primary or fundamental) quantities and then introducing the corresponding units for the basic quantities. The units for some of the derived quantities have been assigned special names and symbols. Some of the important derived units with special names, commonly used in fluid mechanics, in metric gravitational and SI systems of units are as given below:

S. No.	Derived Quantity	System of Units			
		Metric Gravitational		SI	
		Unit	Symbol	Unit	Symbol
(1)	(2)	(3)	(4)	(5)	(6)
1.	Area	square meter	m^2	square meter	m^2
2.	Volume	cubic meter	m^3	cubic meter	m^3
3.	Velocity	meter per second	m/s	meter per second	m/s
4.	Angular velocity	radian per second	rad/s	radian per second	rad/s
5.	Acceleration	meter per second square	m/s^2	meter per second square	m/s^2
6.	Angular acceleration	radian per second square	rad/s^2	radian per second square	rad/s^2
7.	Frequency			hertz	Hz
8.	Discharge	cubic meter per second	m^3/s	cubic meter per second	m^3/s
9.	Mass density (special mass)	metric slug per cubic meter	$ms1/m^3$	kilogram per cubic meter	Kg/m^3

Note:

A. Note that kilogram is written as Kg, not Kg_m, not Kg_f etc. Similarly second as s, not sec or sec. etc. No full stops, No plurals, No dots, No dashes. For example, torque is in N m not N-m etc.

B. The unit of force is Newton with symbol N and there is no such thing as kilogram force in SI units. Just N and its multiples and submultiples. The unit of energy in any from is joule, J = N m.

No horsepower. No metric horsepower. Just watt, W = J/s = N m/s and its multiples and submultiples.

C. Always leave a space between the number and the unit symbol, e.g., 23.2 cm and 2500 N.

For numbers less than unity, zero must be put on the left of the decimal and for larger numbers exceeding five figures one space after every three digits.

❑❑❑

Water Flow Conversion Table

Unit	**m^3/s**	**m^3/day**	**l/s**	**ft^3/s**	**ft^3/day**	**acre feet/ day**	**gal/min**	**gal/day**	**million gallan/ day (mgd)**
1 cubic meter second (cumec)	1	8.64×10^4	10^3	35.31	3.051×10^6	70.05	1.58×10^4	2.282×10^7	22.824
1 cubic/meter day (cumec day)	1.157×10^{-5}	1	0.0116	4.09×10^{-4}	35.31	8.1×10^{-4}	0.1835	264.17	2.64×10^{-4}
1 liter second (l/s)	0.001	86.4	1	0.0353	3051.2	0.070	15.85	2.28×10^4	2.28×10^{-2}
1 cu. foot/second (cusec)	0.0283	2446.6	28.32	1	8.64×10^4	1.984	448.8	6.46×10^5	0.646
1 gallon/minute	6.3×10^{-5}	5.451	0.0631	2.23×10^{-3}	192.5	4.42×10^{-3}	1	1440	1.44×10^{-3}
1 gallon/day	4.3×10^{-8}	3.79×10^{-3}	4.382×10^{-5}	1.55×10^{-6}	0.11337	3.07×10^{-6}	6.94×10^{-4}	1	10^{-6}
1 million gallons/day	4.38×10^{-2}	3785	43.82	1.55	1.337×10^5	3.07	694	10^6	1

Notations

Symbols	Meaning	Units
A	Area of cross section; Projected area of a particle;	m^2
A	dimensionless number = a/D	-
	Major axis of the sediment	-
A_E	Dimensionless height	-
B or b	Bed width of channel	m
B_T or T	Top width of channel	m
C	Coefficient in sediment transport function; Silt factor in the Kennedy's equation, sediment charge or sediment transport divided by water discharge in the Inglis equation, or bed load charge in hundred thousand by weight ; a factor of flow resistance (C = 1.49/n R1/6)	ppm
	Total load concentration (sediment transport) in percent by weight	%
C_a	Suspended sediment concentration at a distance 'a' above the bed	ppm
C_F	Coefficient, CF is unity for laboratory data and 1.268 for field data	-
C_0	Maximum sediment concentration in fraction by volume	ppm
C_m	Sediment discharge concentration in weight per unit volume	ppm
T	Average sediment concentration as defined at appropriate places	ppm
C_s	Sediment concentration by volume or by weight	ppm
C_{2d}	Suspended sediment concentration at a distance '2d' above bed (d is sediment diameter)	ppm
cbl or CBL	Canal bed level	
D,d or h,y	Depth of water flow or full supply depth (F.S.D)	m
D, d	Size of sediment	mm
D_{gr}	Dimensionless particle size	-
d_{50} or d_{si}	Medium size of sediment	mm
d_x	Equivalent diameter for a non uniform granulate	mm
d_m	Effective diameter of the sediment	mm
d_{si}	Mean size of I th size friction of sediment	mm
d_s	The representative size of bed sediment and is usually taken as the median size, d_{50}	mm
d_*	Dimensionless diameter of grain	-
d_{65}	Size of bed sediment for which 65 % of the sediment by weight is finer	mm
d_{35}	Size of sediment for which 35 % of the sediment by weight is finer	mm
E	Specific energy	m
EB	The bed load transport efficiency	-
ES	The suspension efficiency	-

F	Force acting on a particle	Kg/m^2
F_1	Factor in Rubey equation for fall velocity of particle	mm
F_b	Bed factor	-
F_s	Side factor	-
F_r	Froud number	-
F_{gr}	Sediment mobility number	-
F'_{gr}	Value of Fgr at nominal initial motion	-
F	The friction factor of the bed; silt factor	-
f'_b	Darcy-weisbach bed friction factor for the sand grain roughness	-
G	Specific gravity of sediment	Ton/s
G	Gravitational gravity	m/s
G_s	Solid discharge, as total load, by mass	Kg/s
G_{gr}	Dimensionless sediment transport rate	-
G_{sb}	Total bed sediment discharge of the stream in width per unit time	m^3/s
H_t	Height of water flow	m
h_e	Head velocity	m
h_f	Loss of energy	m
h_n	Normal flow depth	m
I_1, I_2	Plasticity index for cohesive sediment (Integral by Einstein)	-
K	Coefficient of Karman constant, K = 0.4; A factor reducing the limiting stress, on the bank particle	-
K	Representative size (mm) of roughness of rippled bed	mm
k_s	Effective grain diameter, (ks = d_{35}) equivalent sand grain roughness of the boundary	mm
L	Length of the dune	m
ΔL	Length of the prism	m
M	Kramer's uniformity coefficient in the transport relations	-
M	Critical velocity ratio	-
N	Transition exponent depending on sediment size	-
N or n	Manning's roughness coefficient	-
n'	Manning's coefficient for plane bed	-
Na	Coefficient of rugosity depend only on grain size of the boundary material of the channel	-
Nv or n_v, zi	Dimensionless exponent in Rouse equation based on sediment size d_{si}	-
P	Exponent; percent by weight corresponding to give size ; wetted perimeter	-
P_e	Transport parameter	-
P_i	Fraction by weight of that fraction of the bed sediment with mean size	-
Q	Discharge if sides were frictionless	m^3/s
q	Water discharge, in cubic meter persecond per unit width	$m^3/s/m$
Q_s or Q_T	Total solid discharge, as total load by volume	m^3/s
Q_{sb}	Solid discharge, as bed load, by volume	m^3/s
Q_{ss}	Solid discharge, as suspended load, by volume	m^3/s
q_{ss}	Rate of suspended load transport in weight per unit width	Kg/s/m
q_{sb}	Rate of bed load transport in weight per unit width of the channel	Kg/s/m
q_{si}	Discharge of bed sediment of mean size d_{si}	$m^3/s/m$
q_{ci}	Critical value of q for initiating motion of sediment of mean size, d_{si}	$m^3/s/m$
q_{sbi}	Discharge of bed load of mean size of sediment, d_{si}	m^3/s
q_T	Rate of total load transport in weight per unit width	$m^3/s/m$

R	Hydraulic radius	m
R'	Hydraulic mean radius of the channel if the bed were unrippled; Hydraulic radius corresponding to grain resistance	m
	Bed Reynolds number	-
S_0	Bed slope	m
S_f	Friction slope (energy line)	m
S_w	Water surface	m
S_a	Average value of Sf	m
S^1	Portion of channel slope due to grain roughness	mm
T	Temperature of water	°C
V	Mean velocity	m/s
V_*	Shear velocity or total bed shear velocity	m/s
V_d	Characteristic velocity	m/s
V_n	Non displacement velocity	m/s
V_p	Detachment velocity	m/s
V_0	Critical velocity	-
V_{cr}	Critical velocity ; average velocity for incipient motion condition	m/s
V_*'	Grain roughness shear velocity	m/s
W	Weight of the water prism	Kg/m^2
X	Variable length of distance, correction term for logarithmic velocity distribution	m
X	Dimensionless factor in Einstein velocity relation; exponent	-
Δx	Length of short channel reach	m
x_1	Distance of the section under consideration from cross regulator	Km
Y	A function of d_{65}/s	-
Y	Coordinate; usually normal to water surface (vertical); exponent; distance from the boundary	m
Z	Actual exponent in suspended sediment distribution equation	-
Z_i	Dimensionless exponent in rouse equation	-
Z_t	Total energy in the flow of the section with reperence to a datum line is the sum of the elevation	m
$\Delta\rho$	Difference between the density of mixture and of the water, = (m-)	Kg/m^3
	Density of fluid (annexure-1)	Kg/m^3
ρ_m	Density of the fluid mixture	Kg/m^3
ρ_s	Density of solid particle	Ton/m^3
ρ_f	Mass density of fluid	Ton/m^3
ρ_s	Specific weight of sediment	Ton/m^3
γ_s	Specific weight of sediment	Kg/m^3
γ_f	Specific weight of fluid, q_f =pg	Kg/m^3
$\Delta\gamma_s$	Difference is specific weight of sediment and fluid	Kg/m^3
τ_0	Average shear stress , Laursen's bed shear stress due to grain resistance; average shear stress corresponding to grain	Kg/m^2
τ'_c	Critical shear stress for particles of size d_{si}	Kg/m^2
τ_b	Effective stress	Kg/m^2
τ_{*c}	Dimensionless critical shear stress	-
τ_{ci}	Critical value of 0 for sediment of size d_{si}	Kg/m^2
τ_s	The shear stress responsible for suspended load transport	Kg/m^2
τ_{oc}	Critical shear stress from shields curve	Kg/m^2

τ_*	Dimensionless shear stress	Kg/m^2
τ_*	Dimensionless shear stress to grains	Kg/m^2
τ_L	Limiting stress	Kg/m^2
α	Energy coefficient, (= 1 for small bed slope)	-
α_t	Coefficient in rough-turbulent equation	-
ν	Kinematic viscosity of fluid	m^2/s
ϕ	Angle of repose for the bed material	0 (degree)
ϕ_T	Total load parameter	-
Φ	Intensity (dimensionless) of solid discharge	-
ϕ^*, ψ^*	Bed-load function	-
Φ^* or Φ	Intensity of transport	-
ψ^*	Intensity of shear	-
ω_o	Fall velocity of particle under ideal condition	m/s
ω_1	Fall velocity of a grain of bed sediment of size d_{si}	m/s
ξ	A fraction indicating the part of the total area of the particle exposed to flow	-
ξ	A function of d_{si}/x	-
tan θ	Friction coefficient	-
ψ	intensity (dimensionless) of shear stress	-
ψ_D	Coefficient with dimension of cubic feet per pound per second	ft^3/lbs/s
Σ_I	Denotes summation for all sets at values of pi, dsi and gci	-
η	Elevation of bed	m
δ	Thickness of boundary layer; factor in Wilson equation for velocity	m
τ_g	Geometric standard diviation of size	-
Δ	Apparent roughness diameter	m
$\Downarrow$	Rouse exponent, $\Downarrow$ = Vss/k V* and Vss, settling velocity	-

Discharge : The discharge Q is the volume of a fluid or solid passing a cross section of a stream per unit time.

Cross-Sectional Area : The cross-sectional area A is the area of a cross section of the flow normal to the direction of flow.

Mean Velocity : The mean velocity V = Q/A is the discharge divided by the area of the water cross section.

Depth of flow : The depth of flow y is the vertical distance from the bed of a stream to the water surface.

Depth of Cross Section : The depth of cross section d is the perpendicular distance from the bed of a stream to the water surface. For an irregular cross section it is the perpendicular distance from a datum line to the water surface. This datum line in general is at the bottom of a rectangle with a wetted area equal to the wetted area of the natural cross section and the water surface of the two cross sections coincides.

Stage : The stage h is the vertical distance from any selected and defined datum to the water surface.

Water Surface width or Top width : The top width T is the width of a stream section at the water surface and it varies with stage in most channels.

Wetted Perimeter : The wetted perimeter P is the length of wetted cross section normal to the direction of flow.

Hydraulic Radius : The hydraulic radius R is the ratio of the cross sectional area to wetted perimeter, R = A/P

Hydraulic Depth or also known as hydraulic mean depth : The hydraulic depth d_h is the ratio of the cross sectional area to the top width of flow, $d_h = A/T$

Water Surface Slope : The slope of the water surface is denoted by S_w.

Slope of the Energy Grade Line : The energy grade line is a graphical representation with respect to a selected datum, of the total head or energy possessed by the fluid. For an open channel the energy gradient is located a distance $V^2 / 2g$ above the free water surface. The slope of the energy grade line is designated by the symbol S_f, or Se.

Bed slope : The bed slope S_b is the longitudinal slope of the channel bed.

□□□

Abbreviations

General Engineering Abbreviations

A.R. = Annual Report
Apr. = April
Am. = American
Ass. = Association
Aug. = August
Bldg. = Building(s)
Bldr. = Builder
Bull. = Bulletin
CBIP = Central Board of Irrigation & Power
CBRI = Central Building Research Institute
CRRI = Central Road Research Institute
CWPC = Central Water & Power Commission
CWPRS = Central Water & Power Research Station
Cem. = Cement
Chart. = Charted
Civ. = Civil
Comis. = Commissions

Found. = Foundation(s)
Gaz. = Gazette
Geoph. = Geophysics
Hlth. = Health
Hwy. = Highway(s)
Hyd. = Hydraulic(s)
IS = Indian Standards
Industry. = Industrial
Inform. = Information
Inst. = Institute
Instn. = Institution(s)
Int. = International
Irrig. = Irrigation
J. = Journal
Jan. = January
Mar. = March
Mater. = Materials
Mech. = Mechanical
Mech. = Mechanics
Oct. = October
P. = Page(s)
Pap. = Paper(s)

Pub. = Publication(s)
Pwr. = Power
Recl. = Reclamation
Rep. = Report(s)
Res. = Research(es)
Rev. = Review
Riv. = River
Rly. = Railway(s)
Sci. = Science
Sci = Scientific
Sep. = September
Ser. = Series
Sfty. = Safety
Coun. = Council
Cu. = Cubic
Curr. = Current
Dec. = December
Dev. = Development
Dig = Digest(s)
Drnge. = Drainage
Dte. = Direcorate
Eastn. = Eastern

Commun. = Communications

Concr. = Concrete

Congr. = Congress

Construc. = construction

Engng. = Engineering

Engr. = Engineer(s)

Enviorn = Environmental

Feb. = February

No., n. = Number

NBO = National Buildings Organisation

Nat. = National

Nov. = November

Soc. = Society

St. = State

Plnr. = Planner(s)

Pract. = Practice

Presr. = Preservers

Princ. = Principal(s)

Proc. = Proceedings

PWD = Public Works Department

Pt. = Part

Stand. = Standard

Stat. = Station

Survr. = Surveyors

Symp. = Symposium

Tech. = Technical

Technol. = Technology

Tim. = Timber

Educ. = Education

Mem. = Memoirs

Memo. = Memorandum

Met. = Meteorology

Mly. = Military

Mon. = Monthly

Monogr. = Monograph

Munic. = Municipal

Transp. = Transport

v. = volume

Val. = Valley

Wks. = Works

Yd. = Yard

❒❒❒

List of Figures

Introduction

Chapter-1

Chapter-2

Chapter-3

Chapter-4

Chapter-5

Chapter-6

Chapter-7

Chapter-7B

Chapter-8

Chapter-9

List of Tables

Chapter-5

Chapter-6

Chapter-7

Chapter-7B

Chapter-8

Chapter-9

Introduction

1. Our Source of Inspiration

There is much that can be learned from a study of the past, for it enables us to view the present in its true perspective and is a basis for speculation as to the future. This is as true of science as of mankind. The science of hydraulics has a great history and one that is well worth recalling, in view of its service to man kind.

Irrigation is the earliest form of hydraulics and has been practiced by mankind from the dawn of history. Archaeological discoveries have shown that canals, dams, and reservoirs existed in Egypt and Mesopotamia as early as 4000 B.C

There have been many contributors, and space permits only a few of the more distinguished to be mentioned. Amplification is provided in the subsequent text where it is felt that knowledge of the detailed historical background would be helpful to the understanding of a particular analysis or empirical formula.

The Greeks contributed much that was of outstanding quality and interest in the fields of philosophy, science, and art. In the 3rd century Archimedes established the principles of buoyancy and flotation. The works of Euclid and Aristotle embraced a wide field, some of which was of relevance to hydraulics.

Where the Greeks were imaginative innovators, the Romans were skilful adaptors. Numerous reclamation and drainage works were undertaken. Large water wheels were constructed for grinding corn. Water supply and sanitation received particular attention; for instance, Rome was supplied by nine aqueducts (canal) and there was a sewerage system. A useful treatise on the methods of water distribution was prepared by Frontinus.

The urge for progress also manifested itself in practical form, for in the 16th and 17th centuries many important hydraulic projects, such as harbours, canals, and reclamation schemes were undertaken. Indeed it was during the mid-17th century that a large part of the fenland of East Anglia was drained, the major credit being due to the Dutch engineer, Vermuyden.

Practicing engineers were also active in experimenting and in their attack on design problems. Pitot devised a simple instrument for determining velocity. To Chezy is due the credit for being the first to put forward a practical formula for channel flow. Du Buat, besides conducting numerous experiments, published an informative treatise on hydraulics.

The progress of hydraulics in the 19th century was along two largely independent paths. On the one hand there was theoretical or classical hydraulics which was often impractical because it involved the rigorous analysis of an ideal fluid behaving without turbulence. On the other hand there was empirical hydraulics which had considerable experimental but little rational basis. In fact, one may comment somewhat cynically that **'in the 19th century fluid dynamicists were divided into mathematicians who explained things that could not be observed and hydraulic engineers who observed things that could not be explained'**. But this is not entirely true, for there were some who by their broad treatment could claim membership of both schools.

The contributions of the hydrodynamicists are not to be belittled for there are many instances where the simplifying assumptions inherent in the concept of an ideal fluid do lead to results that are very near to the truth. Notable researchers in this were Navier, Airy, Saint-Venanat, Stokes, Kelvin, Rayleigh, and Lamb, not all of whom were mathematicians. Also worthy of mention are Hagen and Poiseuille for their experimental investigations of laminar flow.

The exponents of empirical hydraulics were mostly practicing engineers. With the great constructional works stemming from the Industrial Revolution there was an increasing awareness of the need for the formulation of rules and equations which would assist in hydraulic design, particularly with respect to channel and pipe flow. Dracy, Bazin, Ganguillet, Kutter, and Manning were but a few of the engineers who developed workable formulae to fit the field and experimental data that by now were beginning to accumulate.

Many of these formulae have not survived the test of time, and it is fascinating today to turn the pages of some 19th-century treatise on harbour or river engineering and contemplate the simplicity of formulae purporting to describe the most complex sorts of phenomena. One cannot avoid the conclusion that the great engineers of this period wisely placed much more reliance on their own experienced judgment than on the naïve application of some scientifically unproven formula.

In the laboratory, Reynolds is noteworthy for his many experimental achievements and in particular for his investigations on the transition between laminar and turbulent flow, published in 1883. Weisbach, besides undertaking useful experimental work on various aspects of hydraulics, presented a valuable engineering treatise on the dynamics of fluids which was for many years unsurpassed.

At the close of the 19th century there was still an appreciable lacuna between theory and observed behavior. Investigators were probing diligently from the two extremities, the purely theoretical and the strictly empirical, but there was little common ground. It was only later, consequent upon the fundamental physical researches initiated by Prandtl at the beginning of the 20th century that the present unified approach known as fluid mechanics was developed. This seeks to establish the true physical basis of fluid phenomena by a combined process of analytical reasoning and experimentation.

Prandtl's convincing theory, published in 1904, for the behavior of turbulent flow past a solid boundary earned him a chair at Gottingen University. The new concept opened up a rich field for study which he, his associates, and pupils - noteworthy among whom were Blasius, Karman, and Nikurads - proceeded zealously to explore. Turbulence, velocity distribution, form drag, and skin friction drag were all subjected to searching analytical and experimental investigation. Very soon research work in this field was proceeding in many scientific institutions throughout the world, but quite understandably the major effort was directed towards the by now pressing problems of aeronautics, where, because of the streamlined profiles and the absence of a free surface, results of positive practical value were to be quickly forthcoming. Needless to say, the application of the new concept to hydraulics was by no means neglected, and in the first three decades of the 20th century the fundamentals of pipe flow were successfully investigated.

In the wider sense too there was much valuable progress, particularly in dimensional analysis and the related field of model testing. Noteworthy contributors of the early 20th century were Bakhmeteff, Boussinesq, Engels, Forcheimer, Freeman, and Gibson.

Our fleeing journey to the evolution of hydraulic science has been long and laborious. For our present knowledge we are indebted to a multitude of investigators, many unknown, and following diverse callings, but all imbued with a common desire to know more about the mysteries of fluid behavior.

What does the future hold? Suffice it to say that there are many gaps in our knowledge - problems concerning turbulent separation, air-entraining vortices, sediment transport, density currents and interfacial mixing, to name but a few. Although the great fundamental truths may all have been discovered, their application to complex problems affords much scope for the exercise of human intellect. In recent years the number of research workers throughout the world engaged in the study of hydraulic phenomena has greatly increased; also, improved instrumentation techniques (electronic gadgetry) and mathematical tools (computers) have become available, so it would be a pessimist indeed who did not look forward with confidence to a continuing and possibly exciting advancement in the years that lie ahead.

But whilst recognizing the scientific importance of these researches we must not lose sight of the fact that it is engineers who are required to turn the fruits to practical advantage. The facility with which they are able to do so rests on their understanding or present knowledge and their willingness to participate in remedying its inadequacies.

2. Flow Formulae

Huge numbers of flow formulae have been developed by many researchers-as much as 38 or more. A list of 38 formulae is given in table 1 and 2. Research is still continuing to answer many unresolved questions, raised in this text at many places.

Table 1 : List of Flow (Resistance) Formulae

S. No.	Author (Year)	Formula
1.	Chezy (1769)	$V = \frac{C}{\sqrt{g}}\sqrt{gRS} = C'\sqrt{gRS}$; Dimensionless, see table 2 for C.
2.	Manning (1889)	$V = \frac{1}{N}R^{2/3}S^{1/2}$ (Metric) and $V = \frac{1.486}{N}R^{2/3}S^{1/2}$, English Units. For details see Para 1.4, for N see Para 2.0 and table 2.1 to 2.3
3.	Kennedy (1895)	$V_0 = 0.548\ d^{0.64}$ (Metric units); For details see Para 2.3.1, 7.5 $V_0 = 0.84d^{0.64}$ or $V_o = m\ d^n$, (English units) $V = mV_0$; m ranges from o.39 to 0.84 ; and n from 0.52 to 0.73 For more details see chapter 7
4.	Lindley (1919)	$V = 0.95\ d^{0.57}$ $V = 0.57\ B^{0.36}$ $B = 3.8\ d^{1.61}$ English Units
5.	Khannag (1920)	V = 0.0216 RS, English units
6.	Beleida (1921) (see Lacey 1958)	V = 0.02808 RS, English units
7.	Malokal (1921) (see Lacey 1958)	V = 0.046 RS, English Units
8.	Parandtl (1926)	$\frac{V}{V_*} = 8.5 + 5.75\log\frac{y}{k_s}$ (for rough walled pipe), Dimensionless
9.	Lacey (1929-58)	See table 7.14 detailes are given in chapter 7

Contd...

Table 1. Contd...

10.	Bose (1936) (ASCE 1963)	$V = 1.12\ R^{1/2}$ (E) $S = 2.09 \times 10^3\ d^{0.86}/Q^{0.21}$ English units $A = PR\ \ P = 2.8\ Q^{1/2}\ \ R = 0.47\ Q^{0.21}$, English units
11.	Malhotra (1939-40) (see Lacey 1958)	$V = 18.18\ R^{0.632}\ S^{0.343}$ English unit
12.	Blench (1939- 60)	$V = 6\sqrt{F_b F_s Q}\ \ B = \sqrt{F_b \frac{Q}{F_s}}\ \ d = \sqrt{F_s \frac{Q}{F_s}}$ English units $S = F_b^{5/6}\ F_s^{1/12} / \left(1+\frac{C}{233}\right) KQ^{1/6}$ Bed factor $F_b = \frac{V^2}{d}$ Side factor $F_s = \frac{V^2}{B}$ $K=3.63\frac{g}{\sqrt{1/4}}$ B = width of canal
13.	Colbrook and white (1937)	$\frac{1}{\sqrt{f}} = 1.74 - 2\log\left(\frac{k}{r}+\frac{18.7}{R_e\sqrt{f}}\right)$, Dimensionless f = friction factor of Darcy weisbackh $Re = Vr/v$ r = radius of the pipe k = effective roughness of walls
14.	Darcy-Weisbach (1958)	For $\frac{k_s V_*}{\nu} > 70\ \ \frac{1}{\sqrt{f}} = C\log\left(a\frac{R}{k_s}\right)$, Dimensionless For $\frac{k_s V_*}{\nu} > 5\ \ \frac{1}{\sqrt{f}} = C\log\left(R_e\frac{\sqrt{f}}{b}\right)$, Dimensionless For $5 \le \frac{k_s V_*}{\nu} \le 70\ \ \frac{1}{\sqrt{f}} = C\log\left(a\frac{R}{k_s}+R_e\frac{\sqrt{f}}{k_s}\right)$, Dimensionless (For the values of parameters C, a, b, see Table 6.1)
15.	Einstein (1950)	$\frac{V}{V_*^1} = 6.25+5.75\ \log\frac{R'}{k_s}x$, Dimensionless $k = D_{56}\ \ V_*^1 = \sqrt{gR's}\ \ \frac{V}{V_*^{"}}\phi(\varphi');$ $\Psi' = \frac{\rho\ s\text{-}\rho}{\rho}\frac{D_{35}}{SR'}$, Dimensionless
16.	Leliavsky (1955)	$V = TR^{0.85}S^{0.72}T = 147+3.92(z-10)^{0.353}$, English units
17.	Ning Chien (1955)	$\frac{V^2}{R} = C\left(\frac{q_t}{q}\right)^{1/2}\left(R^{1/2}S\right)^{2/3} = C\left(\frac{q_t}{q}\right)^{1/6}$, English units
18.	Inglis (see Lacey, 1958)	$Ws \propto Q^{1/2}\ I^{1/4} / g^{1/4}\ m^{1/4}\ \ A \propto Q^{5/6}\ I^{-1/12}$, English units $S \propto Q^{-1/6}\ I^{5/12}\ g^{1/12}\ m^{5/12}\ \ V = \propto g^{1/2} S / E_m^{1/2}$, English units $V^3/w_s \propto g^{3/2}\ m^{1/2}\ V/gd \propto I^{1/2}$ Inglish no. 1 $I = x\ v_s/(vg)^{1/3}\ E = \frac{\gamma}{w_s} = \frac{d}{R}$

Contd...

Table 1. Contd...

19.	Liu and Hwang (1959)	V = $C_a R_b^x S^y$, English units
20.	Kansoh (1960) (see simons and Albertson, 1963)	V = 0.56 $d^{0.64}$ (E) V = 0.36 $d^{0.64}$, Matric units B = 2.383 $R^{1/2}$ (E) d =0.531 $Q^{0.361}$ for sand beds and cohesive banks d= 0.305 $Q^{0.361}$ for coarse non cohesive materials
21.	Ghalep (see Simons and Albertson, 1963)	V = 284 $d^{0.727}$
22.	Jareki (1960) (see Bureau of Reclamation, 1960)	V_b = 0.465$d^{1/2}$, fine materials V_b = 0.518$d^{1/2}$, coarse materials; V_b = competent bottom velocity
23.	Sethna (1962)	$V = \frac{66.5}{\frac{0.1}{m}} \sqrt{RS}$ bed material moving ; Matric units $V = \frac{2525}{\frac{0.2}{m}} RS$ bed material not moving ; Matric units S = 0.52 $f^{0.6}/R^{0.2}$ (E) $f = \frac{G/L}{8.95m^{0.2}}$; Matric units G/L = gr/it of silt charge
24.	W. Shen (1962)	For $\frac{WD}{v} < 100, \frac{V_*''}{V} = 0.03 + 0.11 \log \frac{\psi'}{v}$, Dimensionless For $\frac{WD}{v} > 100, \frac{V_*''}{V} = 0.064 - 0.0909 \log \frac{\psi'}{v}$ $\Psi' = \frac{\gamma RS}{\gamma_s' D_{50}}$ (see fig. 0.064 variation of v)
25.	Ahmad and Rehman (1963)	S = $K f^{5/3} R^{1/6}$ $K_2 = 0.45x10^{-3}$ to 0.7 x 10^{-3}, English units S = $k_3 f^{5/3}/q^{1/3}$ k_3 = 0.35 - 0.42 f = $k_4 \sqrt{d}$ K_4 = 1.1 - 3.0 $\frac{k_4}{ane}$ =\ 1.9 b = $k_1 Q^{1/2}$ Q = design discharge
26.	Albertson ans Simons (1964)	$\frac{v-V}{V} \frac{C}{\sqrt{g}} = 5.9 \log \frac{y}{y_0} + 2.49$, Dimensionless V = local velocity, y_0 = total depth y = depth corresponding to V.
27.	Strickler Meyer Peter (1923-48)	$N = \frac{D_{50}^{1/6}}{21.1}$ (M) 6.73 $N = \frac{\frac{D^{1/6}}{90}}{26}$ Metric units D_{50} is in meters and represent particale size in which 50 % of bed material by weight is finer

Contd...

Table 1. Contd...

28.	Mostmov (1949)	$C = 22 \log \frac{R_e}{C} - 13.4$ *Smooth flow*, Matric units $C = 22 \log \frac{R_e}{k_s} + \frac{9.5}{R+1.5}$ *Rough flow*, Matric units
29.	Powel (1950)	$\frac{C}{\sqrt{g}} = 7.4 \log \frac{R_e}{C/\sqrt{g}} - 1.13$ Smooth flow, Dimensionless $\frac{C}{\sqrt{g}} = 7.4 \log \frac{R}{\varepsilon} - 1.13$ Rough flow, Dimensionless ε is a coefficient involving the shape of the cross section as well as the size spacing and shape of the roughness element.
30.	Altshul (1952)	$\frac{C}{\sqrt{g}}\ 5.10 \log \left[\frac{R_e}{R\frac{k}{D}+7.0}\right]$, Dimensionless
31.	Lane and Carlson (1953)	$N = \frac{D_{75}}{29.3}$, English units
32.	Sayre and Albertson (1957-63)	$\frac{C}{\sqrt{g}} = 6.06 \log \frac{R_e}{C/\sqrt{g}} - 1.2$ Smooth flow, Dimensionless $\frac{C}{\sqrt{g}} = 6.06 \log \frac{d}{x}$ x is the roughness parameter gicen as a function of the relative size and relative spacing of the baffles.
33.	Koloseus and Davidian (1961)	$\frac{C}{\sqrt{g}} = 5.66 \log \left(\frac{0.56\, v^{0.9} R}{k_s}\right)$, Dimensionless k_s = concentration of roughness elements (vertical roughness elements)
34.	Tracey and Lester (1961)	$\frac{C}{\sqrt{g}} = 5.75 \log \left(\frac{R_e}{C/\sqrt{g}}\right) + 2.38,\ for 0.14 < F_r < 3.96$, Dimensionless
35.	Mirajgavker and charles (1963)	$\frac{C}{\sqrt{g}} = 5.28 \log \frac{d}{x} + 1.72$, Dimensionless
36.	Simons and Richardson (1967)	1- For plane bed $\frac{C}{\sqrt{g}} = 5.9 \log \frac{d}{D_{85}} + 5.44$ 2- For plane bed with motion $\frac{C}{\sqrt{g}} = 7.4 \log \frac{d}{D_{85}}$ 3- for ripples $\frac{C}{\sqrt{g}} = \left(7.66 - \frac{0.3}{V_*}\right) \log d + \frac{0.13}{V_*} + 11$ 4- For dunes and antidunes $\frac{C}{\sqrt{g}} = 7.4 \log \frac{d}{D_{85}} \sqrt{1 - \frac{\Delta SR}{SR}}$ for determination of ΔRS

Contd...

Table 1. Contd...

37. Senturk (1967-73)

1 - for plane bed with sediment motion

$$\frac{V}{V_*} = 6 - 2\log C + 6.5\log\frac{R}{D_{65}},\ \text{Dimensionless}$$

2- for skin friction $\frac{V}{V_*^1} = 6 + 6.5\log\frac{R'}{D_{65}}$, Dimensionless

3- for riplles $\frac{V}{V_*^e} = -52 + 2.5\ \log C + 18\log\frac{R''}{D_{65}}$, Dimensionless

4- For dunes $\frac{V}{V_*^{"}} = 48 - \log C - 11\ \log\frac{R''}{D_{65}}$ Dimensionless

$$C = \frac{(w_{65}D_{65}/v)^2}{S}\frac{D_{65}}{D_{35}}\gamma_s',\ \text{Dimensionless}$$

38. Kutter and Gangulett

$$C = \frac{\frac{1}{n} + \left(23 + \frac{0.00155}{S}\right)}{1 + \left(23 + \frac{0.00155}{S}\right)\frac{n}{\sqrt{R}}}\ \text{in metric system}$$

$$C = \frac{\frac{1.811}{n} + \left(41.66 + \frac{0.00281}{S}\right)}{1 + \left(41.66 + \frac{0.00281}{S}\right)\frac{n}{\sqrt{R}}}\ \text{in F.P.S. system}$$

The difficulty in this formula is the determination of the value of C each time as it changes with the values of bed slope, hydraulic mean depth and coefficient of rugosity. Thus the use of this formula is a bit cumbersome.

Table 2 : Other Equations for resistance coefficients (Besides Equations at serial no. 27 to 31, 33, 34, 35, in table 1)

i)	N. N. Pavlovsky (1925)	: $C = \frac{Rx}{n}$; $x = 2.5\, n^{1/2} - 0.13 - 0.75\ R^{1/2}(n^{1/2} - 0.10)$
ii)	A. E. Bretting (1948)	: $1/f^{1/2} = 2\log(k_s/14.83\,R)$
iii)	J.H. Thijsse (1949)	: $1/f^{1/2} = 2.03\,[\log 12.2\,R/0.282\,\delta + k_s]$
iv)	Powell (1950)	: $C = 42\log R/e$ (Rough Channel)
v)	A.D. Alfshul (1952)	: $1/f^{1/2} = 1.8\log[Re/Re(\Delta/d) + 7]$
vi)	P. Ackers (1958)	: $C = (32g)^{1/2}\log(14.8\,R/k_s)$
vii)	M.L. Albertson (1961)	: $C/g^{1/2} = 6.06\log\left(\frac{d_n}{x} + 2.6\right)$
viii)	H.J. Koloseus and Davidian (1961)	: $1/f^{1/2}\,2\log[(0.56\,\lambda^{0.9}R)/k_s]$

C = Chezy's coefficient ; N or n = Manning's coefficient
R = the hydraulic radius (in meters) ; f = the Darcy-weisbach coefficient
k_s = the equivalent sand grain diameter ; δ = the thickness of the laminar sublayer
k_s = a measure of the roughness ; e = a measure of channel roughness
Re = is the Reynolds number ; Δ = is a linear measure of roughness height
d = is depth of flow ; is a measure of the amplitude of the roughness
d_n = is normal depth ; x = a general roughness parameter
λ = a measure of the concentration of roughness elements
k_s = a measure of the amplitude of the roughness elements

Thus our Journey to the flow through open channel has been very long and still continuing.

Gunther Garbrecht (1961) evaluated 22 Chezy coefficient formulas that have been used in Europe.

3. Basic Hydraulics of Flow

3.1. Classifications of Flows

Water can flow either in closed conduits like pipes or in open channels like rivers or canals. When the flow is in closed conduits, it may be under pressure, and is called as pipe flow or pressure flow. Otherwise even in pipes, when running partly full, or in rivers/canals the flow is called open channel flow.

Either the pressure flow or open channel may be classified as under:

i. Steady and unsteady
ii. Uniform and non uniform
iii. Turbulent and laminar

See figure 1 given below

Steady Flow (velocity At A Point Constant With Time)		**Unsteady Flow (velocity At A Point Changes With Time)**	
Type and Example	Comment	Type and Example	Comment
Uniform 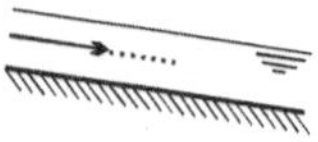	Constant depth at all time, along distance		Depth varies suddenly from time to time (very rare) such as in surges.
Non uniform 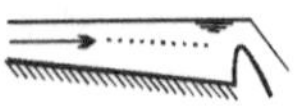 (i) gradually varied	Depth varies slowly with distance such as behind weirs.		Long flood wave
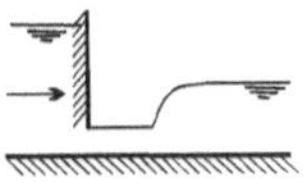 (ii) Rapidly varied	Hydraulic jump causes sudden depth change in a short distance	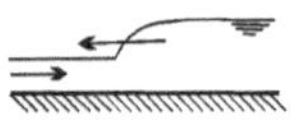	Short surge wave (e.g. tidal bore)

Fig. 1. Flow classification

The flow that occurs in open channels differs from the pipe flow. In a pipe pressure varies from point to point in the flow direction, whereas the free surface pressure in a channel has a constant value, usually atmospheric. Also, since cross-sectional area of flow in a channel is not determined by fixed boundaries, flow depth can vary from section to section without restraint. Flow depth can also vary with time. In actual practice, the flow may be a combination of above types, for example an open channel flow may be steady uniform or steady non-uniform. Different types of flow are further discussed below.

3.2 Steady and Unsteady flow

Steady flow exists if the velocity at a point remains constant with respect to time. ***Conversely, unsteady flow exists if the velocity changes either in magnitude or in direction with respect to time.*** Steady flow is usually much easier to analyze and solve then unsteady flow. In fact, a strictly rigorous solution for unsteady flow is sometimes impossible. Many hydraulic phenomena involve unsteady flow, so that only approximate solutions are possible. Such solutions are usually based on statistical averages or a step analysis for which there is only a small change in velocity over a short increment of time, and for this small time increment the flow is assumed steady.

3.3 Uniform and Non-Uniform

If at a given instant the velocity remains constant with respect to distance along a streamline, the flow is uniform. If there is a change either in magnitude or in direction along the streamline, the flow is non-uniform. In flow around a bend of a pipe or channel, the direction changes with distance, and in flow with changing cross section, the magnitude changes with distance; hence the flow is non-uniform. **In open channels where the velocity changes slowly along the channel, the flow is assumed uniform for an increment of length and is known as gradually varied flow.**

3.4 Turbulent and Laminar Flow

In turbulent flow the fluid particles (small molar masses) move in very irregular paths, causing an exchange of momentum from one portion of fluid to another. The fluid particles can range in size from very small (say a few thousand molecules) to very large (thousands of cubic feet or cubic meter) in a large swirl in a river or in an atmospheric gust. The turbulence sets up greater shear stresses through out the fluid and causes more losses. Also in turbulent flow the losses vary about as $V^{1.7}$ to V^2.

Turbulent flow situations are very common in engineering practice.

In laminar flow fluid particles move along smooth paths in laminas or layers, with one layer gliding smoothly over an adjacent layer. In laminar flow losses are much less than turbulent flow, as they vary as the first power of velocity. Laminar flow is not stable in situations involving high velocity, or large flow passage and may break down in turbulent flow.

3.5 Pressure Flow (or also generally known as pipe flow)

Flow in closed conduits involves a combination of steady or turbulent flow, and flow over smooth or rough boundaries.

At the upstream end of a pipe there is a region of flow development in which the boundary layer is developing and the flow is technically non uniform because the velocity varies from point to point along the streamline. Therefore, the velocity distribution changes from section to section, which means that, in this region of flow development, a standard calibration of measuring devices and the standard loss coefficients for pipe fittings (minor losses) are not applicable. In other words, calibrations must be made and coefficients must be determined in place if reasonable accuracy is to be achieved. Furthermore, in this region the greater boundary shear and any initial separation cause a greater amount of energy loss per unit length of pipe than in the region of fully developed flow downstream. The length of the region of flow development in a pipe of diameter D is approximately 0.06 Re. D for laminar flow and 0.7 $Re.^{0.25\,D}$ for turbulent flow throughout the length from the pipe entrance.

3.6 Open Channel Flow

Flow in open channels has been nature's way of conveying water on the surface of the earth through rivers and streams. These streams have constantly been the subject of study by man, as he has been alternately blessed by the life-giving quality of streams under control and plagued by the destructive power of streams out of control, such as in time of flood.

Open channels include not only those which are completely open overhead, but also closed conduits which are flowing partly full. Examples of such closed conduits which are flowing partly full. Examples of such closed conduits are tunnels, storm severs, sanitary sewers, and various types of pipelines.

Flow in open channels involves a free surface, which is actually an interface between two fluids having different specific weights, such as air and water. Because of the presence of the free surface, flow in open channels has an additional classification to the three types of flow that occur in pressure pipes.

The four classifications of open channel flow are:

1. Uniform or non uniform,
2. Steady or unsteady,
3. Laminar or turbulent, and
4. Tranquil (sub critical) and rapid (super critical); critical.

The fourth type of flow does not occur in pressure pipes.

Uniform flow in open channels has no change with distance in either the magnitude or the direction of the velocity along a streamline. **Otherwise the flow is non uniform, or varied**. Strictly uniform flow rarely exists. For practical purposes, flow in an open channel is generally considered as uniform if the depth of flow is approximately constant in the direction of flow. **The depth of uniform flow is called normal depth.**

Steady flow occurs when the velocity at a point (same point) does not change with time. Otherwise, the flow is unsteady, such as traveling surges and flood waves in an open channel.

Whether laminar flow or turbulent flow exists in an open channel also depends upon the Reynolds number of the flow just as it does in pipes. Like the flow in pipes, turbulent flow may be over either a smooth boundary or a rough boundary, depending on the relative size of the roughness elements as compared with the thickness of the laminar sub layer.

Unlike laminar and turbulent flow, tranquil (sub critical) flow and rapid (super critical) flow occur only with a free surface or interface. The criterion for this classification of flow is the Froude number $Fr = V / \sqrt{gD}$ (Where D = hydraulic mean depth = Area / Top width)

When Fr = 1.0, the flow is critical;

When Fr < 1, the flow is tranquil (i.e. sub critical) and

When Fr > 1, the flow is rapid (i.e. super critical)

Uniform flow in an open channel occurs with either a mild, a critical, or a steep slope, depending on whether the flow is sub critical (tranquil), critical, or rapid (super critical), respectively.

3.7 Specific Energy

The concept of Specific Energy is used only in the open channel flow in the analysis of critical flow and related phenomena.

At any cross-section of a channel the energy with respect to the channel bed is defined as the Specific Energy. If the slope is small,

$$\text{Specific Energy} = E = Y + \frac{V^2}{2g} \qquad \text{Eq. 0.1}$$

Where, Y = depth of flow and V = Velocity of flow.

Assuming a uniform velocity distribution, the specific energy is constant across the section.

Putting V = Q/A ; where Q = discharge and A = cross-sectional area.

$$\text{Then } E = Y + \frac{Q^2}{2gA^2} \qquad \text{Eq. 0.2}$$

The cross-sectional area varies only with the depth of flow, if the geometry of the channel is constant, and therefore for a given discharge the specific energy is a function of depth alone.

When the depth of flow is plotted against specific energy for a given channel section and discharge, a specific energy curve (generally called as energy curve in short) is obtained (figure 0.2). This curve has two limbs Ac and Bc - the limb Ac approaches the horizontal asymptotically towards the right. The limb Bc approaches the line OD as it extends upwards and to the right.

The curve shows that for a given specific energy, there are two possible depths. The low depth (or stage) is called as super critical depth and flow as super critical flow. High stage is called as sub critical depth and flow as sub critical flow. The two depths are known as alternate depths or conjugate depths

At point C the specific energy is minimum. This depth of minimum specific energy is known as "critical depth" and is denoted by Yc. The corresponding flow is known as critical flow. All symbols that have the subscript c refer to a channel section in which flow is critical. Mathematically for critical flow to occur dE/dY should be equal to zero.

Differentiating the Eq. 0.2 with respect to Y and noting that Q is constant,

$$\frac{dE}{dY} = 1 - \frac{Q^2}{gA^3}\frac{dA}{dY}$$

The differential water area d A near the free surface is equal to T dY (see figure 0.2).

$$\text{or } dA / dY = T$$

Therefore $dE/dY = 1 - (Q^2\ T/g\ A^3) = 1 - (V^2\ T/\ g\ A)$ as $(Q^2/A^2 = V)$

$= 1 - (V^2 / g\ D)$ (For wide rectangular channel $A / T = D$, also called as hydraulic mean depth)

For specific energy to be minimum

$d E/d Y = 0 =$ i.e $1 - (V^2/g\ D)$ or $1 - (Q^2\ T/g\ A^3) = 0$

or $V^2 / g\ D = 1$ at critical flow

$$\text{or } \frac{V}{\sqrt{gD}} = 1 \qquad \text{Eq. 0.3}$$

$\frac{V}{\sqrt{gD}}$ is know as Froude number

At a critical flow state, flow is characterized by following conditions.

i. Specific Energy is minimum for a given discharge.

ii. Discharge is maximum for a given specific energy.

iii. Velocity head is equal to one -half the hydraulic mean Depth i.e. Froude No. is unity.

iv. There is a definite discharge - depth relationship independent of channel roughness and flow conditions upstream as well as downstream of flume.

3.8 Critical Depth as a Control Section

In other words critical depth is defined as that depth at which for a given discharge, the specific energy is minimum.

In terms of discharge, the critical depth formula can be written as :-

$$Q^2/g = (A^3 / T)_{crit}$$

Where Q is the discharge, A is the control section area of flow and T is the surface width when depth $Y = Y_{crit}$

Since A, T depends only on the depth y and the channel geometry which in constant, discharge at critical depth is a unique function of depth alone. Such a section is called as a control section.

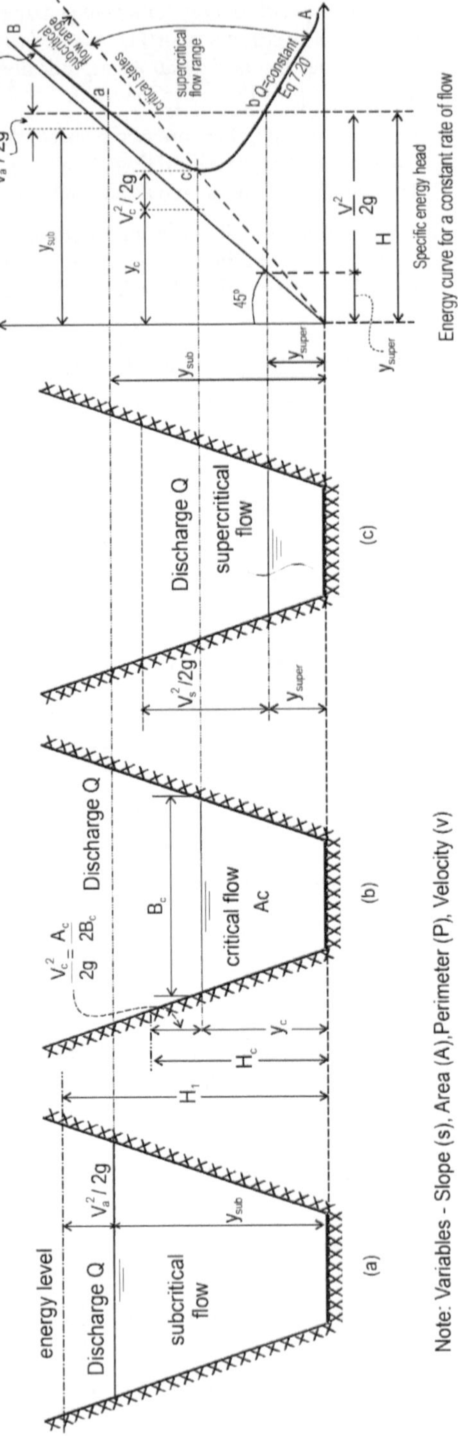

Note: Variables - Slope (s), Area (A), Perimeter (P), Velocity (v)

Figure 2. Alternative water depths for one energy level and a constant flow rate

At the control section, the discharge is given by the equation

$$Q = \sqrt{g A_c / T_c} \qquad \text{Eq. 0.4}$$

In a measuring device such as standing wave flumes, Parshall Flumes, Cut throat flumes and Broad crested weirs of various types, the discharge equation is based on this critical depth concept.

If the control section is rectangular (for instance)

$$Q^2 = g\, b^2\, y_{cr}^{\ 3} \qquad \text{Eq. 0.5}$$

Further $y_{cr} = 2\,H / 3$ where H is the total sill - referenced head at the upstream of the measuring section, and y_{cr} is as measured above the sill.

The discharge equation, therefore, reduces to

$$Q = (2/3)^{3/2}\, C_d\, \sqrt{g}\, b\, H^{3/2} \quad \text{............ For rectangular control sections} \qquad \text{Eq. 0.6}$$

3.9 Fundamental equations of fluid motion

The three fundamental equations used in the solution of flow problems both is pressure flow and open channel flow are based on the principles of conservation of mass, energy, and momentum. These are

(i) Continuity equation

For water, regarded as an incompressible liquid, the principle of conservation of mass leads to the continuity equation,

$$Q = VA = \text{constant or in other words} \qquad \text{Eq. 0.7}$$

$V_1 A_1 = V_2 A_2 = V_3 A_3$ (at various section of the channel /pipe) provided that no water is added or lost in the channel

Where Q = discharge,

V = velocity (assumed constant across the section generally taken as mean v, and

A = cross-sectional area.

(Note: use consistent units on both sides)

(ii) Bernoulli's equation

This is a statement of the principle of conservation of energy applied along a streamline and is given as

$$p / \gamma + \frac{V^2}{2g} + Z = \text{constant or in other words} \qquad \text{Eq. 0.8}$$

$$p_1 / \gamma + \frac{V_1^2}{2g} + Z_1 = p_2 / \gamma + \frac{V_2^2}{2g} + Z_2 \text{ and so on.}$$

where p = pressure,

V = velocity,

Z = elevation above a datum level, and

γ = density of the liquid.

Each expression is in the form of a head of liquid, i.e. in units of length.

where p / γ = pressure head,
$V^2 / 2g$ = velocity head,
Z = gravitation head,

The sum of pressure, velocity, and gravitation heads is known as the total head. The physical significance of these expressions can be appreciated in figure 0.3

Bernoulli's equation 0.8 applies to an in viscid liquid. For a real liquid in which there is frictional loss along the flow path, an expression for frictional head loss must be included. Thus applying Bernoulli's equation to two successive cross-sections along a flow path.

$$p_1 / \gamma + V_1^2 / 2g + Z_1 = p_2 / \gamma + V_2^2 / 2g + Z_2 + h_f \quad \text{Eq. 0.9}$$

where the suffixes 1 and 2 denote the first and second cross-sections and h_f is the frictional head loss.

(iii) The impulse momentum equation

It is derived by the application of Newton's second low of motion. For a given direction (x- direction) it is expressed as:

$$Fx = \frac{\gamma Q}{g}\left[(V_2)_x - (V_1)_x\right]$$

Where Fx is the applied force in the x-direction and $(V_1)_x$ and $(V_2)_x$ are the velocities in the x-direction before and after the change in momentum.

It has been assumed that the distance between the sections where V_1 and V_2 are measured is small and that friction losses are negligible. If this is not the case, F_x includes the force exerted on the flow by friction at the boundary.

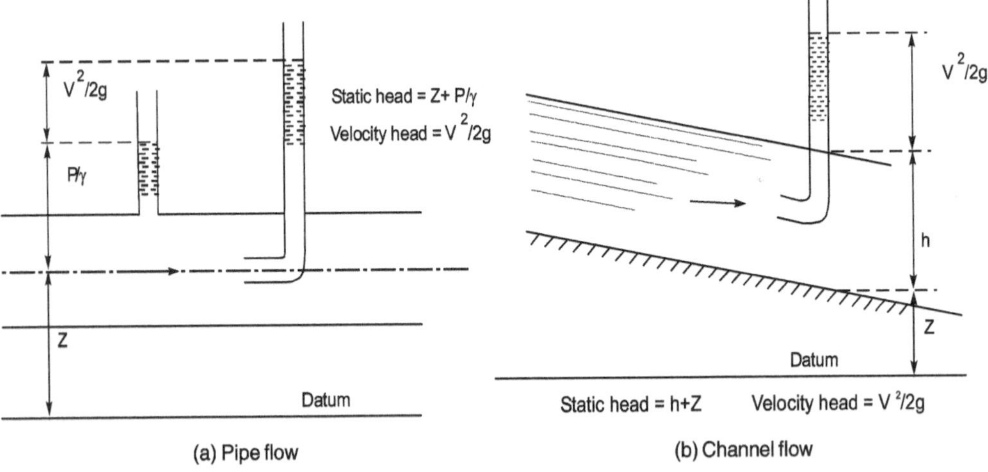

Figure 3 : Static and velocity heads in pipe and channel flow

In the simple treatment of the fundamental equations considered above, the velocity of flow is generally assumed constant across the section. For a more sophisticated approach see Chow (1973) or any other text on engineering hydraulics.

3.10 Flexibility

Because of a changing flow rate, the head upstream of an (irrigation) canal bifurcation usually changes. Depending on the characteristics of the structures in the supply canal and that in the off-take canal such as outlets, the relative distribution of water may change because of the changing head. To describe this relative change of distribution the term **flexibility is used, which has been defined as the ratio of the rate of change of discharge of the off-take or outlet Q_0 to the rate of change of discharge of the continuing supply canal Q_s or :**

$$F = (dQ_s / Q_s) / (dQ_0 / Q_0) \qquad \text{Eq. 0.10}$$

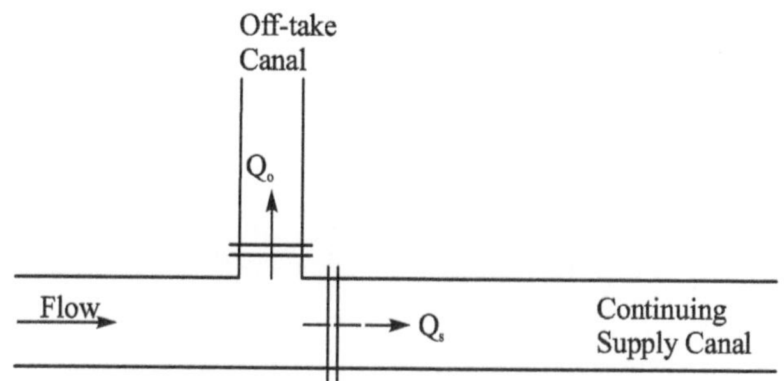

Figure 4 : Canal bifurcation sketch

In general the discharge of a structure or channel can be expressed by the Equation

$$Q = \text{Const } h_1^{\,n} \qquad \text{Eq. 0.11}$$

Hence we can write

$$dQ / dh_1 = \text{Const. } n h_1^{\,n-1} \qquad \text{Eq. 0.12}$$

□□□

Chapter 1

Design of Canals

1.0 General

The design of canal involves following three major aspects:

A Irrigation and drainage canal network planning.

B Determination of discharge / capacities of canals and drainages.

C Determination of shape and size i.e. geometrical dimensions of the proposed canal section, generally known as design of canal section. More appropriately it can be called as design of internal section of canal. It can be further divided in three following parts viz.

i. Design of unlined canals:

ii. Design of lined canals: or a canal may be partly lined and partly unlined.

iii. Choice between the unlined and lined canals. This aspect should be distinguished from the selection of type of lining. However, the scope of choice of lining may include both the aspects.

The design of canal section remains incomplete without the design of its embankments, either in lined or unlined canal, so the next aspect is

iv. Design of canal embankments.

Still to complete the design of a canal one more aspect is required i.e.

v. Canal aesthetics and canal arboriculture.

Without this the design always remains incomplete though the water in canal system can run but the canal net work starts deteriorating at a fast rate. Still this aspect is generally neglected or given thought after a considerable lapse of time, and then it becomes more costly.

All the above (A), (B) and (C) aspects are inter acting, effecting each other substentially and even change the entire system process in a very big way. For example in a dense canal network planning, (i.e. where the length of canal net work per unit area i.e km/ha is more), the capacity of each individual canals may be low, and the system may be very costly and dificult to manage, but may be more easy to the farmers, where as in a low dense canal net work, capacity of individual cansls may be high, the system may cost less and easy to manage, but may be unfavourable to farmers. But, still this is not always so. Canal network planning depends on several other factors besides the effect of above (B) and (C). More details are discussed separately.

Even the above three individual aspects have several factors in themselves which are all interacting and make the design of canals a highly complex process.

Details for the design of **unlined** canals are discussed in chapter 7.

The design of **lined** canals is discussed in this Part. Choice between unlined and lined canals along with the selection of type of lining is discussed separately in chapter 9.

Details about various types of lining are discussed separately.

Details of design of canal embankments are discussed in chapter 8.

This Part deals with only the design of internal over all section of canal, including it's inter relationship with its embankment.

Canal network planning is discussed separately in another publication.

1.1 Approaches of Design of Canal Section

In general there are three methods available for design of internal section of a canal that give the hydraulics of flow, and suitable internal dimension of any canal, may be irrigation, or power or drainage etc. These methods are

I. The method of permissible velocity
II. The method of tractive force and
III. Regime theories (methods)

Out of these, the (I) method of permissible velocity is extensively used throughout the world. Even this approach has several modifications such as limited / given bed slopes or depth. Thus the methodology is no longer of permissible velocity only and becomes method of permissible bed slopes and depth, see table-9.5 and also see para7.8.

In the **first** method, Manning's formula along with the continuity equation Q = VA are used, to find the dimensions of the channel. This approach along with modification / limitations is extensively dealt in this part. A very clear understanding of role of hydraulic radius, side slopes on hydraulic efficiency, b/d ratio, bed slopes, rugosity coefficient are presented in this text. A Master equation (para 4.3) of best sections and trapezoidal sections in general are presented which avoids any trial and error solution.

The second method is of recent origin and is being increasingly advocated by researchers, who have worked on it. In this method, concept of tractive force is applied. Details are discussed in para 7.10, chapter 7.

For details of all above methods see chapter 3, 4, 5 and 7

Classification of Canals and its Design

From hydraulic point of view, design of any channel (irrigation, or drainage or hydropower etc.) may be classified into following two major categories as under:

(i) Rigid boundary (i.e. non erodible or some times also called as non alluvial) channel

The rigid boundary may be a rigid or hard surface lining or canal may pass through hard rock without lining or may pass through semi hard strata, such as soft rock, boulders or say semi flexible boundary. It may further carry

(a) Clear water, such as from reservoirs, or
(b) Sediment (silt) laden water from the source such as in weirs and barrages, or some sediment may enter from the adjoining area through which canal is passing. For example in cutting reaches rain water may erode the side slopes, which may enter the canal, or there may be a land slide, (see typical sections in figure 9.1) or wind blown sand may enter the canal system in the desert (typical example is IGNP).

(ii) Non rigid boundary or generally known as Alluvial channel

These are earthen channels passing through alluvial soils. It may also carry

(a) Clear water as in (i) (a) or (b) sediment-laden water as above in (i) (b).

There may be very few canals to carry clear water or clear water may be for a limited period of a whole year.

All the canals can be designed by the method of permissible velocity. A unique method is given in chapter 7.

1.2 Design of Lined Canals

Lined canals can be constructed of any geometrical shape such as trapezoidal, rectangular, triangular, semi-circular, parabolic, etc. or a combination of above. Geometrical shape to be adopted depends upon the type of lining and its economics, soil characteristics, topography and discharge, which are all interrelated and interacting, see table 9.2.

For example thin linings cannot be adopted in rectangular sections. Or rectangular canal sections are not always economical or feasible in all type of soils. More details about choice of lining and various inter relationships are discussed in other chapters. Many times the cost of lining and its finishing also influences the decision on the canal section to be adopted.

Depending upon the discharge and topography, a combination of above sections can also be easily adopted in a particular reach or in different reaches of the same canal. For example trapezoidal section may be adopted in partial cutting/filling and a rectangular / circular section may be provided in heavy rock cutting. More details are discussed in the chapter of choice of lining.

Different geometrical sections within a canal length may become necessary due to change of soil properties (clay/sandy soil, or rock cutting etc.), topography (heavy filling/cutting) and expected fluctuations in discharge such as perennial/ non-perennial canals, high discharge during Rabi and low discharge in Kharif season, high and low flood diversion; in regulation system i.e. day and night running canals or only day running canals.

1.2.1 Composite Section

A composite section is a combination of several geometrical shapes. Thus in a composite section there is no single uniform geometry. Two or several geometrical sections form a composite section. In nature such sections can be easily seen in soft / hard rocks or Murram (or hard strata). Even in soils, river changes its water flow section according to the variation in discharge, by either silting or erosion or cutting of banks.

Flow through rivers is nature's way of conveying water. This flow varies very greatly in the same river from very small to very large depending upon the rainfall and run-off. The same river carries all the discharges and fulfils the hydraulics of nature. This is so because of the composite nature of the section.

This section also fulfils the requirement of changing or varied discharges, which may occur in different time periods.

A composite section (a mix of two sections, also known as compound section) is shown in figure - 1. 1 can also be adopted for canals of very large discharges and low capacity factors.

These sections are also very good for deep cuttings or heavy fillings, or very vide variation in discharge.

Composite section may be unlined or fully lined or it may be partly lined i.e. inner section may be lined and outer section may be unlined or inner section and or ½ section may be lined. More details of these sections are discussed in para 2.1 in chapter 2.

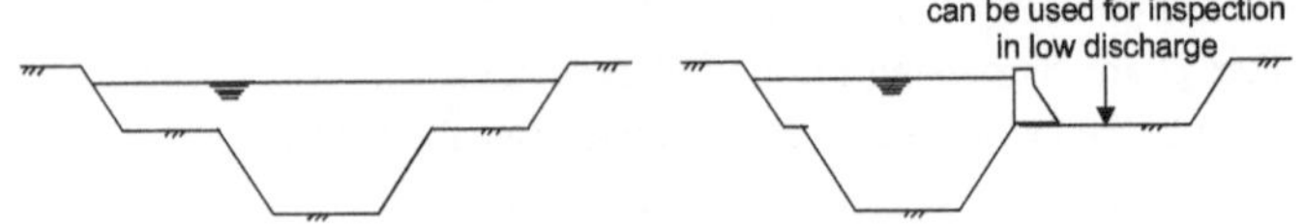

Figure 1.1 : Composite sections

Suitable transitions are provided at any change of canal section for gradual change of shape and smooth flow to avoid excessive head loss. A section can also be abruptly changed at a structure, with permissible/reasonable/acceptable head loss, such as at falls, measuring structure or bridges.

1.3 Hydraulic Design

The geometrical and hydraulic properties of most common lined sections are given in **Table 1.1** More details of geometrical and hydraulic properties of semicircular, Trapezoidal, rectangular, and Mehboob section are given in **Table 1.2.**

Table 1.1 : Geometrical and Hydraulic Properties of common lined channel sections

Section	Area (A)	Wetted Perimeter (P)	Hydraulic Radius R=A/P	Hydraulic Depth D=A/T	Section Factor (Z)
1. Circular $T=(\sin\theta/2)\,d_o$	$1/8\ (\theta-\sin\theta)d_o^2$	$½\ \theta\ d_o$	$¼(1-\sin\theta/\theta)d_o$	$1/8[(\theta-\sin\theta)/(\sin ½\ \theta)]d_o$	$\sqrt{2}/32[(\theta-\sin\theta)^{1.5}/(\sin ½\ \theta)^{0.5}](d_o)^{2.5}$
2. Trapezoidal	$(b+zy)y$	$B+2y\sqrt{1+z^2}$	$(b+zy)y/b+2y\sqrt{1+z^2}$	$(b+zy)y/(b+2zy)$	$[(b+zy)y]^{1.5}/(b+2zy)$
3. Parabolic	$2/3(Ty)$	$T+(8y^2/3T)$	$2T^2y/(3T^2+8y^2)$	$2/3(y)$	$2/9(\sqrt{6}Ty^{1.5}$
4. Round Bottomed triangle $T=2[z(y-r)+r(1+z^2)^{1/2}]$	$(T^2/4z)-(r^2/z)(1-z\cot^{-1}z)$	$(T/z)-(\sqrt{1+z^2})-(2r/z)(1-z\cot^{-1}z)$	A/P	A/T	$A\sqrt{(A/T)}$
5. Rectangle	by	b+2y	by/(b+2y)	Y	$by^{1.5}$
6. Round corne Rectangle	$[(\pi/2)-2]r^2+(b+2r)y$	$(\pi-2)r+b+2y$	$[(\pi/2)-2]r^2+(b+2r)y/(\pi-2)r+b+2y$	$[\{(\pi/2)-2\}r^2/(b+2r)]+y$	$[\{(\pi/2)-2\}r^2+(b+2r)y^2]^{1.5}/\sqrt{b+2r}$
7. Triangle	zy^2	$2y\sqrt{1+z^2}$	$zy/2\sqrt{1+z^2}$	$½\ y$	$(1/\sqrt{2})zy^{2.5}$

1.4 The Manning's Formula

In 1889, an Irish Engineer Robert Manning (1895, chow 1973) in his effort to co-relate and systematise existing data of flow through natural and artificial channels, developed an equation from seven different formulas viz. Chezy (1769), Ganguillet and Kutter (1869), Bazin (1897), and others. This was later modified to its present well known form. This has been verified by many others and later came to be well known as Manning's equation.

$V = 1/N.\ R^{2/3}\ S^{1/2} = R^{2/3}\ S^{1/2}\ /N$ (Metric units) ... Eq. 1.1

or its back conversion gives

$V = 1.486/N.\ R^{2/3}\ S^{1/2} = 1.486\ R^{2/3}\ S^{1/2}\ /N$ (English units).

Where,

V = mean velocity of flow in m/sec in metric system or feet/sec in FPS system.

Equation is not dimensionally homogeneous **(Use consistent units on both sides).**

R = Hydraulic radius in m or ft.

= A/P, A = Water area i.e. area of flow in sq. m. or sq.ft.

P = Wetted Perimeter in m or ft.

S = **Slope of the energy line** (when flow is uniform energy slope gradient may become parallel to the water surface slope and bed of the channel). This is a very common assumption in the design of canals, where flow is assumed as uniform. But in actual practice uniform flow rarely exists. Therefore, after the design of channel, actual water surface profile should be computed, energy grade line plotted, and flow phenomena again checked for different discharges likely to occur say, quarter, half, three fourth and full discharge. This will give actual water surface profile, which will be useful in locating level of outlet, operation and maintenance.

Alternately it is energy (head) loss per unit length, i.e. $S = h_f\ /l$

N = Coefficient of roughness, specifically known as Manning's N, (see Para 2.0 for details.

Owing to its simplicity and satisfactory results, it is most widely used for all open channel flow computations and design.

Several research workers have developed many formulae for the flow through canals and have remained in vogue in different times in different parts of the world. But now Manning's formula is widely used through out the world.

For design of channels another continuity equation (2) is used

$Q = A_1\ V_1 = A_2\ V_2$ ... Eq. (1.2)

Thus equation (1) and (2) are used together and a section is determined by trial and error.

To avoid any trial and error, and to find a unique solution, Suryavanshi (1973) has developed a master equation (4), see para 4.3. For best sections a unique relation is established in para 1.6 and unique solution are presented in para 2.2 and chapter 3, 4 and 5.

1.4.1 Influence of Shape (Form) on the Hydraulic Radius

For a given cross sectional area, the hydraulic radius (R) depends on the shape and proportions of the section.

For the same area there can be several shapes and sizes. A few cross sections shown in Figure 1.2 (a), all have equal areas, but a glance shows that the amounts of rubbing surface (this is different then rugosity which refers to roughness, see para 2.0) differ widely, or the hydraulic radii are different. A wide shallow stream, as in (i) or a narrow deep ditch or gorge, as in (ix), offers great resistance to flow, shape (ix) even more than shape (i). Compact areas of cross section, such as (v) and (vii), offer least resistance. An index of such different conditions is furnished by the value of the hydraulic radius. Dimensionally the hydraulic radius R, is a length, being the cross-sectional area divided by the wetted perimeter.

It is thus obviously the number of units of cross-sectional area of the stream per unit of wetted (rubbing) perimeter. It follows that in a wide stream such as case (i) in Fig. 1.2 (a), where most of the rubbing is at the bottom the hydraulic radius is practically equal to the average depth of the water.

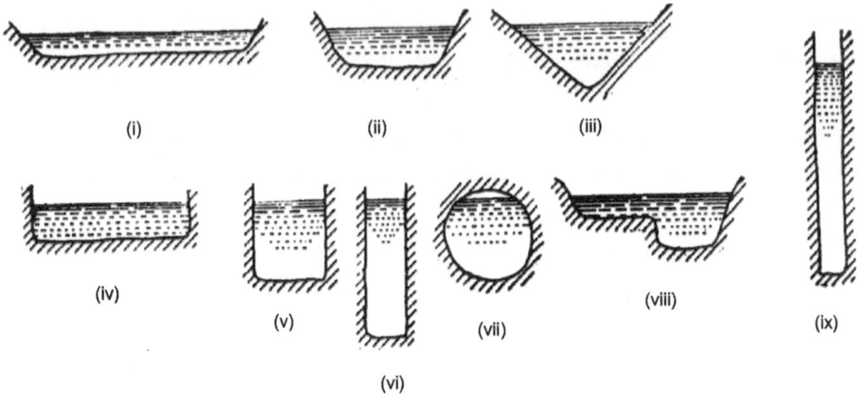

(a) Typical examples of equal cross section area flow.

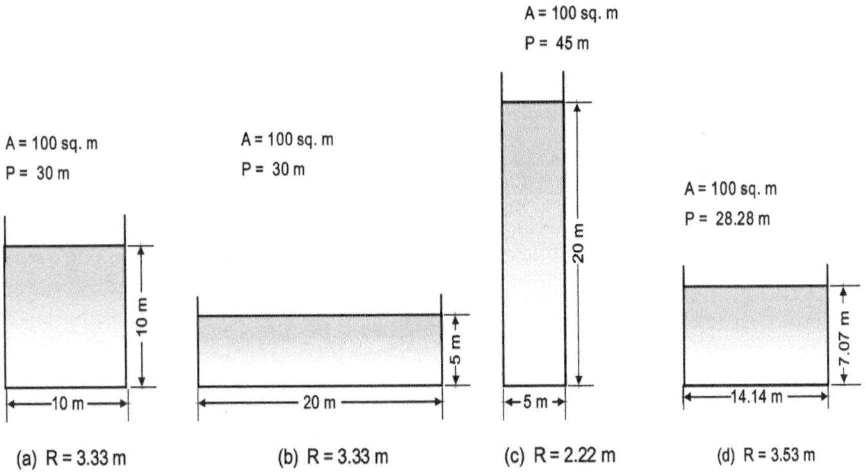

(b) Typical examples of equal area of rectangular sections.

Figure 1.2 : Effect of geometric proportions on hydraulic radius.

Notes:

1. In all above figures, though area of cross- section of flow is same, still all section have different hydraulic radius (R) and consequently different discharges for the same bed slope and rugosity coefficient.
2. R is also influenced by side slope Z, see para 3.3. There are other factors effecting R. These are more explicit in Manning's Master equation, see para 4.3.
3. In a long river all above shapes can be seen, still discharge may be same. This is explained by the energy slope, rugosity coefficient and velocity, which all are changing in a long length.

In fact for most rivers in plains, there are little banks or no steep banks at the water's edge for ordinary stages of the river and the entire rubbing surface is at the bottom as in (i). Therefore for wide channels (many times wider than depth), R = depth (approximately). Hence the alternative expression of hydraulic mean depth as a substitute for hydraulic radius is often used. Note that the value of the hydraulic radius is always less than the depth of water at the deepest part of the cross section.

From the basic formula Eq. (1.1), para 1.4, slope being equal to head loss (h_f) per unit length, it can be written as h_f = S.L = $LV^2 N^2 / R^{2/3}$ $L\left(\frac{P}{A}\right)^{2/3} V^2N^2$. **That means smaller the ratio of the wetted perimeter to the area of cross section of the flowing stream the less is the loss of head by friction; or, reversing the formula, the greater the ratio of the cross-sectional area to the wetted perimeter the greater is the velocity of the water.**

This latter form of the statement is condensed in hydraulics by stating that, slope and roughness remaining the same, the greater the hydraulic radius (or the hydraulic mean depth) the greater is the velocity ; and this is symbolized in the Manning's and other exponential formulas.

As a typical example, consider a rectangular section of a given area 100 sq. meters. It can have several shapes. Four typical shapes are shown in figure 1.2 (b) having the same area, but the hydraulic radius (A/ P) works out different in each case.

Out of the 4 shapes shown above, maximum value of R is 59 per cent in excess of the least value, i.e. as shown in (d) and width is equal to twice the depth.

Velocity, varies as the two - thirds power of R. Hence, for the same cross - sectional area, slope and rugosity coefficient (N), the velocity and discharge will be greatest for that section which has the least wetted perimeter or maximum R. Maximum R is only at d / 2 and it decreases with either increase or decrease in bed width or depth. Also see para 1.6 for most efficient section.

1.4.1.1 Effect of Geometric Proportions on the Hydraulic Radius

Behaviour of R with respect to very wide and shallow channels can be appreciated in figure 1.3, where R versus depth curves are shown.

Consider a rectangular channel of width B, and let depth D increase from very small to very high. Here hydraulic radius R = B D / (B+2D) and it is immediately evident that for D = 0, R =0, and for $D \rightarrow \infty$, $R \rightarrow B/2$, therefore variation of R with D is as shown in figure 1.3 (a).

Now in the same ractangular channel let the width increase for a constant depth. It can be noted that for B = 0, R = 0, and for $B \rightarrow \infty$, $R \rightarrow D$; thus the variation of R with B is as shown in figure 1.3 (b).

For non rectangular sections, the above approximation is a valid if the section is wide and shallow. For a typical trapizoidal section the relation ship as shown in figure 1.3 (c). It has no asymptote since, dR/dD does not approach 0 as D approaches infinity.

Thus R can be maximum up to D (depth of channel) for very wide channels, such as very shallow rivers. In such channels P = (B + 2D) = B, D being very small, can be ignored) and R = A/P = B.D/B = D. These channels are not hydraulically efficient as R is very small because D is small. This is so generally in flat slope Alluvial Rivers. In low stages of river, velocity is very low, and velocity increases in the same river in high stages, (because R increases)

Minimum R can be near to zero (but not zero) for very deep channels, this is so in rocky rivers. In such channels P = B + 2D = 2D (B can ignored being very small) and R = A/P = BD /2D = B/2. Here again R is very small because B is small or say negligible. These channels are also not hydraulically efficient.

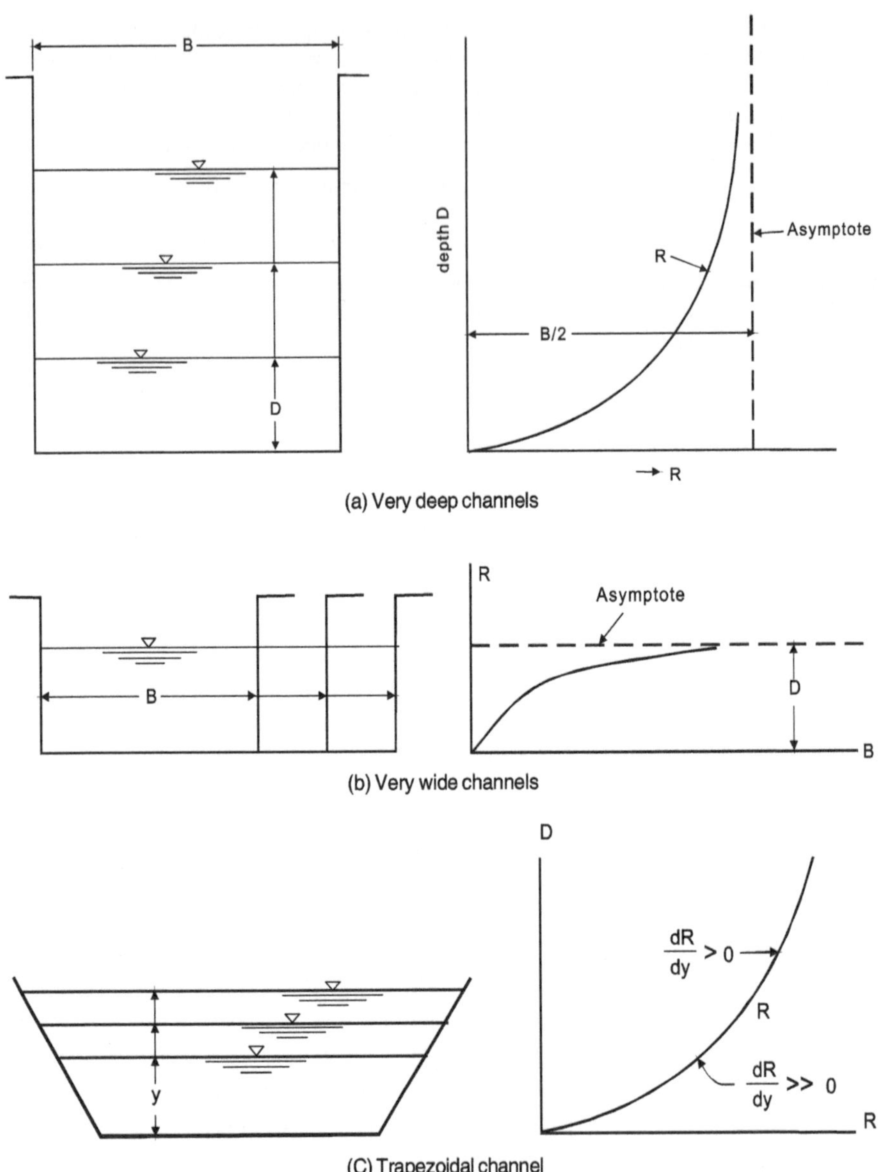

(a) Very deep channels

(b) Very wide channels

(C) Trapezoidal channel

Figure 1.3 Curves of Hydraulic Radius for rectangular and trapezoidal shapes.

Maximum efficiency occurs only at R = d/2. For best section R should be maximum i.e. as much as posible near to d / 2. It holds true both for lined and unlined canals. When R is greater than d/2. (See also para 1.6), the section is no longer most efficient.

For the full circle or semicircle, R = d/4 and, for y = 0, R = 0; thus there is a continous variation of R with y, with maximum value at d/2 as shown in figure 1.4 . Of more engineering significance in such problems is the variation of $AR^{2/3}$ with y since this combination of A and R appears in the Manning's equation. With $AR^{2/3}$ plotted against y (see Figs. 1.4 and ...), another maximum point is noted near top (before top), also see para 4.3. From this it may be concluded that for given S, and n there is a point at which the (uniform) flow rate is also maximum. Engineers usually disregard this in channel design, but it sometimes helps to explain phenomena which occur when channels of convergent walls flow nearly full.

More details of inter relationship of V, R, Q, A, b, d, N, and S are disscussed in chapter 4 and 5.

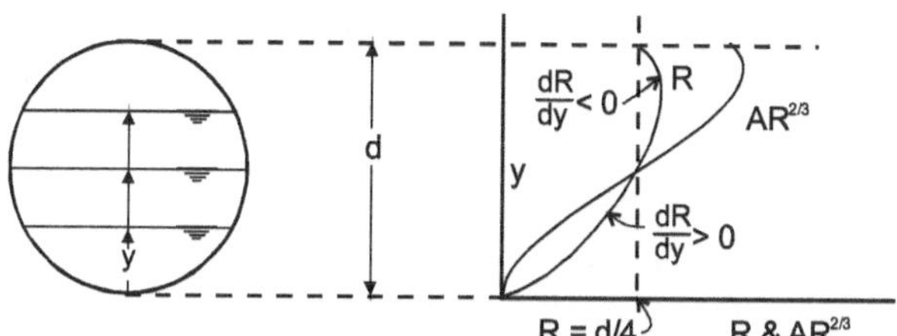

Figure 1.4 Hydraulic radius curve for circular channel

1.4.2 Oddities and Paradoxes of Open Channels

A cross section like (ix) in Fig. 1.2 (a), consists in flow effect of two streams side by side with the same slope, but with the water flowing slower in the shallow portion, a matter of common observation. Such a case should be treated in computations not as a simple single stream, but as a combination of two streams side by side with different velocities and hydraulic radii, but with the same slope. The wetted perimeters include only the contact with the bottom and sides, not water rubbing against water. Also see para 2.1 for more details. More oddities and paradoxes of open channels with the role of R are explained in chapter (5), see para 5.5 and table 5.4.

1.4.3 Laws of Variation

1. *For the Discharge* : when energy Slope and Roughness are constant.

   (a) If the shape of the cross section is fixed but not the size, then R varies directly with d, where d is any linear dimension such as depth or width. Consequently, using the Manning formula, Q varies as $R^{8/3}$ or $R^{2.67}$

   (b) For rivers, many times wider than deepth, R equals d (nearly) where d is the average depth. If the width of the river is nearly constant at different stages then by the Manning's formula Q varies as $R^{5/3}$ or $R^{1.67}$ or $d^{5/3}$.

   (c) For the extreme case of a very narrow channel with depth of water several times the width, R remains nearly constant as the depth changes and Q varies as d.

2. For the Energy Slope when the Discharge and Roughness are Constant.

   (a) When the shape of cross section is fixed then S varies inversely as $R^{1.33}$ or $D^{1.33}$ $\left(\text{i.e.} S \propto \frac{1}{R^{1.33}}\right)$ by the Manning's formula (as for long pipes).

   (b) For wide rivers by the Manning's formula S varies inversely as $R^{3.33}$ or $d^{3.33}$. This law is of great importance in dealing with the deepening of rivers or canals either by dredging or by backing the water up by building a dam downstream.

1.4.4 Open-channel Problems

Among common problems confronting the engineers and their solutions are:

	Given	To find	Solution
I	Q, shape and size of cross section, and N i.e. Q, b, d Z and N	Necessary slope,S	Simple, use Manning's equation. But for complex problems, trials are required, see chapter 4
II	Q, shape but not size of cross section, S, and N (Sometimes the width is known but not the depth, Or vice versa.) i.e. Q, b or d, Z, N, S	Necessary size i. e. b or d	See chapter 3, 4 and 5
III	Shape and size of cross section, S and N i.e. b, d, z , s and N	V and Q	do
IV	Field measurements of R, V and S i.e b, d, z, v, and S	N and Q	do

1.5 Important Terminology (Definitions)

i) Hydraulic Efficient Section or Best Section

Hydraulic efficient section or Best section is a given geometrical shape or section with minimum perimeter through which the given discharge passes (see figure 1.6). All best sections have hydraulic radius as half of depth i. e. R = D/2. But all best sections are not most efficient.

ii) Most Efficient Section (or Optimum Best Section)

Semi circle is the only most efficient section, i. e. optimum best section. All other sections circumscribing the circle are best sections (i.e. hydraulic efficient section, according to that shape), but have hydraulic efficiency smaller than the semi circle.

iii) Optimum (or Most Best) Trapezoidal Section

Trapezoidal section with side slope of 1:0.577 (V : H ; i. e.60° angle with base) is the only most best trapezoidal shape. It is half of a hexagen. All other side slopes have hydraulic efficiency smaller than this. Still the sections are circum- scribing the semi circle and called best section, their relative hydraulic efficiency is indicated in table 3.7. Due to steep side slopes it is generally not adopted and slightly flatter slopes are adopted.

iv) Hydraulic Efficiency

It is the ratio of discharge which will pass in the cross-section of given area to the discharge that will pass in the semi circle of the same area.

Sometimes a better comparison is the area and perimeter of the section to that of semi circle for the same given discharge. It is more convenient and most generally used.

(v) Optimum or Most Economical and Practical Section

It is the section, which has the least cost of land + earth work + lining (if applicable) of the overall canal section including free board, service road, berms, filter toe etc. This section is to be adopted for all practical purposes.

Generally best section result in most economical section in almost all discharge, soil and topographical conditions. However in deep cuttings and heavy fillings a slightly narrow section will be economical. Wider sections are in no way economical, but become essential due to the limitation of depth, and steeper slopes. Also see chapter 6, 7 8 and 9 and table 9.3.

1.5.1 Best Hydraulic Sections

From the Manning's formula it can be known that the discharging capacity (conveyance) of a channel section increases with the increase in hydraulic radius or/and decrease in wetted perimeter.

Therefore, from hydraulic viewpoint, the **channel section having the least wetted perimeter for a given area has the maximum conveyance. Such a section is known as the best hydraulic section.** The geometric elements of the various best hydraulic sections are given in table 1.2 and details are given in Para 1.6.

Table 1.2 : Best hydraulic sections and their comparison for a given area or depth (Arranged in order of their hydraulic efficiency)

S. No.	Section with geometrical proportions	Area(A)	Wetted perimeter (P) in terms Y& A	Hydraulic Radius R=A/P	Hydraulic Conveyance $AR^{2/3}$
1	Semi Circle 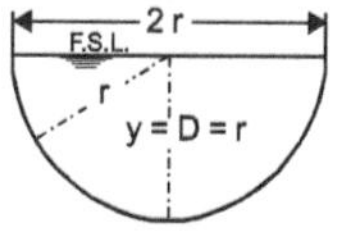	$1.57y^2$	(πy) or $3.14\,y$ or $\sqrt{6.28A}$	$0.5\,y$	$0.987\,y^{8/3}$
2	Mehboob Section 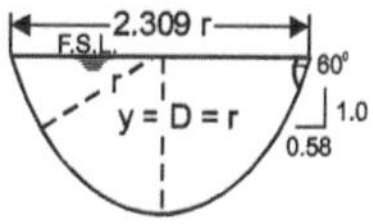	$1.625\,y^2$ (103.5%)*	$3.25y$ (103.5%)*	$y/2$	$1.0237\,y^{8/3}$ (103.72%)*
3	Trapezoidal (Half of Hexagon) 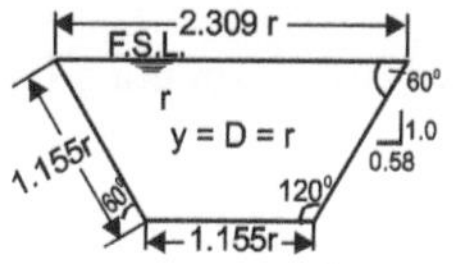	$y^2\sqrt{3}$ or $1.732\,y^2$ (110.3%)*	$2y\sqrt{3}$ or $3.464\,y$ (110.3%)* or $\sqrt{6.93A}$	$0.5\,y$	$1.088\,y^{8/3}$ (110.2%)*
4	Rectangle (Half of a Square) 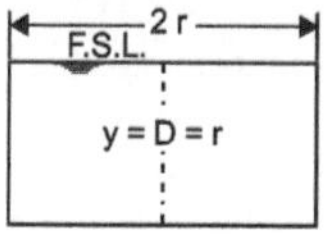	$2y^2$ (127.3%)	$4\,y$ (127.3%)* or $\sqrt{8A}$	$0.5y$	$1.257\,y^{8/3}$ (127.3%)*
5	Triangle (Half of a Square) 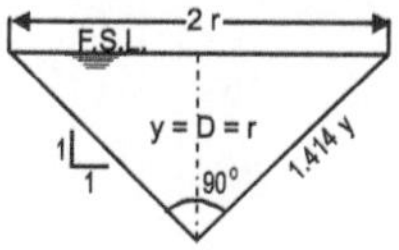	y^2 (63.6%)*	$2\sqrt{2}y$ or $2.829\,y$ (90%)* or $\sqrt{8A}$	$0.3537\,y$ (70.7%)*	$0.5\,Y^{8/3}$ (50.6%)*

Note : * Percentage is with reference to semi Circular Section at serial No. 1 in above table.

The semi circle has the least perimeter among all sections for the same area and hence it is the most hydraulically efficient of all sections. Other sections are listed in the order of their hydraulic efficiency.

All the sections listed above may not always be practical owing to difficulties in construction and availability of materials. In general a channel section should be designed for the best possible hydraulic efficiency with due modifications for practicability and economics of excavation (earthwork / rock cutting) and lining.

1.6 Conditions for Maximum Discharge through a Channel (Hydraulic Efficient Channels)

A channel which gives maximum discharge for a given cross sectional area and slope may be defined as a channel which has a minimum wetted perimeters so that there is a minimum resistance to flow and thus resulting in a maximum discharge. This is also the channel of most economical cross-section or also known as **Best Hydraulic Section.** From the above definitions, it is obvious that while deriving the condition for a channel of most economical cross sections the cross sectional area is assumed to be constant and a relation between depth and breadth of the section is found to give maximum discharge.

A) Rectangular Sections

Consider channels of rectangular section as shown in figure given below.

Area of flow $A = b \cdot d$

and discharge $Q = A \cdot V$

$= AR^{2/3} S^{1/2} / n$

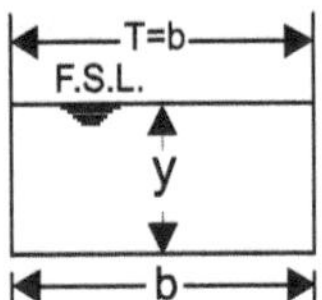

Where,

R = Hydraulic radius

S = Slope of canal

V = Velocity of flow

n = Rugocity coefficient.

P = Wetted Perimeter.

Keeping A, and S constant in the above equation, the discharge will be maximum when R is maximum or wetted perimeter P is minimum as $R = A/P$

For a rectangular channel, we know that

$P = b + 2d, \; = A / d + 2d$

In this equation P will be minimum when $dP / dd = 0$

Differentiating the above equation with respect to d and equating the same to zero i.e.

$dP / dd = - A / d^2 + 2 = 0$

$\therefore A = 2d^2$ or $b \cdot d = 2d^2$

or $b = 2d$ ... (i)

i.e. the bed width is double the depth.

In this case, the hydraulic mean depth, or hydraulic radius

$R = A/p = bd / (b+2d) = 2d \cdot d / (2d+2d)$ (since $b = 2d$)

$= 2d^2 / 4d = d/2$ .... (ii)

Hence for the maximum discharge or maximum velocity, these two conditions (i.e. b = 2d and R = d/2) should be satisfied. This section is circumscribing a semi circular, see figure 1.6.

B) Trapezoidal Section

Consider a channel of trapezoidal cross section ABCD with side slope Z:1 as shown in figure given below.

Area of flow $A = bd + 1 / 2 \cdot 2 \cdot Zd \cdot d$

$= d\,(b + Zd)$ ... (i)

or $A/d = b + Zd$ or $b = A/d - Zd$ .... (i) a

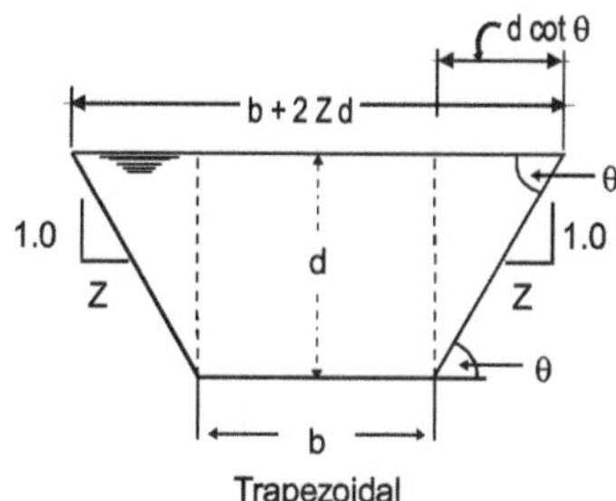

Trapezoidal

and discharge $Q = A \times V = A\ .1/N\ .R^{2/3} S^{1/2}$

$= A\ (A/P)^{2/3}\ S^{1/2} / N$

$= A^{5/3} S^{1/2} / P^{2/3} N$ ... Eq. 2(i)

Keeping A and S constant, in the above equation, the discharge will be maximum when A/P is maximum or the perimeter P is minimum.

We know that $P = b + 2\ \sqrt{d^2 + Z^2 d^2} = b + 2d\ \sqrt{(Z^2 + 1)}$

Substituting the value of b from equation (i)

$P = A/d - Zd + 2d\ \sqrt{(Z^2 + 1)}$

In this equation P is minimum when $dP/dd = 0$.

Differentiating the above equation with respect to d and equating the same to zero

$dP/dd = -A/d^2 - Z + 2\ \sqrt{(Z^2 + 1)}$

or $-A/d^2 - Z + 2\ \sqrt{(Z^2 + 1)} = 0$

or $+A/d^2 + Z = 2\ \sqrt{(Z^2 + 1)}$

Substituting the values of 'A' as $A = d\ (b + Zd)$ We get,

$d\ (b + Zd)\ /\ d^2 + Z = \sqrt{(Z^2 + 1)}$

or $(b+Zd)\ /\ d + Z = 2\ \sqrt{(Z^2 + 1)}$

or $(b+Zd) + Zd = 2\,d\ \sqrt{(Z^2 + 1)}$

or $(b+2\,Zd)/2 = d\ \sqrt{(Z^2 + 1)}$ ... (ii)

i.e. the sloping side is equal to half of the top width.

or $b = 2d\sqrt{Z^2 + 1} - 2\,Zd$ or $b/d = 2\,(\sqrt{Z^2 + 1} - Z)$ ... (ii)a

Thus b/d ratio is fixed

As $P = b + 2d\ \sqrt{(Z^2 + 1)}$ and $A = d\ (b + Zd)$

$R = A/P = d\,(b+Zd) \,/\, [b+2d\sqrt{(Z^2+1)}\,]$

Substituting in the above equations, the value of 2d $\sqrt{(Z^2+1)}$ from Eq. (ii), we get.

$R = d(b+Zd) \,/\, (b+b+2\,Zd) = d\,(b+Zd) \,/\, 2(b+Zd) = d/2$ ... (iii)

Hence for maximum discharge or maximum velocity these two conditions i.e. a fixed $b/d = 2\,(\sqrt{Z^2+1} - Z)$ ***and R = d/2 should be satisfied. This section is also circumscribing the circle. b/d is fixed by side slope Z ; R = d/2 is true for any side slopes as it is independent of side slopes (Z).***

C) Optimum Side Slopes

For optimum side slopes, that makes the area least or most optimum;

Substituting the value of b in the equation for A and P, and solving simultaneously for P, we get P in terms of A , as

$$A = d\,(b + Zd) = d\left[Zd\left(\sqrt{Z^2+1}-Z\right)+Zd\right]$$

$$= d^2\left[2\left(\sqrt{Z^2+1}-Z\right)+Z\right] = d^2\,2\left(\sqrt{Z^2+1}-Z\right)$$

$$\text{or } d = \sqrt{A/2\left(\sqrt{Z^2+1}-Z\right)} \qquad \text{or } d = \frac{\sqrt{A}}{\sqrt{2\left(\sqrt{Z^2+1)}-Z\right)}}$$

$$P = b+2d\,\sqrt{(Z^2+1)} = 2d\left[\sqrt{Z^2+1}-Z\right]+2d\sqrt{Z^2+1}$$

$$= 2d\sqrt{Z^2+1}-2\,Zd+2d\sqrt{Z^2+1} = 4d\sqrt{Z^2+1}-2Zd$$

$$= 2d\left[2\sqrt{Z^2+1}-Z\right] = \frac{2\left(2\sqrt{Z^2+1}-Z\right)\sqrt{A}}{2\sqrt{Z^2+1}-Z}$$

$$= 2\sqrt{\left(2\sqrt{Z^2+1}-Z\right)}\,\sqrt{A}$$

$$P = 2\sqrt{A(2\sqrt{(1+Z^2)} - Z)}$$

Now to find the value of Z that makes P the least, differentiate P with respect to Z and equating dP/dZ equal to zero and solving for Z we get

$$Z = \frac{\sqrt{3}}{3} = \tan 30^\circ = 1/\sqrt{3}$$

This is the most optimum side slope. **This section is half of a hexagon. This section is also circumscribing the circle, see figure 1.6. This R is equal to d/2 the optimum maximum value. It cannot be more than d/2 in any case.**

D) Optimum Values of b and A in Terms of d

Putting the above value of Z in (ii)a, bed width of best section is

or $b=2d\left(\sqrt{1/3+1}-1/\sqrt{3}\right)$

$$b = 2d\left(\frac{2}{\sqrt{3}}-\frac{1}{\sqrt{3}}\right) = 2d/\sqrt{3}=1.155d$$

Putting this value of b and Z in Eq. (i), area of cross section of best section is

$$A = d\left(2d/\sqrt{3}+d/\sqrt{3}\right)$$

$$= d(2d+d)/\sqrt{3} \quad = 3d^2/\sqrt{3} \quad = \sqrt{3}\,d^2 \; = \mathbf{1.732\ d^2}$$

Similarly, $P= b+2d\sqrt{(Z^2+1)} = \frac{2d}{\sqrt{3}} + 2d\sqrt{1/3+1}$

$$= \frac{2d}{\sqrt{3}}+2d\frac{2}{\sqrt{3}} = \frac{2d+4d}{\sqrt{3}} = \frac{6d}{\sqrt{3}}$$

or $P = 2\sqrt{3}\,d \; = \mathbf{3.464\ d}$

The above values of b, A, and P are the most optimum (best) and unique value for optimum side slope. For any other side slopes, b, A and P changes, but there is unique relation ship for efficient hydraulic section, between A, b, d, Z , and P and that can not be altered. See table 1.3 and 3.2.

1.6.1 Some useful derivation of Manning's formula for best sections

It can be stated that, for the efficient section of any shape of more then 2 sides the hydraulic radius will be one-half of the flow depth and the section must be circumscribing a circle. It cannot be inscribed in a semicircle, see Para 1.10

The efficient polygonal section of any specified number of sides can be found to be one which can be circumscribing a semicircle.

The above best/ optimum /or unique values of b, A and P, R have no meaning unless until put in the Manning's equation.

When the value of R as d/2 is put in the Manning's formula and the relationship is a unique relationship as under

$$V = (d/2)^{2/3}(S)^{1/2}/N$$

$$= (1/2)^{2/3}d^{2/3}S^{1/2}/N = 0.63d^{2/3}S^{1/2}/N \qquad \text{... Eq. 1.3}$$

or $S = (VN/0.63d^{2/3})^2 = V^2N^2/0.3969\,d^{4/3}$

This may be called as the Master Equation (1) of Manning's formula for the best sections (or efficient hydraulic section).

For any type of BEST section R = d/2 only and always only, irrespective of any shape, side slope and size. The above equation shows a direct and unique relationship between V, d and S, treating N as constant. For a given velocity or treating V as constant, there is unique depth slope only, and can not be altered. This perimeter is minimum, and this area is also minimum, and b/d ratio is also fixed, and can not be altered, thus no trial and error is required.

This equation can also be written in the form

$$Q = V.\ A = A\ (0.63\ d^{2/3}\ S^{1/2}\ /\ N)$$

(A) Optimum best trapezoidal section

Putting optimum value of A for trapezoidal section of optimum side slope, Z = 0.577

$$Q = \sqrt{3}\,d^2\ (0.63d^{2/3})\ S^{1/2}\ /N$$

$$= 1.732(0.63)\ d^{8/3}\ S^{1/2}\ /\ N \quad = 1.091\ d^{8/3}\ S^{1/2}\ /N$$

$$\text{or } d = \left(\frac{QN}{1.091\ S^{1/2}}\right)^{3/8}$$ - (only for optimum best trapezoidal section i.e. Z = 0.577:1)

Treating Q and N as given or constant, there is a unique relation ship between depth and slope and cannot be altered. For any other side slope there is also a unique relation, see table 2.4. This concept is used in this book for the design of best section.

For a given discharge and bed slope, there is a fixed depth and cannot be changed. Similarly for a given discharge and depth there is fixed slope or minimum slope, and canal cannot be designed on any slope flatter then this, without changing b and d.

Mehboob section and best trapezoidal sections are dealt in chapter III. These sections for high capacity have the problems of large depth for flat slopes.

Large depth again poses the problem of stability of slopes, safety of canal Bank etc. and for only that purpose wide channels are required, and dealt in chapter 4.

A general Master equation and useful derivations of Manning's formula are given in chapter IV.

Other Side Slopes

Any other side slope either flatter or steeper will decrease efficiency; for a given discharge and not for depth see also Para 3.3 and 3.7. Also that section will not be half of a hexagon, see table 3.7

1.6.2 Triangular Section

This section is not in common practice, yet for this case

$$\frac{dP}{dZ} = \frac{d(2d\sqrt{1+Z^2}}{dZ} = \frac{d(2\sqrt{A/Z}\,\sqrt{1+Z^2}}{dZ}$$

$$= 2\sqrt{A}\left[\frac{1}{2\sqrt{(1+Z^2)\ /\ Z}}\left(-\frac{1}{Z^2}+1\right)\right] = 0$$

The corresponding surface width is 2d, so this section can be inscribed in a semicircle. Since the area A is equal to Zd^2 and the perimeter $P = 2d\sqrt{1+Z^2}$ the hydraulic radius is

$$R = \frac{A}{P} = \frac{Zd^2}{2d\sqrt{1+Z^2}} = \frac{d}{2\sqrt{2}}$$

This is not equal to D/2.

Though this section may be called as best triangular section, but is not as optimum as trapezoidal or rectangular. In fact the trapezoidal is a truncated triangle, but circumscribing a circle.

1.6.3 Alternative Approach

Alternatively for any side slopes the same analysis can be done slightly in different way as under. Results are same as above.

The area of cross-section in term of θ, i.e. the angle the sides make with the horizontal can be expressed as

$A = bd + d^2 \cot\theta$ or $b = \frac{A}{d} - d\cot\theta$

And the wetted perimeter is $P = b + 2d \operatorname{cosec}\theta$

Substituting for b in above Equation we get

$$P = \frac{A}{d} - d\cot\theta + 2d\operatorname{cosec}\theta$$

For a given area A, for P to be minimum dp/d (d) = 0 i.e. differentiating above equation with respect to d, treating area to be constant, we get

$$\frac{d(P)}{d(d)} = -\frac{A}{d^2} - \cot\theta + 2\operatorname{cosec}\theta = 0 \quad \text{... (D)}$$

$$\text{or} \quad A = 2d^2 \operatorname{cosec}\theta - d^2 \cot\theta \quad \text{... (E)}$$

Finally, solving for b, the relation comes out as

$$b = 2d(\operatorname{cosec}\theta - \cot\theta) = 2d\left(\frac{1}{\sin\theta} - \frac{\cos\theta}{\sin\theta}\right) = 2d\left(\frac{1-\cos\theta}{\sin\theta}\right)$$

Or b/d = 2 tan θ/2

By making use of this relation and eliminating b, three convenient equations as under are obtained

$$A = \frac{2d^2}{\sin\theta}\left(1 - \frac{\cos\theta}{2}\right) \quad \text{Or} \quad d = \sqrt{\frac{A\sin\theta}{2-\cos\theta}} \quad \text{... Eq.}$$

$$P = \frac{4d}{\sin\theta}\left(1 - \frac{\cos\theta}{2}\right) \text{ and } R = d/2$$

The values of b, A, and P as functions of d and θ can be plotted as in figure 1.5. The graph indicates that of all trapezoidal sections that one having θ = 60° has a smaller wetted perimeter than any other.

The above expression and graph are very useful for any side slopes and are used by many authors in their research works.

The above methods can also be used by adopting different side slopes of the canal on either side of the central line. Different side slopes become essential in typical site conditions, see chapter 7.

However in this text ready reckoning tables for equal side slopes on both sides are given in chapter 3 and 4.

Based on above area formula, all the dimensions i.e. d, b and P etc of best sections as a function of A, are also given in table 1.3. These can be easily used either for a given area or given velocity as A = Q / V.

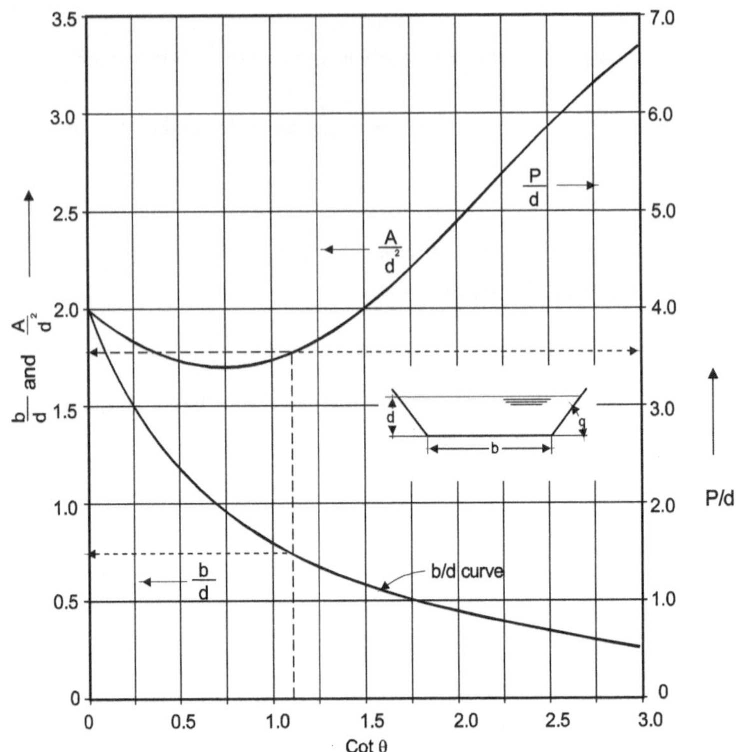

Figure 1.5 The section with the least wetted perimeter

Table 1.3 : Parameters of Best Sections (Circular, Rectangular and Trapezoidal) in Terms of Cross Sectional Area (A)

Side Slope V : H	Angle of slope with horizontal θ	Depth Y or D	Bottom width B	Top Width at water surface	Wetted perimeter P
Semicircular	....	$0.798\sqrt{A}$	....	$1.596\sqrt{A}$	$2.506\sqrt{A}$
Rectangular	900 00	$0.707\sqrt{A}$	$1.414\sqrt{A}$	$1.414\sqrt{A}$	$2.828\sqrt{A}$
1 : 0.5	630 26	$0.759\sqrt{A}$	$0.938\sqrt{A}$	$1.697\sqrt{A}$	$2.645\sqrt{A}$
1 : 1	450 00	$0.740\sqrt{A}$	$0.613\sqrt{A}$	$2.093\sqrt{A}$	$2.705\sqrt{A}$
1 : 1.25	380 40	$0.716\sqrt{A}$	$0.503\sqrt{A}$	$2.293\sqrt{A}$	$2.795\sqrt{A}$
1 : 1.5	330 41	$0.689\sqrt{A}$	$0.418\sqrt{A}$	$2.485\sqrt{A}$	$2.904\sqrt{A}$
1 : 2	260 34	$0.636\sqrt{A}$	$0.300\sqrt{A}$	$2.844\sqrt{A}$	$3.144\sqrt{A}$

1.6.4 Other Methods

Many other authors have tried and found the economical or best section or optimized section solution without trial and error in different ways.

Some of them are Asthana and others (1966), Madhaviani and Ali (1967), Modi and Seth (1968), Deb (1971), Suryavanshi (1972). Swamee et.al (1972), Arya et. al (1995), Atmapoojya (2001).

Recently Chalisgaonkar et al (2002) have given graphical charts for trapezoidal channels with curved ends for different given conditions of discharge, side slopes, rugosity coefficients, bed slopes or depth or B / D ratio and may be very useful to designers. These are based on Master equation of Manning's formula given by Suryavanshi (1972).

Some of above authors have given nomogramical solutions, and some have given graphical solutions. But their hydraulic or economic efficiency is generally not indicated or not known. Also their practical adaptability and suitability to various site and topographical conditions is not given or discussed.

Here in this text tabular unique solution with clarity of the inter relationships of all seven parameters (B, D, Z, V, Q, N and S) is presented with hydraulic efficiency. A comparison of best sections are presented in Para 1.7 to 1.11. The best trapezoidal and Mehboob sections have been found feasible for discharges upto 50 to 80 cumecs. Beyond it wide channels have to be preferred / adopted for limiting depth and slopes (generally flat or mild), see chapter 4. The tabular solution gives a clear open picture of the analysis and dimensional parameters such as top width, length of curved portion, side slope etc in readily available form often, required by construction and operational engineers

Also Master equation of Manning's formula given by Suryavanshi (1972) is given in detail in Chapter 4.

1.7 Comparison of Best Hydraulic Sections (Trapezoidal, Rectangular and Mehboob) with Semi Circle

The geometric and hydraulic properties of best hydraulic sections are given in table 1.2. A comparison of these sections with semi circle is also indicated in the same table in terms of percentages. A good comparison can easily be appreciated in figure 1.6, where these sections are superimposed on each other.

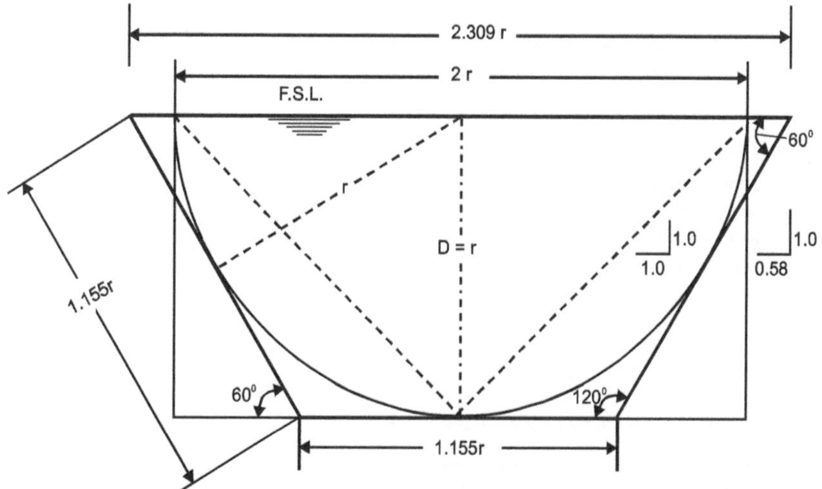

Figure 1.6 Comparison of Best Hydraulic Sections (Semi Circle, Mehboob, Trapezoidal & Rectangular Sections); all best sections are circumscribing the circle

(*actual value is 0.5777 and for easiness it is rounded to 0.58 and same is written at most places in the text)

1. The **best trapezoidal section is one which is circumscribing a semi circle. Compared to the semi circle it has a cross sectional area of 10.3% more, while the perimeter is also 10.3% more,** but the hydraulic radius R is same as that of a semi circle (r/2 or d/2) The hydraulic conveyance **(AR $^{2/3}$) is also 10.3%** more. Thus **it appears as efficient as a semi circle** (But **only as long as hydraulic radius R is equivalent to that of a semi circle).** But this is not a true comparison for a given discharge (see para 1.8)

2. The **best rectangular section** is also the one which **is circumscribing a semi circle.** Compared to the semi circle, **it has an area of 27.3% more**, while the **perimeter is also 27.3% more,** but the hydraulic radius R remains constant as r/2 or d/2. The hydraulic conveyance **($A/R^{2/3}$) is also 27.3% more.** Thus for 27% extra area and extra perimeter, the discharge is also 27% more. Therefore, many times its **hydraulic efficiency is also said to be same as that of a semi circle or** trapezium. But again it is **true only when hydraulic radius R is same that of a semi circle.** Thus it is a false comparison. For true comparison for a given discharge, see para 1.8 .
3. The **best Mehboob section** is also one which **is circumscribing a semi circle**, with tangent sides of the trapezium, (also see Para 3.1 and figure 3.1). It also has the same hydraulic radius **R** as r/2 or d/2 as that of a semi circle, but the **cross sectionals area is 3.5% more** and the **perimeter is also 3.5% more,** as compared to the semi circle. But because, of the hydraulic radius R being same as that of radius of a semi circle. Apparently the hydraulic efficiency or in other words **discharging capacity ($A/R^{2/3}$) is also 3.72% more.** Again for a true comparison for a given discharge, see para 1.8 .
4. Thus **all the three best sections, as long as their geometric proportion are circumscribing a semi circle, can be apparently said to be equally efficient, but not for a given discharge .**
5. In case the depth is increased or decreased then circumscribing a circle, keeping geometrical proportions in the ratio for best hydraulic section, still the section does not remain as efficient as a semi circle. It becomes inefficient in the proportion of either increase or decrease of depth/bed width. This can be well appreciated in the figure 1.9 where best section are inscribed in a semi circle, where depth of flow is less than that in the semi circle. Similar effects can be well appreciated in figure 1.8 where the area is constant or in figure 1.10, where perimeter is constant, and the best hydraulic sections become inefficient as compared to a semi circle.
6. Even the best triangular section is the most inefficient. It has a very high perimeter (90% of semi circle) where as its area is 63.6% of semi circle and its conveyance is only 71.63% of semi circle.

1.8 Comparison of Best Sections for A Given Discharge

Though a comparison of hydraulic efficiency of best sections with semicircle is given above in para 1.7, and all the sections are found practically equally efficient, yet the area of any other section is more than the semicircle. Therefore, for a given discharge, above comparison is not true. Also the comparison of sections for a given area is not true for a given discharge, as the discharge is not directly related to area only, but is also related to the $R^{2/3}$ i.e. $Q \propto AR^{2/3}$ (S and N remaining constant for a particular channel).

A comparison of best sections for a given discharge is shown in figure 1.7 and given in table 1.4 and detailed calculations are also given below.

Table 1.4: Comparison of best sections for a given discharge

Hydraulic properties	Semicircle	Mehboob	Trapezium	Rectangle
(1) Area of flow	**$1.57\ y^2$**	$1.579y^2$	$1.609y^2$	$1.667y^2$
(% of semicircle)	**100**	100.5	102	106
(2) Depth of flow	**y**	0.986y	0.964y	0.913y
(% of semicircle)	**100**	98.4	96.4	91.3
(3) Wetted Perimmeter	**3.14 y**	3.193y	3.34y	3.652y
(% of semicircle)	**100**	101.6	106	116
(4) Hydraulic radius R=A/P	**0.5 y**	0.492y	0.482y	0.456y
(% of semicircle)	**100**	98.4	96.4	91.3
(5) Conveyance $AR^{2/3}$	**$0.987\ y^{8/3}$**	$0.987y^{8/3}$	$0.987y^{8/3}$	$0.987y^{8/3}$
(% of semicircle)	**100**	100	100	100

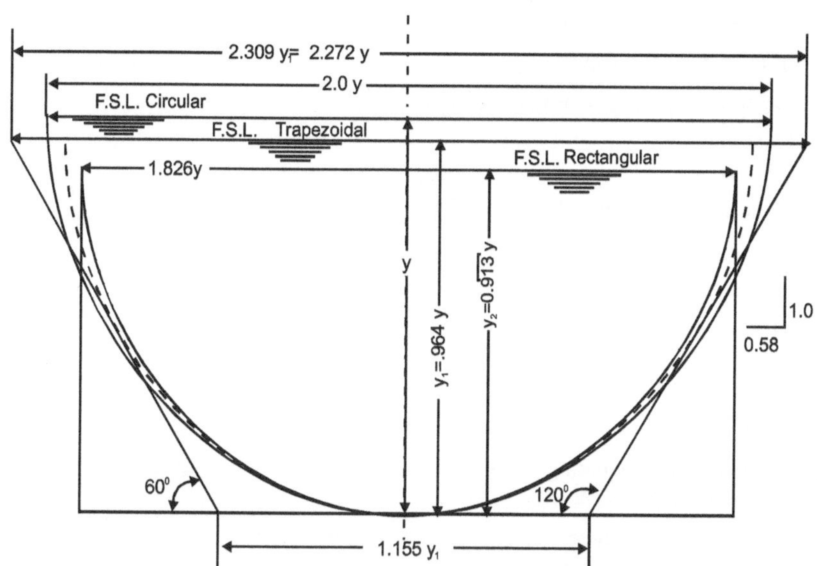

Figure 1.7 Comparison of best sections for a given discharge
Note : For the same FSL bed levels may be different.

The above table shows that for a given discharge, the depth of flow is maximum for semi circle and goes on decreasing with Mehboob, trapezoidal and rectangular sections, as the area is more while circumscribing the circle.

For the given discharge, a Mehboob section has only 0.5 % more area and 2% more perimeter ; the trapezoidal section has 2 % more area and 6% more perimeter, while rectangular section has 6% more area and 16% more perimeter.

Details of calculations

a) Trapezoidal Section

Discharge through semi circle = $0.987\ y^{8/3}\ S^{1/2}/N$ (from table 2.2)

For the same discharge, let y_1 be the depth of flow in the best trapezoidal section, then

Discharge through best trapezoidal section = $1.088\ y_1^{8/3}$

Equating two discharge equations, we get,

$$1.088\ y_1^{8/3} = 0.987\ y^{8/3}$$

or $$y_1^{8/3} = 0.907\ y^{8/3}$$

$$y_1 = (0.907y^{8/3})^{3/8} = 0.964y$$

Wetted Perimeter of Trapezoidal section

$$= 2\sqrt{3}\,y_1 = 2\sqrt{3}\ 3.(0.964y)$$

$$= 3.339y = 3.340y$$

Ratio of this perimeter with that of semi circle perimeter

$$= 3.340y/3.14y$$

$$= 1.063 \text{ or say } 106\%$$

Area of best trapezoid = $\sqrt{3}\,y_1^2$

= $\sqrt{3}$ (0.964y)2 = 1.609 y^2

Ratio of area of Trapezoid with the area of semi circle

= 1.609y^2 / 1.57 y^2

= 1.024 or say 102%

Hydraulic Radius of trapezoid = $y_1/2$ = 0.48 y

Ratio of R with that of a semicircle = (y_1/2) / (y/2) = 0.964 y

= 96.4%

Hence for a given discharge (i.e. the same conveyance) a trapezoidal section has 2% more area where as its perimeter is 6% more than semi circle. Thus it can be said that for the same hydraulic efficiency it is little about 6% costlier.

b) Rectangular Section

For the same discharge equating the conveyance of best rectangular section with the conveyance of semicircle, (from table 1.2) we get

1.257 $y_2^{8/3}$ = 0.987 $y^{8/3}$ (where y_2 is the depth of flow in rectangular section for the same discharge)

y_2 = (0.987 $y^{8/3}$/1.257$)^{3/8}$ = 0.913 y

Wetted perimeter of rectangle = 4y_2

= 4 (0.913 y) = 3.652 y

Ratio of this perimeter with the perimeter of semicircle

= 3.652y/3.14y = 1.162 or 116%.

Area of Rectangle A = 2y_2^2

= 2 x (0.913y$)^2$ = 1.667y^2.

Ratio of this area with the area of semi circle

= 1.667y^2/1.57y^2

= 1.061 or 106%

Ratio of R = (y_2 /2) / (y / 2) = 0.913y/y = 0.913 or 91.3%

Hence for a given discharge a rectangle section has 6% more area whereas its perimeter is 16% more. Thus it can be said that for the same hydraulic efficiency it is costlier by more than 16%.

c) Mehboob Section

For a given discharge equating the conveyance of Mehboob section with the conveyance of semicircle (from table 1.2), we get.

(Conveyance of Mehboob section) = 1.0237 $y_3^{8/3}$ = 0.987 $y^{8/3}$

or y_3 = (0.987 $y^{8/3}$/1.0237$)^{3/8}$ = 0.986 y

Wetted Perimeter of Mehboob section:

P = 3.25y_3

= 3.25 x 0.986 y = 3.204 y

Ratio of this Perimeter with the perimeter of semicircle

= 3.204 y/3.14y = 1.019 or 101.9% or 102%

Area of Mehboob section A = 1.625 y_3^2

= 1.625 x (0.986 y$)^2$ = 1.579 y^2

Ratio of this area with the area of semi circle
$= 1.579y^2/1.57y^2 = 1.005$ or 100.5%

Ratio of $R = y_3/2/y/2 = 0.986\ y/\ y$
$= 0.986$ or 98.6%

Hence for a given discharge a Mehboob Section has 0.5% more area, where as its perimeter is only 2% more. Thus it can be said that for the same hydraulic efficiency it is costlier by little more than 2%.

1.9 Comparison of Best Sections for A Constant Area

A comparison of the best trapezoidal, rectangular and Mehboob section with the semi circle for a given area is shown in the figure 1.8 & Table 1.5

Table 1.5: Comparison of best sections for a given area

Hydraulic properties	Semicircle	Mehboob	Trapezium	Rectangle
(1) Area of flow (% of semicircle)	**$1.57\ y^2$** **100**	$1.57y^2$ 100	$1.57y^2$ 100	$1.57y^2$ 100
(2) Depth of flow (% of semicircle)	**y** **100**	0.983y 98.4	0.953y 95.30	0.886y 88.6
(3) Wetted Perimmeter (% of semicircle)	**3.14 y** **100**	3.193y 101.6	3.297y 105	3.544y 113
(4) Hydraulic radius R=A/P (% of semicircle)	**0.5 y** **100**	0.492y 98.4	0.476y 95	0.443y 88.6
(5) Conveyance $AR^{2/3}$ (% of semicircle)	**$0.987\ y^{8/3}$** **100**	 98.9	 96.6	 92

This table shows that for a given area when compared to a semi circle Mehboob Section is 98.9%, efficient while its perimeter is 1.6% more; A trapezoidal is 96.6% efficient with 5% more perimeter, and a rectangle is only 92% efficient with 13% more perimeter.

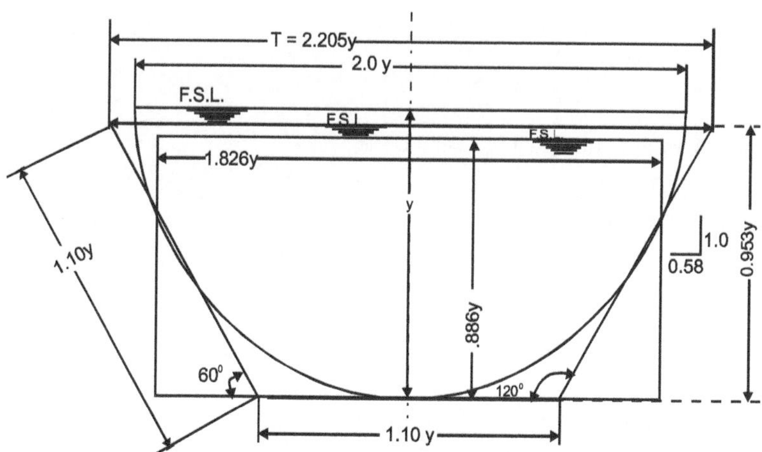

Figure 1.8 : Comparison of best section for a given area

However, this comparison should not be used for a given discharge. For a given discharge i.e. the same hydraulic conveyance as that of semi circle, the percentages are more then as indicated above (see para 1.8). Detailed calculations for given area are explained below.

a) Rectangular Section

For best Rectangular Section B = 2D.

Keeping area A equivalent to that of a semi circle, i.e. equating the two areas.

$A = B \times D = 2D \times D = \pi r^2/2$

or $2D^2 = \pi r^2/2$

or $D = [\pi r^2/4]^{1/2} = ½ r\sqrt{\pi} = 0.886\ r$

Perimeter, $P = B + 2D = 2D + 2D$

$= 4D = 4 \times 0.886\ r = 3.544\ r$

$R = A/P$

$= (\pi r^2/2) / (3.544\ r) = 0.443\ r$

Ratio of P with that of semi circle = 3.544 r/r

= 1.13 or 113%

Ratio of R with that of semi circle = 3.544 r/0.5 r

= 0.0.886 or 88.60%

Ratio of $AR^{2/3}$ with that of a semi circle = $[(\pi r^2/2)(0.443r)^{2/3}] / [(\pi r^2/2)(r/2)^{2/3}]$

$= (0.443 \times 2)^{2/3}$

= 0.92 or 92%

This shows that for a constant area a best rectangular section is only 92% efficient as compared to the semi circle, while its perimeter is 13% more.

b) Trapezoidal Section

For best trapezoidal section

$Z = 0.58$ and $B = 1.1550$

$A = 1.155\ D \times D + ½ \times 2 \times 0.58\ D \times D$

$= 1.155\ D^2 + 0.58\ D^2$

$= 1.735\ D^2$

Keeping area A equivalent to that of a semi circle,

$A = 1.735\ D^2 = \pi r^2/2$

or $D^2 = (\pi r^2/2)(1/1.735) = 0.908\ r^2$

or $D = 0.953\ r$

and $B = 1.155 \times 0.953\ r = 1.10\ r$

Perimeter, $P = 1.155\ d + 2 \sqrt{D^2 + (0.58\ D)^2}$

$= 1.155\ D + 2.31\ D$

$= 3.465\ D = 3.465 \times 0.953\ r = 3.297\ r$

Ratio of P with that of a semi circle = 3.297 r / r

= 1.049 or 105 %

Hydraulic Radius R = A/P

$= (\pi r^2/2)/(3.297\ r) = 0.476\ r$

Ratio of R with that of a semi circle = 0.476 r/0.50 r

= 0.95 or 95%

Ratio of $AR^{2/3}$ with that of a semi circle = $(0.95)^{2/3}$

= 0.966 or 96.6%

This shows that for constant area best trapezoidal section is only 96.6% efficient as compared to the semi circle while its perimeter is 5% more.

c) Mehboob Section

For Mehboob Section area $A = r_1^2\ (\pi/180 + \cot\theta\ )$, *Where* is in degrees, and r_1 is the depth of Mehboob section.

Keeping this area equivalent to that of a semi circle

$A = r_1^2\ (\pi/180 + \cot\theta), \quad = \pi r^2\ /2$

or $r_1^2\ (3.14 \times 60/180 + \cot 60) \quad = \pi r^2\ /2$

or $r_1^2\ (3.14/3 + 0.577) \quad = 3.14\ r^2/2$

or $r_1^2\ (1.624) \quad = 3.14\ r^2/2$

or $r_1^2 \quad = 0.967\ r^2$

or $r_1 \quad = 0.983\ r$

Perimeter P $= 2\ r_1 \cot\ + 2\ r1 \times 2/360$

$= 2\ r_1\ (\cot + /180)$

$= 2 \times 0.983\ r\ (\cot 60 +\ 60/180)$

$= 1.966\ r\ (1.6244) = 3.193\ r$

R = A/P

$= (\pi r^2/2)/(3.193\ r) = 0.492\ r$

Ratio of P with that of semi circle = $3.193\ r/\pi r$

= 1.016 or 101%

Ratio of R with that of semi circle= 0.492 r/0.5 r

= 0.984 or 98.4%

Ratio of $AR^{2/3}$ with that of semi circle

$= (\pi r^2/2)(0.492\ r)^{2/3}\ /(r^2/2)(r\ /2)^{2/3}$

= (0.492)2/3/(1/2)2/3

= 0.6232/0.6299

= 0.989 or 98.9%

This shows that for constant area a best Mehboob Section is only 98.9% efficient as compared to the semicircle, while it's Perimeter is just 1% more.

1.10 Comparison of Best Sections Inscribed in a Semi Circle

A comparison of best section inscribed in a semi circle with the semi circle is shown in figure 1.9 and Table 1.6.

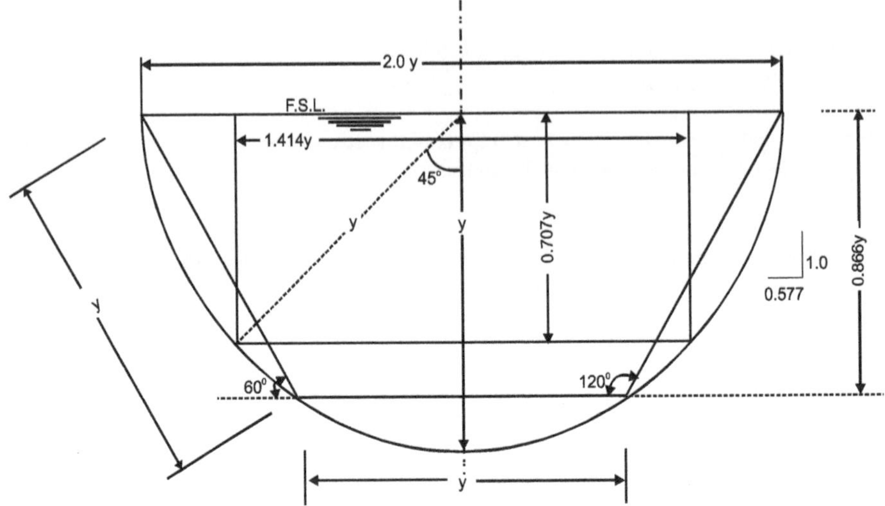

Figure 1.9 : Comparison of Best Sections Inscribed in a Semi Circle

Table 1.6: Comparison of best sections inscribed in a semi circle

Hydraulic properties	Semicircle	Trapezium	Rectangle
(1) Depth of flow (% of semicircle)	**y** **100**	0.866 y 86.6	0.707 y 70.7
(2) Area of flow (% of semicircle)	**3.14 y** **100**	1.2409 y^2 79	y^2 63.66
(3) Wetted Perimmeter (% of semicircle)	**3.14 y** **100**	3.0 y 95.5	2.828 y 90.02
(4) Hydraulic radius R=A/P (% of semicircle)	**0.5 y** **100**	0.4136 y 82.72	0.3536 y 70.72
(5) Conveyance $AR^{2/3}$ (% of semicircle)	**0.987 $y^{8/3}$** **100**	0.6889 $y^{8/3}$ 69.58	0.5 $y^{8/3}$ 50.51

Detailed calculations

a) Rectangular Section

For a rectangular section inscribed in a semi circle D = r sin 45° = 0.707 r

For best hydraulic section B = 2D = 1.414 r

Cross-sectional Area of flow

A = B x D = 1.414 r x 0.707 r = r^2

Ratio of the area with that of semi circle = (r^2) / (π r^2 / 2) = 0.6366 = **63.66%**

Perimeter of the rectangle

$P = B + 2D = 1.414r + 1.414\ r = 2.828\ r$

Ratio of Perimeter with that of semi circle = $(2.828\ r)/\pi r = 0.9002 =$ **90.02%**

R of rectangle = A/P

$= (r^2)/(2.828\ r)$

$= r/2.828 =$ **0.3536 r**

Ratio of R with that of semi circle = $0.3536\ r/0.5\ r = 0.7072 =$ **70.72%**

$A\ R^{2/3}$ for Rectangular section = $r^2(0.3536r)^{2/3} = 0.50\ r^{8/3}$

Ratio of $A\ R^{2/3}$ with that of semi circle = $0.50\ r^{8/3}/0.987\ r^{8/3} = 0.5051 =$ **50.51%**

This shows that a rectangular section inscribed in a semi circle is only 50.51% efficient as compared to semi circle where as its perimeter is 90% of a semi circle and area is only 63.66% of a semi circle.

b) Trapezoidal Section

For a trapezoidal section inscribed in a semi circle,

$B/2 = r \text{ Sin } 30^o = r/2$

or $B = r$ and $D = r \cos 30^o = 0.866\ r$

Cross-sectional area,

$A = BD + 2\ (1/2)\ DD \cos 60^o$

$= BD + D^2/2$

$= r\ (0.866\ r) + [(0.866\ r)^2/2]$

$= 0.866\ r^2 + 0.374\ r^2$

$= 1.2409\ r^2$

Ratio of the area with that of semi circle = $(1.2409r^2)/(\pi\ r^2/2)$

$= 0.790 =$ **79.0%**

Perimeter of the trapezium

$P = B + 2B = 3\ r$

Ratio of Perimeter with that of semi circle = $(3\ r)\ /\ \pi r$

$= 0.9554 =$ **95.5%**

Hydraulic Radius R of trapezium = A/P

$= (1.2409r^2)/(3\ r) =$ **0.41365r**

Ratio of R with that of semi circle = $0.41365\ r/0.5\ r$

$= 0.8272 =$ **82.72%**

$A\ R^{2/3}$ for Trapezoidal section = $(1.2409r^2)\ (0.41365r)^{2/3}$

$= 1.2409 \times 0.5551\ r^{8/3} =$ **0.689 $r^{8/3}$**

Ratio of $A\ R^{2/3}$ with that of semi circle = $0.68888\ r^{8/3}\ /\ 0.98995\ r^{8/3} = 0.6958 =$ **69.58%**

This shows that a trapezoidal section inscribed in a semi circle is only 69.58% efficient as compared to semi circle where as its perimeter is 95.5% of a semi circle and area is only 79% of a semi circle.

1.11 Comparison of Best Sections for Constant Perimeter

A comparison of the best hydraulic sections for a given or constant perimeter with that of a semi circle is shown in figure1.10 and Table 1.7.

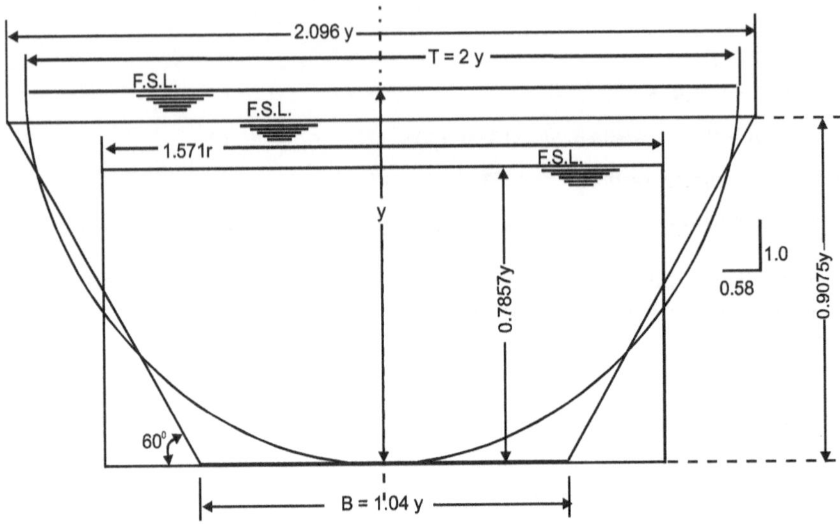

Figure 1.10 : Comparison of Best Shapes for Constant Perimeter

Table 1.7: Comparison of best shapes for constant perimeter

Hydraulic properties	Semicircle	Mehboob	Trapezium	Rectangle
(1) Area of flow		$1.515 y^2$	$1.429 y^2$	$1.2346 y^2$
(% of semicircle)	**100**	96.4	91	78.63
(2) Depth of flow	**y**	0.966 y	0.9075 y	0.785 y
(% of semicircle)	**100**	96.6	90.75	78.5
(3) Wetted Perimmeter	**3.14 y**	3.14 y	3.14 y	3.14 y
(% of semicircle)	**100**	100	100	100
(4) Hydraulic radius R=A/P	**0.5 y**	0.5 y	0.455 y	0.393 y
(% of semicircle)	**100**	100	91	68.6
(5) Conveyance $AR^{2/3}$	**$0.987 y^{8/3}$**	$0.932 y^{8/3}$	$0.845 y^{8/3}$	$0.6627 y^{8/3}$
(% of semicircle)	**100**	94.2	85.6	66.94

Details of calculations

Rectangular Section

For best rectangular section B = 2D

Perimeter of the rectangle

P = B + 2D = 2D+2D = 4D

Equating this perimeter to the perimeter of a semi circle,

$4D = \pi r$

$D = \pi r / 4 = 0.785 r$

Area of rectangle, $A = B \times D = 2D \times D = 2 D^2$
$= 2 (0.7857r)^2 = 1.2346 r^2$

Ratio of area of rectangle with the area of semi circle,
$= (1.2346r^2)/(\pi r^2/2) = (1.2346 \times 2) / \pi$
$= 0.7857 = 78.57\%$

Hydraulic Radius R of rectangular section, $= A / P = 2D^2 /4D = D/2$

$= 0.785 r / 2 = 0.3925 r$

Ratio of R with that of semi circle $= 0.3925 r / 0.5 r = 0.686$ or 68.6%

Conveyance of Rectangular section,
$A R^{2/3} = (1.2346 r^2)(0.393 r)^{2/3}$
$= 0.6627 r^{8/3}$

Ratio of conveyance with that of semi circle $= (0.6627 r^{8/3})/(0.989 r^{8/3})$
$= 0.6694 = 66.94\%$

This shows that for the same perimeter, rectangular section is only 66.94% efficient where as area is 78.63% that of the semi circle.

b) Trapezoidal Section

For best trapezoidal section
$z = 0.58$ and $B = 1.155y$

Perimeter of Trapezium $= 3.464 y$ (see table 2.2)

Keeping perimeter equivalent to that of a semi circle,
$P = 3.46 y = \pi r = 3.14 r$
or $y = 0.9075 r$ and $B = 1.155 \times 0.9075 r$
$B = 1.048 r$

Area of trapezoid,
$A = 1.735 y^2$ (see table 2.2)

Substituting the value of y from above in terms of 'r'
$A = 1.735 \times (0.9075 r)^2 = 1.4288 r^2$

Ratio of A with that of a semi circle $= (1.4288r^2) / (r^2/2)$
$= 0.91 = 91.0\%$

Hydraulic Radius $R = A/P = (1.428r^2)/(\pi r) = 0.455r$

Ratio of R with that of semi circle $= 0.455 r/0.5 r = 0.91 = 91\%$

Conveyance $= A R^{2/3} = (1.4288r^2) (0.455r)^{2/3} = 0.845 r^{8/3}$

Ratio of $A R^{2/3}$ with that of semi circle $= 0.845$ r8/3/$0.987 r^{8/3}$
$= 0.856 = 85.6\%$

This shows that for the same perimeter (constant perimeter), a trapezoidal section is 85.6% efficient where as area is only 91% of a semi circle.

c) Mehboob Section

Let y_3 be the depth of flow for the same perimeter as that of circle

Perimeter of Mehboob Section $= 2 y_3 (\cot \theta + \theta/180)$, where, $\theta = 60^o = 3.25 y^3$

Keeping the perimeter equivalent to that of a semi circle,
$3.248\ y_3 = 3.14\ r$ or $y_3 = 0.966r$

Area of Mehboob Section = $1.625\ y_3^2 = 1.625 \times (0.966r)^2$
$= 1.515\ r^2$

Hydraulic Radius $R = A/P = 0.5\ y_3$
$= 0.5\ .\ (0.966\ r) = 0.483\ r$

Ratio of A with that of a semi circle = $(1.515\ r^2)\ /\ (\pi r^2/2)$
$= 0.964 = 96.40\%$

Conveyance of Mehboob Section $A\ R^{2/3} = (1.515r^2)(1.515\ r^2/3.14r)^{2/3}$
$= 0.932\ r^{8/3}$

Ratio of $A\ R^{2/3}$ with that of semi circle = $0.932\ r^{8/3}/0.989\ r^{8/3}$
$= 0.942 = 94.2\%$

This shows that for a constant perimeter a best Mehboob section is 94.2% efficient as compared to the semi circle where as area is 96.4% that of the semi circle.

1.12 Summary and Conclusion

Of all the flow formulae, Manning's formula due to its simplicity is now widely accepted through the world and extensively used in the design of channels.

In this formula hydraulic radius (R) has a very important role, as discussed in Para 1.4.1.

R is greatly influenced by the shape of the channel, including B/D ratio and side slope (Z).

Semi circle is the most optimum best section. All other (best) sections are circumscribing it. Mehboob section ranks number 2, then trapezoidal (half of a hexagon) at number 3 and rectangular at number 4.

Hydraulic particulars of best sections are given in table 1.2, 1.3 and figure 1.5

A comparison of best sections for a given discharge is given in para 1.8, table 1.4. Mehboob and best trapezium are very close to semicircle.

For maximum velocity R = d /2 (half of flow depth) only for any shape (semicircle, rectangle, trapezoidal with any side slopes or Mehboob) and the section is called as the best hydraulic section or most efficient section, for the given side slopes.

For best section, by substituting d / 2 for R, Manning's formula reduces to a unique relationship.

$V = 0.63\ d^{2/3}\ S^{1/2}\ /\ N$ - in metric units (for R = d/2) this is irrespective of any shape: (or vide slope).

Similarly there is a unique relation ship of discharge and d for each shape.

Thus for design of best section no trial and error is required.

References

Arya, R.S. et al, Design Charts for Earthen and Lined Channels, 1st International Research and Development Conference Water & Energy 2001, Central Board of Irrigation and Power, New Delhi, October, 1995.

Atmapoojya, S. L. and Ingle R.N., Optimal Design for A Canal Cross-Section, Journal of Indian Water Works Association, Mumbai, January-March, 2001.

Chalisgaonkar, R., Saxena, G.C., Mittal A.K., Quick Design of Lined Channels, Journal of Indian Water Resources Society, Vol. 22, No. 1, January 2002.

Chow, V.T, Open - Channel Hydraulic, McGraw Hill 1973.

Chow, Ven Te, A Note on the Manning Formula, Transactions, American Geo-Physical Union, Vol. 36, No. 4, p. 688, August, 1955.

Ishardas, Design of Lined Canals, International Association of Hydraulic Research, Question-1, pp. 76, 1951.

Kala, M.K. Study for Economic Section of Lined Canals; M.E. Thesis University of Roorkee, Roorkee, 1971.

Manning, Robert, on the Flow of Water in Open Channels and Pipes, Transactions, Institution of Civil Engineers of Ireland, Vol. 20, pp. 161-207, Dublin, Supplement, Vol. 24, pp. 179-207, 1891, 1895.

Mirza, Mohamed Din, Note on Canal Cross Section and Lining, Indian Concrete, J. Vol. XVII No. 12, pp - 443 To 445.

Modi, P. N., and Seth, S. M., Design of Most Economical Trapezoidal Section of Open Channels, Irrigation and Power, Journal of Central Board of Irrigation and Power, New Delhi, July, 1968.

Suryavanshi, B.D., The Master Equation for Designing of Canal Section by Manning's Formula, Irrigation and Power, Journal of Central Board of Irrigation and Power, New Delhi, January, 1972.

Swami, P.K., and Bhatia, K.G., Economic Open Channel Section, Irrigation and Power, Journal of Central Board of Irrigation and Power, New Delhi, April, 1972.

❑❑❑

Chapter 2

Practical Aspects of Design of Canals

Theoritical aspects of design has been discussed in chapter 1. Discharging capacity of canal is computed by Manning Equation (1)and continuity Equation (2) (Para 1.4).

In the Manning's equation $V = R^{2/3} S^{1/2} / N$ (Eq. 1, Para 1.4); N and R play most important role. The effect of R has been discussed in para 1.4.1, Chapter 1.

In this chapter the complexities of Manning's n (or N, also known as Manning's rougosity coefficient or simply rougosity coefficient) and some practical aspects of composite sections and velocity in best Mehboob, trapezoidal or rectangular sectioned channels are discussed.

2.0 Manning's Rugosity Coefficient (n)

Value of N depends upon a number of factors such as , surface roughness, vegetation, obstructions, type of lining, age of lining, erosion and damages in lining, curves, discharging capacity , perimeters, depth of flow, velocity etc. These are discussed below and commonly adopted values of n are given in tables 2.1 to 2.3.

The values of N for unlined canals and natural unlined channels are also given in table 2.2 and 2.3. It should be carefully noted that lined canals also behave as unlined canals many times due to silt deposition, damages in lining, or falling of debris from adjoining land and growth of vegetation in the canal. Therefore one should be very judicious in adopting values of N for design purposes.

2.1 Factors Affecting Manning's Roughness Coefficient

In reality, the value of N is highly variable and depends on a number of factors. These factors are not independent but are interdependent.The factors that exert the greatest influence upon the coefficient of roughness in both artificial and natural channels are :

i) Surface Roughness

The surface roughness is represented by the size and shape of the grains of the material forming the wetted perimeter and it produces a retarding effect on the flow. Generally fine grains result in relatively low value of n and coarse grains, a higher value of n.

In alluvial streams where the material is fine grained, such as sand, clay, loam, or silt, the retarding effect is much less than where the material in coarse, such as gravels or boulders. When the material is fine, the value of n is low and relatively unaffected by change in flow stage. When the material consists of gravels and boulders, the value of n is generally high, particularly at low or high stage. Larger boulders usually collect at the bottom of the stream, making the channel bottom rougher than the banks and increasing the value of n at low stages. At high stages, a portion of the energy of flow is used in rolling the boulders downstream, thus again increasing the value of n. However at medium stages value of N may be higher.

ii) Vegetation

Vegetation may be regarded as a kind of surface roughness, but it also markedly reduces the capacity of the channel and retards the flow. Weeds block the way of channel and reduce the flow by as much as 60 to 80%. This effect depends mainly on height, density, distribution, and type of vegetation, and it is very important in designing small drainage channel. Due care is also required in weed prone canals.

Seasonal Change. Owing to the seasonal growth of aquatic plants, grass, weeds, willow, and trees in the channels or on the banks, the value of n may increase in the growing season and diminish in the dormant season.

iii) Obstruction

The presence of logjams, bridge piers, and the like tends to increase n. The amount of increase depends on the nature of the obstructions, their size, shape, number and distribution.

iv) Stage and Discharge

The value of N in most streams decreases with increase in stage and discharge. When the water is shallow, the irregularities of the channel bottom are exposed and their effects become pronounced. However, the value of N may be large at high stages if the banks are rough and grassy.

v) Channel Irregularity

Channel irregularity means irregularities in wetted perimeter and variations in cross section, size, and shape along the channel length. In natural channels, such irregularities are usually introduced by the presence of sand bars, sand waves, ridges and depressions, and holes and humps on the channel bed. Silt deposition occurs in lined canals also.

In rocky reaches presence of boulders and rocks cause a highly irregular shape with lot of holes, humps in the bed and sides. These irregularities definitely introduce roughness in addition to that caused by surface roughness and other factors.

Generally a gradual and uniform change in cross section, size, and shape will not appreciably affect the value of N, but abrupt changes or alternation of small and large sections necessitate the use of a large value of n. In this case, the increase in n may be 0.005 or more. Changes that cause sinuous flow from side to side of the channel produce the same effect.

vi) Channel Alignment

It refers to curvature and meandering. Smooth curvatures with large radius in alignment give a relatively low value of N, whereas sharp curvature with severe meandering gives a higher value of N. On the basis of flume tests, Scobey (1939) suggested that the value of N be increased 0.001 for each 20 degrees of curvature in 30 m (100 ft) of channel. Although it is doubtful whether curvature ever increases N more than 0.002 or 0.003, its effect should not be ignored, for curvature may induce the accumulation of drift and thus indirectly increase the value of N.

The meandering of natural streams, however, may increase the value of N as high as 30%.

vii) Silting and Scouring

Some times silting may change a very irregular channel into a comparatively uniform one and decrease n, whereas scouring may do the reverse and increase n. However, the dominant effect of silting depends on the nature of the material deposited. Uneven deposits such as sand bars and sand waves are channel irregularities and increase the roughness. The amount and uniformity of scouring depend on the material forming the wetted perimeter. Thus, a sandy or gravely bed will be eroded more uniformly than a clay bed. The deposition of silt eroded from the uplands will tend to even out the irregularities in a channel dredged through clay. The energy used in eroding and carrying the material in suspension or rolling it along the bed will also increase the value of n. The effect of scouring is not significant as long as the erosion on channel bed caused by high velocities is progressing evenly and uniformly.

viii) Suspended Material and Bed Load

The suspended material and the bed load, whether moving or not moving consumes energy and cause head loss and increase the channel roughness.

2.1.1 Studies in India and proposed value of n

KERS (1972) has conducted many studies in the field from 1963 to 1972 on evaluation of rugosity coefficient on various types of lined and unlined canals in Karnataka, and have found much higher values then given in the table 2.1 to 2.3. The N- values were considerably different between various Km of the same canals. In actual field, after few years of lining a canal, the roughness increases, due to some silts or flowing weeds etc. Very high maintenance or cleaning several times in a year is not practical, see Para. 9.15.2.

Table 2.1 : Rugosity Coefficients (N) and Commonly Accepted Limiting Velocities for Different Types of Linings

S.No.	Surface characteristics	Value of 'N'		Maximum permissible/ limiting velocity in m/s
		New surface	Old surface	
1.	Concrete lined with surface as Indicated below			2.7
	(a) Formed, no finish	0.013-0.017		
	(b) Trowel, finish	0.012-0.014	0.015 - 0.018	2.7
	(c) Float finish	0.013-0.015		
	(d) Float finish, some gravel on bottom	0.015-0.017		
	(e) Gunite, good section	0.016-0.019		
	(f) Gunite wavy section	0.018-0.022	0.018-0.022	
	(g) P. C. C. Tiles / Slabs		0.018-0.020	
	(h) U.C.R. Random rubble masonry with pointing	0.024-0.026		
2.	Concrete bottom Float finished sides as indicated below			
	(a) Dressed stone masonry		0.019-0.021	
	(b) Course Rubble masonry	0.015-0.017		
	(c) Random stone in mortar (Random rubble masonry)	0.017-0.020	0.020-0.025	
	(d) Random rubble masonry cement plastered	0.016-0.020	0.015-0.017	
	(e) Dry rubble (rip rap), (stone pitching)	0.020-0.030	0.020-0.033	
3.	Boulder ling		0.022-0.027	1.5
4.	Gravel bottom sides as indicated below			
	(a) Formed concrete	0.017-0.020		
	(b) Random stone in mortar	0.020-0.023		
	(c) Dry rubble (rip rap)	0.023-0.033		
5.	Brick	0.014-0.017		
	Burnt clay Brick / Tile	0.018-0.020	0.018-0.020	1.8
6.	Asphalt			
	(a) Smooth	0.013	0.013-0.015	
	(b) Rough	0.016	0.016-0.018	
7.	Wood planned clean	0.011-0.013		

Note : For canals in curves, (other than straight) a small increase in value of N be made to allow for additional loss of energy.

Table 2.2 : Rugosity Coefficients (N) For Unlined Channels.

S.No.	Type of channel and description	Minimum	Normal	Maximum
1.	Excavated or dredged			
	a. Earth, straight and uniform			
	1. Clean, recently completed	0.016	0.018	0.020
	2. Clean, after weathering	0.018	0.022	0.025
	3. Gravel, uniform section, clean	0.022	0.025	0.030
	4. With short grass, few weeds	0.022	0.027	0.033
	b. Earth, winding and sluggish			
	1. No vegetation	0.023	0.025	0.030
	2. Grass, some weeds	0.025	0.030	0.033
	3. Dense weeds or aquatic plants in deep channels	0.030	0.035	0.040
	4. Earth bottom and rubble sides	0.028	0.030	0.035
	5. Stony bottom and weedy banks	0.025	0.035	0.040
	6. Cobble bottom and clean sides	0.030	0.040	0.050
	c. Dragline-excavated or dredged			
	1. No vegetation	0.025	0.028	0.033
	2. Light brush on banks	0.035	0.050	0.060
	d. Rock cuts			
	1. Smooth and uniform	0.025	0.035	0.040
	2. Jagged and irregular	0.035	0.040	0.050
	e. Channels not maintained, weeds and brush uncut			
	1. Dense weeds, high as flow depth	0.050	0.080	0.120
	2. Clean bottom, brush on sides	0.040	0.050	0.080
	3. Same, as above highest stage of flow	0.045	0.070	0.110
	4. Dense brush, high stage	0.080	0.100	0.140
2.	Natural Streams			
	2.1 Minor streams (top width at flood stage<30m (100ft)			
	a. Streams on plain			
	1. Clean, straight, full stage, no rifts or deep pools	0.025	0.030	0.033
	2. Same as above, but more stones and weeds	0.030	0.035	0.040
	3. Clean, winding, some pools and shoals	0.033	0.040	0.045
	4. Same as above, but some weeds and stones	0.035	0.040	0.050
	5. Same as above, lower stages, more ineffective slopes and sections	0.040	0.048	0.055
	6. Same as 4, but more stones	0.045	0.050	0.060
	7. Sluggish reaches, weedy, deep pools	0.050	0.070	0.080
	8. Very weedy reaches, deep pools, or flood ways with heavy stand of timber and underbrush	0.075	0.100	0.150
	b. Mountain streams, no vegetation in channel, banks usually steep, trees and brush along banks submerged at high stages			
	1. Bottom : gravels, cobbles, and few boulders	0.030	0.040	0.050
	2. Bottom : cobbles with large boulders	0.040	0.050	0.070

A long channel may not have same roughness/ smoothness in its entire length. Also some reaches may be lined and some unlined. Also flatter side slopes generally adopted in lined canal encourages silt and weed deposition on sides that increases N. The differences in roughness are partly met by more head loss/ heading of water on upstream and a reasonable value of N is to be considered in a reasonable long length of canal. Therefore a little higher value of N (generally 10 to 20 % higher) is always preferable in the design.

The effect of higher value of n on economics of channel is discussed in Para. 4.5. It does not increase the cost of section proportionate to increase in n, see para 4.5. Also various practices for adoption of N values are discussed in Chapter 6, Table 6.4.They all recommend adoption of higher values based on practicability of maintenance programme.

Table 2.3 : Rugosity Coefficients (N) For Natural Unlined Channels.

S.No.	Type of channel and description	Minimum	Normal	Maximum
2.2	Flood plains			
	a. Pasture, no brush			
	1. Short grass	0.025	0.030	0.035
	2. High grass	0.030	0.035	0.050
	b. Cultivated areas			
	1. No crop	0.020	0.030	0.040
	2. Mature row crops	0.025	0.035	0.045
	3. Mature field crops	0.030	0.040	0.050
	c. Brush			
	1. Scattered brush, heavy weeds	0.035	0.050	0.070
	2. Light brush and trees, in winter	0.035	0.050	0.060
	3. Light brush and trees, in summer	0.040	0.060	0.080
	4. Medium to dense brush, in winter	0.045	0.070	0.110
	5. Medium to dense brush, in summer	0.070	0.100	0.160
	d. Trees			
	1. Dense willows, summer, straight	0.110	0.150	0.200
	2. Cleared land with tree stumps, no sprouts	0.050	0.060	0.080
	3. Same as above, but with heavy growth of sprouts	0.050	0.060	0.080
	4. Heavy stand of timber, a few down trees, little undergrowth, flood stage below branches	0.080	0.100	0.120
	5. Same as above, but with flood stage reaching branches	0.100	0.120	0.160
2.3	Major streams (top width at flood stage > 30m (100 ft). The n values are less than that for minor streams of similar description, because banks offer less effective resistance.			
	a. Regular section with no boulders or brush	0.025	0.042	0.060
	b. Irregular and rough section	0.035	....	0.100

2.2 Composite Section

There are two types of composite section viz.

(i) *Channels of compound section* : These channels have cross section composed of several geometrical shapes or sub sections as shown in figure 2.1 with each sub section having different roughness from the other.

(ii) *Channels of composite roughness* : There are normal singular geometrical shape, but have different values of N for the bed and sides, for example in different lining in bed and sides, having different roughness coefficient.

2.2.1 Channels of compound sections

A typical example of a compound section is a river channel with flood plains as shown figures 2.1 and 2.2. The roughness of the side channels will be different (generally rougher) than that of the main channel. Therefore the mean velocity in the main channel is greater than the mean velocity in the side channels. In such cases the method of analysis is to consider each sub section separately and apply Manning's formula to each subsection. The total discharge is the sum of component discharges computed by the Manning's equation.

Thus in the channel shown in fig. 2.1, assuming that the bed slope is the same for the three sub-areas : the total discharge is

$$Q = \left(\frac{A_1}{n_1} R_1^{2/3} + \frac{A_1}{n_1} R_2^{2/3} + \frac{A_3}{n_3} R_3^{2/3} \right) S_o^{1/2}$$

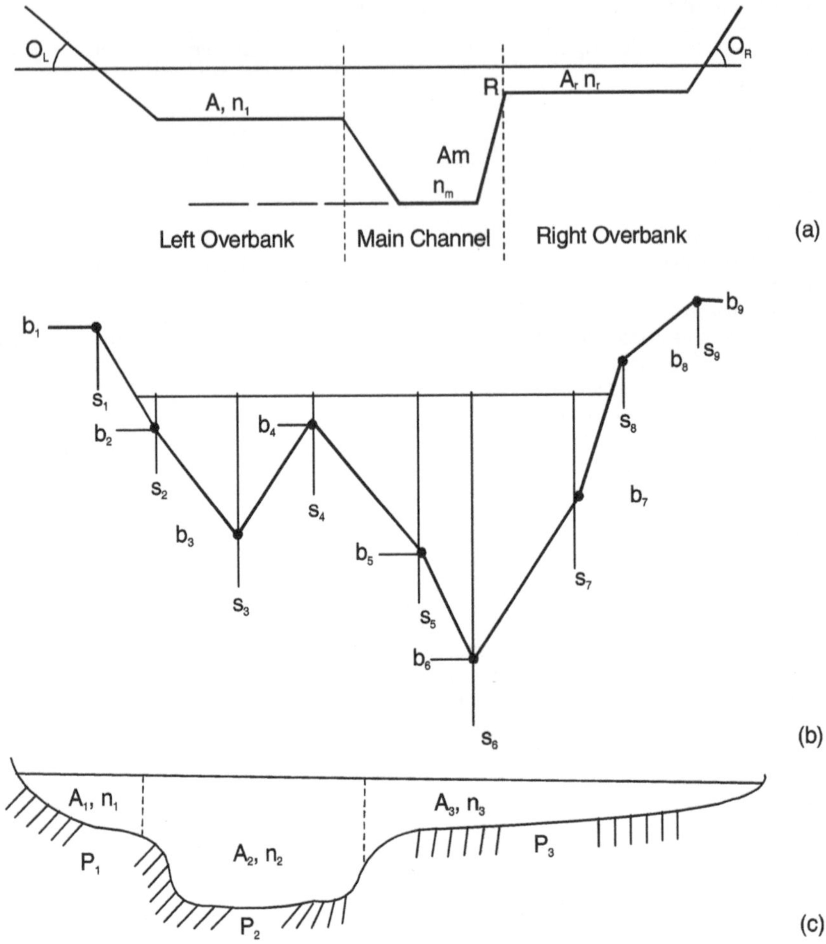

Figure 2.1 : Compound channel sections.

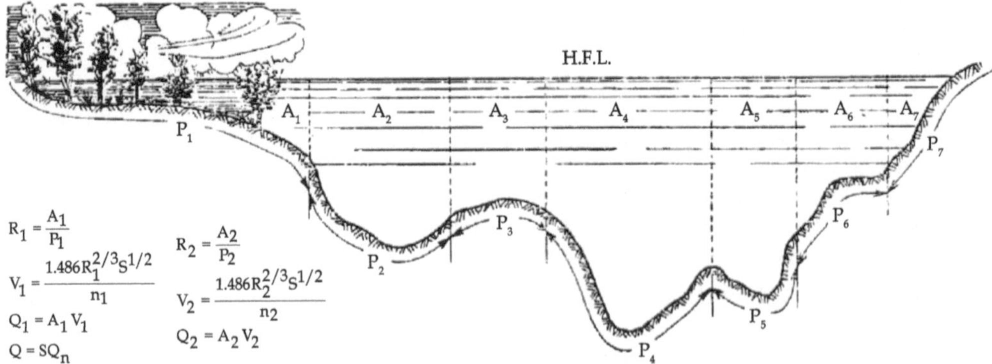

Figure 2.2 : Determination of Equivalent 'R' and 'n' when these vary considerably in different parts of the cross section

With A representing the total cross-sectional area and P the total wetted perimeter, the global value of n given the individual values n_1, n_m, n_r is

$$\frac{1}{n}=\sum_{1}^{r}\frac{1}{n_i}\frac{(A_i/A)^{5/3}}{(P_i/P)^{2/3}}$$

2.2.2 Channels of composite roughness

Many times channels have different values of n , for the bed and sides, see figure below. In applying the Manning's formula to such channels it is necessary to compute an equivalent value of n to be used for the whole section. The water area is 'divided' into N parts having wetted perimeters P_1, P_2.... P_N with associated roughness coefficients n_1, n_2,...,n_N. Horton (1933) assumed that each sub-area has a velocity equal to the mean velocity of the whole section, that i.e., $V = V_2 = V_3 = \ldots\ldots V_N = V$ on the basis of this assumption equivalent n is computed as follows :

$$\text{Thus } n=\left[\frac{\sum_{i=1}^{N} P_i\, n_i^{3/2}}{P}\right]^{2/3} = \frac{(P_1 n_1^{1.5} + P_2 n_2^{1.5} + \ldots\ldots\ldots\ldots + P_n n_N^{1.5})^{2/3}}{P^{2/3}}$$

Where P = total wetted perimeter.

Pavlovskij (1931) and others (1933, 1950) equated the sum of the component resisting forces to the total resisting force and thus found.

$$n=\left[\frac{\sum_{i=1}^{N} P_i\, n_i^{2}}{P}\right]^{1/2} = \frac{(P_1 n_1^{2} + P_2\; n_1^{2} + \ldots\ldots\ldots\ldots + P_N n_N^{2})^{0.5}}{P^{0.5}}$$

Lotter (1933) applied the Manning's equation to sub-areas and equated the sum of the individual discharge equations to the total discharge. Thus the equivalent roughness coefficient is :

$$n=\frac{PR^{5/3}}{\sum_{i=1}^{N}\left(\frac{P_i R_i^{5/3}}{n_i}\right)}=\frac{PR^{5/3}}{\frac{P_1 R_1^{5/3}}{n_1}+\frac{P_2 R_2^{5/3}}{n_2}+\ldots\ldots\frac{P_n R_n^{5/3}}{n_N}}$$

All the above formulae give different discharges, and for practical purposes of canal design a lower value may be adopted.

Example 2.1

A trapezoidal channel has side slopes 1 : 1 and bed slope 1 : 1000. Its bed is 3 m wide composed of sand (n = 0.02) and sides are of concrete (n = 0.014). Determine the discharge when the depth of flow is 2.0 m.

Its discharge should not be worked out by averaging the values of n i.e. (02 + .014) / 2 = .017. Its discharge should be worked out as under.

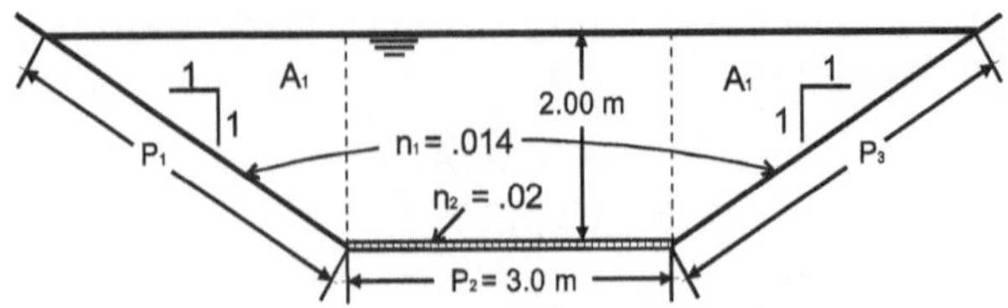

$P_1 (= P_3) = 2.828$ m; $P_2 = 3.0$ m;
$A_1 (= A_3) = (2 \times 2)/2 = 2.0$ m²;
$R_1 (= R_3) = A_1 / P_1 = 0.7072$ m;

Total P = 8.656 m (on solid surface only)
$A_2 = 3 \times 2 = 6.0$ m²; Total A = 10.0 m²
$R_2 = A_2 / P_2 = 2.0$ m; Total R = A/P = 10/8.656 = 1.155 m

Composite roughness is calculated according to various formulae's as under :

i) Horton and Einstein:
$$n = \left[\frac{\sum P_i n_i^{1.5}}{P}\right]^{2/3}$$

$$n = \left[\frac{2(2.828) \times 0.014^{1.5} + 3 \times 0.02^{1.5}}{8.656}\right]^{2/3} = 0.0162$$

ii) Pavlovskij:
$$n = \left[\frac{\sum P_i n_i^2}{P}\right]^{1/2}$$

$$n = \left[\frac{2(2.828) \times 0.014^2 + 3 \times 0.02^2}{8.656}\right]^{1/2} = 0.0163$$

iii) Lotter:
$$n = \frac{P R^{5/3}}{\sum_{i=1}^{N}\left(\frac{P_i R_i^{5/3}}{n_i}\right)} = 0.0157$$

With y = 2.0 m; % difference with least is as under

i) Discharge by Horton's formula = $(A/n)R^{2/3} S_o^{1/2}$ = 21.49 m³/s = 96.93%
ii) Discharge by Pavlovskij 's formula = 21.36 m³/s = 96.34%
iii) Discharge by Lotter 's formula = 22.17 m³/s. = 100%

Example 2.2

The cross-section of the flow in a river during a flood is as shown in figure below.

Wetted perimeter of side channel around the solid boundary only (excluding the interfaces X - X between the main and side channel flows) = 10.24 m.

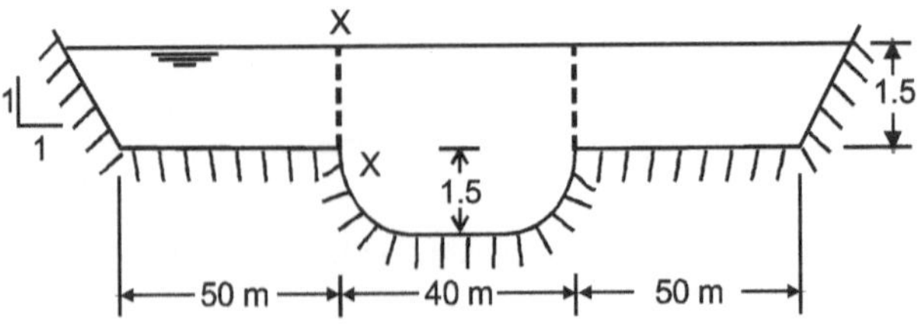

Assuming the roughness coefficients for the side channel and main channel to be 0.04 and 0.03 respectively, estimate the discharge.

Bed slope = 0.005;

Area of main channel (bank full) = 280 m^2

Wetted perimeter of main channel (bank full) = 54 m

Area of flow in side channel = 152.25 m^2

Solution

Area of main channel component = 280 + 40 x (1.5) = 340 m^2

$$\text{Discharge} = \frac{340}{0.03}\left(\frac{340}{54}\right)^{2/3}\sqrt{0.005} + \frac{152.25}{0.04}\left(\frac{152.25}{104.24}\right)^{2/3}\sqrt{0.005} = 3079\, m^3/s.$$

Note that treatment of this problem by the equivalent roughness methods of Horton and Pavlovskij will produce large errors in the computed discharge due to the inherent assumptions. However the Lotter method should produce a similar result to that computed above since it uses basically the method of sub sections.

$$\text{Lotter equivalent roughness } n = \frac{PR^{5/3}}{\sum_{i=1}^{N}\left(\frac{P_i R_i^{5/3}}{n_i}\right)}$$

$$n = 0.0241 \text{ and } Q = \frac{492.25}{0.024}\left(\frac{492.25}{158.242}\right)^{2/3}\sqrt{0.005} = 3077\, m^3/s.$$

2.3 Velocity and Discharge Computation for Best Hydraulic Sections

Velocity of flow through any section of any size and shape is given by the Manning's formula

$$V = R^{2/3} S^{1/2} / N \text{ and } (R = A/P)$$

or $V = [d/2]^{2/3} S^{1/2} / N$ (*where* d=depth of flow, for any best section R=d/2 or r/2 or y/2, irrespective of any size and shape)

$V = 0.63\, d^{2/3} S^{1/2} / N$ Eq 2.2 (i)

Thus for any section velocity is a function of hydraulic radius R and energy slope (S) only, see Para 1.6.2.

For best canal section of any shape and size R is always equal to r/2 or d/2 (half of depth). The velocity is independent of A and P; and is a function of d/2 (half of depth) only, irrespective of size and shape. So velocity does not change with geometry of the section or change of side slopes as long as R is same for any size and shape of river and canal.

But, with the change in geometry of section, R changes, i.e. R > d/2 or < d/2, velocity is no longer independent of A and P; and becomes dependent on A and P.

Velocity of flow for different depth of flow (r) and energy slope (S) and N=0.018 are given in **Table 2.4**. This table is applicable for all kinds of sections with any side slopes. For any other value of N, similar tables can be worked out. The values of N for various finishes of lining are given in **Table 2.1**.

Table 2.4 Velocity of Flow in any Best Sections, Irrespective of Any Side Slope (Z) or Any Shape May Be Rectangular, Trapezoidal, Mehboob. $V = R^{2/3} S^{1/2} / N$. For N = 0.018 $V = (r/2)^{2/3} S^{1/2} / 0.018 = 34.9978\ r^{2/3} S^{1/2}$. In any best section velocity do not change with side slope As velocity is a function of R & S only and R is constant as r/2 for any side slopes

Velocity in m/sec

Bed Slope / Depth r in mt	1 in 1000 v=1.1067 * $r^{2/3}$	1 in 1500 v=0.9036 * $r^{2/3}$	1 in 2000 v=0.7825 * $r^{2/3}$	1 in 2500 v=0.6999 * $r^{2/3}$	1 in 3000 v=0.6389 * $r^{2/3}$	1 in 3500 v=0.5915 * $r^{2/3}$	1 in 4000 v=0.5533 * $r^{2/3}$	1 in 4500 v=0.5217 * $r^{2/3}$	1 in 5000 v=0.4949 * $r^{2/3}$	1 in 5500 v=0.4719 * $r^{2/3}$	1 in 6000 v=0.4518 * $r^{2/3}$	1 in 6500 v=0.4341 * $r^{2/3}$	1 in 7000 v=0.4183 * $r^{2/3}$	1 in 7500 v=0.4041 * $r^{2/3}$	1 in 8000 v=0.3913 * $r^{2/3}$	1 in 8500 v=0.3796 * $r^{2/3}$	1 in 9000 v=0.3689 * $r^{2/3}$	1 in 9500 v=0.3590 * $r^{2/3}$	1 in 10000 v=0.3499 * $r^{2/3}$
1.0	1.1067	0.9036	0.7825	0.6999	0.6389	0.5915	0.5533	0.5217	0.4949	0.4719	0.4518	0.4341	0.4183	0.4041	0.3913	0.3796	0.3689	0.3590	0.3499
1.1	1.1793	0.9629	0.8338	0.7458	0.6808	0.6303	0.5896	0.5559	0.5274	0.5029	0.4814	0.4626	0.4457	0.4306	0.4170	0.4045	0.3931	0.3826	0.3729
1.2	1.2497	1.0204	0.8836	0.7904	0.7215	0.6679	0.6248	0.5891	0.5589	0.5329	0.5102	0.4902	0.4724	0.4563	0.4419	0.42866	0.4166	0.4054	0.3951
1.3	1.3182	1.0763	0.9321	0.8337	0.7610	0.7046	0.6591	0.6214	0.5895	0.5621	0.5382	0.5171	0.4983	0.4813	0.4661	0.45216	0.4394	0.4276	0.4168
1.4	1.3850	1.1308	0.9793	0.8759	0.7996	0.7402	0.6924	0.6529	0.6194	0.5906	0.5654	0.5433	0.5235	0.5057	0.4897	0.47506	0.4617	0.4493	0.4379
1.5	1.4502	1.1841	1.0254	0.9171	0.8372	0.7751	0.7250	0.6836	0.6485	0.6184	0.5920	0.5688	0.5481	0.5295	0.5127	0.49742	0.4834	0.4704	0.4585
1.6	1.5139	1.2361	1.0704	0.9574	0.8740	0.8092	0.7569	0.7137	0.6770	0.6456	0.6181	0.5938	0.5722	0.5528	0.5353	0.51929	0.5046	0.4911	0.4787
1.7	1.5764	1.2871	1.1146	0.9969	0.9101	0.8425	0.7881	0.7431	0.7049	0.6722	0.6435	0.6183	0.5958	0.5756	0.5574	0.5407	0.5255	0.5114	0.4984
1.8	1.6376	1.3371	1.1579	1.0357	0.9454	0.8753	0.8187	0.7720	0.7323	0.6983	0.6685	0.6423	0.6190	0.5980	0.5790	0.5617	0.5459	0.5312	0.5178
1.9	1.6977	1.3862	1.2004	1.0737	0.9801	0.9074	0.8488	0.8003	0.7592	0.7239	0.6931	0.6659	0.6417	0.6199	0.6003	0.58232	0.5659	0.5507	0.5368
2.0	1.7568	1.4344	1.2421	1.1110	1.0142	0.9389	0.8783	0.8281	0.7856	0.7491	0.7172	0.6891	0.6640	0.6415	0.6212	0.60258	0.5856	0.5699	0.5554
2.2	1.8720	1.5285	1.3236	1.1839	1.0807	1.0005	0.9359	0.8825	0.8371	0.7982	0.7642	0.7343	0.7076	0.6836	0.6619	0.64211	0.6240	0.6073	0.5919
2.4	1.9838	1.6198	1.4027	1.2546	1.1453	1.0603	0.9918	0.9352	0.8871	0.8459	0.8099	0.7782	0.7498	0.7244	0.7014	0.68046	0.6613	0.6435	0.6272
2.6	2.0926	1.7085	1.4796	1.3234	1.2080	1.1184	1.0462	0.9864	0.9358	0.8923	0.8543	0.8208	0.7909	0.7641	0.7399	0.71775	0.6975	0.6788	0.6616
2.8	2.1985	1.7951	1.5545	1.3904	1.2692	1.1751	1.0992	1.0364	0.9832	0.9375	0.8975	0.8624	0.8310	0.8028	0.7773	0.7541	0.7328	0.7132	0.6951
3.0	2.3020	1.8796	1.6277	1.4559	1.3290	1.2304	1.1509	1.0852	1.0294	0.9816	0.9398	0.9030	0.8701	0.8406	0.8139	0.7896	0.7673	0.7468	0.7278
3.2	2.4032	1.9622	1.6992	1.5199	1.3874	1.2845	1.2015	1.1329	1.0747	1.0247	0.9811	0.9427	0.9084	0.8775	0.8497	0.82431	0.8011	0.7796	0.7598
3.4	2.5024	2.0431	1.7693	1.5825	1.4446	1.3374	1.2511	1.1796	1.1190	1.0670	1.0216	0.9815	0.9458	0.9137	0.8848	0.85831	0.8341	0.8117	0.7912
3.6	2.5996	2.1225	1.8380	1.6440	1.5007	1.3894	1.2997	1.2254	1.1625	1.1085	1.0612	1.0197	0.9826	0.9492	0.9191	0.89165	0.8665	0.8433	0.8219
3.8	2.6950	2.2004	1.9055	1.7043	1.5558	1.4404	1.3474	1.2704	1.2051	1.1491	1.1002	1.0571	1.0186	0.9840	0.9529	0.92438	0.8983	0.8742	0.8521
4.0	2.7887	2.2769	1.9718	1.7636	1.6099	1.4905	1.3942	1.3146	1.2471	1.1891	1.1385	1.0939	1.0540	1.0183	0.9860	0.95653	0.9296	0.9046	0.8817
4.2	2.8809	2.3522	2.0370	1.8219	1.6632	1.5398	1.4403	1.3581	1.2883	1.2284	1.1761	1.1300	1.0889	1.0519	1.0186	0.98816	0.9603	0.9345	0.9108
4.4	2.9717	2.4263	2.1011	1.8793	1.7155	1.5883	1.4857	1.4008	1.3289	1.2671	1.2132	1.1656	1.1232	1.0851	1.0507	1.01928	0.9906	0.9640	0.9395
4.6	3.0610	2.4993	2.1643	1.9359	1.7671	1.6360	1.5304	1.4430	1.3689	1.3052	1.2496	1.2007	1.1570	1.1177	1.0823	1.04994	1.0203	0.9930	0.9678
4.8	3.1491	2.5712	2.2266	1.9916	1.8180	1.6831	1.5744	1.4845	1.4082	1.3428	1.2856	1.2352	1.1903	1.1499	1.1134	1.08016	1.0497	1.0215	0.9956
5.0	3.2360	2.6421	2.2880	2.0465	1.8682	1.7296	1.6179	1.5255	1.4471	1.3798	1.3211	1.2693	1.2231	1.1816	1.1442	1.10996	1.0787	1.0497	1.0231
5.2	3.3217	2.7121	2.3487	2.1007	1.9176	1.7754	1.6607	1.5659	1.4854	1.4164	1.3561	1.3029	1.2555	1.2129	1.1745	1.13936	1.1072	1.0775	1.0502

Similar tables can be easily prepared for any other value of N.

From Table 2.4 designer can directly read velocity, for any depth of flow for a given slope. Now for a given discharge and slope, the designer need not worry about the shape of the cross section and read the permissible velocity for the given slope for approximate trial depth.

Multiply this velocity with the area (at this depth of the section of choice, say trapezoidal/ rectangular/Mehboob/any other shapes and compute discharge (Q=V.A). If this depth satisfies the requirement the design is O.K.; otherwise try another depth. Normally with one trial the designer is able to design a canal section within few minutes.

Thus no confusion /calculation /trials are required for the design of best canal section of choice of any shape.

2.4 Velocity in Lined/ Unlined Canals

Normally higher velocity is permitted in lined canals, to have a smaller cross-sectional area, and consequently perimeter. This reduces the cost of earth work and lining. Higher velocities also have the advantage of carrying silt. It also retards weed growth. These are put as a strong argument in favour of lining . This may be true for medium discharge canals, but may not be always true or practical, in large canals of big discharges, or canals with flat slopes. Typical bed slopes adopted in practice in some canals are given in table 6.1. Also see, table 9.5, Para 9.12 and 9.13.

To generate higher velocity, steeper bed slope and or higher depths is required. But bed slope of channel depends upon topography. In flat topography steeper bed slopes are not feasible. Even otherwise in irrigation and power canals, head loss is a major problem. Even in drainage channel flat topography restricts bed slope. Therfore it is kept as minimum as possible, so as not to loose command or power. Thus velocities are normally restricted.

Similarly depth of flow is also restricted due to topographic consideration and stability of side slopes. Lower depth of flow also helps in wider channels with smaller velocity, see para 4.1.

Even otherwise, wherever feasible, higher velocity, carrying sediment damage lining by abrasion. Sediment enters any canal systems from the topography through which it is passing or it is brought from the source. Therefore cross - sectional area and perimeter can not be decreased very much as is being claimed in favour of lining.

Permissible/ Limiting Velocities

Even in lined canals, there is a concept of maximum permissible velocity or say permissible or safe or limiting velocity. For unlined canals see chapter 7.

Permissible velocity is the optimum velocity at which water can be safely carried in a channel and which shall result in minimum cross-sectional area , or perimeter or both to minimise the cost. It should be non damaging (or abrasive or non scouring) on lining and also non-silting. Advantages of providing maximum permissible velocity are

1. Minimum cross sectional area/perimeter necessary for minimum cost of construction and maintenance.
2. Minimum wetted perimeter, resulting in reduction of transmission losses.
3. Retard weed growth.

Safe limiting velocity depends upon the type of lining which resists erosion.

Safe limiting velocities for some types of lining generally adopted are as under

i. Boulder lining - 1.5m/s
ii. Burnt clay tile lining - 1.8 m/s
iii. Cement concrete lining - 2.7 m/s

For all other types of lining maximum permissble velocity of 2 m/s is normally adopted. In practice this is also rarely achieved and generally velocity is less than this, see table 9.1. Though normally velocity is restricted, yet in few cases velocity greater than 2 m/s has also been provided.

2.4.1 Kennedy's (1894-95) Critical Velocity Ratio (CVR)

Critical velocity ratio (CVR) is the ratio of actual or designed velocity to the Kennedy's velocity.

$$\text{CVR} = \frac{V}{V_o} = \frac{\text{Designed velocity}}{\text{Velocity given by Kennedy}} = m\,(a\,ratio)$$

Critical velocity (V_o) given by Kennedy's equation is as under

$$V_0 = 0.84\ d^{0.64}\ \text{(FPS units) or } V_0 = 0.548\ d^{0.64}\ \text{(Matric units)}$$

where d = depth of canal

Designed velocity may be compared with the above.

There is nothing like Kennedy's critical velocity ratio in lined canals. Still some minimum velocity is required so that silt (fine sand) is not deposited in the canal bed and continues to move out through outlets/outfalls. Also it should not be high enough to generate scour. Therefore some engineers proposes that Kennedy's critical velocity ratio should be equal to one.

IS 10430 1982 indicated that critical velocity ratio is not applicable in lined canals, but the possibility of silting can not be neglected. This code further suggested that CVR should be aimed higher than unity or by any other method it should be ensured that silting will not take place in lined canal. The concept of CVR higher than one is also included in the revised version of this code (IS 10430 2000). Therfore it would be appropriate to check for CVR.

IGNP lined canals are subject to silting due to wind blown desert sand and therfore Wapcos (1994) recommended for Indra Gandhi Canal Project that 20% high value of Kennedy's CVR or 25% more Lacey's silt factor may be allowed in brick lined canals in cement concrete or cement plaster sections.

It may not be feasible to achieve above CVR in high capacity lined or unlined canals, see Chapter 7 and also para 9.13, table 9.5

In fact above Kennedy's relationship is very near to the Master equation of Manning's formula for best section, where R= d/ 2, see chapter 7. Therefore upto medium dischages only best section, with steepest possible side slope should be adopted.

2.4.2 Hydraulic Sub Critical Velocity

Another important criteria of limiting velocity is that it should be less than the critical velocity i.e. Froude's Number, $F = V/\sqrt{gD_m} < 1$. At F = 1, Velocity is critical, Dm = A/T, (Dm = hydraulic mean depth, A= area of flow and T = Top width), except at rapids to negotiate falls. If the velocity is more than critical velocity (i.e. super critical velocity), than at any obstruction (stones laid by villagers or shurbs or silting etc.) sudden hydraulic jump will be formed, and velocity may become sub critical. This will increase the depth of flow very much and produce waves and may even cause over topping of banks, besides abrasing and damaging the lining. At filling reaches it may cause even breaching of canal, by over topping. This problem normally occurs in low capacity canals, in highly sloping topography, where steep slopes has been feasible. This type of problem was seen on many canals of Gambhiri project chittorgarh, Rajasthan, India. Therefore section should be checked for critical velocity. Alternatinely critical depth is computed with the help of **Table 2.5** and depth of canal little higher than this should be provided.

Table 2.5 : Semi empirical equation for the estimation of critical depth (y_c)

Channel type	Equation for y_c	Remarks
Rectangular (b)	$\left(\frac{Q^2}{gb^2}\right)^{1/3}$	
Trapezoidal (1:2, b)	$0.81\left(\frac{Q^2}{gZ^{0.75}b^{1.25}}\right)^{0.27} - \frac{b}{30Z}$	Range of applicability $0.1<\frac{Q}{b^{2.5}}<0.4$ For $\frac{Q}{b^{2.5}}<0.1$ use equation for rectangular channel
Circular (do)	$\left(\frac{1.01}{d_0^{0.26}}\right)\left(\frac{Q^2}{g}\right)^{0.25}$	Range of applicability $0.02\leq\frac{y_c}{d_0}\leq 0.85$
Triangular (1:2)	$\left(\frac{2Q^2}{gZ^2}\right)^{0.20}$	

2.5 Summary and Conclusion

Factors effecting Manning's roughness coefficient (N) are discussed. Tables for assessment of roughness coefficient, widely accepted are given for lined/unlined and natural rivers. Studies in India on actual operation of canal indicate little higher values than given in the table. Use of little higher value by 20 % is proposed. Channels of compound (general geometrical shape) and composite roughness (i.e. roughness of bed different than that of sides) are explained.

Unique and common velocity table for different bed slopes for all types of best sections (discussed in chapter) is given for quick decisions.

Lastly permissible velocities and their limitations are explained.

References

Ahmed, M. Yassin, Mean Roughness Coefficient in Open Channels with Different Roughness of Bed and Side Walls, No. 27, Verlag Leemann, Zurich, 1954.

Arya, R.S. Et Al, Design Charts for Earthen and Lined Channels, 1st International Research and Development Conference Water & Energy 2001, Central Board of Irrigation and Power, New Delhi, October, 1995.

BIS (Bureau of Indian Standards), Code of Practice for Design of Cross-Section of Lined Canals, IS No 4745 - 1968.

BIS (Bureau of Indian Standards), Code of Practice for Maintenance of Canals Unlined Canals, IS 4839-1968, Part I.

BIS (Bureau of Indian Standards), Criteria for design of lined canals and guidelines for selection of type of lining (first revision), IS 10430: 2000

CBIP, Pub. No. 144: Symposium on Operation and Maintenance of Canal Systems.

Chow, V.T, Open - Channel Hydraulic, Mc. Graw - Hill Book Company Ltd., 1973.

Chow, Ven Te, A Note on the Manning Formula, Transactions, American Geo-Physical Union, Vol. 36, No. 4, p. 688, August, 1955.

Einstein, H.A. and R.B. Banks: Fluid Resistance of Composite Roughness, Tranactions, American Geophysical Union, Vol. 31, No. 4, pp. 603-610, August, 1950.

Engelund, F., Hydraulic Resistance of Alluvial Streams, Journal of the Hydraulics Division, ASCE, Vol. 92, No. HY2, Proc. Paper 4739, pp. 315-327, Mar., 1966.

Framji, K.K., Design Practices of Irrigation Canals in the World, Report of International Commission of Irrigation and Drainage, Central Office, 48 Nyaya Marg, Chanakyapuri, New Delhi - 21, 1972.

George, W. Pickels: Run-Off Investigations in Central Illinois, University of Illinois, Engineering Experiment Station, Bulletin 232, Vol. 29, No. 3, September, 1931.

Horton, Robert E., Separate Roughness Coefficients for Channel Bottom and Sides, Engineering News-Record, Vol. III, No. 22, pp. 652-653, November 30, 1933.

Kaka, N. M., Analysis of Resistance to Flow in Open Channels with Artificial Roughnesses; M.E. Thesis University of Roorkee, 1974.

Lotter, G.K., Considerations on Hydraulic Design of Channels with Different Roughness of Walls, Transactions, All Union Scientific Research Institute of Hydraulic Engineering, Leningard, Vol. 9, pp. 238-241, 1933.

Manning, Robert, on the Flow of Water in Open Channels and Pipes, Transactions, Institution of Civil Engineers of Ireland, Vol. 20, pp. 161-207, Dublin, Supplement, Vol. 24, pp. 179-207, 1891, 1895.

Pavlovskil, N.N., On A Design Formula For Uniform Movement in Channels With Non homogeneous Walls, Transactions, All Union Scientific Research Institute of Hydraulic Engineering, Leningard, Vol. 3, pp. 157-164, 1931.

Sesha, C. R., Pillai, Composite Rugosity Coefficient in Open Channel Flow; Journal of Central Board of Irrigation and Power, March 1962.

Muhlhofer L., Roughness Investigations in A Shaft with Concrete Bottom and Unlined Walls, Munich, Vol. 28, No. 8, p 85-88, 1933.

Tin, U. Kyaw, Study on Friction Factor in Rigid Open Channels; M.E. Thesis University of Roorkee, 1975.

□□□

Chapter 3

Design of Mehboob and Best Trapezoidal Canals

(Medium or Low Capacity Canals)

A canal of approximate discharge of less than 80 cumecs (3000 cusecs) discharge may be called a medium or low capacity canal).

3.1 Mehboob Section

A Mehboob Section is only a modification of best trapezoidal section by making its complete bottom circular as shown in figure 3.1. Also see figure 1.6 and para 1.7. That is why these canals are also known as canals of zero bed width, or trapezoidal canals with curved bottom, or cup shaped canals.

But these cannot be called as triangular canals with curves, as often spoken. In fact it is a truncated triangle circumscribing the semi circle. Trapezium is also a truncated triangle circum scribing a circle.

More accurately it can be said as a combination of semi circular and best trapezoidal section so as to take full advantage of their hydraulic efficiency and practicability.

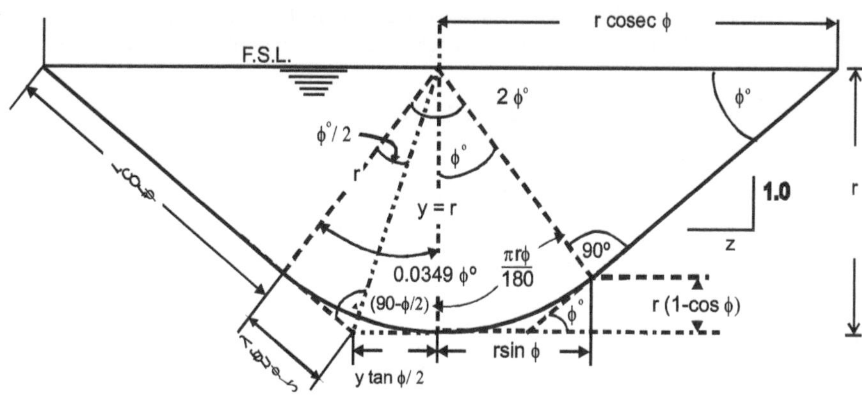

Figure 3.1 : Mehboob section

This is such a beautiful section that the designer can never commit a mistake, even if he wants to commit a mistake. The dimensions are so geometrically fixed that no one can play with any dimension.

The value of this section was realised for the first time by Er S I Mehboob (1943) in his Punjab Irrigation Branch Paper No.260" Lining of Canals". There after this canal section became well known as Mehboob Section after his name. He also stated that the most economical section i.e. the maximum area for the minimum perimeter is when the centre of the arc is at the full supply line of the section. This is also evident in figure 1.6. This is also the condition for all best hydraulic sections. This section is also useful to a higher silt carrying capacity than a wide shallow one.

Bhakra canal manual (1954) proposed this section with a limiting depth of 18 ft. (6m) and maximum velocity of 6ft./s (2m/s), for discharges upto 3000 cusecs (practically 100 cumecs). It was adopted on all Bhankra canal system in the areas of Punjab, Haryana, and Rajasthan, India.

The silt deposit at the off take sites where heading up is done, affects the coefficient of rugosity on a smaller portion of the perimeter and hence cannot greatly affect the velocities. The effect on Lacey's "f" is also correspondingly less. (N is related to silt factor, 'f' by $N = 0.0257\ f^{1/4}$).

The geometric and hydraulic properties of this section are computed as below:

Cross sectional Area of flow

$A = \pi\ r^2\ (2\Phi\ /360) + 2.\ (1/2)\ r.\ r\ \mathrm{Cot}\ \Phi$ (Φ is in degrees)

$= \pi r^2\ (\Phi/180) + r^2\ \mathrm{Cot}\Phi = r^2\ (\pi\Phi/180 + \mathrm{Cot}\Phi) = r^2\ (0.01745\Phi + \cot\Phi)$

Wetted perimeter

$P = 2\pi r\ .\ 2\Phi/360 + 2r\ \mathrm{Cot}\Phi$ (Φ is in degree)

$= 2r\ (\pi\Phi/180 + \mathrm{Cot}\Phi) = r\ (0.0349\ \Phi + 2\ \cot\Phi)$

Hydraulic mean depth

$R = A/P = r^2\ (\pi\Phi/180 + \mathrm{Cot}\Phi)/2r\ (\pi\Phi/180 + \mathrm{Cot}\Phi) = r\ /2$

Thus hydraulic mean depth (R) is a function of depth only irrespective of any side slope, and is always equal to r/2. However minimum area occurs only at Z = 0.577 (1: 0.577, V: H)

$V = R^{2/3}\ S^{1/2}/N = (r/2)^{2/3}\ S^{1/2}/N = 0.63\ r^{2/3}\ S^{1/2}/N$ (in metric system)

The values of area (A), perimeter (P) and hydraulic radius (R) and r of Mehboob Section with different side slopes are given in Table-3.1. Further to facilitate quick designs, computation of these parameters for the different depths (radius) and side slopes are given in Table-.3.1, for ready reference of the designer, and to avoid any computation mistakes. Velocity is read from table 2.1. A multiplication of area A and velocity V gives discharge (Q = A. V). Alternatively r i.e. y (depth of flow) is directly calculated from the equation of discharge (Q) given in the last column of table 3.1. It is derived as under:

$Q = A.\ V. = r^2\ (0.01745\ \varphi + \cot\ \varphi)\ (0.63r^{2/3}\ S^{1/2}\ /N)$

For $\phi = 45^{\circ}$ (or 1:1 side slopes)

$Q = r^2\ (0.01745\ (45) + 1)\ (0.63\ r^{2/3}\ S^{1/2}\ /N) = (1.78525)\ (0.63)\ r^{8/3}\ S^{1/2}/N$

$$\text{or } r^{8/3} = \frac{QN}{1.1247\sqrt{S}} \text{ or } r = =0.957\left(\frac{QN}{\sqrt{S}}\right)^{3/8}$$

Thus for known values of ϕ, depth relation reduces to a very simple form of all known variables, and no trial and error is required.

In the table a comparison with semi circle is also given for the designer's convienience, and to choose the best possible side slope.

Table 3.1 : Mehboob sections with various side slopes (Metric System only)

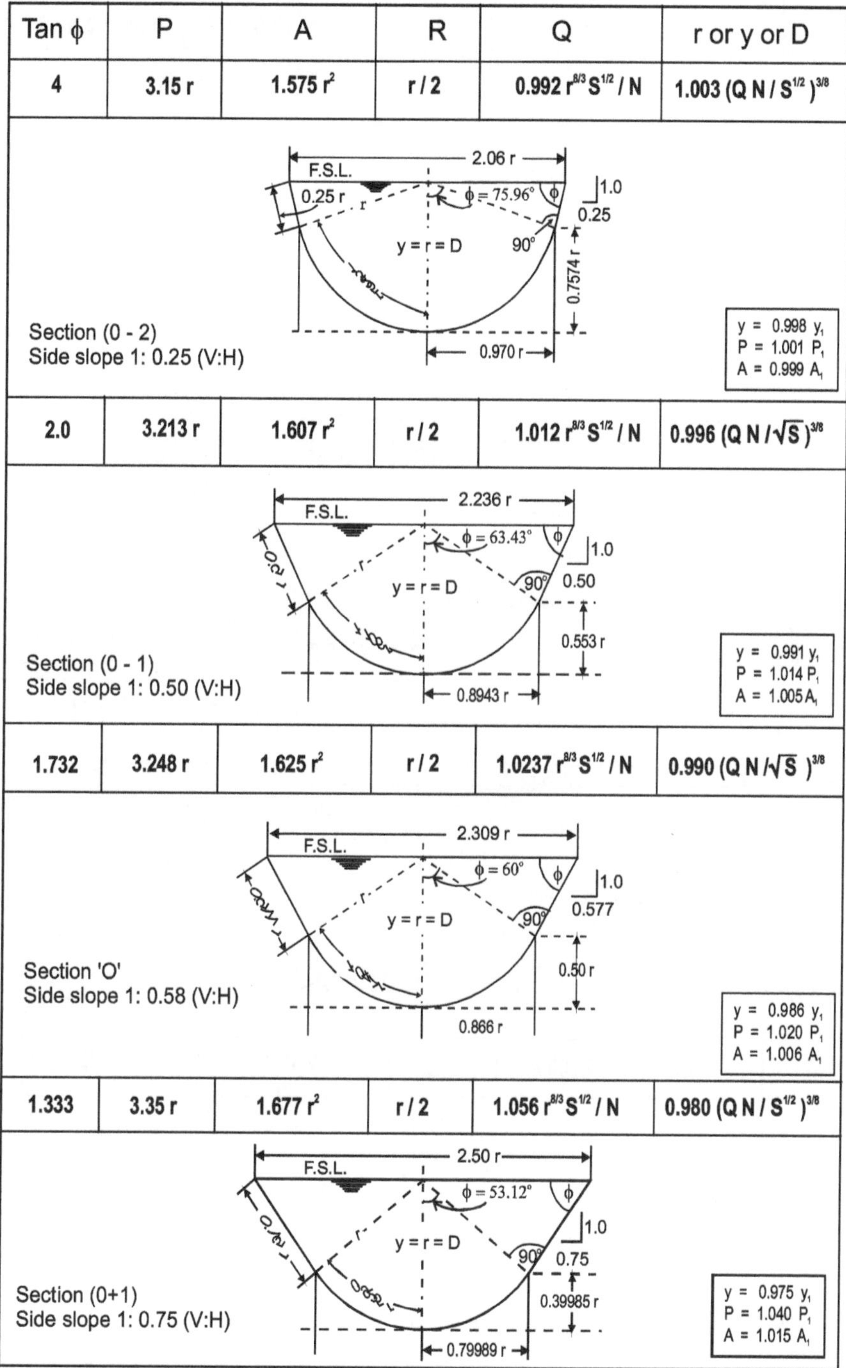

Tan ϕ	P	A	R	Q	r or y or D
4	3.15 r	$1.575\, r^2$	r / 2	$0.992\, r^{8/3} S^{1/2} / N$	$1.003\, (Q N / S^{1/2})^{3/8}$
2.0	3.213 r	$1.607\, r^2$	r / 2	$1.012\, r^{8/3} S^{1/2} / N$	$0.996\, (Q N / \sqrt{S})^{3/8}$
1.732	3.248 r	$1.625\, r^2$	r / 2	$1.0237\, r^{8/3} S^{1/2} / N$	$0.990\, (Q N / \sqrt{S})^{3/8}$
1.333	3.35 r	$1.677\, r^2$	r / 2	$1.056\, r^{8/3} S^{1/2} / N$	$0.980\, (Q N / S^{1/2})^{3/8}$

Contd...

Table 3.1. Contd...

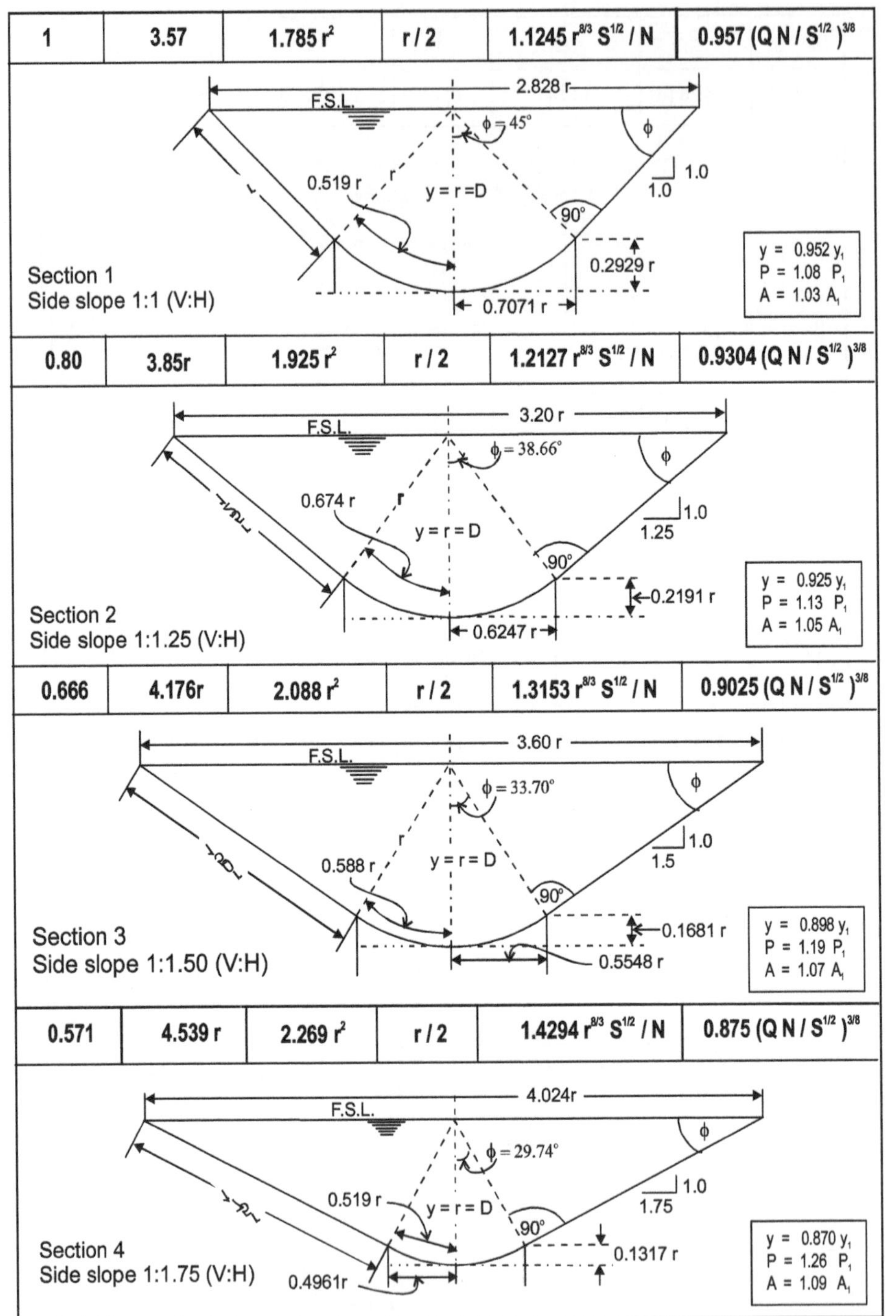

1	3.57	$1.785\ r^2$	r / 2	$1.1245\ r^{8/3}\ S^{1/2} / N$	$0.957\ (Q\ N / S^{1/2})^{3/8}$
Section 1 Side slope 1:1 (V:H)					y = 0.952 y_1 P = 1.08 P_1 A = 1.03 A_1
0.80	3.85r	$1.925\ r^2$	r / 2	$1.2127\ r^{8/3}\ S^{1/2} / N$	$0.9304\ (Q\ N / S^{1/2})^{3/8}$
Section 2 Side slope 1:1.25 (V:H)					y = 0.925 y_1 P = 1.13 P_1 A = 1.05 A_1
0.666	4.176r	$2.088\ r^2$	r / 2	$1.3153\ r^{8/3}\ S^{1/2} / N$	$0.9025\ (Q\ N / S^{1/2})^{3/8}$
Section 3 Side slope 1:1.50 (V:H)					y = 0.898 y_1 P = 1.19 P_1 A = 1.07 A_1
0.571	4.539 r	$2.269\ r^2$	r / 2	$1.4294\ r^{8/3}\ S^{1/2} / N$	$0.875\ (Q\ N / S^{1/2})^{3/8}$
Section 4 Side slope 1:1.75 (V:H)					y = 0.870 y_1 P = 1.26 P_1 A = 1.09 A_1

Contd...

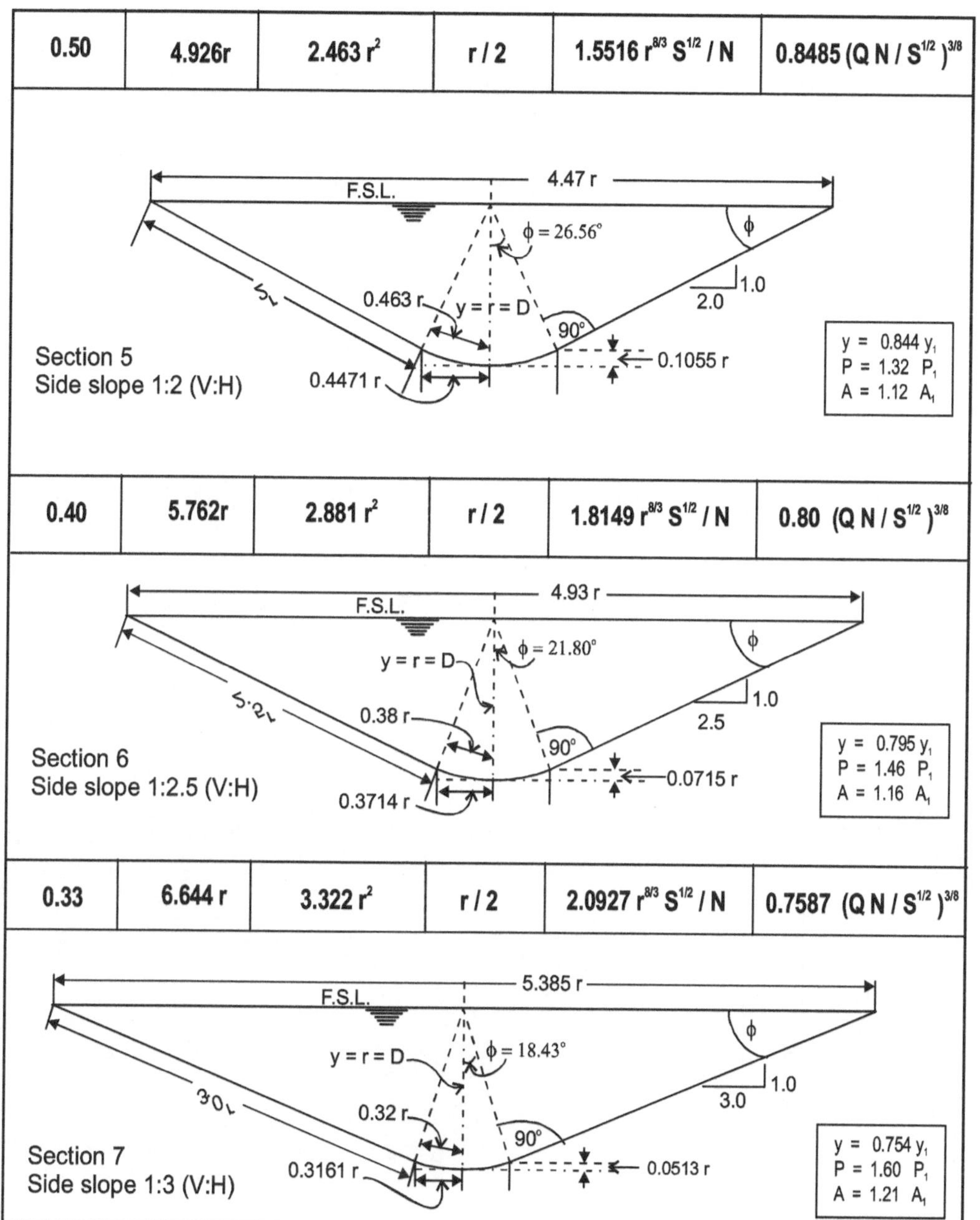

0.50	**4.926r**	**$2.463\ r^2$**	**r / 2**	**$1.5516\ r^{8/3}\ S^{1/2} / N$**	**$0.8485\ (Q\ N / S^{1/2})^{3/8}$**
0.40	**5.762r**	**$2.881\ r^2$**	**r / 2**	**$1.8149\ r^{8/3}\ S^{1/2} / N$**	**$0.80\ (Q\ N / S^{1/2})^{3/8}$**
0.33	**6.644 r**	**$3.322\ r^2$**	**r / 2**	**$2.0927\ r^{8/3}\ S^{1/2} / N$**	**$0.7587\ (Q\ N / S^{1/2})^{3/8}$**

Note : Figures in the right corner box are given for comparison with semi circle. y_1, P_1 and A_1 are the depth of flow, wetted perimeter and area of semi circle. It also helps in comparison of the sections with each other as well as with other section such as Mehboob, Trapezoidal, Tapezoidal with curved ends, Rectangle or circular bottom rectangle.

Table 3.1.1 Cross-sectional area and perimeter of mehboob sections with different side slopes

DEPTH (Radius) r	SECTION - I $\phi = 45^\circ$ 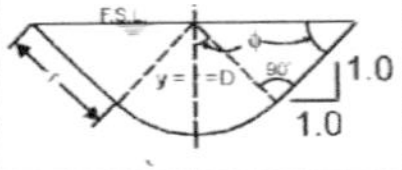		SECTION - II $\phi = 38.66^\circ$ 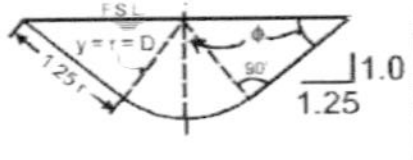		SECTION - III $\phi = 33.69^\circ$ 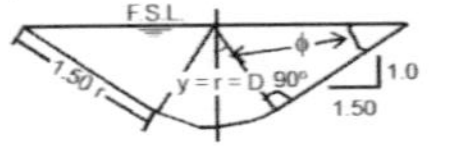		SECTION - IV $\phi = 29.74$ 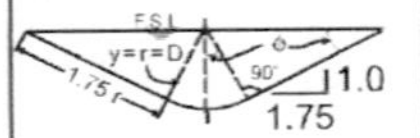		SECTION - V $\phi = 26.56^\circ$ 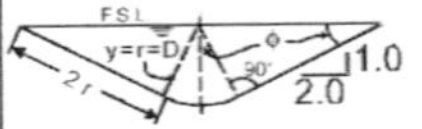		SECTION - VI $\phi = 21.80^\circ$ 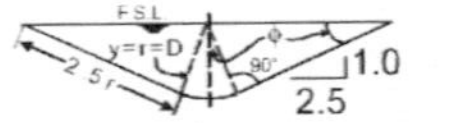	
m	P = 3.57r m	A = 1.785 r^2 m^2	P = 3.85r m	A = 1.925 r^2 m^2	P = 4.176r m	A = 2.088 r^2 m^2	P = 4.539r m	A = 2.269 r^2 m^2	P = 4.926r m	A = 2.463 r^2 m^2	P = 5.762r m	A = 2.881 r^2 m^2
1	2	3	4	5	6	7	2	3	4	5	6	7
1.00	3.570	1.785	3.850	1.925	4.176	2.088	4.539	2.269	4.926	2.463	5.762	2.881
1.10	3.927	2.160	4.235	2.329	4.594	2.526	4.993	2.745	5.419	2.980	6.338	3.486
1.20	4.284	2.570	4.620	2.772	5.011	3.007	5.447	3.267	5.911	3.547	6.914	4.149
1.30	4.641	3.017	5.005	3.253	5.429	3.529	5.901	3.835	6.404	4.162	7.491	4.869
1.40	4.998	3.499	5.390	3.773	5.846	4.092	6.355	4.447	6.896	4.827	8.067	5.647
1.50	5.355	4.016	5.775	4.331	6.264	4.698	6.809	5.105	7.389	5.542	8.643	6.482
1.60	5.712	4.570	6.160	4.928	6.682	5.345	7.262	5.809	7.882	6.305	9.219	7.375
1.70	6.069	5.159	6.545	5.563	7.099	6.034	7.716	6.557	8.374	7.118	9.795	8.326
1.80	6.426	5.783	6.930	6.237	7.517	6.765	8.170	7.352	8.867	7.980	10.372	9.334
1.90	6.783	6.444	7.315	6.949	7.934	7.538	8.624	8.191	9.359	8.891	10.948	10.400
2.00	7.140	7.140	7.700	7.700	8.352	8.352	9.078	9.076	9.852	9.852	11.524	11.524
2.10	7.497	7.872	8.085	8.489	8.770	9.208	9.532	10.006	10.345	10.862	12.100	12.705
2.20	7.854	8.639	8.470	9.317	9.187	10.106	9.986	10.982	10.837	11.921	12.676	13.944
2.30	8.211	9.443	8.855	10.183	9.605	11.046	10.440	12.003	11.330	13.029	13.253	15.240
2.40	8.568	10.282	9.240	11.088	10.022	12.027	10.894	13.069	11.822	14.187	13.829	16.595
2.50	8.925	11.156	9.625	12.031	10.440	13.050	11.348	14.181	12.315	15.394	14.405	18.006
2.60	9.282	12.067	10.010	13.013	10.858	14.115	11.801	15.338	12.808	16.650	14.981	19.476
2.70	9.639	13.013	10.395	14.033	11.275	15.222	12.255	16.541	13.300	17.955	15.557	21.002
2.80	9.996	13.994	10.780	15.092	11.693	16.370	12.709	17.789	13.793	19.310	16.134	22.587
2.90	10.353	15.012	11.165	16.189	12.110	17.560	13.163	19.082	14.285	20.714	16.710	24.229
3.00	10.710	16.065	11.550	17.325	12.528	18.792	13.617	20.421	14.778	22.167	17.286	25.929
3.10	11.067	17.154	11.935	18.499	12.946	20.066	14.071	21.805	15.271	23.669	17.862	27.686
3.20	11.424	18.278	12.320	19.712	13.363	21.381	14.525	23.235	15.763	25.221	18.438	29.501
3.30	11.781	19.439	12.705	20.963	13.781	22.738	14.979	24.709	16.256	26.822	19.015	31.374
3.40	12.138	20.635	13.090	22.253	14.198	24.137	15.433	26.230	16.748	28.472	19.591	33.304
3.50	12.495	21.866	13.475	23.581	14.616	25.578	15.887	27.795	17.241	30.172	20.167	35.292
3.60	12.852	23.134	13.860	24.948	15.034	27.060	16.340	29.406	17.734	31.920	20.743	37.338
3.70	13.209	24.437	14.245	26.353	15.451	28.585	16.794	31.063	18.226	33.718	21.319	39.441
3.80	13.566	25.775	14.630	27.797	15.869	30.151	17.248	32.764	18.719	35.566	21.896	41.602
3.90	13.923	27.150	15.015	29.279	16.286	31.758	17.702	34.511	19.211	37.462	22.472	43.820
4.00	14.280	28.560	15.400	30.800	16.704	33.408	18.156	36.304	19.704	39.408	23.048	46.096

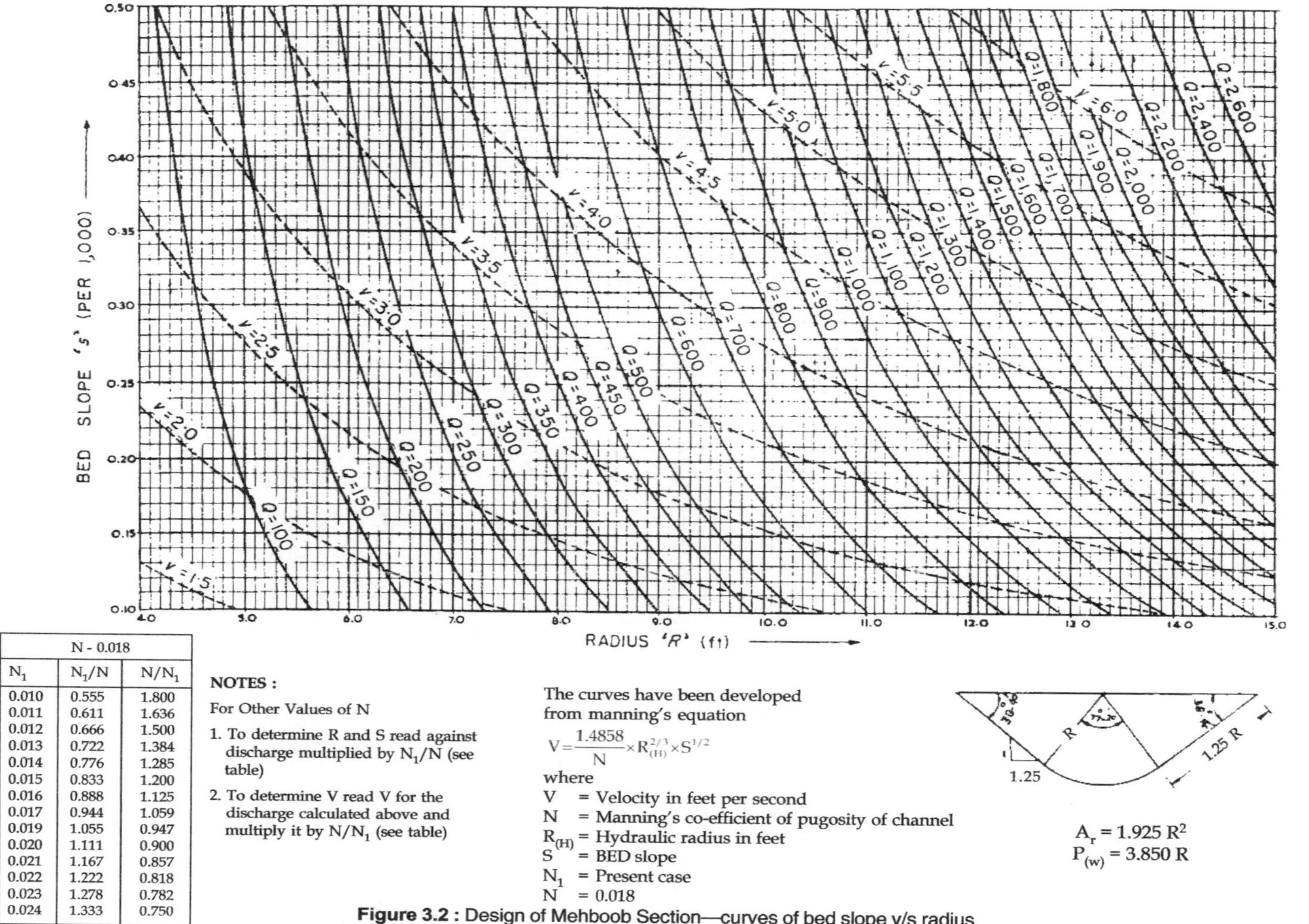

N - 0.018		
N_1	N_1/N	N/N_1
0.010	0.555	1.800
0.011	0.611	1.636
0.012	0.666	1.500
0.013	0.722	1.384
0.014	0.776	1.285
0.015	0.833	1.200
0.016	0.888	1.125
0.017	0.944	1.059
0.019	1.055	0.947
0.020	1.111	0.900
0.021	1.167	0.857
0.022	1.222	0.818
0.023	1.278	0.782
0.024	1.333	0.750

NOTES :

For Other Values of N

1. To determine R and S read against discharge multiplied by N_1/N (see table)
2. To determine V read V for the discharge calculated above and multiply it by N/N_1 (see table)

The curves have been developed from manning's equation

$$V = \frac{1.4858}{N} \times R_{(H)}^{2/3} \times S^{1/2}$$

where

V = Velocity in feet per second
N = Manning's co-efficient of pugosity of channel
$R_{(H)}$ = Hydraulic radius in feet
S = BED slope
N_1 = Present case
N = 0.018

$A_r = 1.925\ R^2$
$P_{(w)} = 3.850\ R$

Figure 3.2 : Design of Mehboob Section—curves of bed slope v/s radius

3.1.1 Limitation of section

These canals are most suited for discharge upto 60 to 80 cumecs. For discharges greater than this, the depth of the section as well as velocity of flow increases. Higher depth may cause the problems of stability of side slopes. Higher velocity may cross the limiting velocity of canal lining material. Therefore depth and velocity are generally restricted.

It may also be desirable to limit the depth in order to avoid excessive leakage, and to reduce the risk of bank failure. Normally a maximum depth of about 5 to 8 m has been preferred. The desirable limit in any particular case depends upon the nature of the ground, the nature of any artificial embankments, and also upon the difference between the canal water-level and the natural water table.

3.2 Choice/Selection between Mehboob and Trapezoidal Sections

The best trapezoidal section apparently appears to have the same hydraulic efficiency as the Mehboob section, see Para 1.7. However for a given area (but not for a given discharge) the Trapezoidal Section is 97.6% efficient of Mehboob Section where as the perimeter is 3.3% more. A very good comparison of semi circle, Mehboob & best Trapezoidal Sections is given in Chapter 1. For a given discharge, trapezoidal section is 1.5% more in area and 3.9% more in perimeter in comparison to Mehboob section, see table 1.4.

Therefore from hydraulic considerations both the sections are practically identical. The only difference is that in Mehboob Section, no one can commit any error or mischief even if one wants to do so, as he cannot play with any of its individual dimensions. Whereas in Trapezoidal Section, there is lot of scope to play with bed width and depth (B/D ratio). However, it should be done to the minimum possible extent of practicability if one is interested in the optimum hydraulically efficient section.

Structurally, Mehboob Section is (i) stronger, (ii) durable than trapezoidal as it provides good foundation to the sides, through it's curved bottom. Also the chances of leakage between joints of the horizontal bed and the sides of trapezoidal are removed in the Mehboob Section, as it has no such joint. The greatest advantage is that is requires less land width because of higher depth and velocity.

Still some prefer trapezoidal section in comparison to Mehboob Section. The arguments placed in its favour with counter arguments are indicated below:

i) *Ease of construction* : Construction of curved bottom is more difficult than the horizontal bed of trapezoidal section. This is a false notion, because of low exposure, to such canals. It is as easy as for trapezoidal as explained below:

a) *Compaction* : In new canals compaction of curved end in not done separately. A provision of lip cutting is made in the compaction of embankments and curved portion is shaped in that. Similarly it is excavated in the compacted bed. Similarly in existing canals provision of lip cutting is made on one side and enough earth is compacted on other side to include lip cutting.

b) *Shaping of earth work* : Some engineers say that earth cannot be easily shaped in the desired curvature. Well this is only an apprehension. Only a wooden frame of desired curvature is required to guide the workers to shape the earth. In actual practice, earth at the bottom in curved shape is more stable than in straight angles.

c) *Lining* : There are some doubts in the minds of many engineers about ease of construction, particularly cement concrete lining. Well it is done in the same way as for sides, only form work is designed accordingly or slip forms are fabricated according to shape. There is no difficulty in brick / tile or any other type of lining of this shape.

ii) *Ease of cleaning the bed* : Tools such as spades can work more easily in horizontal bed than in curved bottom.

But this is not actual problem as spades of different sizes and shapes are available

iii) *Silting of curved bottom* : Some Engineers argue that curved bottom is more quickly or easily silted and then it behaves as a trapezoidal section. Well this argument is not true as the phenomenon of silting is same in both the sections. Rather silting is more in trapezoidal section, because of lesser velocity rather than in Mehboob section (in Mehboob section velocity is higher) More over in partial discharges more silting occurs in trapezoidal channels than in Mehboob canals. Actual experience indicate that silting occurs in parabolic or cup shape. Also see Para. 4.6 and figure 7.9.

iv) *Ease of transitions* : General feeling is that transition; dwarf walls are more difficult in Mehboob canals than in trapezoidal canals. This is a false notion due to low experience. Skilled masons are able to do with strings (thread) and tape measurements.

Well argument (i), (ii), (iii) and (iv) are not as much valid as stressed above.

Several thousands of Km of canal of various discharges ranging from 0.1 to 60 m^3/s (5 to 2000 cusecs) have been constructed in Mehboob section in Indira Gandhi Canal Project in Rajasthan, India. (Year 1955 2001), and many other projects in Haryana (see Para 6.3) and Punjab States of India.

Since the radius is equivalent to full supply depth, curvature is very small and all tools can work in the same way as on a horizontal bed and there is no difficulty of working with any tools. Rather working on small curves is better than in sharp angles. Several thousand Kilometres of small and large channels constructed with curved bottom, in Haryana, Punjab and Rajasthan (India) are working satisfactory. Also see table 6.1.

Similarly all other points are just based on apprehensions with either no or less experience of Mehboob canals in comparison to trapezoidal canals.

Thus Mehboob section is more better, because of greater strength, durability and efficiency and must be preferred.

However for a given discharge the difference in trapezoidal and Mehboob section is also very small so long as the best hydraulic trapezoidal section is adopted, (where $R = d/2$, and in such cases depth is more than bed width). For bed width greater than depth the trapezoidal section is uneconomical and must be avoided.

3.3 Effect of Side Slopes on the Hydraulic Efficiency

The best trapezoidal section is with the side slopes of 1: 0.577 (V: H). But this side slope is not always feasible due to instability of most of soils at this slope. But some rocks may be stable at slopes steeper or even flatter than this (also see Para 9.9). Therefore depending upon the shear strength (angle of repose or shearing resistance or internal friction) of the soil, or rock conditions, and depth of canal, and many other factors, side slope varying from 1 : 0.25 to 1 : 2.5 (Vertical : Horizontal) has been generally adopted in the canals.

With the flattening of the sides, its tangent point with the semicircle shifts down wards, resulting in reduction of the bed width of the canal and increase in its top width. As the bed width decreases, the section approaches to a triangular shape, which has a high perimeter in comparison to its area and is the most inefficient hydraulic section. So with the flattening of side slopes, hydraulic efficiency decreases.

Similarly with steepening of sides slopes, then the optimum side slope of 1:0.577 (V: H), the section approaches a rectangular section which is again not a very efficient section.

A good effect of various side slopes on the hydraulic efficiency can easily be appreciated in figure 3.3(i). In this figure best trapezoidal & Mehboob sections for few side slopes are plotted on 1:1 slope section which is very near to the best optimum section.

These sections still remain best, as long as these are circumscribing the semi circle. With the flattening of side slopes, the cross sectional area of flow and perimeter increases proportionately and so the conveyance. The hydraulic radius (R) remains constant, so apparently it appears that the hydraulic efficiency also remains constant. In other words discharge should increase in the same proportion as the perimeter and cross sectional area increases, but this is not so.

As the bed width decreases, the section approaches to a triangular shape, which has a large perimeter in comparison to its area and is the most inefficient hydraulic section.

A comparison of the Mehboob sections with the various side slopes is shown in table 3.2.

Table 3.2 : Comparison of Hydraulic Conveyance of Mehboob Section for Various Side Slopes, for Constant Depth

S. No.	Side slope V : H	Area		Perimeter		Hydraulic radius R	Conveyance	
		A	% of Item 1 (row)	P	% of Item 1 (row)		$AR^{2/3}$	% of Item 1 (row)
1.	1 : 1	$1.785\ r^2$	100	3.57 r	100	$r/2$	$1.124\ r^{8/3}$	100.00
2.	1 :1.25	$1.925\ r^2$	107.84	3.85r	107.84	-do-	$1.213\ r^{8/3}$	107.91
3.	1 :1.50	$2.088\ r^2$	116.97	4.176 r	116.97	-do-	$1.315\ r^{8/3}$	116.99
4.	1 :1.75	$2.269\ r^2$	127.11	4.539 r	127.14	-do-	$1.429\ r^{8/3}$	127.13
5.	1 : 2	$2.463\ r^2$	137.98	4.926 r	137.98	-do-	$1.552\ r^{8/3}$	138.07
6.	1 : 2.5	$2.881\ r^2$	161.40	5.762 r	161.40	-do-	$1.815\ r^{8/3}$	161.47
7.	1 : 3	$3.332\ r^2$	186.66	6.644 r	186.10	-do-	$2.093\ r^{8/3}$	186.20

Notes: % is with reference to S.No.1. For a given depth, flattening of side slopes increases area (A), so the perimeter (P) and conveyance (i.e. discharge). All increase in the same proportion.

Apparently as it appears and often said that all the sections are equally efficient. That means flattening or steepening of the side slopes should not change the hydraulic efficiency of the sections. But this is not true for a given discharge.

Since with the flattening of sides, area increases, a lower depth of flow has to be adopted for the same discharge. Still this area and perimeter are more as compared to a semi circle. The effect on hydraulic efficiency for Mehboob section for a given discharge is given in table 3.3. A typical comparison of Mehboob section 1 : 1 side slopes and 1 : 2 sides slopes with circular section is shown in figure 3.3 (ii). It may also be carefully noted that as R decreases, velocity also decrease with flattening of side slopes.

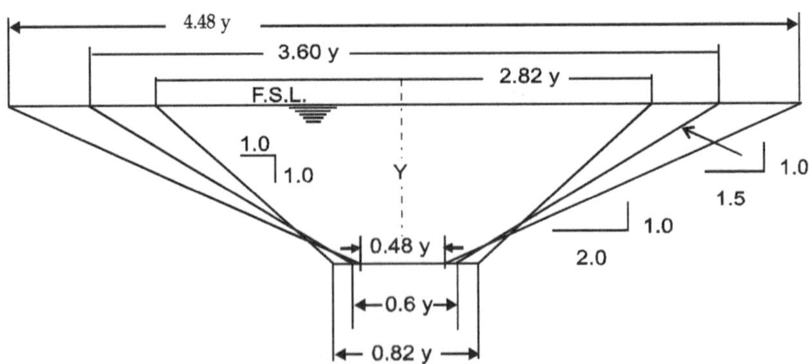

Figure 3.3 : (a) Trapezo ideal

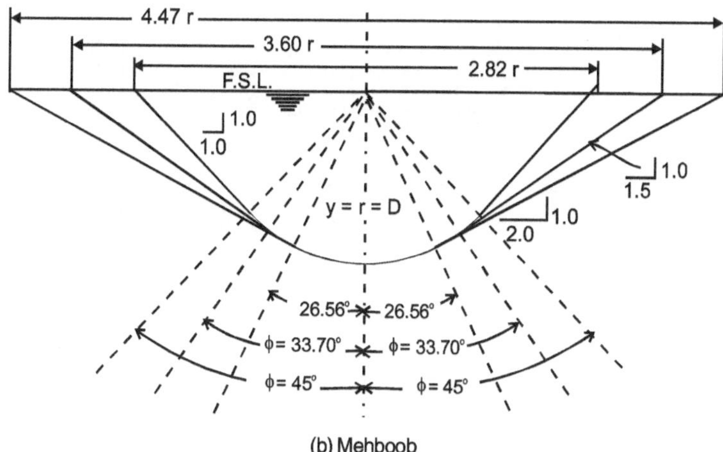

(b) Mehboob

Figure 3.3 (i). Comparison of Trapezoidal and Mehboob sections for various side slopes and constant depth (See Table 3.3)

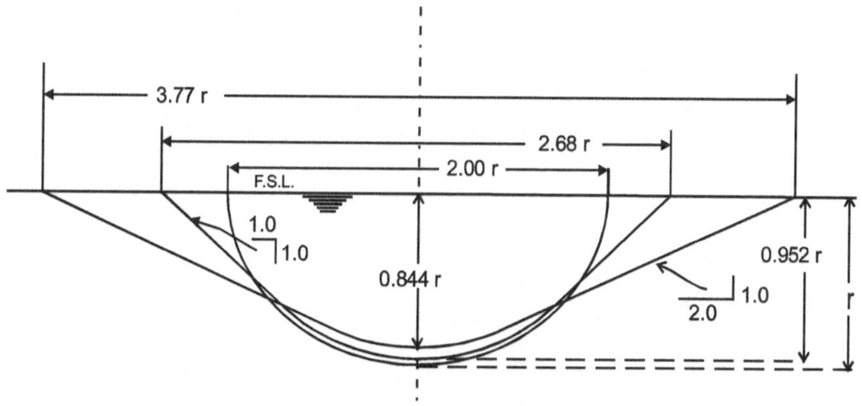

Figure 3.3 (ii). Comparison of Semi Circle and Mehboob sections with two different side slopes (Z) = 1:1 and 1:2 (V:H) for a given discharge (See Table 3.3)

From the above table for a given discharge, a Mehboob section with a side slope of 1:1 has 8.2 % more perimeter as compared to the semi circle and 6% more as compared to Mehboob section of side slope 1:0.58. Its area of cross section is 3% more in comparison to the area of semi circle. Similarly a Mehboob section with a side slope of 1:1.5 has 19.4% more perimeter than that of semi circle and 17.4 % more in comparison to the first Mehboob section that is with side slope of 1:0.58. Thus with flatter side slopes, the section becomes inefficient. This is also evident in table 3.1 and 3.7 where hydraulic efficiency of different side slopes with respect to a semi circle is also given for ready reference for Mehboob and trapezoidal sections.

Similar effects occur in trapezoidal sections and should be appreciated by the designer. Some time a choice may have to be made between flatter slopes versus wider canals with steeper side slopes. In such cases a comparative analysis of both the sections should be made, see para 4.1, 4.4. More detailed and practical aspects of selection of side slopes are discussed in Chapter 4 and 9.

Table 3.3 : Comparison of Mehboob Section for a Given Discharge for Various Side Slopes with Semi Circle (y is depth of flow for a semi circle and y^1 is actual depth of flow in Mehboob section for the same discharge)

S. No.	Side slope V : H	Depth of flow y′	Area		Perimeter		Hydraulic radius		Conveyance	
			A	% of Item 1	P	% of Item 1	R	% of Item 1	$A R^{2/3}$	% of Item 1
1	2	3	4	5	6	7	8	9	10	11
1	Semi circle	y	$1.57y^2$	100	3.14y	100	0.5y	100.0	$0.987\ y^{8/3}$	100
2	1 : 0.58	0.987y	$1.579y^2$	100.57	3.204y	102.03	0.492y	98.4	$0.987\ y^{8/3}$	100
3	1 : 1	0.952y	$1.617y^2$	102.99	3.398y	108.21	0.476y	95.2	$0.987\ y^{8/3}$	100
4	1 : 1.25	0.925y	$1.647y^2$	104.90	3.561y	113.40	0.462y	92.4	$0.987\ y^{8/3}$	100
5	1 : 1.5	0.898y	$1.683y^2$	107.19	3.750y	119.42	0.442y	88.4	$0.987\ y^{8/3}$	100
6	1 : 1.75	0.870y	$1.717y^2$	109.36	3.948y	125.73	0.435y	87.0	$0.987\ y^{8/3}$	100
7	1 : 2	0.844y	$1.754y^2$	111.71	4.157y	132.28	0.422y	84.4	$0.987\ y^{8/3}$	100
8	1 : 2.5	0.795y	$1.820y^2$	115.92	4.580y	145.85	0.397y	79.4	$0.987\ y^{8/3}$	100
9	1 : 3	0.754y	$1.890y^2$	120.38	5.009y	159.52	0.377y	75.4	$0.987\ y^{8/3}$	100

Notes : For a given discharge, as the side slopes flattens, perimeter (P) increases at a faster rate then area (A). Since area of lining increases, hence cost of lining increases.

3.4 Design of Best Trapezoidal Canals

Trapezoidal sections are most common in lined canals, because of its simplicity and practical suitability with almost all type of soils. Its hydraulic efficiency for the best section is also near to the semi circle. The section ranks at number 3 i.e. just next to the mehboob section (which is number 2, as compared to semi circle which is No. 1) see para 1.8. Also the effect of small change of side slopes (which are essential due to the soil conditions) on the hydraulic efficiency may not be high , but become very significant with more flattening of slopes. More details of hydraulic efficiency in comparison to semi circle are discussed in chapter -3 and para 3.3.

Still designers are suggested to adopt Mehboob section (see para 3.1) to the extent possible. They should not be afraid of curved bottom. More details of Mehboob section and curved bottom are discussed in this chapter in para 3.2.

Generally design of trapezoidal channels is done by trial and error. But for best section no trial and error is required. Its dimensions are geometrically fixed as in case of Mehboob section, so that no dimensions can be played with.

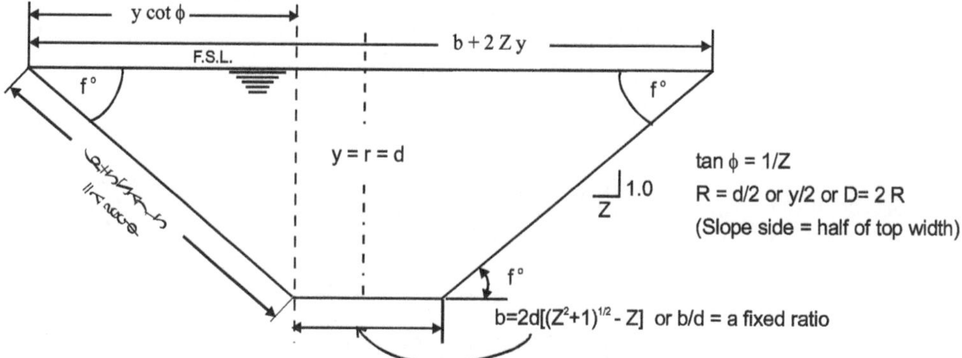

Figure 3.4 : Best trapezoidal section

The optimum best trapezoidal section is with a side slope of 1:0.58. But this slope is not practical and for any other best section of any side slope 1: Z (V:H), following conditions should always be satisfied, from para 1.6.1 (b).

i) $(b + 2Zd)/2 = d(Z^2 + 1)^{1/2}$ or $b = 2d[(Z^2 + 1)^{1/2} - Z]$
   or $b/d = 2[(Z^2 + 1)^{1/2} - Z]$
   i.e. sloping side is equal to half of top width, and

ii) $R = d/2$ or $y/2$
   Cross - sectional area of flow
   $A = bd + Zd^2 = d(b+Zd)$

Minimum wetted perimeter $P = b + b + 2Zd = 2(b + Zd)$

$R = A/P = d/2$

Velocity $V = R^{2/3} S^{1/2}/N = (d/2)^{2/3} S^{1/2}/N = 0.63\, d^{2/3} S^{1/2}/N$

On the basis of above conditions, the geometrical and hydraulic properties of best trapezoidal sections for commonly adopted side slopes are given in Table-3.4, where in, its area (A), perimeter (P), hydraulic radius (R) and conveyance are directly available, in terms of y. All the dimensions have a unique relations with y and no diemention can be played with.

Table 3.4 : Best trapezoidal sections with various side slopes

Metric system only

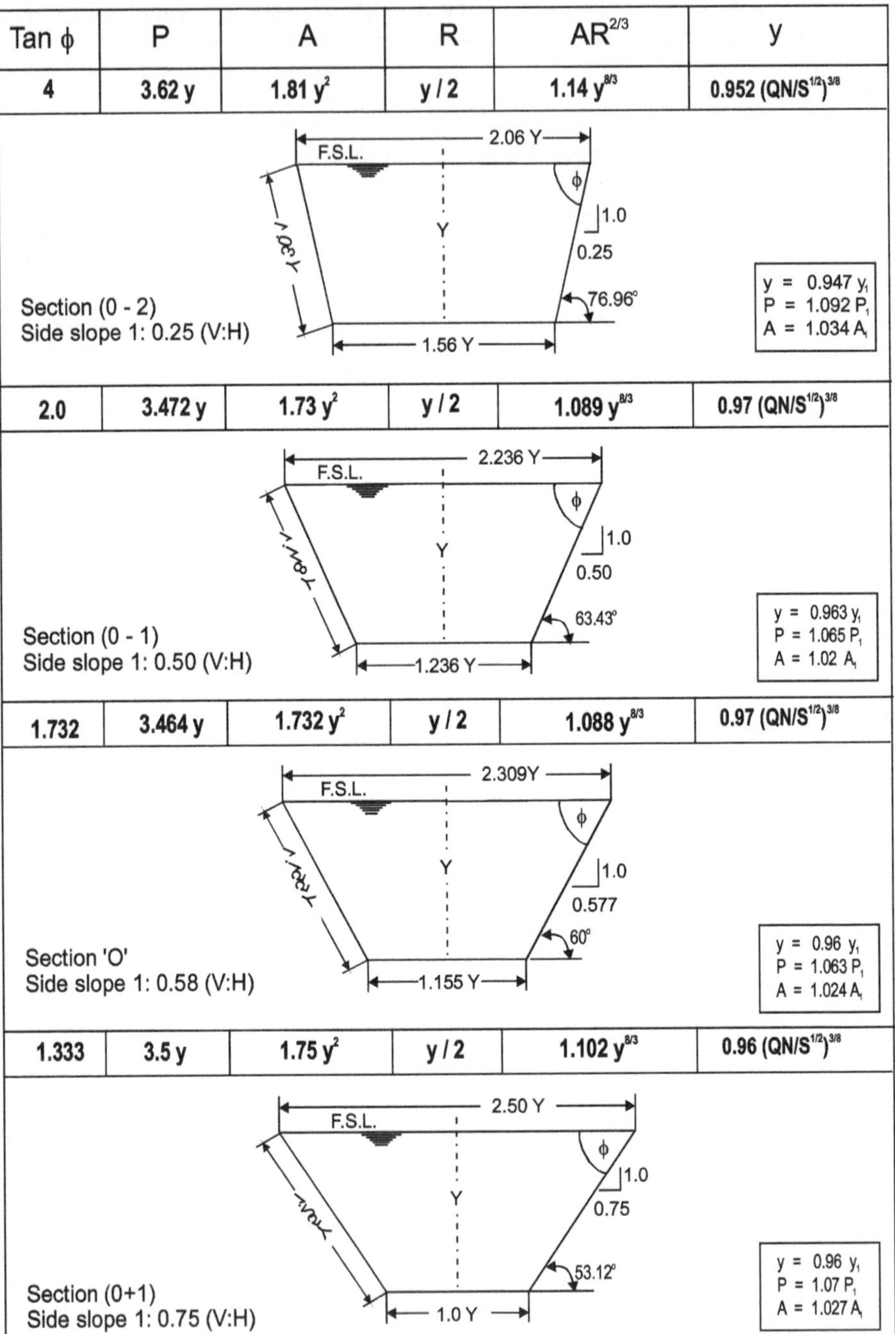

Tan φ	P	A	R	$AR^{2/3}$	y
4	3.62 y	1.81 y^2	y / 2	1.14 $y^{8/3}$	0.952 $(QN/S^{1/2})^{3/8}$
Section (0 - 2) Side slope 1: 0.25 (V:H)					$y = 0.947\ y_1$ $P = 1.092\ P_1$ $A = 1.034\ A_1$
2.0	3.472 y	1.73 y^2	y / 2	1.089 $y^{8/3}$	0.97 $(QN/S^{1/2})^{3/8}$
Section (0 - 1) Side slope 1: 0.50 (V:H)					$y = 0.963\ y_1$ $P = 1.065\ P_1$ $A = 1.02\ A_1$
1.732	3.464 y	1.732 y^2	y / 2	1.088 $y^{8/3}$	0.97 $(QN/S^{1/2})^{3/8}$
Section 'O' Side slope 1: 0.58 (V:H)					$y = 0.96\ y_1$ $P = 1.063\ P_1$ $A = 1.024\ A_1$
1.333	3.5 y	1.75 y^2	y / 2	1.102 $y^{8/3}$	0.96 $(QN/S^{1/2})^{3/8}$
Section (0+1) Side slope 1: 0.75 (V:H)					$y = 0.96\ y_1$ $P = 1.07\ P_1$ $A = 1.027\ A_1$

Contd...

Table 3.4 Contd...

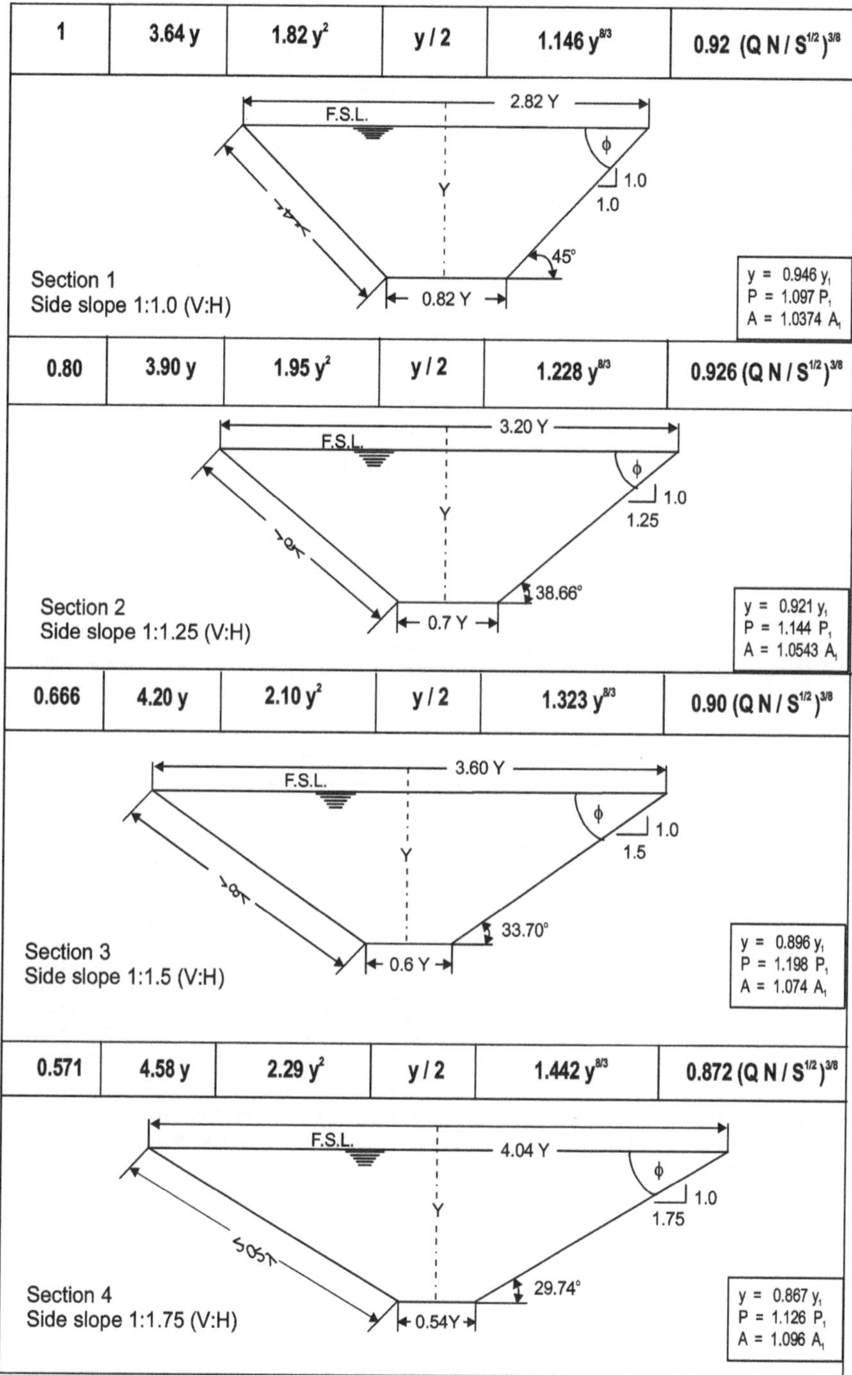

1	**3.64 y**	**1.82 y^2**	**y / 2**	**1.146 $y^{8/3}$**	**0.92 $(Q N / S^{1/2})^{3/8}$**
0.80	**3.90 y**	**1.95 y^2**	**y / 2**	**1.228 $y^{8/3}$**	**0.926 $(Q N / S^{1/2})^{3/8}$**
0.666	**4.20 y**	**2.10 y^2**	**y / 2**	**1.323 $y^{8/3}$**	**0.90 $(Q N / S^{1/2})^{3/8}$**
0.571	**4.58 y**	**2.29 y^2**	**y / 2**	**1.442 $y^{8/3}$**	**0.872 $(Q N / S^{1/2})^{3/8}$**

Contd...

Table 3.4 Contd...

0.50	4.96 y	2.48 y^2	y / 2	1.562 $y^{8/3}$	0.846 $(Q N / S^{1/2})^{3/8}$
0.40	5.36 y	2.68 y^2	y / 2	1.688 $y^{8/3}$	0.822 $(Q N / S^{1/2})^{3/8}$
0.33	5.77 y	2.885 y^2	y / 2	1.818 $y^{8/3}$	0.80 $(Q N / S^{1/2})^{3/8}$

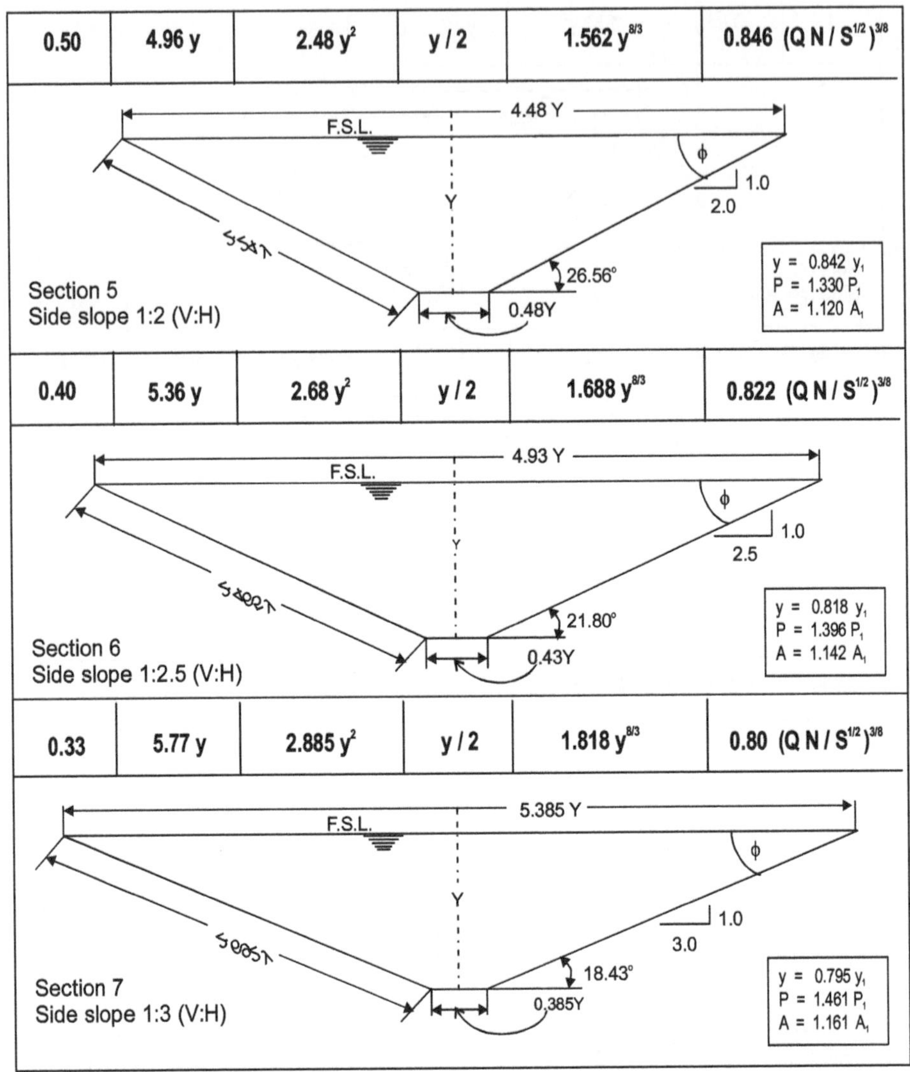

Note : Figures in box are comparison with semi circle for the same discharge.. y_1, P_1 and A_1 are the depth of flow, wetted perimeter and area of semi circle. It also helps in comparison of the sections with each other as well as with other section such as Mehboob, Trapezoidal, Trapezoidal with curved ends, Rectangle or circular bottom rectangle.

Discharge can be computed from the equations given in the table 3.4. For designers convenience , calculations of A and P for various depths and different side slopes are given in Table 3.4.1 From table 3.4.1, area (A) and wetted Perimeter (P) of flow can be directly read from any depth of flow, Velocity is read from table 2.4. A product of A and V gives discharge.

Table - 3.4.1 Cross sectional area and perimeter of best trapezoidal sections with different side slopes

DEPTH	Section I		Section II		Section III		Section IV		Section V		Section VI	
Y = F.S.D.	F.S.L., 1.41 y, 45°, 1.0 / 1.0		F.S.L., 1.6 y, 38.66°, 1.0 / 1.25		F.S.L., 1.8 y, 33.70°, 1.0 / 1.5		F.S.L., 2.02 y, 29.74°, 1.0 / 1.75		F.S.L., 2.24 y, 26.56°, 1.0 / 2.0		F.S.L., 2.465 y, 21.80°, 1.0 / 2.5	
m	P = 3.64 y m	A = 1.82 y^2 m^2	P = 3.90 y m	A = 1.95 y^2 m^2	P = 4.20 y m	A = 2.10 y^2 m^2	P = 4.58 y m	A = 2.29 y^2 m^2	P = 4.96 y m	A = 2.48 y^2 m^2	P = 5.36 y m	A = 2.68 y^2 m^2
1	2	3	4	5	6	7	8	9	10	11	12	13
1.0	3.64	1.82	3.90	1.95	4.20	2.10	4.58	2.29	4.96	2.48	5.36	2.68
1.1	4.00	2.20	4.29	2.36	4.62	2.54	5.04	2.77	5.46	3.00	5.90	3.24
1.2	4.37	2.62	4.68	2.81	5.04	3.02	5.50	3.30	5.95	3.57	6.43	3.86
1.3	4.73	3.08	5.07	3.30	5.46	3.55	5.95	3.87	6.45	4.19	6.97	4.53
1.4	5.10	3.57	5.46	3.82	5.88	4.12	6.41	4.49	6.94	4.86	7.50	5.25
1.5	5.46	4.10	5.85	4.39	6.30	4.73	6.87	5.15	7.44	5.58	8.04	6.03
1.6	5.82	4.66	6.24	4.99	6.72	5.38	7.33	5.86	7.94	6.35	8.58	6.86
1.7	6.19	5.26	6.63	5.64	7.14	6.07	7.79	6.62	8.43	7.17	9.11	7.75
1.8	6.55	5.90	7.02	6.32	7.56	6.80	8.24	7.42	8.93	8.04	9.65	8.68
1.9	6.92	6.57	7.41	7.04	7.98	7.58	8.70	8.27	9.42	8.95	10.18	9.67
2.0	7.28	7.28	7.80	7.80	8.40	8.40	9.16	9.16	9.92	9.92	10.72	10.72
2.2	8.01	8.81	8.58	9.44	9.24	10.16	10.08	11.08	10.91	12.00	11.79	12.97
2.4	8.74	10.48	9.36	11.23	10.08	12.10	10.99	13.19	11.90	14.28	12.86	15.44
2.6	9.46	12.30	10.14	13.18	10.92	14.20	11.91	15.48	12.90	16.76	13.94	18.12
2.8	10.19	14.27	10.92	15.29	11.76	16.46	12.82	17.95	13.89	19.44	15.01	21.01
3.0	10.92	16.38	11.70	17.55	12.60	18.90	13.74	20.61	14.88	22.32	16.08	24.12
3.2	11.65	18.64	12.48	19.97	13.44	21.50	14.66	23.45	15.87	25.40	17.15	27.44
3.4	12.38	21.04	13.26	22.54	14.28	24.28	15.57	26.47	16.86	28.67	18.22	30.98
3.6	13.10	23.59	14.04	25.27	15.12	27.22	16.49	29.68	17.86	32.14	19.30	34.73
3.8	13.83	26.28	14.82	28.16	15.96	30.32	17.40	33.07	18.85	35.81	20.37	38.70
4.0	14.56	29.12	15.60	31.20	16.80	33.60	18.32	36.64	19.84	39.68	21.44	42.88
4.1	14.92	30.59	15.99	32.78	17.22	35.30	18.78	38.49	20.34	41.69	21.98	45.05
4.2	15.29	32.10	16.38	34.40	17.64	37.04	19.24	40.40	20.83	43.75	22.51	47.28

Alternatively depth of flow (y) is directly calculated from the equation given in the last column of table 3.4. It is derived as under:

Q $= A V = d (b + Zd) (0.63\ d^{2/3}\ S^{1/2}/N)$

For m $= 1{:}1;\ b = 2d\ [(1 + 1)^{1/2} - 1] = 2d\ (1.414 - 1) = 0.828d$

and Q $= d\ (b + d)\ (.63\ d^{2/3}\ S^{1/2})\ /N$

$= d\ (0.828\ d + d)\ (0.63\ d^{2/3})\ S^{1/2}\ /N$

$= (1.828\ d^2)\ (.63\ d^{2/3})\ S^{1/2}\ /N$

$= 1.146 d^{8/3}\ S^{1/2}\ /N$

$$\text{or } d^{8/3} = \frac{QN}{1.146\ S^{1/2}}$$

or d or y $= 0.92\ (QN\ /\ \sqrt{S})^{3/8}$

With the help of Table 3.4 a designer can directly choose a particular trapezoidal section for the side slopes of his choice for any given discharge and compute bed width directly, without any trial and error.

All the limitation which apply to Mehboob sections also apply to these sections. Therefore, these sections are also useful only for discharges upto 100 cumecs.

For any other trapezoidal, section, geometric and hydraulic properties are given in Table 3.4 and 3.4.1 with example to read them.

But the designers are advised to adopt bed width as far as possible near to the bed width of best trapezoidal section, to avoid uneconomical sections. Where ever it becomes necessary to increase the bed width, then they should adopt trapezoidal curved ends (See Chapter 4).

3.4.1 Example: 3.1

1. Design hydraulically best sections for a discharge of 12 cumecs, bed slope of 1 in 5000, N = 0.018; for Mehboob and trapezoidal sections take inner side slope of 1:1.5 (V: H) and 1: 2 (V: H); also compare them.
2. Also compare with trapezoidal sections of Z = 1 : 1.5 (V : H), when B = d, B = 2d, and B = 3d

Steps

(i) Assume depth of flow as d and calculate V in terms of d by Manning's Master Eq para 1.6.

(ii) Calculate area of flow A in terms of d

(iii) Multiply V and A, and equate to discharge

(iv) Solve the equation for d.

Solution

(a) All the hydraulically best, circular, rectangular, Mehboob and trapezoidal section have R = d/ 2 and V = 0.63 d2/3 S1/2 / N (see para 1.6.1)

For S = 1 / 5000, and N = 0.018

$$V = 0.63(d^{2/3})\left(\frac{1}{5000}\right)^{1/2} / 0.018 = 0.495 d^{2/3}$$

(i) For circular section

$$Q = V.A = 0.495\ d^{2/3}\left(\frac{\pi}{2}d^2\right) = 12 \text{ or } 0.777\ d^{8/3} = 12 \text{ or } d = 2.79 \text{ m.}$$

Top width = 2d = 2 (2.79) = 5.58 m

Area A = 1.57 $(2.79)^2$ = 12.22 m^2

Perimeter P = πd = 3.14 (2.79) = 8.76 m

Velocity V = 0.495 $d^{2/3}$ = 0.495 $(2.79)^{2/3}$ = 0.98 m/sec

(ii) Rectangular Section

Q = VA = 0.495 $d^2/3$ $(2d^2)$ = 0.99 $d^{8/3}$ or 0.99 $d^{8/3}$ = 12 or d = 2.55 m.

Bed width = 2d = 2 2.55 = 5.10 m

Area A = Bd = 5.10 X 2.55 = 13.0 m^2

Perimeter P = B + 2d = 5.10 + 2 (2.55) = 10.2 m

Velocity V = 0.495 $d^{2/3}$ = 0.495 $(2.55)^2$ = 0.92 m/s

(iii)/1 Best trapezoidal section for side slope of 1 : 1.5 (V : H)

From table 3.4, B = 0.6d, P = 4.20d, A = 2.10 d^2

Q = V.A = 0.495 $d^{2/3}$ (2.10 d^2) = 12 or 1.04 $d^{8/3}$ = 12 or d = 2.50 m

Bottom width B = 0.6 d = 0.6 (2.5) = 1.5 m

Area A = Bd + Zd^2 = 2.10 d^2 = 1.5 (2.5) + 1.5 $(2.5)^2$ = 13.13 m^2

Perimeter P = 4.20d = 4.0 (2.5) = 10.5m

Velocity V = 0.495 $d^{2/3}$ = 0.495 $(2.5)^{2/3}$ = 0.9 m/s

Top width = 3.60d = 3.6 (2.5) = 9.0 m

Hydraulic radius R = d/2 = 2.5/2 = 1.25 m

(iii)/2 Best trapezoidal section for side slope of 1 : 2 (V : H)

From table 3.4, B = 0.48d, P = 4.96d, A = 2.48 d^2

Q = V.A = 0.495 $d^{2/3}$ (2.48 d^2) = 12 or 1.2276 $d^{8/3}$ = 12 or d = 2.35 m

Bottom width B = 0.48 d = 0.48 (2.35) = 1.135 m

Area A = Bd + Zd^2 = 2.48 d^2 = 2.48 $(2.35)^2$ = 13.69 m^2

Perimeter P = 4.96d = 4.96 (2.35) = 11.66 m

Velocity V = 0.495 $d^{2/3}$ = 0.495 $(2.35)^{2/3}$ = 0.87 m/s

Top width = 4.48d = 4.48 (2.35) = 10.53 m

Hydraulic radius R = d/2 = 2.35/2 = 1.18 m

(iv)/1 Mehboob section for side slope of 1 : 1.5 (V : H)

From table 3.1, P = 4.176d, A = 2.088d^2

Q = V.A = 0.495 $d^{2/3}$ (2.088 d^2) = 12 or 1.0335 $d^{8/3}$ = 12 or d = 2.51 m

Area A = 2.088 d^2 = 1.5 (2.51) = 13.15 m^2

Perimeter P = 4.176d = 4.176 (2.51) = 10.48m

Velocity V = 0.495 $d^{2/3}$ = 0.495 $(2.51)^{2/3}$ = 0.9 m/s

Top width = 3.60d = 3.6 (2.51) = 9.04 m

Hydraulic radius R = d/2 = 2.5/2 = 1.25 m

(iv)/2 Mehboob section for side slope of 1 : 2 (V : H)

From table 3.1, P = 4.926 d, A = 2.463 d^2

Q = V.A = 0.495 $d^{2/3}$ (2.463 d^2) = 12

or 1.219 $d^{8/3}$ = 12 or d = 2.36 m

Area A = 2.463 d^2 = 2.463 $(2.36)^2$ = 13.72 m^2

Perimeter P = 4.926d = 4.926 (2.36) = 11.63 m

Velocity V = 0.495 $d^{2/3}$ = 0.495 $(2.36)^{2/3}$ = 0.88 m/s

Top width = 4.47d = 4.47 (2.36) = 10.55 m

Hydraulic radius R = d/2 = 2.36/2 = 1.18 m

(b)/1 Trapezoidal section for side slope of Z = 1 : 1.5 (V : H), B = d

P = d + 3.6d = 4.60d, A = d^2 + 1.5d^2 = 2.50 d^2, R = 0.54 d

V = $(0.54\ d)^{2/3}$ $(1/5000)^{1/2}$ / 0.018 = 0.521 $d^{2/3}$

Q = V.A = 0.521 $d^{2/3}$ (2.50 d^2) = 12

or 1.30 $d^{8/3}$ = 12 or d = 2.31 m

Bottom width B = d = 2.31 m

Area A = Bd + Zd^2 = 2.50 d^2 = 2.45 $(2.31)^2$ = 13.34 m^2

Perimeter P = 4. 60d = 4.60 (2.31) = 10.63 m

Velocity V = 0.521 $d^{2/3}$ = 0.521 $(2.31)^{2/3}$ = 0.91 m/s

Top width = d + 1.5d + 1.5d = 4.00d = 4.0 (2.31) = 9.24 m

Hydraulic radius R = 0.54 d = 0.54 2.31 = 1.25 m

(b)/2 Trapezoidal section for side slope of Z = 1 : 1.5 (V : H), B = 2d

P = 2d + 3.6d = 5.60d, A = 2d2 + 1.5d2 = 3.50 d2 , R = 0.62 d

V = $(0.62\ d)^{2/3}$ $(1/5000)^{1/2}$ /0.018 = 0.571 $d^{2/3}$

Q = V.A = 0.571 $d^{2/3}$ (2.50 d^2) = 12

or 1.998 $d^{8/3}$ = 12 or d = 1.97 m

Bottom width B = 2.0d = 2.0 (1.97) = 3.94 m

Area A = Bd + Zd^2 = 3.50 d2 = 3. 50 (1.97)2 = 13.58 m2

Perimeter P = 5.60d = 5.60 (1.97) = 11.03 m

Velocity V = 0.571 $d^{2/3}$ = 0.571 $(1.97)^{2/3}$ = 0.89 m/s

Top width = 5d = 4.0 (1.97) = 9.85 m

Hydraulic radius R = 0.67 d = 0.54 (1.97) = 1.22 m

(b)/3 Trapezoidal section for side slope of Z = 1 : 1.5 (V : H), B = 3d

P = 3d + 3.6d = 6.60d, A = 3d^2 + 1.5d^2 = 4.50 d^2, R = 0.68 d

V = $(0.680\ d)^{2/3}$ $(1/5000)^{1/2}$ / 0.018 = 0.607$d^{2/3}$

Q = V.A = 0.607 $d^{2/3}$ (4.50 d^2) = 12

or 4.731 $d^{8/3}$ = 12 or d = 1.75 m

Bottom width B = 3.0d = 2.0 (1. 75) = 5.25 m

Area A = Bd + Zd^2 = 4.50 d^2 = 4. 50 $(1.75)^2$ = 13.78 m^2

Perimeter P = 5.60d = 5.60 (1.75) = 11.55 m

Velocity V = 0.607 $d^{2/3}$ = 0.607 $(1.75)^{2/3}$ = 0.88 m/s

Top width = 6d = 6.0 (1.75) = 10.50 m

Hydraulic radius R = 0.68 d = 0.68 (1.75) = 1.19 m

A comparison of all above sections is given in table below. These sections are also plotted in the enclosed figure.

Table For Example 3.1 Comparison of Sections for Q = 12 cumecs, S = 1/5000, Z = 1 : 1.5 and 1: 2, N = 0.018 ; and Trapezoidal Sections B = d, B = 2d, B = 3d

S. No.	Type of Section	Depth of flow (d) (m)	Bed width (B) (m)	Top width (T) (m)	Area of flow (A) (m^2)	Wetted perimeter (P) (m)	Hydraulic Radius (R) (m)	Velocity (V) (m/sec)
1	Semi Circle	2.79	---	5.58	12.22	8.76	1.39	0.98
2	Rectangular	2.55	5.10	5.10	13.00	10.20	1.27	0.92
3	Mehoob							
a.	(Z =1.50)	2.51	___	9.04	13.15	10.48	1.25	0.91
b.	(Z = 2.00)	2.36	_____	10.55	13.72	11.63	1.18	0.88
	% of 3 a	94.02	___	116.70	104.33	110.9	94.92	96.70
	% Variation	-5.98		+16.70	+4.33	+10.9	-5.58	-3.3
4.	Best Trapezoidal							
a.	(Z =1.50)	**2.50**	**1.50**	**9.00**	**13.13**	**10.50**	**1.25**	**0.91**
b.	(Z = 2.00)	2.35	1.13	10.53	13.69	11.66	1.18	0.87
	% of 4 a	94.0	75.33	117.0	104.26	110.04	94.4	95.60
	% Variation	-6.0	-24.67	+17.0	+4.26	10.04	-5.6	**-4.4**
5.	Trapezoidal (Z =1.50)							
a.	B = d	2.31	2.31	9.24	13.34	10.63	1.25	0.91
	% of 4 a	92.4	154.0	102.66	101.6	101.24	100	100
	% Variation	**-7.6**	**+54.0**	**+2.6**	**+1.6**	**+1.24**	**+0.0**	**0.0**
b.	B = 2d	1.97	3.94	9.85	13.58	11.03	1.22	0.89
	% of 4a	78.8	262.7	109.44	103.4	105.04	97.6	97.8
	% Variation	-21.2	+162.7	+9.44	+3.4	+5.04	-2.4	2.2
c.	B = 3d	1.75	5.25	10.50	13.78	11.55	1.19	0.86
	% of 4a	70.0	350.00	116.6	104.95	110.0	952	94.50
	% Variation	-30	+250	+16	4.95	+10.0	-4.8	-5.5

Excess = + and less = -

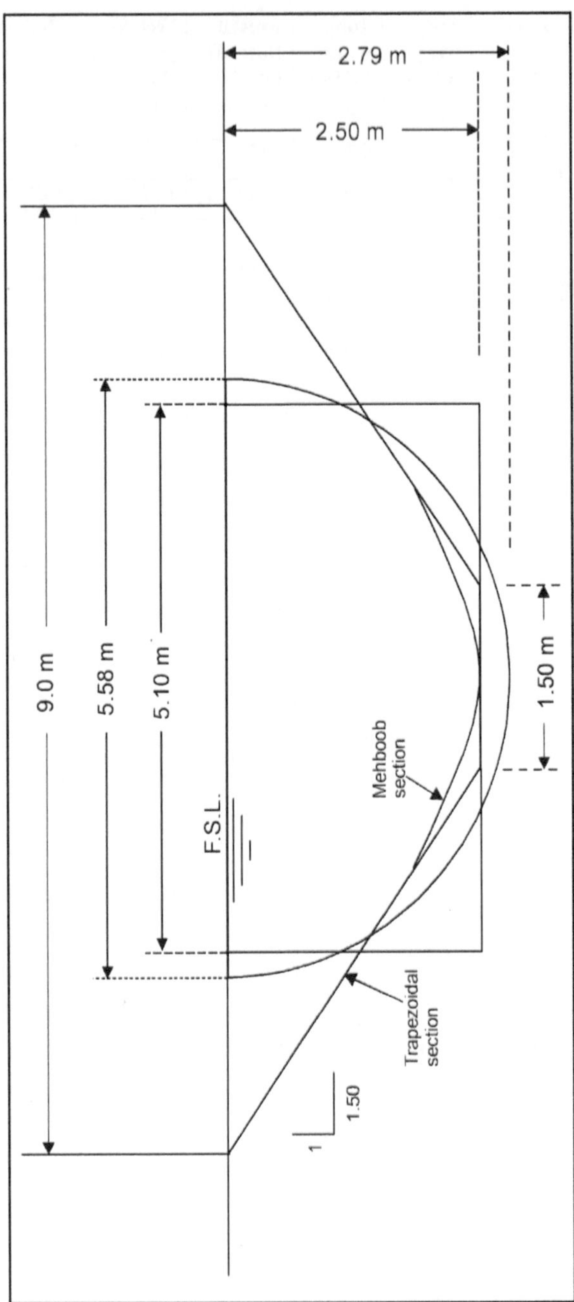

Figure 3.5 (a)for example 3.1 (a) Comparison of inner best sections for a discharge of 12 m^3 / s

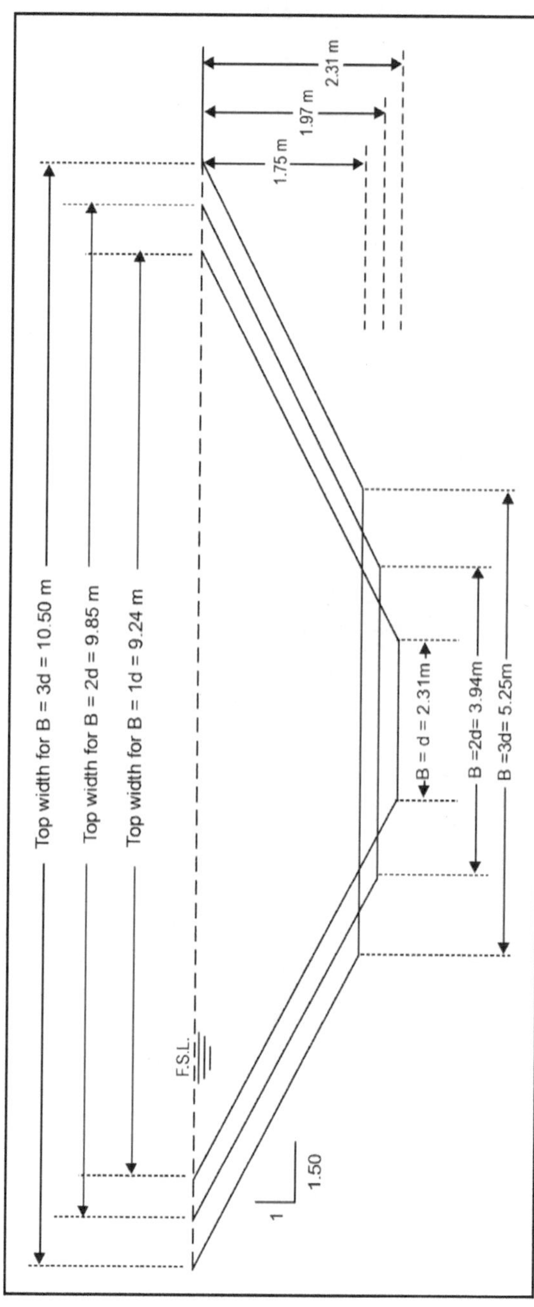

Figure 3.5 (b) for example 3.1 (b) Comparison of inner Trapezoidal Sections for b=d, b=2d and b = 3d, for a given Discharge of Q = 12 cumecs, N = 0.018 and Bed Slope of 1 in 5000 (see example 3.1 and its table)

Note : Since FSD is less than 2.5m, inner side slope of 1:1 or 1:1.25 is more preferable for any type of soil except clay

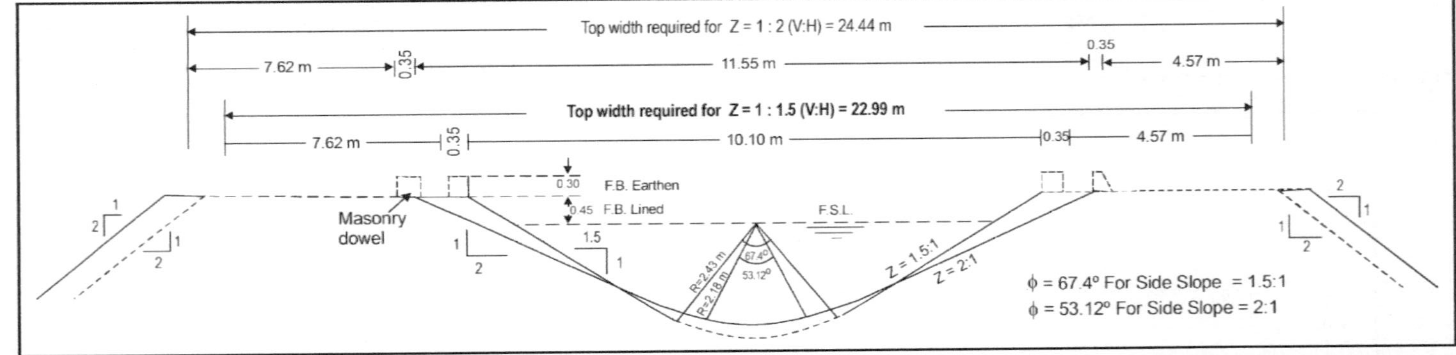

Figure 3.5 (c) for example 3.1 (c) Practical Comparison of Mehboob Sections with Side Slope of 2:1 and 1.5:1 Q = 12 cumecs, N = 0.018 Bed Slope = 1 in 5000, F.B. = 0.50 m

Note : - (1) All dimensions are in meter

(2) Top width for Z = 1 :1.5 (V:H) is less than that for Z = 1:2 (V : H)

Consequently - (i) Cost earth work is reduced in heavy filling or heavy cutting.

(ii) Land width is reduced

Comments / Recommendations

1. This table shows that depth is maximum for circular, top width is higher than rectangular, but less than trapezoidal.
2. Difference in trapezoidal and Mehboob section is nominal; compare 3a with 4a and 3(b) with 4(a)
3. For flatter side slope there is a good difference in all dimension i.e depth, perimeter, top width. It reduces velocity also. Compare sections 3(a) with 3(b) and 4(a) with 4(b) in the table.

   For a side slope from 1:1.5 (V: H) to 1:2 (V: H) top width increases by 17% and perimeter by 10%. Velocity also reduces by 4.4% and decrease in depth is nominal by 6%.

   In contrast to above for B = D and with the side slope of 1:1.5 (V: H), the decrease in depth in by 7.6%. Though increase in bed width is by 54%, yet the top width increases by 2.6% and perimeter by 1.24%. Velocity remains same. Therefore it is economical to increase bed width then side slope.
4. For a depth of 2.5 m, a flatter side slope should be avoided. A slope of upto 1 .5 (V: H) or slightly steeper say 1: 1.25 (V: H) may be reasonable for common soils (except clays).
5. With increase in bed width beyond b = d, there is drastic change in depth, top width, and perimeter. The difference in velocity is nominal. There appears little justification to increase bed width beyond b =d.

   In view of above discussions, Mehbood Section with side slope of 1:1.5 or preferably 1:1.25 (V: H) will be more Economical and therefore recommended.

   The comments/ comparison are more explicit in figure 3.5 of Ex.3.1.

3.5 Summary and Conclusion

Mehboob and best trapezoidal sections are explained in detail. Working tables for different side slopes are provided with comparison of hydraulic efficiency with semi circle.

Mehboob section has been extensively adopted in many parts of India, particularly, Haryana, Punjab and Rajasthan.

There is no difficulty in its construction and it has many advantages over trapezoidal section, see para 3.2.

A clear picture of effect of side slopes on hydraulic efficiency is presented in Para 3.3. It is proposed to adopt side slopes very near to optimum as far as possible within practical limitations. Lastly two typical solved examples are given with procedure to select a most efficient section.

References

Arya, R.S. Et Al, Design Charts For Earthen And Lined Channels, 1st International Research And Development Conference Water & Energy 2001, Central Board Of Irrigation And Power, New Delhi, October, 1995.

Bhakra, Canal Manual Guidelines, General Instructions for the Design of Lined and Unlined Channels of Bhakra Canal System, 1954.

Chow, V.T, Open - Channel Hydraulic, Mc. Graw - Hill Book Company Ltd., 1973.

Ishardas, Design of Lined Canals, International Association of Hydraulic Research, Question-1, P.P. 76, 1951.

□□□

Chapter **4**

Design of Wide Channels

(High Capacity Canals)

A canal of more than 80 cumecs (3000 cusecs) approximetly may be called as a high capacity canal.

It has already been pointed out in chapter 1, that for the best section of any shape, R is always equal to d/2 only, and such depth is the most optimum depth. But such channels are not always practical or most economical due to many other practical reasons, (see para 4.1 and 6.3) and many times shallower depth then this depth is adopted. Typical examples are lining of existing earthen channels, given in para 6.3.

All the best sections have fixed ratio of dimensions very near to the semi circle. For large discharges, radius of semi circle increases that, means depth of canal as well as velocity increases. But higher velocity is not permissible with every type of lining (see table 2.1 and para 2.3). Similarly depth also has limitation of construction, maintenance, stability of slopes, etc. see example–4.2.

Because of limiting velocities and depth, deeper canal section are not practically feasible/possible. Therefore for large discharges, say more than 100 cumecs, Mehboob or best trapizodal sections are just not possible/practical, and recourse has to be taken to trapezoidal section with increased bed width then depth. This section is hydraulically inefficient or uneconomical. Larger the B/D ratio, larger is the inefficiency and section is more costly. (For Mehboob Section, B:D ratio is zero and for best trapizodal section it is much less than 1, see para 3.1, table 3.1 para 3.5 and table 3.4).

Therefore as far as possible B should be kept to the minimum required with the maximum practical permissible depth. Such section can again be said as best possible economical section or optimised section with the given constraints.

Though wide channels in comparison to best sections are un economical, (a typical example is given in para 4.1) but still are unavoidable. This is due to the limitation of bed slopes and depth, see table 2.4. A critical review of para 1.7, figure 1.6; para 1.8, figure 1.7 table 1.3 para 1.9, figure 1.8; para 1.10, and figure 1.9, indicates the economics of perimeter, (area of excavation is not of much importance). With least perimeter i.e. best sections top width is also small, therefore excavation in deep cuttings or embankments in heavy filling are also economical in comparison to wide channels. Also land width required is smaller.

4.1 Effect of Bed Width on Hydraulic Efficiency (Shallow channels versus deep channels)

As with the flattenning of side slopes from the optimum slope of 0.58 : 1 (H : V) efficiency decreases, para 3.3, so in the same way efficiency further decreases with increase in bed width or decrease in depth.

Comparing a trapezoidal section of the best form having side slopes of 1.5 : 1 (H : V), with the semi circle for the same discharge, its area is 6% more then that of the semi circle while its perimeter is 11.9% more. Further a shallower channel having b = 3d, of the same side slopes, bed slope and for the same discharge has 9% greater area (A_3 = 1.09 A_1) and perimeter is 18.7 % more, then that of semi circle.

A comparison of these sections with semi circle is shown in figure. 4.1.

Therefore wherever unavoidable, bed width should be increased to the minimum possible extent, beyond the optimum width. However where comprise is to be made between a canal of steeper side slopes with increse in bed width and decrease in depth versus a canal of flatter sides with smaller bed width and increase in depth, the former may be preffered or for a judicious jugement a comparative analysis of alternatives should be made.

Another important aspect for increasing the bed width and decreasing the depth, is the cost difference in the lining of sides and bed. Many times different types of lining is adopted for sides and bed. Generally to save cost, lining of bed is cheaper than that of sides. A typical example is the lining of Indra Gandhi Main Canal and many important branches on this project. The lining of bed is single tile lining where as that of side is double tile lining.

A typical case of different types of lining in bed and sides is shown in figure 5.2. and chapter 6.

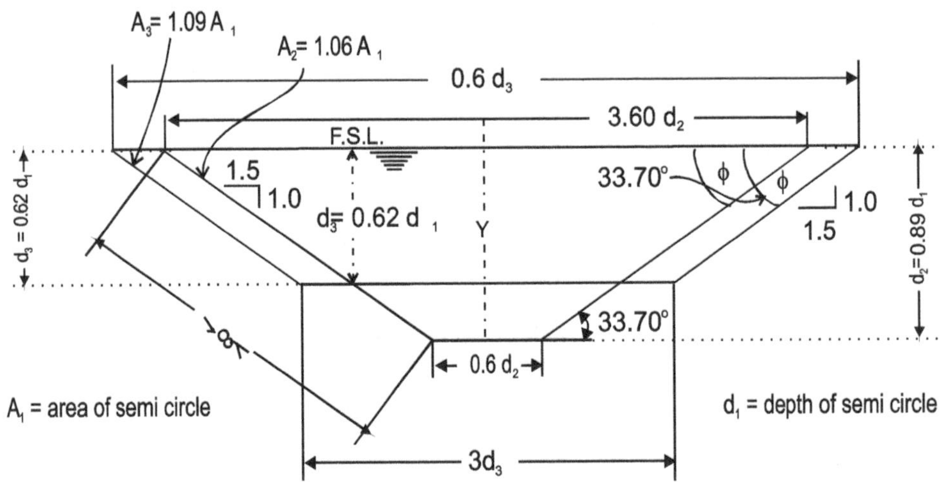

Figure 4.1 Comparison of trapezoidal channel sections of different bed widths and semi circle for a given discharge; (B = 3 D and side slope 1 V : 1.5 H.)

A general rule of thumb may be that the bed width may be kept upto 2.5 times of the depth of channel(i.e. b/d= 2.5), if the cost of bed lining is half of that on sides. Similary if the cost of bed lining is 0.75 times that of sides, then the bed width may be kept as 1.5 times the depth. But in both the cases the cost of excavation/earth work and land width will increase particularly more in heavy cutting/ filling reach. Therefore a critical analysis of various alternatives is required.

The present practice of design such channels is by trial and error. A bed slope is choosen, full supply depth (d) is also limited and decided and by assuming arbitrary some B/D ratio, then A, R, V, and Q are computed. This can be avoided by Manning's Master equation as given in para 4.3.

4.2 Advantages and Disadvantages of a Deep Section (Shallow versus deep section)

The disadvantages of a deep section are (i) greater unit cost of excavation, (ii) difficulty in maintaining side slopes in a deep canal, which may require flatter side slopes, (iii) decrease safety or more danger/risk of breaches and greater damage in case of a breaches.

In deep sections, though unit cost of excavation may be more, yet the over cost of excavation would depend upon the depth of heavy cutting/ depth of heavy filling. In both such situations deep sections are generally cheaper.

The 3rd disadvantage is a wrong notion and depend upon the design of section. The exetent of damage depends on the capacity of canal/discharge at the time of breach.

The disadvantages of a shallow section are (i) greater seepage loss, (ii) more favourable conditions of weeds and plant growth, (iii) greater width of right of way, (iv) greater wetted perimeter, which is disadvantage especially for concrete-lined canals, and (v) higher grade required to produce the same velocity.

An equal fluctuation in the volume of water carried would produce a smaller change in the water level of the shallower section. These advantages are of special value for laterals of the distribution system; for it facilitates the diversion of the water to the land and the regulation of the flow.

Also shallow or wide channels have the advantage of increasing the capacity of just raising the bank, provided sufficient free board is available at the bridges and other structures. However it will increse the head over off taking channels and outlets. Capacity of Upper Ganga Canal has been increased twice without any modifications and raising of structures/Bridges.

4.3 Master Equation

In an attempt to avoid the problem of trial and error, Suryavanshi (1971, 1972, and 1973) in a series of papers, brought out very useful derivations of Manning's formula (also see para 1.6.1). He presented a master equation of Manning's formula. It is derived with the help of continuity equation, for trapezoidal section as under:

$V = R^{2/3}S^{1/2}/N$ -Basic equation, (Metric system) 1

Raising power of equation 1 by cube, we get,

$$V^3 = \frac{R^2 S^{3/2}}{N^3} \quad 1(i) \qquad \text{Eq.4.3.1(i)}$$

and $Q = V.A.$ or $A = \frac{Q}{V}$ - Basic equation 2

For a trapezoidal section, see para 1.6(b)

$A = bd + Z d^2$

Taking $b/d = X$ or $b = X d$ Eq. 4.3.3

$A = X d^2 + Z d^2$ or $A = (X + Z) d^2$ Eq. 4.3. 3 (i)

Also, $A = Q/V$ therefore

$$(X + Z)d^2 = Q/V \text{ or } d = \left[\frac{Q}{V(X+Z)}\right]^{1/2} \qquad \text{Eq. 4.3 .3 (ii)}$$

This is very important relation to include, Q, V, and b/d ratio, along with Z.

Similarly, $P = b + 2d\sqrt{Z^2+1} = Xd + 2d\sqrt{Z^2+1}$

or $P = d(X+2)\sqrt{Z^2+1}$ Eq. 4.3.3 (iii)

$$\therefore R = \frac{A}{P} = \frac{(X+Z)d^2}{(X+2\sqrt{(Z^2+1)})d} \quad \text{Eq. 4.3 .3 (iv)}$$

This relation shows that R being a function of area A, and perimeter P, is also a function of b/d ratio X, side slope Z and depth of flow d.

or $R^2 = \frac{A^2}{P^2} = \frac{A}{P^2}.(\frac{Q}{V})$ replacing A from Eqn. 4.3.2. Eq. 4.3.3 (v)

Thus, because of the relation Q = VA, R is not a mere function of A, P, but is also a function of Q and V. Thus, from 4.3.3 (iii), 3(iv) and 3(v).

$$R^2 = \frac{(X+Z)d^2}{\left(X+2\sqrt{Z^2+1}\right)^2 d^2}\frac{Q}{V} \text{ or } R^2 = \frac{(X+Z)}{\left(X+2\sqrt{Z^2+1}\right)^2}\frac{Q}{V} \quad \text{Eq 4.3.3(v)}$$

Thus R is a function of X, Z, Q and V

Putting this value of R^2 in Eq. 1(i) we get

$$V^3 = \frac{(X+Z)}{\left(X+2\sqrt{Z^2+1}\right)^2}\frac{Q}{V}\frac{S^{3/2}}{N^3}$$

Re-arranging it, we get,

$$\frac{\left(X+2\sqrt{Z^2+1}\right)^2}{(X+Z)} = \frac{S^{3/2}}{N^3}.\frac{Q}{V^4} \quad \text{Eq.4.4}$$

This is called as Master Equation.

The left hand side of this equation is a dimension less number involving X i.e. b/d ratio and side slope (Z), which includes all the parameters of a trapezoidal section. Side slope (Z) is decided on the type of soil. Thus only X is to be determined.

Right hand side of the equations have functions of discharge (Q), slope (S), Rugosity coefficient (N), and velocity (V). Q, S, N are generally known and V is assumed because of limiting velocity. Thus, knowing all other parameters X is uniquely obtained.

After computing X, d is calculated by the Eq. 3(ii) and b from Eq. 3 i.e. b/d ratio. This completes the design of canal.

From this equation, if one wants to change V or S, he will get other values of b, because of the relation 3 (ii). Thus no arbitrary b/d ratio is required.

This is a unique relation to include all the 6 variables of canal design, viz.

(i) Discharge, Q

(ii) Velocity, V

(iii) Rugosity coefficient, N

(iv) Longitudinal slope, S

(v) Side slope Z (horizontal : 1 vertical)

(vi) And b/d ratio, X. In fact X has 2 parameters b and d and thus defines the cross- section completely, alongwith Z, i.e. area A and P. A and P are not independent variables. These have three independent variables i.e. b, d and Z. Again b and d cannot change independently. They have to change simultaneously i.e. b changes d and d changes b. Thus, X takes care of changes in b and d together.

There is no parameter extra and none is missing. Once Q and V are uniquely decided, area A gets fixed. b/d ratio X is uniquely known from the master equation.

In practice Z is fixed from soil characteristics i.e. stability of slopes, and not changed. It leaves only 5 variables. The inter relationship of two variables keeping the other three unchanged (four including Z) can be very well understood by writing the master equation in various forms as under:

1. X relationship with other parameters,

In equation 4, $\dfrac{\left(X+2\sqrt{Z^2+1}\right)^2}{(X+Z)} = \dfrac{S^{3/2}}{N^3}\dfrac{Q}{V^4}$

Though left hand side is quadratic function of X, yet for simplicity X relationship with other parameters can be partially treated as under:

(i) X-V relationship treating Z, S, N, and Q, as constant

$X\, \alpha \dfrac{1}{V^4}$ or $V^4\ \alpha \dfrac{1}{X}$ *or* $V \alpha \dfrac{1}{X^4}$

That means velocity is inversely and greatly affected by b/d ratio (by a power of 4). If X (i.e. b/d ratio) increases, velocity decrease considerably and with decrease in X, velocity increases.

(ii) X-Q relationship treating Z, S, N and V.

X α Q or Q α X most important relation

Discharge is directly proportional to b/d ratio, with increase in X, discharge increases.

(iii) X-S relationship treating Z, N, Q, and V as constant

X α $S^{3/2}$ or $S^{3/2}$ X or S α $(X)^{2/3}$

b/d ratio is effected by slope considerably, X being proportional to 1.5 power of slope. More the b/d ratio, more slope i.e steeper slope is required, or otherwise velocity will decrease.

(iv) X-N relationship treating Z, Q, V and S constant

X α $\frac{1}{N^3}$ or N^3 α 1/X or N α $(1/X)^{1/3}$

That is with increase in N, X decreases slightly, or with decrease in X, N increases slightly.

Thus, X i.e. b/d ratio has a very high relationship with V, Q, S and N and vice versa.

2. N relationship with other parameters

Rewriting the Master equation 4 as under

$$N^3 = \frac{S^{3/2}Q}{V^4}\frac{(X+Z)}{\left(X+2\sqrt{Z^2+1}\right)^2}$$

(v) N-S relationship, treating Q, V, X and Z constant

$N^3 S^{3/2}$ or $N \propto \sqrt{S}$ or $S \alpha N^2$

That means N increases with increase in slope or energy slope increases with increase in N. (It means heading occurs in the upstream and depth decreases in down stream)

(vi) N-V relationship, treating Q, X, S and Z constant

$N^3 \alpha \frac{1}{V^4}$ or $N \alpha (\frac{1}{V^4})^{1/3}$ $N \alpha \frac{1}{V^{4/3}}$ or $V \alpha \frac{1}{N^{3/4}}$

That means N increases with decrease in V or vice versa.

(vii) N-Q relationship, treating S, V, X and Z constant,

$N^3 \alpha Q$ or $N \alpha Q^{1/3}$ or $Q \alpha N^3$

Thus N increases with increase in Q. It also appears as if Q increases with increase in N. But it is not so. Since N is less than 1, its cube powers are much smaller and therefore Q may slightly decrease with increase in N. We should simultaneously see N-V relationship (vi), where velocity decreases with increase in N. So overall effect is that Q decreases.

X-N relationship, treating S, Q, V, Z constant, is indicated in 1(iv) above

3. S-relationship with Q and V.

Re-writing the Master equation 4 as under

$$S^{3/2} = \frac{N^3V^4}{Q}\frac{\left(X+2\sqrt{Z^2+1}\right)^2}{(X+Z)}$$

(viii) S-Q relationship treating, N, V, X, and Z constant,

$S^{3/2} \alpha \frac{1}{Q}$ *or* $S \alpha \frac{1}{Q^{2/3}}$ *or* $Q \alpha \frac{1}{S^{3/2}}$

That means Q is inversely proportional to slope by a power of 3/2.

It may lead to confusion. For example, take a channel of bed slope of 1/1000 i.e. S = 0.001 and another of S_1 = 1/100 i.e. S_1 = 0.01. The inverse of $S^{1.5}$ is less than the inverse of $S_1^{1.5}$, that means discharge increases with increase in slope.

Also see next relationship of S V and V Q relation. Q is a function of V^4, so discharge increases at a fast rate with increase in velocity. See figure 4.4 convex curve of discharge and concave curve of velocity.

(ix) S-V relationship treating N, Q, X, and Z constant

$S^{3/2} \alpha V^4$ or $S \alpha V^{8/3}$ or $V \alpha S^{3/8}$

Velocity increases with 3/8 power of S, because of root power of S,

S-X and S-N relationships are indicated in (1) and (2) above

4. V relationship with other parameters

Re-writing the master equation 4 as under

$$V^4 = \frac{S^{3/2}Q}{N}\frac{(X+Z)}{\left(X+2\sqrt{Z^2+1}\right)^2}$$

(x) V-Q relationship treating S, N, X, and Z as constant
$V^4 \alpha Q$ or $V \alpha Q^{1/4}$ or $Q \alpha V^4$

Thus Q increases at a fast rate with increase in velocity at a much faster rate then velocity it self see figure 4.4, where velocity curve is concave type and discharge curve is convex type.

This relation seems little unpractical or confusing. In any physical model in the laboratory or at site, N, X and Z can be made constant. Actually N cannot be made constant. Also S is not bed slope, but is energy slope 'S' and is beyond control even in laboratory. So is V. This needs further investigations by researchers. Anyway one should not get disappointed by Manning's equation as it is the only equation widely used throughout the world, and well proved by researchers.

Other relations of V-S, V-N, V-X are shown in (1), (2) and (3) above.

Thus, treating Z unchanged in any particular case (this is usual practice) there remains five variables, resulting in 10 inter-relationships. Out of above 10 relations, (i), (iii) and (ix) are often required in design of canal section. Others are useful in understanding the mechanics of channel flow and helps in judicious judgement.

4.3.1 Comparison with common form and some misconceptions

Here, it may be of interest to compare the above relationships with the well known form of Manning's equation and in most common use, i.e.

$$V = \frac{1}{N} R^{2/3} S^{1/2} \qquad \text{Eqn. 1}$$

$$\text{or } N = \frac{1}{V} R^{2/3} S^{1/2} = \frac{1}{V} (\frac{A}{P})^{2/3} S^{1/2} \qquad 1\text{ (ii)}$$

This is the most common form for determination of N in the field or laboratory. V is measured by current meter, A and P are measured by tape and water surface slope (S) measured by levelling instrument. The measurements are refined by making several measurements at many points along the canal, and making adjustments to obtain energy line slope (S).

Also, Q = V. A — Eq. 2
or $Q = AR^{2/3} S^{1/2}/N$ — Eq. 5

$$\text{or } Q = \frac{1}{N} (\frac{A^{5/3}}{P^{2/3}}) S^{1/2} \qquad \text{Eq. 5 (i)}$$

$$\text{or } N = \frac{1}{Q} \frac{A^{5/3}}{P^{2/3}} S^{1/2} \qquad \text{Eq.5 (ii)}$$

Generally engineers take V or Q as inversely proportional to N by treating all other parameters as constant i.e.

$$V \alpha \frac{1}{N} \text{ or } N \alpha \frac{1}{V} \text{ and } Q \alpha \frac{1}{N} \text{ and } N \alpha \frac{1}{Q}$$

But this is not so. These are only mis-conceptions as explained above in sub para 2,of para 4.3 in N-V and N-Q relations. Velocity is inversely proportional to $N^{3/4}$ and $Q \propto N^3$. If it is not so it can not be inferred that N increases with increase in Q. In fact the explicit relationship of Masters equation are hidden in Eq. 5 or 5(i).

In a channel flow phenomena, side slope Z, bed width b, bed slopes, and discharge Q are the only independent variables. Only these can be represented in a physical model in the laboratory or measured in proto type in a natural river or incised canal. Depth d, energy slope(s) and velocity V, gets adjusted according to natural surface roughness existing i.e. coefficient of rugosity N, depending upon b, Q and Z. In the process N also gets adjusted or varies slightly.

However for practical purposes the above general concept of engineers (though not a true concept) may still be taken/used in day to day work as the over all effect of N is to retard velocity and discharge, though not in true inverse proportion.

Similarly there is some misconception about V and Q versus S, see above sub Para 3, S, Q, V relation. Similarly there is some misconception in about V and Q versus S, see above 3, 5, Q, V relation.

Also see para 4.5, where, effect of N on Q is more explicit.

Nomographic solution of Master equation

For very quick designs Suryavanshi (1973) has also presented a nomogram as shown in figure 4.2.

The Z=0 line of the network forms the basic nomogram for Equation (3) with other scales, namely, N, L, S, M, V and Q. For knowing X corresponding to other Z values, one must proceed horizontally (as in an ordinary graph) along the X-Z network. Inversely, for given X and Z values, first come to Z = 0 scale horizontally and then proceed to other scales like N, L, etc., L and M are pivot lines. The left side graduations of V and Q scales should be used together and right side graduations with right side markings. The S scale contains values of S on one side and 1/S values on the other, for convenience of either way of expressions used for this parameter.

For solving Equation 4.3.3(i) scales, V, Q, pivot line A, scale (X+Z) and scale d may be used. Further, last three scales, namely d, X and b form the nomogram for Equation (3). The right side graduations of d and b scales may be used with right side values of V and Q and left side ones with left side values.

For the benefit of designers who use Kennedy's critical velocity criteria (Equation - 2), d scale of the above composite nomogram may be used in conjunction with V scale through pivot points P_L and P_R. This will directly give Vo values. With right side graduations of V and d scales pivot PR may be used and with left side ones, P_L may be used.

A list of pivot lines, which may be used for the inter-relations are mentioned against each. This will greatly facilitate a quick design of the section, once one gets accustomed with the nomogram fully.

Pivot Line	Q	M	P*	L	S	M	d	M
Relation	X-V	X-Q	X-S	X-N	N-Q	S-Q	S-V	V-Q

A beginner in design may try these relations with the help of the nomogram to closely understand the flow phenomenon. It provides a visual feel of the interdependence.

Minimum Slope for a Given Velocity

When a canal section is to be designed for given velocity (V), Q, N and Z then corresponding to minimum perimeter of the cross-section there is minimum bed slope, below which the design becomes impossible. A best discharging section has the following properties:

$$R = d/2, and\ X = 2\left(\sqrt{Z^2+1-Z}\right)$$

The top most line (dotted) in the X-Z network of the nomograph (figure 4.2) represents the above value of X for respective values of Z. As such, in the design if one goes to a point above this line, it indicates that the slope adopted is less than the flattest possible for the assumed velocity. It is necessary in such a case to adopt lower velocity or proceed with desired (new) X-value and find the corresponding velocity, if steeper slope can not be given.

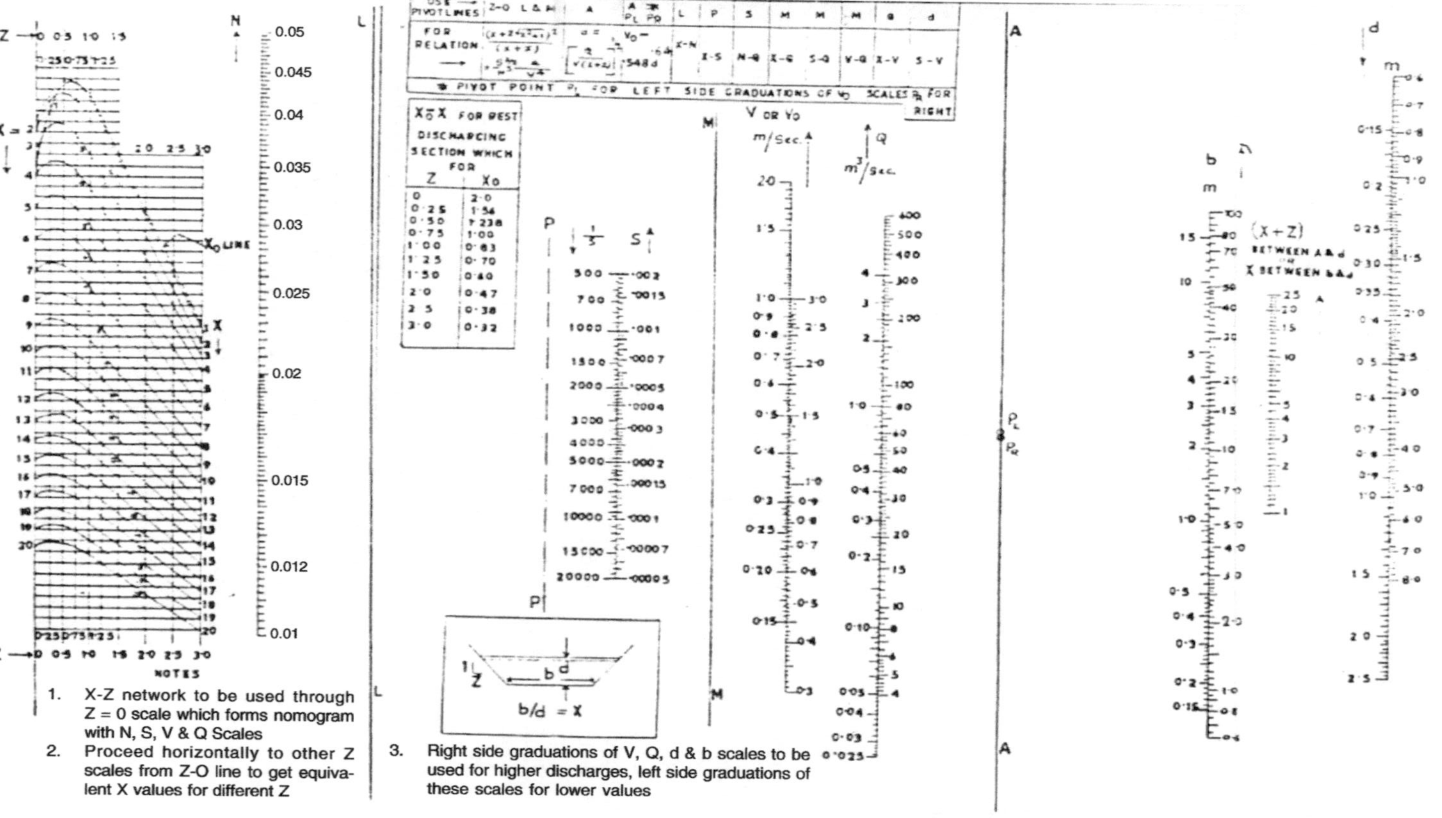

Figure 4.2 A Composite Nomogram for Complete Canal Design

Practical Examples

For getting used to the composite nomogram ten examples are given in table below the bracketed items of each example may be found out from the nomogram for the data given by the unbracketed values in the same line. The bracketed values are answers to the problems found by arithmetical solution of the equations (3), (4) & (5). They will help comparison with the values obtained by the use of nomogram and give confidence to the user.

Table of practical example

S.No.	Q (m^3/s)	Z	N	S	V (m/s)	X	d (m)	b (m)	Vo (m/s)
1	0.04	1	0.035	0.0004	(0.1682)	4	(0.218)	(0.87)	(0.2065)
2	0.1	1.25	0.035	(0.00034)	0.2	0.3	(0.343)	(1.03)	(0.276)
3	3	2	0.025	(0.000422)	0.6	5	(0.845)	(4.225)	(0.492)
4	3.32	0	0.016	1/2000	1	4	(0.91)	(3.64)	(0.516)
5	4	0	0.03	1/2000	0.6	(8.325)	(0.895)	(7.45)	(0.51)
6	15	0.5	0.025	0.00025	0.8	4.72	(1.895)	(8.95)	(0.824)
7	25	1	0.02	0.0002	(0.9641)	5	(2.08)	(10.4)	(0.875)
8	30	1.5	0.0225	1/3750	1	(5)	(2.146)	(10.73)	(0.894)
9	200	2	0.015	(0.000257)	2	8	(3.16)	(25.25)	(1.145)
10	400	1.25	0.018	0.0001	(1.408)	12	(4.63)	(55.5)	(1.46)

4.4 Trapezoidal Sections with Curved Ends

To improve upon the hydraulic efficiency of wide trapezoidal section, the ends are curved as shown in figure 4.3. Normally curves with a radii equal to full supply depth (FSD) are provided at both ends. However curves of higher or lower radius can be provided. Lower radius is not very effective in improvement of efficiency. Also it is not practical in construction (curve being smaller). Higher radius is also not very effective and more over may not be practical. Best radius is equivalent to FSD and is most commonly adopted (similar to Mehboob section). This section was also proposed by Bhakra Canal Mannual (1950) and was adopted in Bhakra Canals, built arround 1950; Indira Gandhi Canals and many other canals in India, see table 6.2.

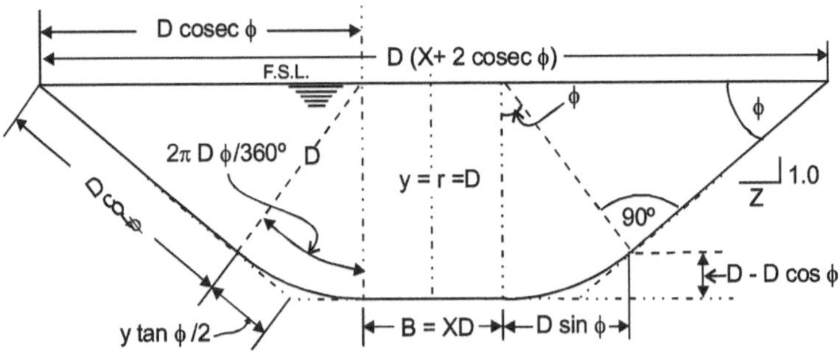

Figure 4.3 Trapezoidal Sections With Curved Ends

The best way to design such section is first design a Mehboob section, and then go on increasing the bed width till the desired depth is reached. Practically only one trial may be required, and it should always be compared with Mehboob section.

The geometrical and hydraulic properties of trapezoidal section with curved ends are computed as under :

Area of flow $A = BD + 2 [\pi D^2 \phi/ 360 + D^2 \text{Cot } \phi/2]$ (ϕ is in degrees)

$= BD + 3.14 D^2 \phi/180 + D^2 \text{Cot } \phi$

Let B = XD, then

$A = XD^2 + 3.14 D^2 \phi/180 D^2 \text{Cot } \phi$

$A = D^2 (X + 3.14 \phi/180+ \text{Cot } \phi)$

Perimeter of flow

$P = B + 2 [D \text{Cot } \phi +2D (3.14 \phi /360)]$ (ϕ is in degrees)

$= B + 2 (D \text{Cot } \phi + 3.14 D \phi /180)$

Since B = XD, therefore

$P = XD + 2D \text{Cot}\phi + 3.14 \phi D/90$

$= D (X + 2 \text{Cot } \phi + 3.14 \phi/90)$

$= D (X + 3.14 \phi/ 90 + 2 \text{Cot}\phi)$

Hydraulic Radius

$R = A/P$

$$= \frac{D^2 (X + 3.14 \phi/180 + \text{Cot}\phi)}{D (X + 3.14 \phi/ 90 + 2 \text{Cot}\phi)} = \frac{D (X + 3.14 \phi/180 + \text{Cot}\phi)}{(X + 3.14 \phi/ 90 + 2 \text{Cot}\phi)}$$

Velocity of flow

$V = R^{2/3} S^{1/2} / N$

$$\text{or } V = \frac{[D^2 (X + 3.14 \phi/180 + \text{Cot}\phi)]^{2/3} S^{1/2}}{[D (X + 3.14 \phi/ 90 + 2 \text{Cot}\phi)]^{2/3} N}$$

$$= \frac{[D (X + 3.14 \phi/180 + \text{Cot}\phi)]^{2/3} S^{1/2}}{[(X + 3.14 \phi/ 90 + 2 \text{Cot}\phi)]^{2/3} N}$$

Discharge

$Q = A.V$

$$= D^2 (X + 3.14 \phi/180 + \text{Cot } \phi) \times \frac{[D (X + 3.14 \phi/180 + \text{Cot}\phi)]^{2/3} S^{1/2}}{[(X + 3.14 \phi/ 90 + 2 \text{Cot}\phi)]^{2/3} N}$$

$$= \frac{(X + 3.14 \phi/180 + \text{Cot } \phi)^{5/3} D^{8/3} S^{1/2}}{[(X + 3.14 \phi/ 90 + 2 \text{Cot } \phi)]^{2/3} N}$$

The above Equations can be further simplified by putting the proper values of ϕ for the side slopes to be adopted. Thus for $\phi = 45^o$, the above expression become as under.

$A = D^2 (X + 3.14 \phi/180 + \text{Cot}\phi) = D^2 (X + 3.14 (45)/180 + 1)$

$= D^2 (X+0.785+1) = D^2 (X+1.785)$

The above expression are further simplified by assuming reasonable value of X. For example taking X= 0.5, and $\phi = 45^o$, we get

$A = 2.285 D^2$; $P = 4.070 D$; $R = 0.561 D$; $V = (0.561D)^{2/3} S^{1/2} / N$

Thus for given values of X and ϕ all the geometric and hydraulic elements can be expressed in very simple form of D. Various canstants k_1, k_2, k_3, k_4, k_5 and k_6 can be used to define these parameters as under :

$A = k_1 D^2$, $P = k_2 D$, $R = k_3 D$, $V = k_4 D^{2/3}$ and $Q = k_5 D^{8/3}$ or $D = k_6 (Q)^{3/8}$

When X = 0, all the above equations get reduced to the form of Mehboob section, see para 3.1.

To facilitate quick designs, A, P, R and Q for various side slopes are given in table 4.1 for different bed width depth ratio of B : D = X = 0.5, 0.75, 1.25, 1.50, 1.75, and 2.00 are given in Table 4.1 to 4.1.4.

Designers are requested to adopt this B:D ratio as minimum as possible or practical, for economical considerations. For an assumed value of X, depth of flow can be directly calculated with the help of table 4.1.

Most economically optimised sections can also be designed quickly by adopting the guide lines for design of best sections given in para 2.14, or by keeping the bed width as near as possible to the dimensions given in Table-3.4. For very large discharge say greater then 100 cumecs (or 3500 cusecs), designers may also consider the possibility of composite section. This may help them to maintain FSL.

4.5 Comparison of Curved End with Trapezoidal Section

Some engineers raise doubts on the success of curved ends. This is only an apprehension based without adequate exposure. Indra Gandhi main canal (formerly known as Rajasthan Main canal). Rajasthan, India with a head discharge of (18500 cusses) and a tail discharge (6000 cusses), 600 Km. long has been constructed with curved ends as discussed above, with clay tile, and P. C. C. block lining. Its performance has been found very good. Similarly many other canals in India have been constructed with curved ends (see table 5.1). All the doubts about curved ends have already been discussed and shed off in para 3.2 under the choice of Mehboob and Trapezoidal section.

Some engineers argue that curved ends make very little difference in perimeter and area in comparison to trapezoidal section, and makes construction difficult.

Well, difference in perimeter depends upon the angle of side slopes. For a most common side slope of 1V:1.5H, the difference between the perimeter and area of best Trapezoidal and Mehboob Section are 4.20 y - 4.176 y = .024 y i.e. practically 2.4 % less and 2.10 y^2 - 2.088 y^2 = .012 y^2 i.e. practically 1.2% less, (bed width and depth of flow remaining constant) see table 3.4 and 3.1.

Though the difference in area may not make much difference, but the difference of 2.4% in perimeter of large and long canals may become some time considerable in economy. This can also be visualised from table 1.3 (Comparison of best section for a given discharge) where in perimeter of trapezoidal section is only 4.4% more than Mehboob section. More over this also effects R, and more discharge can pass through the same section than without curves.

In wide channels these percentages will be slightly different depending upon B:D ratio.

4.6 Effect of Manning's N on Discharging Capacity

The discharge of canal is given by

$Q = AR^{2/3} S^{1/2}/N$

As above, the discharge is inversely proportional to the rugosity coefficient of the channel surfaces, though not truly so, see par 4.3. Still consider as above for practical purposes. Now if the channel surface is smooth N is small (see table 2.1) and for small N, discharge is more than the discharge where N is greater in the channel of the same cross sectional area and slope.

Table 4.1 Trapezoidal Section with Curved Ends

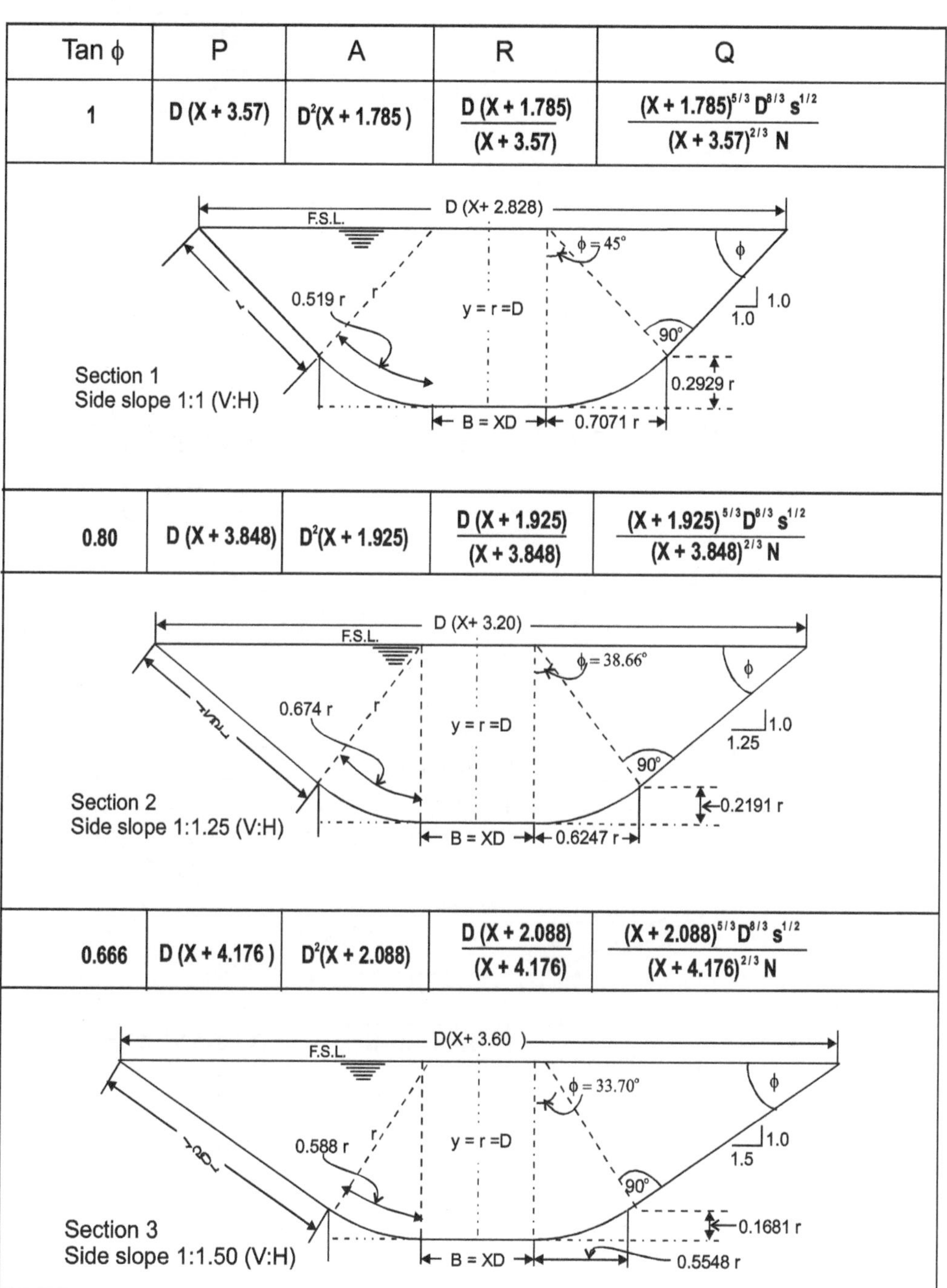

Tan ϕ	P	A	R	Q
1	D (X + 3.57)	D^2(X + 1.785)	$\frac{D(X+1.785)}{(X+3.57)}$	$\frac{(X+1.785)^{5/3} D^{8/3} s^{1/2}}{(X+3.57)^{2/3} N}$
0.80	D (X + 3.848)	D^2(X + 1.925)	$\frac{D(X+1.925)}{(X+3.848)}$	$\frac{(X+1.925)^{5/3} D^{8/3} s^{1/2}}{(X+3.848)^{2/3} N}$
0.666	D (X + 4.176)	D^2(X + 2.088)	$\frac{D(X+2.088)}{(X+4.176)}$	$\frac{(X+2.088)^{5/3} D^{8/3} s^{1/2}}{(X+4.176)^{2/3} N}$

Contd...

Table 4.1 Contd...

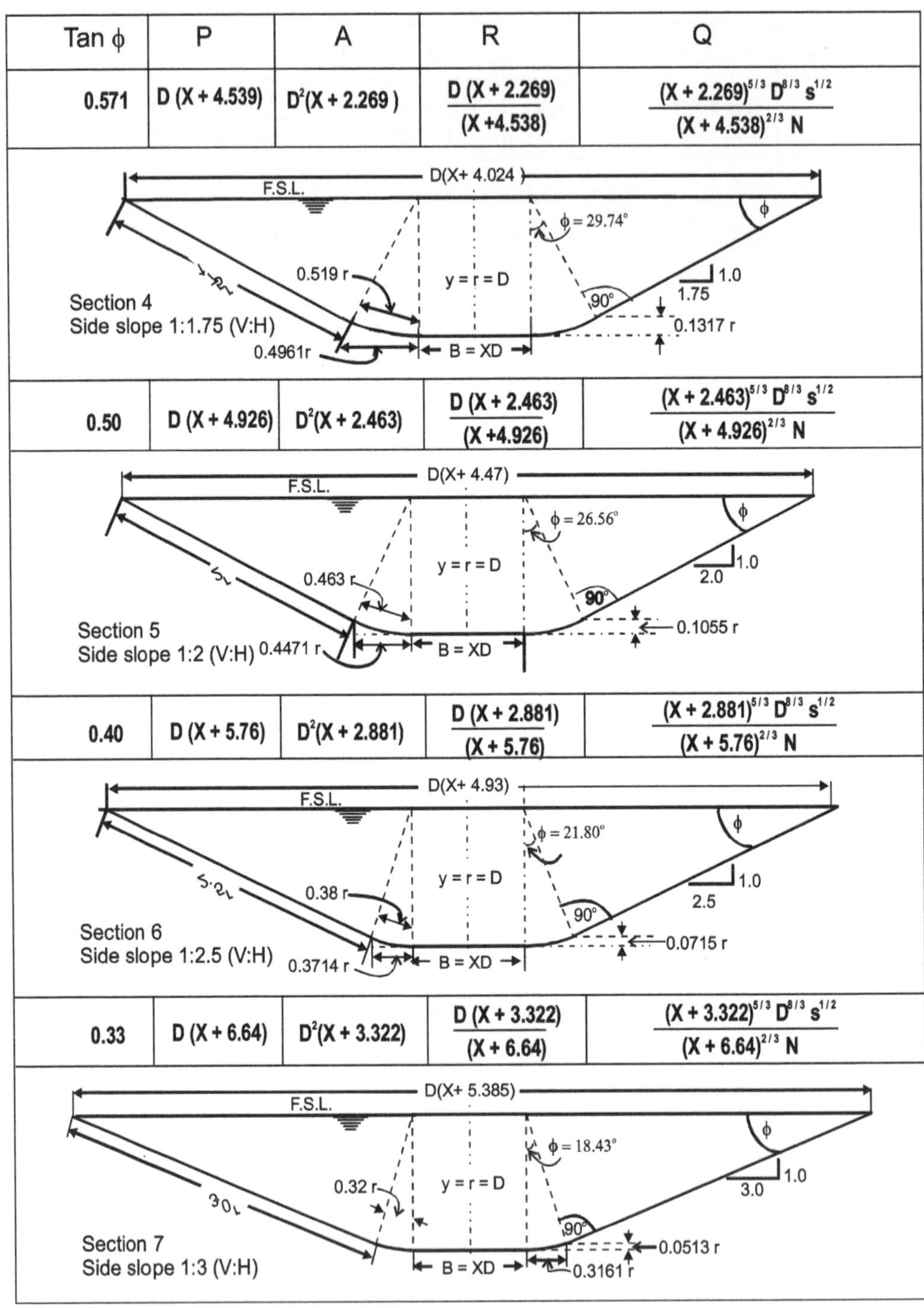

Tan φ	P	A	R	Q
0.571	D (X + 4.539)	D^2(X + 2.269)	$\frac{D(X+2.269)}{(X+4.538)}$	$\frac{(X+2.269)^{5/3} D^{8/3} s^{1/2}}{(X+4.538)^{2/3} N}$
Section 4 Side slope 1:1.75 (V:H)				
0.50	D (X + 4.926)	D^2(X + 2.463)	$\frac{D(X+2.463)}{(X+4.926)}$	$\frac{(X+2.463)^{5/3} D^{8/3} s^{1/2}}{(X+4.926)^{2/3} N}$
Section 5 Side slope 1:2 (V:H)				
0.40	D (X + 5.76)	D^2(X + 2.881)	$\frac{D(X+2.881)}{(X+5.76)}$	$\frac{(X+2.881)^{5/3} D^{8/3} s^{1/2}}{(X+5.76)^{2/3} N}$
Section 6 Side slope 1:2.5 (V:H)				
0.33	D (X + 6.64)	D^2(X + 3.322)	$\frac{D(X+3.322)}{(X+6.64)}$	$\frac{(X+3.322)^{5/3} D^{8/3} s^{1/2}}{(X+6.64)^{2/3} N}$
Section 7 Side slope 1:3 (V:H)				

Table 4.1.1 Geometrical and Hydraulic Properties of Trapezoidal Section With Curved Ends, with Z = 1:1

Section 1 , Side Slope : 1 V : 1 H, or

$\Phi = 45^{O}$ or $\tan\Phi = 1$ or $\cot\Phi = 1$

$\sin\Phi = 0.707$; $\cos\Phi = 0.707$; $\operatorname{cosec}\Phi = 1.414$;

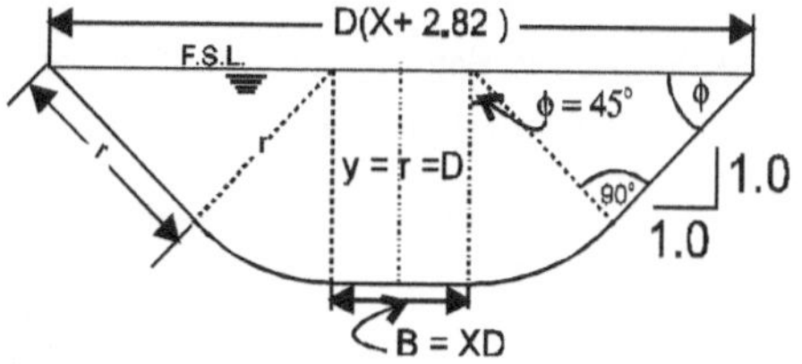

S. No.	X B / D	Area (A)	Perimeter (P)	Hydraulic Radius (R) A/P (col3)/ (col 4)	Velocity (V) (col 5)$^{2/3}$ $S^{1/2}$/N	$AR^{2/3}$ (col3)x (col 5)$^{2/3}$	D see notes below
1	2	3	4	5	6	7	8
1	0.50	2.285 D^2	4.070 D	0.561 D	0.68 $D^{2/3}$ $S^{1/2}$ / N	1.555 $D^{8/3}$	0.847 $(QN)^{3/8}$ / $S^{3/16}$
2	0.75	2.535 D^2	4.320 D	0.586 D	0.70 $D^{2/3}$ $S^{1/2}$ / N	1.775 $D^{8/3}$	0.806 $(QN)^{3/8}$ / $S^{3/16}$
3	1.0	2.785 D^2	4.570 D	0.609 D	0.718 $D^{2/3}$ $S^{1/2}$ / N	2.001 $D^{8/3}$	0.770 $(QN)^{3/8}$ / $S^{3/16}$
4	1.25	3.035 D^2	4.820 D	0.629 D	0.734 $D^{2/3}$ $S^{1/2}$ / N	2.228 $D^{8/3}$	0.740 $(QN)^{3/8}$ / $S^{3/16}$
5	1.50	3.285 D^2	5.070 D	0.648 D	0.7480 $D^{2/3}$ $S^{1/2}$ / N	2.459 $D^{8/3}$	0.714 $(QN)^{3/8}$ / $S^{3/16}$
6	1.75	3.535 D^2	5.320 D	0.664 D	0.761 $D^{2/3}$ $S^{1/2}$ / N	2.691 $D^{8/3}$	0.690 $(QN)^{3/8}$ / $S^{3/16}$
7	2.00	3.785 D^2	5.570 D	0.679 D	0.773 $D^{2/3}$ $S^{1/2}$ / N	2.925 $D^{8/3}$	0.669 $(QN)^{3/8}$ / $S^{3/16}$
8	2.50	4.285 D^2	6.07 D	0.706 D	0.793 $D^{2/3}$ $S^{1/2}$ / N	3.398 $D^{8/3}$	0.632 $(QN)^{3/8}$ / $S^{3/16}$
9	3.0	4.785 D^2	6.57 D	0.728 D	0.809 $D^{2/3}$ $S^{1/2}$ / N	3.871 $D^{8/3}$	0.602 $(QN)^{3/8}$ / $S^{3/16}$
10	3.5	5.285 D^2	7.07 D	0.747 D	0.823 $D^{2/3}$ $S^{1/2}$ / N	4.35 $D^{8/3}$	0.576 $(QN)^{3/8}$ / $S^{3/16}$
11	4.0	5.785 D^2	7.57 D	0.764 D	0.822 $D^{2/3}$ $S^{1/2}$ / N	4.758 $D^{8/3}$	0.557 $(QN)^{3/8}$ / $S^{3/16}$

Explainatory notes : D in column 8 is Calculated from $Q = AV = AR^{2/3} S^{1/2}/N$ or $AR^{2/3} = QN/S^{1/2}$. For first row $1.555\ D^{8/3} = QN/\sqrt{S}$ or $D = (QN)^{3/8}/(1.555)^{3/8}\ S^{3/16}$

Table 4.1.2 Geometrical and Hydraulic Properties of Trapezoidal Section with Curved Ends, with Z = 1:1.25

Section 2 , Side Slope : 1 V : 1.25 H, or

$\Phi = 38.66^\circ$ or $\tan\Phi = 0.80$ or $\cot\Phi = 1.25$

$\sin\Phi = 0.624$; $\cos\Phi = 0.781$; $\mathrm{cosec}\Phi = 1.602$;

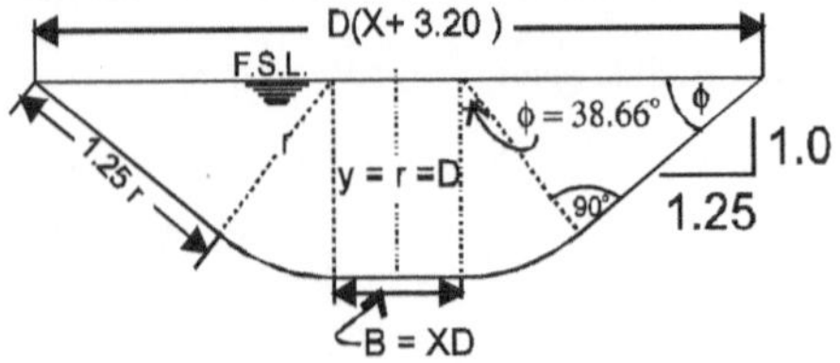

S. No.	X B / D	Area (A)	Perimeter (P)	Hydraulic Radius (R) A/P(col3)/ (col 4)	Velocity (V) (col 5)$^{2/3}$ $S^{1/2}$/N	$AR^{2/3}$ (col3)x (col 5)$^{2/3}$	D see notes below
1	2	3	4	5	6	7	8
1	0.50	2.425 D^2	4.348 D	0.557 D	0.677 $D^{2/3}$ $S^{1/2}$ / N	1.643 $D^{8/3}$	0.830 $(QN)^{3/8}$ / $S^{3/16}$
2	0.75	2.675 D^2	4.598 D	0.581 D	0.696 $D^{2/3}$ $S^{1/2}$ / N	1.864 $D^{8/3}$	0.792 $(QN)^{3/8}$ / $S^{3/16}$
3	1.0	2.925 D^2	4.848 D	0.603 D	0.714 $D^{2/3}$ $S^{1/2}$ / N	2.088 $D^{8/3}$	0.759 $(QN)^{3/8}$ / $S^{3/16}$
4	1.25	3.175 D^2	5.098 D	0.622 D	0.729 $D^{2/3}$ $S^{1/2}$ / N	2.315 $D^{8/3}$	0.730 $(QN)^{3/8}$ / $S^{3/16}$
5	1.50	3.425 D^2	5.348 D	0.640 D	0.743 $D^{2/3}$ $S^{1/2}$ / N	2.544 $D^{8/3}$	0.704 $(QN)^{3/8}$ / $S^{3/16}$
6.	1.75	3.675 D^2	5.598 D	0.656 D	0.755 $D^{2/3}$ $S^{1/2}$ / N	2.775 $D^{8/3}$	0.682 $(QN)^{3/8}$ / $S^{3/16}$
7	2.00	3.925 D^2	5.848 D	0.671 D	0.766 $D^{2/3}$ $S^{1/2}$ / N	3.008 $D^{8/3}$	0.662 $(QN)^{3/8}$ / $S^{3/16}$
8	2.50	4.425 D^2	6.348 D	0.697 D	0.786 $D^{2/3}$ $S^{1/2}$ / N	3.478 $D^{8/3}$	0.627 $(QN)^{3/8}$ / $S^{3/16}$
9	3.0	4.925 D^2	6.848 D	0.719 D	0.803 $D^{2/3}$ $S^{1/2}$ / N	3.953 $D^{8/3}$	0.597 $(QN)^{3/8}$ / $S^{3/16}$
10	3.5	5.425 D^2	7.348 D	0.738 D	0.817 $D^{2/3}$ $S^{1/2}$ / N	4.430 $D^{8/3}$	0.572 $(QN)^{3/8}$ / $S^{3/16}$
11	4.0	5.925 D^2	7.848 D	0.755 D	0.829 $D^{2/3}$ $S^{1/2}$ / N	4.913 $D^{8/3}$	0.550 $(QN)^{3/8}$ / $S^{3/16}$

Explainatory notes : - D is column 8 is calculated from $Q = AV = A R^{2/3} S^{1/2}/N$ or $AR^{2/3} = QN / S^{1/2}$ For first row $1.643 D^{8/3} = QN / S^{1/2}$ or $D = (QN)^{3/8} / (1.643)^{3/8} S^{3/16}$

Table 4.1.3 Geometrical and Hydraulic Properties of Trapezoidal Section With Curved Ends, with Z = 1:1.5

Section 3, Side Slope : 1 V : 1.5 H, or

$\Phi = 33.70^\circ$ or $\tan\Phi = 0.667$ or $\cot\Phi = 1.50$

$\sin\Phi = 0.555$; $\cos\Phi = 0.832$; $\operatorname{cosec}\Phi = 1.802$;

S. No.	X B / D	Area (A)	Perimeter (P)	Hydraulic Radius (R) A/P(col3)/ (col 4)	Velocity (V) (col 5)$^{2/3}$ $S^{1/2}$/N	$AR^{2/3}$ (col3)x (col 5)$^{2/3}$	D see notes below
1	2	3	4	5	6	7	8
1	0.50	2.588 D^2	4.676 D	0.553 D	0.674 $D^{2/3}$ $S^{1/2}$ / N	1.744 $D^{8/3}$	0.812 $(QN)^{3/8}$ / $S^{3/16}$
2	0.75	2.838 D^2	4.926 D	0.576 D	0.692 $D^{2/3}$ $S^{1/2}$ / N	1.965 $D^{8/3}$	0.776 $(QN)^{3/8}$ / $S^{3/16}$
3	1.0	3.088 D^2	5.176 D	0.596 D	0.708 $D^{2/3}$ $S^{1/2}$ / N	2.188 $D^{8/3}$	0.745 $(QN)^{3/8}$ / $S^{3/16}$
4	1.25	3.338 D^2	5.426 D	0.615 D	0.723 $D^{2/3}$ $S^{1/2}$ / N	2.414 $D^{8/3}$	0.718 $(QN)^{3/8}$ / $S^{3/16}$
5	1.50	3.588 D^2	5.676 D	0.632 D	0.736 $D^{2/3}$ $S^{1/2}$ / N	2.642 $D^{8/3}$	0.695 $(QN)^{3/8}$ / $S^{3/16}$
6.	1.75	3.838 D^2	5.926 D	0.647 D	0.748 $D^{2/3}$ $S^{1/2}$ / N	2.873 $D^{8/3}$	0.673 $(QN)^{3/8}$ / $S^{3/16}$
7	2.00	4.088 D^2	6.176 D	0.662 D	0.759 $D^{2/3}$ $S^{1/2}$ / N	3.105 $D^{8/3}$	0.654 $(QN)^{3/8}$ / $S^{3/16}$
8	2.50	4.588 D^2	6.676 D	0.687 D	0.779 $D^{2/3}$ $S^{1/2}$ / N	3.57 $D^{8/3}$	0.620 $(QN)^{3/8}$ / $S^{3/16}$
9	3.0	5.088 D^2	7.176 D	0.709 D	0.795 $D^{2/3}$ $S^{1/2}$ / N	4.045$D^{8/3}$	0.592 $(QN)^{3/8}$ / $S^{3/16}$
10	3.5	5.588 D^2	7.676 D	0.728 D	0.809 $D^{2/3}$ $S^{1/2}$ / N	4.522 $D^{8/3}$	0.568 $(QN)^{3/8}$ / $S^{3/16}$
11	4.0	6.088 D^2	8.176 D	0.744 D	0.821 $D^{2/3}$ $S^{1/2}$ / N	4.999 $D^{8/3}$	0.547 $(QN)^{3/8}$ / $S^{3/16}$

Explainatory notes : - D is column 8 is calculated from $Q = AV = A R^{2/3} S^{1/2}/N$ or $AR^{2/3} = QN / S^{1/2}$ For first row 1.744 $D^{8/3} = QN / S^{1/2}$ or $D = (QN)^{3/8} / (1.744)^{3/8} S^{3/16}$

Table 4.1.4 Geometrical and Hydraulic Properties of Trapezoidal Section with Curved Ends, with Z = 1:2

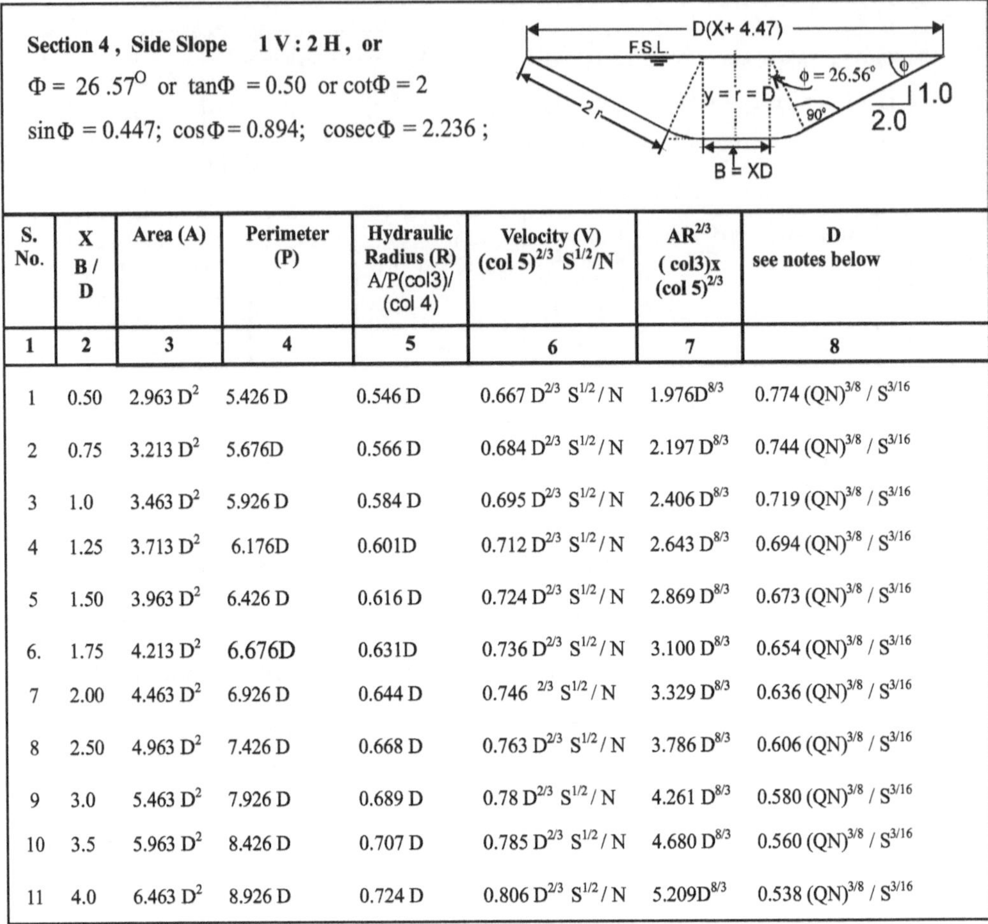

Section 4 , Side Slope 1 V : 2 H , or

$\Phi = 26.57^{O}$ or $\tan\Phi = 0.50$ or $\cot\Phi = 2$

$\sin\Phi = 0.447$; $\cos\Phi = 0.894$; $\mathrm{cosec}\Phi = 2.236$;

S. No.	X B / D	Area (A)	Perimeter (P)	Hydraulic Radius (R) A/P(col3)/ (col 4)	Velocity (V) (col 5)$^{2/3}$ $S^{1/2}$/N	$AR^{2/3}$ (col3)x (col 5)$^{2/3}$	D see notes below
1	2	3	4	5	6	7	8
1	0.50	$2.963\ D^2$	5.426 D	0.546 D	$0.667\ D^{2/3}\ S^{1/2}/N$	$1.976D^{8/3}$	$0.774\ (QN)^{3/8}/S^{3/16}$
2	0.75	$3.213\ D^2$	5.676D	0.566 D	$0.684\ D^{2/3}\ S^{1/2}/N$	$2.197\ D^{8/3}$	$0.744\ (QN)^{3/8}/S^{3/16}$
3	1.0	$3.463\ D^2$	5.926 D	0.584 D	$0.695\ D^{2/3}\ S^{1/2}/N$	$2.406\ D^{8/3}$	$0.719\ (QN)^{3/8}/S^{3/16}$
4	1.25	$3.713\ D^2$	6.176D	0.601D	$0.712\ D^{2/3}\ S^{1/2}/N$	$2.643\ D^{8/3}$	$0.694\ (QN)^{3/8}/S^{3/16}$
5	1.50	$3.963\ D^2$	6.426 D	0.616 D	$0.724\ D^{2/3}\ S^{1/2}/N$	$2.869\ D^{8/3}$	$0.673\ (QN)^{3/8}/S^{3/16}$
6.	1.75	$4.213\ D^2$	6.676D	0.631D	$0.736\ D^{2/3}\ S^{1/2}/N$	$3.100\ D^{8/3}$	$0.654\ (QN)^{3/8}/S^{3/16}$
7	2.00	$4.463\ D^2$	6.926 D	0.644 D	$0.746\ ^{2/3}\ S^{1/2}/N$	$3.329\ D^{8/3}$	$0.636\ (QN)^{3/8}/S^{3/16}$
8	2.50	$4.963\ D^2$	7.426 D	0.668 D	$0.763\ D^{2/3}\ S^{1/2}/N$	$3.786\ D^{8/3}$	$0.606\ (QN)^{3/8}/S^{3/16}$
9	3.0	$5.463\ D^2$	7.926 D	0.689 D	$0.78\ D^{2/3}\ S^{1/2}/N$	$4.261\ D^{8/3}$	$0.580\ (QN)^{3/8}/S^{3/16}$
10	3.5	$5.963\ D^2$	8.426 D	0.707 D	$0.785\ D^{2/3}\ S^{1/2}/N$	$4.680\ D^{8/3}$	$0.560\ (QN)^{3/8}/S^{3/16}$
11	4.0	$6.463\ D^2$	8.926 D	0.724 D	$0.806\ D^{2/3}\ S^{1/2}/N$	$5.209D^{8/3}$	$0.538\ (QN)^{3/8}/S^{3/16}$

Explainatory notes : - D is column 8 is calculated from $Q = AV = A\ R^{2/3}\ S^{1/2}/N$ or $AR^{2/3} = QN / S^{1/2}$ For first row $1.976\ D^{8/3} = QN / S^{1/2}$ or $D = (QN)^{3/8} / (1.976)^{3/8}\ S^{3/16}$

Consider N = 0.02 and 0.03 for the same size and slope of the channel, i.e. it's A, $R^{2/3}$ and $S^{1/2}$ are constant (K). Then the discharge in the smoother channel will be Q_1 = K/0.020 and in the rougher channel it will be Q_2 = K/ 0.030. Then, their ratio

$$or\ Q_1/Q_2 = \frac{K}{.02} \Big/ \frac{K}{.03} = \frac{K}{0.02} \times \frac{0.03}{K} = 3/2 = 1.5 \qquad \text{Eq. (a)}$$

or $Q_1 = 1.5\ Q_2$

It means that

(i) the same channel with smooth surface can carry 50% extra discharge.

(ii) In other words, some Engineers say that the area is reduced by 50% for the same discharge, a very strong argument put in favour of lining of canals.

Though (i) is correct ,as long as R is same in both the cases, but (ii) is true only when the R is same, other wise not as under:

For the same discharge Q, consider two sections of different A_1, N_1, R_1 and A_2, R_2, N_2 but the same slope, then

$$Q = \frac{A_1 R_1^{2/3} S^{1/2}}{N_1} = \frac{A_2 R_2^{2/3} S^{1/2}}{N_2}$$

or ; $Q \propto \frac{A_1 R_1^{2/3}}{N_1} = \frac{A_2 R_2^{2/3}}{N_2}$ slope being same.

Or $Q \propto \frac{A_1 R_1^{2/3} N_2}{N_1} \propto A_2 R_2^{2/3} \propto A_1 R_1^{2/3} N_2 \propto A_2 R_2^{2/3} N_1$

But if R is same in both the alternatives then

$$Q \propto \frac{A_1}{N_1} = \frac{A_2}{N_2} \text{ or } Q \propto \frac{A_1 N_2}{N_1} = \frac{A_2 N_1}{N_2}$$

which is same as Eq. (a)

But in actual practical design R need not be kept same for different N and for a optimum section it should be changed. This is more explicit as below

In the same equation for a given discharge, cross sectional area is expressed as
$AR^{2/3} = Q N / S^{1/2}$ Q and S remaining constant, $AR^{2/3} \propto N$ only

For the best section R = d/2 always; and for Mehboob section with a side slope of 1 : 1, form table 3.1.

$A = 1.785\ y^2$ and $AR^{2/3} = 1.245\ y^{8/3}$
$\therefore 1.245\ y^{8/3} = QN / S^{1/2}$
or $y = (Q N / 1.245\ S^{1/2})^{3/8}$

Treating $Q / 1.245\ S^{1/2}$ as constant (K) for a given discharge and slope, then

$y = K (N)^{3/8}$ -Eq. (b) thus y is proportional to $(N)^{3/8}$
Now for N = .02, and 0.03, we get
$y_1 = K (0 .02)^{3/8}$ and $y_2 = K (0.03)^{3/8}$
or $y_2 / y_1 = (0.03)^{3/8} / (0.02)^{3/8} = (1.5)^{3/8} = (3.3375)^{1/8}$
$= 1.1642$

Similarly ratio of area of flow of rougher and smoother channel is
$A_2/A_1 = 1.785\ y_2^2 / 1.785\ y_1^2 = (y_2/y_1)^2 = 1.3554$

Similarly ratio of perimeter of rougher and smoother channel is
$P_2/P_1 = 3.57\ y_2 / 3.57\ y_1\ (y_2/y_1) = 1.1642$

That means for a given discharge, a rough channel with N = 0.03, Mehboob section with 1 : 1 side slope, has depth of flow (y_2) 16.42% higher than the depth of flow (y_1) in the smoother channel , having N = 0.02. Similarly the cross sectional area is 35.48% higher and perimeter is 16.42% higher only. Or in other words a roughness coefficient of 1.5 times (50% extra), increase the depth by 17% only.

Thus statement (ii) is not true for a realistic canal design. Also see para 4.3 to know as to how the various parameters are related to each others. The cross sectional area makes very little difference in cost in partial filling / cutting reaches, but makes some difference in very heavy cutting /filling reaches, but not as much as 1.5 times. Therefore it is a partial myth that lining reduces the cost of excavation by a reduced cross sectional area due to smooth surface.

Also sepage loss does not increase as much as appriehended.

Also in real practice the value of N increases with age, therefore engineers should not be afraid of using a real or little higher value of N .

Similar comparison can be made with other sections and these should be worked out while working out the economics of smoother channel with rougher channel or choosing the various alternative types of linning.

In actual field even in smooth lined canals, rugosity coefficient increases due to silting and weed growth, there by redusing the discharge, therefore it is always advisable/ preferable to adopt a little higher value of N, then given in table 2.1 to 2.2. Also see table 6.4 which indicates N to be adopted according to maintenance programme.

This aspect is more clear with the following example.

4.6.1 Practical Example 4.1

Design hydraulic best trapezoidal sections for a discharge of 12 cumecs, bed slope of 1 in 5000, inner side slope of 1:1.5 (V:H), for N = 0.01, 0.02 and 0.03 and compare them.

(a) Best trapezoidal section for side slope of 1: 1.5 (V: H), N = 0.01

From table 3.4, B = 0.6d, P = 4.20d, A = 2.10 d^2

$V = (.5d)^{2/3} (1/5000)^{1/2} / 0.01 = .891d^{2/3}$

$Q = V.A = 0.891\ d^{2/3} (2.10\ d^2) = 12$

or $0.871\ d^{8/3} = 12$ or d = 2.01 m

Bottom width B = 0.6 d = 0.6 (2.01) = 1.21 m

Area $A = Bd + Zd^2 = 2.10\ d^2 = 2.1 (2.01)^2 = 13.13\ m^2$

Perimeter P = 4.20d = 4.0 (2.01) = 8.44 m

Velocity $V = 0.891\ d^{2/3} = 0.891 (2.01)^{2/3} = 1.42$ m/s

Top width = 3.60d = 3.6 (2.5) = 9.0 m

Hydraulic radius R = d/2 = 2.01 /2 = 1.00 m

(b) Best trapezoidal section for side slope of 1: 1.5 (V: H), N = 0.02

From table 3.4, B = 0.6d, P = 4.20d, A = 2.10 d2

$V = (.5d)^{2/3} (1/5000)^{1/2} / 0.02 = 0.445d^{2/3}$

$Q = V.A = 0.445\ d^{2/3} (2.10\ d^2) = 12$

or $0.935\ d^{8/3} = 12$ or d = 2.60 m

Bottom width B = 0.6 d = 0.6 (2.6) = 1.56 m

Area $A = Bd + Zd^2 = 2.10\ d^2 = 1.5 (2.6) + 1.5 (2.6)^2 = 14.20m^2$

Perimeter P = 4.20d = 4.0 (2.6) = 10.92m

Velocity $V = 0.445\ d^{2/3} = 0.495 (2.6)^{2/3} = 0.84$ m/s

Top width = 3.60d = 3.6 (2.6) = 9.36 m

Hydraulic radius R = d/2 = 2.6/2 = 1.30 m

(c) Best trapezoidal section for side slope of 1: 1.5 (V: H), N = 0.03

From table 3.4, B = 0.6d, P = 4.20d, A = 2.10 d^2

$V = (.5d)^{2/3} (1/5000)^{1/2} /0.03 = 0.297d^{2/3}$
$Q = V.A = 0.297\ d^{2/3} (2.10\ d^2) = 12$
Or $0.624d^{8/3} = 12$ or $d = 3.03$ m
Bottom width $B = 0.6\ d = 0.6\ (3.03) = 1.82$ m
Area $A = Bd + Zd^2\ 2.10\ (3.03)^2$
Perimeter $P = 4.20d = 4.0\ (3.03) = 12.73$m
Velocity $V = 0.298\ d^{2/3} = 0.297\ (3.03)^{2/3} = 0.9$ m/s
Top width = $3.60d = 3.6\ (3.03) = 10.91$ m
Hydraulic radius $R = d/2 = 3.03\ /2 = 1.51$

These comparisons are given in table and figure enclosed.

Table of Example 4.1 : Comparison of Best trapezoidal Sections for Q = 12 cumecs, S = 1/5000, Z = 1:1.5 (V: H), and N = 0.010, N = 0.018, N = 0.020, N = 0.030

S. No.	Type of Section	Depth of flow (d) (m)	Bed width (B) (m)	Top Width (T) (m)	Area of flow (A) (m^2)	Wetted perimeter (P) (m)	Hydraulic Radius (R) (m)	Velocity (V) (m/s)
1.	Best Trapezoidal N=0.010	2.00	1.21	7.24	8.48	8.44	1.00	1.42
	(%)	100.00	100.00	100.00	100.00	100.00	100.00	100.00
2.	N=0.018 from Ex. 3.1	2.50	1.50	9.00	13.13	10.50	1.25	0.91
	(% of 1)	125.00	123.96	124.30	154.83	124.40	125.00	64.10
3.	N=0.020	2.60	1.56	9.36	14.20	10.92	1.30	0.84
	(% of 1)	130.00	128.90	129.40	167.45	129.40	130.00	59.20
	% variation of 1	30.0	28.90	29.40	67.45	29.40	30.00	- 40.80
	(% of 2)	10400	104.00	104.00	108.14	104.00	104.00	92.30
	% variation of 2	4.0	4.0	4.0	8.14	4.0	4.0	7.7
4.	N=0.030	3.03	1.82	10.91	19.28	12.73	1.51	062
	(% of 1)	151.50	116.6	150.70	222.73	150.83	151.00	43.66
	% variation	51.5	16.6	50.70	122.73	50.83	51.00	- 56.34
	(% of 3)	116.54	166.66	116.56	135.77	11657	161.50	73.81
	% variation of 3	16.54	66.66	16.56	35.77	16.57	61.50	- 26.91

- For less
Above result are plotted in figure of this example.

Remarks

For twice rugosity coefficient (N), i.e 100% increase in N, there is increase in depth by 30%, perimeter by 29.4%, and area by 67.4% only. Similarly for increase in rugosity coefficient by 3 times, increase in depth is 51.5% in perimeter, 50.83% and in area 122.73% only. For increase in N by 50%, depth increases by 16.54%, area by 35.77% and perimeter by 16.57%. Similarly also see the top width which is increasing only in proportions of depth.

In actual field the cost is more a function of top width and perimeter. The role of area and depth is little, in heavy cutting and filling and also partial cutting filling.

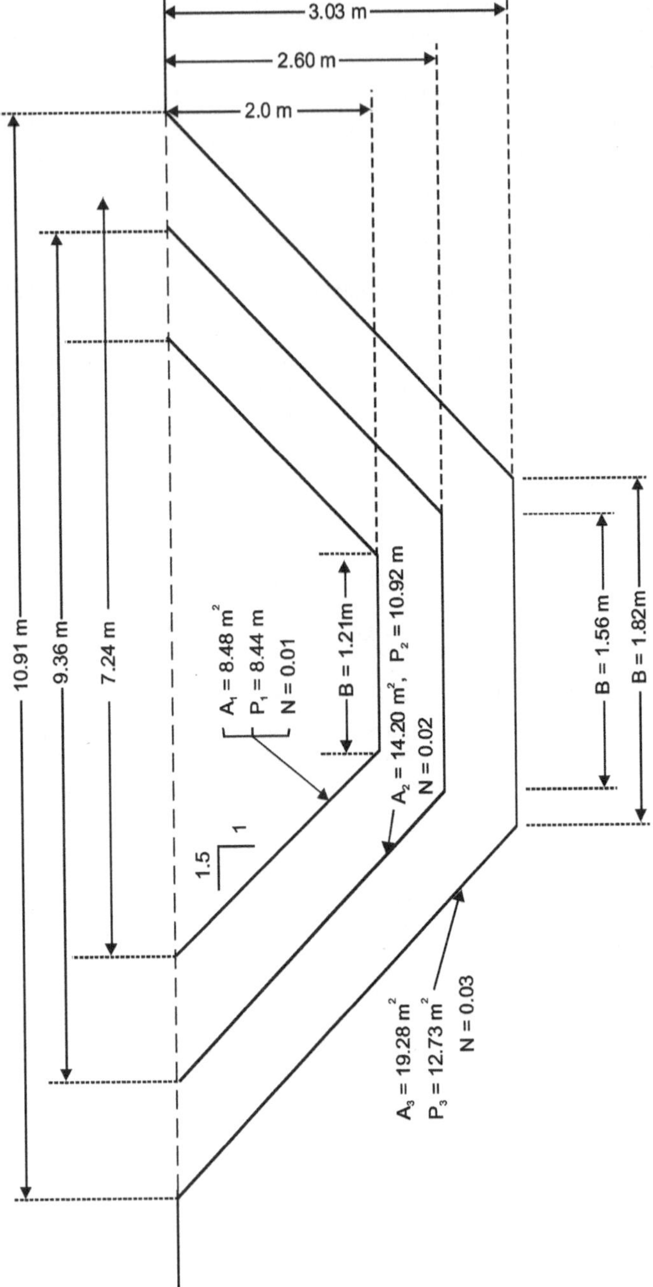

Figure of example 4.1 Comparison of Trapezoidal Sections for a given Discharge of 12 cumecs, Bed Slope 1 in 5000, Side Slopes of 1 : 1.5 and N = 0.01, 0.02 and 0.03 (see table of example 4.1).

Normally in any canal alignment 40% length can be considered in balance cutting/ filling, 25-30% in heavy filling and 30 to 35% in cutting to heavy cutting.

Thus a little higher value of N by 10 to 40% is not increasing the cost by that percentage or substantially.

In actual field even in smooth lined canals, rugosity coefficient increases due to silting and weed growth; there by reducing the discharge. Flat slope in lined canal enhances silt deposition of silt and weed growth. Therefore it is always advisable/ preferable to adopt a little higher value of N, then given in table 2.1 to 2.2.

Also if lining gets damaged, the canal will not be closed for repairs of lining immediately.

Therefore it is proposed to adopt a value of N= 0.018 against N=0.01; or N=0.02 against 0.018 or 0.025 against 0.02

Example 4.2

Propose a suitable lined canal section in sandy soils for a discharge of 280.85 cumecs, for N = 0.017, and canal bed slope of 1/12000.

Solution

For sandy soils a side slope of 1:1 (V: H) is unreasonable for depth beyond 1.2 m and 1: 1.5 or 2 (V: H) appears reasonable which can be further chosen / decided depending upon depth of canal.

(i) To under stand flow phenomena, initially consider a Mehboob section;
From table 3.1, for a side slope of 1: 1, (V: H) and

$$S^{1/2} = \left(\frac{1}{12000}\right)^{1/2} = 0.009128;$$

$$Q = 1.12456r^{8/3}\ S^{1/2}\ /N\ = \frac{1.1245}{0.017}\left(\frac{1}{12000}\right)^{1/2} r^{8/3} = 0.6036\ r^{8/3}$$

$$\text{Or } r = \left(\frac{Q}{0..6036}\right)^{3/8} = \left(\frac{280.85}{0.6036}\right)^{3/8} = 10.0m$$

and maximum velocity $V = 0.63\ r^{2/3}\ S^{1/2}\ /\ N$ from Eq. 2.2(i) (Para 2.2, Chapter 2)
or $V = 0.63\ r^{2/3}\ (1/12000)^{1/2}/0.017 = 0.338\ r^{2/3} = 0.338\ (10.0)^{2/3} = 1.57$ m/s

To clearly understand mechanics of flow computations for various side slopes of 1: 1, 1 : 1.25, 1 : 1.5, 1 : 2, 1 : 2.5 and 1 : 3.0 (V : H) are done with the help of table 3.1 in a tabular form as under.

Comparison of Mehboob Sections for Q = 280.85 cumecs, S = 1 / 12000, N = 0.017; for various side slopes

S. No.	Side Slope (Z) V:H	Depth of flow (d) (m)	Bed width (B) (m)	Top Width (T) (m)	Area of flow (A) (m²)	Wetted perimeter (P) (m)	Hydraulic Radius (R) (m)	Velocity (V) (m/s)
1	2	3	4	5	6	7	8	9
1.	1 : 1	10	0.0	28.28	178.50	35.727	4.996	1.5709
2.	1 : 1.25	9.73	0.0	31.136	182245	37.4605	4.864	1.5416
3.	1 : 1.50	9.438	0.0	33.976	185.990	39.3564	4.726	1.510
4.	1 :1.7.5	9.15	0.0	36.81	189.9664	41.5318	4.573	1.479
5.	1 : 2.0	8.873	0.0	39.662	193.9123	43.7083	4.435	1.449
6.	1 : 2.5	8.366	0.0	41.244	201.6411	48.189	4.184	1.394
7.	1 : 3.0	7.934	0.0	42.724	209.114	52.675	3.97	1.345

From above table it can be inferred that for slope of 1/12000, Section at S.No. (1) gives maximum possible velocity of 1.57 m / s, which is quite reasonable but depth of 10.0 m is little higher. Also slopes of 1:1 (V: H) are unstable for such height in sandy soils. But with flattening of side slopes, velocity and depth will decrease.

Table of computations for Example 4.2 :
Comparison of Sections for Q = 280.85 cumecs, S = 1 / 12000, N = 0.017; and Trapezoidal Sections with curved ends, for various side slopes and B / D ratio

S. No.	B / D ratio (X)	Depth of flow (d) (m)	Bed width (B) (m)	Top Width (T) (m)	Area of flow (A) (m^2)	Wetted perimeter (P) (m)	Hydraulic Radius (R) (m)	Velocity (V) (m/s)
(i) Z = 1 : 1 / V : H,								
1.	0.50	8.86	4.43	29.41	179.46	36.07	4.98	1.56
2.	0.750	8.43	6.32	30.20	180.29	36.43	4.95	1.55
3.	1.00	8.06	8.06	30.79	181.04	36.85	4.91	1.55
4.	1.25	7.74	9.68	31.50	182.02	37.33	4.88	1.54
5.	1.50	7.46	11.19	32.23	182.96	37.84	4.84	1.53
6.	1.75	7.21	12.63	32.95	184.01	38.38	4.79	1.52
7	**2.00**	**6.99**	**13.99**	**33.69**	**185.08**	**38.95**	**4.75**	**1.51**
8.	2.50	6.61	16.53	35.16	187.25	40.13	4.67	1.50
9.	3.00	6.29	18.89	36.61	189.63	41.34	4.58	1.48
10.	3.50	6.03	21.09	38.11	191.89	42.60	4.50	1.46
11.	4.00	5.83	23.31	39.76	196.39	44.11	4.45	1.43
(ii) Z = 1 : 1.5 (V : H)								
1.	0.50	8.49	4.24	34.80	186.50	39.69	4.69	1.51
2.	0.75	8.12	6.09		187.01	39.99	4.68	1.50
3.	1.00	7.79	7.79	35.87	187.73	40.36	4.65	1.49
4.	1.25	7.51	8.39		188.50	40.78	4.62	1.48
5.	1.50	7.26	10.89	35.86	189.36	41.23	4.59	1.48
6.	**1.75**	**7.04**	**12.32**		**190.21**	**41.72**	**4.56**	**1.47**
7	2.00	6.84	13.68	38.293	191.14	42.23	4.53	1.46
8.	2.50	6.49	16.22		193.12	43.31	4.46	1.45
9.	3.00	6.19	18.58		195.06	44.43	4.39	1.44
10.	3.50	5.94	20.79		197.08	45.59	4.32	1.42
11.	4.00	5.72	22.88		199.17	46.76	4.26	1.41
(iii) Z = 1 : 2.0 (V : H)								
1.	0.50	8.39	4.19	41.693	208.64	52.52	3.97	1.346
2.	0.75	8.06	6.05		208.94	52.49	3.98	1.358
3.	1.00	7.76	7.76		208.53	52.46	3.98	1.347
4.	1.25	7.49	9.37		208.51	52.53	3.97	1.347
5.	1.50	7.26	10.89		208.85	52.70	3.96	1.344
6.	**1.75**	**7.05**	**12.33**		**209.18**	**52.92**	**3.95**	**1.342**
7	2.00	6.85	13.71		209.64	53.18	3.94	1.339
8.	2.50	6.52	16.29		210.77	53.83	3.92	1.332
9.	3.00	6.22	18.67		211.67	54.53	3.88	1.326
10.	3.50	5.98	20.92		212.96	65.34	3.85	1.317
11.	4.00	5.75	23.03		214.33	66.20	3.81	1.311

Since the depth is slightly more and a reasonable depth may be near 7 m. For this depth, and side slope of 1: 2(V: H) also appears reasonable. But From above table, Mehboob section with Z = 1: 2 (V: H) gives D = 8.873m. This depth is also higher and therefore it is not a good choice.

Analysis of alternative wide channel

Therefore now consider wide trapezoidal section with curved ends. The calculation of this section for various Z and X values is done with the help of table 4.1.1 to 4.1.4 in a tabular form as above :

Now from the above table, on comparing the sections for a depth of flow of 7m, we get following inferences:

i. For Z=1:1 and X=2; P= 38.95m and V= 1.51m/s
ii. For Z=1:1.5 and X=1.75; P= 41.72m and V= 1.47 m/s
iii. For Z=1:2 and X=1.75; P= 52.92mand V= 1.342m/s

Thus for a proposed depth of 7m, we may very reasonably adopt a side slope of 1: 1.5 and X= 1.75. There is no justification to increase X > 1.75.

If we want to decrease depth, still it is better to adopt a side slope of 1:1.5 and increase X up to 4 only. It gives a section with D = 5.72m and P = 46.76m and V= 1.41m/s. Even Z = 1 and X = 4 gives a section with D = 5.83m and P = 44.11m and V= 1.43 m/s. Similar more trials can be done to satisfy one self and others.

In any way there appears no justification to adopt Z = 2. It encourages deposition of silt on side slopes and gives uneconomical section.

A very critical examination of some sections IGNP are discussed in Chapter 7B

4.7 Flow Characteristics of Channels

Flow characteristics of a channel means as to how velocity and discharge varies in particular channel with reference to depth. It depends upon the geometrical characteristics of that particular channel.

(a) Trapezoidal Canal

In a trapezoidal channel of given side slope (Z) and bed width (b); perimeter (P) and area (A) increases with depth with the following relationships.

$P = b\,d + 2\,d\sqrt{Z^2+1} = d\,(b+2\sqrt{Z^2+1}) = k_1\,d$ (for constant b and Z)

or $P = x\,d^2 + 2\,d\sqrt{(Z^2+1)} = d\,[x + 2\sqrt{(Z^2+1)}] = k_1\,d$

That means P is a straight line function of depth only, treating b and Z as constant.

$A = b\,d + Z\,d^2$ or $X\,d^2 + Z\,d^2$
$= d^2\,(X + Z)$; $(X + Z)$ or b and Z being constant $= k_2\,d^2$

A is a curvilinear function of d^2, it is convex in shape.

An inspection of the Manning's formula indicates that,
$V = R^{2/3}\,S^{1/2}/N = (A/P)^{2/3}\,S^{1/2}/N$

Substituting above values of P and A, (for constant b and Z)
$V = (k_2\,d^2 / k_1\,d)^{2/3}\,S^{1/2} / N = k_3\,d^{2/3}$

S and N are constant for a particular channel, of given dimensions
i.e. velocity is a function of $d^{2/3}$, it increases at slow rate with depth.
and discharge $Q = A\,R^{2/3}\,S^{1/2}/N = k_1\,d^2.\,k_3\,d^{2/3}\,S^{1/2}/N = k_4\,d^{8/3}$

That is Q is function of $d^{8/3}$ for a channel of given dimensions. Thus Q increases faster than P, A and V.

k_1 k_2 k_3 k_4 depends upon the geometry of given section.

Thus in a trapezoidal channel of given cross section, slope and rugosity coefficient N , discharge increases much greater than increase in perimeter and area, being a function of $d^{8/3}$. Also velocity increases with depth, but at slow rate then the area (being a function of $d^{2/3}$). Velocity at higher depth is much more than at low depth. It has concave shape.

The relationships between, perimeter area velocity and discharge with respect to depth for a particular trapezoidal canal, and trapezoidal with curved ends section are shown in figure 4.4 and 7B.5.

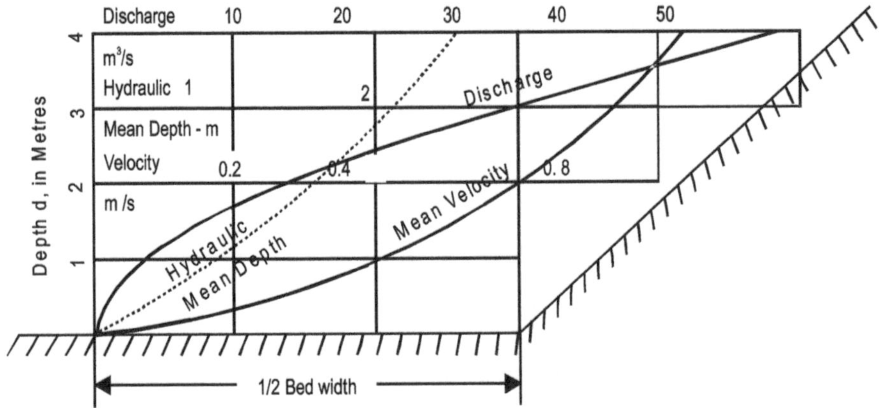

Figure 4.4 Flow characteristics of trapezoidal channel.

Flow Characteristics of all channels are practically identical with slight changes in curvature for k_1, k_2, k_3 and k_4 of that particular channel.

b) Rectangular Section

Similarly in a rectangular given section (b remaining constant) perimeter,

P = b + 2d

= x d + 2d = d (x + 2) = k_1 d (k_1 is constant for constant b)

area A = b.d = x d^2 = k_2 d^2 (k_2 changes with d)

$$\text{Velocity} = R^{2/3} S^{1/2} / N = \left[\frac{bd}{b+2d}\right]^{2/3} S^{1/2} / N$$

$$= (k_2 d^2 / k_1 d)^{2/3} S^{1/2} / N = k_3 d^{2/3}$$

$$\text{Discharge } Q = AR^{2/3} S^{1/2} / N$$

$$= k_2 d (K_3 d^{2/3} S^{1/2} / N) = K_4 d^{8/3} = K_4 d^{2.67}$$

Therefore flow characteristics, practically similar to trapezoidal canals occur in rectangular sections also with slight changes for k1, k2 , k3 and k4 of that particular section. Similar flow characteristics occur in semi circular sections, but flow characteristics of circular sections are different, see para 5.3 and figure 5.6.

Inferences

It can also be seen in nature's way of conveying water in rivers, that river in flood flows much faster than during the lean period of low discharge, reason being velocity increases with depth. This is the reason that during floods rivers have erosion characteristics and same river starts silting during low discharges. Same phenomenon also occurs in canals flowing at partial discharges, but in some canals once silting occurred, it may not get cleaned easily as the maximum velocity at full supply level may not have much tractive force to cause its erosion and transport, where as it occurs in natural streams.

All canals, trapezoidal or Mehboob or rectangular, whether lined or unlined show practically equal signs of siltation at partial discharges. Still Mehboob section will show less siltation than trapezoidal due to cup shape and higher depth at low discharge.

Clear understanding of flow characteristics is also important in locating the cross - regulators, levels of off taking channels and outlets .

Flow characteristic further changes at cross - regulators in the upstream as well as in down stream, predominantly at partial discharge, see chapter 7B.

Cross-regulators have hardly any role at full supply discharge, but their role comes into play at partial discharges. At partial discharges, full supply level is mainained in the up stream, therefore gradually varied flow in the upstream and unstable flow in the down stream occurs for some distance. There after uniform flow down stream full supply is maintained and normal water depth is achieved according to discharge, see chapter 7B.

Flow characteristics are also very important in operation and maintenances of canals.

4.8 Summary and Conclusions

This chapter deals with design of wide channels (other than best section) with reasons and comparative hydraulic efficiencies. Wide channels become necessary to avoid flat side slopes and also to reduce depth. But width should not be unnecessarily increased.

Master Equator of Manning's formula is presented with detailed explanation of inter relation ship of all the 7 parameters of most general trapezoidal section (i.e. Q, V, N, SZ; b and d), in Para 4.3.

A simple nomographic quick solution is presented. It helps in quick decision and adoption of a most economical section without much trial.

Then trapezoidal section with curved ends is presented with very convenient tabular solution.

Effect of rugosity coefficient (N) i.e. smooth and rough surfaces on the discharging capacity of a channels and its economics are discussed. A little higher value say 20 % does not affect economics of canal in that proportion as is anticipated and therefore same are proposed for adoption.

Lastly typical solved examples and flow characteristics of a channel are explained.

References

Arya, R.S. *et al*, Design Charts for Earthen and Lined Channels, 1st International Research and Development Conference Water & Energy 2001, Central Board of Irrigation and Power, New Delhi, October, 1995.

Atmapoojya, S. L. and Ingle R.N., Optimal Design for a Canal Cross-Section, Journal of Indian Water Works Association, Mumbai, January-March, 2001.

Chow, V.T, Open - Channel Hydraulic, Mc. Graw - Hill Book Company Ltd., 1973.

Mehta, A. C., Determination of The Most Economical Supply Depth For Lined Channels, Dissertation, University of Roorkee, Roorkee.

Mirza, Mohamed Din, Note on Canal Cross Section and Lining, Indian Concrete, J. Vol. XVII No. 12, Pp - 443 To 445.

Modi, P. N., And Seth, S. M., Design of Most Economical Trapezoidal Section of Open Channels, Irrigation and Power, Journal of Central Board of Irrigation and Power, New Delhi, July, 1968.

Prasad, V. A Review of Canal Lining in India; M.E. Thesis University of Roorkee, 1975.

Suryavanshi, B.D., The Master Equation for Designing of Canal Section By Manning's Formula, Irrigation and Power, Journal of Central Board of Irrigation and Power, New Delhi, January, 1972.

Swami, P.K., and Bhatia, K.G., Economic Open Channel Section, Irrigation And Power, Journal of Central Board of Irrigation and Power, New Delhi, April, 1972.

□□□

Chapter **5**

Design of Other Canal Sections

A. Rectangular Canals

5.1 Design of Best Rectangular Canals

After the Mehboob and Trapezoidal canals, next best section is rectangle. This section is also generally adopted for **small discharges say upto 10 cumecs (350 cusecs).** This section ranks at No. 4. Yet, this is the only section most commonly adopted for flumes of discharges upto140 cumecs (5000 cusecs). Typical example is Orsang acqueduct, 1 km. long for 1140 cumecs (40000 cusecs) in ractangular sections on Narmada Canal Project, Gujrat, India, built in the years from 1990 - 2000. Such sections are common in power house canals due to limitation of space and very sharp curves. It is free from the effect of change of side slopes as in the trapezoidal canals (see para 3.3) Still some small side slope (batter) is adopted in thick masonry / concrete lining. It adds to structural stability with economy in thickness of wall and as well as hydraulic efficiency. It may still be called as rectangular section.

Details of its hydraulic efficiency in comparison to semi circle, Mehboob and trapeziodal sections are discussed in chapter 1, para 1.8.

Designers are suggested to adopt Mehboob or trapezoidal sections to the extent possible. However at specific site conditions trapezoidal and Mehboob section may not be always feasible or economical such as along hill sides. Then recourse to rectangular section is adopted or some times it is unavoidable, see figue 9.4 (i to v).

For a given discharge its perimeter is 16 % more than semi circle and 10 % more than trapezium. However its area is only 6% more than semi circle and only 4% more than trapezium. But it has smallar depth than semi circle (only 91 .3%) and 95% that of trapezoium, **See figure 1.6**. Also top width is minimum which is 91.3 % that of semi circle i.e. 1.826 y against 2y of semi circle. Thus this section becomes useful in typical site situation, and particularly as a closed section.

This section also has the advantage of lesser requirement of land width then trapezoidal canals, see example 3.1 and figure 5.1. Saving in land width depends upon the depth of cutting above bed and height of filling below bed. More is the height /depth, more is saving, see table 5.1. Also there is some saving in cost of earth work. However lining may be more costly as thin lining can not be adopted. Still this section is very popular for small discharges upto 0.5 cumecs or low depth discharge canals of any discharge. Land saving is more in deeper canals which reduces the increase in width of channel.

This section is also very useful upto 2.0 m depth.

This section is also very useful for closed conduits flowing free, much better than circular which is not a good hydraulic section, see para 5.3.

Table 5.1 : Cost Comparison of Trapezoidal and rectangular section For A Small Canal Sections For 5' Bank Filling / Cutting.

S. No.	Item	Trapezoidal Section	Rectangular section	Difference	%
1	2	3	4	5	6
1.	Earth work				
(i)	5' Bank filling	7375 cft.	6225 cft.	1150 cft.	15.6
(ii)	5' Bank cutting	10375 cft.	9525 cft.	850 cft.	8.19
(iii)	Compaction for 5' bank filling	7750 cft.	5512 cft.	2238 cft.	28.80
2.	Land width for 5' bank filling	23 ft.	21 ft.	2 ft.	8.6

Hydraulic design

For design of best section no trial and error is required. Dimensions of best sections are also geometrically so fixed that it is always circumscribing the semi circle and bed width (B) is equivalent to twice depth (D) ie B = 2D, see para 1.6.1.

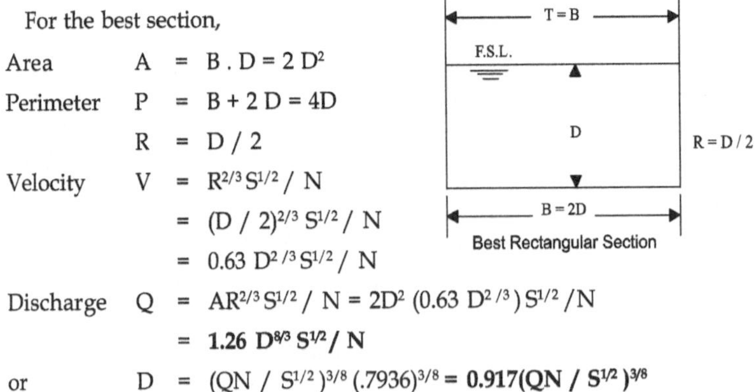

Best Rectangular Section

For the best section,

Area $A = B \cdot D = 2D^2$

Perimeter $P = B + 2D = 4D$

$R = D / 2$

Velocity $V = R^{2/3} S^{1/2} / N$

$= (D / 2)^{2/3} S^{1/2} / N$

$= 0.63\, D^{2/3} S^{1/2} / N$

Discharge $Q = AR^{2/3} S^{1/2} / N = 2D^2 (0.63\, D^{2/3}) S^{1/2} / N$

$= \mathbf{1.26\, D^{8/3} S^{1/2} / N}$

or $D = (QN / S^{1/2})^{3/8} (.7936)^{3/8} = \mathbf{0.917 (QN / S^{1/2})^{3/8}}$

Some depth versus discharges curves for different slopes are given in figure 5.2

For the best section no dimension should be played with. However for an economical cross-section, these dimensions are modified to the extent required. Figure 9.4 shows various types of sections adopted in different topographical situations.

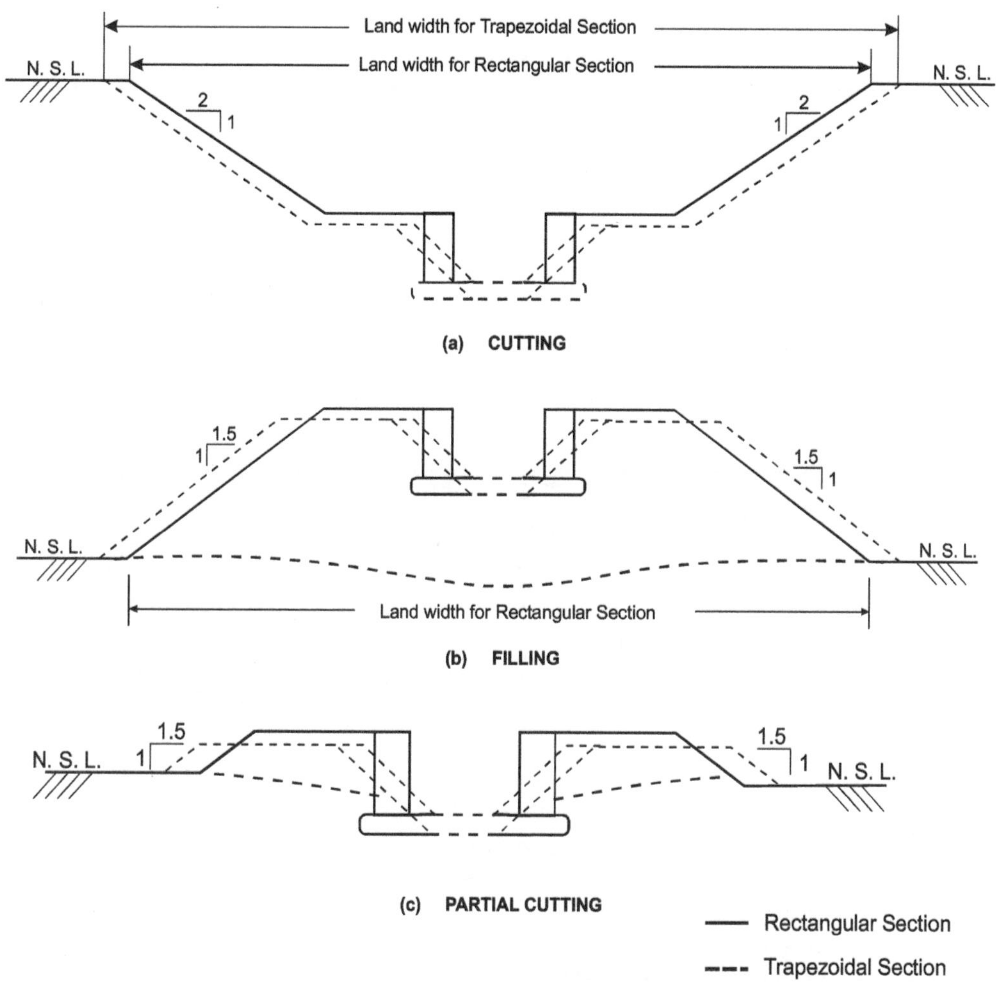

Figure 5.1 Comparison of land width in rectangular and trapezoidal canal sections.

Note : 1 In rectangular section, earthen bank can be further reduced according to the requirement of service road.

2 Land width is further reduced rounded bottom rectangular section and closed conduits.

Efficiency of rectangular section can be increased by rounding the corners or by making its bottom circular. Corners are rounded by suitable radius, See table 1.1. This is not very effective.

5.2 Rounded Bottom Rectangular Section

The best way is by making the bottom circular with a radius equal to $y\sqrt{2}$ (i.e. diagonal of half-best rectangle inscribed in semi circle, see figure 1.8) and this circle becomes its bottom, see figure 5.3 This is simple and very effective. This is the most optimum radius for maximum efficiency of a rectangular section in the same way as Mehboob section (para 3.1) is formed with radius equal to full supply depth.

This section is most suitable to stable rock cutting or hard strata where least earth pressure is exerted on a wall. This is also most suitable to unlined canals in rocks, or thin lining like guniting / shortcrete

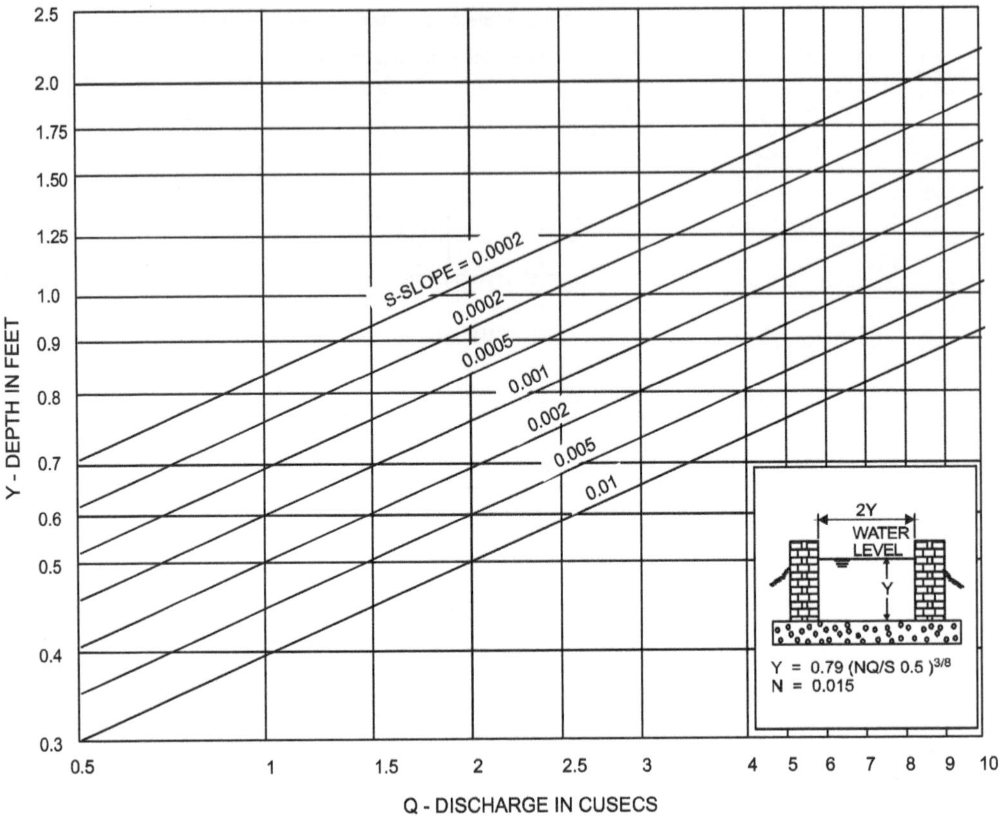

Figure 5.2 Depth V/S Discharge, Curves for Best Rectangular Section

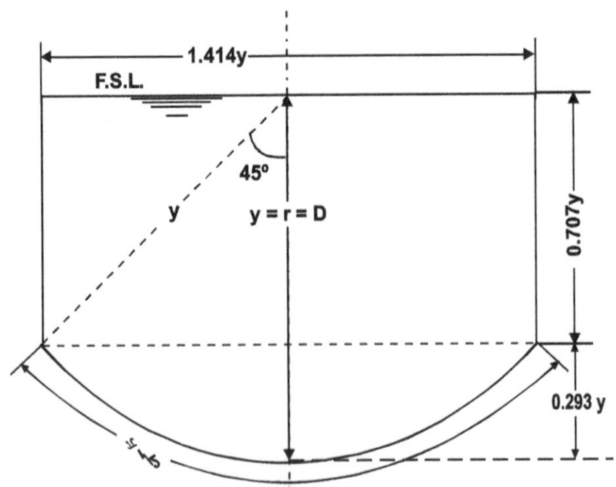

Figure 5.3 : Circular bottom rectangular section

This section has the advantage of adoption of thin lining in bottom and thick section in vertical walls. The thickness of walls is reduced due to the decrease in height of vertical walls and consequently earth pressure acting on walls. Structurally this is next to semi circle and better than rectangle.

For 0.3 to 4 m depth, circular bottom, then simple rectangular and trapezoidal sections will also be reasonably economical. All the advantages of cost economics, higher velocity, non silting, little weed growth, minimum seepage, durability, etc are obtained in these section, see figure, 5.3 and 9.4 for heavy cutting/ filling.

Some may argue about the problems of curves and the joints of curves with vertical walls. This is at all no problem and one should not be afraid of it. This is easily resolved by skills of modifying the dimensions, see figure 5.4.

Its geometric and hydraulic properties are computed as under :

From para 2.2, we get

$A = \pi r^2 (45/180) + 2 (1/2) (.707 r). (.707 r)$
$= \pi r^2 /4 + (.707 r)^2$
$= 0.785 r^2 + .50 r^2 = 1.285 r^2$

$P = 2\pi r (2 \times 45^o / 360) + 2 (.707 r)$
$= 1.57 r + 1.414 r = 2.984 r$

$R = A/P = 1.285 r^2 / 2.98 r = \mathbf{0.431 r}$

Conveyance $A R^{2/3} = 1.285 r^2 (.431 r)^{2/3} = 0.733 r^{8/3}$

Discharge $Q = 0.733 r^{8/3} S^{1/2} / N$

or $r = (Q N / S^{1/2})^{3/8} (1/0.733)^{3/8} = \mathbf{1.123 (Q N / S^{1/2})^{3/8}}$

For the same discharge, the depth of flow will be slightly higher. It can be worked out as under.

5.2.1 Comparison with Semi Circle for a Given Discharge

For a given discharge, let y_2 be the depth of circular bottom rectangular section and y_1 be the depth of semicircular section.

For a given discharge equating the conveyance of circular bottom rectangular section with the conveyance of semicircle, (from table 1.4), we get.

$0.733 y_2^{8/3} = 0.987 y_1^{8/3}$

or $y = (0.987 y^{8/3} / 0.733)^{3/8} = 1.1180 y_1$

Height of vertical wall $= 0.707 y = 0.707 (1.118 y_1) = 0.790 y_1$

top width $= 2 (0.79 y_1) = 1.58 y_1$

Thus for a given discharge its, depth of flow is 11.8% higher than that of semicircle and top width is only 1.58y i.e. 79% of semicircle.

Wetted Perimeter of circular bottom rectangular section:

$P = 2.984 y_2 = 2.984 (1.118 y_1) = \mathbf{3. 336 y_1}$

Ratio of this Perimeter with the perimeter of semicircle

$= 3. 336 y_1 / 3.14 y_1 = 1.062$ or 106.2% or 106%

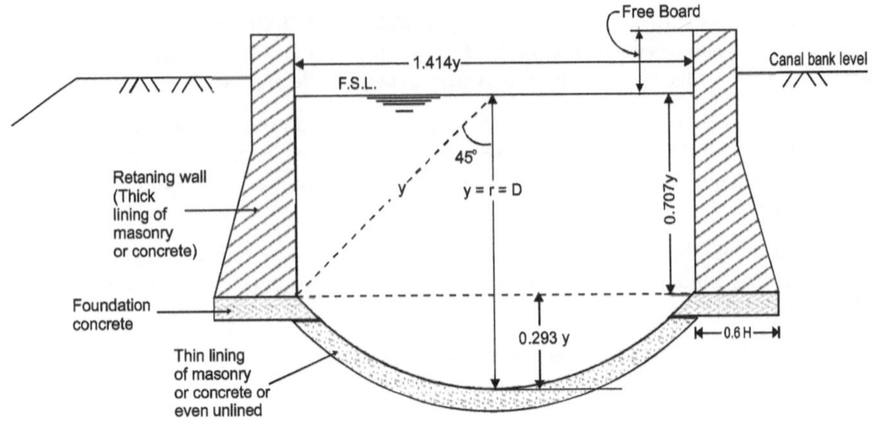

(a) **Circular bottom rectangular section - Bottom lining fully adjusted below walls.**

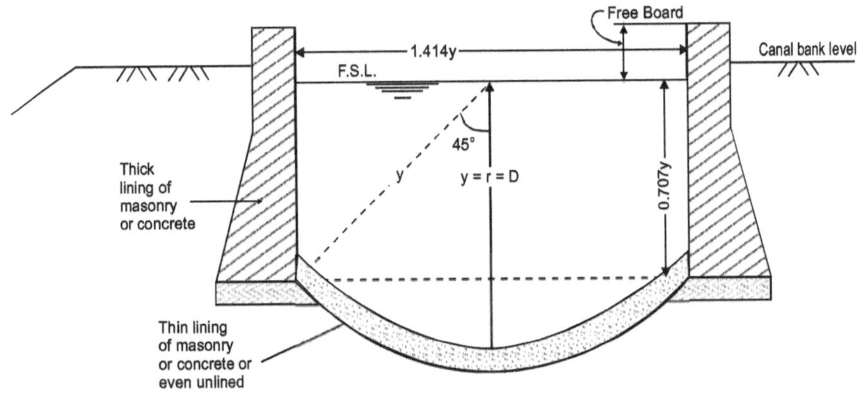

(b) **Bottom lining slightly above (to the extent of thikness of bed lining)**

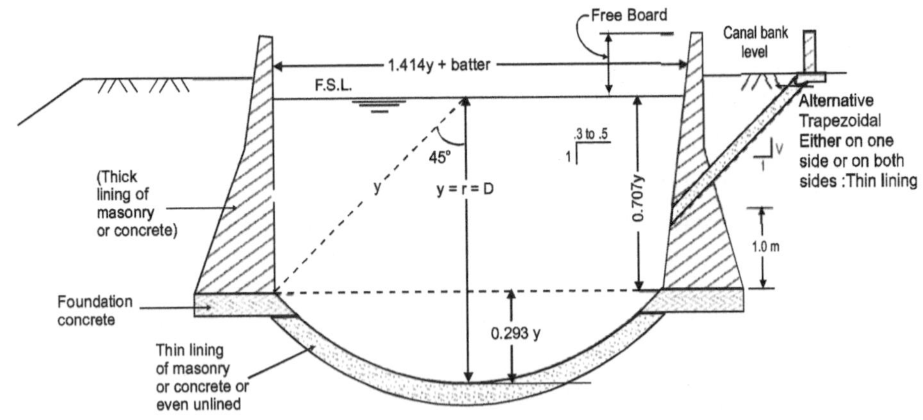

(c) **Circular bottom rectangular - trapezoidal (combind) section - Bottom lining fully adjusted below walls.**

Figure 5.4 : Lined circular bottom rectangular sections

Area of circular bottom rectangular section A = $1.285\ y_2^2$

$= (1.285)\ (1.118\ y_1)^2 = 1.606\ y_1^2$

Ratio of this area with the area of semi circle = 1.606y12 / $1.57y_1^2$ = 1.023 or 102.3%

Ratio of R= $(y_2/\ 2)\ /\ (y_1/2) = (1.118\ y_1/\ 2)\ /\ (y_1\ /\ 2)$

= 1.118 or 111.8 %

Hence for a given discharge a circular bottom rectangular section has an area of 2.3% more, where as its perimeter is only 6% more than the equivalent semi circle. Thus it is very near to best trapezium; where as its top width is only 79% of semicircle. Thus hydraulically it is the best section as a close conduit. In comparison to best rectangle, its area is 3.7% less and perimeter is 10% less.

A comparison of these sections is shown in figure 5.4 and table 5.1.

This section is very useful and economical for parallel canals of hydel powerhouses, as well as for flumes, syphons, and closed conduits with flat or circular roofs.

Vig (1981) has shown that even in very small discharge canals (water courses) curved bed is economical and better.

B. Circular and Other Sections

5.3 Circular Section

Though a semi circular section is the best hydraulic section, yet the circular section is the worst hydraulic section, see figure 5.6. When running full its area is 15% more than the equivalent semi circular channel. This is so, because the perimeter increases more than the area as the depth of flow increases above d_0 / 2, i.e. above simi circle.

Still, it is structurally the best section, and therefore most suited as a closed conduct, either as a tunnel or pipe running at more than 0.75 to 0.9.do or may be running under pressure. The channel flow at full depth is unstable and becomes soon a pressure flow.

The geometric elements and free flow characteristics of a circular section are shown in figure 5.6 . in the form of corresponding ratios when the section is flowing full. The dimensionless curves for $AR^{2/3}/A_0R_0^{2/3}$, $R^{2/3}/R_0^{2/3}$, V/V_0, Q/Q_0 are also shown in figure 5.6 (b). The subscript zero indicates full flow condition.

In these curves the value of N is taken as constant, (practically N varies slightly with depth, see para 2.0)

The velocity curve shows that a maximum value occur at 0.81 do (it is also obtained by equating the first derivative of $R^{2/3}$ (with respect to y) to zero, as $V \propto R^{2/3}$ (for a constant N and S). Then it goes on decreasing till the circle is full, and reaches the same valus as at 0.0 to 0.5 do.

The discharge curve shows that maximum discharge occurs at 0.938 do (it can also be obtained by equating to zero the first derivative of $AR^{2/3}$ with respect to y, as $Q \propto AR^{2/3}$ (for a constant N and S). There after it shows a decreasing trend till circle is full, reaches the same value as at 0.82 d_0.

The hydraulic elements of a circular section for the ratio of depth to diameter are given in Table 5.3. It may be carefully seen that both R and $AR^{2/3}$ have a maximum value at a depth less than full.

The maximum R occurs when y/d = 0.813, and $AR^{2/3}$ when y/d = 0.938. The existence of a maximum value for V and Q is characteristic not only of circular sections but also of others such as egg shaped or horse shoe section as shown in Figures 5.7 and 5.8.

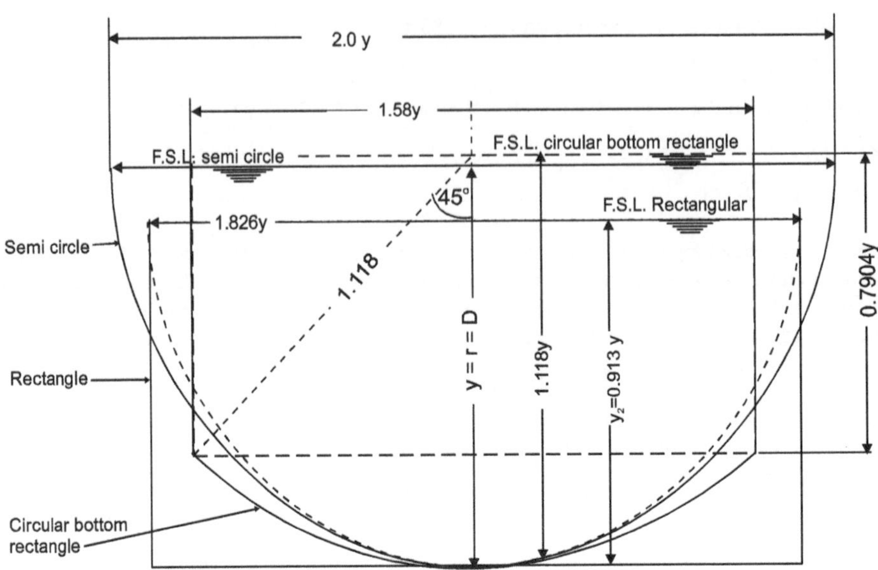

Figure 5.5 : Comparison of circular bottom rectangular section with semi-circle
Note: Bed level has been kept same in above figure. Alternatively FSL can be kept same.

Table 5.2 Comparison of best rectangular and circular bottom rectangular sections for a given discharge

Hydraulic properties	Semicircle	Rectangle	Circular bottom rectangular
(1) Area of flow $1.57\,y^2$	$1.57\,y^2$	$1.667y^2$	1.606y
(% of semicircle)	100	106	102.3
(2) Depth of flow y	y	0.913y	1.118 y
(% of semicircle)	100	91.3	111.8
(3) Wetted Perimmeter	3.14 y	3.652y	3.336 y
(% of semicirlce)	100	116	106
(4) Hydraulic radius			
R=A/P	0.5 y	0.456y	0.559 y
(% of semicirlce)	100	91.3	111.8
(5) Conveyance $AR^{2/3}$	$0.987\,y^{8/3}$	$0.987y^{8/3}$	$0.987y^{8/3}$
(% of semicirlce)	100	100	100

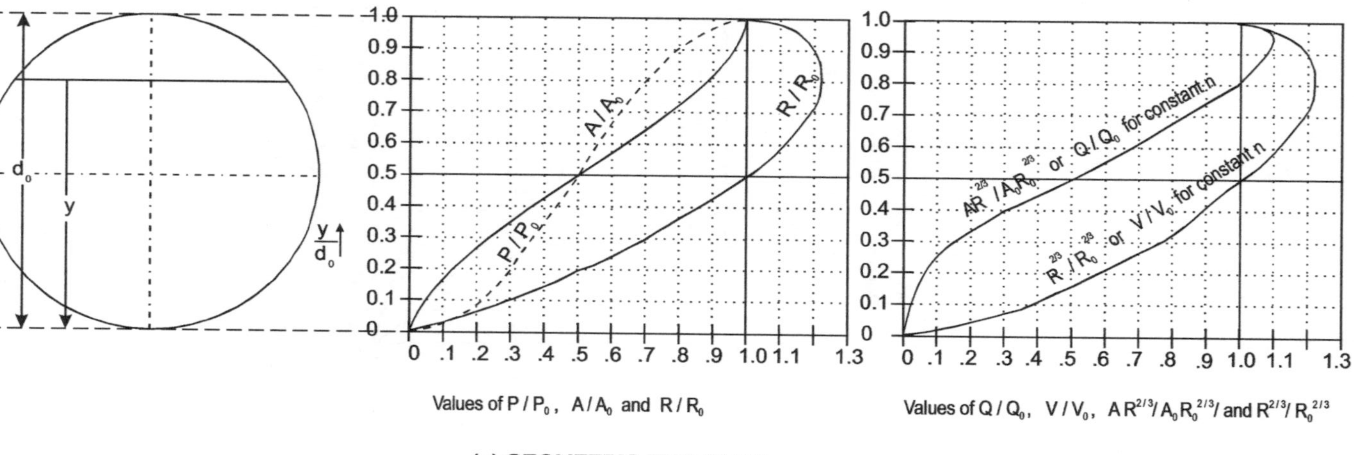

Subscript "0" indicates the full flow condition

Q_0, V_0, A_0, R_0, d_0 are the discharge, velocity, area, hydraulic radius and dia of pipe at full depth

$$P_0 = \pi\ d_0 = 3.1416\ d_0, \quad A_0 = \frac{\pi}{4} d_0^2 = 0.7854\ d_0^2, \quad R_0 = \frac{1}{4} d_0 = 0.25 d_0$$

Figure 5.6 : Geometric and hydraulic properties of a circular section

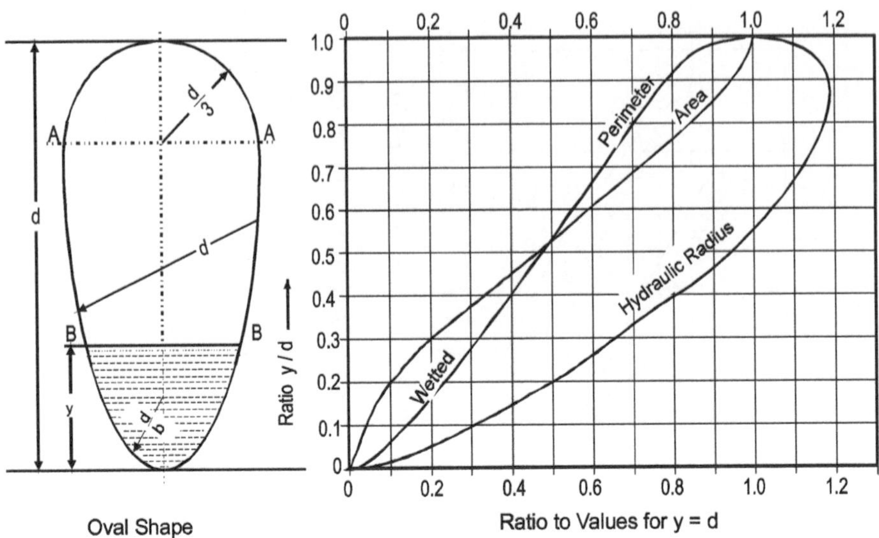

For the full section A = 0.5101 d^2, P = 2.643 d, R = 0.1930 d, and A $(R)^{2/3}$ = 0.170 $d^{8/3}$

Figure 5.7 : The Hydraulic elements of an egg-shaped cross section.

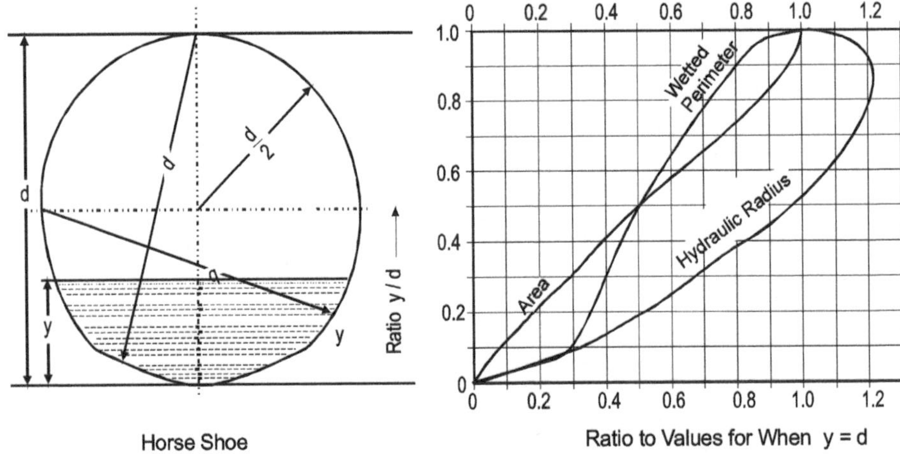

For the full section A = 0.8167 d^2, P = 3.261 d, R = 0.2505 d, and A $(R)^{2/3}$ = 0.324 $d^{8/3}$

Figure 5.8 : The Hydraulic elements of a horseshoe section.

5.4 Sewers - Oval shapes

Sewers are laid with as flat a slope as possible, not because there is much danger of erosion, but to minimise the cost of deep excavation.The minimum slope should give the necessary velocity i.e. about 0.6 to 1.2 m/s (2 to 4 ft./s) to prevent deposition of mud, grit, etc., when the least discharge (dry weather flow) is passing. Because of the relation between depth, discharge, and velocity, para 4.6, it follows that the velocity when the sewer is carrying its full supply may be unnecessarily high. To some extent this tendency may be counteracted by building the sewer of the ovoid, cross-section shown in Figure 5.7; although the discharge corresponding to the level BB is only 20 percent of that corresponding to level AA, yet the velocity corresponding to level BB is 75 percent of that corresponding to level AA.

Here is another instance of controlling the transport of a suspended load by adjustments to the channel shape. The advantages of the ovoid sewer are shared, in a smaller degree but at smaller cost, by a U-shaped section, having a semi-circular bottom, verical sides and a flat top.

Table 5.3 : Hydraulic Elements of a Circular Section

$\frac{y}{d_0}$	Cosθ	Area (A/d_0^2)	Wetted Perimeter (P/d_0)	Hydraulic Radius (R/d_0)	$(R/d_0)^{2/3}$	$AR^{2/3}/d_0^{8/3}$
1	2	3	4	5	6	7
0.1	0.80	0.0409	0.6434	0.0635	0.1592	0.0065
0.2	0.60	0.1118	0.9273	0.1206	0.2441	0.0273
0.3	0.40	0.1982	1.1593	0.1709	0.3080	0.0611
0.4	0.20	0.2934	1.3694	0.2142	0.3580	0.1050
0.5	0.00	0.3927	1.5708	0.2500	0.3968	0.1558
0.6	-0.20	0.4920	1.7722	0.2776	0.4255	0.2093
0.7	-0.40	0.5872	1.9823	0.2962	0.4443	0.2609
0.8	-0.60	0.6736	2.2143	0.3042	0.4523	0.3047
0.9	-0.80	0.7445	2.4981	0.2980	0.4461	0.3321
1.0	-1.0	0.7854	3.1416	0.2500	0.3968	0.3116

5.5 Oddities and Paradoxes of Open Channels

Some odities of open channel flow are discussed in para 1.4. More detial for close shapes are discussed below :

For a culvert or sewer of square cross section and flowing almost full , but not quite full, so that the water surface does not touch the flat top as in figure 1.2(a) v, then the value of the hydraulic radius R is (A/P) = $d^2/3d = d/3$. But if, for any reason, the water level rises a trifle so that the square is "plump" full, the flowing water rubs against the top, and R = $d^2/4d = d/4$, a decrease of 25 per cent from d/3, involving a decrease in velocity and discharge of some 17 per cent. Such a reduction of velocity and discharge will result if the head cannot change rapidly. An example of this is a rail / road culvert in flat country with large ponds of water on both sides of the track. For a square sewer, the water would rise in manholes until the hydraulic slope had increased over 40 per cent to maintain the same velocity with the greater rubbing surface, i.e., with the smaller hydraulic radius.

Likewise, interesting and practically important comparisons may be made for circular section when flowing not quite full, for instance, with a depth of water = 0.93D. In this case the area A is 97 per cent of the full circle value, but the wetted perimeter is only 83 per cent of the full circumference. Hence the hydraulic radius R is 17 per cent larger than for he filled circle. This results in a velocity about 11 per cent greater and a discharge about 9 per cent greater than when the circular section runs full.

Of course, if a culvert or sewer runs full "under pressure," strictly open-channel conditions no longer prevail, with surface slope parallel to the bed of the waterway, but the hydraulic slope is steeper (or sometimes gentler) than the grade of the sewer. Such hydraulic slope is indicated by the heights to which water rises in manholes of the sewer.

Hydraulic elements for a few of the common types of artificial open channels are shown in table 5.4. Such values are usually calculated as the need arises. This table is not so much for reference but only for an opportunity for becoming familiar, by checking, with the calculations.

Table 5.4 : Comparative Sample of Hydraulic Elements of closed sections

Shape of cross section	Depth of (d) water	Width of free surface (T)	Area of cross section (A)	Wetted perimeter (P)	Hydraulic radius, R= A/P	Remarks
1	2	3	4	5	6	7
1. Circle.........	**D**	**Zero**	**3.14r²**	**6.28r**	**r/2 = D/4**	
1.1 Circle.........	0.9D	1.2r (0. 6D)	2.98r²	5.00r	0.596r	A is 5.2 % and P is 20.5% less than full circle.
1.2 Circle.........	0.75D	1.73r	2.53r²	4.19r	0.604r	A is 19.6% and P 33.3% less than full circle.
1.3 Circle.........	0.5D	2r =D	1.57r²	3.14r	r/2 = D/4	R same as full circle.
1.4 Circle.........	0.25D	1.73r	0.614r²	2.09r	0.293r	
2.1 Square (Water touching roof)	D=2 r	Zero	d² = 4 r²	4d = 8r	d/4 = r/2	R same as for inscribed circle.
2.2 Square, Just full	D=2 r	D=2 r	d² = 4 r²	3d = 6r	d/3 = 2 r/3	A is same as in 2.1 but P is 75% and R=133% of 2.1
2.3Half Square,	D/2 = r	D=2r	d²/2 2r²	2d = 4r	d/4 = r/2	R same as for full square. A is 50% and P is also 50% of 2.1
Rectangle.........	D	B	Bd	b +2d	Bd/(b 2d)	
V-Flume, 60°	D=2r	1.154d	0.577d²	2.309d	d/4	R same as for inscribed semicircle.
V-Flume,90°	D	2d	d²	2.83d	d/2.83	

Notes: D = Diameter of circle; d = depth of flow ; r = radius of circle

For a given slope and roughness, velocity and discharge are different.

5.6 Comparison of discharging capacities of closed sections with reference to circle

A comparison of discharging capacities of closed conduits with reference to circle is given below in table 5.5 and shown in figure in 5.9. The numbers in the shape in figure 5.9 indicates the hydraulic efficiency with reference to 100 in the circle. This table shows relative elements of conduits when flowing full and at approximate point of maximum discharge. A square section is hydraulically more efficient then semi circle. Its hydraulic radius is also maximum. Hydraulic efficiency of rectangular section is further improved by making its bottom circular as in figure 5.3 to 5.5.

For hydraulic efficiency, r of equivalent semi circle for the given discharge should be worked out and then A, P, should be compared, see Para 1.6. For closed conduits it is a general practice to compare with circle.

5.7 Equivalent diameter of irregular shape or non-circular sections

For irregular-shaped ducts, the term equivalent diameter, de, is used in place of the diameter (d). It is de = 4A/P =4R

For a rectangular duct of sides a and b, de = 4ab/2(a +b) = 2ab/(a +b). For such a duct with one side very small compared to the other, i.e., with b« a, then b can be neglected in the denominator, so that de = 2ab/a = 2b or twice the minimum dimension. For an annulus of inner and outer diameters d_1 and d_2, de = $(d_2^2 - d_1^2)/(d_2 - d_1) = d_2 - d_1$, i.e., twice the annular width.

The italicized words in the definition should be carefully noted, as sometimes a duct carrying a liquid is not completely filled or is said not to run full. A is the flow area, not the duct area, and P is the perimeter of the duct in contact with or wetted by the liquid, i.e., not including the free surface.

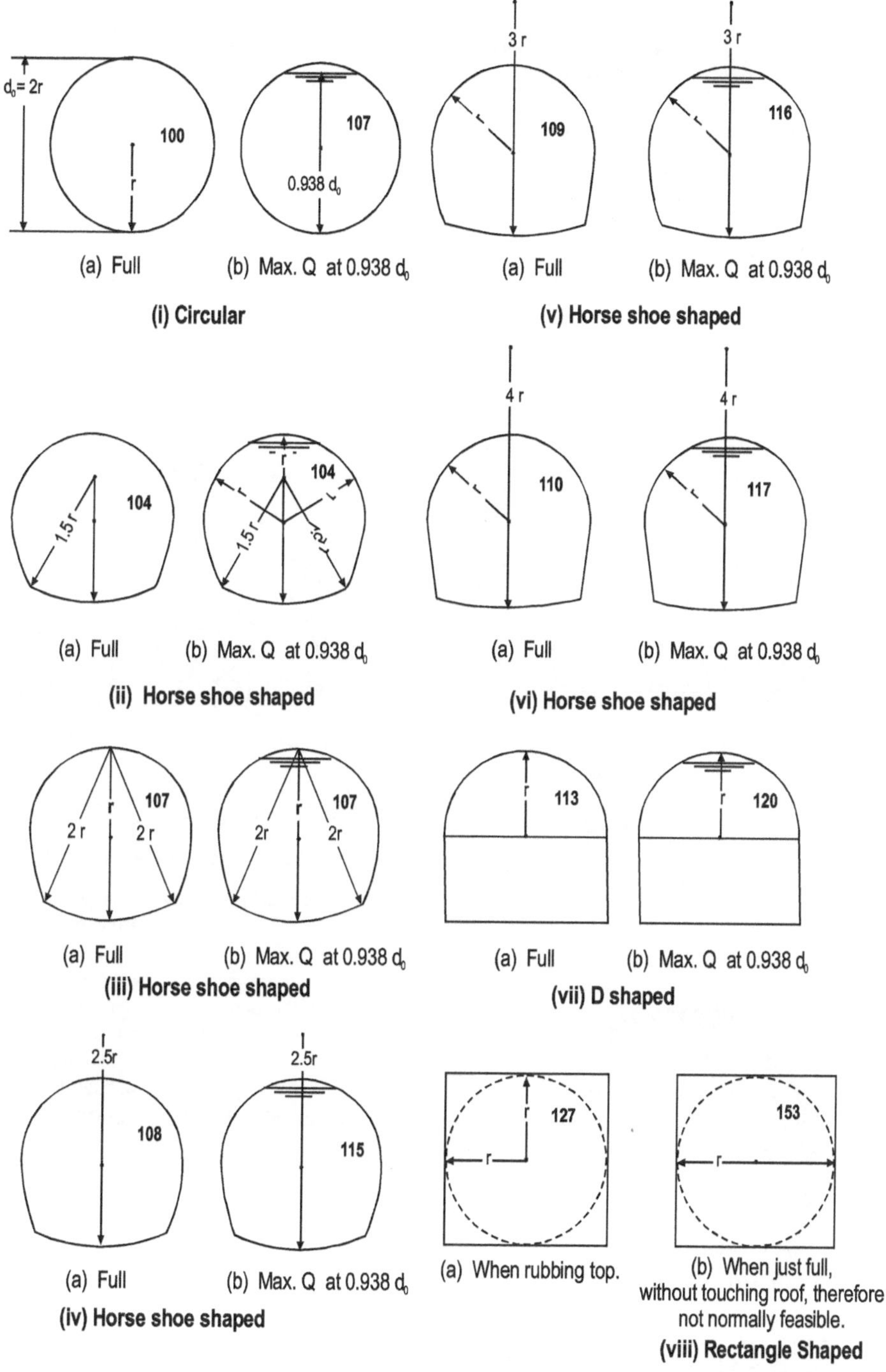

Figure 5.9 : Relative Discharging Capacities of Sections (aqueducts) with Reference to Circle

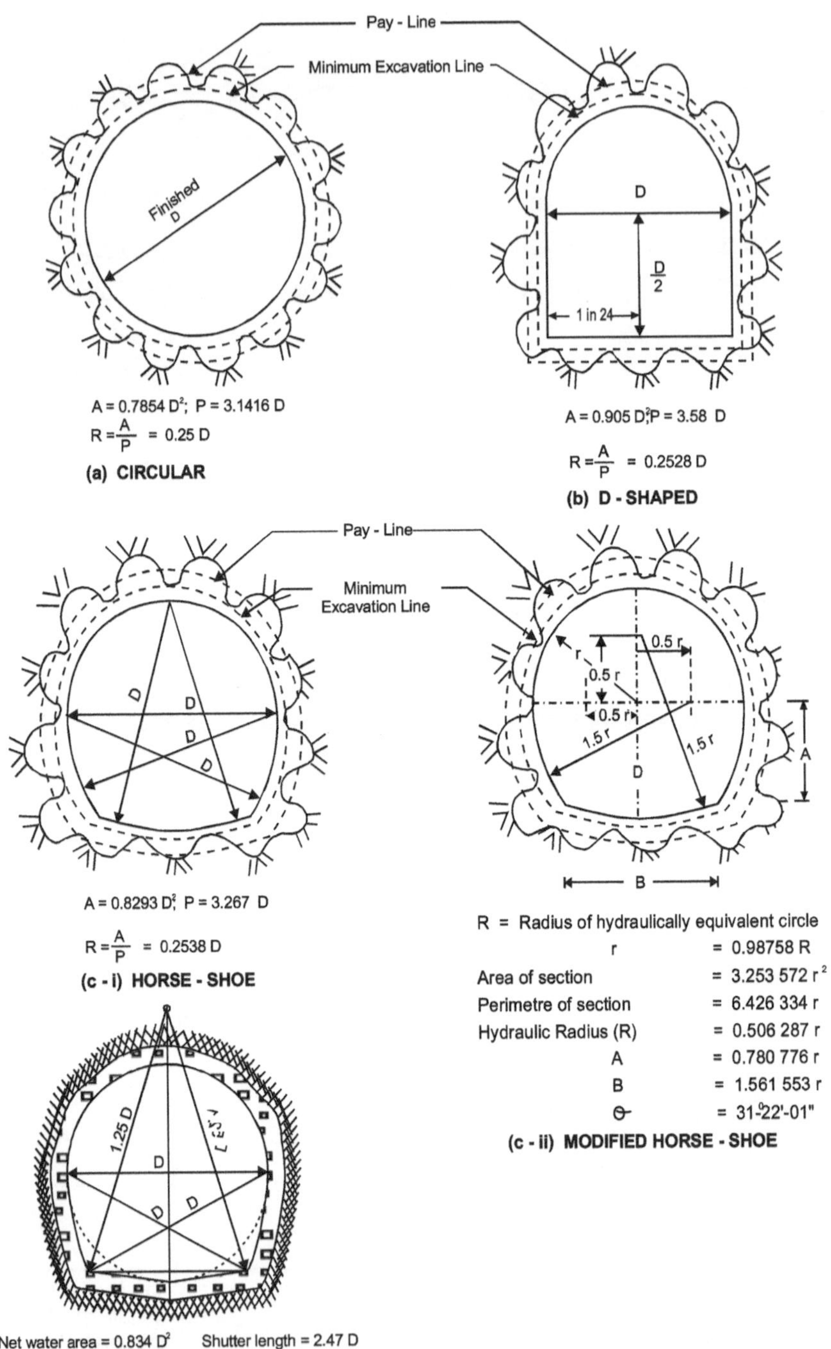

Figure 5.10 : Geometrical Properties of Some Tunnels

Table 5.5 : Comparison of discharging capacities of closed sections with reference to circle.

Section	Relative Elements of Conduits when Flowing Full.				At Approximate Point of Maximum Discharge			
	Area	Wetted perimeter	Mean hydraulic radius	Velocity	Area of flow	Wetted perimeter	Mean hydraulic radius	Velocity
i) Circle	100.0	100.0	100.0	100.0	975.0	841.0	116.0	109.8
ii) r_1=1.5r	103.4	102.3	101.1	100.7	100.9	864.0	116.8	110.3
iii) r_1=2.0r	105.7	104.0	101.8	101.1	103.2	881.0	117.2	110.6
iv) r_1=2.5r	107.1	105.4	101.8	101.1	104.6	895.0	116.9	110.4
v) r_1=3r	107.8	106.3	101.6	101.0	105.3	904.0	116.5	110.1
vi) r_1=4r	108.9	107.6	101.4	100.9	106.4	917.0	116.0	109.8
vii) D-shaped (1/2 square in bottom	113.6	113.6	100.0	100.0	111.1	97.7	113.7	108.3
viii) square	127.3	127.3	100.0	100.0	127.3	95.5	133.3	119.9

The value de has been useful in correlating performances of ducts that are not circular. There are some discrepancies, particularly with ducts having sharp corners, which may give rise to significant secondary flow patterns causing additional loss, but these occur most often with turbulent flow, and the correlation for laminar flow is satisfactory.

5.8 Tunnel Sections

Tunnels are generally made circular or horse shoe sections, or D- shaped section. A few shapes are shown in figure 5.10 and 9.7, 9.8 with some details on geological aspects. Also see para 9.7.

5.9 Summary and Conclusion

In this chapter rectangular section with modifications of circular bottom, circular section, elliptical, horseshoe and other tunnel sections are presented. The circular bottom rectangular section is very efficient in comparison to best rectangular, see table 5.2. It is also very close to optimum best trapezoidal section and same is recommended for adoption. Simple tabular solutions are given for quick designs.

References

King, H.W., & Brater, E.F., Hand Book of Hydraulics, 1963.

Vig, S. N., Use of Curved Bed in Lined Water Curses; Journal of Central Board of Irrigation And Power, Oct. 1981

□□□

Chapter 6

Experiences from Some Canals

6.0 General

No lessons can be drawn without the experiences of the past. In fact, much of engineering, rather all sciences have developed in bits and pieces from the experiences of the past. The engineers by and large have never become wiser in one day. They have become wiser slowly and slowly with their own experiences and from the experiences of past engineering works. Wisdom lies to gain wisdom from experiences/draw lessons from the marvellous engineering works constructed so far. It helps us in a direction for further development of technological skills.

Here an effort is made to draw lessons from experiences of the canals constructed in the past. There are several aspects of such experiences. A few important aspects of studies are:

(i) Internal hydraulic sections of the canal with respect to discharge, topography, stability of side slopes, type of lining, operation and maintenance problems and their solutions etc.

(ii) Design of canals banks (embankment) with respect to discharge, depth of flow, depth of cutting / filling, below/above natural ground level, type of soil, topography, operation and maintenance, drainage of banks etc.

(iii) Canals aesthetics and arboriculture

(iv) Other utility/service of canal such as multi-objectives of hydropower, irrigation, drinking/industrial water supply, social needs of recreation, forests etc.

(v) Environmental issues of the canal such as water logging, spread of water borne diseases (malaria and others) and weed growth.

(vi) Methods of planning and design of cross drainage works such as aqueducts, siphon and super passage etc. special treatment given to them at specific site conditions.

(vii) Sediment inflow into the canal from adjoining areas, (topography) and exclusion techniques. For example in IGNP area, wind blown desert sand enters, too much in cutting reaches. Similarly in hilly regions rainwater enters and brings lot of debris. Similarly rock falling, slides in cutting reaches are serious problems.

(viii) Hydraulic behaviour of canal-during full supply, partial discharges, excess supplies. This may not be uniform flow as is assumed, but may be gradually varied flow.

(ix) Operation and maintenance problems of canals.

(x) Modernization.

The above are some of the issues of the studies and there may be many more issues. In fact a detailed study of the issues over many projects may bring out good examples of the treatments of the past and directions/guidance for the future. Here an attempt is made to discuss only the internal sections of the canals.

6.1 Review of built Canals

Details of some lined canals in USA are given in table 6.1 and that in India are given in table 6.2 and 6.3.

To have a better and quick feeling of discharge versus cross section and lined perimeter, data of some Indian canals have been plotted on the same horizontal and vertical scale and shown in **figure 6.1.** Also a uniform scale is adopted in plotting these sections from Section 1 to 5 and again a different but uniform scale for section 6 to 27. For their comparison, discharge (Q), area (A), perimeter (P), hydraulic radius (R) and bed slope (S) are given in the diagrams. A quick comparison / economics of lined perimeter can easily be felt from these diagrams and detailed analysis given in the tables. Detailed geometric and hydraulic properties and analysis are given in table 6.2 and 6.3.

With the details of most efficient hydraulic canal section given in chapter I, a quick review of some existing/constructed lined canal section may be useful to the reader.

Detailed cross sections of lined canals of varying discharges from 1000 m^3/s to 4 cumecs (m^3/s) in heavy cutting to filling for some Projects in India are given in figure 6.2 these are self explanatory and show the details of embankments also. A minute comparison and examination of internal section shows that there is no uniform practice and there is a very wide variation in depth, B/D ratio, internal side slopes, rugosity coefficients etc.

Also canal systems of some projects are discussed in Para 6.3 to 6.13

6.1.1 Significance of Comparison

(i) Comparison aims at understanding the quality of a subject and it's systematizing

(ii) Comparison enables us to see resemblance and difference, and to point out universality and individuality

(iii) Comparison helps validation and explains variation in existing theories and practices

(iv) Lastly comparison helps in development/modification of existing theories, practical limitation and use.

Such an elementary comparison serves to focus attention to the end results and accordingly to promote further research into the practical aspects of the subject, with a view to more economical, efficient operation and maintenance i.e. trouble free design practices.

6.1.2 Comparison of Sections in table 6.1 to 6.3

A comparison of sections in table 6.1 indicates that though the data of bed slope are not available, yet Table 6.1 shows that there is wide variation in B/D ratio and it does not have a uniform trend of increase with increase in discharge. For a discharge of 6000 cusecs (Gravity Main), B/D ratio is 1.52 and Z = 1:0.75 (V: H) only and depth is increased up to 21.0 ft. This section is economical in heavy filling/heavy cutting or partial filling/cutting in comparison to another section for 4600 cusecs (Delta -Mendota) which has a B/D = 2.9 and D = 16.56 ft.

Table 6.1 : Cross Sections of Lined Canals of Some Projects in USA

Canal	Project	Discharge (cusecs)	Side Slope (H:V)	Depth of flow,ft.	Bottom width, ft.	Ratio, width to depth (B/D)	Mean velocity ft /s	Free board ft.
Lined Earth Sections								
Contra Costa	Central Valley	26	1.25 : 1	1.84	3.0	1.63	2.33	1.16
South Branch	Yakima	220	1.25 : 1	3.80	5.0	1.32	5.94	1.00
Ridge	Yakima	600	1.25 : 1	7.46	7.0	0.94	4.93	1.04
Heart Mountain	Shoshone	914	1.25 : 1	7.51	8.0	1.06	7.00	1.49
Kittitas Main	Yakima	1,320	1.25 : 1	8.99	11.0	1.22	6.63	1.25
Black Canyon	Boise	1,089	1.25 : 1	9.39	12.0	1.28	4.86	1.61
Ridge	Yakima	2,200	1.25 : 1	11.20	14.0	1.25	7.02	1.80
All - American	All - American	2,800	1.66 : 1	10.64	22.0	2.70	6.62	2.25
Delta - Mendota	Central Valley	4,600	1.5 : 1	16.56	48.0	2.9	—	—
Lined Rock Sections								
Kittitas Main	Yakima	1,275	0.5 : 1	9.00	14.00	1.65	7.65	1.25
Main (Gravity Extension)	Minidoka	2,695	0.25 : 1	13.80	20.00	1.45	8.2	1.50
Gravity Main	**Gila**	**6,000**	**0.75: 1**	**21.07**	**32.00**	**1.52**	**5.96**	**—**

Similarly, a careful examination and comparison of canal section in table 6.2 and figure 6.1 indicates, that some canals have a B/D ratio of nearly 1:1, where as some have a high ratio of as much as 1:20 or more. In all these canals, values of Bed slope range from 1 in 7000 to 1 in 12,000 for high capacity canals, and 1 in 5000 to 1 in 8000 for medium and low capacity canals.

Though data on N are not available, yet it generally ranges in between 0.015 to 0.02. Velocity in all canals ranges between 1 to 2 m/s. It is only B/D ratio which is the most pivotal in the economics of the canal section.

Table 6.2 and figure 6.1 show that Sunder Nagar Hydel channel for a discharge of 254 cumecs has a B/D ratio of 1.54, near to 1. The detail in figure 6.1(b), 6.2 and in Para 6.10 show that this canal section is more economical and carries/transport most of the silt to a balancing reservoir at the end of canal, where as the Gandak canal for a discharge of 241 cumecs, has a B/D ratio of 10.9, much more wide and a low depth canal is more costly both in construction and maintenance, (also see chapter) Also operation of this canal is likely to be more costly due to low velocity and more flexibility in excess discharge accommodation due to large width.

Table 6.2 : Details of Lined Canals Sections of Some Projects, India

S.No (1)	Name of Project (2)	Head Discharge Q (m^3/s) (3)	Channel Bed Slope S (4)	Value of N (5)	Velocity V(m/s) (6)	Side Slopes (V : H) (7)	Bed Width B (m) (8)	Depth D (m) (9)	B/D Ratio X (10)	Wetted Perimeter P (m) (11)	Area A (m^2) (12)	Hydraulic Radius R=A/P (m) (13)
(A) High Capacity Canals												
1	Narmada Main Canal, Gujarat, (C. C.)	1132.83	1/12500 (0.00008)	NA	1.6892	1:1.5	75.60	7.60	9.95	103.00	661.20	6.42
2	Rajasthan Feeder Canal, Punjab, (D.T.L.)	523.90	1/12000	0.017	1.425	1:1.5	79.25	4.40	18.01	97.63	389.10	4.40
3	Western Jamuna Canal Haryana, (Partly D.T.L.)	454.00	1/6250 (0.00016)	.014	2.87	1:1.5	50.00	2.91	17.18	60.50	158.20	2.61
4	Nangal Hydel Channel, Punjab, (C.C.)	354.00	1/10000 (0.0001)	NA	2.190	1:1.25	24.38	6.28	3.88	42.80	196.06	4.58
5	Sundernagar Hydel Channel Beas Project (H.P)	254.85	1/6666 (0.00015)	0.018	1.890	1:1.5	9.44	6.13	1.54	31.51	118.71	3.77
6	Gandak canal, (Bihar)	241.44	NA	NA	1.500	1:1.5	41.405	3.80	10.90	55.06	179.98	3.27
7	Western Kosi Canal, Bihar	236.70	1/8000 (0.000125)	NA	1.51	1:1.5	.06	3.66	9.58	50.34	156.28	3.10
8	Satluj Yamuna Link Canal, Haryana	212.00	1/10000 (0.0001)	NA	1.68	1:1.5	14.02	5.48	2.56	33.78	126.09	3.73
9	Mahi Right Bank Canal, Gujarat	198.10	1/7000	0.018	1.520	1:1.5	16.46	6.23	3.34	32.24	111.52	3.46
10	Bhakra Canals, Haryana	192.00	1/8000 (0.000125)	NA	2.19	1:1.5	12.30	4.57	2.69	28.78	87.56	3.04
(B) Medium and Low Capacity Canals												
11	Jawaharlal Nehru Lift Canal, Haryana	90.08	1/5000 (0.0002)	NA	1.42	1:1.5	20.25	2.62	7.73	29.71	63.44	2.14
12	Augmentation Canal, Haryana	87.60	1/5000 (0.0002)	NA	1.34	1:1.5	13.40	3.35	4.00	25.48	65.16	2.56
13	Dhansiri Main Canal, Assam	56.60	1/1500 (0.000667)	NA	2.08	1:1.5	14.44	1.61	8.97	20.24	27.14	1.34
14	Loharu Lift Canal, Haryana	39.65	1/7407 (0.000135)	NA	0.968	1:1.5	21.17	1.72	12.31	27.38	40.94	1.50
15	Ukai Left Bank Canal, Gujarat	35.00	1/8000 (0.000125)	NA	1.06	1:1.5	6.50	3.00	2.17	17.32	33.00	1.91

Contd...

Table 6.2. Contd...

(1)	(2)	(3)	(4)	(5)	(6)	(7)	(8)	(9)	(10)	(11)	(12)	(13)
16	Damanganga Right Bank Canal , Gujarat	34.78	1/1000 (0.001)	NA	1.59	1:1.5	4.50	2.60	1.70	15.87	21.87	1.38
17	Bardikarai main canal	30.01	NA	NA	1.19	1:1	15.70	1.38	11.38	20.00	25.07	1.254
18	Karjan Reservoir Left Bank Canal, Gujarat	28.30	1/3000 (0.000333)	NA	1.32	1:1.5	3.89	2.70	1.44	13.62	21.44	1.57
19	Jui Lift Canal, Haryana	21.14	1/6060 (0.000165)	NA	0.93	1:1.5	10.20	1.77	5.76	16.57	22.74	1.37
20	Salauli Irr. Project, Goa Damn & Diu	14.00	1/5000 (0.0002)	NA	NA	1:1.33	5.95	1.65	3.61	11.44	13.44	1.17

Notes : NA = Not available, Incomplete due to non availability of data. C.C. = cement concrete, DTL = Double tile lining H.P. = Himachal Pradesh, India.

Table 6.3 : Data of Some Irrigation Canal Projects, India.

S.No	Details of System	Name of irrigation projects Sukhi project canals (Gujarat) Left bank / right bank	Sipu Right Main Canal (Gujarat)
(1)	(2)	(3)	(4)
I - General			
1	Location of headworks	Dist. Vadoadr	Dist . Banaskatha
2	Year of commencement/ Completion		
	(i) Storage works	**1978 / 1982**	**1981 / 86**
	(ii) Barrage works		
	(iii) Canal works	1979 / 1982	1981 \ 86
II - Canals			
1	Length of the main canal	3.1 km / 10.3 km.	22.85 km.
2	Design discharge	3.66m³/s / 12.82 m³/s	7.668 m³/s
	Length of Branch Canals		12.15 K_m
3	Length of Distributaries	19.0 km / 27.5 km	39.5 km
4	Canal Section		
5	Unlined/Lined	Lined	Lined
6	Type of lining	Precast concrete Blocks	C.C. (1:3:6) 7.5 cm thick
7	Water surface width & depth at design discharge	6.65m / 9.8 m 1.45 / 2.3 m	6.96 m& 7.86 1.82 &1.82
8	Side slopes	1.5:1 / 1.5:1	1.5.:1
9	Bed width	2.30 m / 2.90 m	0 and 1.30 m
10	Bed slope	1 in 5000 / 1 in 5000	1 in 3000
11	Free board	0.60 m / 0.60 m	0.60 m
III - Irrigation Statistics			
1	Gross command area	31525 ha	32091 ha
	(i) By lift irrigation		
	(ii) By flow irrigation		
2	Total cultivable command area	20660 ha	24338 ha
3	Gross area irrigated during last 5 years		
4	Other approved uses of the canal (Navigation, waters ply, hydro power etc)		

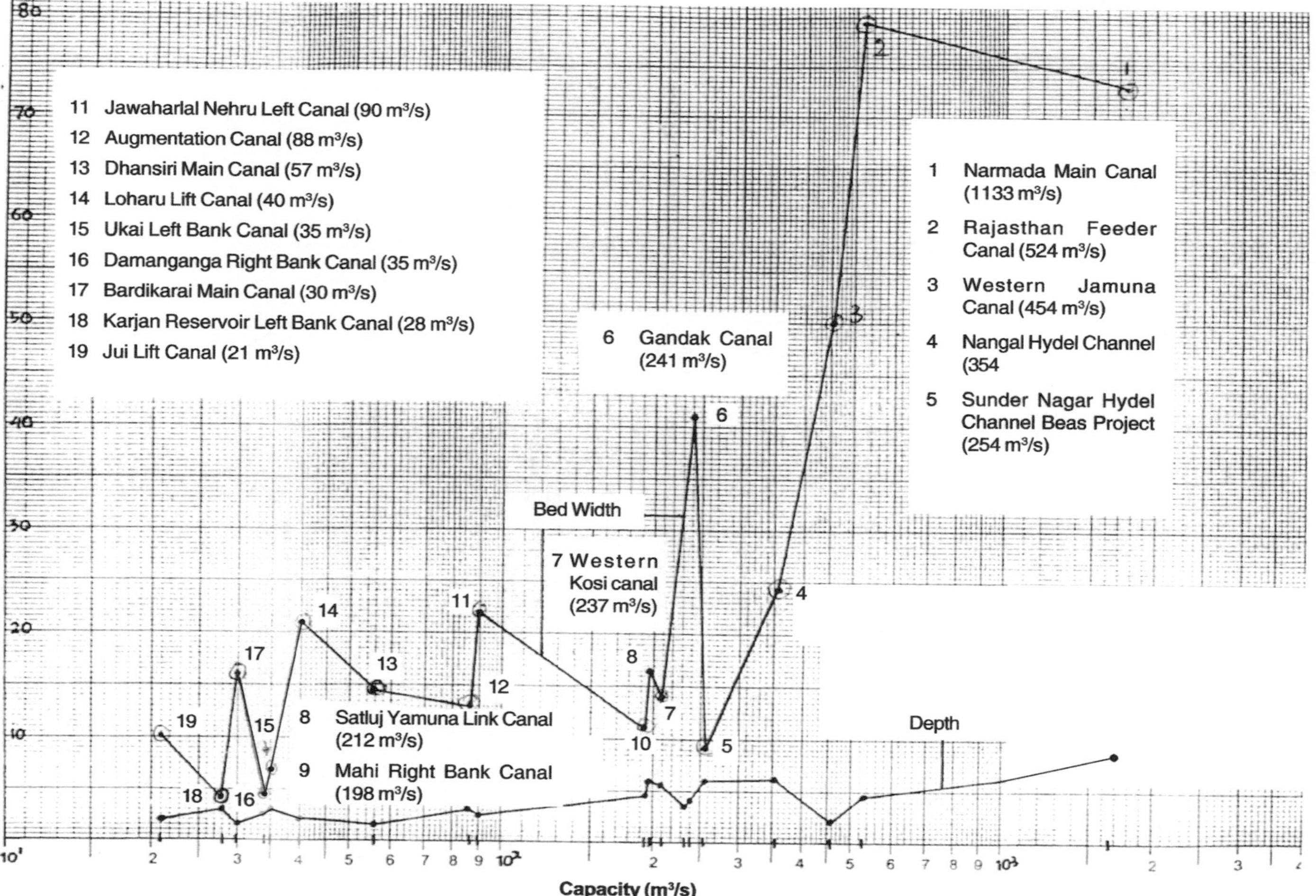

Figure 6.1 (a) : Base width and water depth versus capacity of some lined canal in India

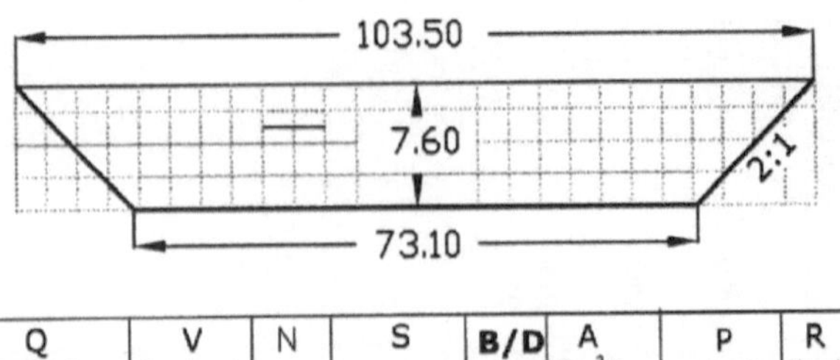

Q (m^3/s)	V (m/ s)	N	S	**B/D**	A (m^2)	P (m)	R (m)
1132.83	1.7	NA	1/12500	**9.62**	671.10	103:00	6.27

1. Narmada Main Canal, Gujarat

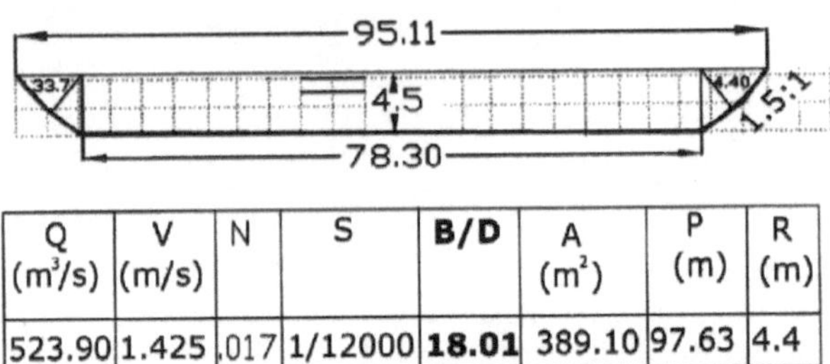

Q (m^3/s)	V (m/s)	N	S	**B/D**	A (m^2)	P (m)	R (m)
523.90	1.425	.017	1/12000	**18.01**	389.10	97.63	4.4

2. Rajasthan Feeder Canal, Punjab

Note : Narmada canal is economical then Rajasthan feeder, also see figure 6.32

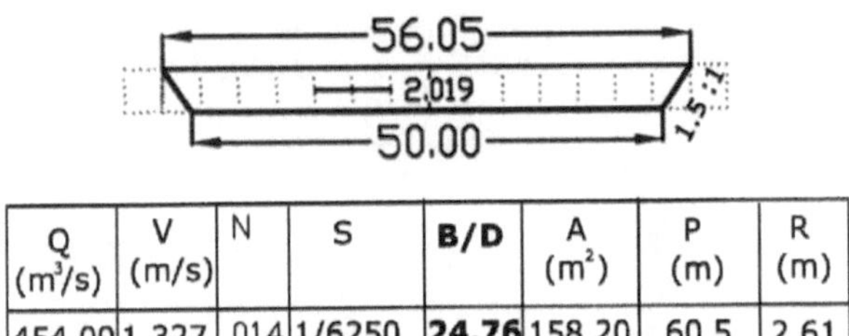

Q (m^3/s)	V (m/s)	N	S	**B/D**	A (m^2)	P (m)	R (m)
454.00	1.327	.014	1/6250	**24.76**	158.20	60.5	2.61

3. Western Jamuna Canal, Haryana

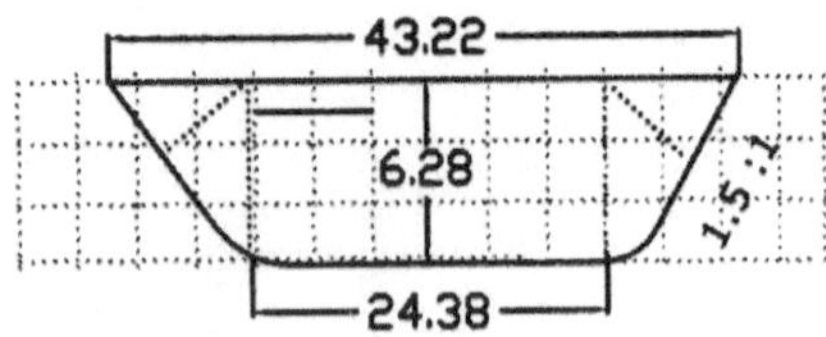

Q (m^3/s)	V (m/s)	N	S	**B/D**	A (m^2)	P (m)	R (m)
354.00	2.190	NA	1/10000	**3.88**	196.06	42.80	4.58

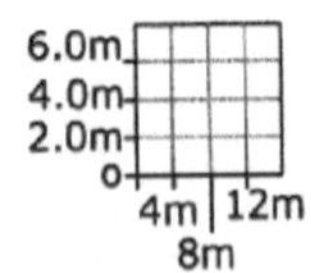

4. Nangal Hydel Channel Punjab

Figure 6.1 (b) : Comparison of Hydraulic Sections of some lined Canals in India

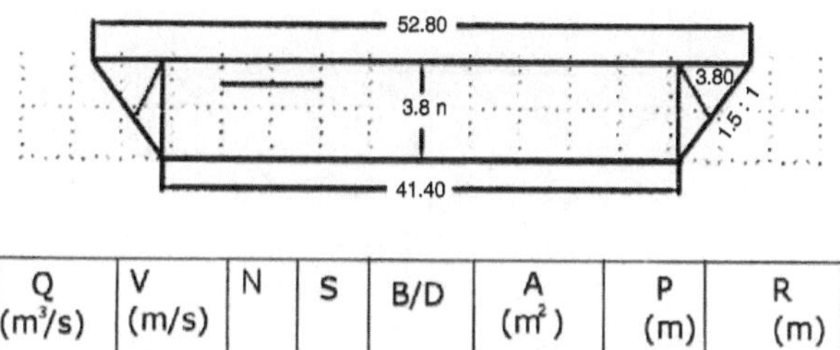

Q (m^3/s)	V (m/s)	N	S	B/D	A (m^2)	P (m)	R (m)
241.44	1.500	NA	NA	10.90	179.98	55.06	3.27

5. Gandak Canal Lining, Bihar

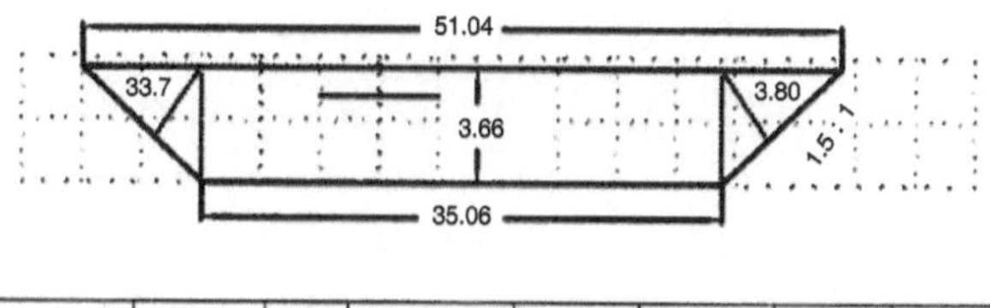

Q (m^3/s)	V (m/s)	N	S	B/D	A (M^2)	P (m)	R (m)
236.70	1.515	NA	1/8000	9.58	156.28	50.34	3.10

6. Western Kosi Canal, Bihar

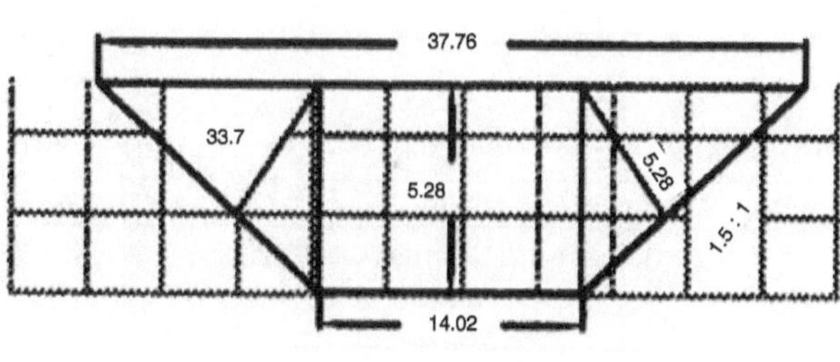

Q (m^3/s)	V (m/s)	N	S	B/D	A (m^2)	P (m)	R (m)
212.00	1.681	NA	1/10000	2.56	126.09	33.78	3.73

7. Satluj Yamuna Link, Canal Haryana

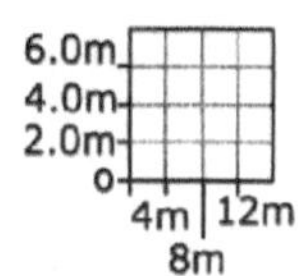

For details see table 6.2 and figure 6.2

Figure 6.1 (b) : Comparison of Hydraulic Sections of some lined Canals in India

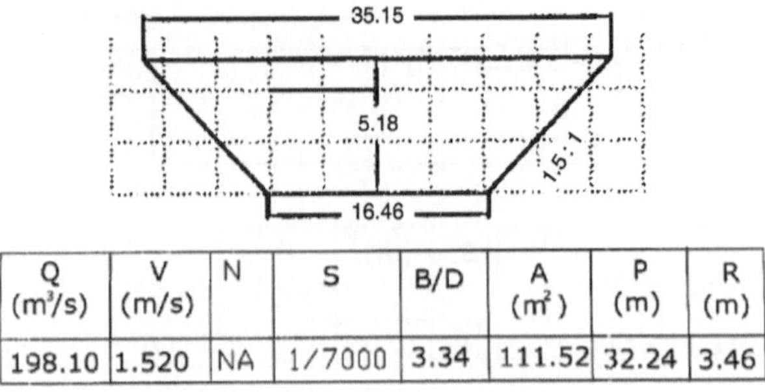

Q (m^3/s)	V (m/s)	N	S	B/D	A (m^2)	P (m)	R (m)
198.10	1.520	NA	1/7000	3.34	111.52	32.24	3.46

8. Mani Right Bank Canal, Gujarat

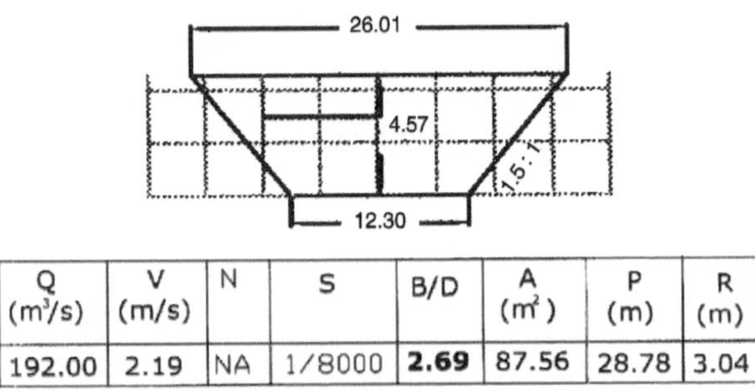

Q (m^3/s)	V (m/s)	N	S	B/D	A (m^2)	P (m)	R (m)
192.00	2.19	NA	1/8000	**2.69**	87.56	28.78	3.04

9. Bhakra Canal, Haryana

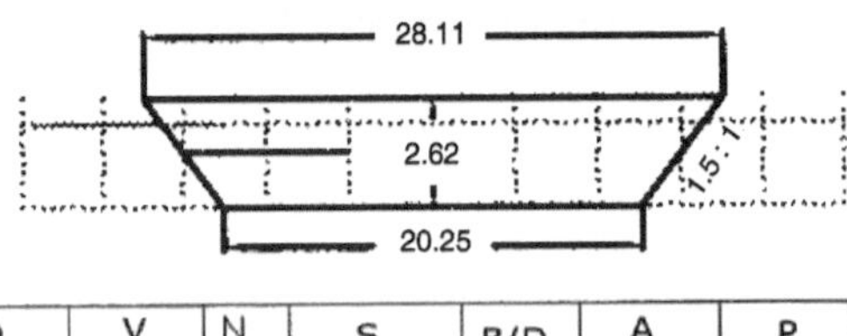

Q (m^3/s)	V (m/s)	N	S	B/D	A (m^2)	P (m)	R (m)
90.08	1.420	NA	1/5000	**7.73**	63.44	29.71	2.14

10. Jawaharlal Nehru Lift Canal, Haryana

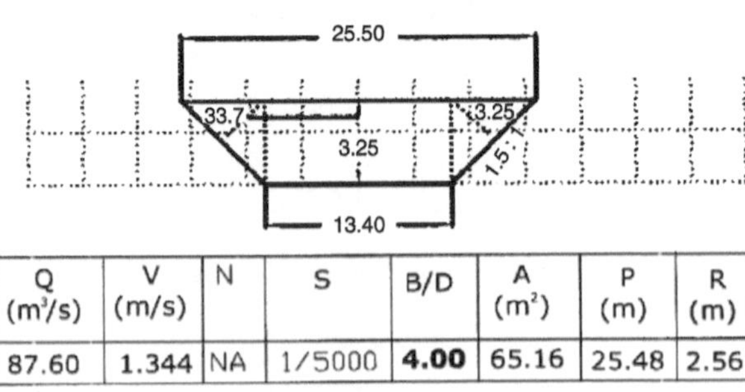

Q (m^3/s)	V (m/s)	N	S	B/D	A (m^2)	P (m)	R (m)
87.60	1.344	NA	1/5000	**4.00**	65.16	25.48	2.56

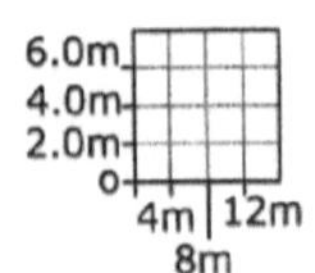

11. Augmentation Canal, Haryana

Figure 6.1 (b) : Comparison of Hydraulic Sections of some lined Canals in India

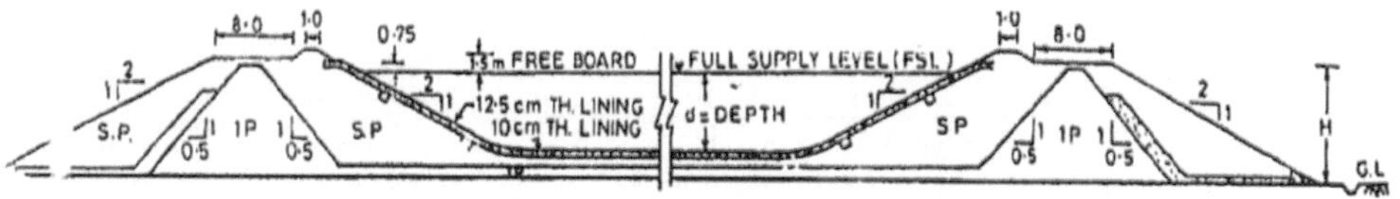

(a) Section of Canal in Case of Natural Ground below Canal Bed

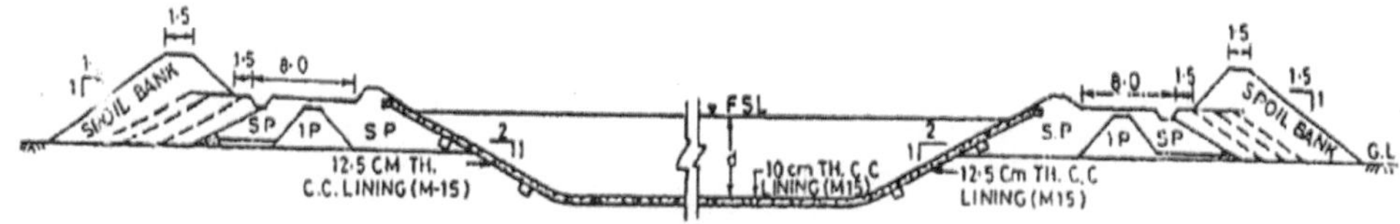

(b) Section of Canal in Case of Natural Ground between Canal Bed & Full Supply Level

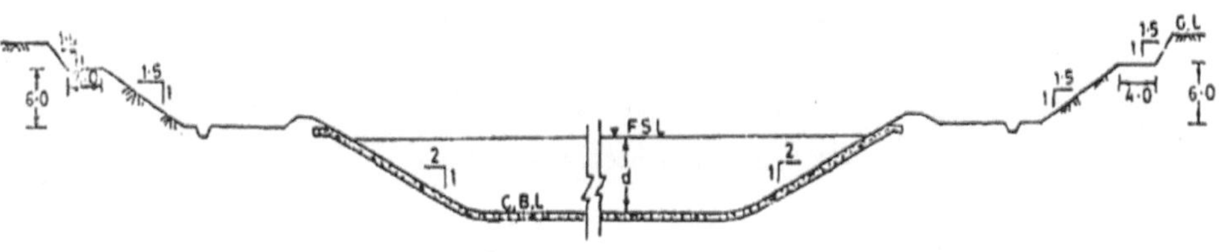

(c) Section of Canal where Natural Ground Level is Above F.S.L.

Q (m³/s)	V (m/s)	N	S	B/D	A (m²)	P (m)	R (m)
1132.8	1.6892	0.017	1/12500	75.6/7.60=9.95	661.20	103.0	6.42

Also see table 6.12 for more details

1. Typical section of Narmada Main Canal (also see figure 6.32 for more details)

Note : The comparison in all these figures be seen with respect to B/D ratio and comparison of internal section shown in table 6.1 and figure 6.1 (a) and (b) and all over sections be compared for land width, cost earthwork and lining perimeter and velocity for silt transport.

2. Regasthan feeder canal. see figure 6.6.

Figure 6.2 : Comparison of over all Canal sections of some lined Canals in India contd...

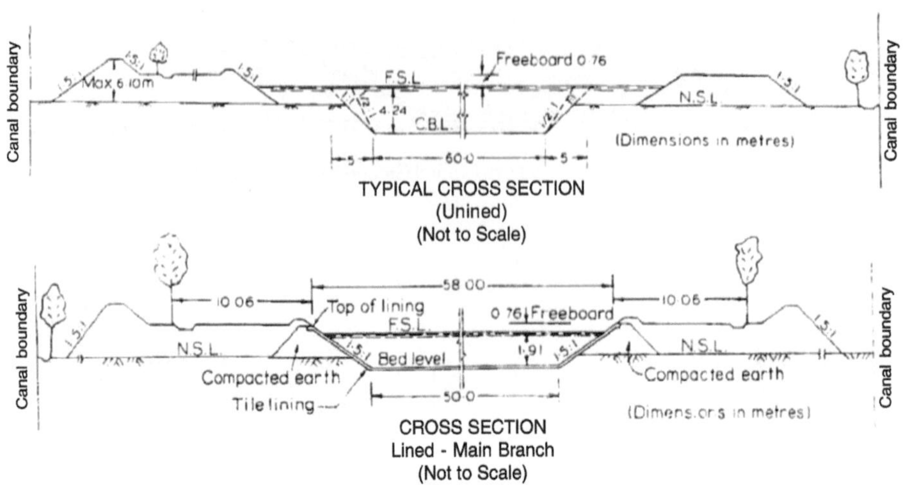

TYPICAL CROSS SECTION
(Unined)
(Not to Scale)

CROSS SECTION
Lined - Main Branch
(Not to Scale)

Q (m³/s)	V (m/s)	N	S	B/D	A (m²)	P (m)	R (m)
454.0	2.87	0.014	1.6250	50.0/2.91=17.18	158.0	60.5	2.61

3. Western Jamuna Canal Haryana

(100 years old. Remodelled in 1956-60)

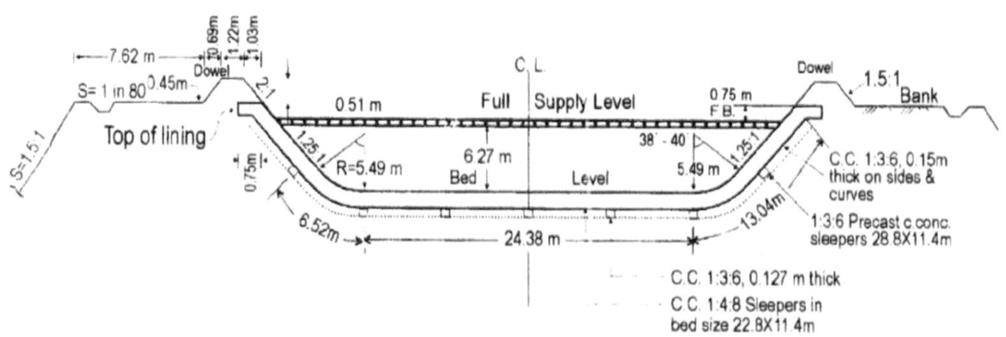

Q (m³/s)	V (m/s)	N	S	B/D	A (m²)	P (m)	R (m)
354.0	2.19		1/10000	24.38/6.28=3.88	190.06	42.8	4.58

Also see Para 6.9 for more details

4. Nangal Hydel Channel (also see figure 6.11)

Figure 6.2 : Comparison of over all Canal sections of some lined Canals in India contd...

Q (m³/s)	V (m/s)	N	S	B/D	A (m²)	P (m)	R (m)
254.85	1.890	0.018	1/6666	1.54	118.71	31.51	3.77

5. Sundernagar Hydel Channel Beas Project, Himachal Pradesh (see figure 6.12)

Q (m³/s)	V (m/s)	N	S	B/D	A (m²)	P (m)	R (m)
241.44	1.500	NA	NA	10.90	179.98	55.06	3.27

6. Gandak Canal Lining, Bihar

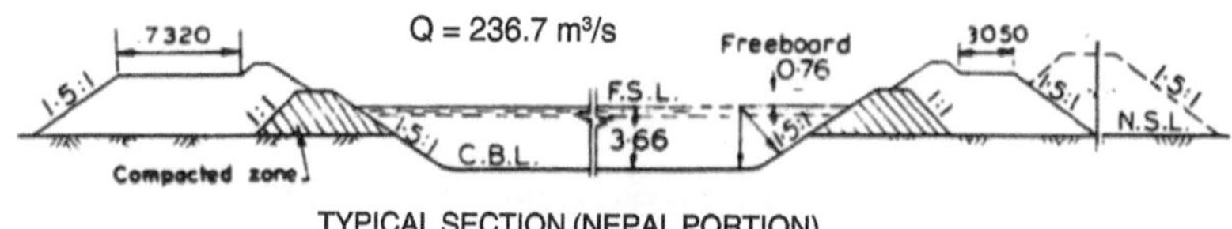

TYPICAL SECTION (NEPAL PORTION)
(CUTTING UP TO 4.42m)

Boulder lining in head reach. STL in bed and DTL on Sides

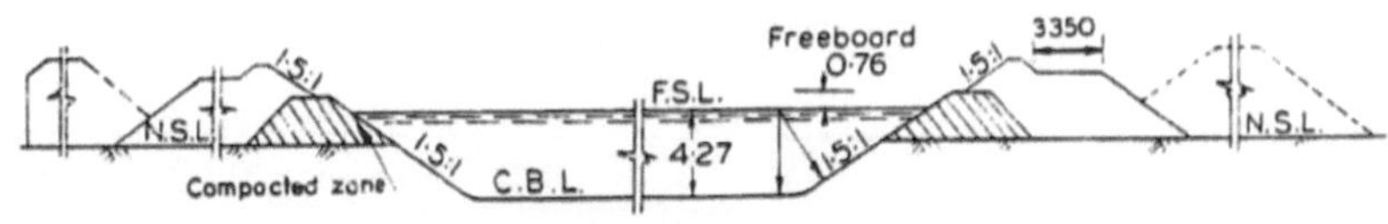

TYPICAL SECTION (NEPAL PORTION)
(CUTTING UP TO 5.03m)

Q (m³/s)	V (m/s)	N	S	B/D	A (m²)	P (m)	R (m)
236.70	1.515	NA	1/8000	9.58	156.28	50.34	3.10

7. Western Kosi Canal, Bihar (Year 1974)

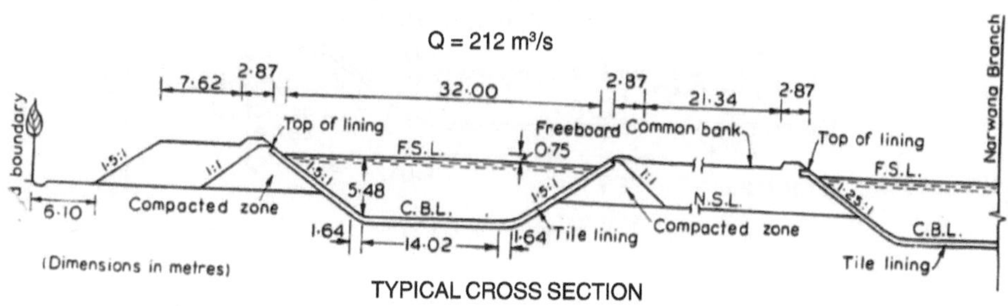

TYPICAL CROSS SECTION

Q (m³/s)	V (m/s)	N	S	B/D	A (m²)	P (m)	R (m)
212.00	1.681	NA	1/10000	2.56	126.09	33.78	3.73

8. Satluj Yamuna Link Canal, Haryana

Figure 6.2 : Comparison of over all Canal sections of some lined Canals in India contd...

Q (m³/s)	V (m/s)	N	S	B/D	A (m²)	P (m)	R (m)
198.10	1.520	NA	1/7000	3.34	111.52	32.24	3.46

9. Mahi Right Bank Canal, Gujarat

Q (m³/s)	V (m/s)	N	S	B/D	A (m²)	P (m)	R (m)
192.00	2.19	NA	1/8000	2.69	87.56	28.78	3.04

10. Bhakra Canal, Haryana (year 1974-82)

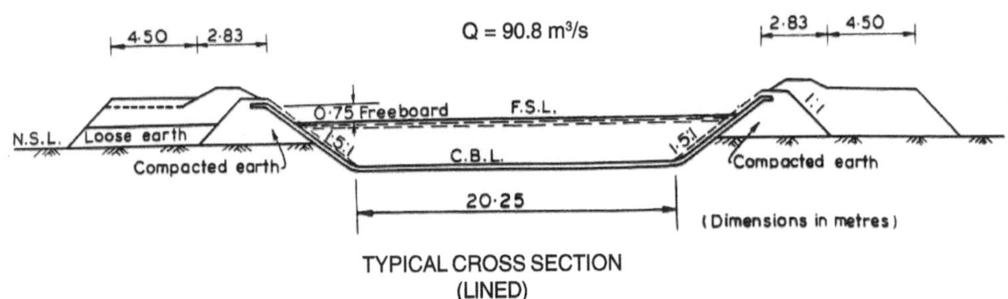

TYPICAL CROSS SECTION
(LINED)

Q (m³/s)	V (m/s)	N	S	B/D	A (m²)	P (m)	R (m)
90.08	1.420	na	1/5000	7.73	63.44	29.71	2.14

11. Jawaharlal Nehru Lift Canal, Haryana (year 1974-82)

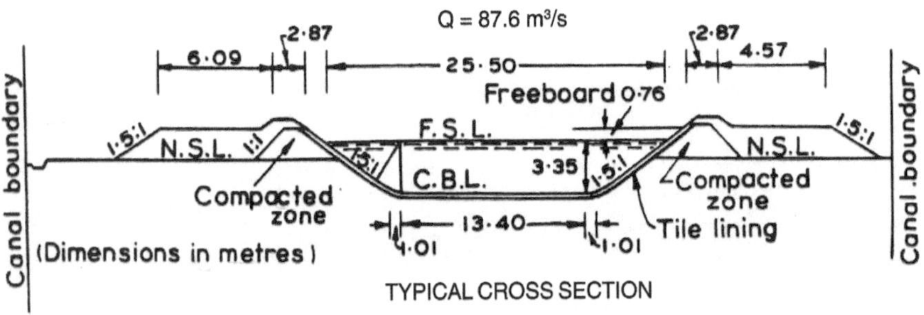

TYPICAL CROSS SECTION

Q (m³/s)	V (m/s)	N	S	B/D	A (m²)	P (m)	R (m)
87.60	1.344	NA	1/5000	4.00	65.16	25.48	2.56

12. Augmentation Canal, Haryana (year 1971-72)

Figure 6.2 : Comparison of over all Canal sections of some lined Canals in India contd...

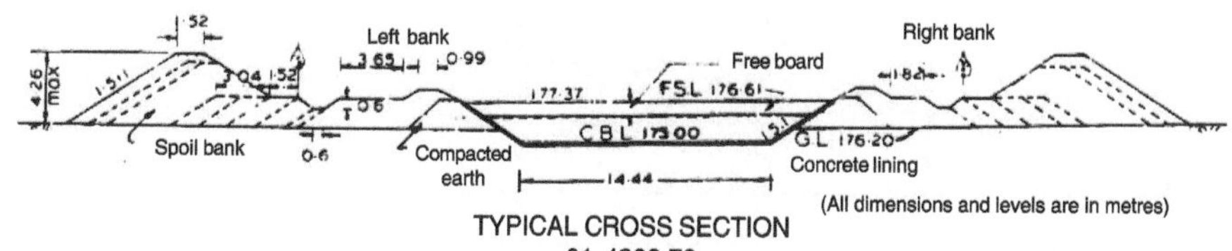

TYPICAL CROSS SECTION
01 4206-70

Q (m³/s)	V (m/s)	N	S	B/D	A (m²)	P (m)	R (m)
56.60	2.085	NA	1/1500	8.97	27.14	20.24	1.34

13. Dhansiri Main Canal, Assam (Year 1975-80)

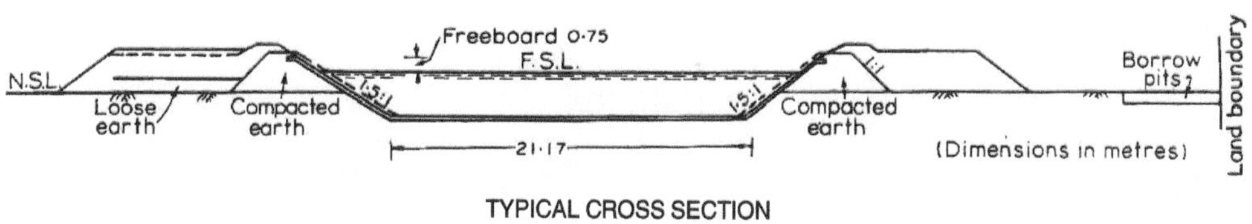

TYPICAL CROSS SECTION
(Lined)

Q (m³/s)	V (m/s)	N	S	B/D	A (m²)	P (m)	R (m)
39.65	0.968	NA	1/7407	12.31	40.94	27.38	1.50

14. Lohuru Lift Canal, Haryana (Year 1970-1979)

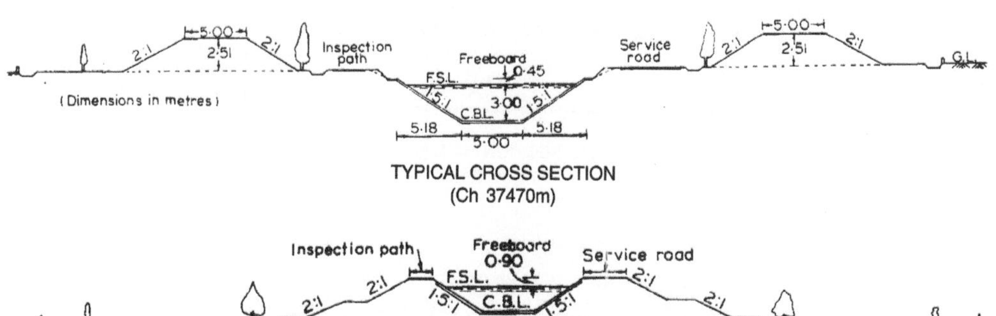

TYPICAL CROSS SECTION
(Ch 37470m)

TYPICAL CROSS SECTION
IN FULL BANKING

Q (m³/s)	V (m/s)	N	S	B/D	A (m²)	P (m)	R (m)
35.00	1.061	NA	1/8000	2.17	33.00	17.32	1.91

15. Ukai Left Bank Canal, Gujarat

Figure 6.2 : Comparison of over all Canal sections of some lined Canals in India contd...

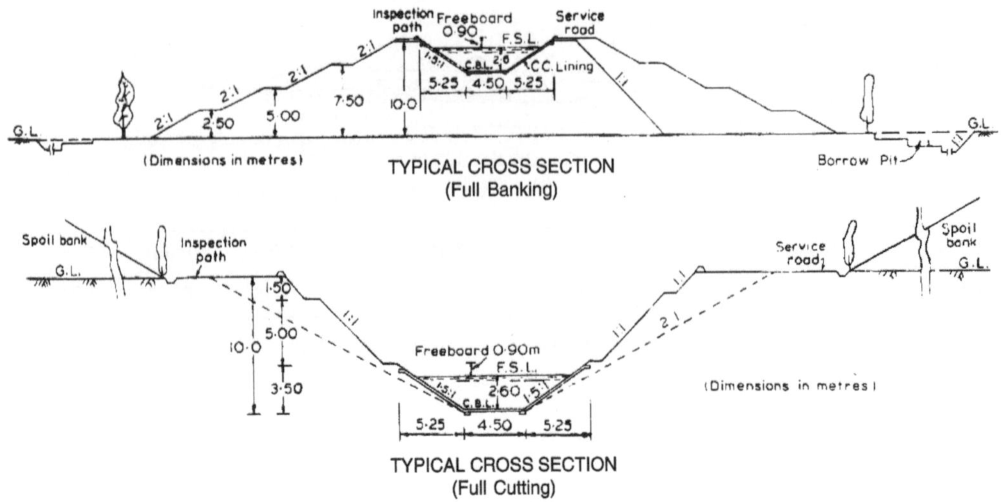

TYPICAL CROSS SECTION
(Full Banking)

TYPICAL CROSS SECTION
(Full Cutting)

Q (m^3/s)	V (m/s)	N	S	B/D	A (m^2)	P (m)	R (m)
34.78	1.59	NA	1/1000	1.70	21.87	15.87	1.38

16. Damanganga Right Bank Canal, Gurjarat

Q = 30.0 m^3/s

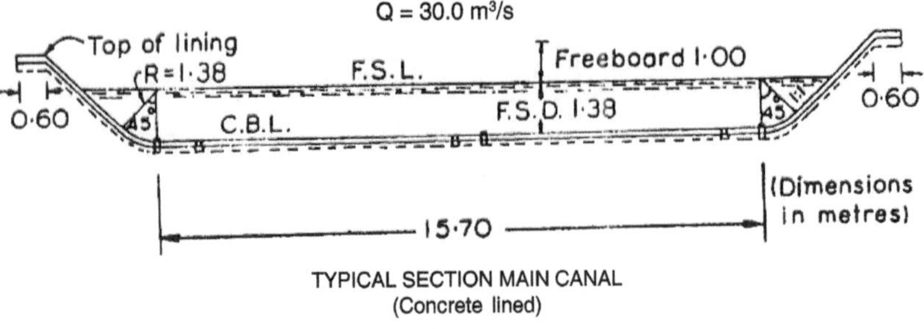

TYPICAL SECTION MAIN CANAL
(Concrete lined)

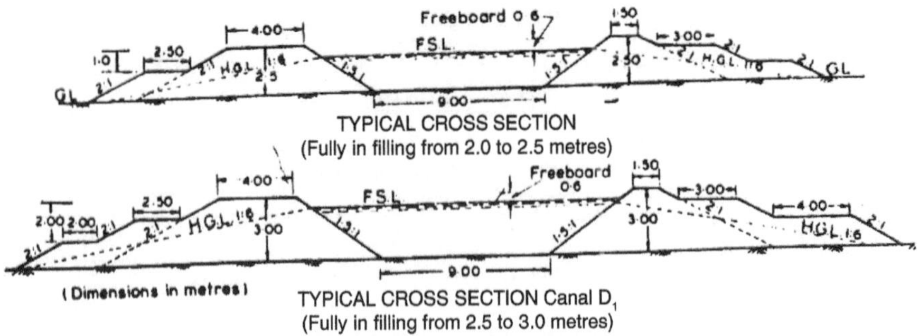

TYPICAL CROSS SECTION
(Fully in filling from 2.0 to 2.5 metres)

TYPICAL CROSS SECTION Canal D_1
(Fully in filling from 2.5 to 3.0 metres)

17. Damanganga Right Bank Canal, Gujarat (1976-80)

Figure 6.2 : Comparison of over all Canal sections of some lined Canals in India contd...

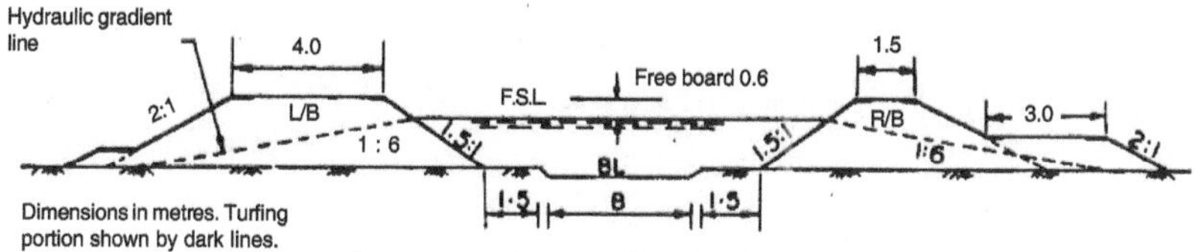

TYPICAL CROSS SECTION D_2 Canal System
Depth of cutting 0.2m
Height of filling above 1.8m below 2.0m

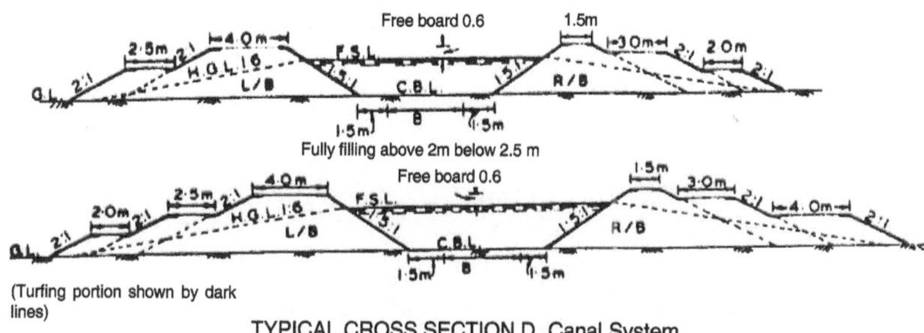

TYPICAL CROSS SECTION D_2 Canal System
(Fully filling above 2.5m below 3.0m)

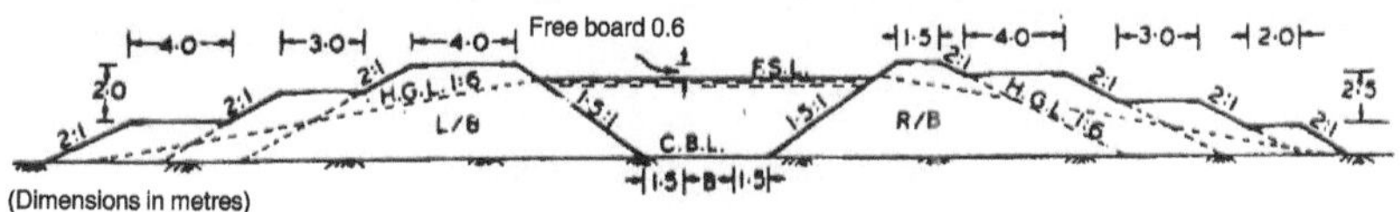

TYPICAL CROSS SECTION D_3 Canal System
(Fully filling above 3.0m below 3.5m)

Q (m^3/s)	V (m/s)	N	S	B/D	A (m^2)	P (m)	R (m)
30.01	1.19	NA	NA	11.38	25.07	20.00	1.25

17. Damanganga Right Bank Canal, Gujarat (1976-80)

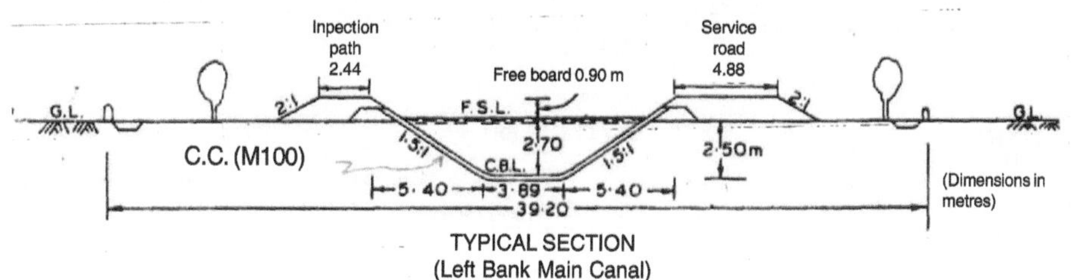

TYPICAL SECTION
(Left Bank Main Canal)

Q (m^3/s)	V (m/s)	N	S	B/D	A (m^2)	P (m)	R (m)
28.30	1.370	NA	1/3000	1.44	21.44	13.62	157

18. Karjan Reservoir Left Bank Canal, Gujarat (1979-82)

Figure 6.2 : Comparison of over all Canal sections of some lined Canals in India contd...

Q = 21.14 m³/s

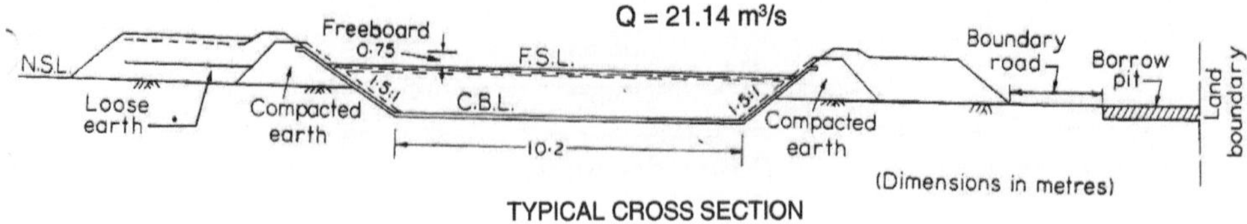

TYPICAL CROSS SECTION

Q (m³/s)	V (m/s)	N	S	B/D	A (m²)	P (m)	R (m)
21.14	0.93	NA	1/6060	5.76	22.74	16.57	1.37

19. Jui Lift Canal, Haryana

Q = 24.16 m³/s

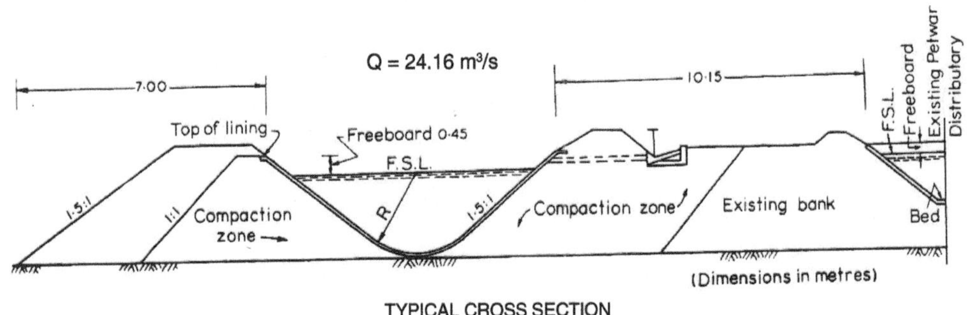

TYPICAL CROSS SECTION

20. Sewani Lift Canal, Haryana

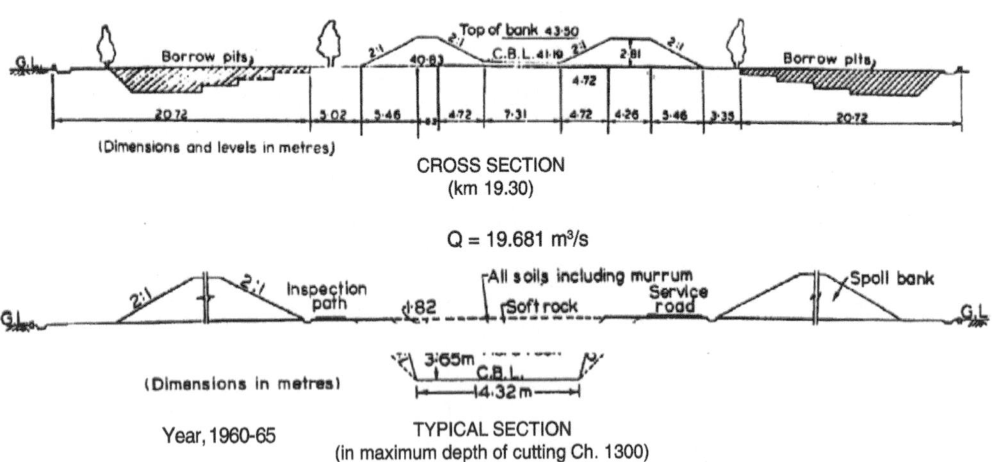

CROSS SECTION
(km 19.30)

Q = 19.681 m³/s

Year, 1960-65

TYPICAL SECTION
(in maximum depth of cutting Ch. 1300)

21. Shetrunji Right Bank Main Canal, Gujarat

Figure 6.2 : Comparison of over all Canal sections of some lined Canals in India contd...

Q = 12.09 m³/s

(Dimensions in metres)

Year 1960-65

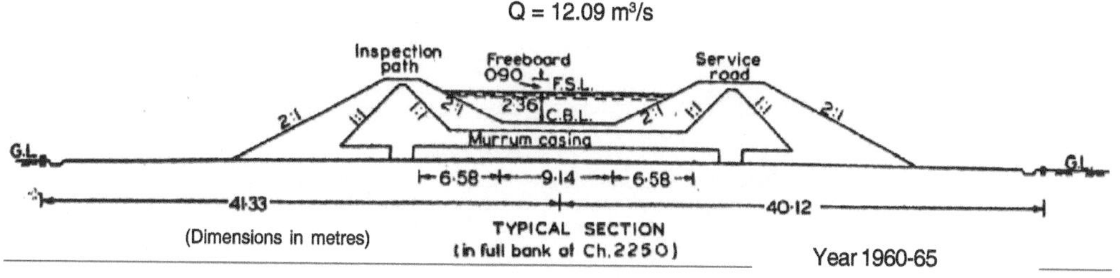

22. Shetrunji Left Bank Main Canal, Gujarat

Q = 14.0 m³/s

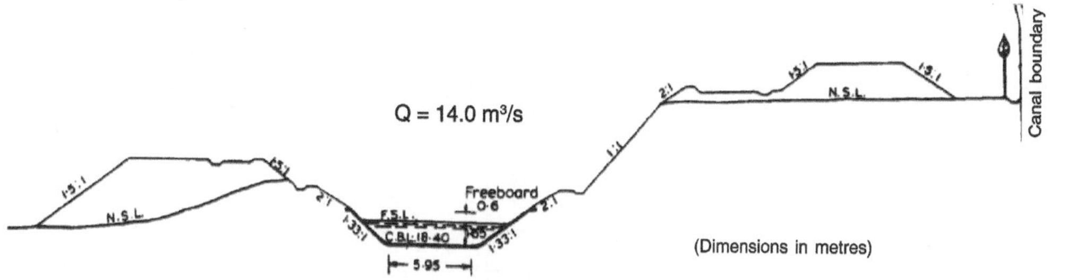

(Dimensions in metres)

(a) Cross Section at R.D. 360
(Typical N.S. Above F.S.L.)

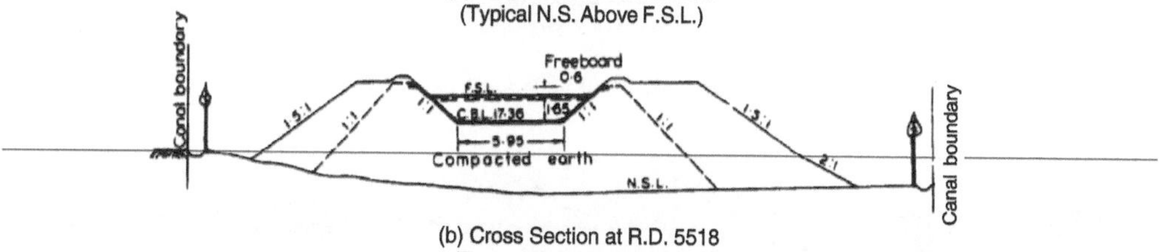

(b) Cross Section at R.D. 5518
(Typical N.S.L. Below BED)

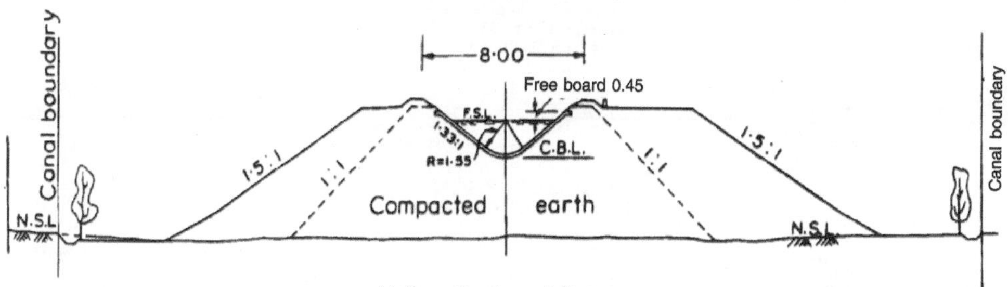

(c) Cross Section at R.D. 24150
(Typical N.S.L. Below BED)

(Dimensions in metres)

23. Saluali Irrigation Project, Goa Dam & Diu Year 1977-83

Figure 6.2 : Comparison of over all Canal sections of some lined Canals in India contd...

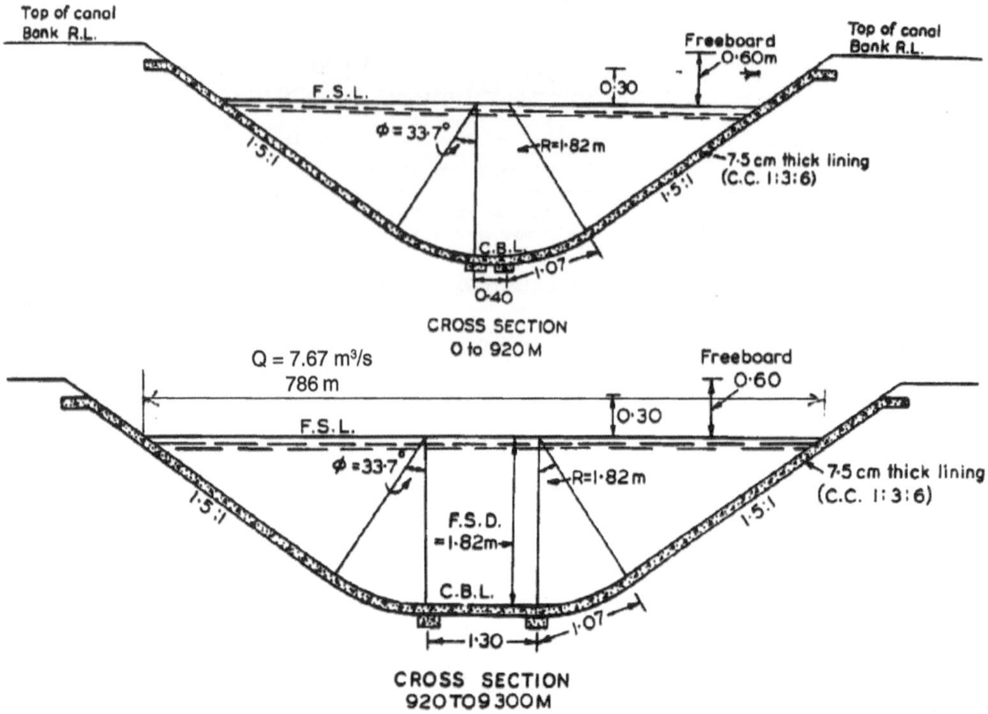

Top of canal Bank R.L.

Freeboard 0·60m

Top of canal Bank R.L.

F.S.L.

0·30

ϕ = 33·7°

R=1·82 m

1·5:1

7·5 cm thick lining (C.C. 1:3:6)

1·5:1

C.B.L.

1·07

0·40

CROSS SECTION
0 to 920 M

Q = 7.67 m³/s

786 m

Freeboard 0·60

0·30

F.S.L.

ϕ = 33·7°

R=1·82 m

7·5 cm thick lining (C.C. 1:3:6)

1·5:1

1·5:1

F.S.D. =1·82m

C.B.L.

1·30

1·07

CROSS SECTION
920 TO 9300 M

24. Sipu Right Main Canal, Gujaraat, Year - 1981-86

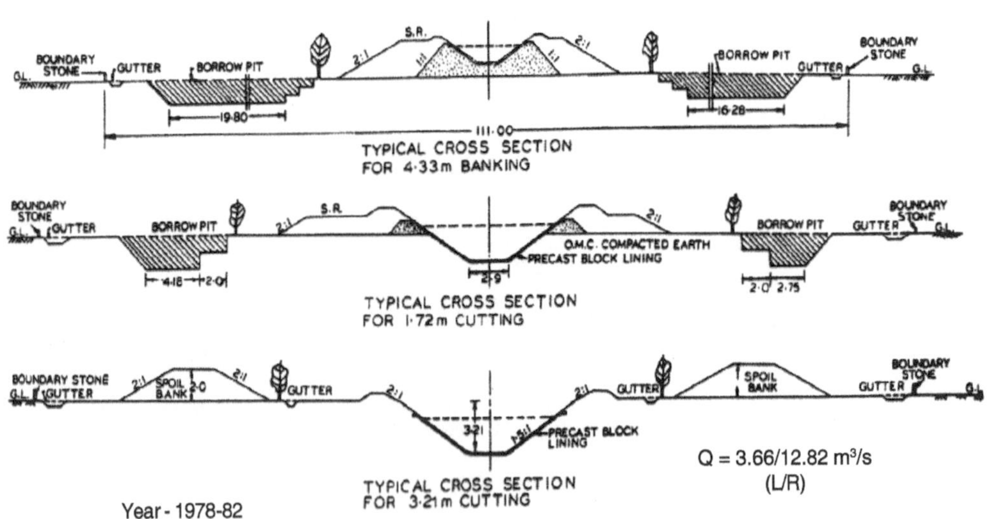

25. Sukhi Project Canal, Gujaraat, Year - 1981-86

Figure 6.2 : Comparison of over all Canal sections of some lined Canals in India contd...

Similarly Rajasthan feeder in Punjab, for a discharge of 523 cumecs is a very vide canal and is not very economical, even when compared to other reaches of the same canal. See Para 6.6

Narmada canal for a discharge of 1132 cumecs may perhaps have limitation of depth, also see Para 6.16, and is costly in comparison to many other high capacity canals.

However there are many high capacity canals in the world, which have larger depth, see Para 6.17 and figure 6.30.

Reader can now himself compare them with each other and best sections and feel how best or economical theses may be when compared to best sections. However, he should not get disappointed but should also refer to other points for most economical cross section given in Para 6.3 and Chapter 9. He should also look to the various limitations of knowledge, depth of canal, type of soil/rocks, water table in the area, type of lining in bed and sides, methods and the technology of construction, such as by manual labour or by machines etc. There he may not get straight forward answers, but may have to work hard to develop several analysis in order to find an optimum economical feasible solution.

However the discussions in this chapter and that in chapter 7, 8 and 9 help the designer to make quick analysis and judgement.

6.2 Development of Design Practices

Design of economic, low maintenance/trouble free channel section had been the concern of every engineer; from the day canal irrigation came to practice. Slowly and slowly many engineers developed theories and guidelines for such designs, initially for unlined canals and subsequently for lined canals. With more stress on lining, the emphasis also shifted to the least perimeter for a given area. While reviewing the canals and their designed sections, it will also be of interest to review the various theories / methodologies / practices / knowledges on which their design might have been based.

Though it is difficult to compile the practices developed in a somewhat historical chronological order, yet an attempt is made here as under:

6.2.1 Books on Canal Design/Engineering

There are no specialised and Practical books only on the canal design. General books on irrigation Engineering includes canal design. The only book on canal that appeared is "Design Practices of Irrigation Canals" by Framji (1972), published by ICID. However there are some publication by CBIP, USBR and other manuals, guide lines and codes generally in use in that country and are referred in this text. Some major books, which appeared in the past, along with the name of the author and the year of publications, are listed in the following table in a chronological order.

List of Books on Irrigation Engineering-

1	Neville	(1875)	8	Ellis	(1950)
2	Mullis	(1890)	9	Sharma	(1944,3rd Ed.1959,3vol, 840 Pages+22 Plates)
3	Flynn	(1892)	10	Khushalani	(1954, 4th Ed.1971, 6 vol. 1500 Pages)
4	Wilson	(1897)			
5	Clibborn	(1901)		**Only Canal books**	
6	Buckley	(1893, 2nd Ed. 1905)	11	Chitale	(1966, 400 Pages)
7	Etcheverry	(1915, 3 vol., 1015 Pages)	12	Framji	(1972, 300 Pages)

The complete titles of these publications are given in the list of references.

Chitale (1966) prepared a detailed account of almost all Research on Hydraulics of stable channel, (published by C.W.P.C.)

The above list does not include all the books published world wide, as same is very difficult and may be very long. During last forty years, many books on irrigation engineering for under graduate courses have come up, that deal the subject just in very short, generally may not be sufficient for any major field work.

The references in the above books are usually very important sources for greater details. Still most publishers' stress the replacement of old references with more recent ones. The reader should be aware of the fact that the author of the reference may be the compiler rather than the originator of a certain theory, and some times may either miss or could not appreciate the concept of the originator.

Many times the concept may not be well explained in / by theory and therefore get missed by the compiler. Also all text books have limitation of text (pages), scope of the work and the contents of the University courses. Even the practical books have the limitation of the experience of the author, time, cost, financial support, market, cost and not the least on the involvement of university teachers with practical aspects (now greatly declining).

The present system of engineering education in India is only theoretical in utter disregard of field experience. It is not on the lines of Medical education, where the teaching is from the practicing Doctors. Also there is little financial support either by All India Council of Technical education (ACTE) or University Grants Commission (UGC), India to field engineers to write their practical field experience. The result is purely theoretical teachings and production of theoretical engineers finding difficult to fit in the field and over all technical skills are declining. A sincere effort is needed by the teachers and field engineers to improve their technical skills, also see chapter 7.

Almost all textbooks in hydraulics indicated the semicircle, as the best hydraulic section. Neville (1875) proposed a graphical method for the best trapezoidal section by extending full supply line out side the semicircle, and drawing three tangents one in the bottom and one an either side for the proposed side slope, similar to one shown in figure 1.6. Buckley (1905) and then Etcheverry (1915) have highlighted the importance of best hydraulic section circumscribing a circle (see Para 1.6) and gave the dimensions in terms of area (A), see table-1.3. These engineers found many difficulties in using the same in field, as all the time they were trying to apply to unlined canals.

6.3 Design Practices

6.3.1 U. S. Reclamation Service (USRS - Etcheverry 1915) Practice

The early U.S. Reclamation Service (Etcheverry, 1915) gave a simple empirical rule for proportioning the irrigation channel section as under:

$D = 0.5\sqrt{A}$; Where D = full supply depth of water in feet; A = water area in ft^2.

For a trapezoidal section it can be seen that this rule may also be expressed by a simple formula

$X = 4 - Z$

Where X is the width-depth ratio (B / D) and Z is the horizontal projection of the side slope corresponding to 1 ft vertical.

The depth of water should not exceed 8 to 10 feet except in uncommonly large canals. The ratio of bottom width to depth usually ranges from 2:1 to 6:1. The side slopes of a canal depend upon the stability of the material in which it is constructed. Inside slopes of 1.5:1 or 2:1 (horizontal to vertical) are practically standard for earth canals under ordinary conditions; on side-hill locations the inside slope of the uphill bank may be made steeper, if the material will stand, to avoid excessive excavation. Outer bank slopes may be made 1.5:1 under ordinary cut conditions.

6.3.2 Indian Practice

Engineers in India, (Molesworth, 1871) have earlier used an empirical formula

$$D = \sqrt{A/3} = 0.577\sqrt{A}$$

This is equivalent to X = 3 − Z for trapezoidal sections;

(D, A, X, Z have same meaning as above)

Comments

This is much near to the best section (see table- 1.3.) then the earlier USRS practice.

6.3.3 USBR Practice (1952)

Because of the many variable factors involved in the design of a lined canal, selection of the ratio of base width to water depth depends largely upon judgment. As a guide in selecting this ratio, curves showing the relation between the base width or depth and the flow capacity, for large number of lined canals designed by the Bureau, are given in Figure 6.3. The curves indicate average conditions for the various capacities shown. Side slopes of lined canals might, theoretically, vary from vertical to very flat. **The steeper slopes require forming, and the flatter slopes require more lining material**. The majority of Bureau-designed sections have used side slopes of 1-1/4:1 or 1-1/2:1. In smaller canals, it is questionable whether the small saving in lining material obtained by use of 1-1/4:1 slope offsets the possible necessity of forming and the general problem of slope instability. In the design of larger canals in suitable material, the situation may merit close study. In general, the Bureau is considering the standardization of a 1-1/2:1 slope for the usual sizes of lined canals involved. One advantage of this selected slope is that it will usually be sufficiently flat to allow the practicable use of any type of lining or lining treatment now contemplated

In unlined canals, the velocity of flow should be such as to prevent cutting of the canal prism or deposition of silt. The maximum velocity allowable to prevent cutting or the minimum allowable to prevent silt deposition will, of course, depend upon soil characteristics and natural factors, but general limits can be set down from experience. These limits are, under ordinary conditions:

	Maximum	**Minimum**
In large earth canals	3.5 ft/s	2.0 ft/s (fps)
In small earth laterals	2.5 ft/s	1.5 ft/s

The Kennedy (1895) formula for silt-laden water is

$V_s = CD^{0.64}$ (FPS units); Where Vs = velocity for non silt and non scour in ft/ sec.

D = depth of water in feet; C = coefficient for various soil conditions.

Values of the coefficient C are as follows:

For fine, light, sandy soil	0.84
For coarser, light, sandy soil	0.92
For sandy, loamy silt	1.01
For coarse silt or hard soil debris	1.09

A suggested modification of the Kennedy formula for clear water is, $V_s = CD^{0.5}$

Kennedy's formula in details is discussed in Para 7.10.

Curves for depth to allowable velocity in earthen canals are given in figure 6.4.

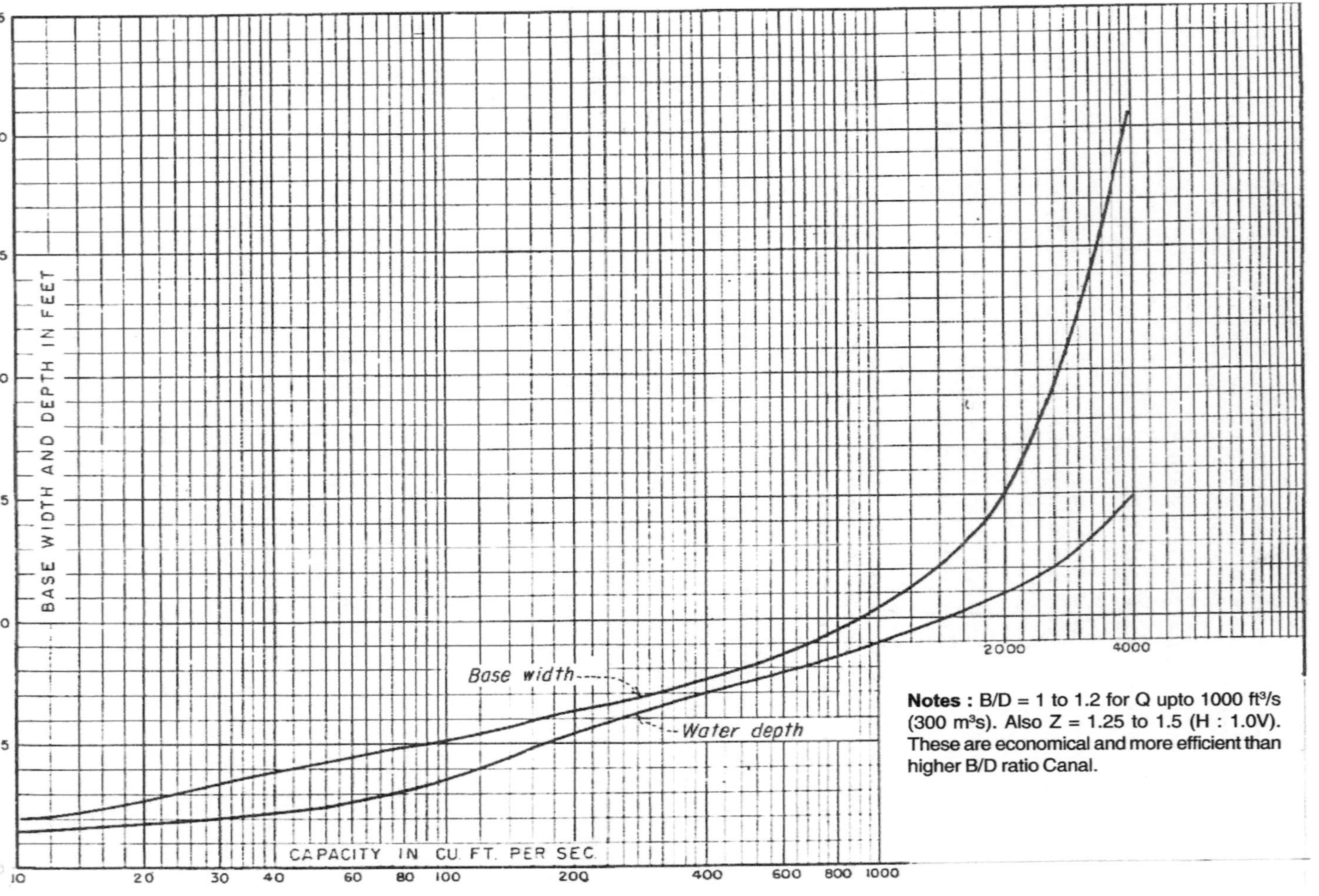

Notes : B/D = 1 to 1.2 for Q upto 1000 ft³/s (300 m³s). Also Z = 1.25 to 1.5 (H : 1.0V). These are economical and more efficient than higher B/D ratio Canal.

Figure 6.3 : Base Widths and Water Depth of Lined Canals, USA, USBR (1952)

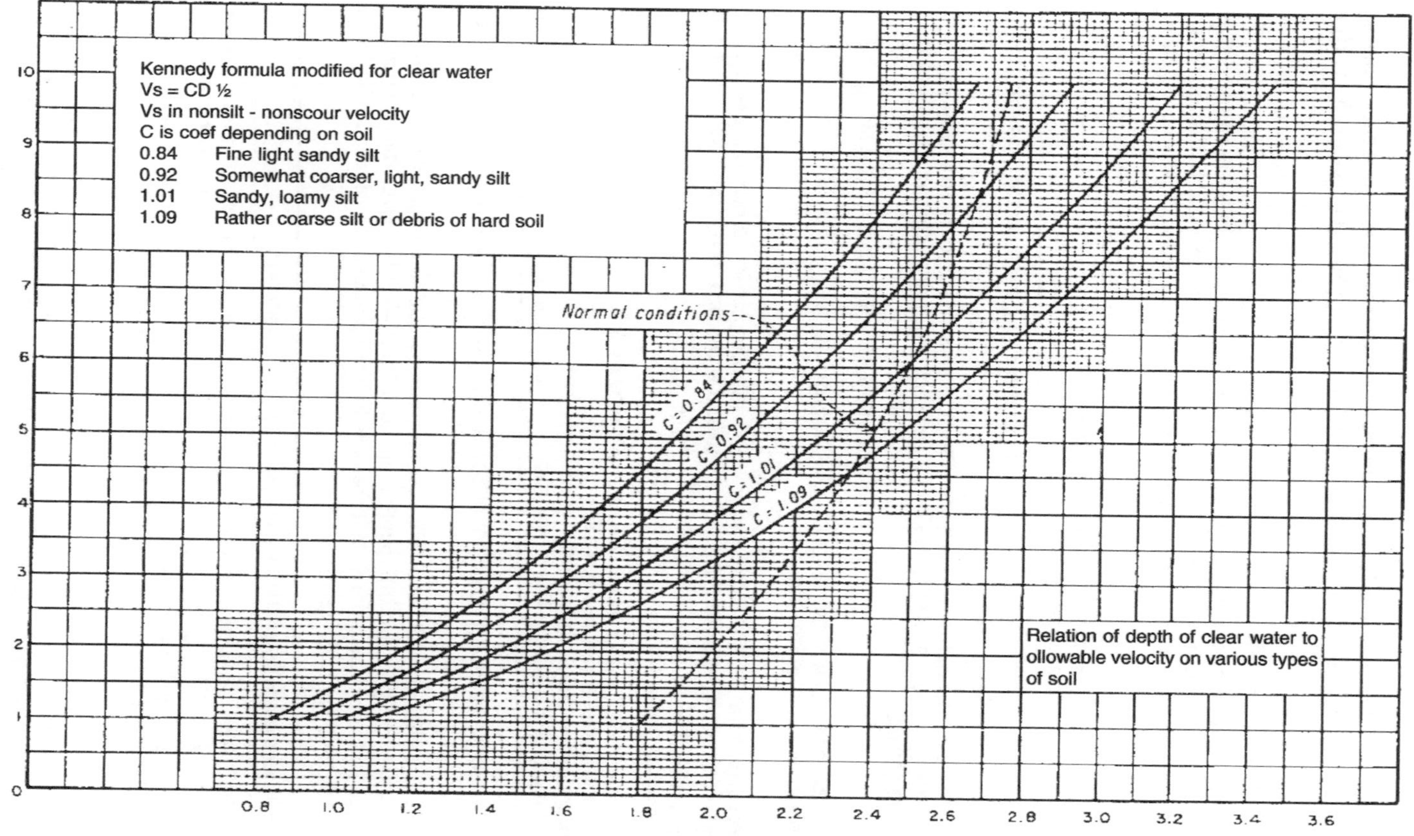

Figure 6.4 (i) : Earth Canals Relation to Depth to Allowable Velocity, USBR (1952)

Comments

The curves shows a B/D ratio of nearly 1:1 for discharges upto 1000 cusecs (nearly 30 cumecs), which is near to best trapezoidal section of side slope Z = 1:1; though the side slope of 1:1.25: (V: H) is little flatter. Also see comments in Para 6.3.4.

6.3.4 Bhakra Canal Manual Guidelines (1954)

The central design organization of Bhakra Project, (India), prepared and issued "General instructions for the design of lined and unlined channels of Bhakra canal system" (1954). Its important recommendations for unlined/lined canals are given in table 6.4. That for lined canals and comments on it are as under:

	Recommended guide lines for *Lined canals* Discharge up to 3000 cusecs (80 cumecs approximately)	Comments
(i)	Mehboob section up to 18 ft(5.5m) depth with a maximum velocity of 6 ft/sec (1.8 m/s)	Most efficient and stable section See Para 3.1 to 3.2
(ii)	Side slopes of 1:1 (V: H) for radius (i.e. depth) less than 12' (3.7 m) and 1:1.25 (V: H) for radius greater than 12' (3.7 m).	This gives maximum hydraulic efficiency and reasonable stable side slopes for common sandy loam soil, See Para 3.3, 7.9, 7.13, and 9.9, 9.13
(iii)	Steepest possible bed slopes, to get maximum permissible velocity and minimum value of Lacey's silt factor, f' = 1.2	See Para 7.13
(iv)	For discharges more than $90m^3/s$ (3000 cusecs) trapezoidal section with curved ends.	B/D ratio not specified, Adoption of higher B/D ratio gives uneconomical section, see Para 4.1 to 4.4.
(v)	Graphs connecting Q, slope, radius, f, V and N=0.018, are given in the manual	No comments
(vi)	It also further lays down the guidelines for banks, free boards, dowel, outer side slopes, hydraulic gradient, boundary road, spoils, borrow pits curves, command water allowance, losses etc.	See chapter 8

6.3.5 Central Water Commission (CWC) /Central Board of Irrigation and Power (CBIP) Guidelines, 1968 (1984)

CWC modified its earlier guide lines of 1968, and incorporated the guide lines of CBIP (1970) and BIS (IS 4745, 1968). These are given in CBIP publication No. 171 (1984). It has given the guidelines for the design of unlined and lined canals. Important abstracts for unlined canals are given in table 6.4 and for lined canals are as under:

	CWC Recommended guide lines– for lined canals	Comments
(i)	Manning's $V=R^{2/3}S^{1/2}/N$ has been proposed, besides the use of formulae for unlined channels.	Formulae of unlined canal normally not required. It may give higher bed width.
(ii)	Coefficient of Rugosity (Manning's N) given by IS-4745-1968 was recommended. These are similar to one given in table 2.1 to 2.3, and revised IS 10430: 2000.	Little higher values are recommended, see Para 2 and 4.5.
(iii)	Side Slopes: In order to keep down the thickness of lining on sides it is usually made to rest on slopes corresponding to the angle of	Slopes recommended are most suitable and economical for ordinary soils see Para 3.6 and 7.9

Contd...

repose of the natural soil so that the slopes shall be such that no earth pressure is exerted on the back of lining. The slopes steeper than these have to be designed as gravity sections and require greater thickness and are consequently uneconomical. The side slopes 1:1 to 1.25:1, the flatter being for deeper canals have been found convenient to work. Generally, a slope 1:1 for depth less than 3.66 m and 1.25:1 slope for more than 3.66 m is adopted. In filling compaction at optimum moisture content is done and stable slopes for such compacted banks are adopted.

(iv) Internal Section:
It has been seen that the most economical section, i.e., having maximum area with minimum perimeter, is the one with a curved bed with its centre at the F.S.L. and the sides tangential to the curve. This type of section, however, does not give any horizontal portion in the bed and is suitable for small channels only carrying discharges up to 60 cumecs. For discharges exceeding 60 cumecs, it becomes necessary to provide horizontal bed to keep depth within reasonable limits. The bed is however, joined to the side slopes in a curve of radius equal to F.S.D.

(i) This section is same, as Mehboob section.
(ii) Given in table 3.1, Para 3.1. This section is same as Section No 1 and 2. given in table 4.1, Para 4.3. B/D ratio not specified,

(v) Full Supply Depth:
Full supply depth is more a matter of assumption. It is however, guided by the following considerations:

(a) Having fixed the alignments and FSL from the command point of view, the depth may be such as to give balancing depth as far as possible.

In above no B/D ratio is proposed and left at the discretion of designer or his Experience.

(b) Velocity may not exceed 2 m/sec.

(c) If the water is clear from silt and sand, considerable velocities can be allowed in lined section. This is, however, not possible in actual practice. In canals taking of from rivers, the slope is further limited from considerations of command. In view of this velocity is usually limited to 2 m/s and the critical velocity ratio should be aimed at higher than unity though critical velocity ratios are not applicable to lined canals but the possibility of silting cannot be neglected.

Silting occurs and higher critical velocity ratio is desirable see para 7.4

(d) The bed of the channel may be higher than the sub-soil water level as far as possible, so as to avoid pumping during constructions and uplift pressures after closure of channel.

No comments

(e) In case of main and branch canals the depth should be such that it gives bed width enough to provide convenient working space within it, consistent with maximum economy in the cost of excavations, land and masonry works.

Since discharge is generally higher, so sufficient bed width is achieved by optimum B/D ratio, see chapter 4.

(f) It also further laid down the details of bank width, free board, berms and dowels drainage and pressure release arrangements, see chapter 8.

No comments

6.3.6 Indian Standard Codes (1968, 1982 and 2000)

IS 4745 = 1968, code of practice for design of cross section of lined canals, now with drawn and superseded by IS 10430 - 2000

IS 10430-1982 -Do- (now revised)

IS 10430-2000

IS 10430 - 2000 is the latest and most up to-date guidelines in India on the subject. It replaces its earlier IS 10430-1982. Some abstracts from it are given in the text at relevant place along with comments/ modifications proposed. It recommends cup shaped section (referred as Mehboob section in the text) only up to only 3 cumecs, and for other sections no B/D ratio is proposed. However, velocities have been limited.

Comments

This code in the present form does not explain the complexities of economical section and much is left at the discretion of designer. Therefore Bhakra manual (1954) is better than IS 10430-2000 code.

In the Bhakra Mannual, Mehboob or best trapezoidal section is recommended up to 80 or say even up to 100 m^3 /s, with a side slope 1:1 for depth up to 1.5 m and 1:1.25 for depth up to 2.5 m and for more discharge, most economical B/D ratio is proposed as discussed in chapter 4. Instead of flattening of side slopes it is advisable to increase bed width.

6.3.7 Other Practices in India and Different Countries

There are varied Hydraulic design Practices about the internal section i.e. rugosity coefficient (N), side slope (Z), B/D ratio (X), bed slope (S_0), and velocity (V) in different parts of India and many other countries. These five parameters are generally assumed on the experience, therefore tabulated in a comparative form in table 6.4 for easy reference and guidence.

Inferences

The above comparison indicates that the experience and practices vary greatly from country to country and within the country also. The important findings can be summarised as under:

(i) About Rugosity coefficient 'N'

a) Some countries adopt N, with due consideration of weed growth and algae formation and N is based on maintenance programme. This value is little higher, then the values based on the type of surface.

b) Some countries adopt N simply on type of surface, lower value for smoother surface and higher for rough surface.

c) Some countries also give due consideration to discharge, depth and size of canal. For higher discharge, smaller value than for lower discharge is adopted. But over all values are higher than that for the type of surface.

d) Some countries adopt a fixed value, irrespective of size and surface.

(ii) About internal side slope 'Z' (1.0 V: H)

a) Some countries, base it on the type of soil / rock

b) Some countries adopt it according to canal section in cutting, or filling or both. Different values for cutting then in filling.

c) Some countries adopt straight fixed values between 1.0 to 2.0 and even upto 2.5. Uniform value is adopted for machine construction to reduce cost of form work. Some countries base it on discharge, steeper slope for small canals and flatter slope for big / larger canals.

d) Some countries base it as 0.5 for unlined canals i.e. according to Kennedy's theory.

e) Some countries adopt best section for low to Medium discharges, both for lined / unlined canals. (See Para 1.5)

f) Generally Z is flatter for lined canals, then in unlined canals.

(iii) About B/D ratio (X)

a) Some countries prefer shallow canals i.e. wide channels.

b) Some adopt width according to requirement of machines / construction equipment.

c) Some countries adopt small values 1 to 2.

d) Some countries adopt B/D from economical consideration. Higher depths are preferred for high capacity canals and try to limit X within 2 to 5.

e) Experience curves for lined canals of USA, show X nearly equal to 1.0 for Q upto 100 m^3/s and are followed by many countries.

(iv) About S_0: (Also Para 9)

a) Generally most countries prefer according to topography.

b) Fixed on command consideration.

c) Fixed on Limitation of maximum permissible velocity consideration.

d) Considerable velocity of 2 to 3 m/s has also been permitted, in some countries.

(v) About velocity: (Also see table 7.8, Para 2 and Para 7.9)

For unlined canals

a) Some countries adopt their own experience of Maximum permissible velocity or non - scouring velocity.

b) Kennedy's velocity has been preferred by many countries. Some countries have modified coefficient of Kennedy's formula.

c) Velocities are kept less than hydraulically super critical velocity, i.e. $V < Vc$ (sub critical).

d) Many countries increase velocity by a laying a protective cover of gravel stones or any hard material. Thus section is economised. This also reduces silting.

e) Few countries adopt velocities given by Scooby and others (table 7.8). These velocities are little lower and some countries adopt their own practice.

f) Few countries also adopt Limiting (Critical). Tractive force approach, (Para 7.10)

g) Very few countries adopt Lacey's formula or concepts.

h) There is more stress on non scouring velocity, and less stress about non silting velocity.

i) For lined canals there is less restriction on velocity. Velocity of 2 to 3 m/s has also been permitted in some countries.

Generally unlined / lined canals were designed with Chezy's, Kutter's or Manning's formula, with assumption of all above parameters.

All the above 5 parameters, actually 6 (as X as has 2, B and D), are inter woven. The intrinsic relationship is explained only by Manning's Master Equation, (Para 4.4) and their effect on each other can be seen by giving a fixed value to one. These are re summarised as below for convenience.

(i) Higher 'N' does not increase cost proportionately (Para 4.5) and should be preferred based on maintenance programme.

(ii) Role of higher Z than (0.6 or 1) is to reduce velocity. It encourages silting on sides and increase over all cost of section.

(iii) X higher than 1 also reduces velocity, and encourages silting and needs steeper slope.

(iv) For flatter slopes, lower B / D ratio gives little higher velocity than that for higher B/D ratio.

Table 6.4 : Hydraulic Design Practices of Canals in Various Countries

Rugosity Coefficient	Side Slope	Bed width to depth ratio	Longitudinal Bed Slope	Velocity
N	1: Z (V : H)	B/D = X	S_0	V
(1)	(2)	(3)	(4)	(5)
1. Australia: (i) Victoria: State Rivers and Water Supply Commission: Lined/unlined canals are designed by Manning's formula				
Unlined Canals- Between 0.022 and 0.025 in earlier designs but in current designs this has been increased to 0.0275, to take into account the effect of weed growth causing reduction in capacity of canal.	**For Q = 60 ft³/s** (1.70 m³/s), Z > 1: 2 For Q > 60 ft³/s (1.70 m³/s), Z >1: 2.5	Between 2 and 5 for most of the canals and vary from 1.5 to 7 for remodeled channels.	Selected as far as possible to balance the earthwork. Topography of the country is generally flat with slopes of 0.05/1,000 (1/20,000) to 0.5/1,000. (1/ 2,000)	**Unlined canal** - Normally 1 ft/s (0.3 m/s) to 1.5 ft/s (0.46 m/s) with maximum limit of 2 ft/s (0.61 m/s).
1. Australia: (ii) New South Wales-Water Conservation and Irrigation Commission: Lined/unlined canals are designed by Manning's formula, lining is exceptional.				
For Q < 70 m3/s, unlined – canals, N = 0.03 For Large canals of Q > 2,500 ft³/s (70.79 m³/s)	**Unlined** canals - Z=2 and lined canals Z= : 1.5	**Unlined** canals-generally fixed so as to give more or less semi-circular section. **In large canals,** shape is modified according to the	For irrigation channels slope is based on the natural ground level as well as the desirable minimum water levels For drainage channels	Maximum permissible velocities are adopted as given below: Irrigation canals — V ft/s (m/s) 1. (a) Smaller channels (25 ft³/s or 0.71 m³/s) — 1.7 (0.52) Bigger channels (2500 ft³/s or 71 m³/s) and depth of flow > 8 ft (2.44m) — 2.5 (0.76) 2. Drainage canals — 1.3 (0.40) 3. lined canals — 8.0 (2.44)

Contd...

Table 6.4. Contd...

N	Z	X	S_0	V
(1)	(2)	(3)	(4)	(5)
For **drainage** channels N= 0.035 Higher N provides some margin / flexibility in operation and maintenance Programme. Drainage channels are also checked for non-scouring velocity with N=0.025. **Lined canals** **N** = 0.014. (Concrete Lined).		requirements of construction equipment. For typical sections, viz., 6x3 (ft) to 100 x10 (ft) [1.83 x0.91 (m) to 30.48 x3.05 (m)], B/D vary between 2 to 10. **Lined canals-** B/D ratio is fixed very near to 1.	Slope is based on desirable permissible maximum water levels.	These values, however, vary in specific cases. The velocity allowed in drainage channel is less since they traverse mostly in the more silty and erodible soils. **Lined canals**, velocities are either clearly sub-critical or super-critical.
1. Australia: (iii) Queensland-Irrigation and Water Supply Commission: Lined/unlined canals are designed by Manning's formula.				
Discharge m^3/s — N **Unlined Canal** < 1.42 — 0.040 1.70 to 2.83 — 0.035 3.96 and above (In between a graded value of N is used) — 0.030 High values of N are adopted to allow for loss in capacity due to excessive weed growth.	**Unlined** and Clay-lined channels Z = 2.0 Concrete-lined channels Z = 1.5	**Unlined** small (D upto 5 ft)=2 to 4D Med. (D>5<10 ft)=4 to 6D Large (D>10 ft)= 6 to 8 D Minimum bed width is 8.0 ft (2.44 m) **Lined channels** small (D = 5 ft)=1.5 to 2D	Selected to best suit ground slope and Earthwork balance, but slope is reduced if necessary, by use of drops, to restrict the velocity of flow to the allowable maximum.	**Earth channels, maximum** - 2.0 ft/s (0.61 m/s), **Concrete-lined** 8.0 ft/s (2.44 m/s). In normal lined channels, V is kept below critical Velocity 'V_c' by providing depth of flow at least 1.15 times the critical depth 'y_c'.

Contd...

Table 6.4. Contd...

N	Z	X	S_0	V
(1)	(2)	(3)	(4)	(5)
Lined canals: For manually-or mechanically - placed concrete lining 0.015 For pneumatically-placed mortar lined canals 0.018		Med. (D > 5 < 10 ft) = 2 to 3D Large (D > 10 ft) = 3 to 4D Minimum bed width 2.0 ft (0.61 m)		
2. Bulgaria: canals are generally lined. Designed by Chezy's or Manning's formula				
Unlined Chezy's formula is used Coefficient C is determined by Pavlovski's formula, $[C= (R)^{Y}/N]$ $Y = 2.5\sqrt{N} - 0.13\text{-}0.75$ $\sqrt{R}\,(\sqrt{N} - 0.1)$ **Large canals**, N = 0.025 small canals, N = 0.030 If resistance is excessive due to growth of vegetation. N= 0.035 It is, however, ensured that	Adopted in accordance with the ground features. Usually, 1: 1.5 but for some canals, made as steep as 1:1. **Lined canals:** Z = 1.50 or 1.25. Usually for some small canals, it can even be as steep as 1.0. Steeper permissible side	Minimum bed width 'B_{min}' is adopted according to construction method. In manual work 'B_{min}' = 0.3 m (1 ft); For excavation by machines 'Bmin = 1 m (3.3 ft) and depend upon the dimensions of the machines. B/D is chosen for obtaining hydraulically Most efficient trapezoidal cross section, (to gives shortest wetted perimeter),	Calculated to be most economically favorable. For such gradient, the mean annual expenses (running plus maintenance) for the canal and the annual losses due to areas being left out of command should be minimum. Besides, gradient should also result in meeting requirements about maximum and minimum velocities.	The maximum permissible water velocity '$v_{(perm)\,max}$' is chosen so as to avoid bed scour and bank erosion. Unlined canals usually >/ 0.8 m/s (2.5 ft/s). Minimum permissible velocity '$v_{(perm)min}$' is chosen to Avoid silting of the canal well as freezing of

Contd...

Table 6.4. Contd...

N	Z	X	S_0	V
(1)	(2)	(3)	(4)	(5)
the canals are kept in good condition by regular cleaning, to keep them clear from partial landslides, vegetation and other obstructions. **Lined canal**: As above, For concrete linings without cement coating, N = 0.014.	Slope is so chosen as to make it possible to place the concrete without shuttering.	the formula: B/ D = $2\left(\sqrt{1+Z^2}-Z\right)$ is used However, not strictly observed to fulfill other requirements, such as economic feasibility of the canal project depending on the capital invested, its maintenance cost during operation. **Lined canal:** The ratio B/D is chosen to obtain hydraulically most efficient trapezoidal cross section, formula B/ D = 2 $\left(\sqrt{1+Z^2}-Z\right)$ For Z =1.5, B=0.61 D For Z = 1.25, B=0.70 D for Z = 1.00, B=0.83 D for Z= 1, B = 0.83 D This condition is, however, not strictly observed where other requirements are also to be fulfilled.	It is worked out to achieve maximum economy. For such a gradient, the mean annual expenses (running and Maintenance costs) and the annual losses due to areas left out of command should be minimum. Besides, gradient should also confirm to the requirements related to the maximum and minimum velocities.	water if the canal runs during the winter, for instance, for filling a reservoir. This velocity is usually < 0.3 to 0.5 m//s (1.0 to 1.5 ft/s). '$V_{(perm)\ max}$' should be such as not to cause wearing out of the lining material. In practice such high velocity is rarely reached. **Lined canal** As above. In the case of a canal to fill a reservoir, velocity < 0.5 m/s (1.6 ft/s).

Contd...

Table 6.4. Contd...

N	Z	X	S_0	V
(1)	(2)	(3)	(4)	(5)
3. Canada: Lined/unlined canals are designed by Manning's formula				
Unlined canals: For all sizes of canals, N = 0.025 is adopted in Kutter's formula or Manning's formula. This value may be conservative but ensures full capacity flow with some weed growth.	For good soil Z = 1 : 2. For lighter soil, Flatter side slopes of 1:3 are used. Outside slopes may be 1:3 for lighter soils or when clay lining is used they could be steeper. Recent tendency is to use flatter slopes for ease of maintenance.	B/D is provided as 2 for smaller canals to 8 for Large capacities.		Water generally supplied from storage reservoirs, it is mostly silt free. Canals are designed for higher velocities to discourage algae and weed growth and to prevent silting. Excessive velocities, which could cause erosion, are however avoided. Kennedy IS modified formula as below is used $V=CD^{0.4} + 1.0$ ft/s C =0.5 for lightest soils, 0.6 for light soils and silts, 0.7 for fair sandy soils, 0.8 for good soils, sand and clay, 0.9 for best heavy clay soils.
4. Colombia: Lined/unlined canals are designed by Manning's formula				
Unlined canals: N = 0.027	Z = 1: 1.5 is generally adopted.	Adopted according to USBR standards. It usually ranges from 2 for small canals with capacity about 31 ft^3/s (0.90 m^3/s) to 8 for canals with capacities of about 10,000 ft^3/s (283 m^3/s).		As per USBR practice, see Para 6.3.3

Contd...

Table 6.4. Contd...

<table>
<tr><th>N</th><th>Z</th><th>X</th><th>S_0</th><th>V</th></tr>
<tr><td>(1)</td><td>(2)</td><td>(3)</td><td>(4)</td><td>(5)</td></tr>
<tr><td colspan="5">5. Cyprus: Mostly Unlined canals upto 50 cusecs (1.35 cumecs) and maximum 30 cusecs (0.90 cumecs) for lined canals.</td></tr>
<tr><td>Unlined canals: mostly constructed by farmers long ago and still working satisfactory.
Lined canals: 0.014 is assumed in Manning's formula as well as in the Bazin's formula.</td><td></td><td></td><td></td><td>In the small concrete canals used for diversion in hilly areas the velocity permitted is 10 ft/s (3.05) m/s) and in large canals used for distribution it is about 6 ft/s (1.82 m/s).</td></tr>
<tr><td colspan="5">6. France: Lined/unlined canals are designed by Manning's formula</td></tr>
<tr><td>Unlined canals:
Depending upon the nature of boundary material, value of Rugosity Coefficient 'N' in Manning's Formula are as under:<table><tr><td>Nature of wetted perimeter</td><td>Characteristics of the canals</td><td>N</td></tr><tr><td>Concrete</td><td>Entirely lined with concrete and straight reaches</td><td>0.013 to 0.014</td></tr></table></td><td><table><tr><td>Nature of soil</td><td>Side slope (Horizontal : vertical)</td></tr><tr><td>Ordinary soil</td><td>1.5 : 1.0</td></tr><tr><td>Clayey – sandy soil</td><td>2.0 : 1.0</td></tr><tr><td>Highly sandy soil</td><td>3.0 : 1.0</td></tr><tr><td>Uncemented soil</td><td>1.0 : 1.0</td></tr><tr><td>Conglomerates, hard clay</td><td>0.75 : 1.0</td></tr><tr><td>More or les weathered rock</td><td>0.5 : 1.0</td></tr><tr><td>Rock</td><td>0.25 : 1.0</td></tr></table></td><td>No fixed criteria. Several formulas given in Para 3.6, and also Para 9 are used.
Generally arbitrary judgment.</td><td>Worked out from Kennedy's formula for D = R.</td><td>As a guide maximum velocities given by Scobey (Table 7.8) in addition the formula as under:
$V = 6.5\,Q\,D^{4/9}\,(\gamma - 1)^{0.5}$
D = grain diameter in mm γ = specific weight of grain
Derived from the experimental data at the University of Iowa is also used. Velocity given by Kennedy formula is adopted as minimum permissible.</td></tr>
</table>

Contd...

Table 6.4. Contd...

	N		Z	X	S_0	V
	(1)		(2)	(3)	(4)	(5)
Concrete	Entirely lined with concrete cross section containing curves	0.015 to 0.017	The canals in earth, both lined and unlined, are provided with trapezoidal section whereas the reaches through rock-cut are sometimes provided with rectangular section. Side slopes are provided depending upon the nature of the soil, type of cross section (cut, fill or mixed) and mode of construction. Z = 1.5 H in filling and Z = 1.0 H in cutting are adopted. Table C gives side slopes generally assumed according to the nature of soil. Reinforced concrete flumes are generally made up of semi-circular, trapezoidal or rectangular sections.			
Concrete and shot Crete	With concreted bed, shot creted slope on rock	0.020 to 0.025				
Concrete and rock	Consisting of a concreted bed, bed wall and rocky slope	0.020 to 0.025				
Concrete and rock	Consisting of an unlined rocky bed, a bed wall and a rocky slope	0.025 to 0.030				
Rock	Rocky canal (Main canal, important main laterals and distributaries)	0.030 to 0.035				
Rock	Rocky canal of small cross section	0.035 to 0.040				
Earth	Earthen canal	0.020				

Contd...

Table 6.4. Contd...

N	Z	X	S_0	V
(1)	(2)	(3)	(4)	(5)
7. GREECE: Lined/unlined canals are designed by Chazy's or Manning's formula and checked for Kennedy's formula				
Kutter's formula is adopted.	For **Unlined canals** both in cutting as well as in embankment, same values of side slopes are adopted which vary from Z = 1.0 H to 1.5 H. **Lined canals:** The shapes of the section vary from rectangular to trapezoidal. If the reinforced concrete lining is used, Z both in cutting as well as in embankment is between 2.0H to 3H. If the lining material is other than reinforced concrete, Z in embankment is between 1.25 H and 1.5 H.			The maximum limiting velocity is normally between 0.8 m/s (2.6 ft/s) to 1.0 m/s (3.3 ft/s) depending upon nature of soil (excepting rock-cut) and minimum at 0.3 m/s (0.98 ft/s). Tractive force concept is used. Similar to unlined canals, limiting values are set for maximum and minimum velocities in case of lined canals also. The maximum limiting velocity is between 1.5 m/s (4.92 ft/s) and 2.0 m/s (6.56 ft/s) whereas the minimum limit is 0.20 m/s (0.66 ft/s).

Contd...

Table 6.4. Contd...

<table>
<tr><th>N</th><th>Z</th><th>X</th><th>S_0</th><th>V</th></tr>
<tr><td>(1)</td><td>(2)</td><td>(3)</td><td>(4)</td><td>(5)</td></tr>
<tr><td colspan="5">8. India: (i) Andhra Pradesh: Lined and unlined canals are designed by Manning's formula. Also see figure 6.19 for canal section.</td></tr>
<tr>
<td>No allowance in N is made for weed growth, velocities are considered to be sufficiently high to prevent weed growth.
Unlined canals:
N is adopted on the basis of discharge as under:
<table>
<tr><th>Type of canal</th><th>N</th></tr>
<tr><td>1. Small canals Q=50 ft³/s (1.42 m³/s) and less</td><td>0.025</td></tr>
<tr><td>2. Larger canals in good condition, Q= 50 ft³/s to 3,000 ft³/s</td><td>0.0225</td></tr>
<tr><td>3. Canals in rock-cutting</td><td>0.035</td></tr>
<tr><td>Lined canals</td><td>0.018</td></tr>
</table></td>
<td>Z = 0.5 H is assumed in design, but section is constructed in different slopes, depending upon the nature of soil as below:
(a) In filling
Height of bank above ground level
(i) Upto 10 ft. (3.05) m = 1:1.5
(ii) Above 10 ft. (3.05) m = 1:2
(b) In cutting
Black cotton, kankar, gravel soft disintegrated rock = 1:1 to 1.5
Hard disintegrated soil=1:0.25
Hard rock = 1:0.5</td>
<td>Central water power Commission, India 1950, guidelines are generally followed, see Para 6.3.5</td>
<td></td>
<td>Unlined canals:
Maximum Permissible Velocities are adopted according to the type of soil as below
<table>
<tr><th>Type of soil</th><th>Velocity, ft/s(m/s)</th></tr>
<tr><td>1. Black cotton soil</td><td>3.0 (0.91)</td></tr>
<tr><td>2. Gravel</td><td>4.0 (1.22)</td></tr>
<tr><td>3. Disintegrated rock</td><td>5.0 (1.52)</td></tr>
</table>
Generally in normal sections, permissible velocity is considered in the range of 2 to 4 ft/s (0.6.1 to 1.22 m/s)
Lined Canals:
Maximum velocity is not allowed to exceed 6 ft/s (1.83 m/s). In flumed sections and in constrictions, allowed to vary within 5 to 10 ft/s (1.52 to 3.05 m/s), and in tunnels 10 to 20 ft/s (3.05 to 6.10 m/s). The velocity adopted is such that it does not allow weed growth in channels.</td>
</tr>
</table>

Contd...

Table 6.4. Contd...

N	Z	X	S_0	V
(1)	(2)	(3)	(4)	(5)
8. India: (ii) Bihar, (a) River Valley Project Department - Gandak Project: A combination of Lacey's Kennedy and Manning's formula is adopted. Also see figure 6.20, 6.21.				
Unlined canals: For discharge >1,000 ft^3/s (28.32 m^3/s), N is adopted based on field observations / experience. For less discharge, N is adopted as under: Q, ft3/s (m3/s) — N up to 5 (0.14) — 0.030 5 to 50 (0.14 to 1.42) — 0.025 50 to 500 (1.42 to 14.16) — 0.0225 Above 500 (14.16) — 0.020	The general practice is to assume that after berm formation, steeper side slopes will be realized than adopted at the time of construction. In design for canals in loamy soil, Z = 0.5 H and for canals in sandy soil, Z = 1.0 H. During construction in filing in loamy soil, Z =1.0 H and in cutting, Z = 1.5 H in sandy soil Z = 2.0 H.	For discharge up to 350 f^3/s (9.91 m^3/s). CWPC Manual guidelines are adopted. For higher discharge, practice is to limit the full supply depth to 10.5 ft (3.20 m).		Kennedy's Critical velocity ratio V/Vc is assumed to decrease from 1.1 in head reach to 0.96 in the tail reach. For branches and distributaries, it is generally kept the same as that of the parent channel at the off take point.
8. India: (ii) Bihar (b) River Valley Project Department - Kosi Project: A combination of Lacey's Kennedy and Manning's formula is adopted.				
Unlined canals: Adopted according to discharges as under: Q, ft³/s (m³/s) — N Less than 5 (0.14) — 0.03 5 to 25 (0.14 to 0.71) — 0.025 25 to 1,000 (0.71 to 28.32) — 0.0225 Above 1,000 (28.32) — 0.020	In design cutting, Z=1.0 H but in actual construction Z = 1.5 H In embankment, Z=2.0 H for discharge above 150 ft^3/s (4.25 m^3/s) and 1.5 H : for discharge range of 5 to 150 ft^3/s (0.14 to 4.25 m^3/s) On the land or outer side slope of Z = 2.5 H is adopted for the discharge range of 300 ft^3/s (8.50 m^3/s) and above Z = 2.0 H and 1.0 H is for discharge range of 5 to 300 ft^3/s (0.14 o 8.50 m^3/s)	Adopted as per CWPC manual.		In the head reach, V/V_c is generally kept the same as the V/V_c adopted in the particular reach of the parent channel from which the sub-branch or distributaries takes off. The V/V_c is assumed to decrease from head to tail reaches. At tail end it should not be less than 0.8.

Contd...

Table 6.4. Contd...

N	Z	X	S_0	V
(1)	(2)	(3)	(4)	(5)
8. India: (ii) Bihar (c) Damodar Valley Corporation: A combination of Lacey's Kennedy and Manning's formula is adopted.				
Unlined canals: Q — N < 500 ft³/s (14.16 m³/s) — 0.0225 500 ft³/s (14.6 m³/s) and above — 0.020	Generally Z=1.0 H in cutting except in case of main irrigation-cum-navigation canal, where a flatter slope Z=2.0 H is provided with a view to minimize the effect of wave wash. When the canal is in embankment, Z of 1.5 H: 1.0 V is adopted on both sides of the bank except, in some cases where the outside slope has been provided at 4.0 H: 1.0 V from the point where the cover over the hydraulic gradient is 2 ft (0.61 m) up to the ground level.	For the discharge range up to about 125 ft³/s (3.54 m³/s) - A fixed B/D ratio of 3 (minimum) is maintained. This is increased thereafter for the increase in discharge.		The maximum permissible velocity is generally less than 2 ft/s (0.61 m/s) though so metimes 4 ft/s (1.2 m/s) is also allowed in main canals. Kennedy's critical velocity ratio V/V_0 = 0.9 to 1.1 is adopted. When the velocity is 1.5 ft/sec (0.46 m/s) or less, V/V_0 or less, V/V_0 > 1 : 1

Contd...

Table 6.4. Contd...

N	Z	X	S_0	V
(1)	(2)	(3)	(4)	(5)
8. India: (ii) Bihar (d) Irrigation Department: A combination of Lacey's Kennedy and Manning's formula is adopted. Also see figure 6.26 (i)				
In earthen channels and in self-silted bed N is taken as 0.0225. Lined canals: Lining material — N Brick in cement mortar — 0.01212 to 0.017 Cement mortar surface — 0.011 to 0.015 Concrete — 0.012 to 0.018	The side slopes are adopted according to the stability of the bank materials on basis of experience. The slope for design purposes is taken as 0.5 H : 1.0 V, and for construction. For clay soil = 1.0 H : 1.0 V is adopted. For loamy soil = 1.5 H : 1.0 V, and for sandy soil = 2.0 H : 1.0 V or 3.0 H : 1.0 V For lined canals: Z = 1 : 1.5		For channels taking off from reservoirs, the slope as, far as possible, is kept not flatter than that obtained for f = 0.80 or 0.75. When the country permits of a steeper slope, the slope provided should not be steeper than that obtained for f = 0.9 or 1.0. f = Lacey's silt factor	Generally Kennedy's or Lacey's formulas are used. (Para 7.12 and 7.13)
8. India: (ii) Bihar (e) State Electricity Board: Both for unlined and lined canals. Manning's formula is adopted. Also see figure 6.15 and 6.16.				
Unlined canals: 0.022 to 0.025 **Lined canals** = 0.018 Only Manning's formula is used both unlined and lined canals.	Generally slope of 1.25 H : 1.0 V is adopted.	Bed width is arbitrarily chosen.	The slope is chosen arbitrarily to give the permissible velocity. Generally as flat a slope as possible is provided.	Maximum velocity of 2.5 ft/s (0.76 m/s) is allowed in unlined earthen channels, 6 ft/s (1.83 m/s) in lined channels. In rock-cutting. However, maximum velocity is taken to be about 5 ft/s (1.52 m/s).

Contd...

Table 6.4. Contd...

N	Z	X	S_0	V
(1)	(2)	(3)	(4)	(5)
8. India: (iii) Gujarat: Unlined canals are designed by Kennedy and Manning's formula, see Para 7.10 to 7.12. Also see figure 6.17 & 6.22.				
N, is taken according to discharge ranges as below,	Except in hard strata, such as soft rock or hard rock, Z assumed as 0.5 H : 1.0 V in design, but actual excavation or embankment is done according to stable side slopes permissible for the type of soil as below:	Generally B/D taken to be not less than 1.5 for unlined canals. In canals of Hathamati and Kakrapar projects these values were taken to vary from 2 to 11, according to capacities of canal.	The bed slope of canal is chosen taking into consideration slope of the country as below:	Maximum permissible velocity in unlined canals is adopted according to the nature of soil as below

Column (1):

Discharge ft³/s (m³/s)	N
Greater than 500 (14.16)	0.0225
500 to 100 (14.16 to 2.83)	0.025
100 to 20 (2.83 to 0.57)	0.0275
20 and less (0.57 and less)	0.030
In case of rock cutting	
For soft rock	0.035 to 0.040
For hard rock	0.050

Column (2):

Type of soil	Z (H : V)	
	Cutting	Banking
Rock	0.25 : 1.0 to 0.5 : 1.0	2.0 : 1.0
Murum or friable soft rock	0.75 : 1.0 to 1.0 : 1.0	2.0 : 1.0
Compact clay, BC soil etc.	1.0 : 1.0	2.0 : 1.3
Silt loam or sand	1.5 : 1.0	2.5 : 1.0
Loose sand and sandy loam	2.0 : 1.0	2.5 : 1.0

Column (4):

Canal	S	
	Flat country	Country with steep fall
Main canal	1/10000 to 1/4000	1/5000 to 1/3000
Branches, Distributaries, etc.	1/6000 to 1/3000	1/4000 to 1/2500

Column (5):

Type of soil	Non-scouring maximum velocity ft/s (m/s)
1. Very light loss soil	1.00 –1.75 (0.30-0.53)
2. Coarse sand or light sandy soil	1.75-2.25 (0.53-0.69)
3. average, sandy soil	2.00-2.05 (0.61-0.62)
4. Sandy loam	2.05-3.00 (0.62-0.91)
5. Average loam or soil	2.75-3.25 (0.84.0.99)
6. Firm loam clay, loamy soft clay	3.00-3.75 (0.91-1.14)
7. Soft rock	4.00-8.00 (1.22-2.4)
8. Hard rock	8.00-15.00 (2.44-4.57)
9. Very hard rock or cement concrete	15.00-25.00 (4.57-7.62)

The actual velocities in contact with bed material are taken as 4/5"(80%) of the velocities given above.
Kennedy's CVR is also kept between 1.1 to 0.9 from head to tail

Contd...

Table 6.4. Contd...

<table>
<tr><th>N</th><th>Z</th><th>X</th><th>S_0</th><th>V</th></tr>
<tr><td>(1)</td><td>(2)</td><td>(3)</td><td>(4)</td><td>(5)</td></tr>
<tr><td colspan="5">Lined canals: Design procedure is practically similar to that for unlined canals. Typical cross sections of lined canals are shown in figure 6.17</td></tr>
<tr><td>N are chosen according to the type of lining as below
<table><tr><td>Type of lining</td><td>N</td></tr><tr><td>Brick masonry lining or sandwich brick</td><td>0.018</td></tr><tr><td>2. concrete lining (lime concrete or cement concrete or precast c.c. slab)</td><td>0.017</td></tr><tr><td>Stone masonry lining</td><td>0.020</td></tr></table></td><td></td><td>The junction of canal bed and sides of lined canals is provided with a curve of radius equal to
(i) Full supply depth if the side slopes are 1.0 horizontal :n 1.0 vertical or flatter, and
(ii) 1/2 full supply depth if the side slope is steeper than 1.0 horizontal : 1.0 vertical.</td><td></td><td>The velocity of 5 ft/s (1.52 m/s) to 8 ft/s (2.44 m/s) is permitted in the lined canals.</td></tr>
</table>

Contd...

Table 6.4. Contd...

N	Z	X	S_0	V
(1)	(2)	(3)	(4)	(5)
8. India: (iv) Jammu and Kashmir: Unlined canals are designed by Kutter's or Manning's and Kennedy's formula see Para 7.10 to 12 **Lined canals:** Design procedure is practically similar, except N, V and S are slightly modified.				
Unlined canals: N = 0.0225 **Lined canals:** Cement concrete = 0.018	In design Z = 0.5 is assumed, whereas in construction kept at 1.0 H when the section is in cutting and at 1.5 H when the section partial cutting and filling or in complete filling. When the canal is in conglomerate rock, Z = 0.25 H is adopted.		Suitable bed slope is assumed depending upon the discharge in the canal. For the discharge between 400 ft^3/s (11.33 m^3/s) and 150 ft^3/s (4.25 m^3/s) slope of 1 in 5,000 is assumed whereas for small discharge up to 20 ft^3/s (0.57 m^3/s), slope of 1 in 2,500 is adopted.	CVR is derived from Kennedy's formula and is taken to vary between 1.1 at head to 0.9 at tail/low discharge. Lined canals: Maximum permissible V is 5 to 6 ft/s (1.52 to 1.83 m/s)

Contd...

Table 6.4. Contd...

N	Z	X	S_0	V
(1)	(2)	(3)	(4)	(5)
8. India (v) Kerala: Both for unlined/lined canals, Manning's formula is used. Kennedy's and Lacey's formula are considered as not applicable as soils are gravelly and laterite.				
Unlined canals = 0.025 and In rock-cuttings = 0.035. **Lined canals:** N is assumed according to the type of lining. For concrete lining N=0.017 to 0.018 is generally used.	The canal is usually designed with a trapezoidal section. Z adopted are 0.25 H to 1.0 H in cutting and 1.25 H to 2.0 H in filling.		The bed slope is fixed considering the actual fall available from the source to the lowest and farthest ayacut to be served. Generally slope of 1 in 2,000 to 1 in 5,000 is provided for unlined canals.	The velocity is usually limited between 2 ft/s (0.61 m/s) to 3 ft/s (0.91 m/s) in earth channels and 6 ft/s (1.83 m/s) in concrete lined or rock-cut channels.

Contd...

Table 6.4. Contd...

N	Z	X	S_0	V
(1)	(2)	(3)	(4)	(5)
8. India: (vi) Madhya Pradesh: Unlined and lined canals are designed by Manning's formula, by assuming maximum permissible velocity.				
N is assumed as 0.0225 for all channels in average condition irrespective of discharge.	Adopted according to type of soil as under: Type of soil — Z (H : V) Light sandy soil — 2.0 : 1.0 Black cotton soil, Matasi soil and such other soils — 2.0 : 1.0 Sandy soil — 2.0 : 1.0 Murum — 1.5 : 1.0 rock — 0.5 : 1.0	This ratio is so chosen that the hydraulic mean radius (R) approximates to ½ full supply depths, i.e. best section, see chapter 3	As far as possible general ground slope.	This is fixed according to the type of soil through which the canal is passing and depth of flow, as under: Type of soil — For full supply depth maximum permissible velocity ft/s (m/s): Less than 2 (0.51) / 2 to 4 (0.61 to 1.22) / More than 4(1.22) Light loamy soil — 1.5 (0.46) / 1.75 (0.53) / 2 (0.61) Black cotton soil, kankar, Matasi soils and other such similar soils — 2.0 (0.61) / 2.50 (0.76) / 3.0 (0.91) Murum and other similar hard strata — 3.0(0.91) / 3.0 (0.99) / 2.5 (0.76)
8. India: (vii) Maharashtra: Unlined and lined canals are designed by Manning's formula. Canals are generally from reservoirs and are silt free.				
Unlined canals: N in all types of strata is adopted except rock is given as under: Design Discharge ft^3/s (m^3/s) — N Up to 20 (0.57) — 0.030 20 to 99 (0.57 to 2.82) — 0.0275 100 to 499 (2.83 to 14.15) — 0.025 >500 (14.16) — 0.0225 In rock-cutting N varies from 0.035 to 0.40 depending upon the roughness of the excavated	In design Z are assumed as 0.5 H. In actual construction, Z as given below is adopted.	For unlined canals the ratio is generally kept between 4 to 5, greater ratio being	The slope is fixed to obtain the desired permissible velocity. It is fixed by trial and	The V/V_c ratio is more indicative than being an essential requirement, since the canals in Maharashtra draw their supplies from storage reservoirs and hence are relatively silt free. This ratio is allowed to range between 0.9 to 1.1 though there are examples of canals having been designed with as low a ratio as 0.55.

Contd...

Table 6.4. Contd...

<table>
<tr><th>N</th><th>Z</th><th>X</th><th>S₀</th><th>V</th></tr>
<tr><td>(1)</td><td>(2)</td><td>(3)</td><td>(4)</td><td>(5)</td></tr>
<tr>
<td>surface. In extreme cases of cutting in hard rock, where the surface is rough with sufficient protrusions, a value of N = 0.050 is also adopted.
Lined canals:
N is based on type of surface as under:
<table>
<tr><th>Type of lining</th><th>N</th></tr>
<tr><td>1. Precast cement concrete (P.C.C.) lining and Shahabad lining</td><td>0.016 to 0.011</td></tr>
<tr><td>2. Uncoursed rubble masonry</td><td>0.020</td></tr>
<tr><td>3. Dry rubble pitching</td><td>0.0250</td></tr>
</table>
</td>
<td>
<table>
<tr><th>Soil</th><th>Z</th></tr>
<tr><td>Granular soil</td><td>1.5 : 1.0</td></tr>
<tr><td>Impervious clayey soils like black cotton soil</td><td>2.0 : 1.0</td></tr>
<tr><td>Loose sand and sandy loam</td><td>2.0 : 1.0</td></tr>
<tr><td>Silt loam compact sand and compact clay</td><td>1.5 : 1.0</td></tr>
<tr><td>Murum and friable soft rock</td><td>0.75 : 1.0 to 1.0 : 1.0</td></tr>
<tr><td>hard rock</td><td>0.2 : 1.0 to 0.5 : 1.0</td></tr>
</table>
</td>
<td>adopted when the discharge increases. Economic considerations, however, some times, dictate greater B/D ratio, even up to 10.</td>
<td>error keeping in view the factor of economy and to afford maximum irrigation. Generally slope in unlined canals is provided as 1 in 500 to 1 in 5,000 depending upon capacity of the canal and slope of the country.</td>
<td>Lined canals:
Maximum velocity permitted ranges between 4 to 8 ft/s (1.2 to 2.4 m/s)</td>
</tr>
</table>

Contd...

Table 6.4. Contd...

<table>
<tr><th>N</th><th>Z</th><th>X</th><th>S_0</th><th>V</th></tr>
<tr><td>(1)</td><td>(2)</td><td>(3)</td><td>(4)</td><td>(5)</td></tr>
<tr><td colspan="5">8. India (viii) Mysore (Karnataka):Both unlined and lined canals are designed by Manning's formula.</td></tr>
<tr>
<td>N is adopted according to type strata
<table>
<tr><td>Material</td><td>N</td></tr>
<tr><td>Normal sections in earthen cuttings</td><td>0.025</td></tr>
<tr><td>Hard murum and boulders</td><td>0.030</td></tr>
<tr><td>Soft rock are rugged hard rock cuttings</td><td>0.35 to 0.040</td></tr>
<tr><td>Lined canals</td><td>0.015 to 0.018</td></tr>
</table>
</td>
<td>In design Z = 0.5 H and in construction as under:
<table>
<tr><td>Type of material</td><td>Side slope Z</td></tr>
<tr><td>soil, soft murum</td><td>1.0:1.0 to 1.5:1.0</td></tr>
<tr><td>Hard murum</td><td>0.5:1.0 to 1.0:1.0</td></tr>
<tr><td>Hard rock</td><td>0.5:1.0 to 1.125:1.0</td></tr>
</table>
In lined canals, above Z is adopted in design also.</td>
<td></td>
<td>Adopted according to the topography as under for various projects:
<table>
<tr><td>Name of project</td><td>Longitudinal slope ft/mile (m/Km)</td></tr>
<tr><td>Ghataprabha</td><td>0.75(0.14) 3.0 or 1/7142</td></tr>
<tr><td>Tungabadra</td><td>0.05 (0.09) or 1/11111</td></tr>
<tr><td>Malaprabha</td><td>0.67 (0.13) or 1/7692</td></tr>
<tr><td>Bhadra</td><td>0.5 (0.09) for main canal or 1/11111
2.0 (0.38) for distributaries or (1/5263 to 1/2632)</td></tr>
<tr><td>Other projects</td><td>1.0 to 2.0 (0.19 to 0.38)</td></tr>
</table>
</td>
<td>Adopted according to nature of soil and full supply depth and generally as below:
<table>
<tr><td>Project</td><td>V ft/s (m/s)</td></tr>
<tr><td>Ghataprabha</td><td>2.5 to 8 (0.8 to 2.4)</td></tr>
<tr><td>Bhadra</td><td>3.0 to 5.5 (0.9 to 1.7)</td></tr>
<tr><td colspan="2">Kabini and other projects</td></tr>
<tr><td>Loose sand</td><td>1 to 1.5 (0.3 to 0.5)</td></tr>
<tr><td>Light sandy soil</td><td>1.5 to 2 (0.5 to .6)</td></tr>
<tr><td>Ordinary soil</td><td>2.0 to 3.0 (0.6 to 0.9)</td></tr>
<tr><td>Hard soil</td><td>3.0 to 4.0 (0.9 to 1.2)</td></tr>
<tr><td>Rocky soil</td><td>5 to 6
(1.5 to 1.8)</td></tr>
<tr><td>Lined - Tungabadra</td><td>4.5 (1.4)</td></tr>
</table>
</td>
</tr>
<tr><td colspan="5">8. India: (ix) Orissa: Most of the canals are unlined, near the coast. Canals were designed by Lacey's formula with f = 0.85 to 1.0. Due to difficulty in assumption of f, later on designed by Kennedy's formula, with values under:</td></tr>
<tr><td>Unlined canals:
N = 0.0225</td><td>Z = 0.5</td><td>Designed tables prepared X is taken from experience.</td><td></td><td>Kennedy's critical velocities and his tables and charts are followed:</td></tr>
</table>

Contd...

Table 6.4. Contd...

N	Z	X	S_0	V
(1)	(2)	(3)	(4)	(5)
8. India (x) Punjab: Most of the canals are from Bhakra reservoir, release water is silt free and due to slit excluders and ejectors.Main and Branch canals are lined, remaining are unlined. Bhakra canal Manual is followed. See Para 6.3.4.				
Unlined canals: Lacey formula is followed with f = 0.9 in head and 0.8 in tail. **Lined canals:** N = 0.018 actual value varies between 0.012 to 0.016 but to allow to weed growth 0.018 is adopted.	**Unlined canals:** Z = 0.5 H in design and in construction as under: Filling, Z = 1.5 H, Cutting, Z = 1.0 H Lined canals: Z = 1 for D < 12 ft. and 1.25 for > 12 ft.,provided angle of repose of soil is not flatter than 1.5. On Bhakra canal Z = 1.25 was adopted and subsequently changed to 1.5	Unlined canals: Lacey's chart are followed: Lined canals: Mebhoob section is followed;	Lacey's slope for f = 0.8 to 0.9 adopted. Lined canals: Steepest possible practical slope upto maximum velocity is adopted	Lacey's critical velocities and his tables and charts are followed: Lined canals: Maximum permissible velocity of 6 ft./s (1.83 m/s) are adopted.
8. India: (xi) Rajasthan, Unlined canals: (a) Bhakra Canal System: Above procedure is followed, being adjoining and extension is of same canals, with some modification in 'f' = 0.8 at the head and 0.7 at the tail. In sandy soil f = 0.9 to avoid sand deposition.				
	Unlined canals: Generally Z = 0.5 in cutting.			
Lined canals: Type of lining – Value of N Concrete lining – 0.016 Brick lining – 0.018	Z = 1.25	Mehboob section, see chapter 3		For discharge < 25 cusecs (0.71 cumecs) = 3 ft./s (0.91 m/s)

Contd...

Table 6.4. Contd...

N	Z	X	S_0	V
(1)	(2)	(3)	(4)	(5)
8. India: (xi) Rajasthan, Unlined canals: (b) Ganga Canal System For unlined canals - Kennedy's formula was adopted. All other data are same as in Bhakra Canal System. For lined canals – As for Bhakra Canal System. But Z = 1.0 upto a height of 2 ft (0.6 m) above bed level and Z = 1.5 further up.				
8. India: (xi) Rajasthan: (c) Chambal Project: Design of lined and unlined canals: Design of unlined and lined canals was treated silt free and deigned by Manning's formula.				
Adopted according to type of soil as under: Nature of soil | N 1. Ordinary soil and discharge more than 500 ft^3/s ($14.16 m^3/s$) | 0.020 2. Ordinary soil and discharge less than 500 ft^3/s ($14.16 m^3/s$) | 0.025 3. Unlined rock-cuts | 0.035 4. Lined section | 0.018 5. Plastered masonry | 0.017	In cutting, Z = 1.0 H In filling Z = 1.5 H		In order to bring maximum area in the command a flat gradient as adopted. Main canal | 1 in 12,500 Branch canal | 1 in 5,000 Distributaries | 1 in 2,500	Maximum velocity was assumed on the basis of material for prevention of erosion of banks. Minimum velocity is not allowed to be less than Lacey's formula with f= 0.8 so as to prevent silt deposition. Type of canal | Maximum velocity ft/s (m/s) Ordinary earthen section | 3.0 (0.91) Unlined rock-cut | 7.0 (2.13) Lined earthen section | 5.0 (1.52) Lined masonry section | 7.0 (2.13)

Contd...

Table 6.4. Contd...

<table>
<tr><th>N</th><th>Z</th><th>X</th><th>S_0</th><th>V</th></tr>
<tr><td>(1)</td><td>(2)</td><td>(3)</td><td>(4)</td><td>(5)</td></tr>
<tr><td colspan="5">8. India: (xi) Rajasthan: (d) Rajasthan Canal Project, now Indira Gandhi canal: Also see Para 6.6 and chapter 7B.</td></tr>
<tr><td>Unlined canals: smaller canals for discharges up to 300 cusecs (8.5 cumecs) were designed by Garret's Table, Kennedy's curves, and Chezy's formula: N = .0225
Higher discharge canals were designed with Lacey's curves.
f = 0.8 as the head and 0.6 at the tail.
Lined canals: designed by Manning's formula, N = .017 for tile lining</td><td>Unlined canals: only medium to low: in design Z = 0.5 H and in construction Z = 1.5 H
lined canals: For high discharge Z = 2.0 H and for medium to low Z = 1.5H</td><td>Unlined 2 to 10.
Lined canals: Mebhoob section is followed;</td><td>Flat so as not to loose command.</td><td></td></tr>
<tr><td colspan="5">8. India: (xi) Rajasthan: (e) (Minor Irrigation Projects): Most of the canals are small discharge canals. Both lined and unlined canals are generally designed by Manning's formula. Sometimes Kennedy's formula is also adopted.</td></tr>
<tr><td>Adopted according to strata
<table>
<tr><th>Nature of soil</th><th>Value of N</th></tr>
<tr><td>Ordinary or sandy soil</td><td>0.025</td></tr>
<tr><td>Hard soil</td><td>0.0225</td></tr>
<tr><td>Hard rock or stratified soft disintegrated rock</td><td>0.030</td></tr>
<tr><td>Concrete lining, or plastered masonry</td><td>0.018</td></tr>
</table></td>
<td><table>
<tr><th>Type of soil</th><th>Side slope (H : V)</th></tr>
<tr><td>Ordinary soil</td><td>1.25 : 1.0 to 1.5 to 1.0</td></tr>
<tr><td>Hard soil</td><td>1.0 to 1.0</td></tr>
<tr><td>Rocky strata</td><td>1.25 to 1.0 to 0.1 to 1.0 depending upon nature of strata</td></tr>
<tr><td>Lined canal</td><td>1.0 to 1.0 to 1.5 to 1.0</td></tr>
</table></td>
<td></td>
<td>Slope is fixed depending upon soil conditions and command consideration., as below
<table>
<tr><th>Nature of soil</th><th>Longitudinal slope</th></tr>
<tr><td>Ordinary soil</td><td>1/3000 to 1/5000</td></tr>
<tr><td>Hard soil</td><td>1/2500 to 1/5000</td></tr>
<tr><td>Hard rock</td><td>1/500</td></tr>
<tr><td>Soft rock</td><td>1/1000</td></tr>
<tr><td>Lined section</td><td>1/400 to 1/500</td></tr>
</table></td>
<td>It is based on the bank material to with stand the cover.
<table>
<tr><th>Type of soil</th><th>Maximum velocity ft/s (m/s)</th></tr>
<tr><td>1. Ordinary soil</td><td>1.5 to 2.0 (0.46 to 0.61)</td></tr>
<tr><td>2. Hard soil</td><td>2.5 (0.76)</td></tr>
<tr><td>3. Soft disintegrated rock</td><td>5.0 (1.52)</td></tr>
<tr><td>4. Hard rock</td><td>10.0 (3.05)</td></tr>
<tr><td>5. Cement concrete lining or masonry with plaster</td><td>3.7 to 4.99 (1.13 to 1.52)</td></tr>
</table></td></tr>
</table>

Contd...

Table 6.4. Contd...

<table>
<tr><th>N</th><th>Z</th><th>X</th><th>S₀</th><th>V</th></tr>
<tr><td>(1)</td><td>(2)</td><td>(3)</td><td>(4)</td><td>(5)</td></tr>
<tr><td colspan="5">8. India: (xii) Tamil Nadu: Unlined and lined canals are designed by Kutter's or Manning's formula.</td></tr>
<tr><td>Unlined canals:
N is adopted according to discharge as under:
<table><tr><td>Discharge</td><td>N</td></tr><tr><td>Upto 50 ft³/s (1.42 m³/s)</td><td>0.025</td></tr><tr><td>> 50 and <1000 > (1.42 and < 28.32) ft³/s (m³/s)</td><td>0.0225</td></tr><tr><td>> 1000 ft³/s (28.32 m³/s)</td><td>0.02</td></tr></table>Lined canal:
Generally concrete lining = .015 to .017</td><td>For design, Z = 0.5 H but for construction as blow:
<table><tr><td>Filling soil</td><td>1.5 : 1</td></tr><tr><td>Cutting</td><td>1.0 : 1</td></tr><tr><td>Rock</td><td>0.5 : 1</td></tr></table></td><td>1.25 to 5 according to discharge.
Lined canal
1.25 to 1.5</td><td>According to topography and command consideration.</td><td>For ordinary soil 3 ft/s (0.9 m/s) Kennedy's CVR 1.1 at head to 0.9 at tail. For lined canal 4 to 5 ft/s (1.2 to 1.5 m/s)</td></tr>
</table>

Contd...

Table 6.4. Contd...

<table>
<tr><th>N</th><th>Z</th><th>X</th><th>S_0</th><th>V</th></tr>
<tr><td>(1)</td><td>(2)</td><td>(3)</td><td>(4)</td><td>(5)</td></tr>
<tr><td colspan="5">8. India: (xiii) Uttar Pradesh: For unlined canals, there are 3 practices (i) In alluvial soil Kutter's and Kennedy's formula, or (ii) Lacey's formula and (iii) On Gandak Project Lindley's formula is used (see chapter 7)</td></tr>
<tr>
<td>N depend upon type of surface, discharge and condition. Generally values as below are adopted.
<table>
<tr><th>Q</th><th>N</th></tr>
<tr><td>Up to 500 ft³/s (14.16 m³/s)</td><td>0.0225</td></tr>
<tr><td>Above 500 ft³/s (14.16 m³/s)</td><td>0.020</td></tr>
</table></td>
<td>Section is designed for Z = 0.5 H, but constructed to = 1 to 2 as below:
<table>
<tr><th>Material</th><th>Angle of response</th><th>Z (H : V)</th></tr>
<tr><td>Firm clay well drained</td><td>45°</td><td>1.0 : 1.0</td></tr>
<tr><td>Clay loam, average sand loam</td><td>30° to 33°</td><td>1.33 : 1.0 to 1.5 : 1.0</td></tr>
<tr><td>Sandy soil or gravel</td><td>33° to 26°</td><td>1.5 : 1.0 to 2.0 : 1.0</td></tr>
</table></td>
<td></td>
<td>Assumed as Lacey's regime slope and the terrain slope. Generally between 1/8333 to 1/2630.</td>
<td>Kennedy's CVR is generally kept between 0.9 to 1.3</td>
</tr>
<tr><td colspan="5">Design of Lined canal: designed by Manning's formula</td></tr>
<tr>
<td>Assumed N is on the type of surface/lining as below:
<table>
<tr><th>Lining material</th><th colspan="4">N for channel condition</th></tr>
<tr><th></th><th>Perfect</th><th>Good</th><th>Fair</th><th>Poor</th></tr>
<tr><td>Brick in cement mortar</td><td>0.012</td><td>0.013 to 0.015</td><td>0.015 to 0.017</td><td>0.017 to 0.018</td></tr>
<tr><td>Cement plastered surface</td><td>0.011</td><td>0.012 to 0.013</td><td>0.013 to 0.015</td><td>0.015 to 0.017</td></tr>
<tr><td>Concrete lining</td><td>0.012</td><td>0.014</td><td>0.016</td><td>0.017</td></tr>
<tr><td>Precast concrete tiles</td><td>---</td><td>0.012</td><td>0.014</td><td>0.016</td></tr>
</table></td>
<td>Adopted on above according to type of surface</td>
<td></td>
<td>Adopted according to country slope, subject to maximum permissible velocity.</td>
<td>Maximum permissible V according to type of lining is as under:
Minimum V is also kept according to Lacey/Kennedy formula.
<table>
<tr><th>Material</th><th>Maximum permissible velocity ft/s (m/s)</th></tr>
<tr><td>Concrete tiles</td><td>7.0 (2.13)</td></tr>
<tr><td>Burnt clay tiles</td><td>6.0 (1.83)</td></tr>
</table></td>
</tr>
</table>

Contd...

Table 6.4. Contd...

<table>
<tr><th>N</th><th>Z</th><th>X</th><th>S_0</th><th>V</th></tr>
<tr><td>(1)</td><td>(2)</td><td>(3)</td><td>(4)</td><td>(5)</td></tr>
<tr><td colspan="5">8. India: (xiv) West Bengal: Unlined canals are designed by Kennedy's, chezy' s/ Kutter's formula</td></tr>
<tr>
<td>N is based on type of soil, rock and anticipation of weed growth as below
<table>
<tr><td>Discharge ft³/s (m³/s)</td><td>N</td></tr>
<tr><td>Above 500 (14.16)</td><td>0.020</td></tr>
<tr><td>500 to 250 (14.16 to 7.08)</td><td>0.0225</td></tr>
<tr><td>250 and below (7.08 and below)</td><td>0.025</td></tr>
</table></td>
<td>Z both in design and construction, are chosen depending upon the discharge and the type of section as given below
<table>
<tr><td rowspan="2">Type of section</td><td colspan="3">Z for Q ft³/s (m³/s)</td></tr>
<tr><td>More than 50 (1.42)</td><td>50 to 5 (1.42 to 0.14)</td><td>5 and less (0.15 and less)</td></tr>
<tr><td>Partial banking</td><td>2.0 : 1.0</td><td>2.0 : 1.0</td><td>1.0 : 1.0</td></tr>
<tr><td>Full banking</td><td>2.0 : 1.0</td><td>1.0 : 1.0</td><td>1.0 : 1.0</td></tr>
<tr><td>Full cutting</td><td>1.0 : 1.0</td><td>---</td><td>---</td></tr>
</table></td>
<td></td>
<td>Slope for main canals and branches is kept between 0.115 to i. e. 1/8695 and 0.200 per thousand to 1/5000 and for minors and distributaries between 1/6666 and 1/2000 per thousand depending upon the country slope so as to reduce the number and height of falls. The loss of command area and maximum permissible velocity are also other important considerations.</td>
<td>In the head reach the velocity is limited to 3.5 ft/s (1.07 m/s) and in the tail reach it is 3.0 ft/s (0.91 m/s).
Kennedy's Critical Velocity is assumed to vary from 1.1 at head to 0.9 at tail reach.</td>
</tr>
<tr><td colspan="5">Lined canal (are designed by Manning's formula)</td></tr>
<tr><td>N is generally taken as 0.016</td><td>Generally slope of 1.25 horizontal : 1.0 vertical is assumed.</td><td></td><td></td><td>Velocity of 5.0 to 6.5 ft/s (1.52 to 1.98 m/s) is assumed for concrete lining with smooth sand cement plaster.</td></tr>
</table>

Contd...

Table 6.4. Contd...

<table>
<tr><th>N</th><th>Z</th><th>X</th><th>S_0</th><th>V</th></tr>
<tr><td>(1)</td><td>(2)</td><td>(3)</td><td>(4)</td><td>(5)</td></tr>
<tr><td colspan="5">8. India: (xv) Central water and power commission, now CWC: Also see Para 6.3.5</td></tr>
<tr>
<td>Unlined canals – Taking off from a reservoir: clear water expected. Designed by Manning's and Kennedy's formula
<table>
<tr><td>Discharge ft³/s (m³/s)</td><td>N</td></tr>
<tr><td>Upto 5 (0.14)</td><td>0.030</td></tr>
<tr><td>5 to 50 (0.14 to 1.42)</td><td>0.025</td></tr>
<tr><td>50 to 500 (1.42 to 14.16)</td><td>0.0225</td></tr>
<tr><td>Above 500 (above 14.16)</td><td>0.020</td></tr>
</table>
Rocky cuts – value of n equal to 0.035 to 0.050 is assumed when rock portion at least 6 inches</td>
<td>Depend on the nature of soil in which the channel is to be excavated and on the depth of cutting or height of embankment.
<table>
<tr><td>Nature of soil</td><td>Cutting</td><td>Embankment</td></tr>
<tr><td>Hard clay or gravel</td><td>0.75 : 1.0</td><td>1.5 : 1.0</td></tr>
<tr><td>Soft clay and alluvial soils</td><td>1.0 : 1.0</td><td>2.0 : 1.0</td></tr>
<tr><td>Sandy loam</td><td>1.5 : 1.0</td><td>2.0 : 1.0</td></tr>
<tr><td>Light sand</td><td>2.0 : 1.0</td><td>2.0 : 1.0 to 3.0 : 1.0</td></tr>
<tr><td>Hard rock</td><td>0.125 : 1.0 to 0.25 : 1.0</td><td>---</td></tr>
<tr><td>Soft rock</td><td>0.25 : 1.0 to 0.5 : 1.0</td><td>---</td></tr>
</table></td>
<td>CVR is not a dominant factor and the channel may be designed for any velocity which the soil can stand without scouring.
The Kennedy's depth formula gives the maximum depth that can be allowed in an earthen channel.</td>
<td>It is fixed arbitrarily from experience that a steeper slope is required for a smaller discharge at the tail end which is gradually made flatter as the discharge increases. It may generally approximate the ground slope unless command considerations demand flatter slope. In the latter case slope is determined mainly by the command</td>
<td>Large value of V/V_C is taken at the head reach which decreases to 0.8 in the tail reach. Main canal taking off from river may have to be designed for V/V_C of 1.1 at head depending upon location of head and quantity and grade of silt entering the canal.</td>
</tr>
</table>

Contd...

Table 6.4. Contd...

N	Z	X	S_0	V
(1)	(2)	(3)	(4)	(5)
(1.52 cm) above the excavated bed level is left out in working out cross - sectional area. If no portion above bed level is left out, N equal to 0.050 to 0.080 is assumed.	The slopes given above are suitable for average height up to 10 ft (3.05 m). For greater heights it is preferable to provide ledges 5 ft (1.52 m) to 7.5 ft (2.29 m) wide at intervals of 10 ft (1.05 m) instead of flattening the slopes. The latter method is, however, slightly economical.	As for example, a channel situated in a soil which can stand mean scouring velocity of 3.5 ft/s cannot have full supply depth of more than 9.3 ft	levels required at the head of the off taking channels and also by the level required at the head of outlets. The slope is also controlled to some extent by CVR.	
Unlined Channels following in Alluvium: These are designed by using Lacey's formulae, see Para 7.13				
A fairly correct value of 'f' may be arrived at by averaging the values obtained from various equations: $f = 0.75\ V^2/R$ $f = 2.0\ V^{2/5}/A^{0.5}$ $f = 1.176\ D^{0.5}m$ D_m = Mean diameter of bed material	In design Z is generally assumed as 0.5 H : 1.0 V. But in construction slopes are adopted as given in above table.	Help of Lacey's diagram is taken.	Help of Lacey's diagram is taken.	Silt factor: A fairly correct value is arrived at by averaging the values obtained from various 4 equations given in table 7.9.

Contd...

Table 6.4. Contd...

N	Z	X	S_0	V
(1)	(2)	(3)	(4)	(5)
Lined canals: Generally designed by Manning's and Kennedy formula, also see Para 6.3.5				
N have not yet (1960) been established but following values are tentatively suggested: Pneumatically applied mortar 0.016 Asphalt 0.015 Concrete 0.014 Values for different type of surfaces generally adopted in India are Nature of surface / Good / Fair / Bad 1. Cement mortar surface: 0.012 / 0.013 / 0.015 2. Brickwork in cement mortar: 0.013 / 0.015 / 0.017 3. Concrete lining: 0.014 / 0.016 / 0.018 4. Stone work: 0.018 / 0.020 / 0.0225 For design purposes, the values given in the last column are generally adopted.	Generally of 1.0 H : 1.0 V for a radius depth less than 12 ft (3.66 m) and 1.25 horizontal : 1.0 vertical slope for more than 12 ft (3.66 m) is adopted.	For Q upto 60 m^3/s Mehboob section and for larger Q. X at the discretion of designs.		Maximum limit 2 m/s
9. Japan: Lined and unlined canals both are designed by Manning's formula				
N is based on the type of surface/material, as per table 6.4.1	Side slope is adopted according to side conditions.	Unlined canal:		

Contd...

Table 6.4. Contd...

<table>
<tr><th>N</th><th>Z</th><th>X</th><th>S_0</th><th>V</th></tr>
<tr><td>(1)</td><td>(2)</td><td>(3)</td><td>(4)</td><td>(5)</td></tr>
<tr>
<td></td>
<td>
Unlined Canals
<table>
<tr><td>S. No.</td><td>Soil Condition</td><td>Side slope H : V</td></tr>
<tr><td>(1)</td><td>Sound rock (farm)</td><td>0.25 : 1</td></tr>
<tr><td>(2)</td><td>Fissured weathered rock</td><td>0.5 : 1</td></tr>
<tr><td>(3)</td><td>Hardpan of gravelly clay soil</td><td>0.5 : 1 to 1.0 : 1</td></tr>
<tr><td>(4)</td><td>Stiff gravelly clay</td><td>1.0 : 1 to 1.5 : 1</td></tr>
<tr><td>(5)</td><td>Sandy clay loam</td><td>1.5 : 1 to 2.0 : 1</td></tr>
<tr><td>(6)</td><td>Gravelly sandy loam</td><td>2.0 : 1</td></tr>
<tr><td>(7)</td><td>Sandy loam loose clay</td><td>3.0 : 1</td></tr>
</table>
Lined canals
<table>
<tr><td>(1)</td><td>10 m³/s and above (Large-scale canal) (359 ft³/s and above)</td><td>1.5 : 1</td></tr>
<tr><td>(2)</td><td>2 m³/s to 10 m³/s (Medium - scale canal) (70.6 to 353 ft³/s)</td><td>0.25 : 1</td></tr>
<tr><td>(3)</td><td>2 m³/s and less (small – scale canal) (70.6 ft³/s and less)</td><td>1.0 : 1</td></tr>
</table>
Retaining wall type canals
<table>
<tr><td>(1)</td><td>Reinforced concrete flume</td><td>0 : 1</td></tr>
<tr><td>(2)</td><td>Cantilever wall</td><td>0 : 1</td></tr>
<tr><td>(3)</td><td>Buttress wall</td><td>0 : 1 to 1.0 : 1</td></tr>
<tr><td>(4)</td><td>Concrete gravity wall</td><td>0 : 1 to 0.5 : 1</td></tr>
<tr><td>(5)</td><td>Concrete block</td><td>0.3 : 1 to 1 : 1</td></tr>
<tr><td>(6)</td><td>Fabricated canal</td><td>0 : 1 to 1.0 : 1</td></tr>
</table>
</td>
<td>Adopted according to maximum permissible velocities and economics. Lined canals – standard values between 2 to 1 are adopted. Studies indicates that 80 % of lined canals has B/D = 1.4 to 3.3</td>
<td>Taken on the basis of topography, economics and permissible velocity.</td>
<td>
Unlined Canals
<table>
<tr><td>Kind of material</td><td>Velocity m/s (ft/s)</td></tr>
<tr><td>Fine sand</td><td>0.45 (1.48)</td></tr>
<tr><td>Sandy loam</td><td>0.60 (1.97)</td></tr>
<tr><td>Loam</td><td>0.70 (2.30)</td></tr>
<tr><td>Clayey with sand</td><td>0.90 (2.95)</td></tr>
<tr><td>Clay</td><td>1.00 (3.28)</td></tr>
<tr><td>Clay with sand</td><td>1.20 (3.94)</td></tr>
<tr><td>Soft rock</td><td>2.00 (6.56)</td></tr>
<tr><td>Medium sound rock</td><td>2.50 (8.20)</td></tr>
<tr><td>Sound rock</td><td>3.00 (9.84)</td></tr>
</table>
Lined canals
<table>
<tr><td>Kind of materials</td><td>Velocity m/s (ft/s)</td></tr>
<tr><td>Thick concrete</td><td>3.00 (9.84)</td></tr>
<tr><td>Thin concrete</td><td>1.50 (4.92)</td></tr>
<tr><td>Asphalt</td><td>1.00 (3.28)</td></tr>
<tr><td>Dry masonry block [stay length, 30 cm (11.8 inches) below]</td><td>1.50 (4.92)</td></tr>
<tr><td>Dry masonry block [stay length, 30 cm (11.8 inches) over]</td><td>2.00 (6.56)</td></tr>
<tr><td>Dry masonry block</td><td>2.50 (8.20)</td></tr>
</table>
</td>
</tr>
</table>

Contd...

Table 6.4. Contd...

N	Z	X	S_0	V
(1)	(2)	(3)	(4)	(5)
10. Malaysia: Most of the canals are unlined, only few canals of 4 to 10 cusecs (0.11 to 0.3) are lined. Canals are designed by Manning's formula.				
Assumed on the basis of experience a depth of flow.	Z = 1.5 H to 3 H depending upon type of soil.		Slope is fixed so as to have a working head of 0.3 m in the command area.	Following permissible velocities are adopted. Sandy loam: 2 ft/s (0.61 m/s) Clay loam: 3 ft/s (0.91 m/s) Stiff loam: 4 ft/s (1.22 m/s)
11. New Zealand: Generally USSBR practice is followed.				
12. Pakistan: Unlined canals are designed by Lacey's formula (Para 7.13). Lined canal carrying silt laden water is designed by combination of Lacey's and Manning's formula. Designs charts are followed, along with CHOP (Canal and head works data observation programme, 1962 - 63) Also see figure 6.14.				
13. Poland: Generally silt free water, no canal fluctuations, Manning's formula adopted.				
14. Portugal: Generally lined canals, Manning's formula is followed, for concrete N = 0.014, Z = 1.0 H, S = 1/10000 to 1/5000, R = D/2 (best section)				
15. Rhodesia: Mostly lined canals. Generally designed by Manning's formula.				
N = .016 and also on algae formation.	Z = 1.25 and steeper in rocks	According to most economical section		USBR standards are followed.
16. Spain: Mostly lined canals. Generally designed by Bazin's or Manning's formula. Mostly circular sections are adopted. In rocky semi circular section is used. N is based on maintenance programme and varies from .014 to 0.02				
17. Turkey: Lined and unlined canals are designed by Manning's formula. Unlined canals are also checked for regime theories. USBR practice is followed				

Contd...

Table 6.4. Contd...

N	Z	X	S_0, X and V
(1)	(2)	(3)	(4)

18. U.A.R: Unlined canals are designed by Hassib El-Difrawy, Molseworth and Yenidonia which were evolved on the basis of data of stable canals in Egypt. Lined canals are designed as most economical section i.e. R = D/2

(1)

Type of canal	N
Canals in very good condition	0.020 to 0.022
Ordinary Canals	0.0225
Canals affected by weed growth	0.0286 to 0.033

(2) Adopted according to type of soil as below:

Type of soil	Z = H : IV
Clayey soil	1.0 : 1.0
Sandy soil	1.5 to 2.0 : 1.0

(3) Adopted according to the Molseworth and Yenidonia as under

$$y = \left(0.1\frac{S_0}{2} + 4\right)\sqrt{B}$$

When water depth is more than 1.62 m (5.31 ft), and

$$y = \frac{(S+8)^2 B}{650}$$

When water depth is less than 1.62 m (5.31 ft), where S is in cm/Km.

(4) S is taken from Prevalent diagram

Permissible velocity is taken from Eq.

V = (0.0033 D + 0.0042) /N

Typical values are given in table below.

B	D	B/D	S		Velocity
m (ft)	m (ft)		Cm/Km	ft/mile	V
1.0 (3.28)	0.70 (2.30)	1.43	15 (0.79)	1/6666	0.26 (0.85)
1.6 (5.25)	0.90 (2.95)	1.78	11 (0.58)	1/9090	0.28 (0.92)
2.5 (8.20)	1.20 (3.94)	2.08	10 (0.53)	1/10000	0.32 (1.05)
3.0 (9.84)	1.40 (4.59)	2.14	9 (0.48)	1/1111	0.35 (1.15)
4.0 (13.12)	1.60 (5.25)	2.50	8 (0.42)	1/1250	0.37 (1.21)
5.0 (16.40)	5.80 (5.91)	2.78	8 (0.42)	1/1250	0.40 (1.31)
7.0 (22.97)	2.00 (6.56)	3.50	7 (0.37)	1/1428	0.43 (1.41)
10.0 (32.81)	2.50 (8.20)	4.00	7 (0.37)	1/1428	0.50 (1.64)
18.0 (59.06)	3.00 (9.84)	6.00	6 (0.32)	1/1666	0.56 (1.84)
25.0 (82.02)	3.50 (11.48)	7.14	6 (0.32)	1/1666	0.63 (2.07)
33.0 (108.27)	4.00 (13.12)	8.25	6 (0.32)	1/1666	0.69 (2.26)
50.0 (164.04)	5.00 (16.40)	10.00	6 (0.32)	1/1666	0.82 (2.69)

Contd...

Table 6.4. Contd...

N	Z	X	S_0	V
(1)	(2)	(3)	(4)	(5)
19. U.S.A.: Lined/unlined canals are designed with the help of Manning's formula. In unlined canals non silting and non scouring velocity is estimated by modified Kennedy's formula, See Para 6.3.3. Also see figure 6.18 for All American canal.				
Unlined canals: Discharge ft^3/s (m^3/s) — N Up to 100 (2.83) — 0.025 More than 100 (2.83) — 0.0225 or 0.020 depending upon type of material along the canal. For uniform cannel section covered with sand and gravel, the Manning's 'N' is determined by the Strickler equation, $N = 0.0342 D_{50}^{1/6}$ Where D_{50} equals the size in feet for which 50 per cent of bed material by weight is finer.	Z = 1.5 to 2 are adopted.	For small canals X = 2, For large canals, X = upto 8.		Maximum permissible velocity is determined by modified Kennedy's formula, see Para 6.3.3 Generally velocities of 1.0 ft/s to 3.5 ft/s (0.3 m/s to 1.0 m/s) are permitted depending upon the material and size of the canal. Smaller the canal lesser the velocity permitted. Tractive force (τ) limits are also sometimes used in conjunction with the allowable velocities. Allowable maximum τ is generally assumed between 0.05 to 0.07 lb/ft^2 (0.24 and 0.34 kg/m^2). Figures 6.4 (ii), (iii) gives relation between τ and plasticity index and maximum τ against ratio of B/D.

Contd...

Table 6.4. Contd...

<table>
<tr><th colspan="3">N</th><th>Z</th><th>X</th><th>S_0</th><th>V</th></tr>
<tr><td colspan="3">(1)</td><td>(2)</td><td>(3)</td><td>(4)</td><td>(5)</td></tr>
<tr><td colspan="3">Lined canals: These are divided into three groups – hard surface, buried membrane and earth linings. Hard surface linings include cement concrete, shotcrete, asphalt concrete, exposed prefabricated asphalt, brick, stone linings; precast concrete linings exposed plastic, soil-cement linings.</td><td rowspan="14"></td><td rowspan="14">B/D ratio varies from 1 for small canals to about 3 for large canals. In Burried membrane this ratio is 4.</td><td rowspan="14"></td><td rowspan="14">Maximum velocity permitted 8 ft/s (2.5 m/s). Mathematical check using N of 0.003 less than design N is applied to make certain that depth does not approach critical depth y_c.
Maximum velocity permitted in buried-membrane lined canals is about 2/3 of that permitted in unlined earth canals in similar material.
The permissible velocities in earth-lined canals vary with the type of lining and material and usually range from 1 to 4 ft/s (0.03 to 1.2 m/s).</td></tr>
<tr><td>Type of lining</td><td>R</td><td>N</td></tr>
<tr><td colspan="3">1. Straight and clean reach</td></tr>
<tr><td rowspan="2">(a) cement concrete</td><td>< 4 ft (1.22 m)</td><td>0.014</td></tr>
<tr><td>> 4 ft (1.22 m)</td><td>Larger than 0.014, to be determined from the equivalent sand grain roughness of 0.005 ft in the Colebrook-White equation</td></tr>
<tr><td>(b) Shot Crete (smoothed with steel-edged screed and rebound removed</td><td>Approaching to 20 ft (6 m)</td><td>0.017</td></tr>
<tr><td>(c) Shot Crete (average)</td><td>---</td><td>0.016</td></tr>
<tr><td>(d) Asphalted concrete (machine placed)</td><td>---</td><td>0.017</td></tr>
<tr><td>(e) Exposed prefabricated asphalt</td><td>---</td><td>0.014</td></tr>
<tr><td>(f) Soil cement</td><td>---</td><td>0.015 or 0.016</td></tr>
<tr><td colspan="3">2. Sinuous reach and aquatic growth</td></tr>
<tr><td>(g) All above linings</td><td>---</td><td>More than above values</td></tr>
<tr><td colspan="3">3. For discharge less than 100 ft³/s (2.83 m³/s)</td></tr>
<tr><td>(h) Buried membrane and compacted earth lining</td><td>---</td><td>0.025</td></tr>
</table>

Contd...

Table 6.4. Contd...

<table>
<tr><th>N</th><th>Z</th><th>X</th><th>S_0</th><th>V</th></tr>
<tr><td>(1)</td><td>(2)</td><td>(3)</td><td>(4)</td><td>(5)</td></tr>
<tr><td colspan="5">20. U.S.S.R: Unlined/lined channels are designed in accordance to all union Construction Standards and Regulations and The Norms and Technical Conditions of Irrigation System Design. Canal cross sections for discharge less than 10 m^3/s (353 ft^3/s) are chosen from standard sections without design calculations. For higher discharges, the section and bed gradient are found using the Chezy's formula. The velocity worked out by the Chezy's formula is also checked up to verify that it is non - scouring and non - silting.</td></tr>
<tr><td>Unlined canals:
N is taken according to discharge and type of material/surface as given in table 6.4.1 (ii).</td><td>In cutting:
Z is decided according to geology/hydrogeology
For D < 5 m. Z is adopted according to material as under:
<table>
<tr><th>Material</th><th colspan="2">Z (H : 1.0 V)</th></tr>
<tr><td></td><td>Under water surface</td><td>Above water surface</td></tr>
<tr><td>Non-weathered rock</td><td>0.1</td><td>0.0</td></tr>
<tr><td>Dusty sand</td><td>3.5</td><td>2.5</td></tr>
<tr><td>Pebbles, gravel clay and heavy loam</td><td>1.0 to 1.5</td><td>0.5 to 1.0</td></tr>
<tr><td>Coarse and medium sands, light loam and sandy loam</td><td>1.25 to 2.25</td><td>1.0 to 1.5</td></tr>
</table>
For D > 5 m. Z is adopted according to material.</td><td></td><td></td><td>In the head reach 20% higher than the silting velocity and lower than the scouring velocity is provided. Design velocity is obtained by combination of R and S.
Non scoring velocity: Approximately computed by the formula
$V_C = CQ^{0.2}$, C = 033 to .55 depending on the fall velocity of sediment. V_C is not allowed to be < 0.3 m/s except for clear water where V_C = 0.2 m/s. It is adopted on the basis of sediment size and flow depth, as given in table 6.4.1 (i) and (ii).</td></tr>
</table>

Contd...

Table 6.4. Contd...

<table>
<tr><th>N</th><th>Z</th><th>X</th><th>S_0</th><th>V</th></tr>
<tr><td>(1)</td><td>(2)</td><td>(3)</td><td>(4)</td><td>(5)</td></tr>
<tr>
<td></td>
<td>
<table>
<tr><td>Soil type</td><td colspan="4">Z according to discharge</td></tr>
<tr><td></td><td>> 10 m^3/s</td><td>< 10 > 2 m^3/s</td><td>< 2 > 0.5 m^3/s</td><td>< 0.5 m^3/s</td></tr>
<tr><td>Clay, heavy and medium loam</td><td>1.25</td><td>1.0</td><td>1.0</td><td>1.0</td></tr>
<tr><td>Light loam</td><td>1.5</td><td>1.25</td><td>1.25</td><td>1.0</td></tr>
<tr><td>Sandy loam</td><td>1.75</td><td>1.5</td><td>1.50</td><td>1.25</td></tr>
<tr><td>sand</td><td>2.25</td><td>2.0</td><td>1.75</td><td>1.5</td></tr>
</table>
In filling, Z is adopted according to discharge as under:
<table>
<tr><td>Lining</td><td>Z (H : 1.0 V)</td></tr>
<tr><td>Concrete and reinforced concrete</td><td>≥ 1.0</td></tr>
<tr><td>Clay, loam, turf, laminated turf, watertight screens</td><td>≥ 1.25</td></tr>
<tr><td>Rip – rap, gravel – fill, asphalt lining</td><td>≥ 1.50</td></tr>
</table>
</td>
<td>In preliminary design of canal cross section, depth 'D' is estimated by the formula**.
$D = C_1.(Q)^{1/3}$
Wherein C_1 is a coefficient varying between 0.7 and 1.0 with an average value of 0.85 and Q is the discharge in m^3/s.The ratio B/D of the canal bed with 'B' to water depth 'D' is kept in accordance with $B/D = 3(Q)^{1/4} - Z$ wherein Z is side slope.
The canal cross section, which gives maximum discharge capacity, is found by
$B/D = 2\sqrt{(1+Z)} - 2Z$ The canal cross section which causes minimum water seepage is given by Kostiakov's formula
$\frac{B}{D} = \frac{2C_2\sqrt{(1+Z^2)} - 2Z}{1 - C_3}$
Lined canals:
Typical standard designs have been worked out of assembly lining made of non - prestressed and prestressed reinforced concrete slabs as well as cast - in - situ concrete and reinforced concrete lining in single and double layers. These designs are guided by the requirement of doing work by machines for level leveling the foundation under lining, for concreting and for cutting the joints.</td>
<td></td>
<td>Verification of canal velocity being non – scouring is made for discharge higher than normal, Q_{norm} i.e., normal canal discharge, as given below:
$Q = 1.2 + 1.3\ Q_{norm}$: when $Q_{norm} < 1\ m^3/s$ (35.3 ft^3/s)
$Q = 1.15 + 1.20\ Q_{norm}$: when $Q_{norm} < 1 < 10\ m^3/s$ (35.3 – 353 ft^3/s)
$Q = 1.10 + 11.5\ Q_{norm}$: when $Q_{norm} < 10\ m^3/s$ (353 ft^4/s)
The channels of the interfarm irrigation network as well as distributing channels working in turn at maximum discharge are designed for Q_{norm}
The working of the canals at low supplies is checked for prevention of silting, efficiency of cross regulators, etc., on basis of Q_{min} i.e., minimum discharge. For large canals with a discharge mopre than 50 m^3/s (1,765 ft^3/s), the admissible non - scouring velocity is determined by special investigators</td>
</tr>
</table>

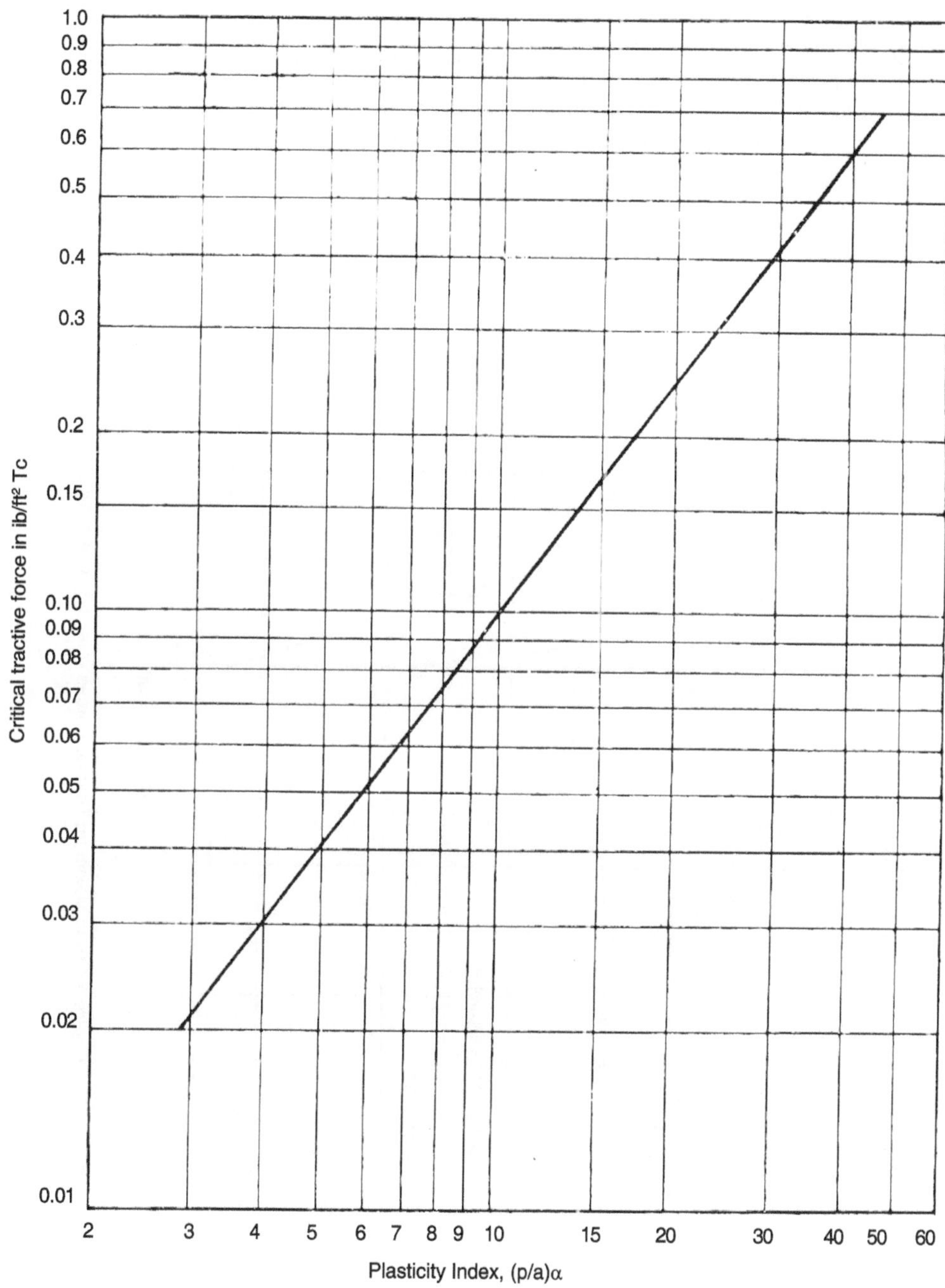

Figure 6.4 (ii) : Critical Tractive Force 'τ_0' for channel design, USA

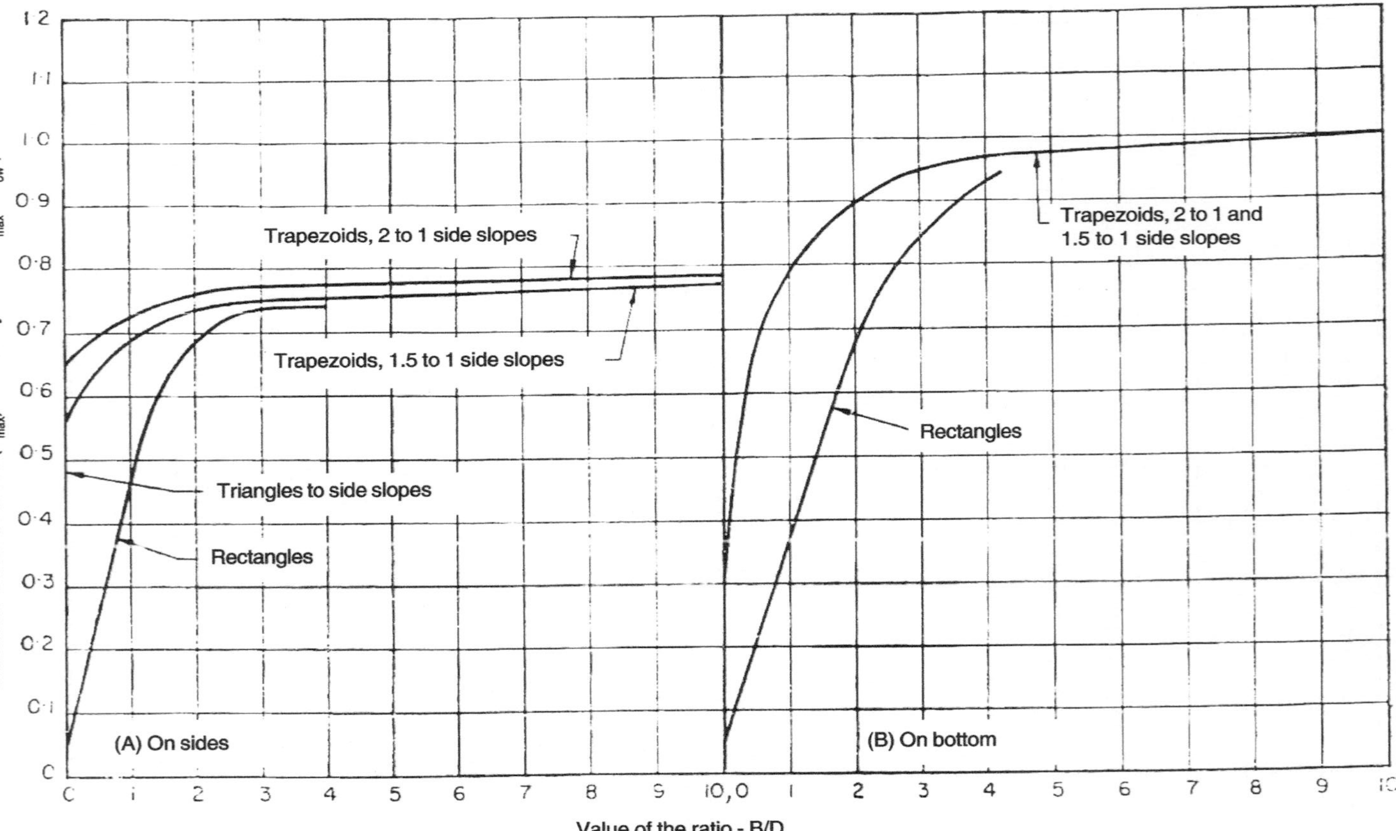

Figure 6.4 (iii) : Maximum Tractive Forces versus B/D ratio in a channel, USA

Table 6.4.1 (i) : Coefficient of roughness 'N', Japan

S. No.	Canal materials and its condition	Coefficient of roughness		
		Minimum values	Normal values	Maximum values
Unlined canal				
1.	Earth, straight and uniform			
	(a) Clean, recently completed	0.016	0.018	0.020
	(b) Clean after weathering	0.018	0.022	0.025
	(c) Gravel, clean	0.022	0.025	0.030
	(d) With short grass, few weeds	0.022	0.027	0.033
2.	Earth, winding and sluggish			
	(a) No vegetation	0.023	0.025	0.030
	(b) Grass, some weeds	0.025	0.030	0.033
	(c) Dense weeds or aquatic plants	0.030	0.035	0.040
	(d) Earth bottom and rubble sides	0.028	0.030	0.035
	(e) Stony bottom and clean sides	0.025	0.035	0.040
	(f)Cobble bottom and clean sides	0.030	0.040	0.050
3.	Dragline - excavated or dredged			
	(a) No vegetation	0.025	0.028	0.033
	(b) Light brush on banks	0.030	0.050	0.060
4.	Rock - cut			
	(a) Smooth and uniform	0.025	0.035	0.040
	(b) Jagged and irregular	0.035	0.040	0.050
Lined canal				
1.	Smooth steel surface (unpainted)	0.011	0.012	0.014
2.	Smooth steel surface (painted)	0.012	0.013	0.017
3.	Steel plate (corrugated)	0.021	0.025	0.030
4.	Cement (motor)	0.011	0.013	0.015
5.	Concrete; travelling type steel from (trowel finis h)	0.013	0.014	0.015
	(a) Fixing type steel form	0.012	0.014	0.015
	(b) Steel made panel type form	0.012	0.015	0.016
	(c) Smooth wood form	0.012	0.015	0.016
	(d) Rough wood form	0.015	0.017	0.020
	(e) Gunite (good section)	0.016	0.019	0.023
	(f) Gunite (wavy section)	0.018	0.022	0.025
	(g) Ready - made flume	0.012	0.014	0.016
6.	Concrete block (smooth surface)	0.014	0.016	0.017
7.	Concrete block (irregular surface)	0.015	0.017	0.018
8.	Earth Lining	—	0.025	—
9.	Asphalt (smooth)	—	0.014	—
10.	Aspha lt (rough)	—	0.017	—
11.	Masonry (dry rubble)	0.023	0.032	0.035
12.	Vegetative lining (sodding)	0.030	0.040	0.050

Table 6.4.1 (ii) : Admissible non — scouring Mean flow velocities (m/s), USSR.

Table 6.4.1 (ii) (a) : For sandy and coarse fragmented ground

Average diameter of ground particles, mm ($D_{s(pt)}$)	At average flow depth, m			
	0.5	1	3	5
	Non - scoring velocity m/s			
0.25	0.37	0.39	0.41	0.45
0.37	0.38	0.41	0.46	0.48
0.50	0.41	0.44	0.50	0.52
0.75	0.47	0.51	0.57	0.59
1.00	0.51	0.55	0.62	0.65
2.00	0.64	0.70	0.79	0.83
2.50	0.69	0.75	0.86	0.90
3.00	0.73	0.80	0.91	0.96
5.00	0.87	0.96	0.10	1.17
10.00	1.10	0.23	1.42	1.51
15.00	1.26	1.42	1.65	1.76
20.00	1.37	1.55	1.84	1.96
25.00	1.48	1.65	1.98	2.12
30.0	1.56	1.76	2.10	2.26
40.00	1.68	1.93	2.32	2.50
75.00	2.01	2.35	2.89	3.14
100.00	2.15	2.54	3.14	3.46
150.00	2.35	2.84	3.62	3.96
200.00	2.47	3.03	3.92	4.31
300.00	2.90	3.32	4.40	4.94

Table 6.4.1 (ii) (b) : For clay ground

Design specific grou nd cohesion* (Coh_{dgn})spec	At average flow depth, m			
	0.5	1	3	5
	Non - scoring velocity m/s			
0.005	0.39	0.43	0.49	0.52
0.010	0.44	0.48	0.55	0.58
0.020	0.52	0.57	0.65	0.69
0.030	0.59	0.64	0.74	0.78
0.040	0.65	0.72	0.81	0.86
0.050	0.71	0.77	0.89	0.98
0.075	0.83	0.91	1.04	1.10
0.100	0.96	1.04	1.20	1.27
0.125	1.03	1.13	1.30	1.37
0.150	1.13	1.23	1.41	1.49
0.175	1.21	1.33	1.52	1.60
0.200	1.28	1.40	1.60	1.69
0.225	1.36	1.48	1.70	1.80
0.250	1.42	1.55	1.78	1.88
0.300	1.54	1.69	1.94	2.04
0.350	1.67	1.83	2.09	2.21
0.400	1.79	1.96	2.25	2.38
0.450	1.88	2.06	2.35	2.49
0.500	1.99	2.17	2.50	2.63
0.600	2.16	2.38	2.72	2.88

Note: Average diameter of particles of the given ground is taken from the formula:

$$\overline{D}s(pt) = \frac{\sum D_{s(pt)_1} \cdot \%W_{s(pt)_1}}{\sum \%W_{s(pt)_1}}^{*}$$

Where, $D_{s(pt)_1}$ and $\%W_{s(pt)_1}$ are respectively average size of particles and percentage content of each fraction by weight.

Table 6.4.1 (iii) : Admissible nonscouring Mean Flow velocities (m/s) for stabilized channels (Protection on earth), USSR.

Protection	Design mark of concrete of grout with regard to compressive strength	At flow depth, m			
		0.5	1	3	5
		Velocity m/s			
(1)	(2)	(3)	(4)	(5)	(6)
(i) Concrete lining (no sandy and pebbly sediments in the streamflow)	50	9.6	10.6	12.3	13.0
	75	11.2	12.4	14.3	15.2
	100	12.5	13.8	16.0	17.0
	150	14.0	15.6	18.0	19.1
	200	15.6	17.3	20.0	21.2
	300	19.2	21.2	24.6	26.1
(ii) Stone protection (no sandy and pebbly sediments in the streamflow)	150 – 50	7.4	8.7	10.7	11.6
	25	6.3	7.4	9.1	9.8
	10	4.3	5.0	6.2	6.7
(iii) Gabions (0.5 × 0.5 × 1 m and more)	—	4.7	5.5	6.8	7.3
(iv) Rip – rap in netting	—	3.0	3.5	4.0	4.4
(v) Single – layer protection on rubble or clay layer (10 – 15 cm) with a cover of clay, silt, straw or hay : on made compacted ground with the size of stones					
(a) 15 – 20 cm	—	2.4	2.8	3.5	3.8
(b) 20 – 30 cm	—	2.8	3.3	4.0	4.4
(vi) One settled or thoroughly compacted ground with the size of stones					
(a) 15 – 20 cm	—	2.6	3.0	3.7	4.0
(b) 20 – 30 cm	—	3.6	3.6	4.5	4.9
(vii) Double – layer protection on rubble layer with the size of stones					
(a) 15 – 20 cm	—	3.0	3.5	4.3	4.7
(b) 20 – 30 cm	—	3.1	3.7	4.7	5.1

Table 6.4.1 (iv) : Values of Roughness factor 'N' for channels, USSR

Characteristics of canals	N	
	Irrigation	Catchment - out - fall (Drainage channel)
(1)	(2)	(3)
I. In earth channels		
(i) Q > 25 m³/s		
(a) In cohesive and sandy ground	0.0200	0.0225
(b) In gravelly –pebbly ground	0.0225	0.0250
(ii) Q = 1 to 25 m³/s		
(a) In cohesive and sandy ground	0.0225	0.0250
(b) In gravelly –pebbly ground	0.0250	0.0275
(iii) Q < 1 m³/s	0.0250	0.0275
(iv) Channels of permanent network of periodical action	0.0275	—
(v) Temporary irrigation ditch	0.030	—
II. In rock base		
(i) Thoroughly treated ground surface	0.020 to 0.025	
(ii) Satisfactorily treated ground surface without benches	0.030 to 0.035	
(iii) Satisfactorily treated ground surface, with benches	0.040 to 0.045	
III. Lined channel		
(i) Concrete lining (well finished)	0.013	0.014
(ii) Concrete lining (rough)	0.015	0.017
(iii) Prefabricated R.C. flumes Cobble stone	0.012	0.014
(iv) Board lining	0.013	0.017
(v) Brickwork	0.013	0.013
(vi) Rubble masonry on cement mortar	0.017	0.03
(vii) Asphalt - concrete lining	0.013	0.16

(v) It may be preferable to increase B / D ratio than to increase Z.

(vi) Kennedy's velocity is perhaps non scouring and non silting, also very near to maximum velocity for the most optimum (or most economical) section, see Para 7.15.

(vii) USSR practice lays stress both on non - scouring and non silting velocity, and appears much better.

(viii) For silt laden water, or where silt concentration changes considerably in a year (typically in canals off taking from weirs), hard surface protection of stones, gravels appear a good choice.

6.4 Lining of Existing Earthen Channels

In order to reduce seepage losses, (for losses see Para 9.15.2) and many other advantages being claimed, (though actually may not be available, see Para 9.15, also long term-evaluation normally not available), linings are being done/ proposed in the existing channels. Generally, there are following four schools of thoughts for lining the existing earthen channels:

(i) Modify / reduce the existing canal cross section so as to adopt an optimised lining section for minimum perimeter. This will reduce the cost of lining, but increases cost of earth work and few more aspects of structures as discussed in subsequent Para.

(ii) To provide the lining in the existing section without much altering or reducing the section. Only slight modifications of the section is carried out to remove the irregularities (uneven ness), weeds and suite to lined section, but by and large the existing section is maintained, see figure 6.5(i)

Besides above, two thoughts, there are a 3rd and 4th thought also. These are

(iii) To construct a parallel canal with one bank common. The existing channel is abandoned or used as a drain on one side. This has been adopted in some cases when the canal is perennial and no time is available for lining. Also all the parameters of canal i.e., discharge, full supply levels, bed slopes, side slopes and the entire section has been changed considerably, which may not fit in the existing/re-modelling of section, see figure 6.5 (ii)and (iii)

(iv) The parallel canal may be little away also, with out any common bank.

In the (i), with the optimisation of the perimeter, either the Mehboob section or best trapezoidal or trapezoidal with curved ends or any other economical section is adopted. The section is adjusted on one side, making use of one bank. Excavation of deep section is utilized to make or fill the gap of other side. The lining of such section may be same throughout its perimeter, but if the bed width is more, than an economical lining in the bed is generally provided.

Besides above, there are many other practical problems of adjustments of structures, such as bridges, Head Regulators, Cross-Regulators, Escapes cross drainage works and canal outlets. Larger the discharge and size of structures, larger are the complexities of adjustments. Besides it, the number of structures is also important. A typical case of re-modelling of Mahi canal is discussed in Para 6.8.

Still there is a third problem of working time, but only in perennial canals. Lining in such canals may be done only in some closures of limited period. This depends upon the magnitude of discharge, area being served, crops being grown, and other requirements of water say hydropower, drinking water etc.

Many-times the problem of availability of construction time is resolved by making suitable temporary arrangement of water diversion and isolating reaches for construction. This practice was adopted in lining of constructed earthen channels of IGNP. But this is very costly. In this approach two cross bund are made at suitable distances across the existing channel than water is diverted by making a loop canal towards left or right side depending upon easiness, availability of land and soil, and restored in the original channel after some distance. When lining is completed cross bunds are removed. Thus overall cost becomes very heavy.

In the (ii) concept, generally the depth of the existing earthen canal is low and bed width is more. Therefore, generally a cheaper lining is adopted for the bed. Normally the cost of remodelling of channel section i.e. earth work is less but difficult to achieve, see figure 6.5 (i). A typical section of lining of Borsad Branch (Gujarat) is shown in figure 6.5(iv). The problem of construction time remains same as above.

Some times for low depth and wide channels a rectangular section is also preferred, which is easy in construction, requiring little or no earth shaping and less time of construction.

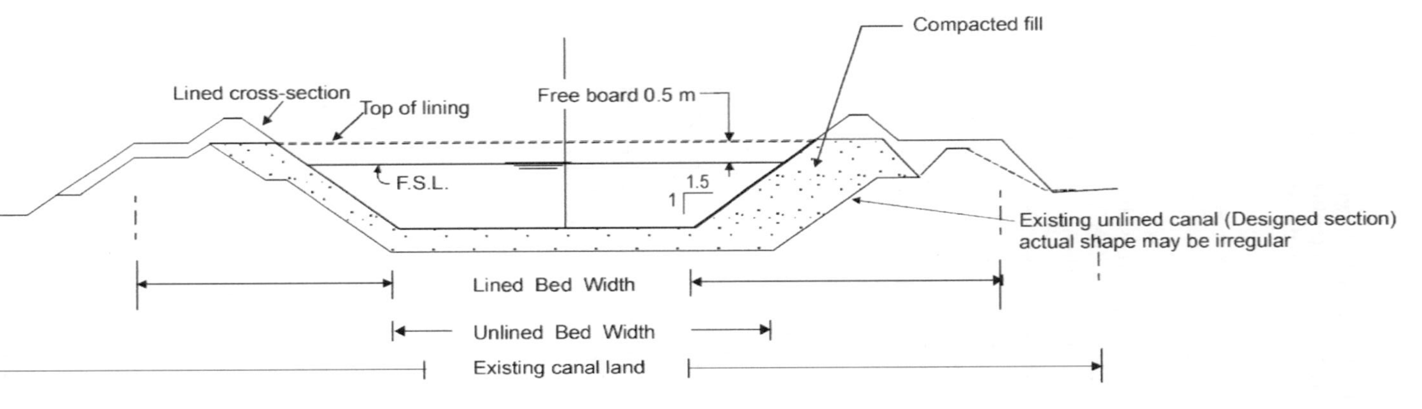

Figure 6.5 (i) : Lining within the existing Earthen Channel Section.

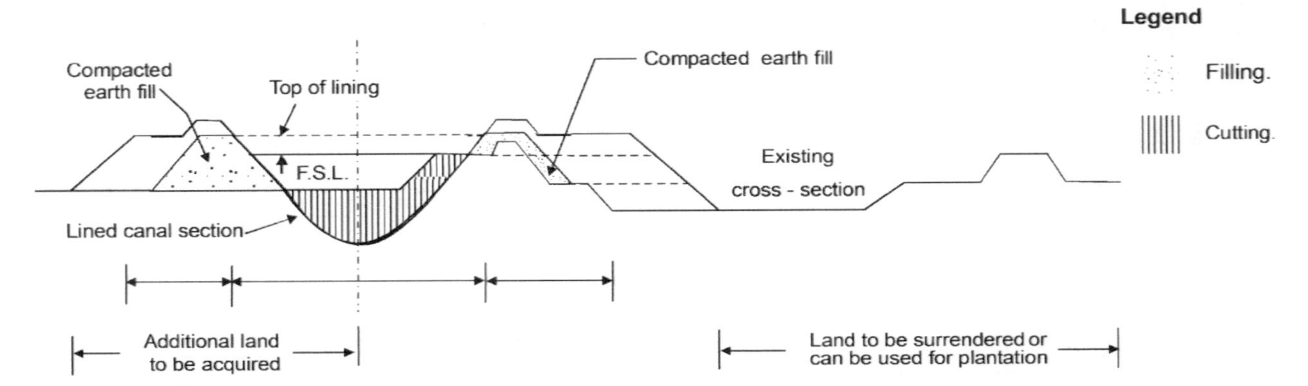

Figure 6.5 (ii) :Construction of Parallel Lined Canal

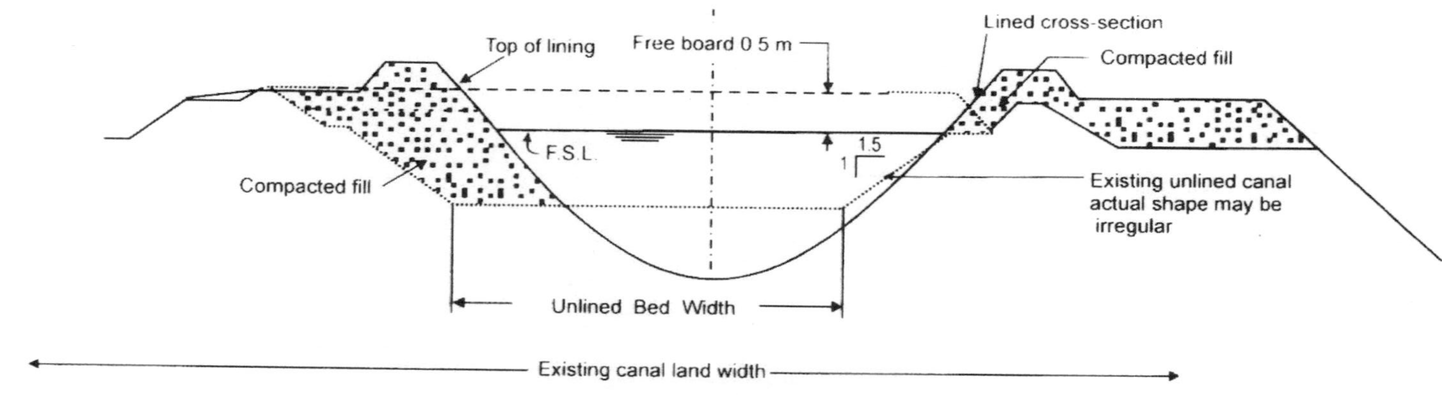

(iii) :Fitting of Mehboob Section inside the Existing Earthen Channel

Notes :

(i) Mehboob section can be fitted on either side depending upon minimum earthwork. However longtudinal curves should be avoided.

(ii) On one side there will be cutting and other side cutting is utilized in filling or on bank.

(iii) Alternatively best trapezoidal section can be adopted in place of Mehboob section. Both are practically identical

Figure 6.5 : Lining of Existing Earthen Channel

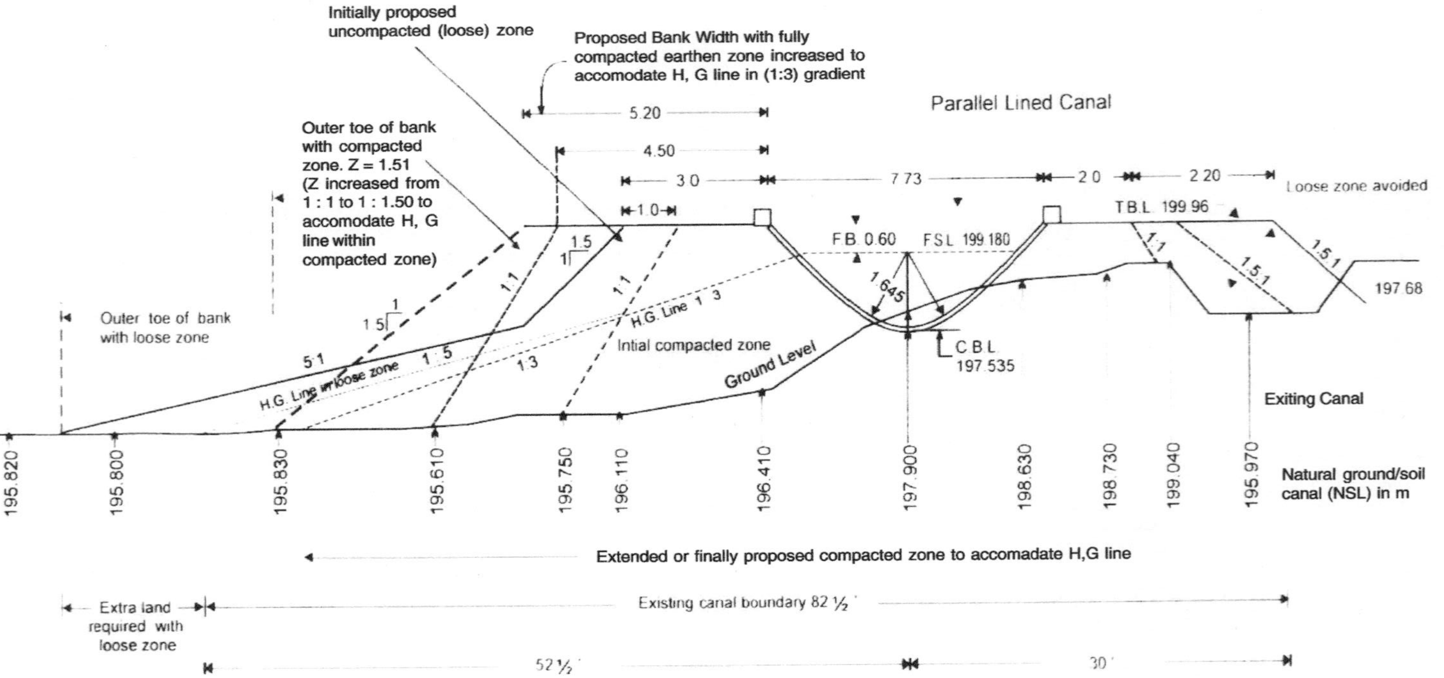

Notes :

1. Because of heavy filling, parallel canal section is costly
2. With fully compacted zone over all cost of earth work and land decreases

Figure 6.5 (iv) :Construction of common bank Parallel Lined Canal X-Sec. at Km. 2.500 of Jasana distributory of Noha Project, Hanumangarh, Rajasthan, India

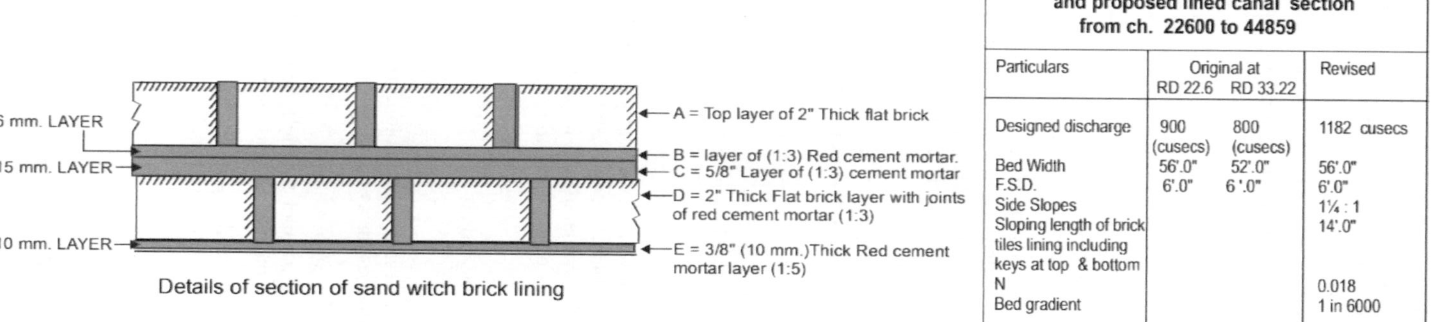

Existing section of unlined canal and proposed lined canal section from ch. 22600 to 44859

Particulars	Original at RD 22.6	Original at RD 33.22	Revised
Designed discharge	900 (cusecs)	800 (cusecs)	1182 cusecs
Bed Width	56'.0"	52'.0"	56'.0"
F.S.D.	6'.0"	6 '.0"	6'.0"
Side Slopes			1¼ : 1
Sloping length of brick tiles lining including keys at top & bottom			14'.0"
N			0.018
Bed gradient			1 in 6000

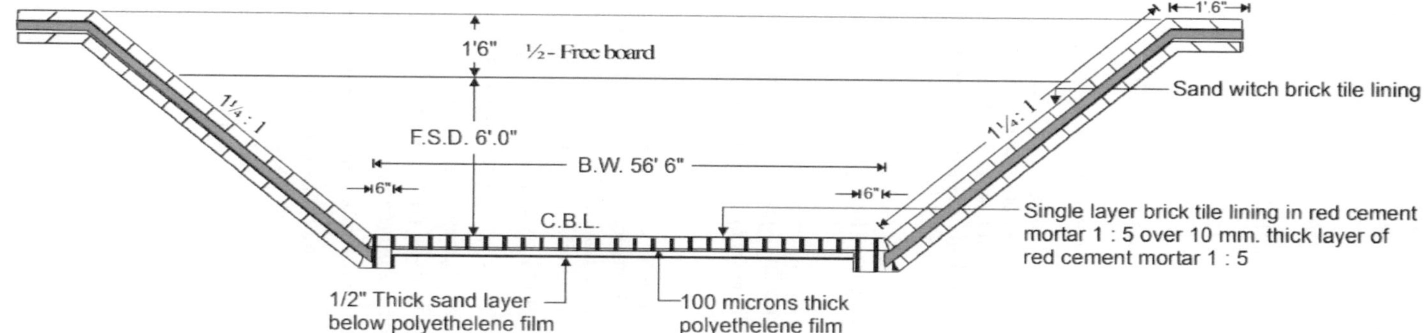

Figure 6.5 (v) Typical cross section of composite lining-sandwiched brick tile lining in slopes & single tile lining (in polyethylene) in bed. Design of Boarsad branch From Ch. 22600 To 44859

Because of a low cost lining in bed, in the (ii) proposal, the over all cost of both the alternatives (i) and (ii) may be practically same. In this case, there are practically little problems of adjustment of existing structures, as the full supply levels and depth of the canal remains practically the same. But in this case various aspects of modernisation may not be resolved.

Alternative (iii) is favoured when there are large number of problems of adjustment of existing structures in the proposed new lined canal section, cross bunds and water diversion for irrigation and closing the flow in the section to be lined during construction.

Alternative (iv) is adopted when sufficient land and earth is not available by the side of the existing canal or it is not possible to acquire the same. It is also adopted when the alignments also needs a change. In this all the problems of modernisation are easily resolved, but it may be very costly. A typical example is Parallel Upper Ganga Canal (PUGC) from Hardwar to Roorkee, constructed as a lined canal in 1990-2002, parallel to 150 years old earthen Upper Ganga canal (UGC), with a common bank in some length and without common bank in other portions, see Para 6.5.2.

Therefore all the pros and cones, of all the alternatives needs to be examined in detail and work out over all cost along with the feasibility of construction and maintenance of irrigation system. Never-the-less, the decision maker should never be afraid with the quantum of the studies required and should have such alternatives worked out before taking any decision. In fact such studies should be appreciated and encouraged, as it makes the decision process easy.

In working out the economics of lining of existing channels, cost of all above aspects should be included in the over all cost of lining versus unlined canals. For economies of lining see Para 9.15.1 and 9.15.2.

6.4.1 Merits and demerits of lining during canal closures

The merits and demerits of lining during canal closures may be summarised as under.

Merits

(i) The available period of canal closure depends upon the water availability at the source of water and regulation pattern Some canals may be Kharif or Rabi and good closure time may be available others may me perennial and little or no time may be available, still in large projects branches, distributaries run on rotation of 15 days on and 15 days off or some other suitable periods.

(ii) Earth work is reduced to a very small proportion;

(iii) The extent of compaction of earth work is reduced substantially;

(iv) The work can be commenced speedily;

(v) Remodelling of bridges; falls and other masonry works is mostly avoided because the depth of water (FSD) and the existing bed levels are maintained as far as possible;

(vi) Outlets can be retained after suitable adjustments;

(vii) There is no change in alignment, whereas in the case of parallel channels the alignment has to be shifted and curved links are required to be provided at the sites of bridges/falls.

(viii) Fresh acquisition of agricultural land is not required resulting in economy in cost. The land acquisition proceedings which could impede the progress of the project are avoided.

Demerits

(i) When the lining has to be done during a closure period of 16 days, it is a difficult task and is in the nature of rush job because of the constraint of time;

(ii) When canals run by rotation from first week of November to about June, then only about 120 days of closure is available for lining activities in a year;

(iii) Once the channel is run, the lining is covered by a layer of silt / sand and is not exposed to inspection by sight;

(iv) The lined section of the existing channel is generally determined in such a manner that the depth is maintained more or less at the same level and the bed width is tightened on one side or both sides. This results in wider section which is not only uneconomical but is prone to silting and berm formation. The wetted perimeter is 25% to 50% more than the economic section resulting in increase in cost and seepage from lined channels;

(v) Sometimes the command can be improved substantially by eliminating falls, but this methodology does not suit the requirements of earth work in bed.

By and large, construction of parallel channel has been favoured excepting where the land acquisition may be too problematic or the channel may be in very high filling.

Haryana,(India) Irrigation Department Practice:

In Haryana Irrigation Phase-I Project 2900 Km of branch-canals, distributaries and minors were lined from 1978 to 1983 at a cost of about Rs. 830 millions. This programme was continued as Haryana Irrigation Phase-II Project, to further line 2859 Km of channels. Thus, about 5800 Km of channels were lined by 1987. In all total 29.94 million sq m surface area of channels were lined, upto 1985.

Generally Cup shaped (Mehboob) section has been provided being the most economical, as it helps to keep the channel free from silt and berm formation. The seepage losses with cup shaped section are minimum.

Lining and section has been reported as very successful and effective Single (1985). It has reduced seepage losses by 10 to 15%.

6.5 Twin canals

Many times another canal is planned adjoining an existing canal. A typical example is Rajasthan Feeder canal (Now also known as Indira Gandhi canal feeder) along Ferozepore feeder, both taking off from Harike barrage in Punjab, India. These are unavoidable due to following reasons.

Initially one canal is planned from a source. Additional canal may be planned subsequently, with increased water supply at the source or revised hydrology or agreements. Enhancement of canal capacity may not be acceptable due to

(a) Administrative and political reasons

(b) May serve two different areas under different administration

(c) Water delivery/accounting may be difficult.

(d) Some times may be technically very costly, or uneconomical, generally for high discharges

Twin canals can be said as similar to parallel canals. There may be lot of operational differences, consequently there are lot of differences in structural design of embankments, which have to take care of different operating/stress conditions. These are

(i) That one canal is running and the other may be closed, causing backwater pressures.

(ii) Also one may be lined and other may unlined,

(iii) Their depth of flow, full supply and bed levels may be different.

Two typical cases are discussed below.

6.5.1 Rajasthan Feeder Canal, India

The Rajasthan Feeder Canal of 525 m^3/s at head flows through Punjab State, alongside the Ferozepore Feeder up to mile 11 (18 Km) and then along side the Sirhind Feeder. The canal enters the Rajasthan territory near mile 111 (180 km) in the vicinity of village Jandwala in district Hissar.

The sections of the Rajasthan feeder are as under :

(i) First 3 miles (5 Km) is unlined section due to high sub soil water table. It is 332 ft. wide and 14.0 ft. deep, with a common bank of Ferozepore feeder.

(ii) Next 31 miles lined section is 260 ft x 14.4 ft; partly along Ferozepore feeder and then along Sirhind feeder. The cross-section along Sirhind feeder as shown in figure 6.6

(iii) Next 2 miles (3 km) is transition length from a section of 260' x 14.4' to 125' x 21.0', (ft).

(iv) Last 98 miles (155 km) is lined section 125' x 21.0', (ft) with burnt clay tiles.

Two typical cross-sections are shown in figure 6.6.

In order to bring maximum area under command the feeder has been designed with a slope of 1 in 12000. A value of N = 0.017 has been adopted in the design by Manning's formula.

The various alternatives which would effect common bank in different ways as given below, needs tackling although in a common way, i.e., providing a suitable, adequate and efficient drainage system for the common bank when:

1. Both channels are running at full supply.
2. Both channels are dry.
3. Either one is dry.
4. The draw down conditions when the closure of a channel is to take place.

Truly speaking, the major portion of the lining remains vulnerable to the back pressure of water.

Summing up the problems of this common bank, it can be stated that:-

The common bank is to be made safe against the settlement by providing suitable adequate and efficient drainage system. This can be done either by compacting whole of the common bank at conditions much more than optimum moisture; or

Keeping the water up-till full supply levels in both the channels by striking temporary bunds in between and lining the channels afterwards; Or

Providing some artificially mechanical system of drainage by creating sump wells at reasonable and suitable distances and installing pumps there on. But this is very costly and also very difficult in maintenance.

More elaborate drainage arrangements has been adopted on the Parallel Upper Ganga Canal, as discussed in next Para 6.5.2

6.5.2 Parallel Upper Ganga Canal (PUGC), India

A Parallel upper Ganga Canal (Known PUGC), to carry a discharge of 10,000 cusecs from Hardwar to Roorkee U.P. India has been constructed from the years 1990-2000. It is a lined canal parallel to 150 years old upper Ganga Canal. (U.G.C. See Para 6.12) This Canal has a common bank with the old UGC, which has been retained and is likely to work. Thus this common bank also have the same problem as discussed in para 6.5.1. Much more and sophisticated drainage arrangements has been provided on this canal as shown in figure 6.7(i) to (v). The cross section of the two canals from centre to centre is shown in figure 6.7(i) (a) A plan of the two canals is shown in figure 6.7(ii) The surface water is drained in to a catch water drain on the bank of UGC, which is at a little lower level than that of PUGC. The seepage water is drained in a fitter trench shown in 6.7(i) (b). Also cross filters are provided as shown in figure 6.7(i) (c). Elaborate drainage arrangement behind lining is shown in figure 6.7(iii) to (v).

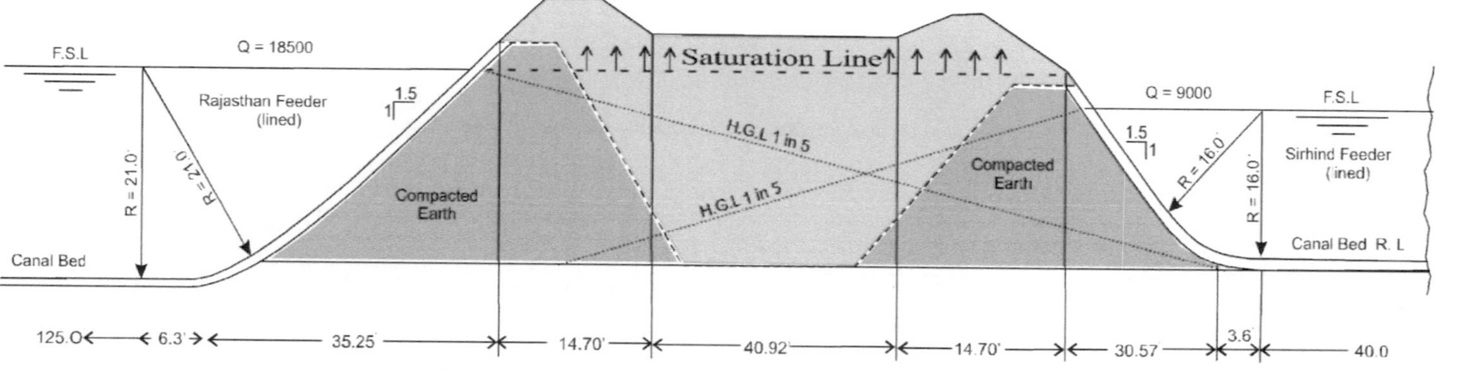

(i) Detials of Cross Section Showing Saturation of Common Bank At R. D. 376906 of Rajasthan Feeder

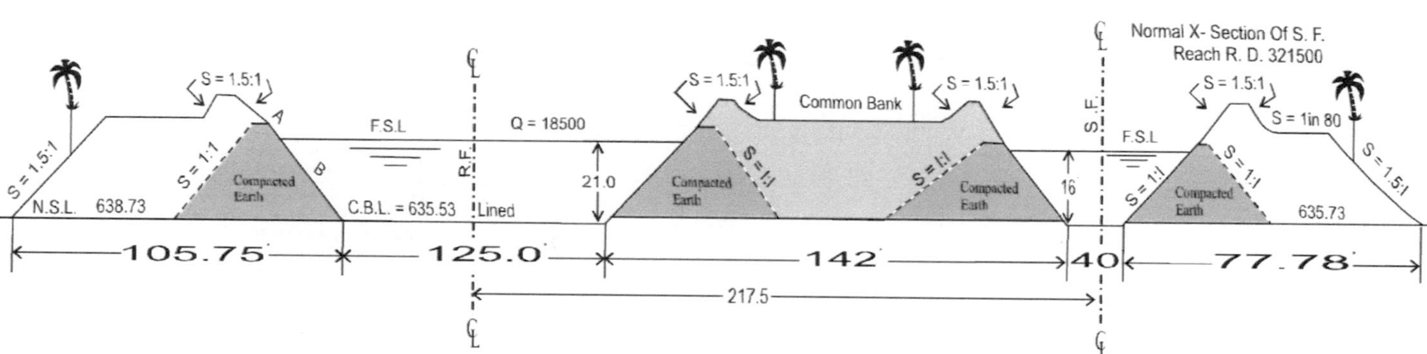

(ii) Cross Section With Common Bank At R. D. 376906 of Rajasthan Feeder
Notes : for surface Drainage of common bank see figure 8.3.

Figure 6.6 : Cross- Section of Common Bank of Two High Capacity Canals of Rajasthan Feeder & Sirhind Feeder.

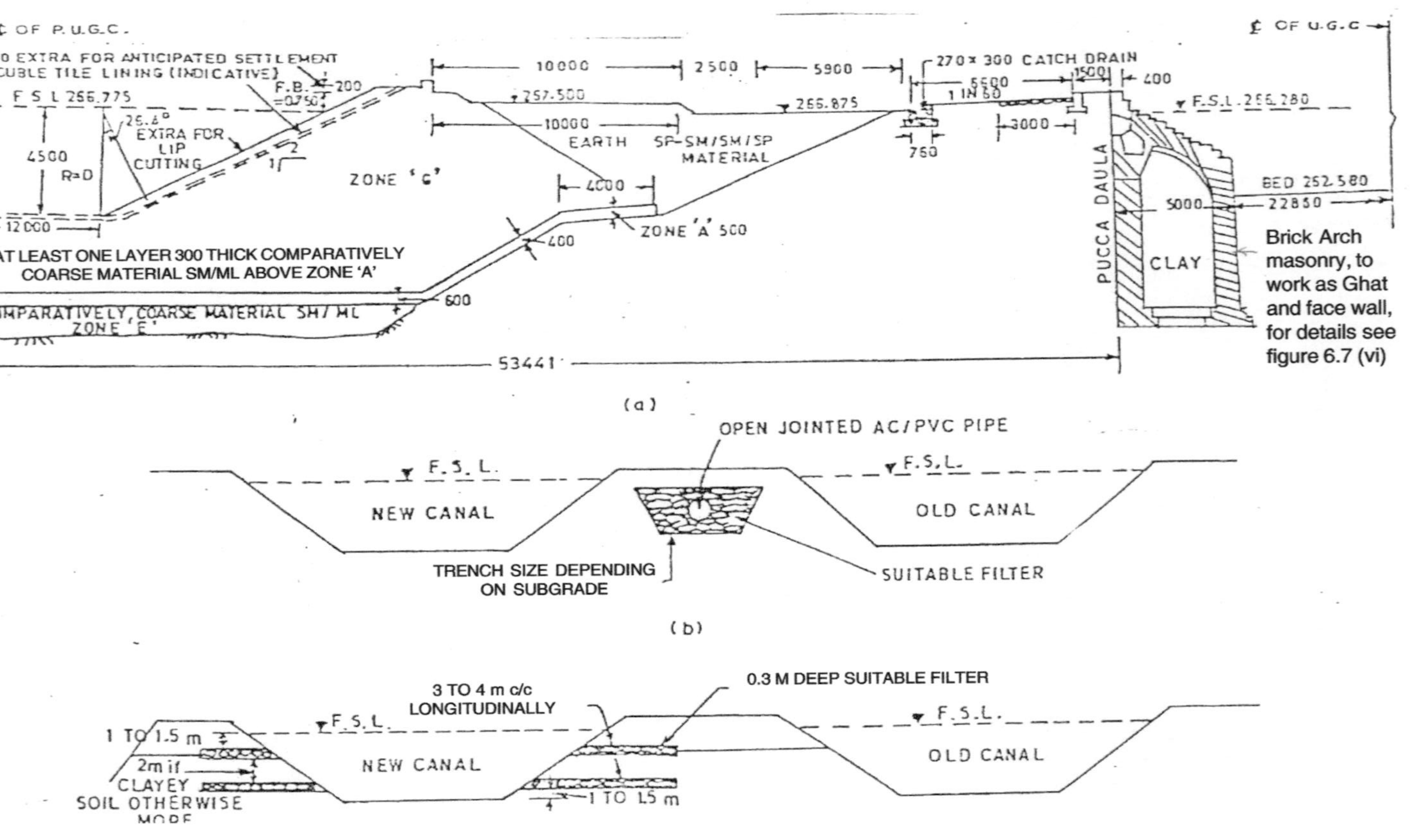

Figure 6.7 (i) Drainage Arrangement below Common Bank in PUGC

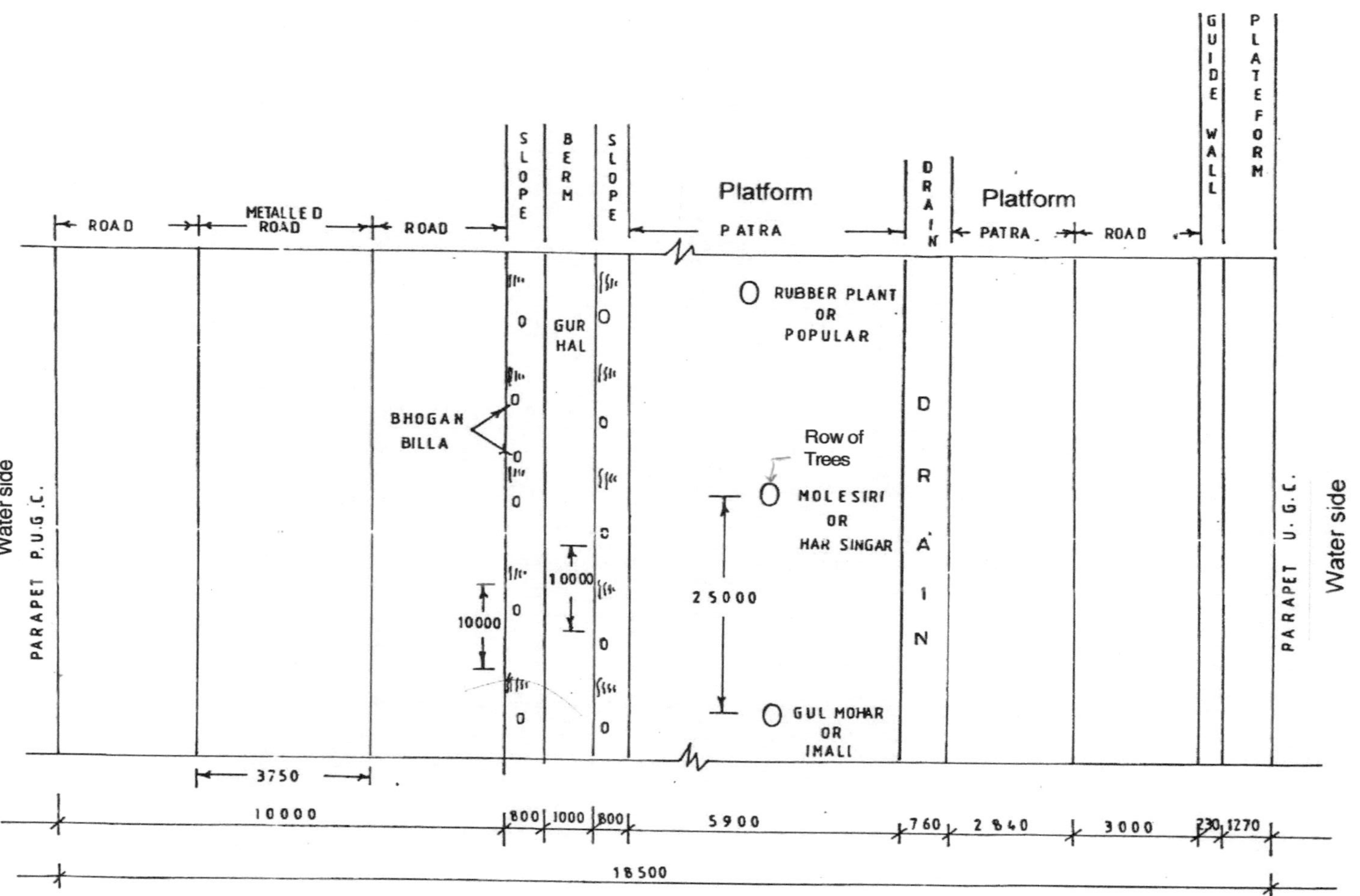

Figure 6.7 (ii) Plan of common bank between Parapets of PUGC (Parallel Upper Ganga Canal Line) and UGC (Upper Ganga Canal), India

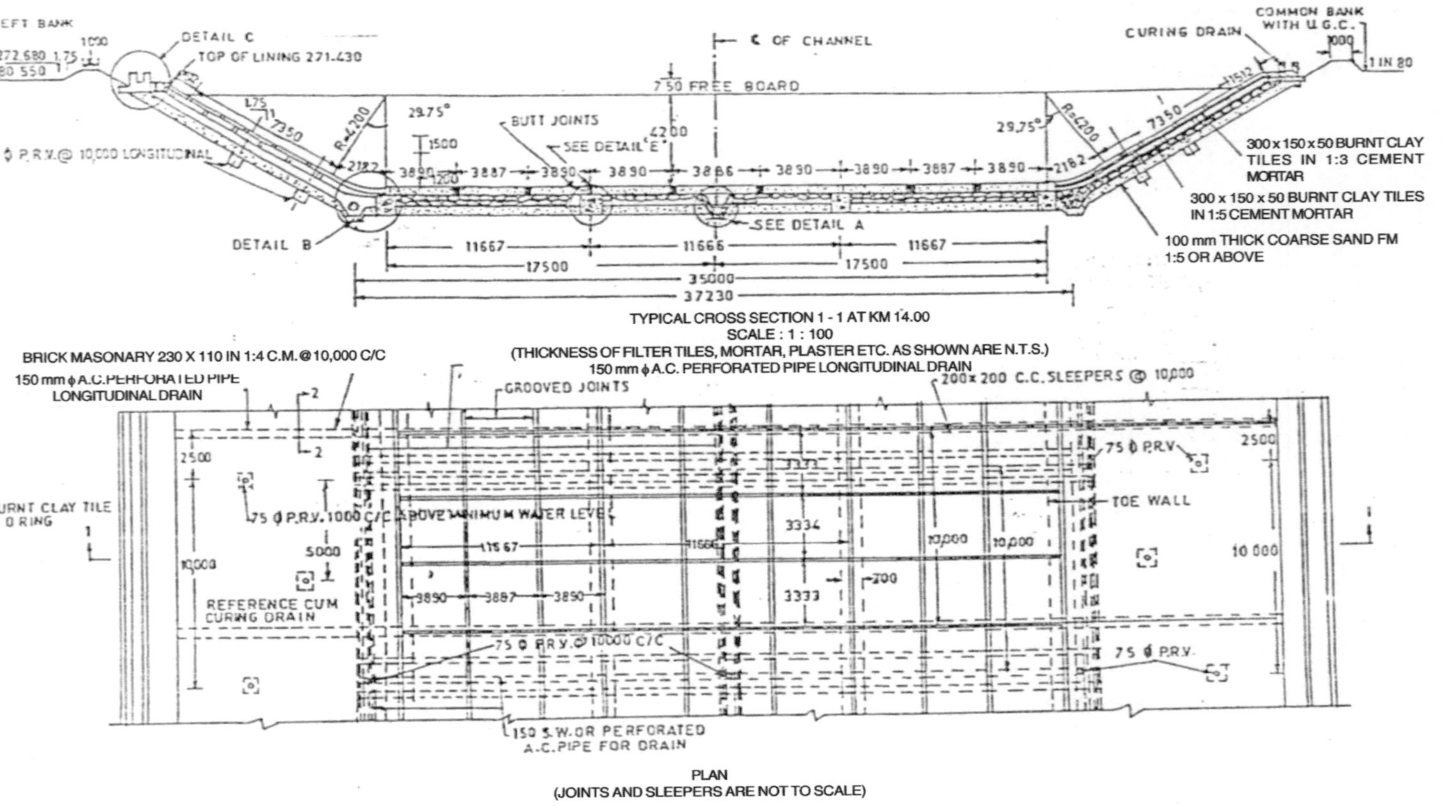

Figure 6.7 (iii) Cross Section and Plan of PUGC

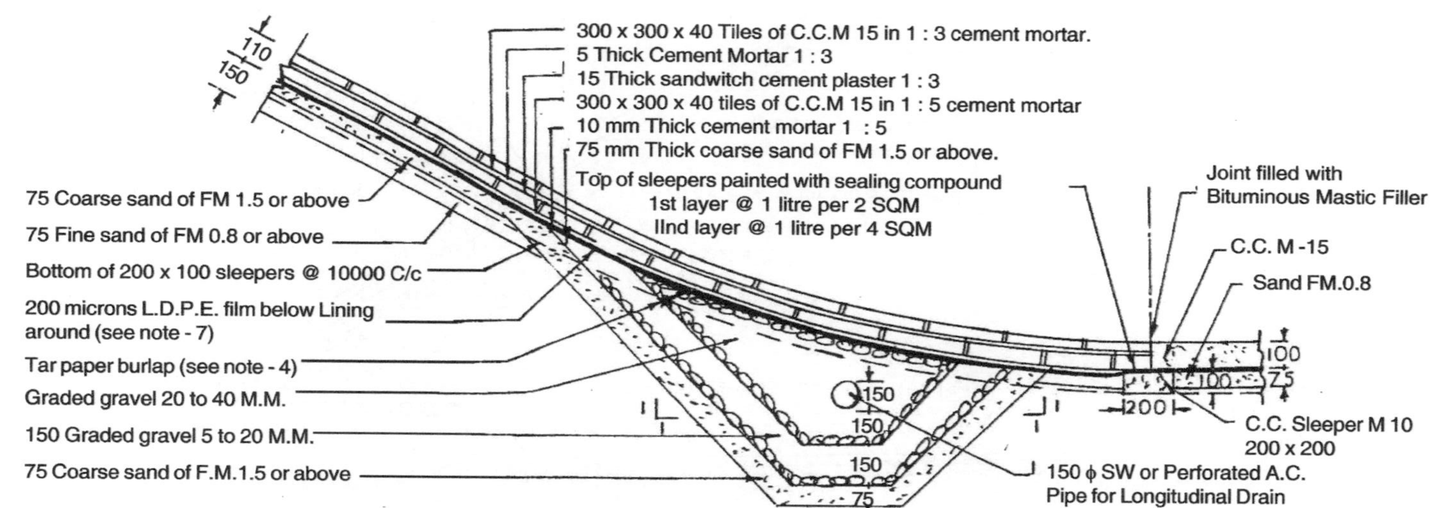

Figure 6.7 (iv) Detail of Lining and drainage arrangement in PUGC from Km. 6 to 10

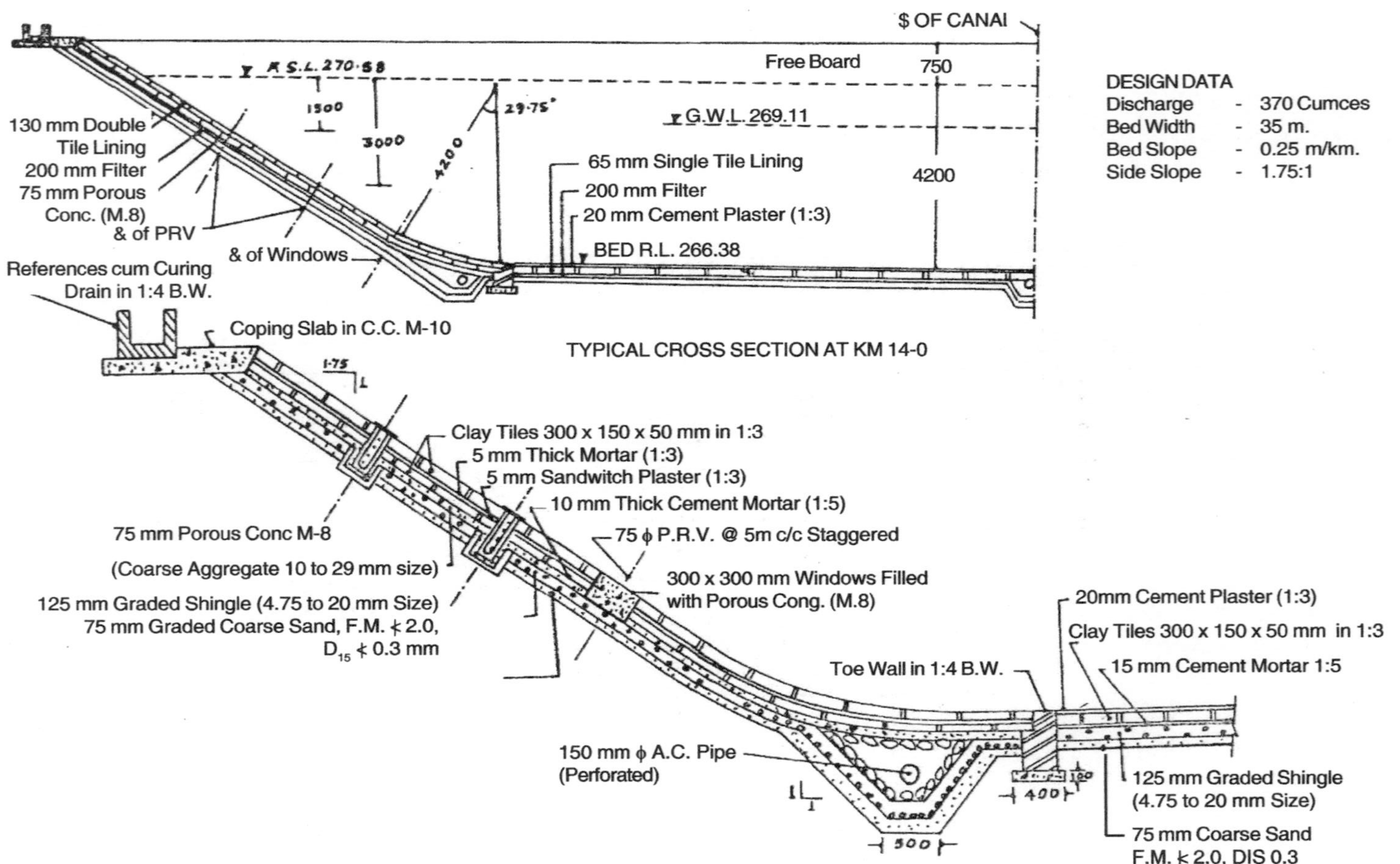

Figure 6.7 (v) Details of Lining and drainage arrangement in Bed and Side Slope of PUGC

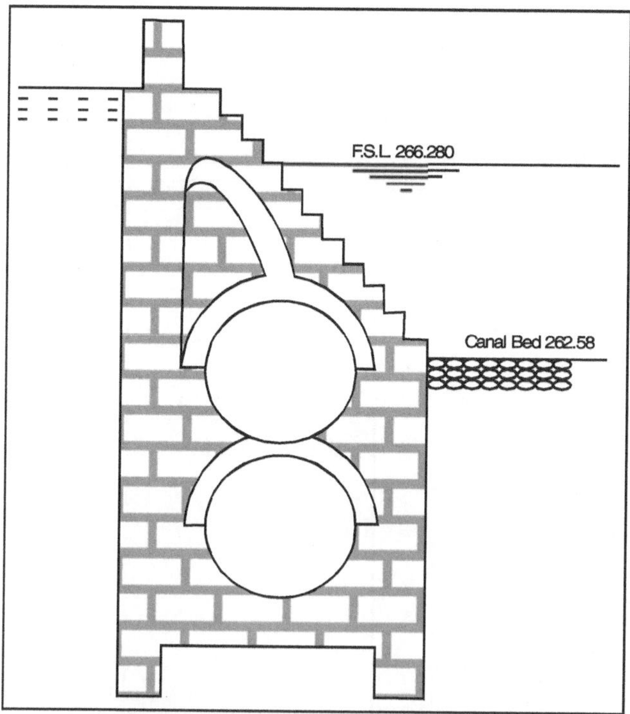

(a) Original design, the principal of which was that the revetment should as far as stability was concerned, be entirely independent at the pressure of the earth, either from the embankment on out side, or from the canal on the inside. The carrying out of this design, however, depended on the resources in kankar and brick; its cubic content was considerable.

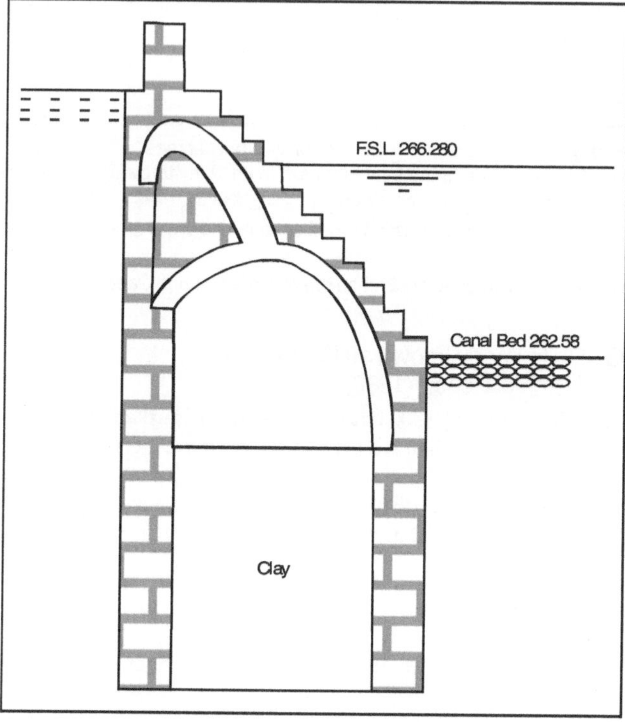

(b) Modified design, the front and rear walls were tied together by a transom or cross wall at every 15', this cross wall extending to a height of feet. In this section the decrease of masonry over that in fig.1 was a very great. Its stability, however, depended on the earth work, & on the contingencies of thrust, arising from the earth both on the canal and embankment sides. Consequences arising from this, which occurred during the rains of 1848, led me to modify the section by attempting to obtain more satisfactory equilibrium of the parts, than it had at that time. It was difficult to do this without falling back into the evil of increasing the cubic content, an evil which the above design had got rid off.

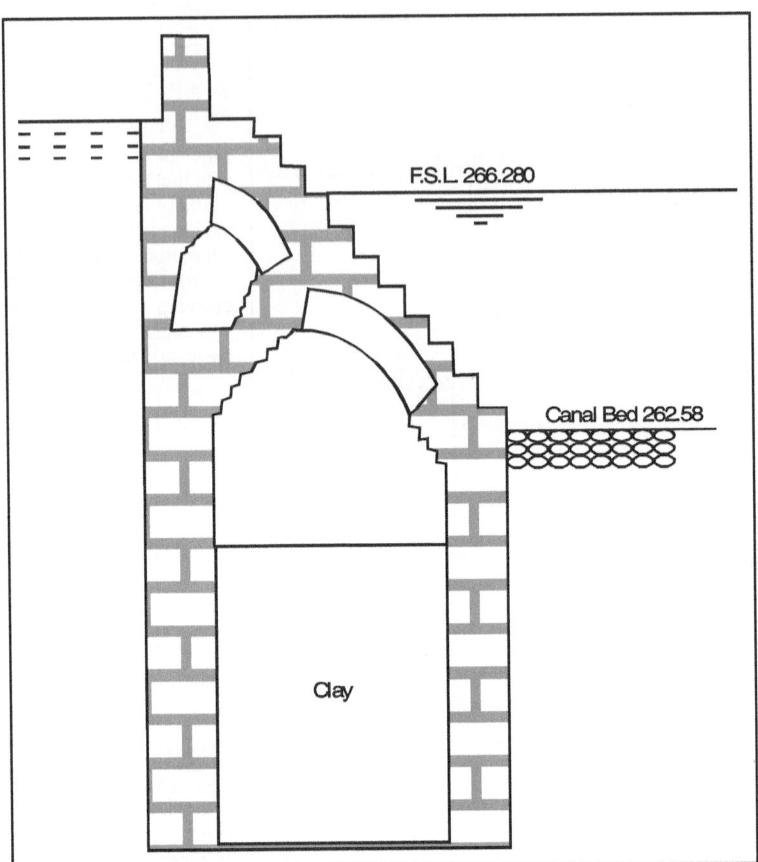

(c) Shows the remedy that was applied in the above case. It increases the cubic content, but decreases the amount of arch-work, & in principal it relives the back wall from a portion of the thrust that exit in fig. 2, by throwing the weight more forward, and by dividing the action, both of weight and thrust, more equally throughout the miss.

The height varies with that of revetment the cross-wall being in all cases carried up to the height of springing of the arch of figure (ii)

Figure 6.7 (vi) : Revetments (Face Wall or Masonry Cut Off) of Solani Embankment, Upper Ganga Canal, Meerut

6.6 Indira Gandhi Nahar Project (IGNP), formerly known as Rajasthan Canal Project (RCP) India: Revised Project Report (1986)

This is a gigantic canal project to carry 524 cumecs (18500 cusecs) water from Harika Barrage in a 204 Km long feeder canal in Punjab, to the vast Great Indian Desert known as Thar Desert, in western Rajasthan. The canal network is spread in an area of about 60 Km wide by 1000 km long belt. It consists of 204 Km of feeder, 450 Km main canal, 8000 Km of distribution networks and several thousand Km of lined water courses, to spread over a gross command area of 2.5 M ha and provide irrigation to a culturable command of 1.55 M ha.

The project was conceived by the great Indian Engineer Kanwar Sen Jain, around the year 1940 and construction was started in the year 1958. Since then the project has gone under considerable modifications/changes/revision after revision, several times. It is still (year 2008) under construction near tail areas.

Planning, design and construction of canal system is managed by a high powered Canal Board with many advisory, technical committees consisting of eminent engineers/consultants.

6.6.1 Main Canal

It is a contour canal with distribution network and irrigation on the right side only. Few lift schemes are provided on the left.

The main canal though initially thought as a unlined canal was subsequently designed and constructed as a lined canal.

The main canal passes through sandy desert soils. High cuttings of about 20 m above bed level to heavy bed filling of more than 4 m (from bed level i.e. 12 m to top bank level) are encountered in the alignment of the canal. Banks are of homogenous earth of sandy desert soil.

The over all cross-sections of heavy cutting and filling are sown in figure 7B.3 and 7B.4. Also see Para 8.10 for details.

No materials of construction except the desert sand are available all along the canal, or even within 100 Km. There are no rivers or stone hillocks nearby. However, there are some stone hillocks near Ratangarh at a distance of 200 to 300 Km from main canal and in tail areas after 450 Km of canal (near Mohangarh) at a distance of about 50 Km from tail. Even the coarse sand (locally known as Bajri) required for cement mortar is available at a distance of 200 to 300 km away from canal in deep quarries of Shivbari (Bikaner) and Bap (Phalodi). However, clay soil for manufacture of tiles/bricks is available in small pockets in depression in between sand dunes; at distances varying from 5 to 100 Km. The section of lined canal and type of lining was decided after detailed deliberation and discussions in a symposium, RCB (1950, 1960).

6.6.2 Internal Section

The internal hydraulic section of lined feeder (canal) and details of few more sections are given in Table 6.5 and Figure 6.8. The depth of the canal has been limited to 6.5 m (21 ft.), in head reaches for stability of sandy soils, operational problems and easiness in construction. It gradually decreases in tail. Internal side slopes of 1:2 (V:H) were considered stable for sandy soils and provided for depth from 6.5 m to 5 m, through out the entire length of 450 Km of main canal.

Bed slope has also been restricted to 1 in 12000, because of long length of canal, practically 650 km, and to have sufficient command. Even with this flat slope, the drop in water level is 54 m from head to tail. It is uniform from head to tail in 450 km length. Thus velocities are also very much limited.

There are only two ways to increase the velocity, one is to increase the slope and second is to increase R (as per Manning's equation). By limiting depth and energy slope and internal side slope, velocity is very much limited in the entire length from 1.5 m/s to 1.2 m/s. Bed width varies, practically from 11 times the depth at head to 2 times the depth at tail. These appears little justification for such large B/D ratio at Head, see Para 6.1.1, figure 6.1 to 6.2 and chapter 7B.

For the design of canal a value of N = 0.017 for burnt clay tile lining was adopted. It was assumed that with the flow of water the tile surface will get smoothened. Also a thin layer of moss will get deposited on the surface of tile that will reduce the Manning's rugosity coefficient. Therefore a value of N = 0.017 was considered reasonable. But in actual practice this has not been found true and silt deposition and weed growth has been noticed more details are discussed in subsequent Paras and chapter 7B.

6.6.3 Lining

Single title lining in bed and double tile lining on sides has been adopted. Burnt clay tile lining up to 365 Km and thereafter P.C.C. block lining is adopted from 365 Km to 450 Km (tail). Details of tile lining are as under:

Table 6.5 : Details of Indira Gandhi Canal Sections, India. (Double Tile Lined (DTL) and PCC Block Lining)
Bed Slope 1 in 12000, n = 0.017, Trapezoidal Section with Curved Ends, Side Slopes 2 : 1 (H : V)

S. No	Location of Canal	Discharge Q (m^3/s)	Velocity V (m/s)	Bed Width B (m)	Depth D (m)	B/D ratio X (5/6)	Wetted Perimeter P (m)	Area A (sq.m)	Hyd-raulic Radius R=A/P (m)
1	2	3	4	5	6	7	8	9	10
1.	Km 30 (RD 100)	447	(4.64) 1.529	33.83	6.40	5.29	70.72	317.56	4.49
2.	Km 122 (RD 400)	358	(4.58) 1.469	22.25	6.40	3.48	59.14	243.43	4.12
3.	Km 134 (RD 440)	353	(4.45) 1.466	21.95	6.40	3.43	58.82	241.38	4.10
4.	Km 146 (RD 475)	317	(4.46) 1.43	18.60	6.40	2.91	55.09	216.77	3.93
5.	Km 170 (RD 560)	308	(4.45) 1.43	18.29	6.34	2.88	54.78	214.77	3.92
6.*	Km 189 (RD 620)	280	(4.35) 1.41	15.85	6.34	2.50	52.37	199.49	3.81
7.	Km 216 (RD 710)	267	(4.35) 1.395	15.24	6.25	2.49	51.25	191.52	3.74
8.*	Km 250 (RD 820)	237	(4.35) 1.360	12.50	6.25	2.00	48.50	174.34	3.59
9.*	Km 293 (RD 962)	178	(4.06) 1.263	12.19	5.42	2.25	47.04	163.55	3.48
10.	Km 341 (RD 1120)	166	(4.07) 1.24	12.19	5.33	2.29	42.91	135.03	3.15
11.*	Km 384 (RD 1260)	151	(3.98) 1.217	10.67 v	5.27	2.02	41.00	124.51	3.04
12.*	Km 445 (RD 1400)	136	(3.88) 1.18	10.36 v	5.06	2.05	39.52	115.55	2.92

Source: IGNB-Revised Project estimate, (1990) 1 RD = 1000 ft.
* See figure 6.8 for sections; + PCC Block Lining from Km 384 (RD 1260) to tail.

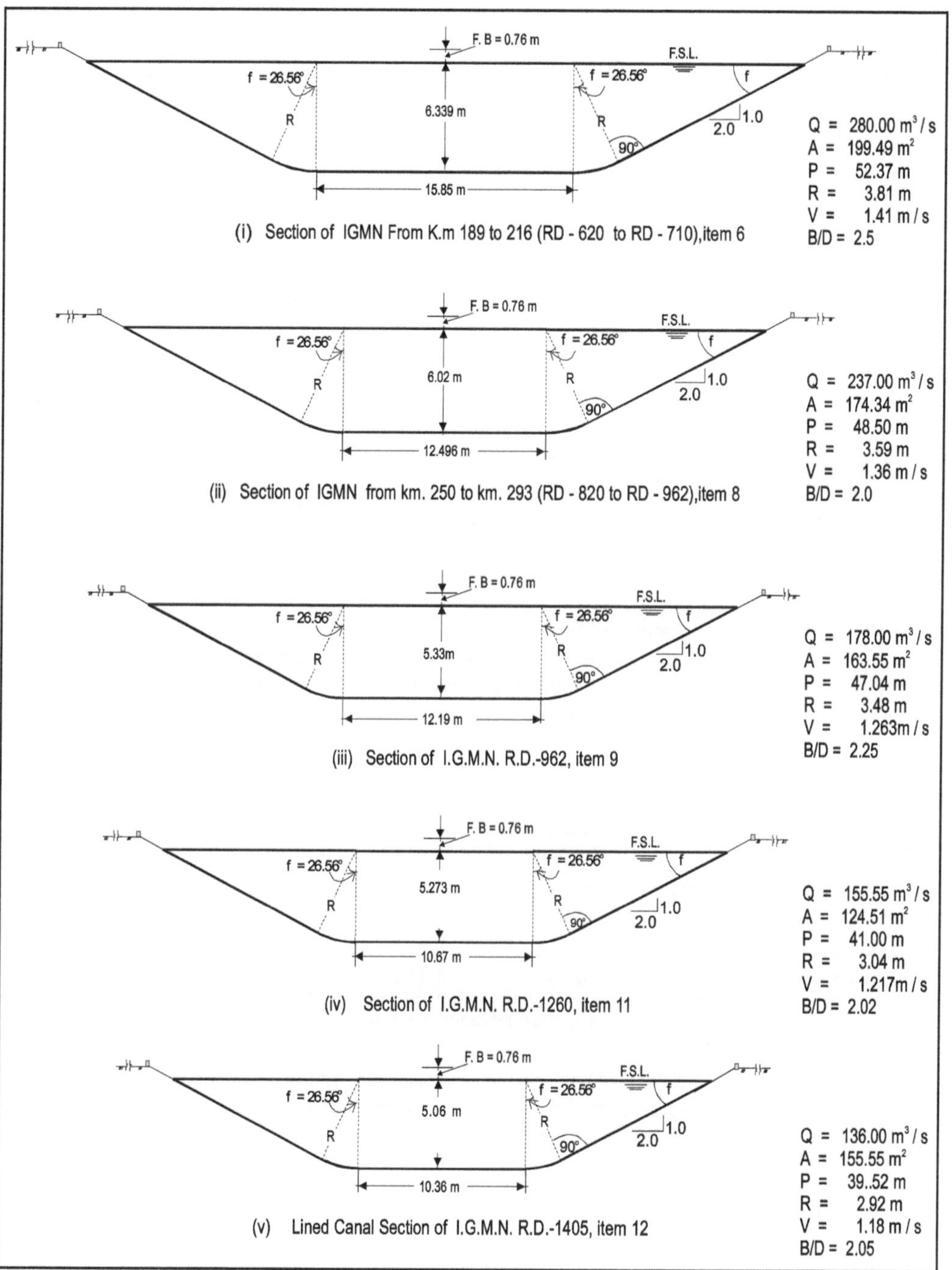

Figure 6.8 : Canal Sections of Indira Gandhi Main Canal (IGMN) Between K_m. 189 to 445.

Notes : Also see table 6.5; Double tile lining on sides and single tile in bed; N =0.017

(A) Burnt clay tile lining in bed and sides

(a) Bed - Single clay tile lining

(i) Base coat 9 mm (3/8") thick, 1:5 cement mortar plaster over compacted and dressed sub grade.

(ii) 50 mm thick 305 x 152 x 50 mm (12" x 6" x 2") burnt clay tiles laid in 6 mm thick 1:5 cement mortar on base coat

(iii) 19 mm (3/4") thick 1:3 (Cement Mortar) plaster over tiles.
Total thickness of lining = 9 + 6 + 50 + 9 = 74 mm (3")

(b) Side slopes - double tile lining.

(i) Base coat 9 mm (3/8") thick, 1:5 cement mortar plaster over compacted and dressed sub grade.

(ii) 50 mm thick 305 x 152 x 50 mm (12" x 6" x 2") burnt clay tiles laid in 6 mm thick, 1:5 cement mortars on base coat.

(iii) 15 mm (5/8") thick 1:3 (Cement Mortar) plaster over tiles, called as sandwich plaster

(iv) 2nd layer of tiles laid in 1:3 (Cement Mortar) 6 mm thick over sandwich plaster, total thickness of lining = 127 mm (5¼ inches)

Thus it can be seen that lining of bed is cheaper than sides. The ratio of cost may be around 40:60, between bed and sides.

P.C.C. block lining from Km 365 to Km 450 (tail) has been adopted

6.6.4 Lining of Branches, Distributaries and Minors

Initially all branches, their distributaries and minors were planned, designed and constructed as unlined channels. The area through which these are passing varied from sandy dunes of desert to hard clay, soils of all sorts and many times Kankar, gypsum from soft to very hard (locally known as Dhandla, a weak stone).

These channels were designed by Kennedy's formula, with the help of Garrat's Diagram, and the following stipulation about Manning's N. (RCP 1962)

Discharge	Manning's N	Corresponding value of Silt factor (f)
(i) Up to 5 cusecs (0.11 cumecs)	.03	3.16 (gravel)
(ii) 5 to 50 cusecs (1.35 cumecs)	.025	1.52 (coarse sand)
(iii) 51 to 500 cusecs (1.35 to 1.4 cumecs)	.0225	1.00 (standard silt)
(iv) Above 500 cusecs (1.4 cumecs)	.02	0.62 (fine silt)

(Lacey related his silt factor 'f' to Manning's N, the coefficient of rugosity by the expression, $N = .0225\, f^{1/4}$) .

Also a bed width depth ratio (B: D) for unlined channels was adopted as per CWPC (1960) practice. According to it, B: D ratio varies from 2.9 to 7.9 for discharge range from 10 cusecs (0.25 cumecs) to 2000 cusecs (60 cumecs). That means most of the channels had width more than 5 to 9 times the depth. Also critical velocity ratio (V / Vo) was kept near unity and velocity was restricted to 2.5 ft/sec.

Thus, the channels are very wide enough even for discharges less than 3000 cusecs, with very flat slopes. Typical cross-section of unlined canals is shown in figure 6.19 to 6.30.

Later on, around the year 1970, it was decided to line these earthen channels and design/construct future canals as lined channels.

Different practices of lining of internal earthen section of existing channels were adopted see Para 6.4. Generally same existing channels were lined. Also capacity of the canal system was increased.

Most of new channels were designed and constructed as Mehboob section and lined with single or double tile lining. But near tail areas, the design was further changed to suit to different type of lining in bed and sides, and use LDPE/PVC film.

Most of the channels sections are very wide and shallow with low velocities, having silt deposits, weed growth, and consequently have low discharging capacity to tail, high head loss and rise in full supply levels in head reaches.

6.6.5 Performance

So far no systematic procedures and methodology has been developed for monitoring the following issues of performance.

(i) Hydraulic flow characteristics - head losses, rise in FSLs, discharges

(ii) Silt deposition/ weed growth and their effect

(iii) Seepage and its effect

(iv) Lining performance

(v) Cost of operation and maintenance.

(vi) Monitoring damages to canal banks inner and outer slopes, surface and seepage drains, structures/ repairs and maintenance, Also see Para 8.10 for breaches of main canal banks.

(vii) Maintaining/ monitoring proper water accounts of each canal minor wise and distributory wise; and each individual farmer, (just like bank account).

(viii) Maintaining/ Monitoring proper revenue accounts- minor wise and distributory wise.

How ever some scattered data have been collected on occasional basis. These data are lying scattered. These have so far not fully analyzed and utilized to prepare any operational manual or direction for future.

How ever in this text an attempt is made to review the hydraulic design of main canal and likely performance in respect of item (i) above and the detailed methodology is discussed in chapter 7B. Researchers and managers should compile the scattered data and focus attention to all issues as pointed out above in (i) to (viii) and throw light on the operational behaviour versus designed behaviour.

6.7 Canal systems of Sidhmukh and Nohar Projects, Rajasthan, India

The canal network of these two project, initially designed as a lined canal and built in 1991-2000

The canal systems are adjoining the area of Bikaner canal (localy known as Gang canal, built around 1927) Bhakra canal (built around 1950) and Indira Gandhi Nahar (canal) Project (IGNP) portions built around 1960-65), all in Rajasthan. Therefore the design practices of these three projects with some modifications were adopted on this project. The canal network designs have following important features.

S. No.	Discharge range m^3/s	Type of section	Internal side slope (V:H)	Bed width	Remarks
1.	<0.5	Trapezoidal	1:1	0.30	
2.	>0.5&< 1.5	Mehboob	1:1.5 and 1:1.25*	0.0	
3.	>1.5&< 14	Do	1:1.5	0.0	

* Earlier the side slope was 1: 1.5 and subsequently changed.

The side slope of 1:1 and 1:1.25 are more efficient. A discharge criterion is adopted in deciding the type of section and internal side slopes. A better criterion would be depth of flow, as the discharging capacity varies with bed slope.

Table 6.6 : Details of over all Cross Section of Sidhmukh and Nohar Projects, Rajasthan, India.

S. No	Discharge (cumecs)	Full cutting			Partial cutting Filling				Fully filling				Type of Section	Inner Side slope Of lining H: V	Free Board(m)	Type of Lining	Bed Width (m)	Minimum cover On h. G. Line (m)	Slope of H. G. Line.		Internal Angle. (a) Mehboob in section
		Bank Width (m)		Cutting Slope H: V	Bank Width (m)		Side Slope H: V		Bank Width (m)		Side Slope H: V								In Comp. Zone H: V	In Loose Zone H: V	
		L	R		L	R	Loose Zone	Comp Zone	L	R	Loose Zone	Comp Zone									
1	2	3	4	5	6	7	8	9	10	11	12	13	14	15	16	17	18	19	20	21	22
1.	Up to 0. 5	1.00	1.00	1:1	1.00	1.00	-	1. 5:1	1.00	1.00	-	1. 5:1	Trapez oidal Section	1:1	0.45	S.T. Lin-Ing	0. 30	0. 30	3:1	5:1	-
2.	0. 5 to 1. 50	5.20	2.75	1:1	5.20	2.75	1.5:1	1:1	5. 20	2.75	1. 5:1	1:1	Mehboob Section	1.25:1	0.45	D.T. Lin-Ing	0. 00	0. 30	3:1	5:1	38°- 66′
3.	1. 50 to 14	5.20	3.00	1:1	5.20	3.00	1.5:1	1:1	5.20	3.00	1. 5:1	1:1	Mehboob Section	1.5:1	0. 60	D.T. Lin-Ing	0.00	0.60	3:1	5:1	33°- 40′

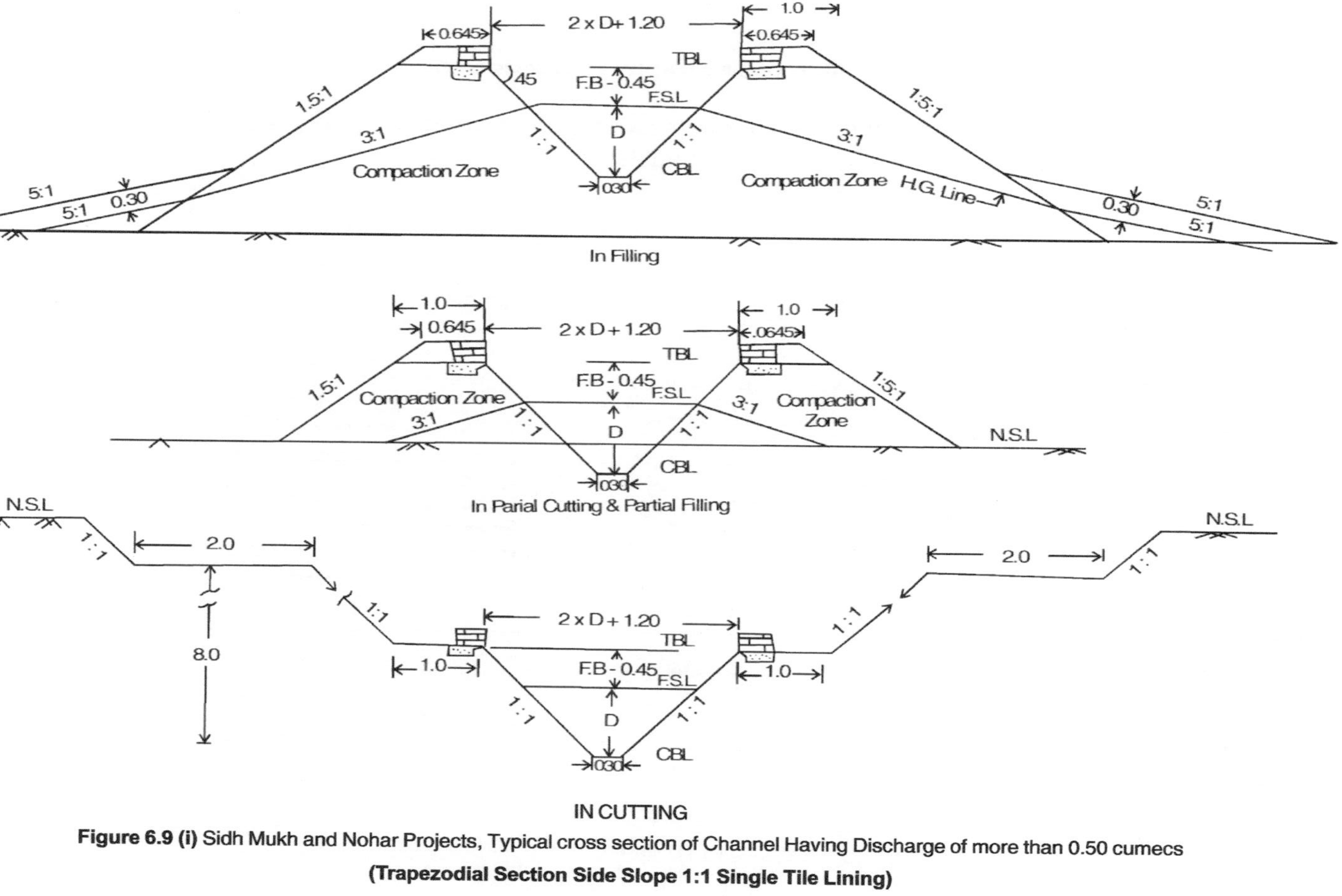

Figure 6.9 (i) Sidh Mukh and Nohar Projects, Typical cross section of Channel Having Discharge of more than 0.50 cumecs

(Trapezodial Section Side Slope 1:1 Single Tile Lining)

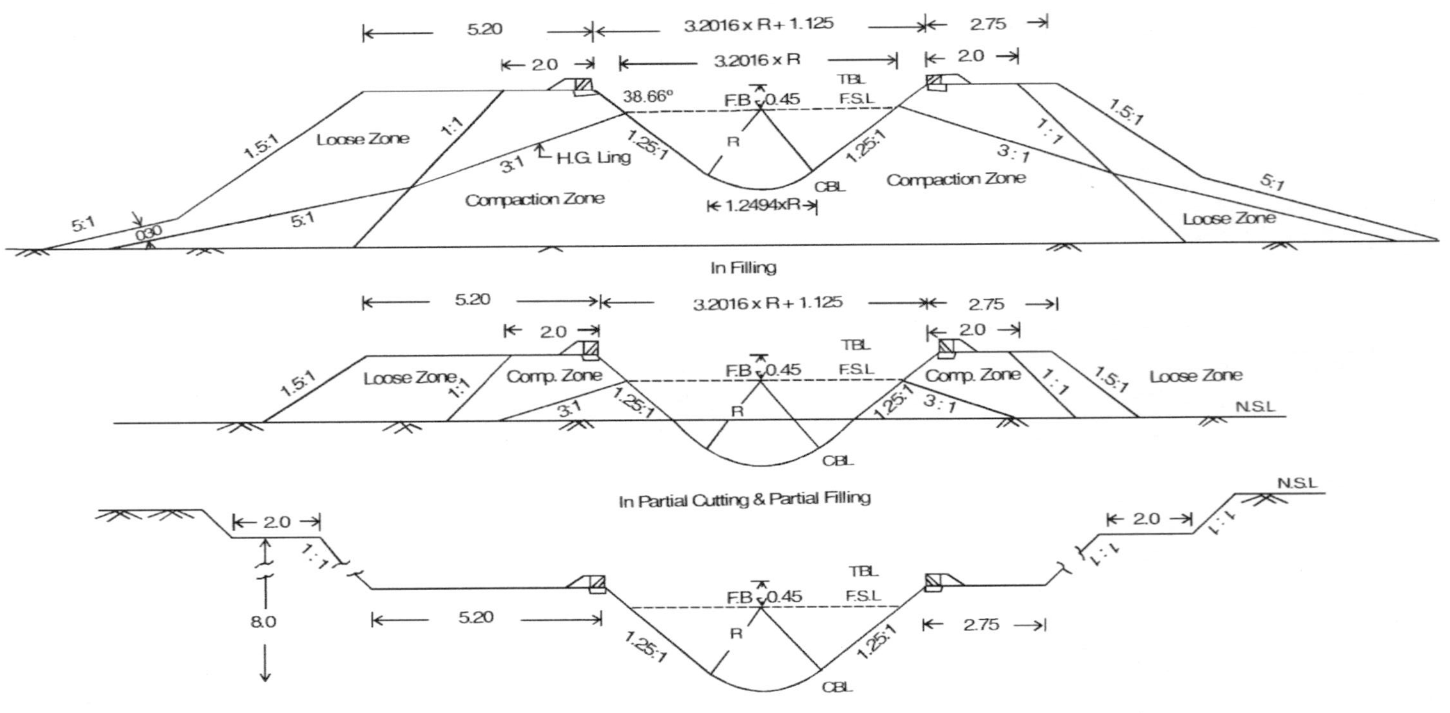

Figure 6.9 (ii) Sidh Mukh and Nohar Projects, Typical cross section of Channel Having Discharge of more than 0.50 cumecs but less than 1.50 cumecs

(Side Slope 1:1.25 (V.H) double Tile Lining)

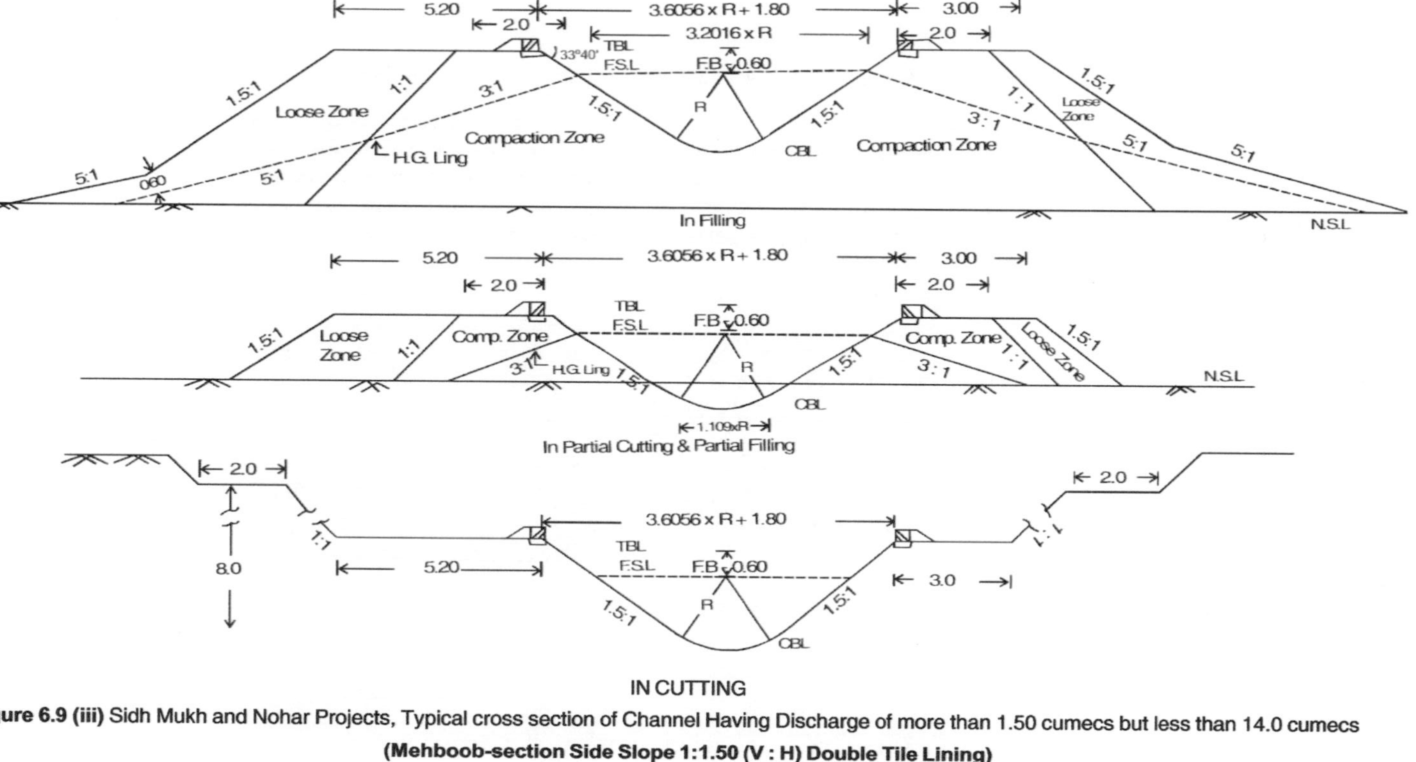

Figure 6.9 (iii) Sidh Mukh and Nohar Projects, Typical cross section of Channel Having Discharge of more than 1.50 cumecs but less than 14.0 cumecs
(Mehboob-section Side Slope 1:1.50 (V : H) Double Tile Lining)

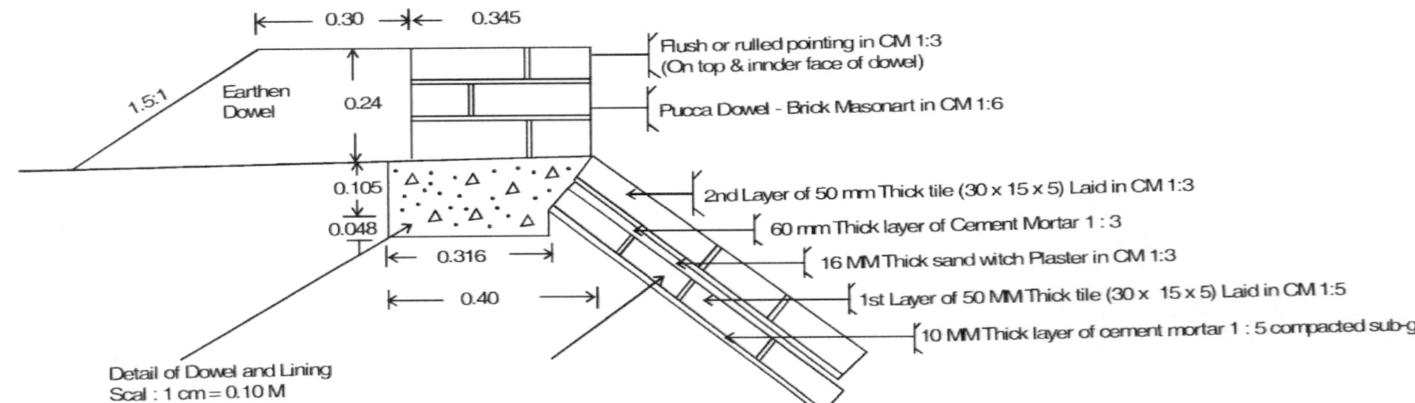

Note :

1. 0.30 M Earthen Cover on H.G. Line should be provide in Partial Cutting & Partial Filling section if required.
2. 15M wide clear berm shall beleft between Borrow Area or Spoil bank & Outer toe of the Canal Embankment.
3. Earth obtained from cutting shall be utilize in Adjoining filling portion up to 500 meter lead.
4. Spoil banks shall not be more than 3 meter in Weight when disposal is done manually and 5 meter by other means
5. Depth of Borrow pits shell not exceed than 1.0 meter.

Figure 6.9 (iv) Sidhmukh and Nohar Projects, Details of Dowel and Lining

The soils of embankment are generally sandy soils and no drainage arrangement behind lining is provided.

Single tile lining has been adopted up to a discharge of 0.5m^3/s and double tile lining for discharge greater then 0.5 m^3/s. For the tile surface value of N is taken as 0.018. The value is adopted according to the general practice in the area.

Typical canal sections along with details of lining are shown in figure 6.9.

Details of embankments, free board, dowel etc. are also given in the table 6.6.

No provision for drainage of rain water on banks and its disposal was made in the initial stages but subsequently with the experience of rain water erosion, the section was suitably modified.

Dust storms as in IGNP also occur in this area and the performance depends on silting and weed growth which is yet to be ascertained.

6.8 Mahi Bajaj Sagar Project, Banswara, Rajasthan, India

Remodelling of Right Main Canal, (RMC)

Initially lined canal was constructed in the year 1975-84 and remodelling done in the year 1988-2000. (Revised Project Report and Estimates, 1987)

RMC off takes from Kagdi - pick - up - weir. The canal is 74 km long. On the canal ,there are two Power Houses - i) at KM 29.3 for a capacity of 2 X 400 KW and known as Ghatol Power House; and ii) At KM 63.81 for 165 KW known as Ganera Power House. It is a contour canal and passes along hill toes. Right bank is on the hill side. Left bank is in part filling . The hill side consists of hard to soft rock with lots of criss - cross joints, yet generally in stable dip. The left side consists all types of mixed soils may be pebbles and boulder mixed, rock over burden.

At some places it contains clayey and finer materials also.

It was initially designed for discharge of 12.30 cumecs (434 cusecs) at head and constructed in the years 1975-84. This discharge was for C.C.A. of 35,000 Ha.

Subsequently around the year 1987 it was decided to increase the area from 35,000 to 61,500 Ha. and the discharge was also increased from 12.28 cumecs (434 cusecs) to 30 cumecs (1060 cusecs) more than double. The area was increased on Narwali distributory and Sagwara Canal off taking from Narwali distributory. Therefore to accommodate the increase in discharge modification of constructed canal, upto Km 29 and Narwali disributory (off taking at Km 29) was started in the year 1988.

The Right Main canal was originally lined throught out its length. Modification has been carried out by slightly increasing the FSD from 2.02 m to 3.30 m and by increasing the bed width in some reaches. This increase in bed width was done in the reaches where the original side slopes was 1 / 2 : 1 and 1 / 8 : 1 (H:V). The increase in bed width was done by dimantling the existing lining and excavation of the bank, on one side, generally in the left bank, being in filling. Typical cross-sections of normal and heavy cuttings are given in figure 6.10 (i) to (iv).

Details of canal sections initially designed and subsequently modified are given in table 6.8.

The linning consists of cement concrete lining. 100 mm thick on sides and bed.

Besides this, modification of all the structures such as bridges, cross drainage works, head regulators of off taking channels, gated out lets etc. was planned. Most of the existing structures were discarded and abondoned. Their R.C.C. slabs were also made redundent. New structures constructed in their place. The number of structures and cost involved is very high.

Table 6.7 : Details of Initial and Remodelling Design of Right Main Canal of Mahi Project, Banswara, Rajasthan, India

MDDL of Kagdi Reservoir = 230.75 m; Original FSL Of Canal At Head = 230.25 m; Modified FSL Of Canal At Head = 231.53 m																			
		Original design									Modified design								
S. No.	Reach	Dis-charge	Bed width	FSD	B/D	Side slopes	Area	Wetted Perimeter	Hydraulic Radius	Vel. In	Dis-charge	Bed width	FSD	B/D	Side slopes	Area in sq m	Wetted Perimeter	Hydraulic Radius	Vel. In m/sec.
		Q	B	D	X	Z	A	P	R	V	Q	B	D	X	Z	A	P	R	V
	(m)	(cumecs)	(m)	(m)		(H:V)	(m^2)	(m)	(m)	(m/s)	(cumecs)	(m)	(m)		(V:H)		(m)	(m)	
1	2	3	4	5	6	7	8	9	10	11	12	13	14	15	16	17	18	19	20
	Bed Slope 1 in 5000																		
1	0 to 1020	12.28	4.5	2.02	2.23	1:1	13.17	10.21	1.29	0.93	30	4.5	3.3	1.36	1:1	25.74	13.83	1.86	1.15
2	1020 to 1190	12.28	5.45	2.02	2.70	1/2:1	13.05	9.97	1.31	0.94	30	6.2	3.3	1.88	1/2:1	25.90	13.58	1.91	1.143
3	1190 to 3560	12.28	4.5	2.02	2.23	1:1	13.17	10.21	1.29	0.93	30	4.5	3.3	1.36	1:1	25.74	13.83	1.86	1.15
4	3560 to3740	12.28	6.3	2.02	3.12	1/8:1	13.23	10.37	1.28	0.926	30	6.2	3.3	1.88	1/2:1	25.90	13.58	1.91	1.143
5	3740 to 5520	12.28	4.5	2.02	2.23	1:1	13.17	10.21	1.29	0.93	30	4.5	3.3	1.36	1:1	25.74	13.83	1.86	1.15
6	5520 to 5700	12.28	5.45	2.02	2.70	1/2:1	13.05	9.97	1.31	0.94	30	6.2	3.3	1.88	1/2:1	25.90	13.58	1.91	1.143
7	5700 to 6060	12.28	4.5	2.02	2.23	1:1	13.17	10.21	1.29	0.93	30	4.5	3.3	1.36	1:1	25.74	13.83	1.86	1.15
8	6060 to 6230	12.28	5.45	2.02	2.70	1/2:1	13.05	9.97	1.31	0.94	30	6.2	3.3	1.88	1/2:1	25.9	13.58	1.91	1.143
9	6230 to 9060	12.28	4.5	2.02	2.23	1:1	13.17	10.21	1.29	0.93	30	4.5	3.3	1.36	1:1	25.74	13.83	1.86	1.15
	At Km. 90.60, Narwali Distriburory off takes																		
10	9060 to 10830	11.42	4.5	1.94	2.32	1:1	12.49	9.99	1.25	0.915	29.14	4.5	3.24	1.39	1:1	25.08	13.67	1.83	1.14
11	10830 to 11240	11.42	5.4	1.94	2.78	1/2:1	12.36	9.74	1.27	0.922	29.14	6	3.24	1.85	1/2:1	24.69	13.24	1.86	1.14
12	11240 to 13290	11.42	4.5	1.94	2.32	1:1	12.49	9.99	1.25	0.915	29.14	4.5	3.24	1.39	1:1	25.08	13.67	1.83	1.14
13	13290 to 13740	10.87	5.4	1.89	2.86	1/2:1	11.99	9.69	1.24	0.9	28.52	6	3.19	1.88	1/2:1	24.23	13.19	1.84	1.128
14	13740 to 15360	10.87	4.5	1.89	2.38	1:1	12.08	10.75	1.12	0.9	28.52	4.5	3.19	1.41	1:1	24.53	13.52	1.81	1.138
15	15360 to 15540	10.87	5.35	1.81	2.96	1/2:1	11.32	9.40	1.20	0.89	27.63	5.9	3.15	1.87	1/2:1	23.55	12.95	1.82	1.12
16	15540 to 19470	10.87	4.5	1.81	2.49	1:1	11.42	9.62	1.19	0.88	27.63	4.5	3.15	1.43	1:1	24.1	13.41	1.80	1.158
17	19470 to 19670	9.2	4.5	1.72	2.62	1:1	10.7	9.36	1.14	0.86	26.25	4.5	3.05	1.48	1:1	23.03	13.13	1.75	1.115
	Bed Slope 1 in 2000																		
18	19670 to 20150	9.2	3	3.05	0.98	1/8:1	10.31	9.15	1.13	0.95	26.25	7	3.05	2.30	1/8:1	22.51	13.15	1.71	1.41
19	20150 to 20350	9.2	2.7	2.95	0.92	vertical	7.97	8.60	0.93	1.18	26.25	2.7	3.05	0.89	vertical	7.97	8.60	0.93	1.18
												5.2	3.05	1.70	1/8:1	17.02	11.35	1.50	1.55
20	20350 to 21000	9.2	3	3.05	0.98	1/8:1	10.31	9.15	1.13	0.95	26.25	7	3.05	2.30	1/8:1	22.51	13.15	1.71	1.41
	Bed Slope 1 in 5000																		
21	21000 to 24030	9.2	4.5	1.72	2.62	1:1	10.7	9.36	1.14	0.86	26.25	4.5	3.05	1.48	1:1	24.4	13.13	1.86	1.115
22	24030 to 24150	9.2	5.3	1.72	3.08	1/2:1	10.59	9.15	1.16	0.87	26.25	5.8	3.05	1.90	1/2:1	22.34	12.62	1.77	1.114
23	24150 to 29350	9.2	4.5	1.72	2.62	1:1	10.7	9.36	1.14	0.86	26.25	4.5	3.05	1.48	1:1	24.4	13.13	1.86	1.11

Notes: This table indicates that velocities in the original design were near 0.9 to 0.95 m/sec. practically same for a unlined canal in rocks or murram etc. Velocities in revised section are nearly canal section due to .... To 1.5 m/sec, again not too much to justify a reduced.

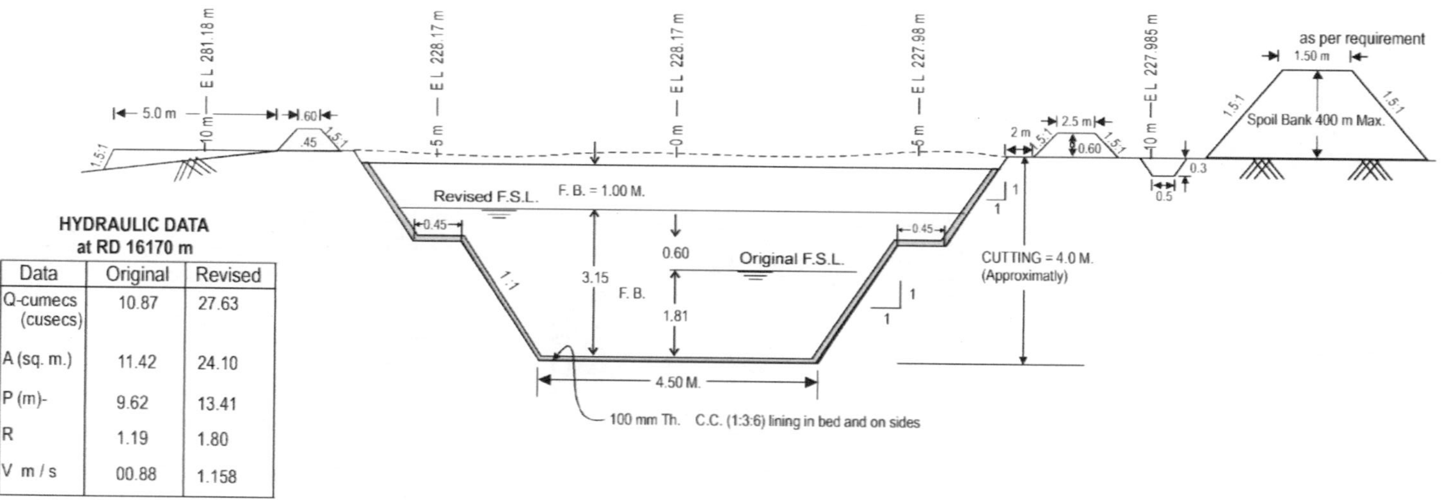

HYDRAULIC DATA at RD 16170 m

Data	Original	Revised
Q-cumecs (cusecs)	10.87	27.63
A (sq. m.)	11.42	24.10
P (m)-	9.62	13.41
R	1.19	1.80
V m / s	00.88	1.158

(i) Original and modified typical cross section in full cutting at RD 16170 m of RMC - Mahi Bajaj Sagar project
Note : Only FSL is raised and no other modification.
Modified section is more efficient

Figure 6.10 Modification of Canal Section for increasing capacity of existing lined canal of Mahi Bajaj Sagar Project, Rajasthan, India

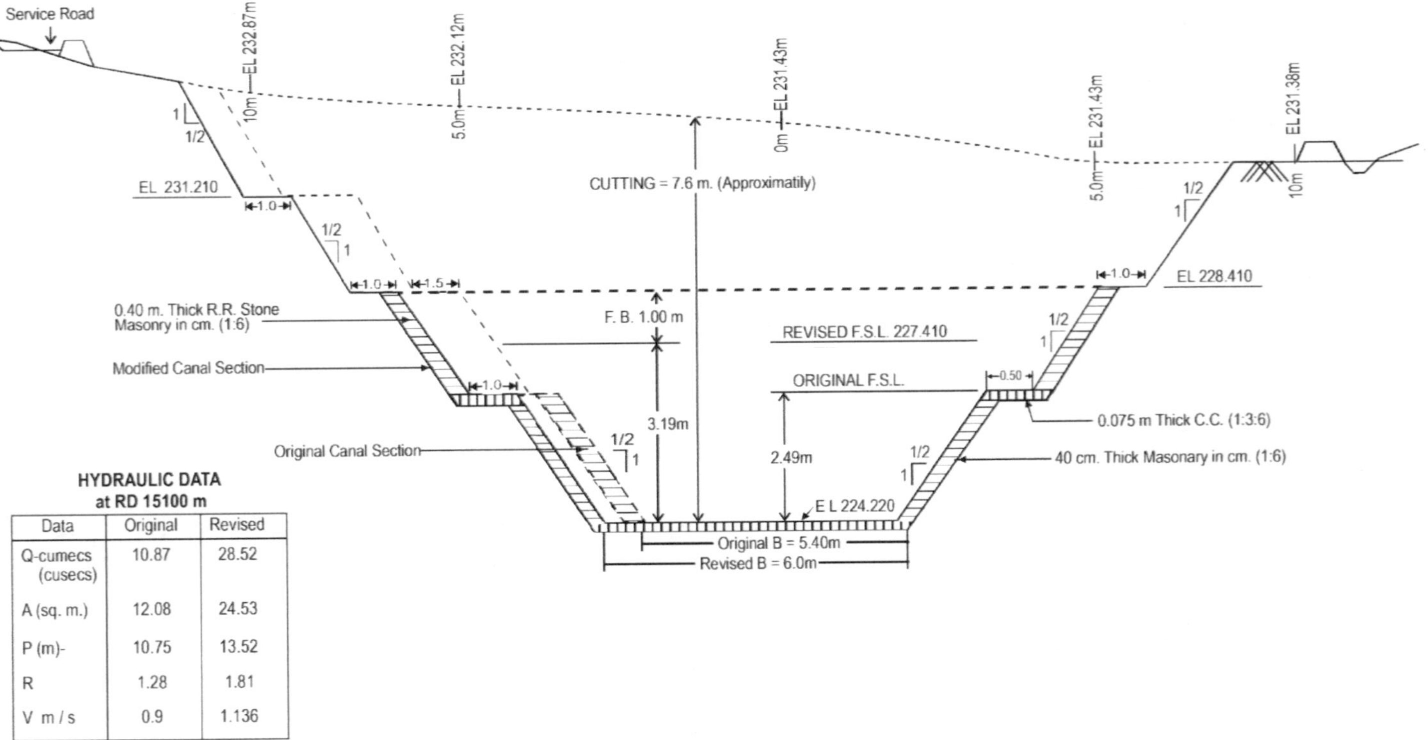

HYDRAULIC DATA at RD 15100 m

Data	Original	Revised
Q-cumecs (cusecs)	10.87	28.52
A (sq. m.)	12.08	24.53
P (m)-	10.75	13.52
R	1.28	1.81
V m / s	0.9	1.136

(ii) Original and modified typical cross section for heavy cutting at RD 15100 m of RMC - Mahi Bajaj Sagar project

Figure 6.10 Modification of Canal Section for increasing capacity of existing lined canal of Mahi Bajaj Sagar Project, Rajasthan, India

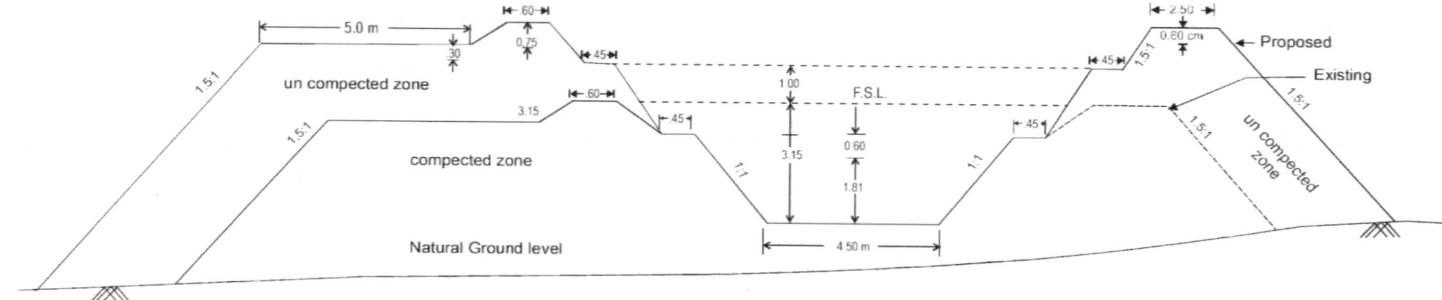

HYDRAULIC DATA
at RD 15100 m

Data	Original	Revised
Q-cumecs (cusecs)	10.87	28.52
A (sq. m.)	12.08	24.53
P (m)-	10.75	13.52
R	1.28	1.81
V m / s	0.9	1.136

(iii) Typical original and Modified Cross Section in Full FIlling at R.D. 17610 M. of R.M.C

Notes :

1. Similar modifications has been adopted in partial with of filling section at RD 15620 m.
2. The earthwork for raising the bank of canal is to be laid after removal of the Grass, picking. The old surfaces and earthwork is laid in layers. The lining is to be laid after 2 rainy seasons so as the allow enough time for natural settlement of earthwork.
3. The Lined section is addopted.
4. In bed 10 cm. thick C.C. (1:3:6) lining is proposed.
5. On side 10 cm. thick C.C. (1:3:6) lining / 5 cm. thick P.C.C. slab lining is to be proposed.
6. Only FSL is raised and there is no other modification.

Figure 6.10 Modification of Canal Section for increasing capacity of existing lined canal of Mahi Bajaj Sagar Project, Rajasthan, India

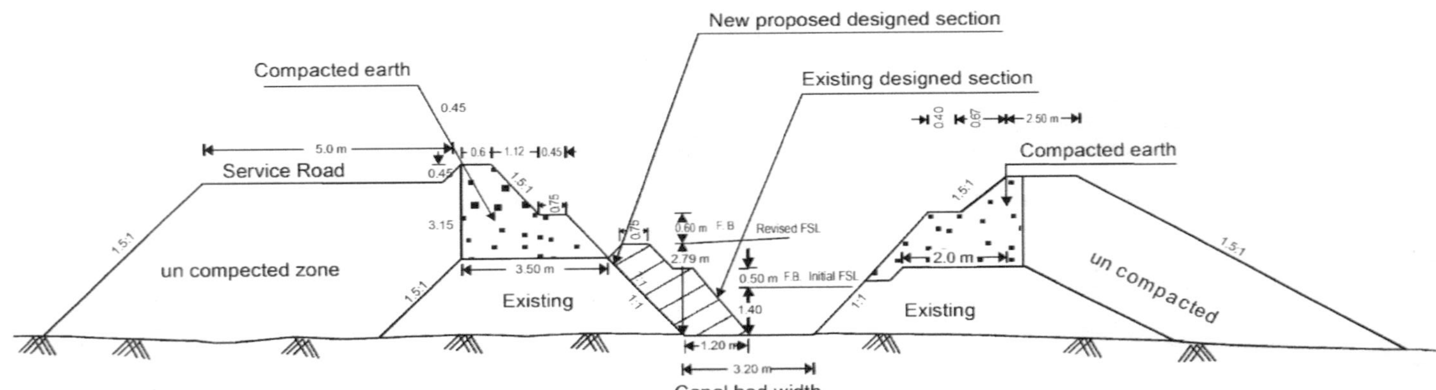

(iv) Original and modified typical cross section in cutting (shallow) from RD 0 to 3280 m of Narwali Distributory - Mahi Bajaj Sagar Project

Note :
1. All dimension are in meters except otherwise mentioned.
2. FSL and bed width was modified.

Figure 6.10 Modification of Canal Section for increasing capacity of existing lined canal of Mahi Bajaj Sagar Project, Rajasthan, India

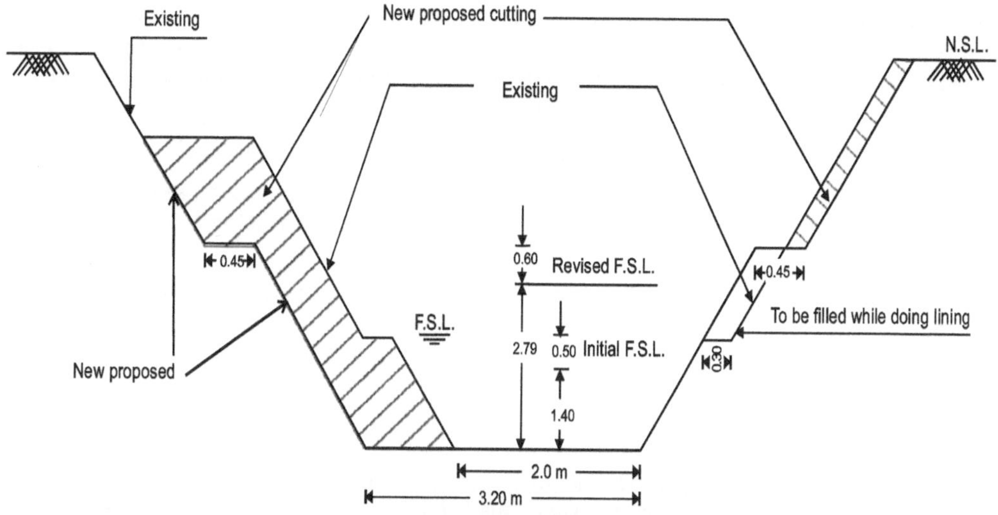

(v) Typical Remodelled Cross-Section in Cutting (Shallow) 7 Narwali Distributory

Figure 6.10 Modification of Canal Section for increasing capacity of existing lined canal of Mahi Bajaj Sagar Project, Rajasthan, India

Performance

Performanance of linning may be satistactory at few places but was unsatistactomy at some places. Failure of many panels at many location was seen in the year 1999 or even before. Generally failaure is by back water pressure, with intial signs of cracks and subsequently panels slipped down.

This occurred generally on the hill side i.e. right bank. This may be due to back water pressure of rainfall or ground water flow.

Also at the junction of horizontal bed and sides there is erosion of lining or cement concrete coming out. However it has helped in release of back water pressure and reduction of subsequent failure of lining.

There is no record of failures. Maintenance / repairs / replacement done. It was done as and when funds were available.

There is no monitoring of seepage losses. The farmers generally reported heavy seepage losses from the joints and fissures of rocks. Water can be seen in the fields and natural drains, when ever the canal is running. Soon after closure this water disappears.

Also, in modified section there is a horizontal 1' berm before raising of lining, see figure- 6.10. It was provided for a working space. However in rainy season during canal closure trees, bushes grow on the plat form and their roots penetrate behind lining and subsequently damage lining. It should be avoided in raising of sides of the section.

The water delivered to the canal, being from a 2nd Reserveir is practically silt free.

This canal is a non perennial canal, generally used in Rabi season only. In Karif season there is a good rainfall and normally no irrigation is required except a pre sowing irrigation and a last irrigation at maturity.

How ever there is zaid crop and sugar cane sowing some times in the end of Febuary or early March, and water is required for it.. But the requirement is very small. Requirement being small, canal has to flow for a small discharge and to feed the outlets, farmers put obstructions to raise water level.

Thus designing the canal section for a smallar discharge in the beginning and to increase capacity subsequently is generally costlier than to design the canal with higher discharge initially. Irrespective of the fact that smaller discharge is required in the beginning, yet some extra margin is always desirable. It does not mean to design initially for a very high discharge. It may be generally good to design the main canals for 10 to 20% higher discharge to catre for unforseen contigencis, such as high crop water requirement, growth of vegitation in the canal , weeds and a higher value of N against the value taken in the initial design, or to supply water on demand. Also such canals are helpful for temporary storage and to feed tails.

Capacity could also be increased by adopting a compound section and without dismentalling the initial lining, which is practically a new one.

6.9 Nangal Hydel Channel, Punjab, India

Nangal Hydel Channel with a capacity of 354 cumecs (12500 cusecs) offtakes on the left side of Nangal Pond.

On Nangal Hydel Channel, there are two power houses one at Ganguwal and the other at Kotla, each generating 77 MW. The supply in Nangal Hydel Channel after being used for power generation, feeds the Bhakra canal irrigation system of Punjab, Haryana and Rajasthan States. Besides, it also meets drinking water requirements of many villages, towns and cities including Delhi.

Head to Ganguwal Power House at 18288 m (60000 ft.)

In this reach the canal passes along the Shivalik foot-hills having gravel strata. The channel has been taken on such a contour that the left bank is generally in cutting while the right bank is in filling. In order to effect economy the cutting and filling have been balanced as far as possible. In this reach, there are two major cross drainage works - Donala Aqueduct and Debitwali Aqueduct and 21 No. other cross drainage works.

Ganguwal Power House to Kotla Power House

In this reach, the canal crosses the most undulating terrain. There are both deep ravines and high mounds. There are about 20 number heavy cutting and 13 Number heavy filling reaches in a short length of about 10 dm. Maximum cutting is of the order of 25 m between 26030 m and 26182 m. while maximum filling at Mendhwan Aqueduct at 25864 m (84854 ft) is 41 m (134 ft). The typical cross-section of channel in heavy cutting and heavy filling are shown in Figures 6.11

There are 8 numbers of cross drainage works in this reach, the important ones being Lamlehri and Mendhwan Aqueducts.

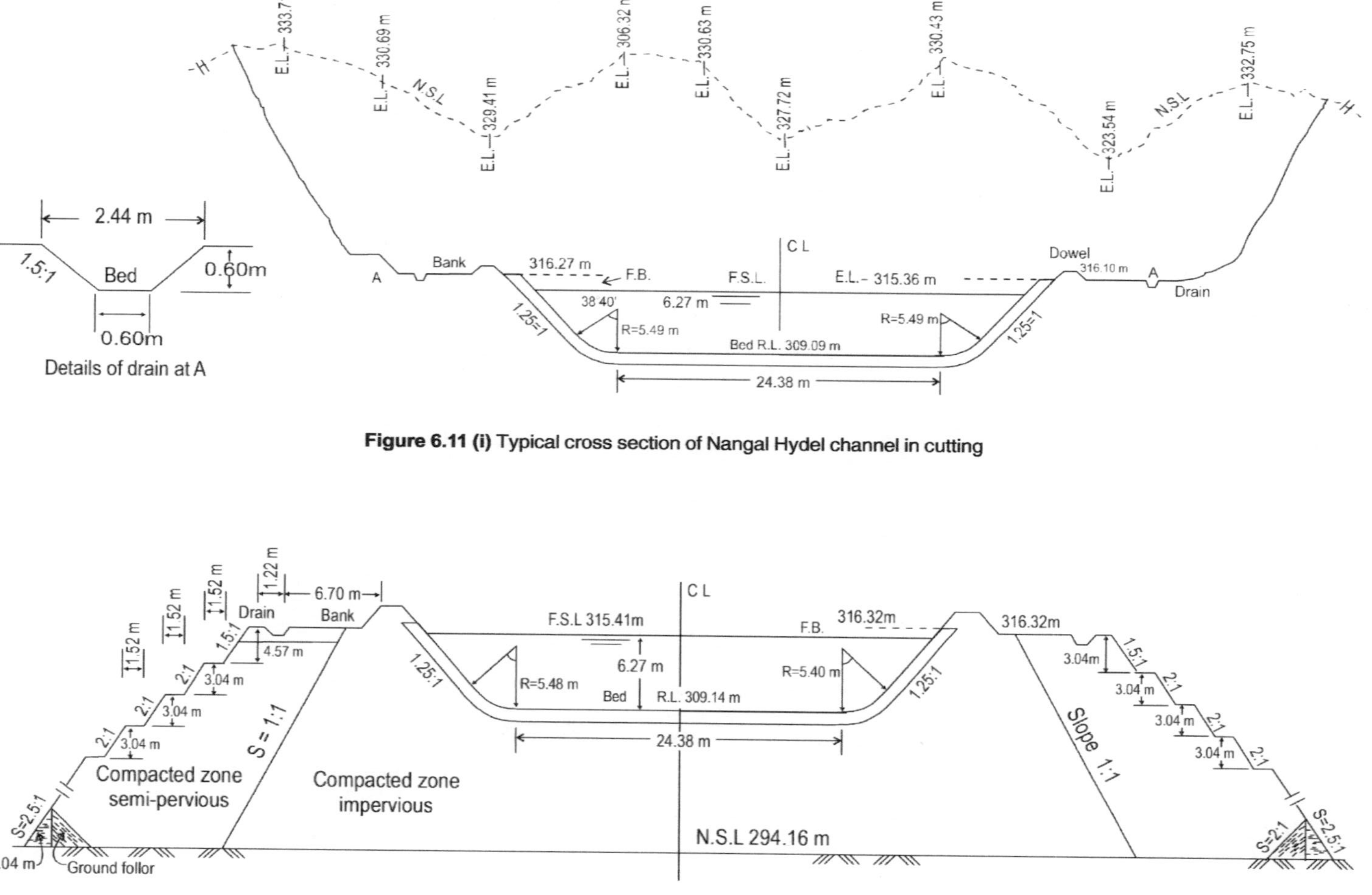

Figure 6.11 (i) Typical cross section of Nangal Hydel channel in cutting

Figure 6.11 (ii) : Typical cross-section of Nangal Hydel Channel in filling

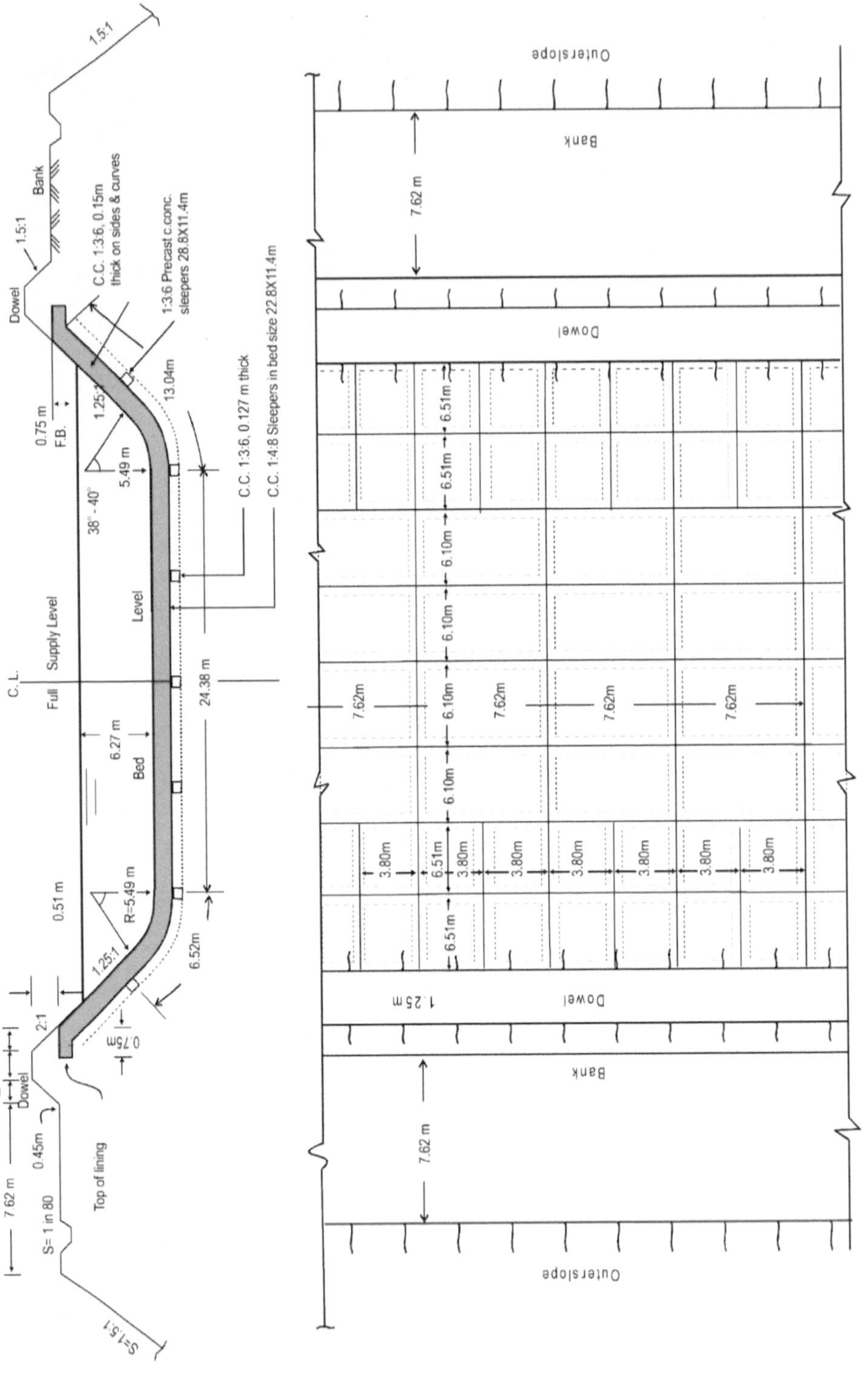

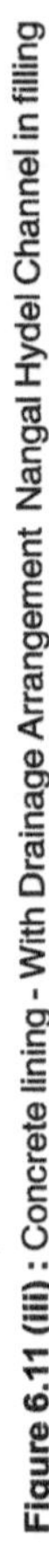

Figure 6.11 (iii) : Concrete lining - With Drainage Arrangement Nangal Hydel Channel in filling

Kotla Power House to Tail

In this reach the canal generally passes through plains. There are two regulators one at Nakian and the other at Lohand. The regulation of supplies in the channel for meeting demands of irrigation both on Bhakra Main Line and Sirhind Canal system is done from these regulators. There are 27 number cross drainage works in this reach, the major ones being Lotan Syphon, Kundlu Syphon and Sirsa Adueduct.

Lining and Construction details

(a) Bed

(i) 1: 3: 6 cement concrete, 127 mm thick in 7.6 m x 6.1 m slab. The bed slabs rest on cement concrete bed sleepers of 1: 4: 8 mix of 229 mm X 114 mm laid in situ.

(ii) 203 mm thick cement concrete in reaches below Kotla Power House where sub-soil water-level was high.

(iii) Two layers of well-burnt tiles 50 mm thick with a 15.9 mm layer of 1 : 3 cement plaster sandwiched in between the first layer is laid on 9.5 mm thick layer of 1 :5 cement mortar and the other on 6.3 mm thick layer of 1:3 cement mortar.

(b) Sides

(i) 1:3:6 cement concrete 150 mm thick. The side slabs rest on precast cement concrete bed blocks 229 mm X 114 mm and of about 0.60 m length.

(ii) 229 mm thick 1: 3: 6 cement concrete in the reach below Kotla Power House where sub-soil water-level was high.

(iii) Two layers of well-burnt tiles 50 mm thick with a 15.9 mm layer of 1:3 cement plaster sandwiched in between. The first layer is laid on 9.5 mm thick layer of 1:5 cement mortars and other on 6.3 mm thick layer of 1:3 cement mortars.

(c) Joints

9.5 m x 38 mm expansion joints were provided.

Pressure release arrangements

(a) Bed

(i) 75 mm thick filter was used in bed where the strata was porous and water-table was high along with longitudinal drains.

(ii) Flap valves operating automatically at 101.6 mm differential head at 152.4 m interval in the centre and on either side.

(b) Sides

(i) 114 mm thick filter was used on sides along with transverse drains in reaches of porous strata and high water table.

(ii) Same as item (ii) of bed.

(c) Joints

9.5m x 38 mm expansion joints were provided.

Performance

Lining in Continuous running Reaches: Ever since the canal was commissioned, it has not been possible to close it up to km 33 due to requirements of water at Kotla and Ganguwal Power houses. The continuous running in this reach has given rise to the following problems:

a) Failure and collapse of slabs have been caused due to unequal settlement in the backfill or due to creation of cavities underneath arising from movement of lines through joints.

b) The bulging of slabs has been caused due to excess hydrostatic pressure as a result of inadequate drainage against high water table outside the canal.

c) Due to continuous running of the canals the joints could not be attended during the last several years in which the hot pour compound has been washed out.

d) Some leakage has been observed through the canal embankment and at the cross-drainages during the past few years. Although the source of the leakage can only be located on closure of canal, it appears o be a result of gradual saturation of the banks and the subsoil due to continuous seepage through joints and cracks.

Lining in Heavy Cutting Reaches : Some major damages have occurred due to sub-soil water in the reaches of heavy cutting especially downstream of Km 33. The reaches, where proper drainage arrangements were initially provided, gave less trouble. General rise in water table has taken place in the vicinity of the canal. Serious damage has resulted due to closure for annual repairs downstream of Lohand Regulator. Experience on this canal has indicated that concrete lining has given better overall performance in high water-table reaches than tile lining.

Lining in Filling Reaches: there have been practically little damages in filing or partly filling reaches. In such cases major cause of damage was the entry of rainwater into the sub grade behind the lining on he slopes.

6.10. Sunder Nagar Hydel Channel-Beas Sutlej Link (BSL) Project, Himachal Pradesh, India

The Sundernagar Hydel channel is 11.8 Km long concrete lined hydel channel (Beas Design Organization Design Memorandum and Project Reports, & 1965). It off takes from the exit portal of Pandoh Baggi Tunnel and outfalls into Sunder Nagar, Balancing Reservoir, before the inlet portal of Sundernagar Satluj Tunnel across the Suketi Valley, near Sundernagar Township in Himachal Pradesh.

The channel has a full supply discharge of 254.85 cumecs (9000 cusec) at head and 238.68 cumecs (8250 cusecs at tail. It is an open channel, concrete lined throughout its length, running along the periphery of hillocks of Suketi Valley.

The alignment of the channel has been demarcated on the basis of map prepared by Survey of India on a scale of 4" to a mile (1:15, 840) and 10 feet contour intervals. Subsequently finalized after marking proposal(s) on the ground with a detailed survey to a scale of 1" to 200 feet (1:2400) and 10 ft contour intervals in a strip of 1000 ft. width along the channel alignment.

The sub strata were also adequately investigated by Geological Investigation.

In general the following considerations were taken into account in fixing the alignment:

(i) To provide the entire hydraulic section in cutting. This being preferred to a fill section in view of seismicity of the area and nature of the terrain.

(ii) To keep the depth of cutting to the minimum.

(iii) To provide right angle crossing with natural drainages.

(iv) To avoid steep hill slopes.

According to practice in Punjab the minimum radius of curvature for the canal alignments is as follows:

Capacity of Channel	Minimum Radius
1500 to 2499 cusecs	1000 ft.
2500 to 4999 cusecs	1500 ft.
5000 to 7499 cusecs	2000 ft.
7500 to 9999 cusecs	2500 ft.
10000 and above	3000 ft

Considering the topography of the terrain through which the Hydel Channel passes, and to avoid excessive cuttings and fillings, a minimum of 2000 ft radius has been adopted.

Hydraulic Section

The hydraulic section of the channel has been designed for a water surface slope of 1 in 6666. The side slopes for the channel section has been kept as 1 vertical to 1.25 to 1.5 horizontal. The flatter slope was adopted keeping in view that the channel is lying in an area of high seismicity.

The hydraulic section has been designed on the assumption that Manning's co-efficient of rugosity 'N' for a concrete lined channel would be 0.016. In this connection reference is invited to the U.S.B.R. design standard No. 3 "Canals and Related Structures" wherein under Para 1.19 it is stated that a value of 'N' for Portland cement concrete lining of 0.014 may be used for design with the proviso, the value must be adjusted with the size of the channel. The value of 0.014 is considered satisfactory for hydraulic radius less than 5 but the value should be increased to 0.017 when the hydraulic radius approaches 20. On this consideration the design for the Sundernagar and Hydel Channel has been proposed for value of 'N'=0.016, the hydraulic radius (R) for the proposed section being expected to be 12 m.

On the basis of alternative studies for various combinations of bed width and depth, a bed width of 31 ft, with the corresponding depth as below has been adopted.

For 9000 cusecs	d = 21.1 ft
For 8250 cusecs	d = 19.2 ft
For 7500 cusecs	d = 18.3 ft

A depth much greater than 20 ft. was not favoured and bed width has been kept constant in the entire length of the canal for facility for construction. The section adopted is shown in figure 6.6 (a) and (b). The maximum velocity for 9,000 cusecs discharge is 6.2 ft/s.

Free Board

A free board of 2.5 feet above designed water level to top of concrete lining has been provided. In the head reach upstream of the first escape at RD 5,500 the free board above designed full supply level has been increased to 4 feet. This has been done to provide some margin in the event of the actual coefficient of rugosity for the tunnel being less than 0.016 assumed in design.

All masonry works along the channel are designed for a free board of 2.5 feet for a discharge of 9000 cusecs, which may be the designed capacity. This would facilitate remodeling without any interference with the masonry works.

Typical Cross Sections

Typical cross sections adopted for the Hydel Channel were proposed for the following conditions:

a. Section in fill to a maximum of 25 feet below bank level.
b. Section in cut to a maximum of 50 feet above channel bed level.
c. Section in cut involving excavation more than 50 feet above bed level.

The maximum cut anticipated along the channel lining is about 120 feet.

In those portions of the channel where depth of fill was in excess of 25 feet below bank level, the adequacy of slopes was tested for the soil properties to be used.

Two typical sections for heavy cutting and filling are shown in figure 6.12.

Besides the typical cross-sections, Beas Design organization has also prepared detailed plans showing contours, curves, L- sections on the same scale and same drawing sheets, and detailed cross - sections at suitable intervals. The embankments have been designed as earthen dams based on detailed soil analysis of borrow areas, and stability of slopes. Zoned section with chimney drain and filter toe has been adopted in heavy fillings (embankment). Adequate drainage has also been provided behind lining.

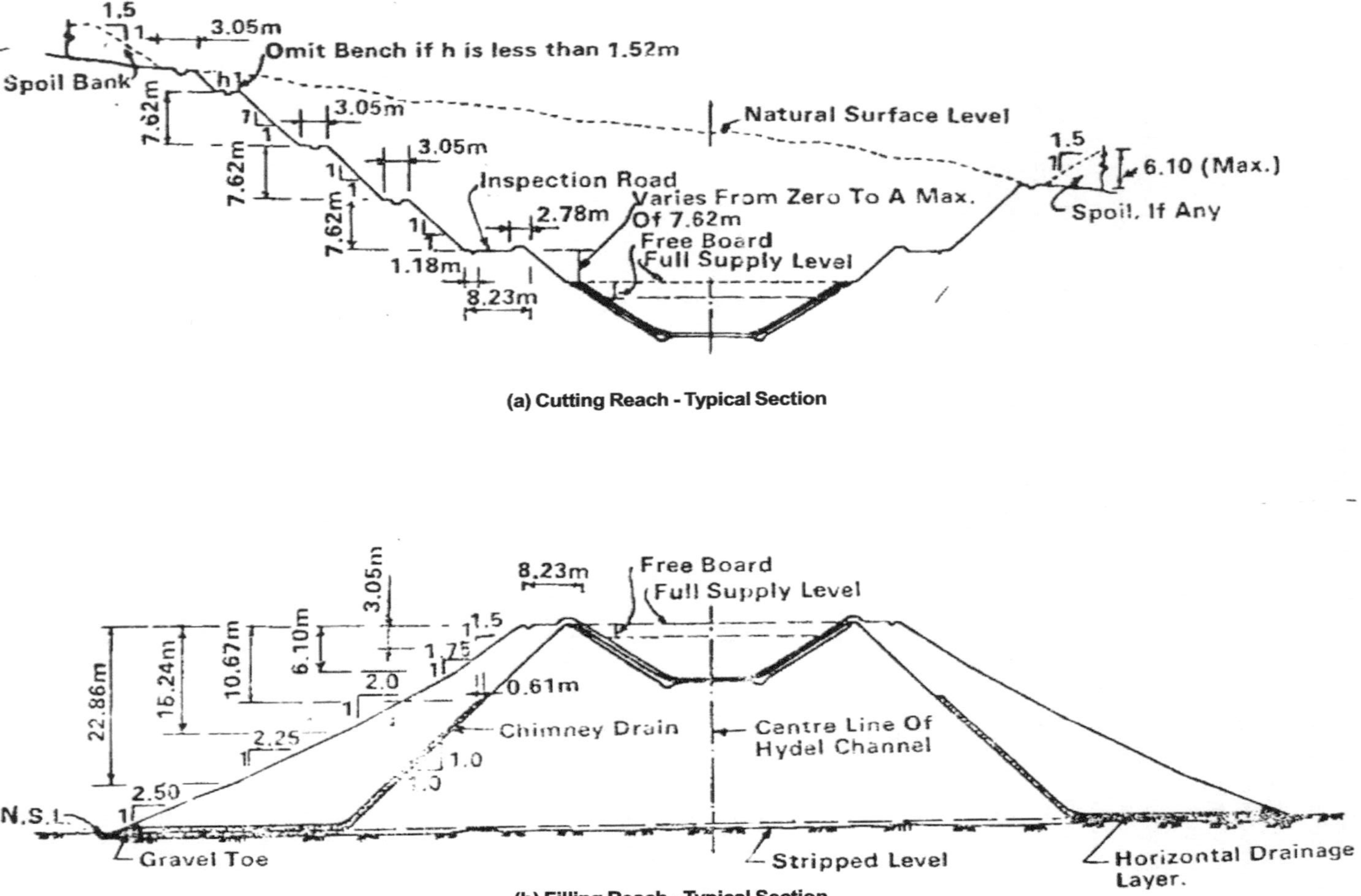

(a) Cutting Reach - Typical Section

(b) Filling Reach - Typical Section

Figure 6.12 Sunder Nagar Hydel Channel, Beas Sutlej Link (BSL) Project

Typical section for cutting and filling section are shown in figure 6.12. It is interesting to note that service road or **Top Bank** level is not parallel to FSL i. e. at free board. It has a smooth gradient to reduce the cost of heavy cuttings.

6.11 Performance of other Lined Canals

6.11.1 Yamuna Power Channel, U.P. India

This channel has been constructed in the years 1963-65 to carry a discharge of 198.1 m^3/s. The channel has a velocity of 1.775 m/s and has been lined in a length of 13.6 Km. with following details.

Sub-grade: It consists of fine and coarse consolidated sand with gravel and boulders with permeability, K = 600 to 7,500 m/yr.

Type of lining : Concrete lining in bed. Normal Reaches - 100 mm thick concrete laid in situ in 1: 3 colgrout. In reaches where ground water-table is high - 150 mm thick concrete lining was adopted.

Side lining consists of Brick and concrete tile lining.

Joints : 12.5 mm wide expansion joints have been provided at 10.06 m interval. The joints are on brick sleepers and filled by joint filler developed at U.P.I.R.I. Roorkee, using maxphalt.

Expansion joints of 20 mm width were provided in the lining at every 10 meter interval.

Pressure release arrangements

(a) *Bed:* Normal reaches-No pressure release arrangement has been provided. Reaches where ground water-table is high 30 cm diameter perforated cement concrete pipes at either junction of the side and bed and one at the centre of the bed fitted with pressure release valves at 5.18 m c/c.

(b) *Sides:* Reaches where ground water-table is high (i) The first layer of the filter consists of 150 mm thick layer of the material ranging from 2 mm to 50 mm. (ii) The next 150 mm thick layer consists of the material ranging from 37.5 mm to 50 mm in size. Special drain was constructed at the berm level for curing arrangement and dewatering.

Operation : In the interest of power generation, the power channel runs throughout the year. The power channel was specially closed in the years 1972, 1978 and 1984 for detailed inspection of the damages and to carryout repairs. It has been noticed that in the following reaches some panels of side lining of the channel were found damaged in almost every closure.

(i) In 00-1 Km on the hillside bank and just d/s of the curve.

(ii) In 8.00 to 9.5 Km on both banks.

The concrete lining in the bed was damaged in small patch in a length of almost 30 meter at Km 13.40 Km, as was noticed in the year 1984. It has also been noticed that the damages to channel lining are frequent during rainy season.

The damage to side lining in deep cutting reaches might have occurred due to:

(i) Bulging of tiles due to backpressure generated by high spring level or otherwise and absence of arrangement for release of back pressures. This damage is further extended by the forces of continuous flow in channel especially in the downstream of curve.

(ii) Loosening of the joints in the lining due to continuous water seepage from the earthen embankment into channel due to high spring level. This loosened mortar is further sucked by the flow in the channel thus damaging the lining.

The damage in the concrete lining in bed was observed only in the year 1984 and that too at only one place. The seepage of water from toe of the embankments and pits of 0.3 m to 1.0 m in the damaged area was seen. It is probable that some cracks in the bad lining had developed giving way to seepage of canal water. Such damaged panels were repaired with modified lining; in which 300 mm boulder pitching was done over 250 mm thick filter. Panels were formed along the length and slope of the channel to lay boulder pitching. Seventy percent area of the boulder pitching was grouted and 30% area in patches at the bottom was left ungrouted to release back pressures developed due to high spring level. This lining was first experimented in 1978. In the closure of 1984 these panels of modified lining were found satisfactory except dislodging of some boulders. On casual closure of the channel it was seen that these panels had again been damaged due to unloading of boulders.

The channel is running since 1965. The channel lining has behaved well generally except in the deep cutting reaches. Special care should be taken while providing lining in the channels in deep cutting reaches. Suitable changes are proposed as wire crating of the ungrouted boulder pitching. Suitable pressure release valves and drainage arrangement along with proper filter is also recommended for adoption in high spring areas along channel alignment.

Proper drainage arrangement is required at the outer toe of the embankment to avoid damages in the channel bed lining. A boulder pitching along with suitable filter is recommended for that purpose.

6.11.2 Rishikesh-Chilla Power Channel

The following types of lining have been provided in the power channel:

(i) Boulder Pitching lining
(ii) *In-situ* concrete Lining
(iii) Cement Concrete Tile Lining

The power channel was first time run in July, 1980. After running continuously for more than 4 years it was closed for 4 weeks for inspection and necessary repairs in Feb. 1985. In all 39 panels of c.c. tile lining were found partially damaged. These were mostly in deep cutting reaches where the ground water level on the back was quite high. The rest of the lining was found in good condition. The damage to panels was apparently on account of high back water pressure which could not be released through pressure release valves when the water level in power channel got suddenly reduced during monsoon due to choking of head regulator trash rack. It was observed that in most of the pressure release valves, the rubber flapper had damaged and silt laden water from the channels had entered into the pipe and had virtually choked it. Silt was coming out of some of the valves when the channel was closed indicating release of back water pressure through them. A number of valves had to be cleaned manually by inserting mild steel bars and when silt was thus removed, back water pressure could release. Some of the valves were found damaged and had almost dislodged from their respective place.

It was apparent that in the damaged panels the pressure release valves had not discharged their function properly. The damaged panels were repaired in the same manner as original tile lining was done and instead of installation of new valves, pockets of 1 meter x 1 meter at 3 to 4 meters center to center staggered in elevation, were left in the lining as shown in Fig. 6. These pockets were filled up with filter material and boulder pitching flushing with the lining surface was done. This has been found to work very well. The panels in which such arrangement was tried have not damaged again.

Canal closure of 3 days was taken in Feb. 1986 and 10 other panels found damaged were repaired in the manner detailed above including arrangement for the release of back water pressure. Seven days closure of the channel was taken during March, 1987 and 14 other panels were found damaged. The same procedure for repair and pressure release arrangement was provided.

The lining of power channel using compressed cement concrete tiles has been quite successful. These have performed well except in reaches where back water pressure had developed and could not be effectively released in case of sudden draw-down in the channel. The functioning of pressure release valves has not been very effective.

The details of drainage arrangement on this canal, Khara Power Canal and Eastern Ganga Canal are given in table 6.9. The subsoil water table in these canals was between canal FSL and bed.

Table 6.8. Under Drainage Provision in Some lined Canals

S.N.	Name of Canal	Type of Sub Grade	Sub Soil Water Level	Provision of Filter	Pressure Release Arrangements
1.	Chilla Power Channel (U.P.)	K<10^{4} cm/ sec	Between F.S.L. And Canal Bed	On Slope 75 mm. Coarse Sand **150 mm. Graded Filter (1to 50)** 225 To 300 mm> Graded Filter (37 To 100) 75 mm Porous Concrete	On Slope 75 mm Dia Prvs @ 5 m c/c In Two Rows. 3 m And 6.5m Belo w F.S.L. In Bed 3 Nos Iso mm Dia A.C. Perforated Pipe Longitudinal Drains, 150 mm Dia A.C. Perforated Pipe Transverse Drains @ 10 m c/c 75 mm dia prvs @ 10 m c/c.
2.	Khara Power Channel (U.P.)	K<10^{4} cm/ sec	Between F.S.L. And Canal Bed	On Slope 75 mm Coarse Sand 150 mm Graded Shingle 75 mm Porous Concrete In Bed 75mm Porous Concrete	On Slope 75 mm Dia Prvs @ 2.5mc/c In Two Rows, 2 m and 4 m below F.S.L. In Bed 3 Nos, 150 mm Dia A.C. Perforated Pipe Long Drains, 150 m.m. Dia A.C. Perforated Pipe Transverse Drains@10 m c/c 75 mm dia prvs @5 m c/c
3.	Eastern Ganga Canal (U.P.)	K<10^{4} cm/sec	Between F.S.L. And Canal Bed	On Slope 75 mm Coarse Sand 225 mm Graded Shingle 5 To 20 75 mm Porous Concrete In Bed 75 mm Porous Concrete	On Slope 75 mm dia prvs @ 2.5 m c/c Longitudinally 2.5 m below F.S.L. In Bed 3 nos, 150 mm dia a.c. perforated Pipe Long Drains. 75 mm dia prvs @5 mc/c on Drains.

6.12 Brief Description of Some More Lined Canals

(i) Typical cross-section of Sadul Branch, Bhakra Canal, Rajasthan, India constructed around 1950 is given in figure 6.13. The canal section is Mehboob lined with double tile lining. It has a depth of 13.3 ft and inner side slopes of 1 : 1.25 (V : H) was adopted. Soils of the embankment are sandy soils and no drainage arrangement was adopted. No damage of lining has been reported and performance can be considered as satisfactory.

(ii) Typical cross-section of existing lined canals in Pakistan, are shown in figure 6.14 (i) to (iii). These canals were lined, much earlier, when lining of canals was taken up in India around, 1950.

(iii) Typical cross-section of lined canals, adopted in Power (Hydel) canals in Bihar, India is given in figure 6.15 and 6.16.

(iv) General typical lined canal cross-sections, adopted in Gujrat, India are given in figure 6.17.

It may be seen that, generally a tangential curve of radius equal to full supply depth have been adopted is most of the practices, Details about advantage / disadvantage are discussed in Para 4.4.

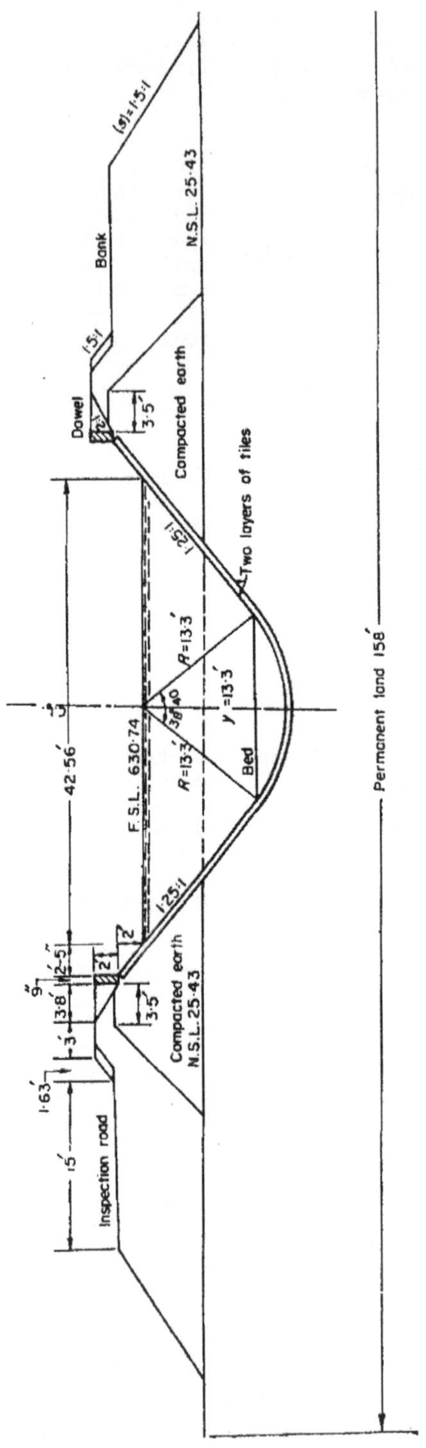

Figure 6.13 Typical cross section of Sadul Branch, Bhakra Canal (Rajasthan), India

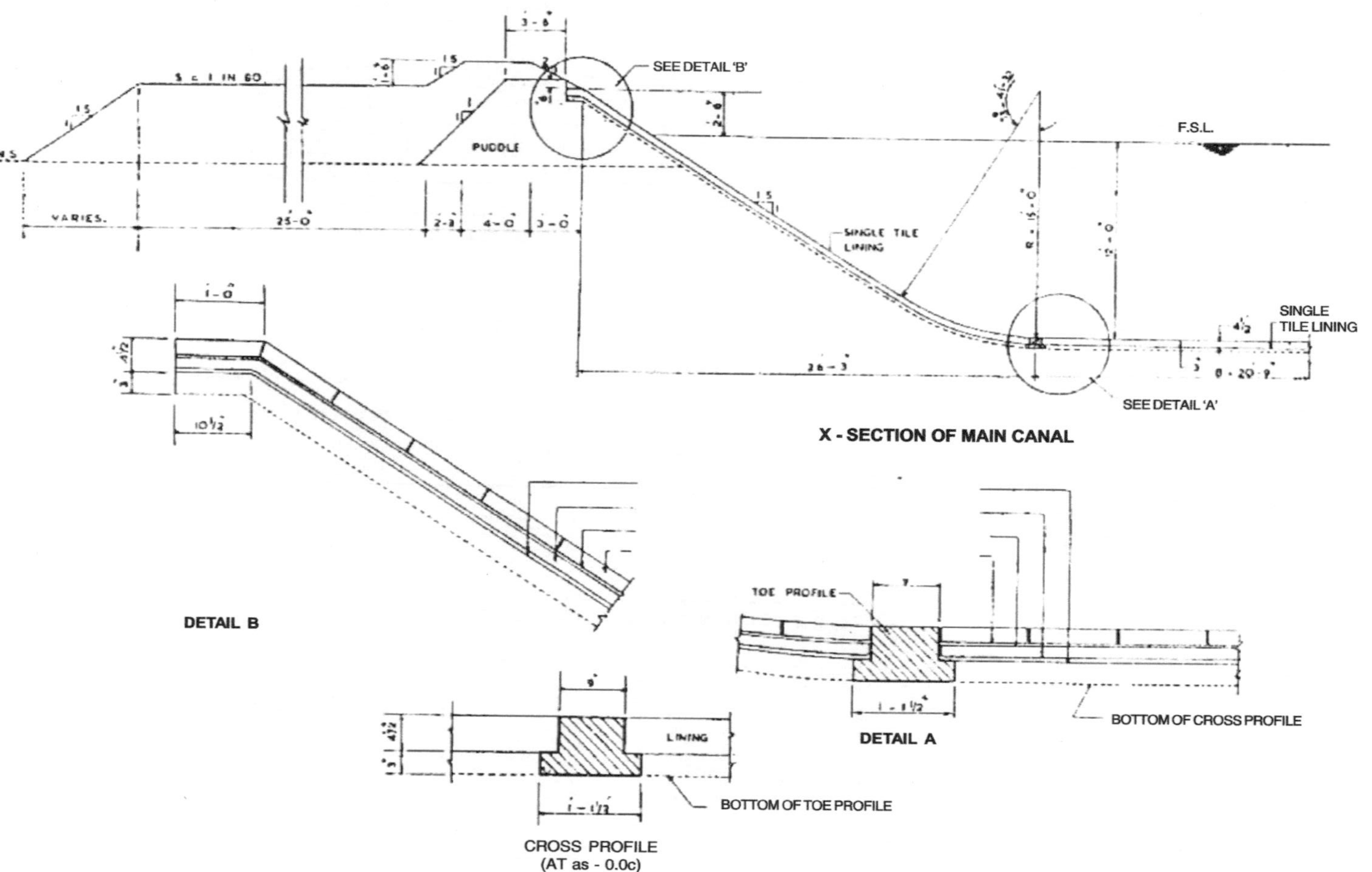

Figure 6.14 (i) Typical Cross Section of (Thal Main Line Lower) Lined Canal, Pakistan

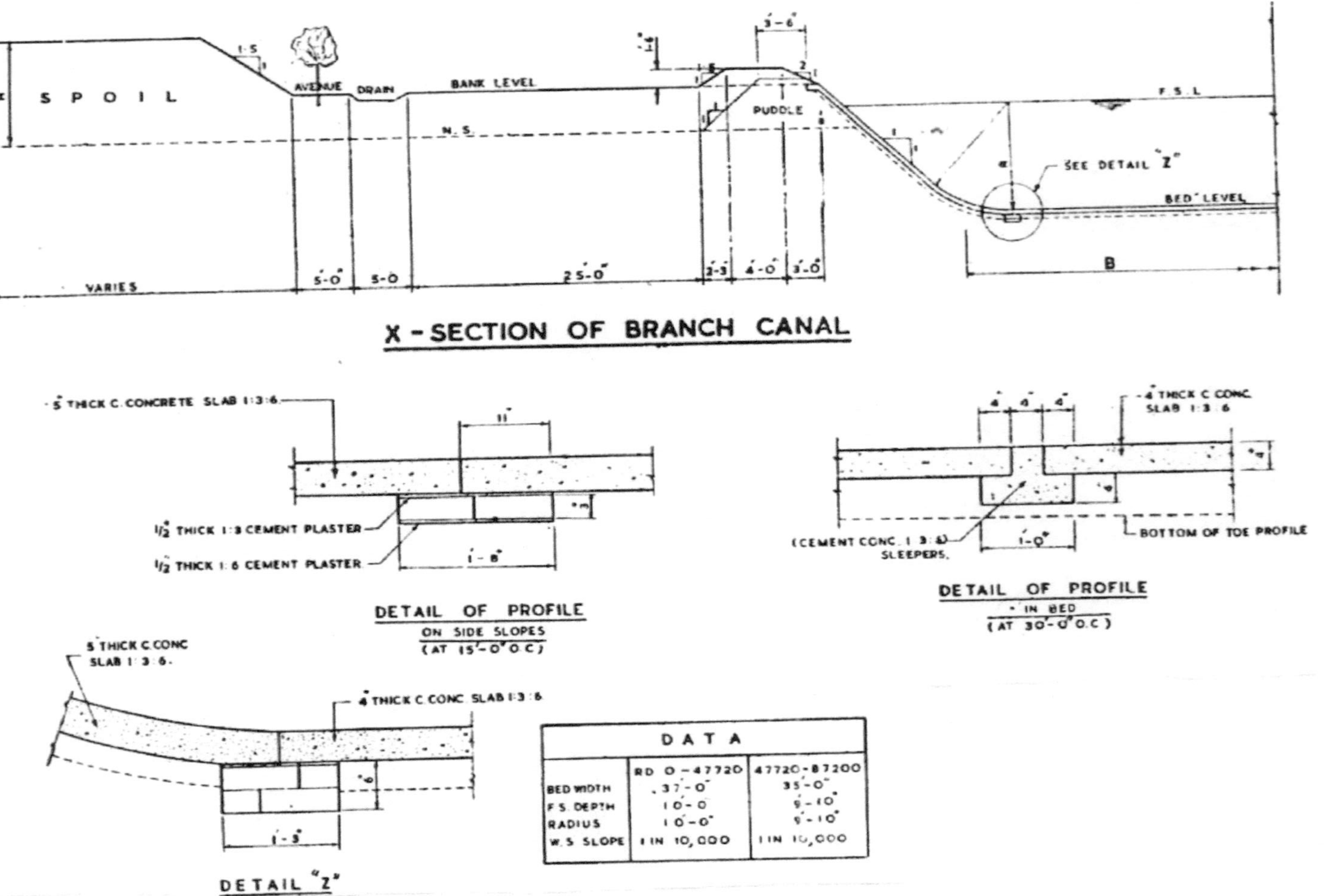

Figure 6.14 (ii) Typical Cross Section of (Khizar Branch) Lined Canal, Pakistan

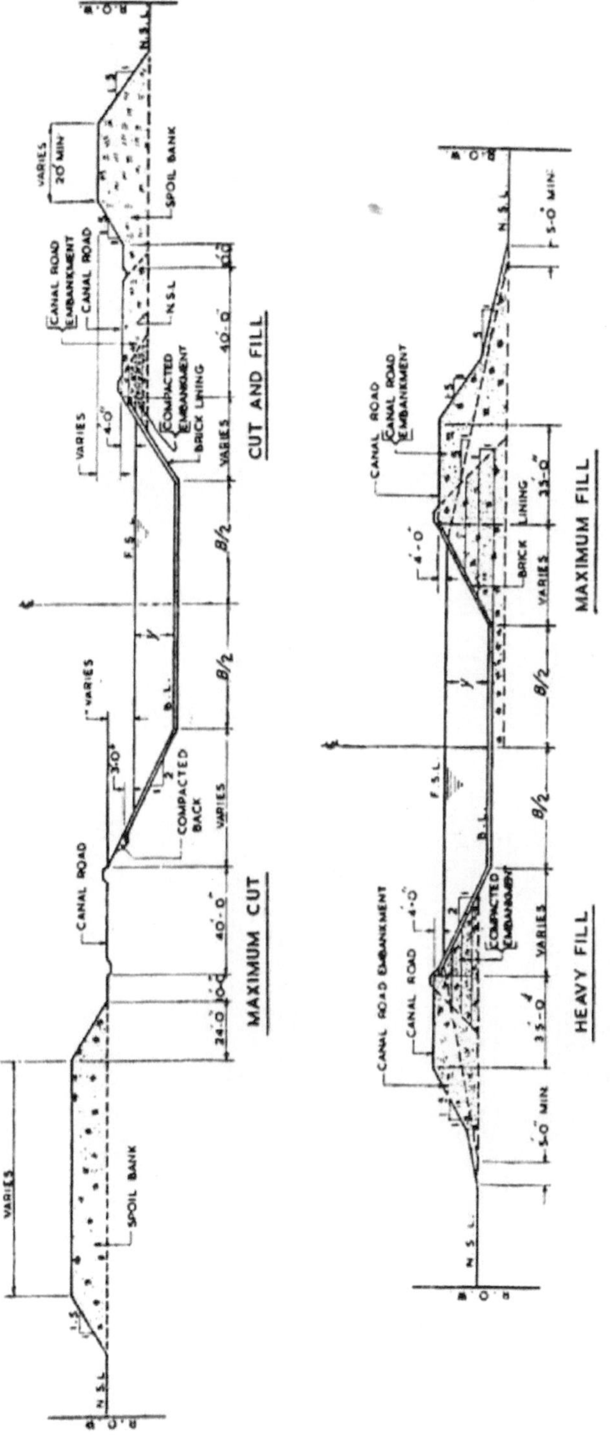

Figure 6.14 (iii) Typical Cross Section of (Sidhnai - Mailsi-Bahawal Link) Lined Canal, Pakistan

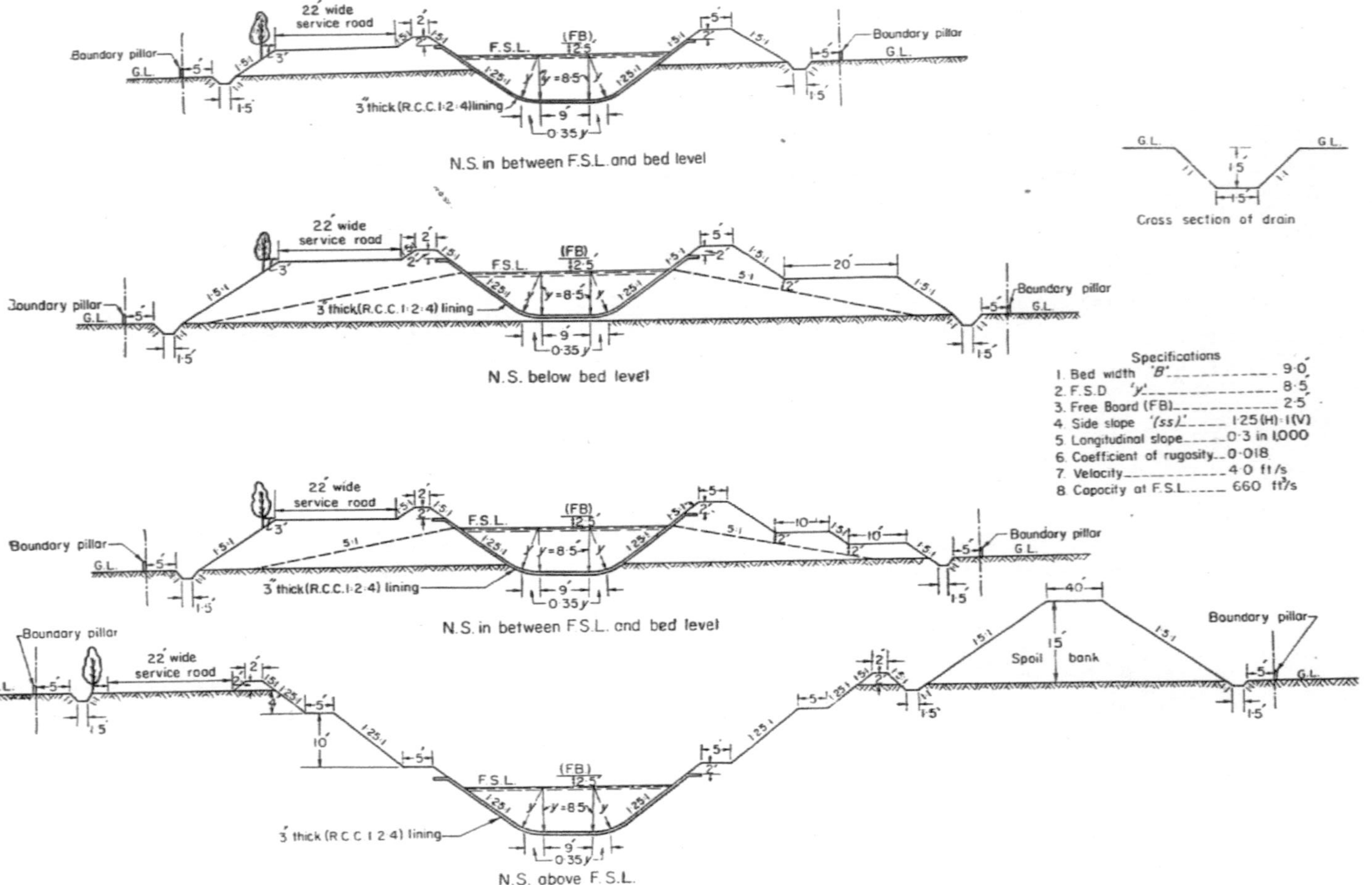

Figure 6.15 Bihar State Electricity Board, India, Typical Cross Sections of Basia Power Channel, South Koel Hydroelectic Project

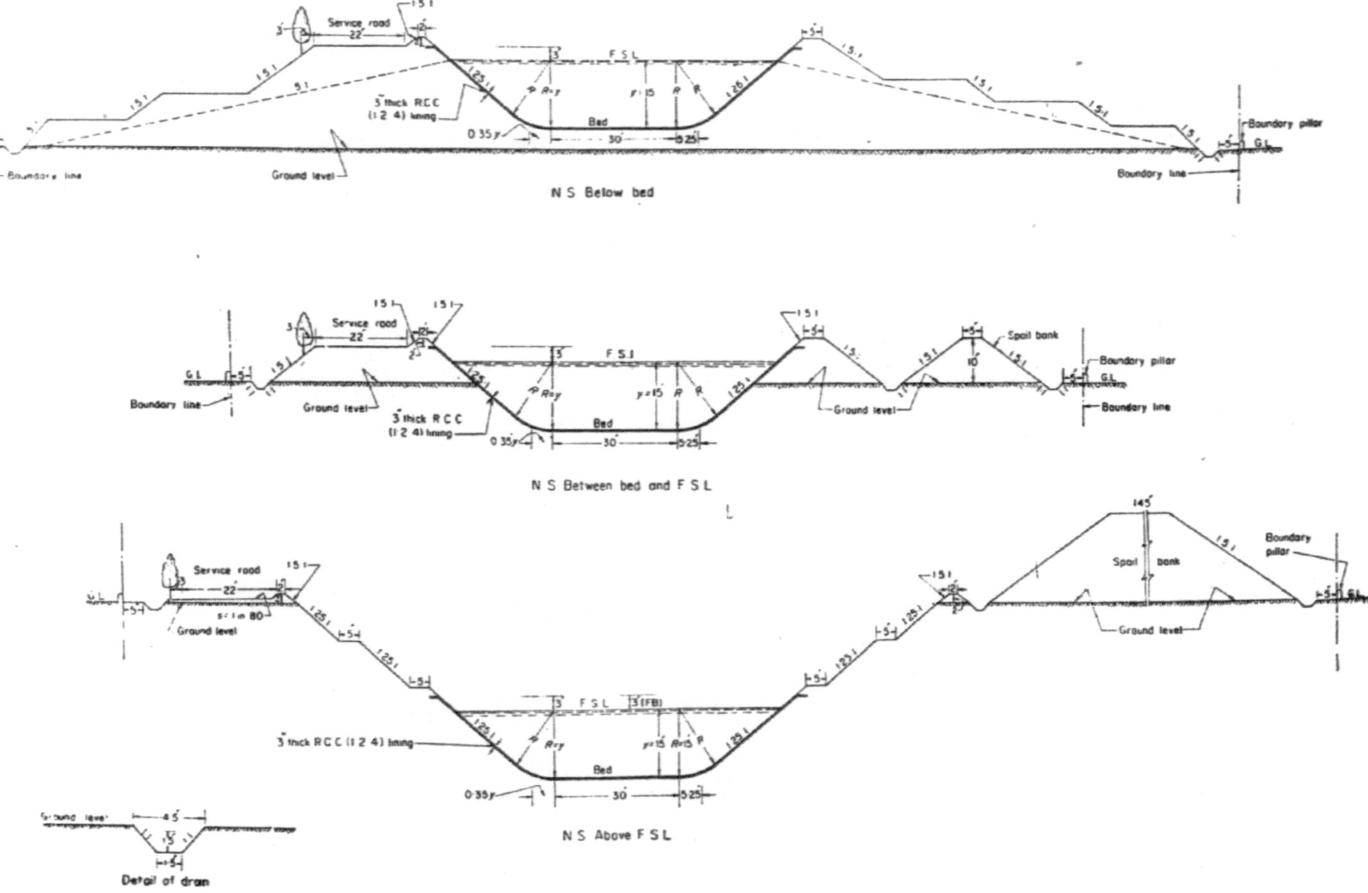

Figure 6.16 Bihar State Electricity Board, India, Typical Cross Sections, Subernrekha Hydro-Electric Project

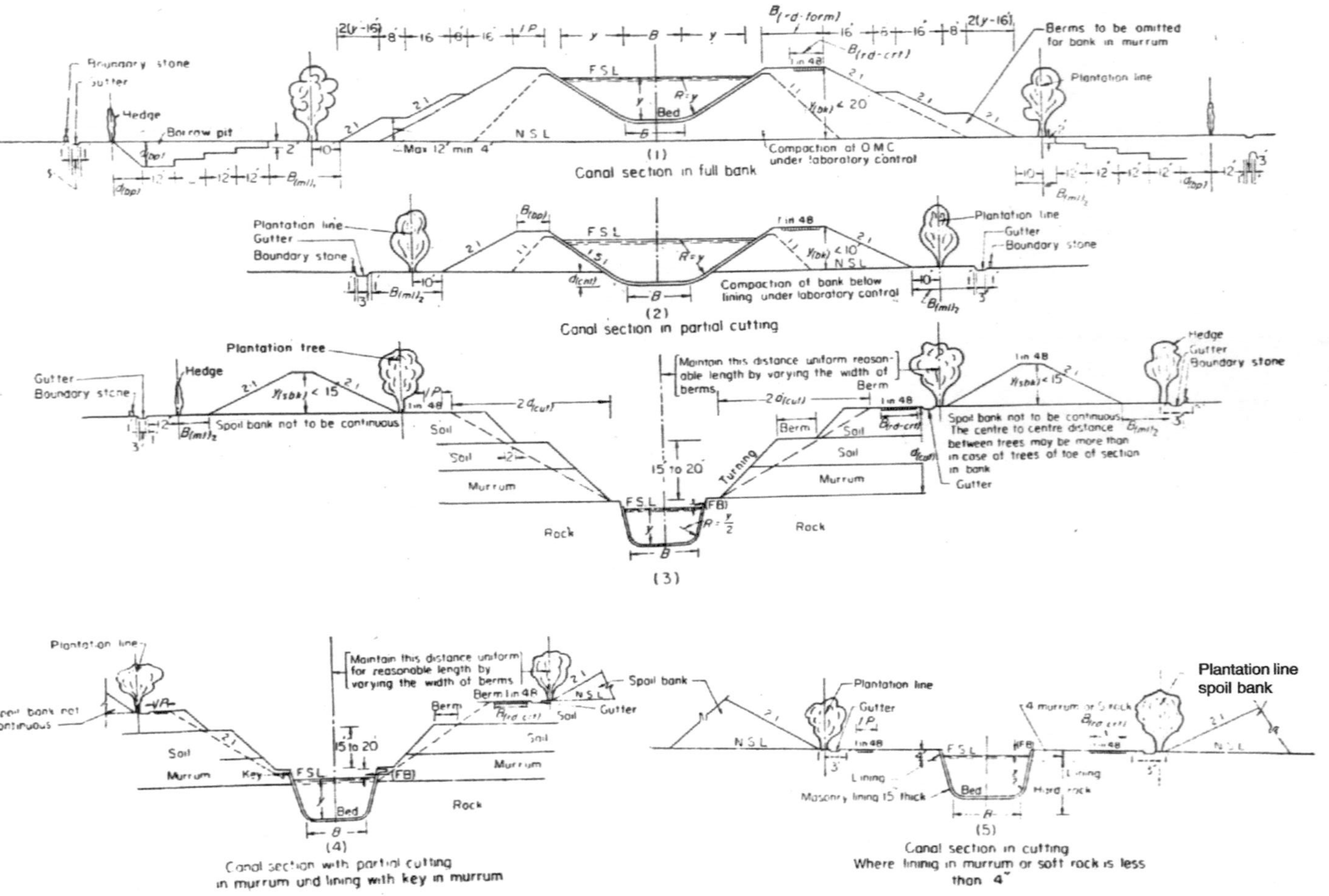

Figure 6.17 Typical Cross Sections for Lined Canals, Gujarat, India

6.13 Upper Ganga Canal (UGC), India

This canal is off taking from Hardwar, initially designed to carry 6600 cusecs, (190 m^3/s) at its head, and subsequently the discharge was raised to 10,000 cusecs, (290 m^3/s) without any modifications in banks and structures. All the structures had sufficient free board and high strength in brick and lime mortar, to pass 50% additional discharge.

It may be interesting to know the design details of this 150 years old canal, constructed in the years 1845-54 by Colonel Cautley.

The high embankment of unlined canal near Roorkee (about 30 Km from Head) has a positive cut off in brick masonry arches in lime mortar, on the face and also works as Ghats (bathing place) as shown in figure 6.7 (i) and more details shown in figure 6.7(vi) and further discussed in the notes give below the same figure (a), (b), and (c). For most of the works modified alternative (c) was adopted.

The Brick arch masonry in lime mortar works as positive cut off for the earthen embankment placed behind it, as well as face wall as Ghats, very useful to the local people for bathing and collection of water. The brick arch masonry was so designed that it stood the earth pressure, when the canal is empty. Also structurally it was independent of earth pressure/water pressure from either side, a very novel and economical design of positive cut off.

Section of canal

The dimensions of the main canal and branches were worked out by Dubuat's formula which is as follows:

$$V = \frac{306.55(-\sqrt{R} - .10325)}{-\sqrt{b} - HypLog - \sqrt{b+1.6}} - 0.31(-\sqrt{R} - .10325)$$

Where V = Velocity of flow, R = Hydraulic mean radius,

and 1/b = the denominator of the fraction expressing the slope of bed.

Channel dimensions of the main canal so obtained were :

Mileage (1)	Bed slope(2)	Bed width ft (3)	Water depth ft (4)	Side slopes (5)
0	18"/mile (1/3520)	140	10	1:1 side slope was adopted for the entire length of the canal.
50	18"/mile (1/3520)	130	9	
110	18"/mile (1/3520)	108	9	
180	15"/mile (1/3520)	90	7.5	
250	14"/mile (1/3520)	80	7.0	
281	12"/mile (1/3520)	75	6.0	
Terminus	12"/mile (1/3520)	25	4.0	

S84(*d*)

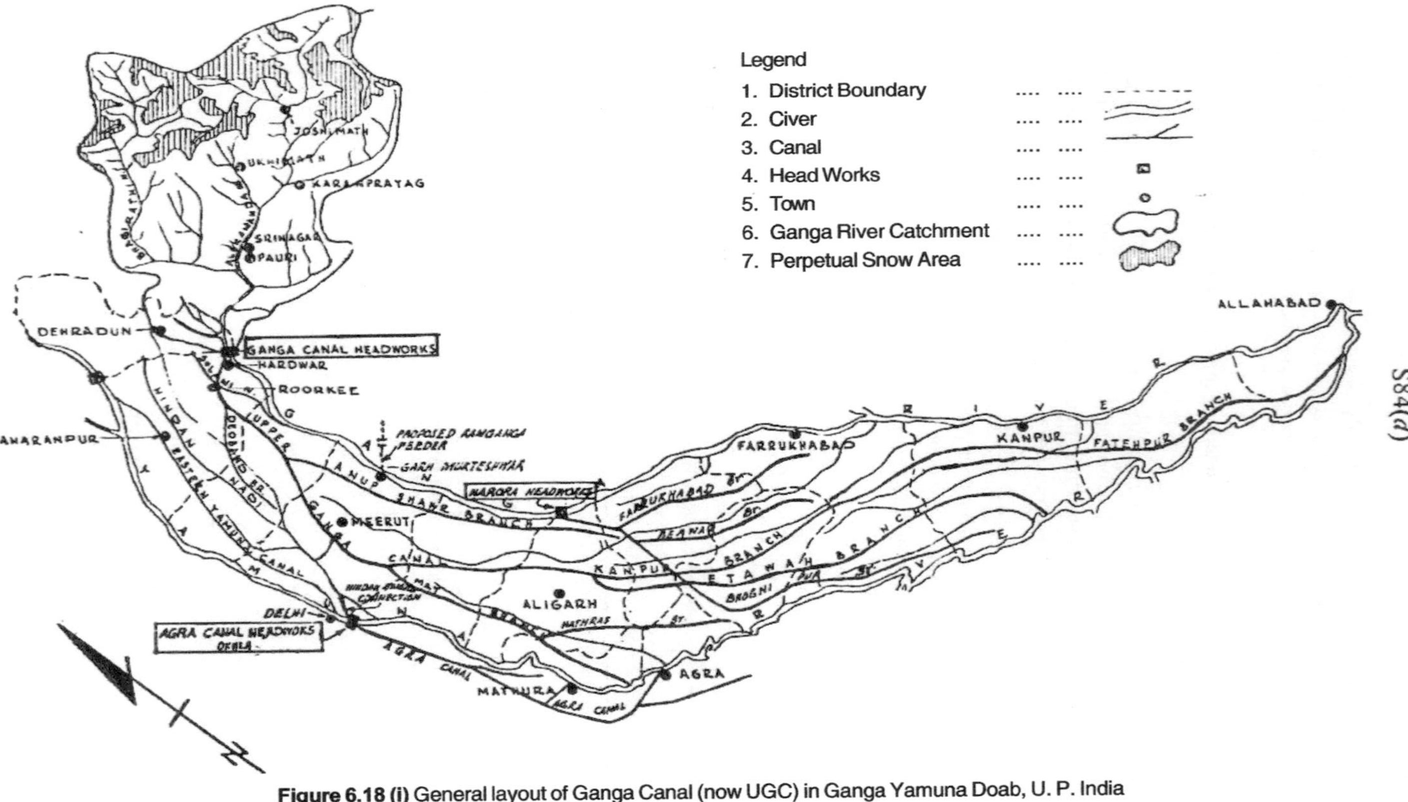

Figure 6.18 (i) General layout of Ganga Canal (now UGC) in Ganga Yamuna Doab, U. P. India

S84(a)

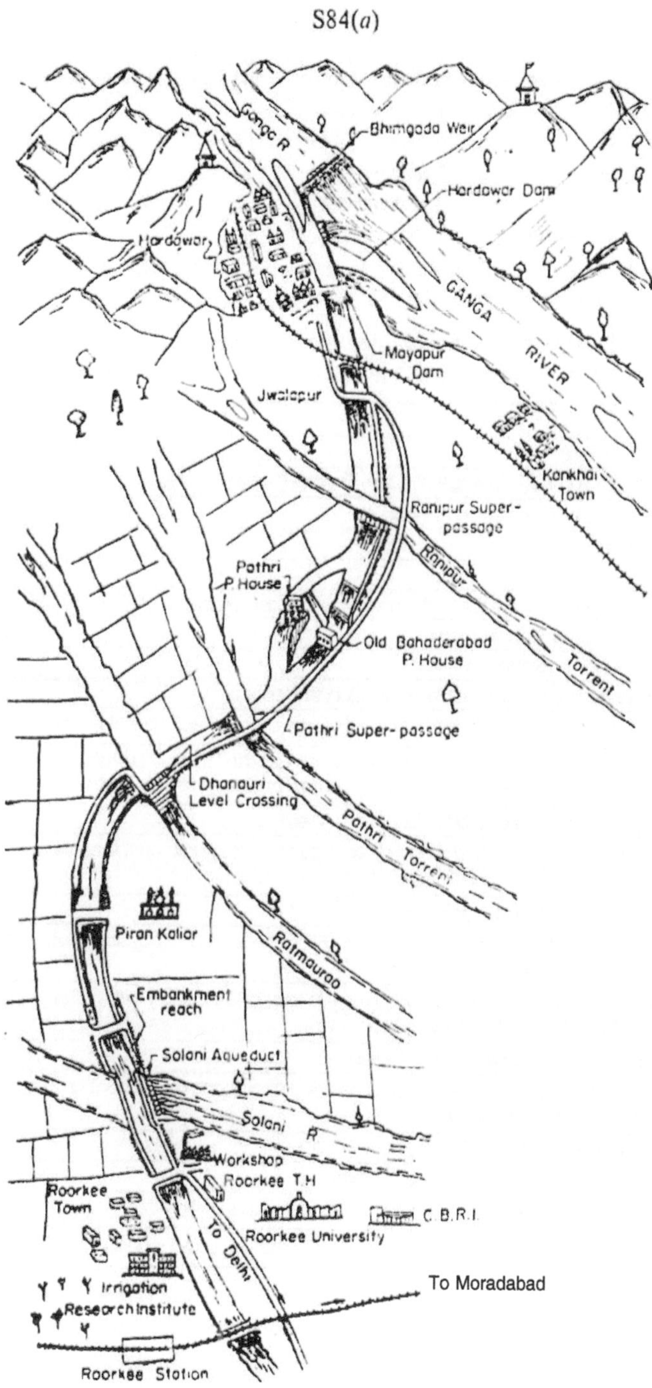

Figure 6.18 (ii) Sketch map showing cross drainage works on Upper Ganga Canal in its first 20 miles, U. P. India

The bed slopes are quite steep and velocities in main canal in head reach are of the order of about 4 to 5 ft/s (1 to 1.5 m/s).

However the permissible velocities in the design of Ganga Canal system were:-

(i) In light sandy soil 1.5 to 2 feet per sec.
(ii) In sandy loam 2.5 feet per sec.
(iii) In ordinary firm loam 3 feet per sec.
(iv) In stiff clay or kankar soil 4 feet per sec.
(v) In shingle bed and boulders. 5 feet per sec.

Lot of slit enters this canal, being off taking from a weir/barrage at the foot hills of the river. The typical silt problem is discussed in Para 7.20.

Historical Events of Upper Ganga Canal (earlier known as Ganga Canal)

A brief historical event of this canal which got a great name and fame in those years is given below.

S. No.	Events	Year
1	First Feasibility report	1837-38
2	Establishment of Thomson College of Civil Engineering at Roorkee	1845
3	Establishment of Govt. workshop at Roorkee	Ist Jan, 1848
4	Commissioning of the Project	8th April, 1854
5	Start of irrigation	1 May, 1855
6	First major remodeling of the canal (to improve initial deficiencies)	1868
7	Construction of Lower Ganges Canal	1874-1878
8	Start of Osrabandi (warabandi)	1880
9	Heavy damage to head-works at Hardwar (Gohna Floods)	1894
10	Construction of first Hydel power station on the system at as well as in the state	1912-1915
11	Settlement with British Government for leaving free-gaps in Bhimgoda Weir and Mayapur Dam for passing on uninterrupted flow of Ganga to the Ghats	1917
12	Addition of permanent Head-Works (Bhimgoda Weir)	1917-1922
13	Major damages to the head-Works and other important structures on the canal (Damages at Dhanauri Cross Drainage Works)	1924
14	Addition of subsequent Hydel Power Stations	1930-1955
15	Increase in the head discharge of the canal (6750 Cusecs. to 8500 Cusecs) with out any modification of banks and structures.	1938
16	Further increase in head-discharge (8500 cusecs to 10500 Cusecs) with out any modification of banks and structures.	1951-1952
17	Century celebrations of upper Ganga Canal	1954-1955
18	Silting in the head reach of canal requiring desilting operations on account of Alakananda Calamity (land slide)	1970
19	Addition of silt Ejecting drain (Slit Ejector at Km 22)	1975-1977
20	Addition of New Head-Works (Bhimgoda Barrage)	1978-1984
21	Strengthening pathri super passage (major remodeling near Km 16)	1986-1988
22	Commissing of parallel canal (PUGC) upto down stream of solani aqueduct	Nov.2001

A general layout of this canal is shown figure 6.18 (i), and sketch map of first 20 miles, showing cross-drainage works in given in figure 6.18 (ii). There are many torrents (rivers) crossing the canal. The original L-Section and subsequent modification carried out for power houses is shown in figure 6.18 (iii).

The silt problem of this canal is discussed in Para 7.20.

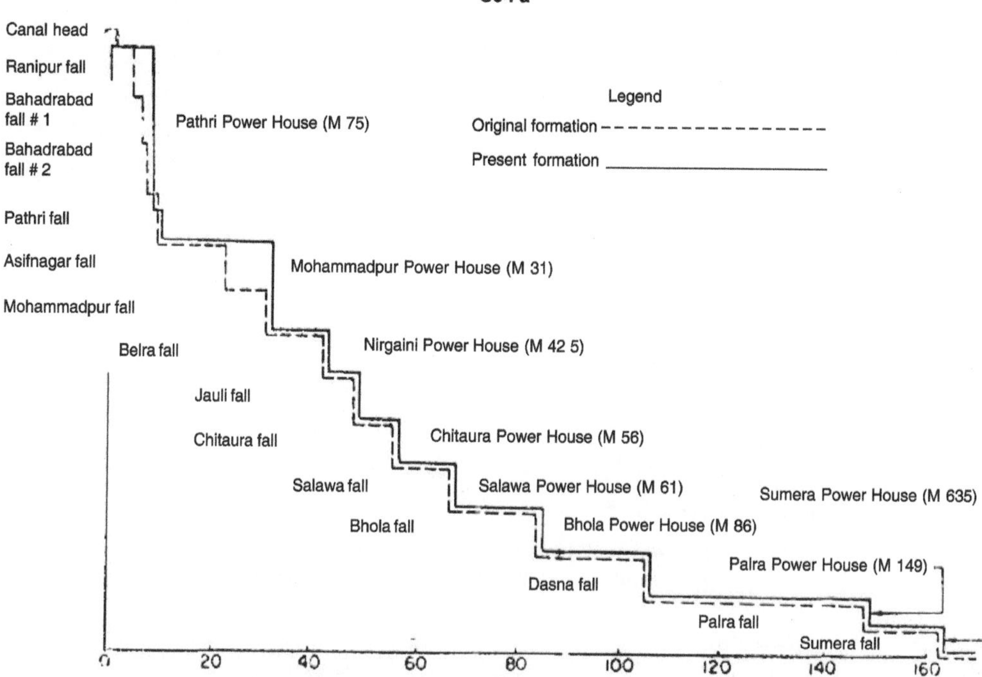

Figure 6.18 (iii) Falls on Ganga Canal as originally provided and as modified for hydro power generation, U. P. India

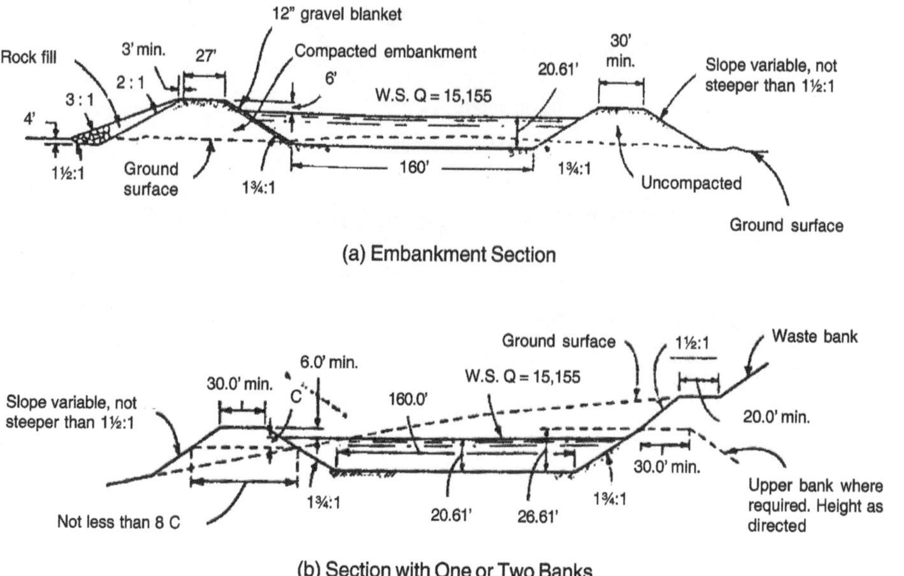

Figure 6.18 (iii) Typical Cross Section of All American Canal, USA

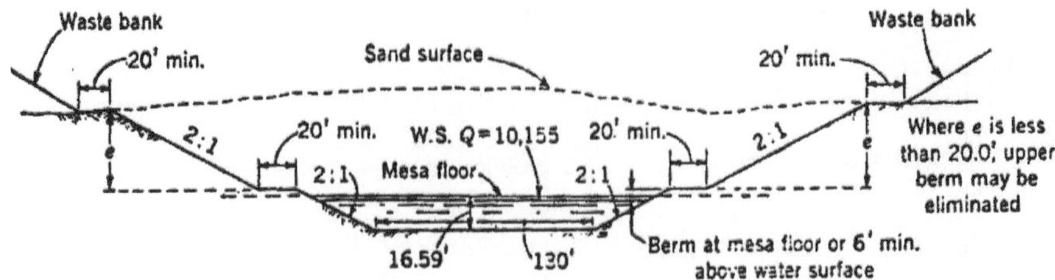

(c) Earth Section (Through cut. Mesa Floor 6' or less above water surface)

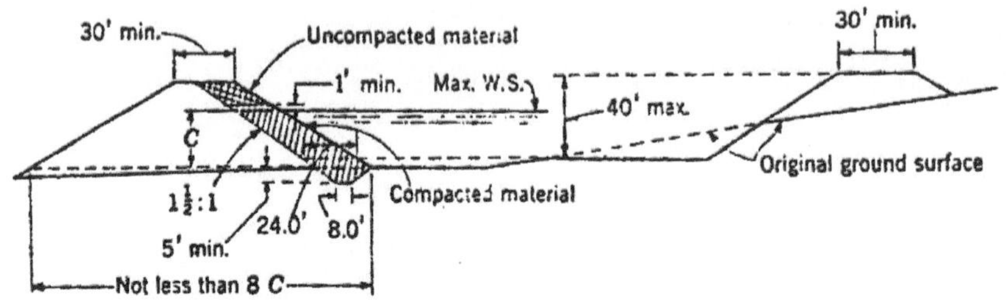

(d) Section with Compacted Bank Slope (Ground surface below bottom grade)
Note : (a) to (d) earth section of All American Canal, USA

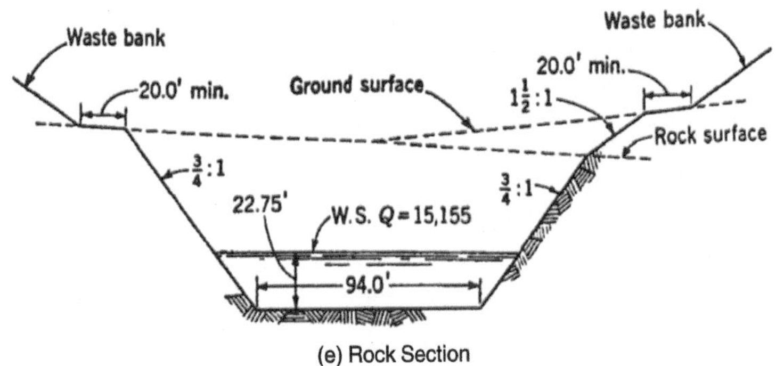

(e) Rock Section

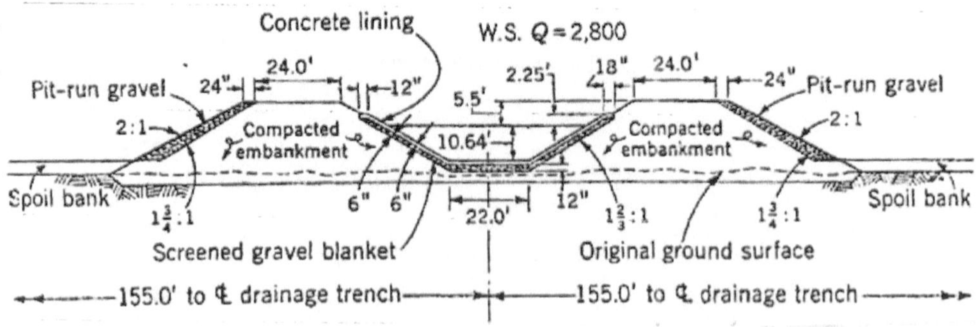

(f) Concrete-Lined Section
Note : (e) and (f) Rock and Concrete lined sections used in the All-American Canal

Figure 6.18 (iv) Typical Cross Section of All American Canal, USA

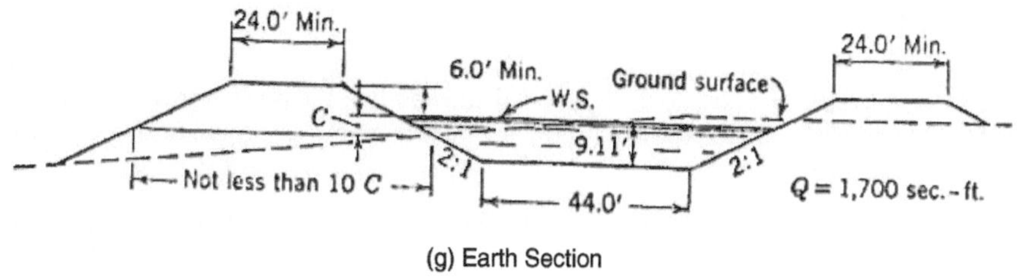

(g) Earth Section

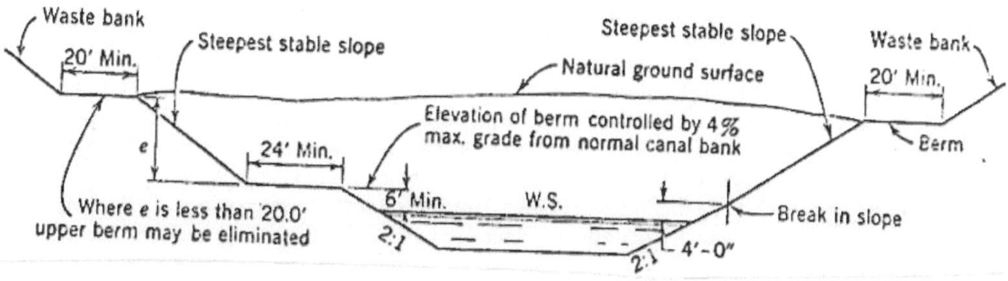

(h) Typical Section in Through Cut

Note : (g) to (h) Earth sections Coachella Canal a branch of All American Canal Project.

Details of Sections of All American Canal

S. No.	Q (ft/s)	B/D	A (ft²)	Z	P (ft)	R (ft)
(a)	15155	160/20.61 = 7.76	4040.95	1.75	243.08	16.62
(b)	15155	160/20.61 = 7.76	4040.95	1.75	243.08	16.624
(c)	10155	130/16.59 = 7.84	2707.16	2.0	204.19	13.26
(d)	----	----		----		
(e)	15155	94/22.75 = 4.13	2526.67	0.75	150.87	16.75
(f)	2800	22/10.64 = 2.07	423.14	1.67	63.42	6.67
Coachella Branch Canal						
(g)	----	44/9.11 = 4.8	566.82	2.00	84.74	6.69
(h)	---	---		2.00		

Figure 6.18 (iv) Typical Cross Section of All American Canal, USA

6.14 The All American Canal (U.S.A.)

The All-American Canal, 120 Km (80 miles) long carries Colorado river water from imperial Dam near Yuma Arizona to imperial Valley. Built by the Bureau of Reclamation during 1934 to 1940 was designed for maximum discharge of 433 cumecs (15,155 cusecs) at the head to 79.5 cumecs (2,600 cusecs) at the tail end.

It involved practically all problems of canal design. Some sections pass through earth formations, other through rock strata. One section west of Yuma required deep excavation through sand dunes. Some sections are wholly in excavations, some partly in excavation and some entirely between earth embankments. One embankment section near the lower end was lined with reinforced concrete.

At certain sections of the canal along sloping ground, inner slopes of the downhill bank were lined with thick layers of compacted earth. At other locations where piping might take place, cutoff walls of sheet piling were driven beneath the lower banks. One 13 mile section has 4-inch blankets of loose clay along the sides.

Observation during early operations of the canal showed that the plans were effective in reducing seepage, and that troublesome quantities of sand were not blown into the canal through the sand-dune sections.

Six typical sections of the All-American canal are shown figure 6.18. Section in (a) is an embankment section where one bank was compacted and one uncompacted; section in (b) shows a hillside section, mostly in excavation, where compaction of the downhill bank was not specified; a sand-dune section through deep cut where berms were provided along both banks is shown is (c); section (d) is a hillside section where a thick layer of compacted material was placed along the water face of the downhill bank.

Section (e) shows a deep rock cut planned for maximum depth of flow of 22.75 feet. The Section (f) shows the concrete-lined channel between compacted embankments near the downstream end of the canal. Lining is provided in exceptional high filling reaches. The inner slopes of the embankments at the lined section were provided with a 6-inch blanket of screened gravel and covered with a ½ - inch coating of Gunite before placing the concrete linings. The outer slopes were provided with thick blankets of pit-run gravel. Drains were installed under the lined section at 250-foot spacing, and drainage trenches were excavated along both banks.

Typical cross sections of Coachella Canal, main branch of the All American Canal are shown in figure 6 at (g) to (h). It carries water 145 miles north along the east side of the Imperial Valley.

In the design a mean velocity of 3.7 feet per second was used for earth sections, 160 feet wide at the bottom and 20.6 feet deep, with side slopes of 1.75: 1. Canals carry some fine silt in suspension. All of the heavier sediments are removed by the desilting works at Imperial Dam

As compared to other canals in Para 6.1 to 6.12, the section is more economical with a B/D ratio of 7.8, and side slope of 1:1.75 (V: H) even for depth of 20.6 feet. The canal is lined only is filling reaches.

In the long canal, width, depth, and side slopes are not constant, even for the same discharge. These change with topography for economical consideration, similar to Main canal (Para 6.8)

6.15 Some Unlined Canals and Their Cross-Section

6.15.1 Nagarjunsagar Project, Right Main Canal, Andhra Pradesh, India

This canal has been designed and constructed around the year 1955-65 as a two stage canal. In the first stage the canal has to pass a discharge of 11000 cusecs and (311.5 cumecs) and the second stage or ultimate discharge is 17000 cusecs (481.1 cumecs). Typical cross sections of this canal showing two stage design and construction are shown in figure 6.19 (i) to (iii) with notes to these figures as under:

Notes to figure 6.19 (i)

(1) Section (1) will have to be formed for the first phase. Section (1) and (2) may be formed if surplus earth is available and full ultimate section may be formed if more surplus earth is available from canal excavation.

(2) When the ultimate section is formed, platforms at regular intervals will have to be provided for avenue trees.

(3) For greater heights of embankments i.e. more than 30 ft., 5 ft. berms and 2.5 : 1 slopes may be provided at convenient intervals so as to ensure 2 ft. soil cover over hydraulic gradient of 4 : 1 before it is met with rock toe.

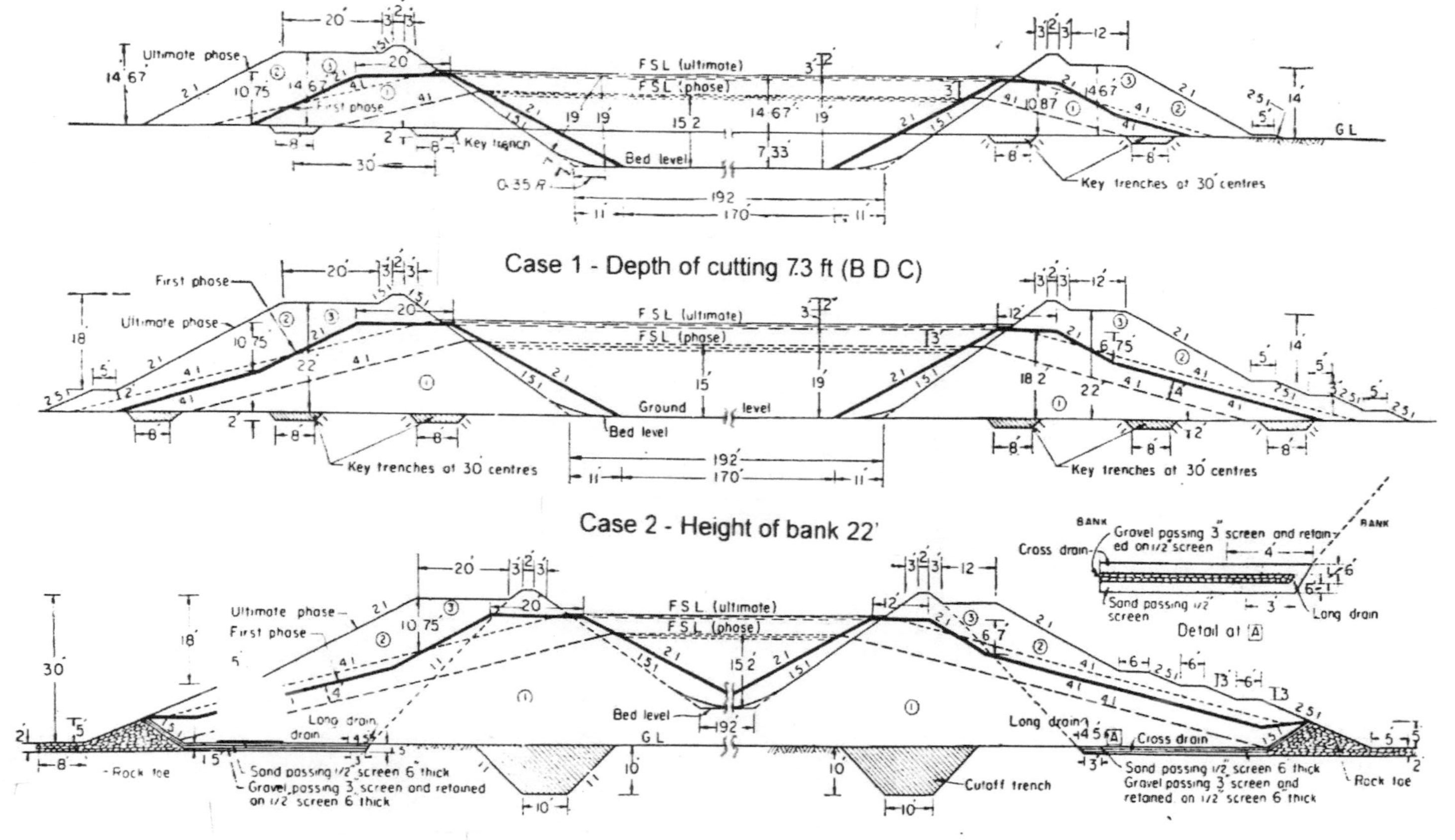

Note : Canal designed and Constructed around the year 1955-65 as Two phase construction.

Figure 6.19 (i) Andhra Pradesh, Nagarjunsagar Project, Right main Canal, (M, 30/3 to 41/4) - Typical Cross Sections

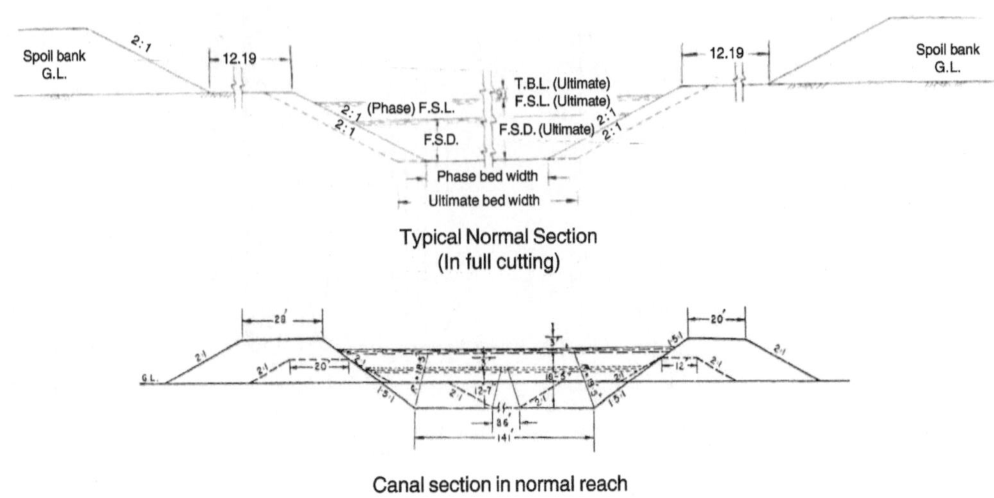

Typical Normal Section
(In full cutting)

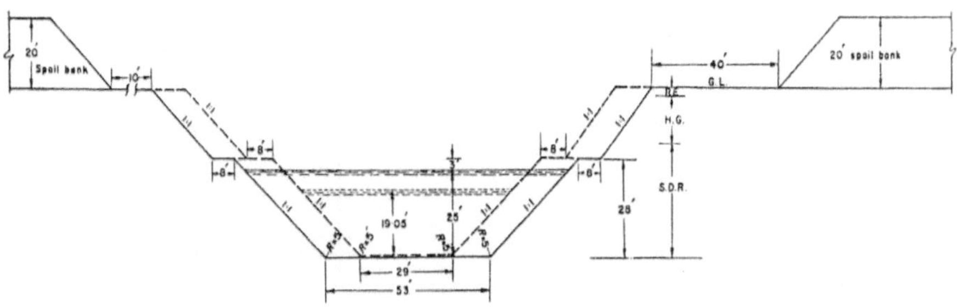

Canal section in normal reach

Canal section in flume from M 57 - 4 - 318 to 58 - 1 - 626

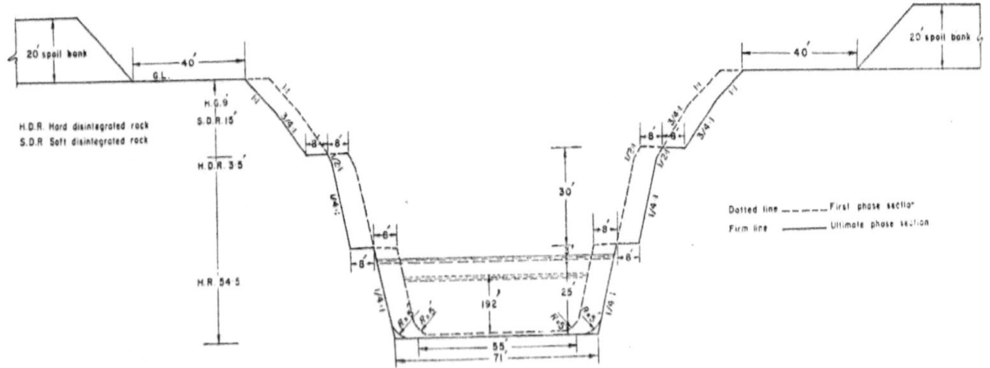

Canal section in flume reach M 60 - 2 - 33 to 62 - 4 - 413

Figure 6.19 (ii) Nagarjunsagar Project, Right Canal, Typical Cross Sections, Andhra Pradesh

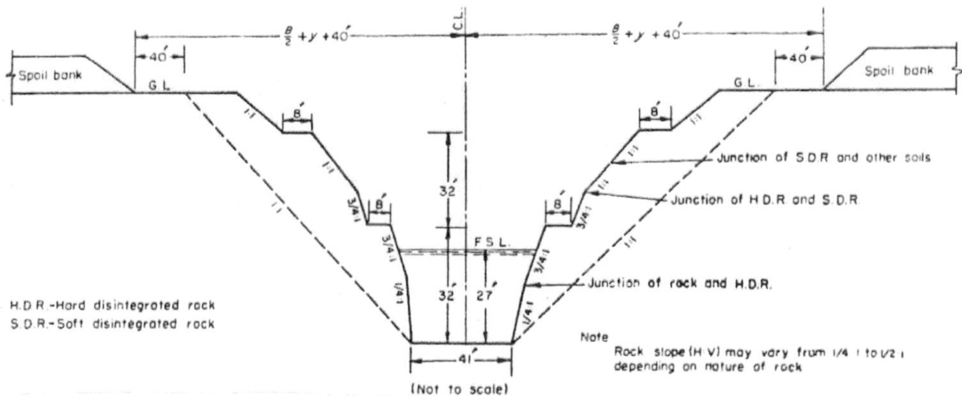

(a) Section of flume where depth of rock-cut is less than Full Supply Depth (F.S.D) – Nagajunasagar Project (not to scale)

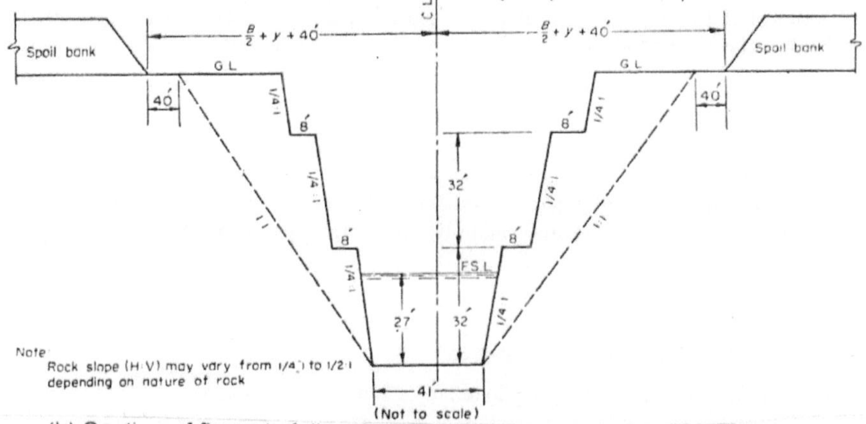

(b) Section of flume is full rock-cutting Nagarjunsagar Project (not to scale)

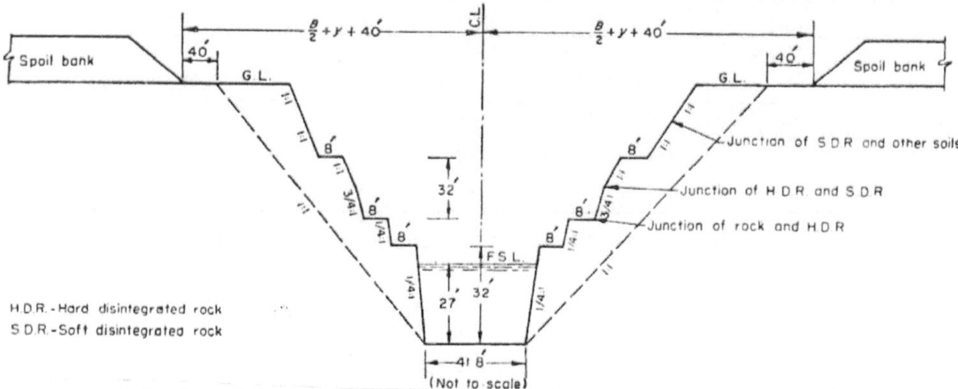

(c) Section of flume where depth of rock-cut in more than Full Supply (F.S.D) – Nagarjunsagar Project

Note : Rock slopes H : V may vary from 1.4 to 1 to 1/2 to 1 depending on the nature of rock (not to scale).

Figure 6.19 (iii) Nagarjunsagar Project, Right Canal, Andhra Pradesh

(4) When there is rock at the ground level itself, filter drain will have to be formed above ground level.

(5) The thickness of filter drain may be increased to 3 ft. consisting of 2 ft. layers of sand 1 ft. each passing ½" screen and with sand witch layer of 1 ft. thick of gravel passing 3" screen and retained on ½" screen, if the height of bank is more than 40 feet.

(6) The Transverse filter drains are to be spaced at 100 ft. intervals or as decided during execution.

(7) Rock toes with filter drains will have to be provided when the height of bank for ultimate discharge exceeds 25 feet. This may be at 1/3 F S D or to the impervious soil level whichever is less. Cut-off trench is to be provided only if the soils are porous and if trench is needed for the same.

(8) No cut-off is required for the following cases

i) Banks of 12 ft. height if there is at least 1 ft. thick of impervious soil below the base of bank.

ii) Banks of 12 ft. - 15 ft. height if there is at least 2 ft. thick of impervious soil below the base of bank.

iii) Banks of 15 ft. - 2 ft. height if there is at least 3 ft. thick of impervious soil below the base of bank and canal bed level.

iv) Banks of 25 ft. - 30 ft. height if there is at least 4 ft. thick of impervious soil below the base of bank.

6.15.2 Canals of Gandak and other projects, Bihar, (India)

Typical cross - section of canal system of Gandak Project are shown in figure 6.20 and 6.21 and that of other projects in figure 6.22. Other details of free board, top width of bank etc. are given in notes as under:

2. Hydraulic Gradient '$s_{(Hbk)}$'

The Hydraulic Gradient line may be taken as:

(a) 1 in 4 for clay; (b) 1 in 5 for clayey loam; (c) 1 in 6 for sandy loam

3. Side Slopes (H : V): as under

Soil and Design Side Slope	Construction	
	In cutting	In banking both inner** and outer slopes
(a)	(b)	(c)
Loamy soil, 1/2 :1	1 : 1	1.5 : 1
Sandy soil, 1 : 1	1.5 : 1	2 : 1

**Where the height of banks is more than 20 ft (6.56 m), it shall have designed sections.

4. When Pushtas (berms) are provided, the top level of pushta shall be a least 1.0 ft (0.33 m) below F.S.L. and top width of pushta shall be 6.0 ft (1.97)
5. The minimum top width of spoil bank shall be 5 ft (1.64 m) and maximum height 15 ft (4.9 m). The intermediate height and top width shall be fixed such that the land width required is minimum.
6. The crest of the banks shall have transverse slope of about 1 in 60 away from the canal side.
7. Land acquisition may be reduced where ever it is possible.

   More notes are given in figure 6.22

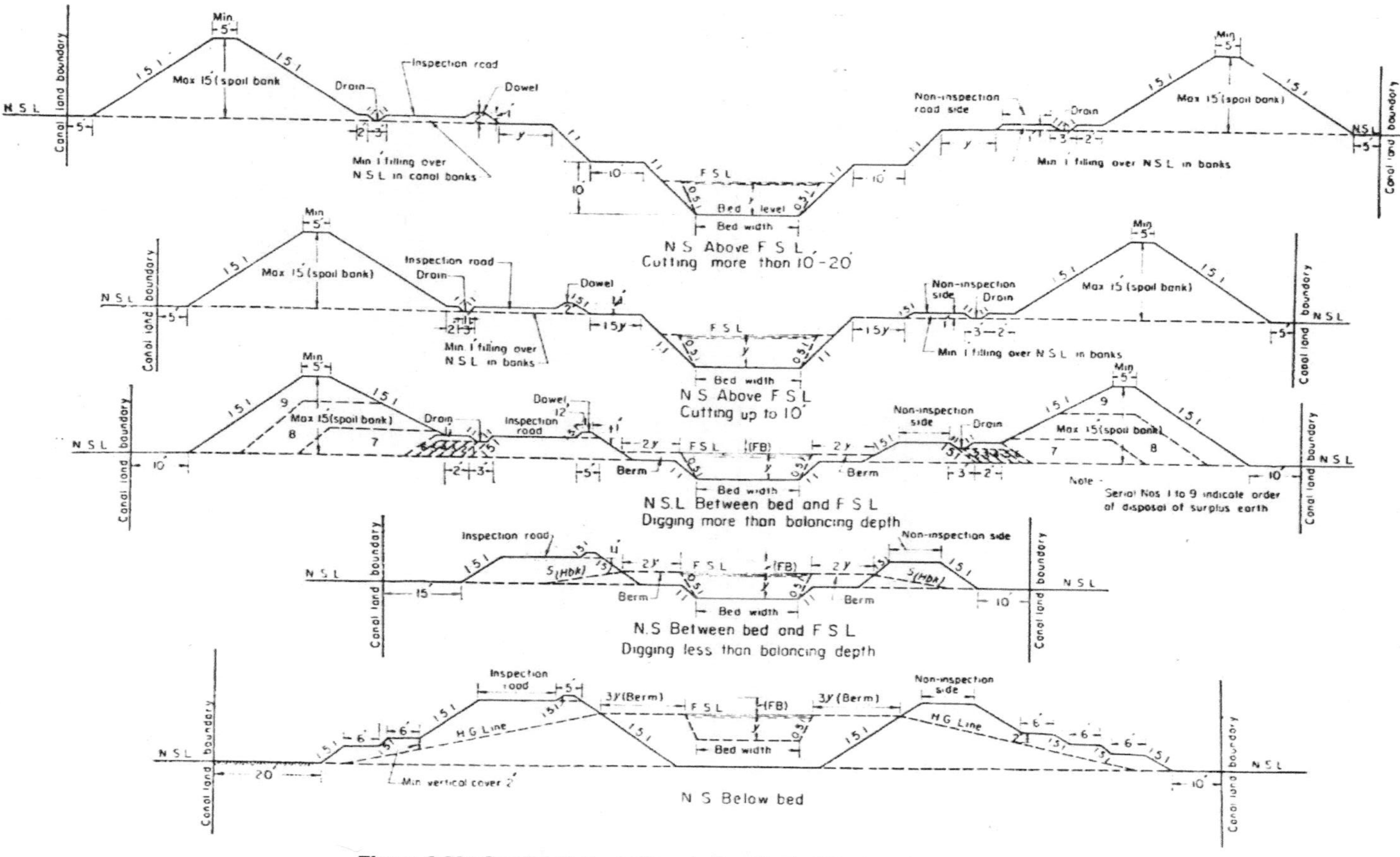

Figure 6.20 : Gandak Project, Bihar, India – Typical Earthen Canal Sections

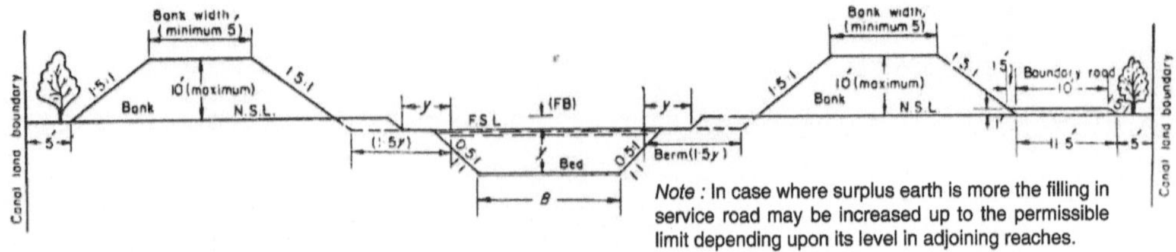

N.S. above F.S.L.

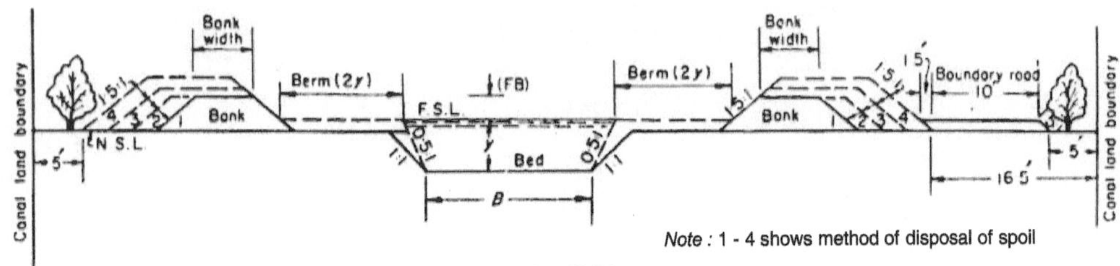

N.S. between bed and F.S.L.
Digging less than balancing depth

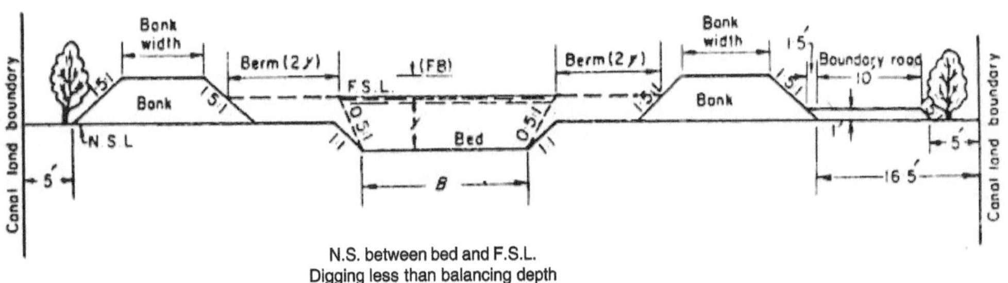

N.S. between bed and F.S.L.
Digging less than balancing depth

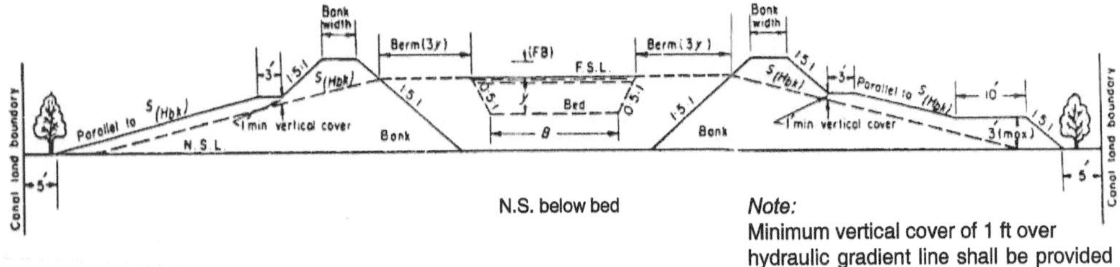

N.S. below bed

Note:
Minimum vertical cover of 1 ft over hydraulic gradient line shall be provided

Canal Discharge

	up to 49 ft³/s	50 to 199 ft³/s	200 to 350 ft³/s
Top width of bank in feet	4	5	6
Free board in feet	1.5	1.5	2.0

Figure 6.21 : Gandak Project, Bihar, Typical Cross Section for distributaries and minors (Q up to 350 ft³/s or 9.91 m³/s)

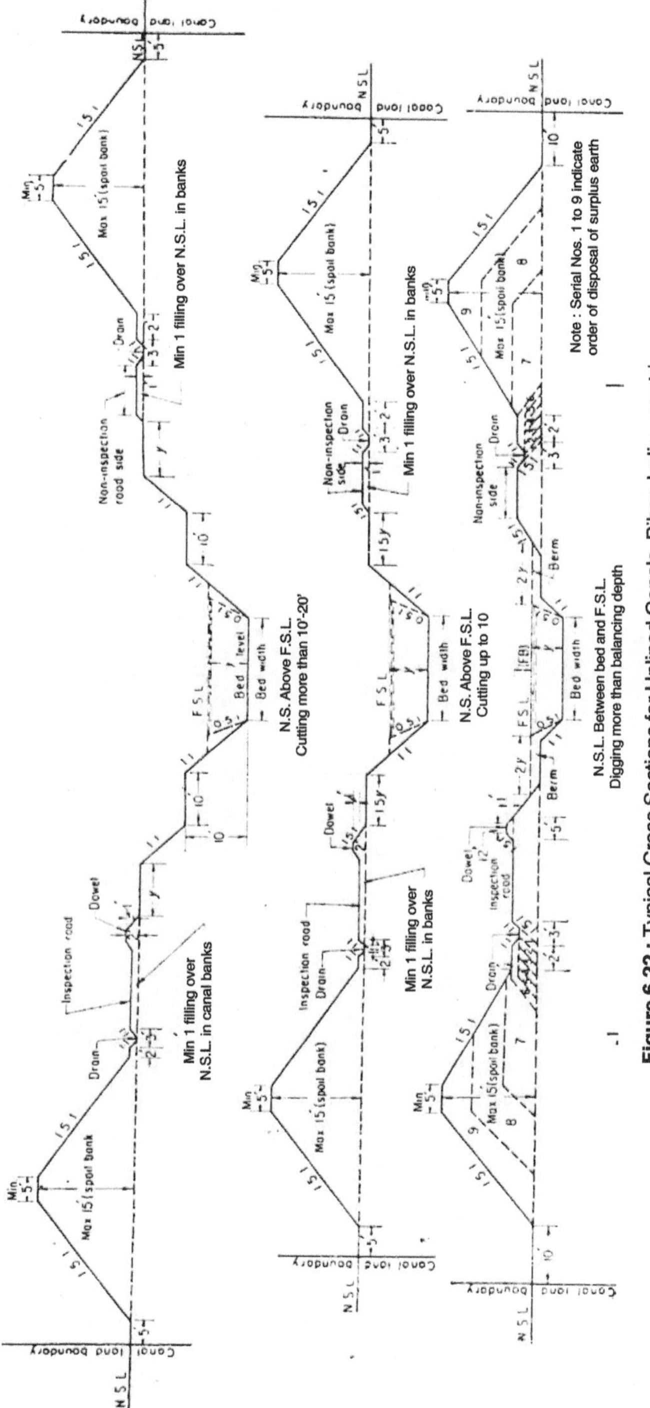

Figure 6.22 : Typical Cross Sections for Unlined Canals, Bihar, India contd...

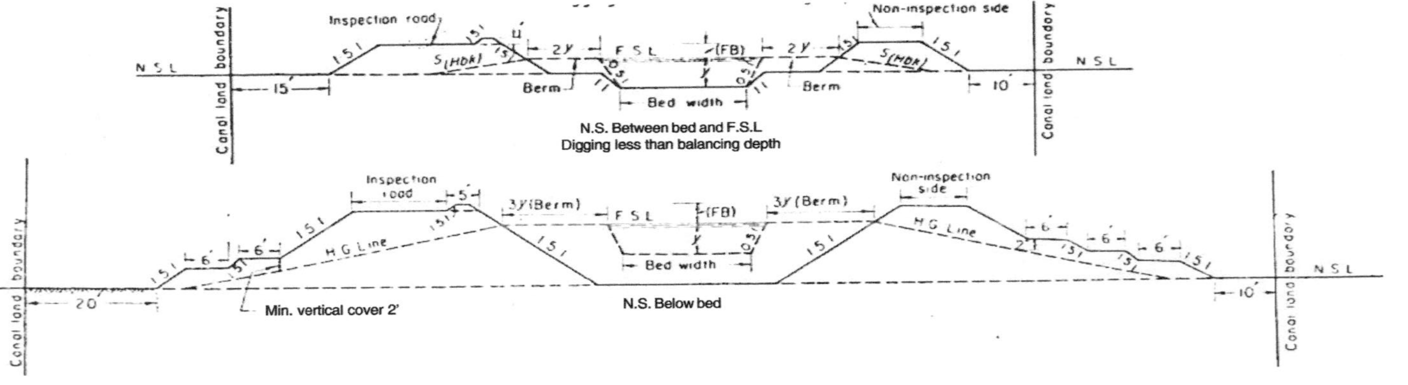

Figure 6.22 : Typical Cross Sections for Unlined Canals, Bihar, India

Notes (particulars) to figure 6.22 :

Discharge in ft³/s (m³/s)	Free Board in ft (m)		Top widths of banks in ft (m)*		Minimum vertical cover over hydraulic gradient line in ft (m)
	For complete cutting	For complete filling or partial cutting	Inspection side excluding dowel	Non-inspection side (no dowel)	
(a) 350-500 (9.05-14.15)	2.0 (0.65)	2.0 (0.65)	15.0 (4.9)	8.0 (2.6)	1.0 (0.33)
(b) 500-2,500 (14.15-70.75)	2.5 (0.82)	2.5 (0.82)	15.0 (4.9)	10.0 (3.3)	1.5 (0.50)
(c) 2,500-7,500 (70.75-212.25)	2.5 (0.82)	3.0 (0.98)	15.0 (4.9)	10.0 (3.3)	2.0 (0.66)
(d) (7,500 and above) (212.25 and above)	3.0 (0.98)	3.5 (1.15)	20.0 (6.5)	10.0 (3.3)	2.0 (0.66)

*Whenever surplus earth is available from digging, both bank widths shall be increased accordingly.

Figure 6.22 : Typical Cross Sections for Unlined Canals, Bihar, India

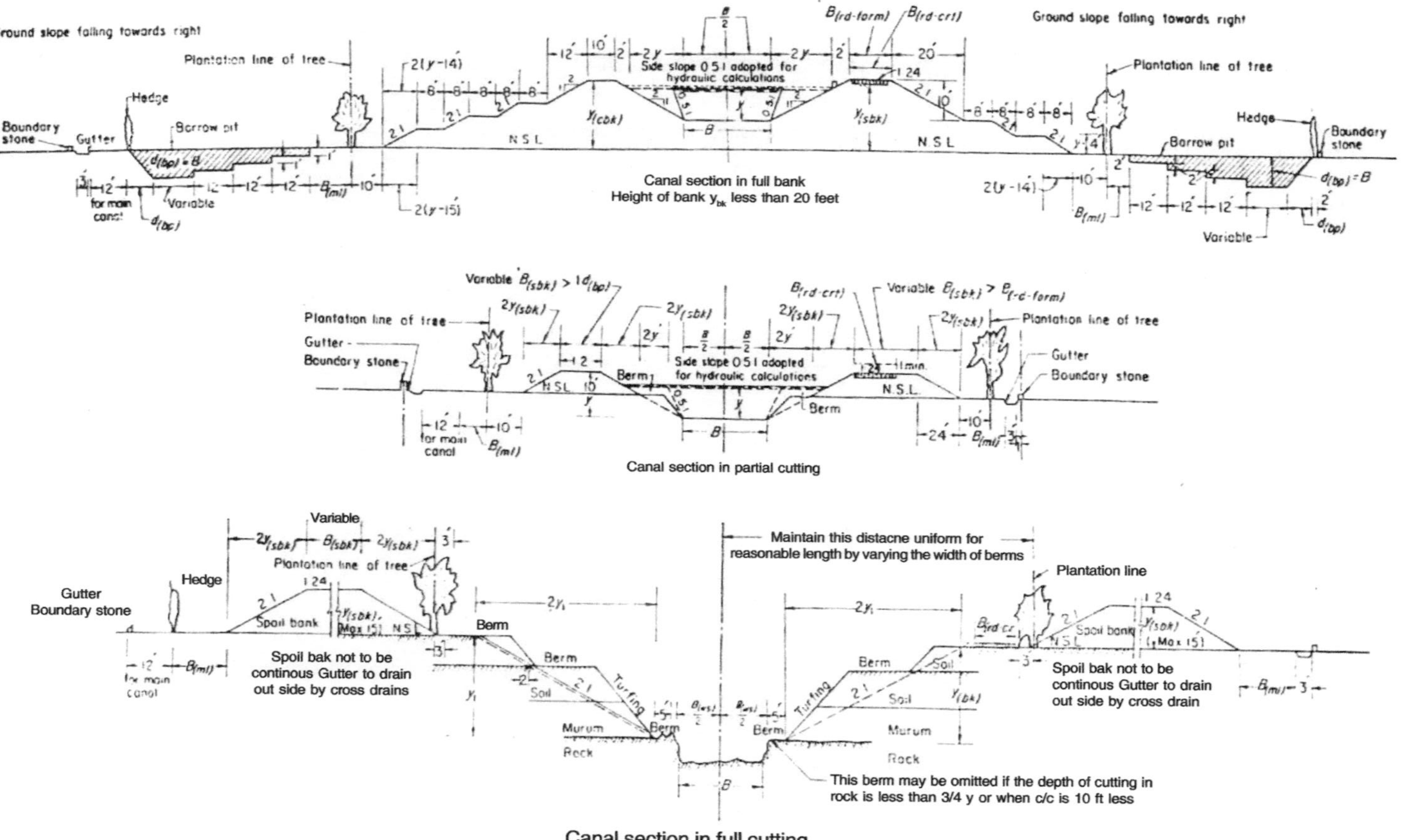

Figure 6.23 : Typical Cross Sections for Unlined Canals, Gujrat

Note : For Canals in filling having discharge from 25 to 250 cusecs (0.7 to 7.0 cumecs), the service road is taken outside at ground level

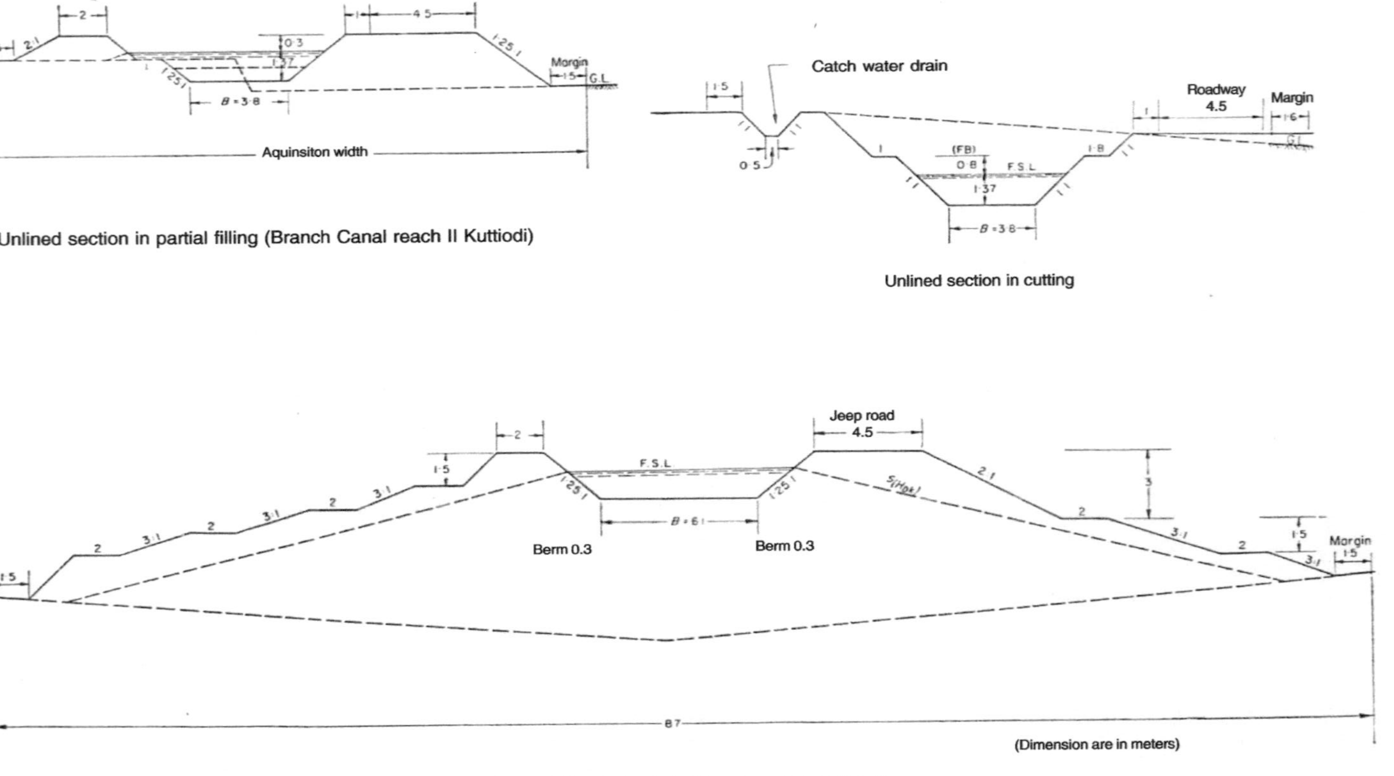

Unlined section in partial filling (Branch Canal reach II Kuttiodi)

Unlined section in cutting

section in filling (L.B. Main Canal reach II Kuttiadi)

Figure 6.24 : Kerala, India, Typical Cross Sections of Unlined Canals

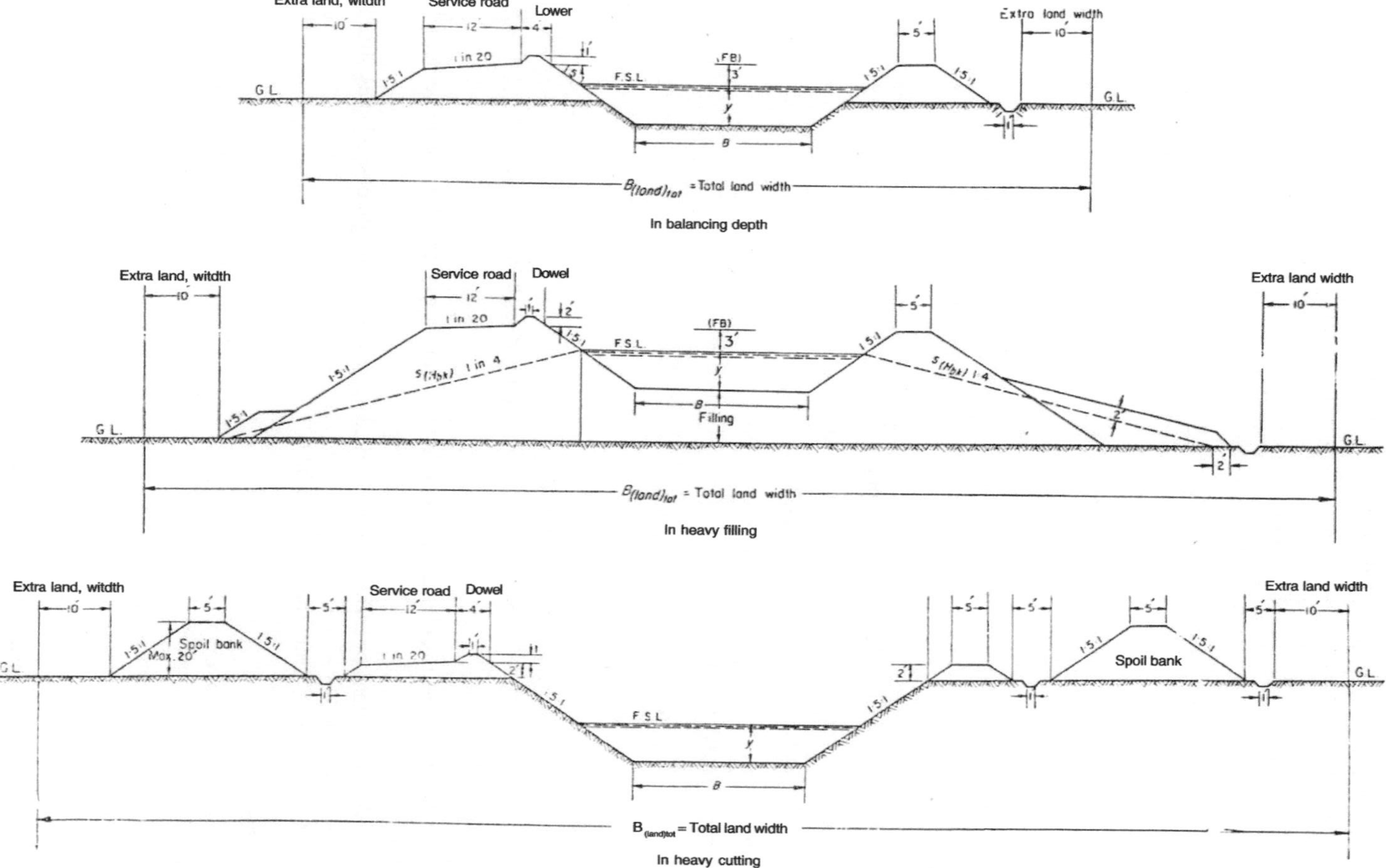

Figure 6.25 : Madhya Pradesh, Typical Sections Earthen Irrigation Canals
Discharge between 100 ft³/s and 350 ft³/s

6.15.3. Unlined canals in Gujrat, Kerala and Madhya Pradesh

Typical cross - sections are shown in figure 6.23, 6.24 and 625 respectively.

6.15.4. Bhakra Nangal Punjab, Canal System

Typical cross- sections are given in figure 6.26 (i) and (ii), Details of design methods are discussed in Para 6.3.4 and table 6.4. Here notes to these figures are also given as below:

Notes of figure 6.26 (i)

Width of road

(i) The width of patrol (service) bank on left shall be 25 ft. (7.62 m) including dowel.

(ii) The width of non-patrol bank on right shall be

(iii) For channels with discharge 500 ft^3/s (14.2 m^3/s) or above = 12 ft (3.7 m)

(iv) For channels with discharge 350 ft^3/s to 499 ft^3/s (9.91 m^3/s to 14.13 m^3/s) = 8ft (2.5 m)

(v) There should be only one boundary road on right 15 ft (4.6 m) wide with 5 ft (1.5 m) additional land for avenues

II Height of bank above F.S.L.	2.5 ft (0.76 m)
III Width of dowel at top	1.25 ft (0.381 m)
IV Width of dowel at base	5.0 ft (1.52 m)
V Height of dowel	1.25 ft (0.381 m)
VI Hydraulic gradient (H: G)	

1. The H.G. line should have a cover of one foot
2. Top level of Pushta 10 ft (3.05 m) below F.S.L.
3. Top width of Pushta 6.0 ft (1.83 m)
4. The slope of H G line to be adopted should be

   (i) Good soil 1 in 4 (ii) Average soil 1 in 5 (iii) Bad soil 1 in 6

VII Spoil Banks

(i) Maximum height of spoil bank shall be kept 20 ft (6.1 m) only where work is to be done by machinery.

Where work is to be done by manual labour or donkeys, maximum height of spoil bank shall be reduced to 12 ft (3.66 m).

(ii) Minimum width of spoil bank shall be 5 ft (1.52 m).

6.15.5. West Bengal

Typical cross - sections of canal section in West Bengal are shown in figure 6.27 (i) along with, notes to canal sections are as under:

Provision of inspection path along canals should be as follows:

1. Main and branch canals on one side over crest or berm 15′ wide.
2. Distributory and minors up to 5 ft^3/s: 15′ wide path along ground level on borrow pit side.
3. The size of ridge bund should be 1′ crest and 1′ height having side slope towards canal side the same as that of canal slope and 1 : 1 on the outer side.

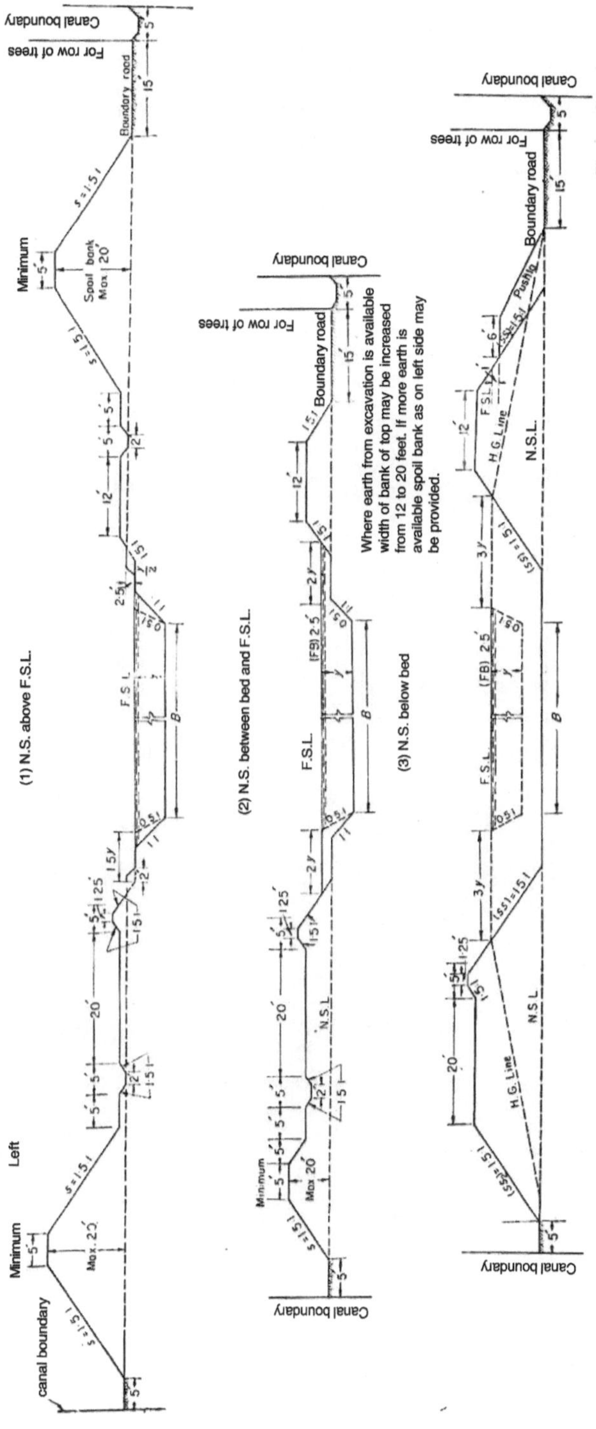

Figure 6.26 (i) : Bhakra Nangal, Punjab, Typical Cross Sections for Earthen Branch Canals

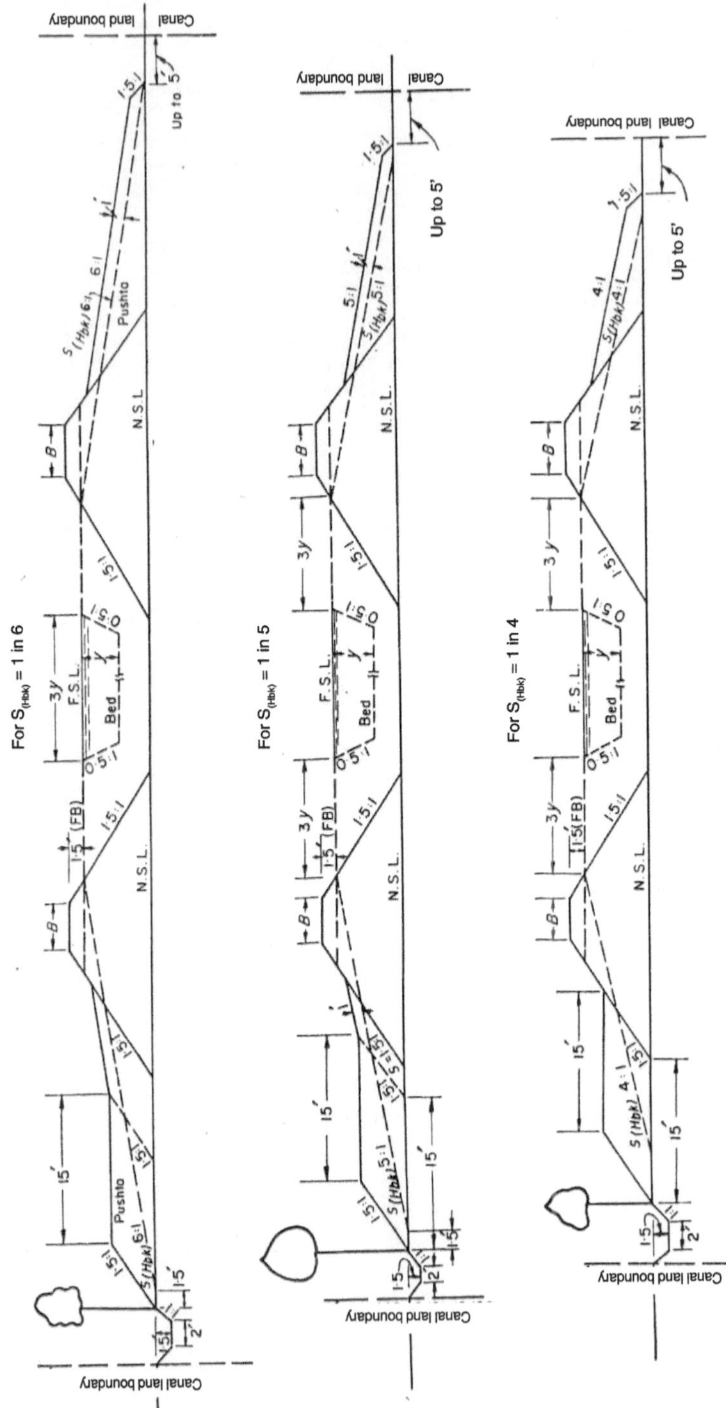

Figure 6.26 (ii) : The Bhakra Nangal Project - Typical Earthen Canal Cross Sections for Distributaries when bed level is above natural surface

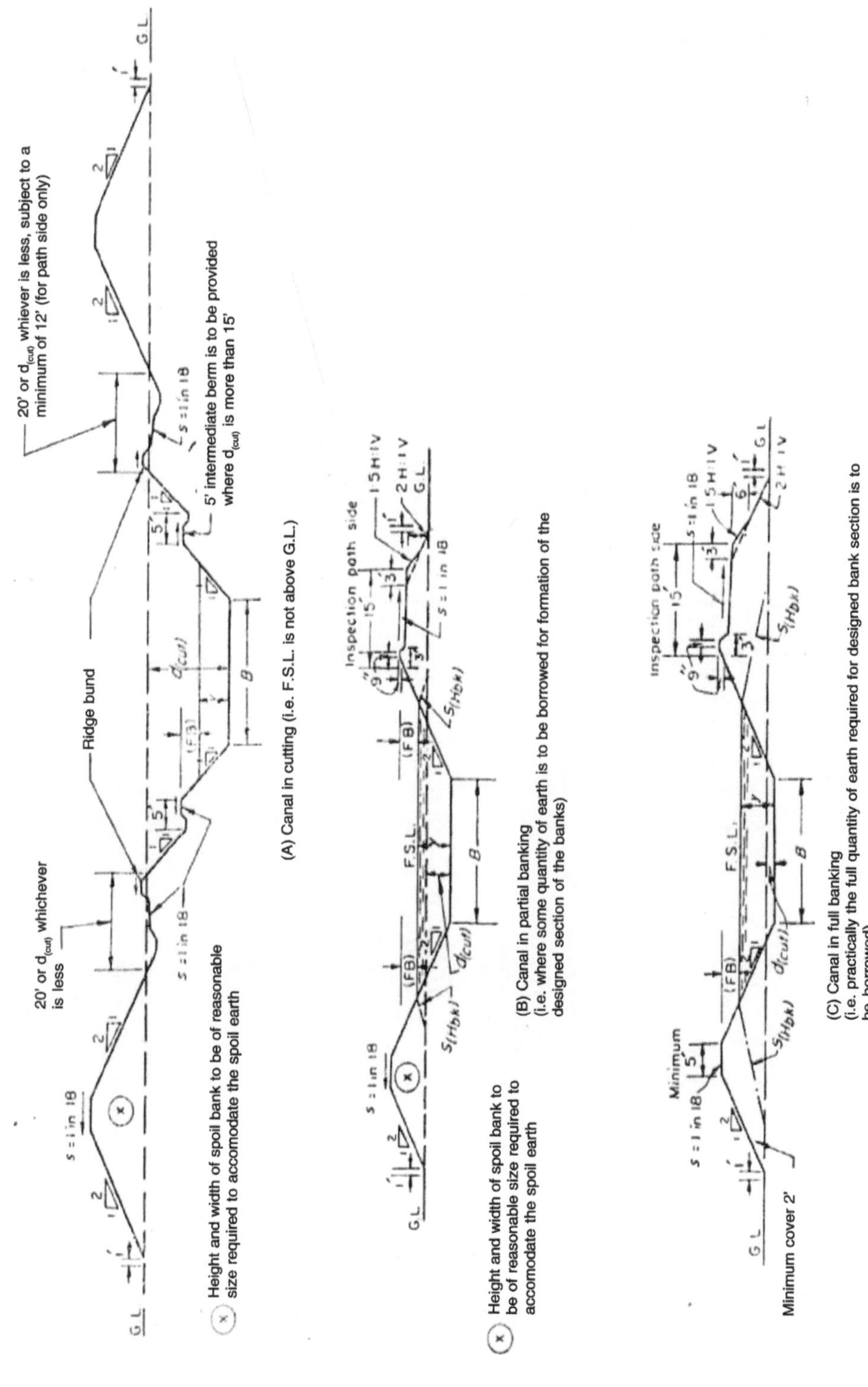

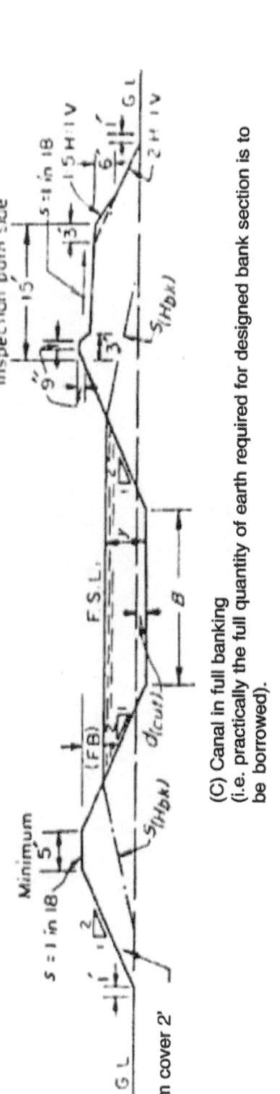

20' or $d_{(cut)}$ whiever is less, subject to a minimum of 12' (for path side only)
Ridge bund
20' or $d_{(cut)}$ whichever is less
5' intermediate berm is to be provided where $d_{(cut)}$ is more than 15'
s = 1 in 18
G.L
Height and width of spoil bank to be of reasonable size required to accomodate the spoil earth
(A) Canal in cutting (i.e. F.S.L. is not above G.L.)
Inspection path side
1.5H:1V
2H:1V
F.S.L.
(FB)
B
Height and width of spoil bank to be of reasonable size required to accomodate the spoil earth
(B) Canal in partial banking (i.e. where some quantity of earth is to be borrowed for formation of the designed section of the banks)
Minimum cover 2'
Minimum 5'
(C) Canal in full banking (i.e. practically the full quantity of earth required for designed bank section is to be borrowed).

Figure 6.27 (i) : Typical Cross Sections for Unlined Canals, West Bengal Project

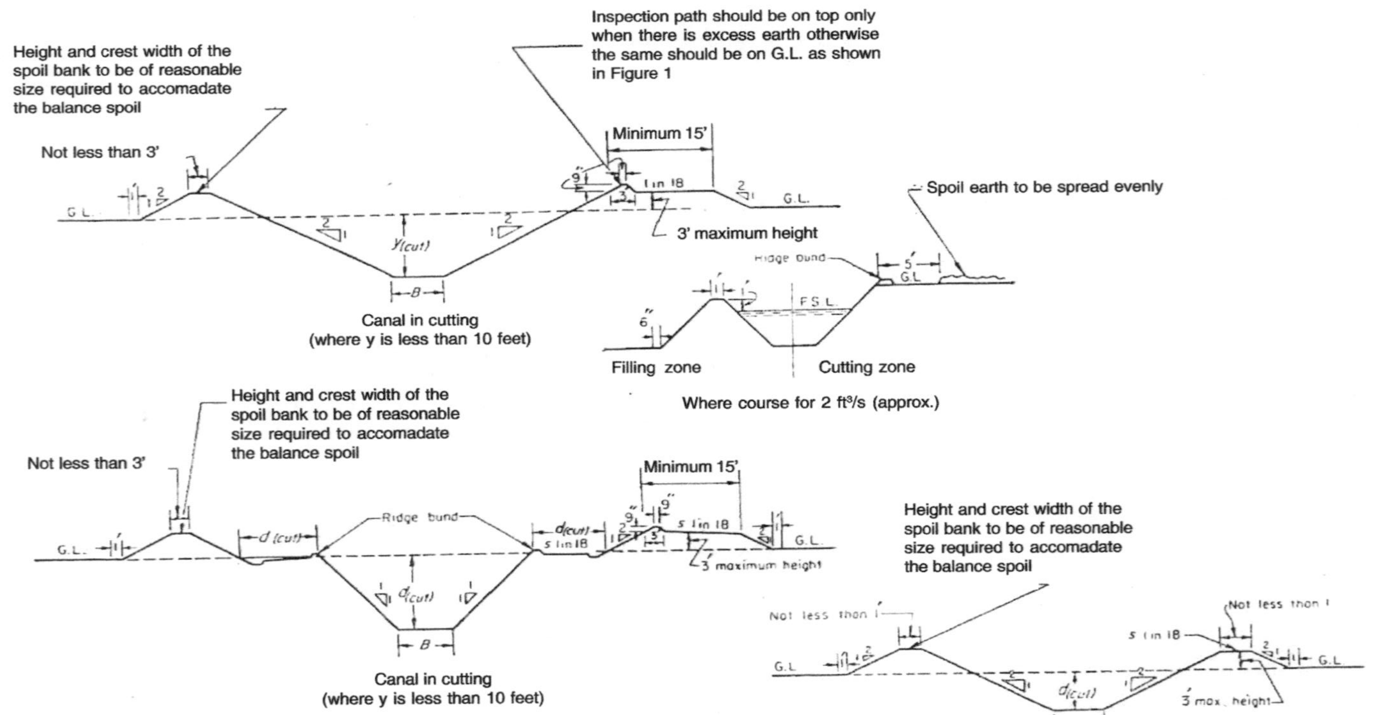

Note :

1. d (cut) = Depth of cutting
2. The size of ridge bund should be 1' crest and 1' height having side slope towards canal side the same a that of canal slope and 1 : 1 on the canal side.

Figure 6.27 (ii): Typical Cross Sections of distributaries (Q less than 50 ft³/s), minors and watercourses, West Bengal

4. Top soil is to be placed at centre and course material to be placed on the sides in case where heavy excavation is necessary. Bank should be so formed as to have a wide area for growing trees and for that purpose same top soil may be necessary on the crest.
5. Hydraulic gradient line is to be allowed 1 in 4 for ordinary soil and 1 in 5 for porous soil. Minimum cover is 2′.
6. Free board should be maintained as follows:

   i) Free board 3.5′ for discharge above 1,000 ft^3/s
   ii) Free board 3′ for discharge between 1,000 and 500 ft^3/s
   iii) Free board 2.5′ for discharge between 500 and 100 ft^3/s
   iv) Free board 2′ for discharge between 100 and 50 ft^3/s
   v) Free board 1.5′ for discharge between 50 and 5 ft^3/s
   vi) Free board 1′ for discharge less than 5 ft^3/s

6.16 Thick Compacted Earth Linings

Due to high cost of Hard surface linings (cement concrete, brick / tile, masonry etc) earth and Burried Membrane linings are have also been used extensively in many countries.

A very simple, cheap, very effective and novel method developed by Er. Evans (1900) and now modified is shown in figure 6.28 (i). It consist of laying unburnt clay dry bricks (known as Katcha in India) in bottom as two layers, and one layer on sides as shown in the figure. With water it becomes impervious mud or clay lining. Thickness can be increased by laying more layers. This is very effective and easy in construction and noise compaction required.

In USA : Thick compacted earth linings normally have a 3 to 8 ft (0.91 to 2.44 m) thickness on the canal sides, measured horizontally and a 12 to 44 inches (0.30 to 0.61 m) bottom thickness of compacted select material. However, any compacted section 12 inches (0.31 m) thick or more is considered to be thick compacted earth lining. Figure 6.28 (ii) shows a typical earth lined section. Thin compacted earth linings usually have a 6 to 12 inches (15.2 cm to 30.5 cm) layer of compacted cohesive soil. A protective cover of 6 to 12 inches (15.2 cm to 30.5 cm) of coarse soil or gravel is also laid over it. Loosely placed earth lining generally consists of a loose earth blanket of selected fine-grained soils dumped into the canal and spread over the bottom and sides to a thickness of about 12 inches (30.5 cm). Bentonite soil mixtures usually consist of a sandy soil and betonite mixed together and compacted. The thickness varies with local conditions.

6.16.1 Buried Membrane Lining

Buried membrane lining is normally installed only to reduce water loss by seepage. A cover must be provided to protect the membrane from exposure to the elements and from injury by turbulent water, stock, plant growth and maintenance equipment. The depth of cover depends on cover material, size of canal, water velocity and canal side slopes. The canal bottom width should be about 4 times the water depth or greater and the side slopes 2.0 horizontal to 1.0 vertical or flatter. Long radius curves are desirable at the intersection of the side slopes and the bottom to improve stability and to more nearly approach the final shape of the canal section after it has been in operation. When rounded uniformly graded gravels or sands have been used for cover material, 2.0 horizontal to 1.0 vertical side slopes have proved to be too steep. There is also danger that the cover material may slough down the bank during placing. Such sloughing may carry the membrane down the slope with it causing it to crack or tear. Flatter side slopes and placing the cover material on the canal bottom and lower portions of the side slopes first may prevent damaging the membrane lining. At structures, the membrane should be carefully bonded to the cutoff and should be lowered a sufficient distance away from the structure to provide space for extending riprap protection where required. Laboratory test of the available cover

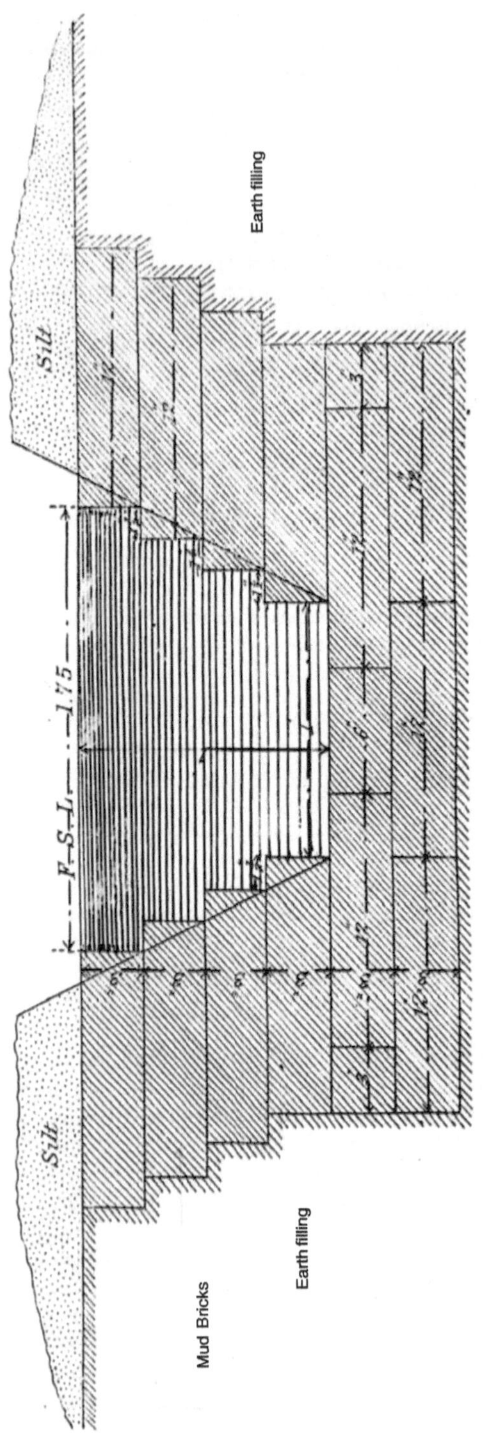

Note :
Designed and constructed by Er. A.C. Evans, Executive Engineering, Aligarh Dn. Ganges Canal, India, Dated : 18-01-1900

Figure 6.28 (i) : Canal Section with Mud/ (Kacha) Brick Lining Suitable Upto 2 m Depth

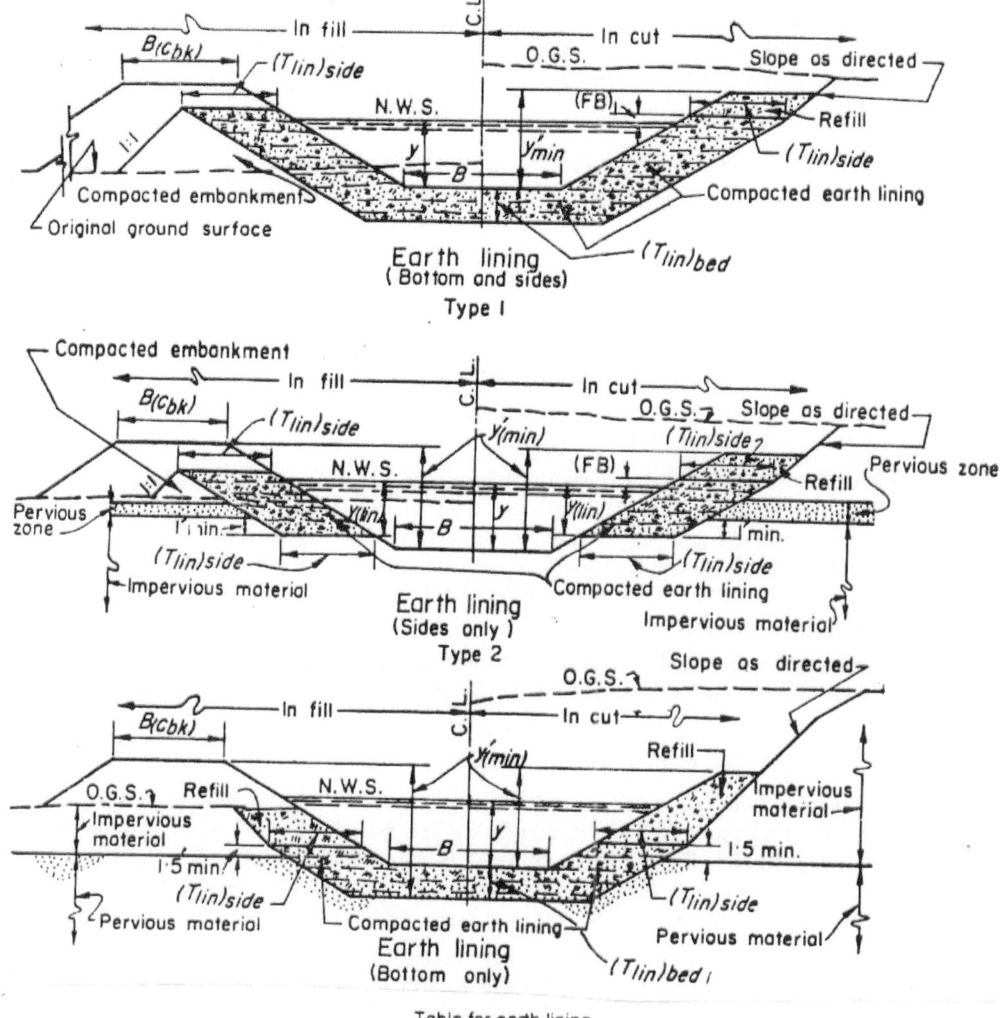

Table for earth lining

y or $y_{(lin)}$	$T_{(lin)}$ bed	$T_{(lin)}$ side
2.0' max.	1.0'	3.0'
4.0' max.	1.5'	4.0'
6.0' max.	2.0'	6.0'
Over 6.0'	2.0'	8.0'

Notation

B(cbk) = Top width of bank
T(lin) bed = Thickness of Lining on bed
T(lin) side = Thickness of Lining on sides
y or y (lin) = Depth of flow of lined canal
FB = Free Board

Note :

If lining material requires a protective cover of gravel or nprap to prevent scour or erosion, excavation shall be extended to provide for the designated thickness of lining plus the gravel or riprap cover.

If a previous zone containing mostly coarse material is encountered and requires a graded filter to prevent loss of lining material, excavation shall be lowered on sides or bottom to provide for the designated thickness of lining plus the filter.

For Free Board see chapter 8

Figure 6.28 (ii) : Typical Thick Earth - Lined Canal Sections, U.S.A

material are desirable in determining the side slopes and mixtures of available materials to be used for best results. Crushed rock or angular cover material should never be placed directly on the membrane because of the danger of puncturing it. Earth, gravelly material and gravel have been used for cover over membrane lining. A small amount of lean clay will add stability especially to gravelly materials and gravel. Details of buried membrane linings are shown in Figure 6.29.

6.17 Some Other Examples of high capacity and short length canals.

i) The *Shannon* headrace canal, built between 1924 and 1929, is 7 ½ miles long and was designed to carry 500 cumecs (18,000 cusecs) at a maximum velocity of 4.9 ft. per sec. It is also used for navigation (The Engineer, December 1927, March, 1928).

   In sections where the bank and bed material is insufficiently water-tight, the canal is lined with a layer of clay, 0.60m (2 ft.) thick. As a protection against erosion the canal is lined with a 0.40m (16-in.) layer of broken stone, and those portions of the bank slopes subject to wave action are further protected by a concrete slab, 12 cm (4.7 in.) thick placed by screeding. The cross-section, which is trapezoidal, has a base width of 31 m (103 ft.) Where protected by concrete the bank slope is 1 in 2; (V:H) elsewhere it is 1 in 3. (V:H).

ii) The *Beauharnois* headrace canal, *(Annon,1931),* begun in1929, is 25 Km (15 ½ miles) long and is carried between dykes for its entire length. For the initial installations only part of the ultimate channel was excavated below original ground, but the dykes were built 3000 ft. apart to allow the ultimate diversion of 623.6 cumecs (22,000 cusecs) at a maximum velocity of 0.68 m/s (2.25 ft per sec). As additional water diversion was required the channel was dredged accordingly. The canal side slopes are at 1 in 3 (V:H) and are lined with rip-rap as a protection against wave action.

   It also intended that the canal will ultimately be part of the St. Lawrence seaway, when a 180 m (600 ft.) width of the canal will be deepened to 8.0 m (27 ft.) to form a shipping channel.

iii) The *Tummel* headrace canal, (Annon, 1934) built during 1932-33, is 4.5 Km (2 ¾ miles) long and carries 65.2 cumecs (2300 cusecs) at a maximum velocity of (6.5 ft. per sec).

   The canal is lined with reinforced concrete varying generally in thickness from a minimum of 5 cm (2 in.) over rock points to 12 cm (5 in.) over cut ground, and 18 cm (7 in) .over made ground. Where the cut is in rock, the cross-section, which is trapezoidal, has a base width of 31 ft. 6 in. and side slopes of 1 in 0.25 (V : H). Where the cut is in soft ground, the base width is 18 ft. and the side slopes 1:1.5 (V : H).

iv) The *Kendoon* canal (J.I. GE.1937 - 1938), built between 1935 and 1936, is 1/2 mile long and was designed to carry 1900 cusecs at a maximum velocity of 7 ft. per sec.

   The canal, of trapezoidal section, has a bottom width varying from 7 ft. to 10 ft and was lined throughout its length with 6-in. concrete slabs cast in place, the side slopes having a batter of 1 : 1.5 (V : H). At some points in the embankment where leakage occurred, a further waterproof lining was laid consisting of a layer of felt and bitumen covered by a 4-in. reinforced concrete slab. In other places protection was provided by a continuous sheet -pile core.

v) The *Ottmarsheim* canal, built between 1948 and 1952, is 9 miles long and was designed to carry 41,000 cusecs at maximum velocity of 4 ft. per sec. It is also used for navigation.

   The cross-section, is trapezoidal, with a base width of 262 ft. with bank slopes of 1:3 (V:H). The canal is lined with cast in situ concrete slabs over all faces above phreatic level. In the downstream section of the canal the slabs were precast using a vacuum process. Where a concrete invert is provided, it is covered by a bed of gracel 3 ft .or more thick to allow vessels to anchor.

vi) The *Donzere* canal (Water Power Jan. 1953*),* built between 1948 and 1952, is 17 miles long, designed to carry 54,000 cusecs at a maximum velocity of 4.25 ft per sec. It is also used for navigation.

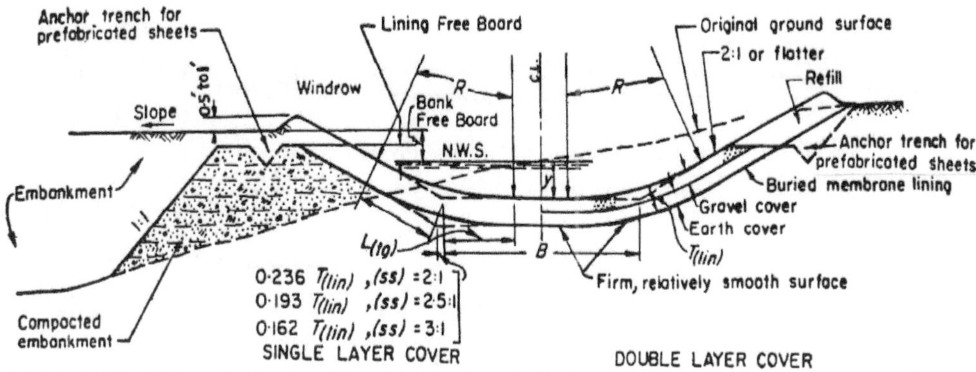

Transverse section of waterway $T_{(lin)}$ = 10" + Y/12: where $T_{(lin)}$ is in inches and y is water depth in inches. (12" minimum)

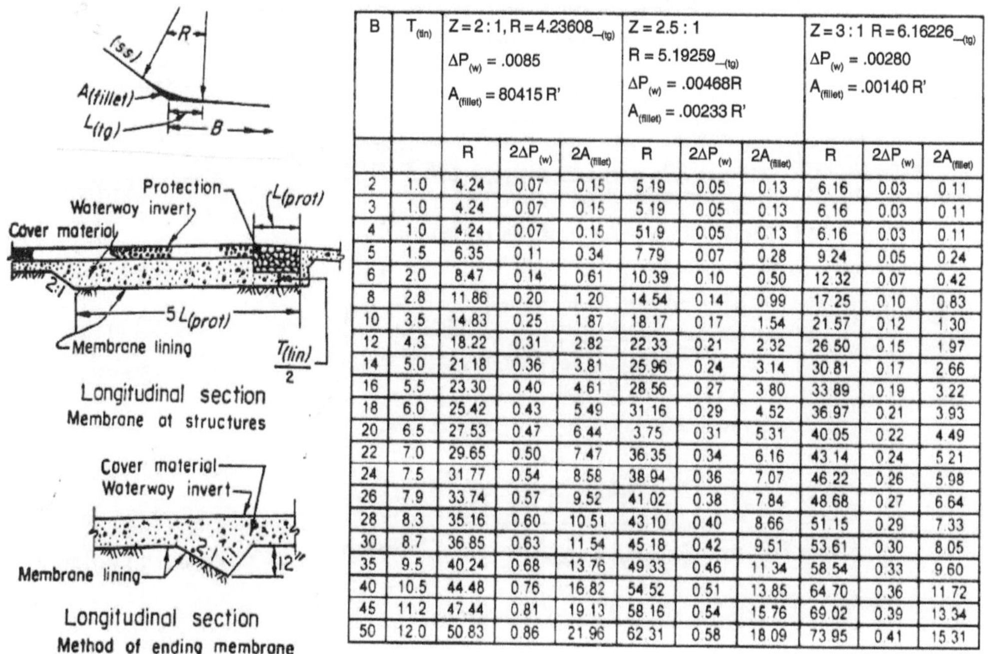

B	$T_{(lin)}$	Z = 2 : 1, R = $4.23608_{-(tg)}$; $\Delta P_{(w)}$ = .0085; $A_{(fillet)}$ = 80415 R'			Z = 2.5 : 1; R = $5.19259_{-(tg)}$; $\Delta P_{(w)}$ = .00468R; $A_{(fillet)}$ = .00233 R'			Z = 3 : 1 R = $6.16226_{-(tg)}$; $\Delta P_{(w)}$ = .00280; $A_{(fillet)}$ = .00140 R'		
		R	$2\Delta P_{(w)}$	$2A_{(fillet)}$	R	$2\Delta P_{(w)}$	$2A_{(fillet)}$	R	$2\Delta P_{(w)}$	$2A_{(fillet)}$
2	1.0	4.24	0.07	0.15	5.19	0.05	0.13	6.16	0.03	0.11
3	1.0	4.24	0.07	0.15	5.19	0.05	0.13	6.16	0.03	0.11
4	1.0	4.24	0.07	0.15	51.9	0.05	0.13	6.16	0.03	0.11
5	1.5	6.35	0.11	0.34	7.79	0.07	0.28	9.24	0.05	0.24
6	2.0	8.47	0.14	0.61	10.39	0.10	0.50	12.32	0.07	0.42
8	2.8	11.86	0.20	1.20	14.54	0.14	0.99	17.25	0.10	0.83
10	3.5	14.83	0.25	1.87	18.17	0.17	1.54	21.57	0.12	1.30
12	4.3	18.22	0.31	2.82	22.33	0.21	2.32	26.50	0.15	1.97
14	5.0	21.18	0.36	3.81	25.96	0.24	3.14	30.81	0.17	2.66
16	5.5	23.30	0.40	4.61	28.56	0.27	3.80	33.89	0.19	3.22
18	6.0	25.42	0.43	5.49	31.16	0.29	4.52	36.97	0.21	3.93
20	6.5	27.53	0.47	6.44	3.75	0.31	5.31	40.05	0.22	4.49
22	7.0	29.65	0.50	7.47	36.35	0.34	6.16	43.14	0.24	5.21
24	7.5	31.77	0.54	8.58	38.94	0.36	7.07	46.22	0.26	5.98
26	7.9	33.74	0.57	9.52	41.02	0.38	7.84	48.68	0.27	6.64
28	8.3	35.16	0.60	10.51	43.10	0.40	8.66	51.15	0.29	7.33
30	8.7	36.85	0.63	11.54	45.18	0.42	9.51	53.61	0.30	8.05
35	9.5	40.24	0.68	13.76	49.33	0.46	11.34	58.54	0.33	9.60
40	10.5	44.48	0.76	16.82	54.52	0.51	13.85	64.70	0.36	11.72
45	11.2	47.44	0.81	19.13	58.16	0.54	15.76	69.02	0.39	13.34
50	12.0	50.83	0.86	21.96	62.31	0.58	18.09	73.95	0.41	15.31

Note :

Notation are same as in figure 6.28

The graduation type and thickness of cover material is dependent on tractive forces and velocities in the section and the type of material available in the area.

DPiwi is the difference between 21 (fg) and the arc length

To obtain the wetted perimeter for a section subtract 2DP (w) from the wetted perimeter for a trapezoidal section of width B. A (fillet) is the area of the fillet

To obtain the area for a section subtract 2 A (fillet) from the area of a trapezoidal section of width B.

For Free Board see chaper 8

Figure 6.29 Typical Buried Membrane Lining in U.S.A

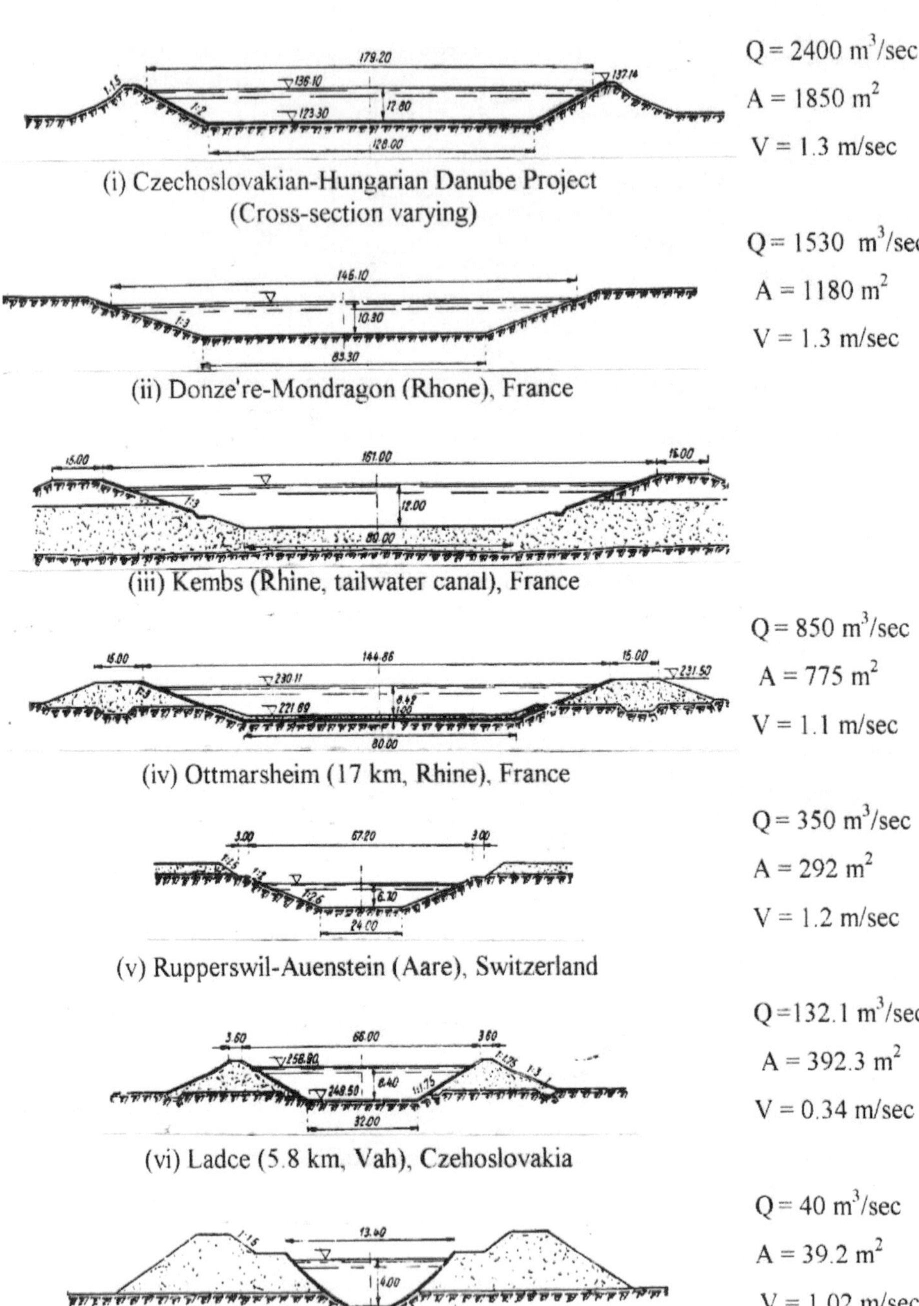

(i) Czechoslovakian-Hungarian Danube Project (Cross-section varying)

(ii) Donze're-Mondragon (Rhone), France

(iii) Kembs (Rhine, tailwater canal), France

(iv) Ottmarsheim (17 km, Rhine), France

(v) Rupperswil-Auenstein (Aare), Switzerland

(vi) Ladce (5.8 km, Vah), Czehoslovakia

(vii) Kesznyeten (4.8 km, Hernad), Hungary

Figure 6.30 Normal Cross-Sections and main Hydraulic Data of some power Canals

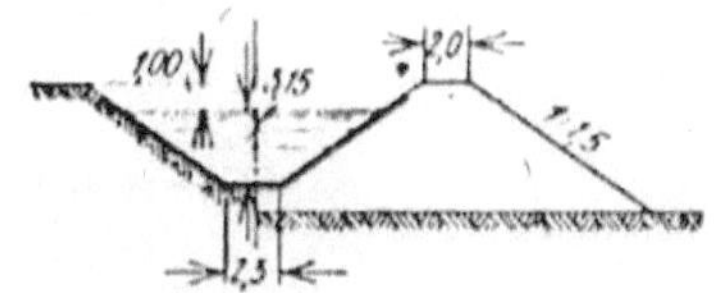

Q (m³/s)	V (m/s)	S	B/D	A (m²)	Z	P (m)	R (m)
40	2	1/3333	2.3/3.15 = 0.73	20	1.25	12.38	1.61

(a) Walchensee

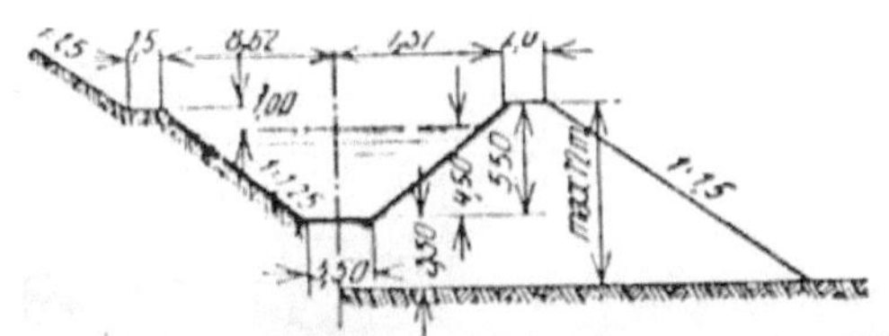

Q (m³/s)	V (m/s)	S	B/D	A (m²)	Z	P (m)	R (m)
60-70	1.3-1.5	1/5555	3.5/4.5 = 0.77	46	1.25	17.91	2.57

(b) A/z Burghausen

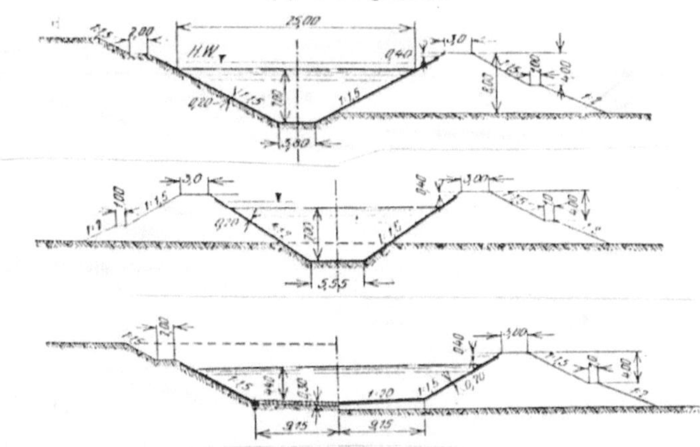

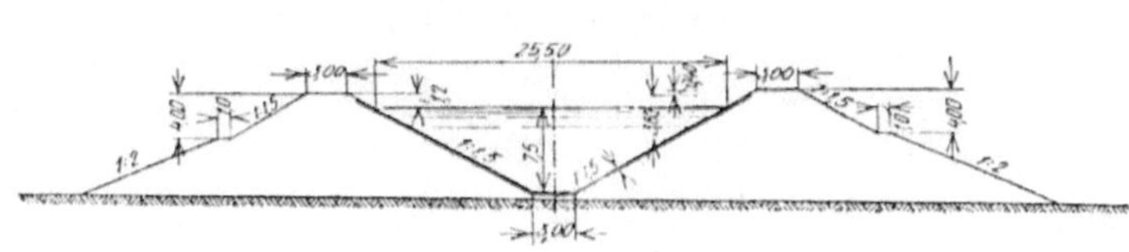

Q (m³/s)	V (m/s)	S	B/D	A (m²)	Z	P (m)	R (m)
150	1.5	1/8333	3.80/7.00 = 0.54	100	1.5	29.04	3.44
150	1.3	1/8333	5.55/7.00 = 0.79	110	1.5	30.79	3.57
125	1.15	1/5882	18.30/4.40= 4.15	110	1.5	34.16	3.22
125	1.15	1/1538	3/7.5 = 0.40	107	1.5	30.04	3.56

(c) Mittlere Isar

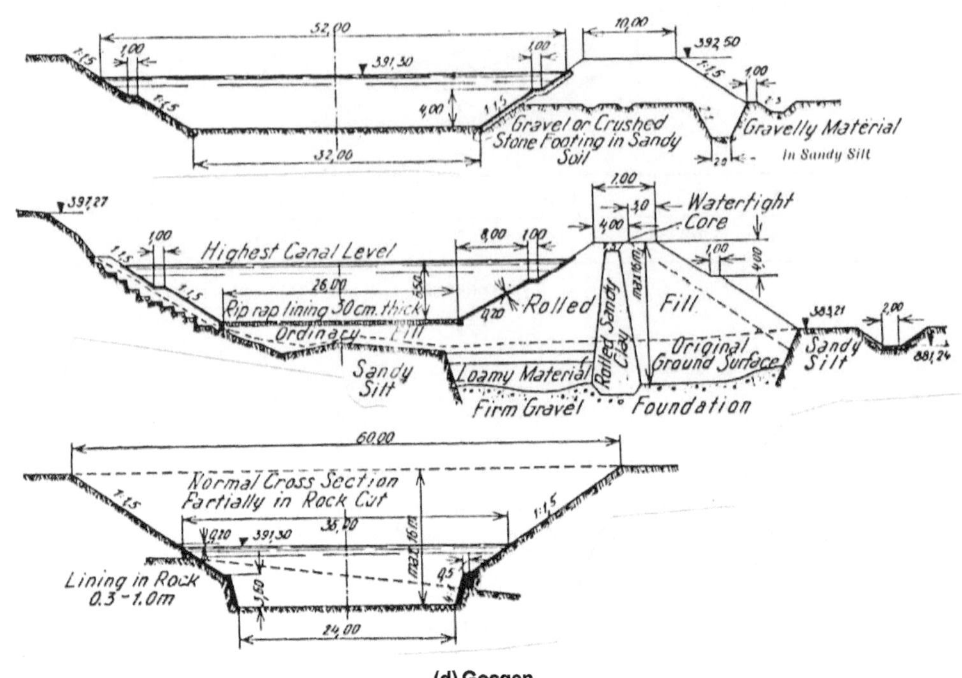

(d) Gosgen

Q (m³/s)	V (m/s)	S	B/D	A (m²)	Z	P (m)	R (m)
350	1.4	1/7695	32/	250	1.5		
350	1.40	1/7692	26/6.5 = 4	250	1.5	49.44	
350	1.80	1/4348	24/	190	1.5		

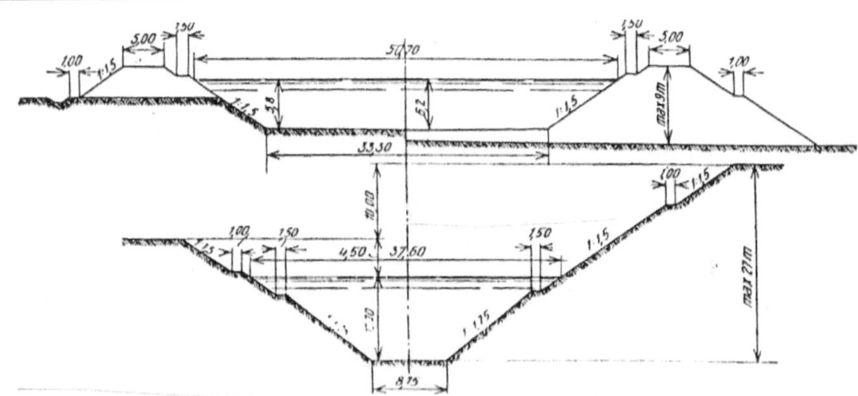

Q (m³/s)	V (m/s)	S	B/D	A (m²)	Z	P (m)	R (m)
300	1.40	1/6289	33.5/5.2 = 6.40	250	1.5	52.04	4.80
300	1.40	1/6289	8.75/10 = 0.87	250	1.75	49.061	4.80

(e) Mittlere Inn

Notations : A = Area of flow, V = Mean veolocity, S = Hydraulic slope

Figure 6.31 Typical canal cross sections, a) Walchensee, b) Alz Burghauscn c) Mittlere Isar, d) Gosgen, e) Mittler Inn.

The canal, is trapezoidal with, 274 ft. width at the base. The bank slope is normally 1:3 (V : H), but in areas where there is a considerable thickness of silt the slope has been flattened to 1 in 5 and lined with a 6-ft. layer of remixed alluvium obtained from the excavation. Except for portions of the slope subject to wave action, the canal is unlined, the alluvium being rendered watertight by deposition of silt from the canal water. A bituminous concrete revetment is provided as a protection against wave action.

vii) *Butterbridge* aqueducts (about 1 ½ miles) of the Loch Sloy scheme, completed in 1952, is an open channel designed to carry 4.5 cumecs (160 cusecs) at a maximum velocity of 2.7 m/s (9 ft. per sec).

The bottom width of the channel is 3 ft. with side slopes of 1 in 1 and a height of 3 ft. 6 in. Except for a short section which is pitched, it is lined throughout with concrete 5 in. thick, placed *in situ*.

viii) The T*arraleah* canal was constructed by the Hydro-Electric Commission of Tasmania. It is a concrete-lined canal carrying 25.5 cumecs (900 cusecs). At side slopes steeper than 1 : 2.5 (V : H) the canal was replaced by a reinforced concrete flume. The *Bronte* canal, constructed by the same Authority during 1951 - 53, is unlined and has a capacity of 46.8 cumecs (1650 cusecs). It has a bottom width of (30 ft). and side slope of 1: 2 (V:H). The subsoil consisted of shale or mudstone, covered by dolerite clay. At one end only the dolerite clay was present, and two lengths of canal were formed as reinforced-concrete flume.

ix) The *Roanoke* canal opened in 1955 is the tailrace for the installation constructed by the Virginia Electric and Power Co. at Roanoke Rapids, North Carolina. The canal is 8000 ft. long by 80 ft. wide, and depth varies form 31 to 56 ft. in granite.

x) Masony (1987) has given some canal sections as shown in figure 6.30 very high to medium capacity power canals in different countries. Discharge ranges from 2400 m^3/s to 40 m^3/s. Their dimensions, area of flow and velocity are shown in figure 6.30. The lining of the tail race canal of the Donzere Mondragon power plants, shown in figure 6.30 (ii) serves neither the purpose of seepage prevention nor those of reducing friction, but are intended to protect against wave action. The lining, does not extend down to the bottom, but protects the slopes with in 5.5m wide range of water level fluctuations. The lining itself consists of concrete slabs laid side by side, without sealing.

xi) The canal of the Arnoldstein development near villach, Carinthia, has a trough cross-section shown in figure 6.31 and is entirely in fill. It has a reinforced concrete lining 12 cm. (4. 75 in.) thick that has proved entirely satisfactory.

Where the canal extends down to rock, the slope is broken by berms, as indicated in figure 6. The thickness of the lining in such a canal depends on the conditionand character of the rock.

If the canal passes through valuable or very narrowly confined land, vertical or nearly vertical sides may be economically justified.

a) Walchensee
b) Alz Burghausen
c) Mittlere Isar
d) Gosgen
e) Mittlere Inn

6.18 Narmada Main Canal

The Narmada main canal, planned as a contour canal, will perhaps be the biggest lined irrigation canal in the world. It will be 460 Km long up to Gujarat- Rajasthan border having discharging capacity of 1133 cumecs (40,000 cusecs) at its head tapering down to 71 cumecs (2500 cumecs) at the Gujarat-Rajasthan border. The section of the canal is 73.1(B)x 7.6 m (FSD) at its off take. The bed gradient of the canal is 1 in 12,500 up to Mahi River crossing (144.5 Km) and flatter, 1 in 15000 beyond that up to

Rajasthan border to suit higher FSL of 44.96 m (147.5 ft. required by Rajasthan. The canal system will be operated during Kharif and Rabi seasons and will remain closed for about 2.5 months of hot weather season due to limited availability of water.

The canal water is expected to be silt free being taking off from a reservoir, or may have little silt in Kharif.

The main canal passes through different types of soils/rock strata and has varying depth of cutting and height of banking. Since the main canal is contour canal it crosses numerous rivers and streams. Very high banks are, therefore, required to be provided in the approach reaches. Similarly, at some places the canal passes through deep cutting. The maximum height of embankment is of the order of 19 m and the maximum depth depth of cutting is of 16 to 19 m. Even where the canal is in partial cutting / partial banking its total height above canal bed level is as high as 9.1 m (7.6 m fully supply depth + 1.5 m free board). Hence it was decided to carry out systematic slope stability analysis of canal banks and foundation on the same lines as followed for design of earth dam.

The canal bank sections, both in cutting and filling (banking) are designed as per the guidelines given in the IS: 7894 - 1975 "Code of practice for stability analysis of earth dams". The canal sections in banking are zoned section with either upstream impervious zone or with central impervious core and shells of semipervious type of material, as shown in figure 6.32. Detailed study was carried out to decide whether to provide zoned section or homogeneous section (for all heights). Keeping in view different soil strata met in cutting and the construction methods, homogeneous bank was not found feasible. Zoned section was adopted. When the height of banking is less than 5 m, neither horizontal filter nor rock-toe cum toe-drain is provided. Horizontal filter and rock-toe cum toe-drain are provided for zoned section up to 9.1 m height and for height of banking more than 9.1 m inclined filter in addition to the horizontal filter and rock-toe cum toe-drain is provided.

Detailed study carried out to decide whether the impervious zone should be located in the centre or on the upstream just behind the lining. Generally, a zoned section with impervious material placed just behind the lining is found unsafe under sudden draw-down condition. Hence, zoned section with central core and outer shells of semi pervious material has been generally adopted so far.

Stability analysis of canal bank

(i) Narmada main canal in its first phase (i.e. 144.5 Km) passes through seismic zone no. III. Hence as per IS: 1893 -1975 " criteria for earth-quake resistant design of structures", basic horizontal seismic coefficient of 0.04 and important factor of 2.0 is considered. Thus, horizontal seismic coefficient of 0.08 was considered in the analysis.

(ii) Effective stress method using Swedish slip circle analysis is followed.

(iii) The canal bank section is checked for construction, sudden draw-down and operating/steady seepage condition. The earthquake effect is combined with the operating /steady seepage conditions.

After carrying out stability analysis for different types of soils and different heights of bank/depth of cutting, a general decision was taken to provide inner side (water side) slope of 2:1 with suitable adjustment in zoning of embankments. The outer side slope would vary according to slope stability requirements.

Internal section

The entire canal is cement concrete lined with details shown in figure 6.32(ii).

The Internal section of the canal upto Km 460 Rajasthan border are given in table 6.9, with details of B/D ratio (X), P, A, R, V, Z and S. X varies from 9.5 at Head to 3.3 at the tail. This is very high.

Highest depth of the canal is 7.6m at the head and gradually reduced to 4.4m at the tail. For all the various heights of canal embankment, a uniform, Z = 2 : 1 (H : V) is adopted in the entire length of 460 Km.

Table 6.9 : Details of Narmada Main Canal

Notes: Bed gradient: From ch. 0 to 144.5 Km. = 1:12,500: From ch. 144.5 to Guj-Raj Border =1: 15,500
Inner side slope Z = 2:1 (H: V) except for reach from ch. 17.250 to 21.390 Km, where Z = 2.25:1 (H: V)

S. No.	Canal Reach Km	Head-Design Discharge Q (m^3/s)	Bed Width B (m)	Full Supply Depth D (m)	B/D Ratio X	Wetted Perimeter P (m)	Area A (m^2)	Hydraulic Radius R=A/P (m)	Velocity V (m/s)	Details of off taking canals		
										Name	Chain of off taking	Discharge (m^3/s)
(1)	(2)	(3)	(4)	(5)	(6)	(7)	(8)	(9)	(10)	(11)	(12)	(13)
1.	0.00–17.25	1133.55	73.10	7.6	9.61	107.08	671.1	6.266	1.69	Wadia	9.931	6.947
2.	17.25 – 21.39	1133.55	72.10	7.6	9.48	109.52	677.9	6.189	1.67	Tilakwada	17.871	5.934
3.	21.39 – 48.06	1133.55	73.10	7.6	9.61	107.08	671.1	6.266	1.69	Mandwa Angari Timbi Sankheda	25.263 32.694 38.694 45.109	10.711 9.980 1.680 5.069
4.	48.08 – 51.55	1129.25	72.80	7.6	9.57	106.78	668.8	6.262	1.69			
5.	51.55 – 67.15	1126.38	72.60	7.6	9.55	106.58	667.3	6.260	1.69	Miyagam	62.916	143.480
6.	67.15 – 82.04	1070.51	68.70	7.6	9.03	102.68	637.6	6.209	1.68	Gojali Vadodara	70.236 81.834	7.000 76.040
7.	82.04 – 95.23	1037.60	66.40	7.6	8.73	100.38	620.16	6.177	1.67	Dena	88.755	6.858
8.	95.23 – 105.34	1033.31	66.10	7.6	8.69	100.08	617.8	6.173	1.67	Dumad Sakarda	100.026 102.953	5.240 21.780
9.	105.34-129.30	1021.87	65.30	7.6	8.59	99.28	611.8	6.161	1.67	Zmkha Nahara Desar	106.905 111.653 126.645	2.340 3.850 5.400

Contd...

Table 6.9 Contd...

(1)	(2)	(3)	(4)	(5)	(6)	(7)	(8)	(9)	(10)	(11)	(12)	(13)
10.	129.30–144.50	1009.01	64.40	7.6	8.47	98.38	604.9	6.149	1.67			
11.	144.50–184.92	1003.80	72.00	7.6	9.47	105.99	662.7	6.253	1.51	Thasra Dabhali Vajeval Sanali Vasanadarat	148.485 156.456 166.462 171.961 179.150	1.970 2.460 2.760 5.060 3.180
12.	184.92–196.19	974.38	75.20	7.3	10.30	107.84	655.5	6.078	1.48	Mahemdabad	187.074	9.918
13.	196.19–210.93	944.24	72.70	7.3	9.95	105.34	637.3	6.049	1.48	Ghodasar	202.368	15.720
14.	210.93–229.92	927.37	71.30	7.3	9.76	103.94	627.1	6.033	1.48	Vehlal Daskroi	212.546 223.667	18.890 13.553
15.	229.92–246.48	881.63	67.50	7.3	9.24	100.14	599.3	5.984	1.47	Dholka	246.286	56.544
16.	246.48–263.26	822.76	62.60	7.3	8.57	95.24	563.6	5.917	1.46	Sanand Saurashira	258.632 263.000	23.006 319.100
17.	263.26–277.80	583.57	53.70	6.5	8.26	82.76	433.6	5.238	1.35	Viramgham-I Viramgam-II	267.063 277.166	8.706 14.213
18.	277.80–293.50	558.89	51.20	6.5	7.88	80.26	417.3	5.199	1.34	Goraiya Kharaghoda	290.605 292.398	22.568 19.507
19.	293.50–310.25	513.57	46.60	6.5	7.17	75.66	387.4	5.120	1.32	Zinzuwada	301.041	56.360
20.	310.25–314.70	454.69	40.60	6.5	6.25	69.66	348.4	5.001	1.30	Dodiwads	312.764	2.035

Contd...

Table 6.9 Contd...

(1)	(2)	(3)	(4)	(5)	(6)	(7)	(8)	(9)	(10)	(11)	(12)	(13)
21.	314.70–321.11	454.69	40.60	6.5	6.25	69.66	348.4	5.001	1.30	Mithigharial	316.478	1.295
22	321.11-333.59	454.70	40.60	6.5	6.245	69.67	348.4	5.000	1.30	Bolera	326.389	18.244
23	333.59-348.57	431.24	38.20	6.5	5.878	67.27	332.8	4.947	1.29	Rajapura	344.772	21.736
24.	348.57–356.42	407.83	35.80	6.5	5.51	64.87	317.2	4.890	1.28	Amarapura	354.000	19.382
25.	356.42–362.01	387.41	33.70	6.5	5.18	62.77	303.5	4.836	1.27			
26.	362.01–375.88	387.41	33.70	6.5	5.18	62.77	303.5	4.836	1.27	Radhanpur	374.486	37.750
27.	375.88–388.16	346.95	33.90	6.1	5.56	61.18	281.2	4.596	1.23	Kachchh	385.814	170.000
28.	388.16–406.16	169.45	20.30	5.2	3.90	43.55	159.6	3.665	1.06	Vejpur	405.722	18.400
29.	406.16–424.47	149.72	17.30	5.2	3.32	40.55	144.0	3.552	1.04	Madka Malsan	417.792 423.732	12.600 20.243
30.	424.47–442.45	114.77	14.10	4.9	2.88	36.01	117.1	3.252	0.98	Dhima	440.060	17.731
31.	442.45–458.32	95.70	14.60	4.4	3.32	34.27	102.9	3.004	0.93	Gadasisar	453.837	20.188

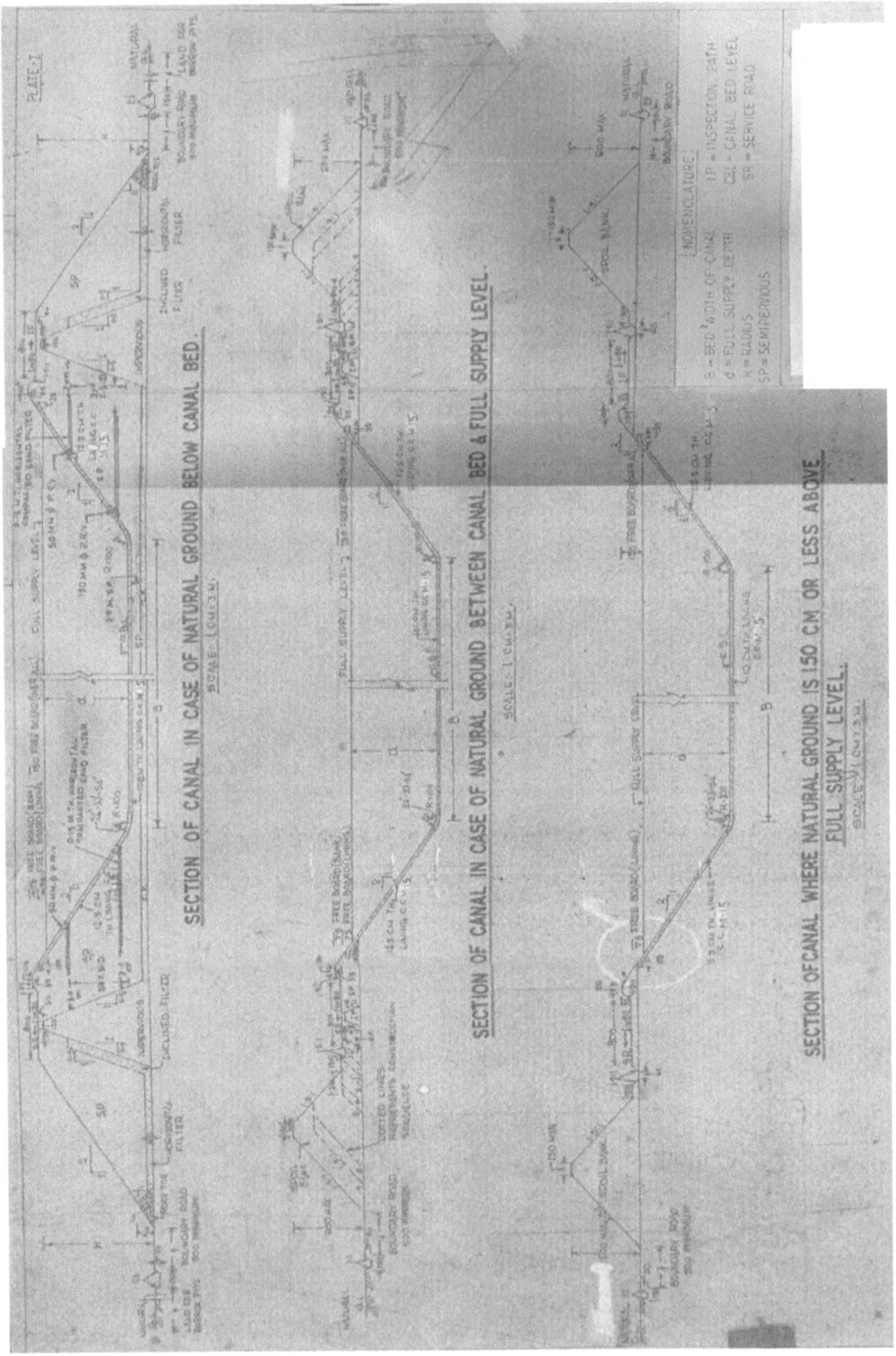

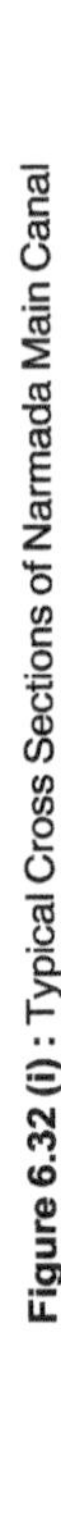

Figure 6.32 (i) : Typical Cross Sections of Narmada Main Canal

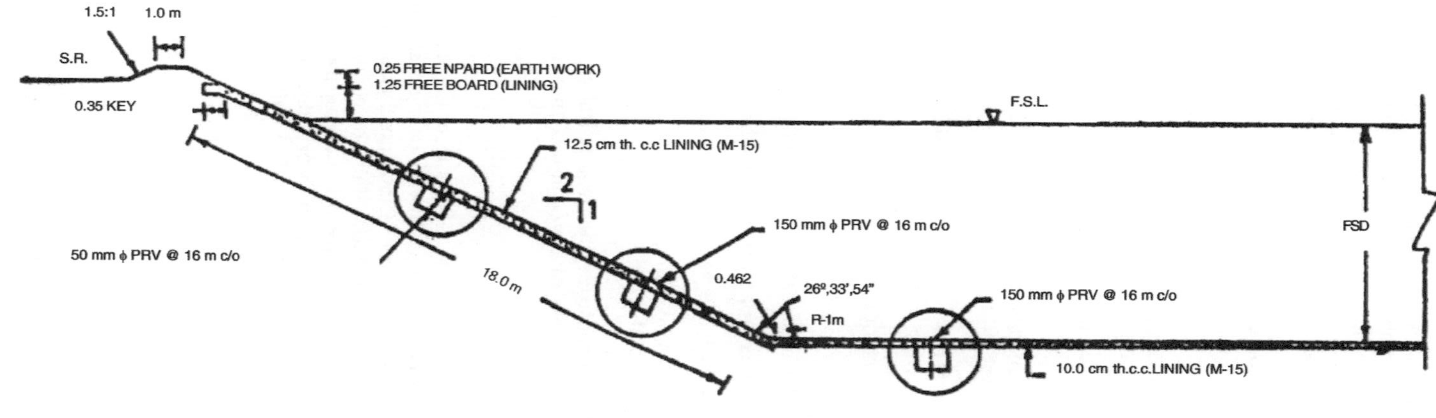

Note :

1. Longitudinal Joints and Transverse joint are Provided @ 4m c/c
2. RCC Lining 20 cm thick in bed and 25 cm thick on side slope is provided near the structures

Fig. 6.32 (ii) Details of Lining in Bed and on side Slope in Narmada main Canal.

The canal section is costly in comparison to many other lined canal section, typically, Nangal and Sunder Nagar hydel canal, see figure 6.1. Also when compared to the All American canal and other canals in figure 6.30 and 6.31, the section is costly, because of a higher B: D ratio, and restricting the depth. The velocities in this lined canal are also low particularly in the tail reaches. Higher velocity can be achieved, even for a very flat slope by increasing the depth of flow and side slope Z. A steeper side slope and smaller B : D could be adopted in the tail areas, to achieve higher FSL.

One high capacity canal versus two parallel low capacity canals

Two alternatives, namely, one single canal of full design capacity versus two parallel canals to be constructed in stages were examined in view of the following considerations :

(i) The smaller capacity canal would give irrigation benefit at relatively early stage since such a canal can be completed in shorter time, later on another parallel canal can be constructed.

(ii) Part of the financial and physical inputs can be diverted towards the development of distribution network in the command area.

(iii) Early development of irrgation could generate more resources to the State which could be ploughed back towards the completion of the rest of the system in later years.

The study indicated that the single canal of large capacity would be better than two parallel canals of smaller capacities of account of the following factors:

(i) The overall increase in the base total cost for doubling the capacity would be much less than 50%.

(ii) Water loss in conveyence would be more with two canal system.

(iii) Land acquisition requirements will be appreciably larger compared to a single canal.

(iv) Operation of two canals would be more complex particularly when canal automation is envisaged.

(v) Construction of second parallel canal will face serious dewatering problems.

(vi) Unbalanced hydrostatic pressure on canal lining when one canal is closed and the other is operating. Also see Para 6.5, figure 6.6 abd 6.7 (i)

In view of the above, it was decided to construct one canal of full capacity instead of two stage construction. Feasibility of navigation along the main canal was also examined and is was been concluded that the proposal is techno-economicaliy not feasible.

6.19 Summary and Conclusions

In this chapter an attempt has been made to draw lessons from the experiences of many canals constructed so far, books, codes and guide lines in vague in various organisations and countries. This study indicates that the canal design has been a matter of individual experience and judgment and differs greatly in practice, from country to country and within the country also. A comparison of practices is made in the text at appropriate location. Inferences of comparison of practices is given in Para 6.3.7. Based on this experience, and sediment transport theory (discussed in chapter 7) a unified procedure of design of lined/unlined canal is proposed in Para 7.20. Suitable design guide lines to suit all possible cases are given in table 9.2 and necessary details in chapters 9.

The major findings of this comparison for both lined and unlined canals can be summarized as under:

(i) B/D ratio of 1 gives a very economical section for Q up to $50m^3/s$.

(ii) B/D ratio between 1 to 2 may be reasonable for Q > $50m^3/s$ and < $500m^3/s$.

(iii) A maximum B/D ratio between 3 to 4, and only in exceptional cases up to 5 may be reasonable for Q > $500m^3/s$.

(iv) With increase in D, the height of embankment does not increase, (as the FSL remains same) but the height of bed filling decreases, over all width of section decreases. Thus cost of land and earth work reduces.

(v) Inner side slope Z = 1: 1 for depth up to 2.0 m has generally been preferred and appears stable for most of the soils. It is also gives higher velocity and economical section.

(vi) Z = 1.25 to 1.5 for depth > 2.0m up to 5m has been recommended by many manuals and generally adopted.

(vii) Z > 1.5, low depth and wide channels has been a matter of individual judgment. These are generally costly, encourages silt deposition on sides.

(viii) All earthen stable channels has shown Z, practically near vertical (or 0.5 : 1.0) and also recommend by most regime theories.

(ix) Earth work of unlined canals is generally less costly/ economical than earth work of lined canals of same capacity. The unlined canals, normally has internal berms. Full bed filling not done, (done only partly in patches) borrows pits in bed in cutting are permitted. Has sufficient provision for future silting and future strengthening of canals banks.

(x) Zoned section with positive cut off has been adopted for high filling.

(xi) The unlined all American Canal (Para 6.14) for a discharge of 433 m^3/s (15,155 cusecs) with a B/D = 7.8 and Z = 1:1.75, for 20.6 ft. depth of flow is more economical in comparison to Narmada lined canal and many other lined/unlined canals.

(xii) Sunder Nagar Hydel Channel lined canal for Q = 255 Cumecs (9000 cumecs) (Para 6.10) with Z = 1.25 to 1.5, and B/D = 1.54 is very economical even in high cutting/fillings and has a zoned section.

(xiii) Even for same Q, B, D and Z section change according to topography and economics, (due to other parameters, S, V, N) similar to modified section of RMC, Mahi Project (Para 6.8). This section is very economical.

References

Anon, Les Voies Navigables, Travaur (Paris), June 1952.

Anon, Lots of Digging Downstream, Eng. News Rec., 22nd July 1954.

Anon, The Beauharnois Canal and Hydro-Electric Power Scheme, The Engineer, 13th Feb. 1931.

Anon, The Tummel Development, The Engineer, 6th And 13th July 1934.

Bhakara, Canal Manual Guidelines, General Instructions For The Design of Lined and Unlined Channels of Bhakra Canal System, 1954.

BIS (Bureau of Indian Standards), Code of Practice For Design of Cross-Section of Lined Canals, IS No 4745 – 1968, now with drawn.

Buckley, Irrigation Department of Egypt, Irrigation Practice, P. 207.

Buckley, Irrigation Pocket Book, 1893, 2nd Ed. 1905.

CBIP, (Central Board of Irrigation & Power), Manual on Current Practices in Canal Design in India, Malcha Marg, Chanakyapuri New Delhi, June 1968.

CBIP, Design Practice for Unlined Incised Canals, Technical Report -7 May 1970.

CBIP, Manual on Canal Linings, 1975.

CBIP, Pub. No.187, Major Canal in India, Malcha Marg, Chanakyapuri New Delhi, 1986.

CBIP, Pub. No. 144: Symposium on Operation and Maintenance of Canal Systems.

Chitlae, Hydraulics of Stable Channel, Published By CWPC, 1966.

Clement, M., Ottmarsheim Development, Travaur (Paris), Jan. 1951.

Clibbron, 1901 J., Roorkee Treatise. Irrigation Works in India, Thomson Press, Roorkee, 1901.

Ellis, Col. W.M, Manual-Irrigation Engineering, College of Engineering, Madras, 1950.

Etcheverry, B.A., Irrigation Practice and Engineering, Vol. II, Conveyance of Water, Mc Graw-Hill Book Company, Inc., New York, 1st Ed., P. 122, 1915.

Flynn, P.J. Irrigation Canals and Other Irrigation Works, George Spaulding & Co., San Francisco, Cal., 1892.

Fortier, S. Conveyance of Water In Irrigation Canals, Flumes and Pipes Water Supply Paper No. 43, U.S. Geological Survey, Washington, D.C., 1901.

Framji, K.K., Design Practices of Irrigation Canals in The World, Report of International Commission of Irrigation and Drainage, Central office, 48 Nyaya Marg, Chanakyapuri, New Delhi - 21, 1972.

Fraser, D.D., The Construction of the Butterbridge Diversion Aqueducts, Loch Sloy Hydro-Electric Project, Jour. Inst. Water Eng., May 1952.

Guilford, L. Molesworth, Pocket-Book of Engineering Formulae (Useful Formulae and Memoranda) For Civil and Mechanical Engineers, E & F.N. Spon, London, 7th Ed., P. 176, 1871.

Houck, Irrigation Engineering, Vol II., 1951.

Hudson, W. and Hunter, J.K., The Galloway Hydro-Electric Development With Special Reference to the Constructional Works, Jour. Inst. C.E., Vol. 8, 1937-38.

Irrigation Department, of Various States In India and other Countries, Project and Revised Project Estimates And Design Notes.

Jaeger, C., The Rhone Development, Water Power, Jan. 1953.

Khushalani, K.B., Irrigation Design and Practice, 4th Ed. 1971, 6 Volumes, 1954.

Krishna, Iyenger, M. B., High Level Canal in Tungbhadra Project Salient Features Vol. XVII, No.4, P. 10-20, April 1967.

Molesworth, Guilford L., Pocket-Book of Engineering Formulae (Useful Formulae and Memoranda) for Civil and Mechanical Engineers, P. 176, E & F.N. Spon, London, 7th Ed., 1871.

Mullins, J., Irrigation Manual, Spon E and F.N. 1890.

Murthy, K. Satya Narayana, Design of Canal Sections in Nagarjuna Sagar Project Vol. XV, No. 5, P.-219-229, May 1970.

Murti, N. G. K., Irrigation Canal System, Construction Cost Economics; Journal of Central Board of Irrigation and Power, Oct. 1981

Nayak, P.R. F., The Tungabhadra Left Bank Canal Vol. IX, No. 2, P. 18-22 Feb. 1959.

Neville, J., Hydraulic tables, third edition, Lock Wood and Co. London, 1875.

Parker, P.M., The Control of Water, Second Edition, 1932.

Rajasthan Canal Board, Canal Lining, Sanchai Bhawan, Bhawani Singh Road, Jaipur.

Sharda, C.B., Lining of Canals With Special Reference of Rajasthan Canal M.E. Dissertation, WRDTC, University of Roorkee, 1968.

Sharma, K.R., Irrigation Engineering, 3rd Ed., 3 Volumes, D.V. Sharma, Publishing Company Ltd., 1944, 1959., (For M/S. Omex Trade Links Jullundur City At India Printers, Training College Road, Jullundur City).

Shetty, D. B., A Critical Review of Different Types of Canal Linings with Special Reference To Tungabhadra High Level Canal; M.E. Thesis University of Roorkee, 1971.

Single, A.M, C.E Haryana - Canal Modernisation in Haryana And Is Benefits, Bhagirath, C.W.C., Ministry of Irrigation, Sewa Bhawan, R.K. Puram, New Delhi. -110066, Vol xxxii No2, April, 1985.

Sitarama Sastry, M. and L. R. Kapur, Nagarjuna Sagar Left Canal Vol. XIX, No. 10, P. 433-438, Oct. 1969.

USBR, (United States Bureau of Reclamation), Linings for Irrigation Canals, July, 1952.

USBR, Canals and Related Structures Design Supplement No.3.

USBR, Lining of Irrigation Canals, Pp - 149, 1963.

Venkata Rao, M. Phase Construction of Nagarjuna Sagar Right Canal Vol.xix, No.11, P. 478-456, Nov. 1969.

Wilson, Irrigation Engineering, Wollis R. and W.W.G.Yeh, 1987.

□□□

Chapter 7

Sediment Transport and Canal Design Theories

7.1 Terminology

A number of common terms, frequently used in hydraulics and sediment transport literature are defined as under for easiness, and ready reference to readers.

Density : The density ρ ($Kg\text{-}sec^2/m^4$ or $lb\text{-}sec^2/ft^4$) of a fluid or solid is the mass that it possesses per unit volume. Both the density of the water-sediment mixture and the density of sediment are important variables.

Specific Weight : The specific weight γ (Kg/m^3, T/m^3, lb/ft^3) is the weight per unit volume. It is related to the density by $\gamma = \rho\, g$.

Where, g is the gravitational acceleration in (m/sec^2, ft/sec^2).

Specific Gravity: The specific gravity S is the ratio of the specific weight of a fluid, solid or fluid-solid mixture to the specific weight of water at 4°C or 39.2°F.

Viscosity: Viscosity is the property of a fluid that resists relative motion and deformation in the fluid and causes internal shear. Therefore, viscosity is a property exhibited only under dynamic conditions. According to Newton, the shear (τ) at a point within a fluid is proportional to the velocity gradient du/dy at that point

$$\text{or } \tau = \mu \frac{du}{dy}$$

Where μ in $Kg\text{-}sec/m^2$ or $lb\text{-}sec/ft^2$, is the dynamic viscosity. When divided by the density ρ, it is the kinematic viscosity $\nu = \mu/\rho$ in m^2/sec or ft^2/sec. Under ordinary conditions of pressure, viscosity varies only with temperature. The viscosity of a liquid decreases with increasing temperature, the reverse is true for gases.

Surface Energy : Surface energy, in lb/ft or Kg/m, also known as surface tension is caused by cohesion and adhesion of molecules. Cohesion is the force of attraction between molecules of the same kind, and adhesion is the force of attraction between molecules of different kinds. Surface energy causes the spherical shape of water droplets and their internal pressure, as well as certain types of movement of water and other fluids through porous media. The surface tension of liquids decreases as their temperature rises.

Drag Force : The drag force F_d is the force component exerted by a moving fluid on any object submerged in the fluid. The direction of the force is the same as that of the free stream of fluid.

Lift force : The lift force F_L is the force component exerted on a body submerged in a moving turbulent fluid. The force acts in the direction normal to the free stream of fluid.

Shear force : The shear force is the shear developed on the wetted area of the channel and it acts in the direction of flow. This force per unit wetted area is called the shear stress τ_ρ and can be expressed as $\tau_\rho = \gamma RS_b$, where γ is the specific weight, R is the hydraulic radius, and S_b is the bed slope.

Shear Velocity : The shear velocity V_* is defined as $V_* = \sqrt{\tau_0/\rho}$

Reynolds Number : The Reynolds number Re is

$$R_e = \frac{\rho V^2/L}{\mu V/L^2} = \frac{\rho VL}{\mu} = \frac{VL}{\nu}$$

where V is the velocity, L is the length and ν is the kinematic viscosity. The Reynolds number relates the inertia forces to the viscous forces and is usually involved wherever viscosity is important, such as in slow movement of fluid in small passages or around small objects.

Froude Number : The Froude number Fr is

$$F_r = \left(\frac{\rho V^2/L}{\gamma}\right)^{1/2} = \frac{V}{\sqrt{gL}} = \frac{V}{\sqrt{gd_h}}$$

Where g is the gravitational acceleration, d_h is hydraulic mean depth = Area/ Top width. The Froude number relates the inertia forces to the gravitational effects and is important wherever the gravity effect is dominating, such as with water waves, flow in open channels, sedimentation in lakes and reservoirs, salt-water intrusions, and the mixing of air masses of different specific weights.

Weber Number : The Weber number W_e is

$$W_e = \frac{\rho V^2/L}{\sigma/L^2} = \frac{\rho V^2 L}{\sigma}$$

where σ is the surface energy. The Weber number relates the inertia forces to the forces of surface energy and is important, especially in connection with formation of water droplets and movement of water in capillaries and interstices of porous media such as soils.

Relative Roughness : The relative roughness is k/R where k is the height of the roughness element and R is the hydraulic radius. The inverse of the relative roughness is encountered in resistance formulas.

Sedimentation Parameter : The sedimentation parameter (G) is

$$G = \frac{\rho\nu^2}{\gamma_s D_s^3}$$

where γ_s is the submerged specific weight of the particle with diameter D_s. This dimensionless number occurs in sediment flow formulas and is a determining factor considering threshold conditions and in the computation of the fall velocity of solid particles.

Particle Reynolds number : the particle Reynolds number is related to the fall velocity of the particle

$$\frac{\omega D_s}{\nu}$$

Where ω is the fall velocity of the solid particle with diameter D_s. This dimensionless number is extensively used in the study of the dynamic properties of submerged solid particles and in sediment transport relations.

The Mobility Index : The mobility index is $\frac{V_*}{\omega}$

This dimensionless number is useful to predict bed forms.

The Entrainment function : The entrainment function is similar to the Froude number but involves the grain size. It is $1/\Psi = \tau_0 / \gamma_1' D_s$

This dimensionless number is basic to the study of beginning of motion and sediment transport.

Other dimensionless numbers such as the Boussinesq number, the Mach number, etc., are commonly used in fluid mechanics. But these numbers have no substantial relations to sediment problems.

Temperature taken as a parameter is important because a change in temperature changes the value of the viscosity which alters the value of Reynolds number. For example, the variation in $\frac{\omega D_s}{\nu}$ relates to the formation of bed forms. Note that the change of water temperature in streams during a 24-hour cycle can exceed 40° F (22° C).

Bed Layer : A flow layer, several grain diameters thick (usually taken as two grain diameters thick) immediately above the bed.

Bed Load : Sediment that moves by saltation (jumping or hopping), rolling, or sliding in the bed layer, or along the bed of a stream.

It is also divided in two parts

(a) *Contact Load:* Sediment particles that roll or slide along the bed in almost continuous contact with the streambed.

(b) *Saltation Load:* The sediment bouncing and hopping along the stream bed by the energy and turbulence of the flow and by other moving particles.

Bed-Load Discharge (or Bed load or Sediment): The quantity of bed load passing a cross section of a stream in a unit of time.

Bed Material (or Bed Sediment): The sediment mixture of which the streambed is composed.

Bed-Material discharge (or Bed-Material Load): That part of the total sediment discharge which is composed of grain sizes found in the bed and it is equal to the transport capability of the flow.

Density of Water-Sediment Mixture: The bulk density which is the mass per unit volume including both water and sediment.

Discharge-Weighted Concentration: The dry weight of sediment in a unit volume of stream discharge, or the ratio of the discharge of dry weight of sediment to the discharge by weight of water sediment mixture.

Fine-Material Load (or Wash Load): That part of the total sediment load that is composed of particle sizes finer than those represented in the bed. Normally the fine-sediment load is finer than 0.062 mm for a sand-bed channel. The silts, clays and sand could be considered as wash load in coarse gravel and cobble bed channels.

Load (or Sediment Load): The sediment that is being moved by a stream.

Sediment (or Fluvial Sediment): Fragmental material that originates from weathering of rocks and is transported by, suspended in, or deposited by water.

Sediment Concentration (by Weight or by Volume): The quantity of sediment relative to the quantity of transporting or suspending fluid, or fluid-sediment mixture. The quantity may be by weight or by volume.

Sediment Discharge (or Sediment Load): The quantity of sediment that is carried past any cross section of a stream in a unit of time. The discharge may be limited to certain sizes of sediment or to discharge through a specific part of the cross section.

Sediment Yield: The total sediment outflow from a watershed or a drainage area at a point of reference and in a specified period of time. This is equal to the sediment discharge from the drainage area.

Spatial Concentration: The dry weight of sediment per unit volume of water-sediment mixture in place, or the ratio of the dry weight of sediment to the total weight of water-sediment mixture in a sample or unit volume of the mixture.

Velocity-Weighted Sediment Concentration: The dry weight of sediment discharged through a cross section during unit time.

Suspended Load (or Suspended Sediment): Sediment that is supported by the upward components of turbulent currents and that stays in suspension for an appreciable length of time.

Suspended-Sediment discharge (or Suspended Load): The quantity of suspended sediment passing through a stream cross section above the bed layer in a unit of time.

Total Sediment Discharge: The total sediment discharge of stream. It is the sum of the suspended-sediment discharge and the bed-load discharge.

Total Sediment Load (or Total Load): The total sediment load of stream. It is the sum of the suspended load and the bed load, or the sum of the bed-material load and the wash load.

Wash-Load Discharge (Wash Load): That part of the total sediment discharge which is composed of particle sizes finer than those represented in the bed and is determined by available bank and upslope supply rate.

Armoring of the Bed: It is the natural equilibrium of the bed reached after the fine material is transported away by the flow resulting in the development of a thin layer of coarser particles at the bed surface.

7.1.1 Tractive Force (Shear stress)

When water flows in a channel, a force is developed that acts in the direction of flow on the channel bed. This force which is simply the pull of water on the area of wetted perimeters (i.e. perimeter x length) is known as the tractive force, or sheer stress on the boundary material.

This approach is based on the consideration of equilibrium of a sediment particle resting on the bed under the action of

(i) Drag
(ii) Lift force
(iii) Tractive force caused by the following fluid and the submerged weight of the particle.

Considering steady uniform flow in a rectangular channel and consider equilibrium of a water prism abcd under various forces acting on it Fig. 7.7 Since there is no acceleration of the fluid, the summation of all the forces acting in the direction of flow must be zero.

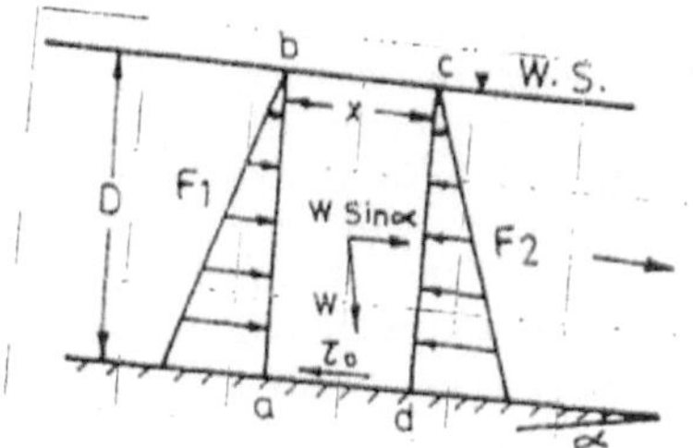

Figure 7.1 Forces Acting on a Water Prism

Hence $\sum F = F_1 + w \sin\alpha - F_2 - \tau = 0$

Where

F_1, F_2 = hydrostatic forces

W = weight of the water prism per unit area of wetted perimeter

τ = average shear stress of the boundary since the depth of flow is the same at section ab and cd

$F_1 = F_2$ and $\tau = \dfrac{W \sin\theta}{\text{area of wetted perimeter}}$

Where

Area of wetted perimeter = P. ΔL; ΔL = length of the prism; P = area of wetted perimeter.

$W = A\Delta L\gamma_f$; γf = specific weight of fluid $\simeq 1$ (for water), A = area of prism;

Therefore,

$$\tau = \frac{A\Delta L\gamma_f \sin\theta}{P.\Delta L}$$

Replacing A/P by R; Where R = hydraulic radius

we get

$\tau = \gamma_f\ R \sin\theta$; for small θ, $\sin\theta = \tan\theta = S_0$ (channel slope).

Or $\tau = \gamma_f\ R.S_o$ - Eq. 7.1 (Kg/m^2 in metric units and lbs/ sq. ft in FPS units)

τ is not dimensionally similar so use consistent units on both sides.

For wide channels R = D, therefore $\tau = \gamma_f$ DS or γRS

γ = Specific Weight of sediment laden water

S = Bed slope of the channel

(Use consistent units on both sides)

The force exerted by water on the channel bed will have the same magnitude but will act in the direction of flow. This shear stress can be directly related to the velocity distribution near the boundary and the viscosity of the fluid.

This tractive force is used in the estimation of sediment transport see table 7.3 to 7.5 and also used in the design of channels known as tractive force theory, see Para 7.10.

7.2 Silt inflow in Canals, its Effect and Control

Generally most of canals, whether lined or unlined carry some silt. The silt in flow in a canal may be more in case of out flow from weirs/barrages or low dams or diversion schemes, particularly when situated in hills or hill toes. The silt inflow is less in canals taking off from reservoirs. In reservoirs much of the silt is deposited in it and clear water passes down in the canal. However, some silt may also pass in the canal depending upon the inflow-release pattern and time, location of canal out let and sill level, water level in the reservoir during operation, for example in monsoon season. In non monsoon season it may be silt free.

Silt inflow in a canal varies according to the silt inflow in the river, its discharge and the diversion flow. Rivers receive huge quantity of sediment along with water due to erosion of drainage basin. Silt may also enter into a canal from the topography, through which the canal is passing, such as wind blown sand and rain cuts on the cut slopes or the storm water inflow in the canal at inlets. All canals of IGNP, Rajasthan, India, are subject to wind blown sand/silt into the canal. Typical examples of storm water inflow into the canal are the inlets of Upper Ganga Canal, U.P. and India. In some canals, failures of inner slopes of banks have also caused silts and debris. This is also a seasonal and occasional phenomenon and is not uniform over time.

Silt is very harmful in power canals. It damages and erodes the blades of the turbines. In irrigation canals silt is useful when transported to the fields. It has high agricultural productivity and is beneficial to the crops. But this silt becomes harmful in irrigation canal also, when deposited in it. It blocks/reduces water way, carrying capacity/discharges and may encourage weed growth on sides. When at high velocity it has an abrasion effect on lining or damages it. In earthen channels the bed and banks are eroded. The phenomenon varies from project to project, velocity and nature of banks. Engineers have made use of Low velocity to help in deposition of silt for strengthening the earthen canal banks, Sharma (1955).

An analysis of sediment inflow in river and canals is very essential for proper design and regulation.

The objectives of a canal design, operation and maintenance are:

(i) Exclude entry of silt, debris (or sediment) in the canal, as much as possible, i.e. provide Silt Excluders at the source.

(ii) Whatever has entered may be ejected from a canal to the extent possible, so provide Silt Ejectors, at appropriate locations. Pass out some portion of water according to silt inflow.

(iii) Also entrap the maximum silt or as much as possible in Silt traps/Silt tanks or de-silting chambers and flush out or eject at suitable locations. It may be called as a modified Silt Ejector.

(iv) And lastly channel design should be such that it is non-silting and non-scouring i.e. carries the silt and passes out to the distribution system and to fields and does not settle or erode bed. In case of lined canals it should not damage lining, but may have silt cleansing velocity. Some silt can be permitted only in irrigation unlined canals but that should not be deposited in bed. It should be made use to strengthen the banks of canal.

(v) However power canals should be silt free.

Since the sediment concentration change with time in a year and available water or discharge also change, therefore such design may not be feasible. The only alternative is to reduce silt in flow and provide silt traps and eject the silt, and a balanced design can be attempted, see Para 7.11.

7.3 Sediment and Its Transport in a River/canal

7.3.1 Sediment generation (Production)

Sediment is produced by soil erosion. Three forces are predominant in soil erosion (i) wind (ii) water and (iii) gravity. Generally water and gravity work together but also joined by wind. These are discussed in brief as under

7.3.1.1 Forces/Variables Causing Sediment Generation

(i) Wind (velocity, duration and direction), causes soil erosion. Typical example is desert or sandy soils. Substantial portion reaches in rivers and canal

(ii) Rainfall (intensity and duration) disturb top crust, Soil eroded joins rivulets (nallah). River/canal banks, slopes, are more easily eroded.

(iii) Geophysical and topographical characteristics, such as type of soil, unstable mountains and steep slope, land slides, size, and shape of particles density hard and soft rocks see figure 8.4 to 8.6.

(iv) Ground water movement causing instability of mountain slopes, land slides, etc.

(v) Vegetative cover, with holding soil and reducing wind, rain and runoff soil erosion.

(vi) Discharge and velocity, channel geometry and energy/bed slopes. Velocity can take care of energy slope (S). These change with change in sediment concentration.

Discharge in river varies with rainfall- runoff process.

Magnitude of forces and so sediment quantity and quality varies with time at a particular cross-section and all along the river/canal. in a year depending upon stability of bank slopes, the rainfall, topography, landslides, stream flow characteristics etc. All the processes and forces causing river bank failures, shown in figure 8.3 to 8.7 and Para 9.9 are applicable to the banks of every small tributary/ nallah or main river, itself or even to the canal banks.

Since the quantity and quality, size etc. of sediment available depends on the geophysical characteristics of topography/land, it cannot be easily related to discharge only. It is not a mere function of discharge. All the three are independent variables and also related to each other to the extent of erosion within the channel and transporting capacity of river.

The sediments at the foot hills of each mountain are different then at the plains and then near the sea.

7.3.2 River Regions and its sediment flow characteristics

For a better understanding and clarity of silt problem, the course of a river can be broadly divided into following four parts.

(i) The mountains

(ii) The sub-mountainous or boulder tract

(iii) The trough

(iv) (a) The alluvial, (b) The deltaic,

All these stages can be seen in all the rivers.

The slope of the river decreases from the source to the sea. There may not be marked points of division into these four regions but a trained eye, can easily distinguish them as under :

(i) *The Torrent*: In the mountainous regions the river is called a torrent. It has a rocky bed, narrow/ gorges and many abrupt falls. The stream is rapid, deep, narrow, comparatively straight, clear and sparkling.

(ii) The course of the torrent in mountainous region can be divided into two sections:

a) The zone of erosion. In this there goes on constant erosion of bed, by cutting back and increasing of slope.

b) The zone of deposition. It is the zone where the eroded material is deposited in the upper reach in fan-shaped layers, and the country is raised by the yearly deposits just as in the deltas. The deposited stuff goes on gradually restricting the flow of water and in the course of years may impede the discharge so that on the occasion of probably an unusual flood the stream forces its passage through the ground made by it. It thus cuts a channel which in flood will carry forward the shingle and sand to a lower reach below which it then attacks in a similar manner. Some writers divide the course of the torrent into four parts on a similar basis as the course of the river.

(ii) Sub-mountainous Regions. In the sub-mountainous region the river still has a steep slope. The width is more, but it is interspersed by high lands with many hill torrents affluent.

(iii) Trough Region. In the trough portion the river on account of its work during ages has cut into the soil and has high land on the sides of the valley, the valley may be 5 to 16 Km wide and in it the river follows more or less a tortuous course. The river in the course of many centuries has varied its location throughout the valley leaving steep scarps on either side of the boundary of its wanderings. The high land on each bank, which may be 3 to 15m, above the general level of these valleys, is called 'Bhangar land', whereas the lower land where it wanders is called 'Khadir'. The khadir lands are very fertile. But the 'Bhangar' lands though formed of good soil once brought down by the rivers from the mountains are either covered with grass or brushwood or are entirely waste lands. The Bhangar lands form the 'Doabs'. The word 'doab' means "between two waters (or two rivers), and it originated in the Punjab (India).

(iv) Alluvial Region. This stage of the river, as said above, is often divided into two parts, the alluvial proper and the deltaic.

Table 7.1 gives at a glance the principal characteristics of the different reaches of the river. Figure 7.2 shows schematically the course of a river.

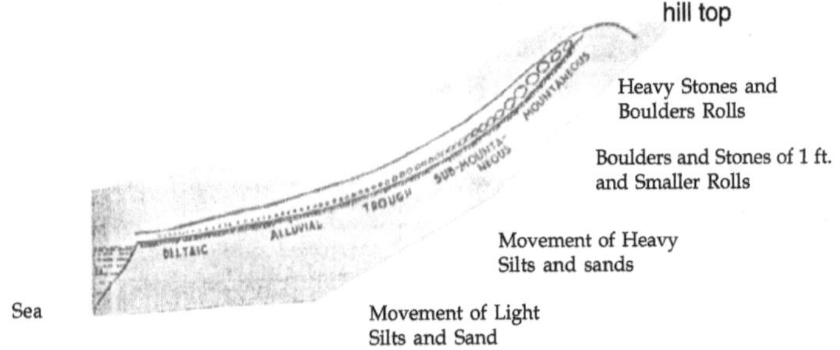

Figure 7.2 River Regions and sediment transport (See table 7.1).

Table 7.1 Characteristics of River Regions (also see figure 7.2)

Parameter	Regions				
	Regions			Deltaic	
Regions	Mountainous	Sub-mountainous	Trough	Alluvial	Deltaic proper
1. Bed slope ft. per mile. m. per Km ratio V:H	Over 600 to 15 114 to 3 1/9 to 1/333	20 to 5 4 to 1 1/250 to 1/1000	5 to 2 1 to 0.4 1/1000 to 1/2550	2 to 0.5 0.4 to 0.1 1/2500 to 1/10000	0.5 to 0.25 0.1 to 0.05 1/10000 to 1/20000
2. Flood velocity ft/s m/s	15 to 45 5 to 15	5 to 30 1.5 to 9	3 to 5 1 to 1.3	1 to 4 0.3 to 1.2	1 to 3 0.3 to 0.9
3. Course (Section)	Narrow, Fixed and Deep	Well defined	Meander ing with invalley	Wide, meandering & shallow	Very wide, Shallow & changing
4. Material	Heavy stones	Boulders from About 1 ft. (0.3m.) diameter to Shingle ½ in. (13 mm.) diameter	Heavy silt and sand	Light silt and sand	Very light silt andsand

The characteristics of both these parts alluvial and deltaic are the same. The difference is only of degree. In the alluvial portion the river has usually a single high water channel in greater portions of its length, though in some portions there may be high islands, which of course get sub-merged during very high floods. In the low water season the river may have more than one channel. Among them usually one is very active. The entire river course is confined within a few kilometers, say 3 to 10. In the deltaic region the river has more than one channel at considerable distances apart. Often two or three may be equally active both in the high and low water. Other principal characteristics of the alluvial and deltaic tracts are as under:

(a) Characteristics of Alluvial Tracts

The river has generally a tortuous course. At the discharge of the river is not constant but varies considerably in the ratio of even 1:100, no fixed course is possible. The maximum flood width may be 20 times the width of the river during the low water season. The velocity and the silt charge also change very greatly. They may be several times more during flood. The country generally slopes away from the river.

(b) Characteristics of Deltaic Tracts

(i) *Absence of tributaries.* As the country is flat and the rainfall is less than on the mountains no tributaries join the rivers in the deltas. The little rainfall that occurs as in the Indus delta is carried by surface drains directly into the river.

(ii) *Its constantly changing course.* The river runs in the soft soil causing it to erode on one side and thus introducing curvature into the course. The erosion goes on at the outer side of the curve and deposition takes place on the inner. Each curve needs a reflex curve on the downstream. Thus the river is made to take to a sinuous path. When the sinuosity increases considerably the river shortens its course by cutting across the chord. The course thus constantly goes on changing.

(iii) *Tendency to sub-divide.* Due to the flattening of slopes near the sea, the river is not able to push water into the sea and has therefore a constant tendency to seek lower levels and thus go on sub-dividing, see Fig. (7.8).

(iv) *Sector shape.* As the river sub-divides, it more or less forms a set of radiating channels which when entering the sea encroach upon it in the form of an arc of a circle. It is the shape of the tract near the sea which gives the name to this tract as delta.

(v) *Land slopes away from the river.* Due to a constant heavy deposit in the bed of the river and on the berms the river channel goes on continuously rising in elevation leaving the surrounding country at a lower level. The land, therefore, slopes away from the river. The portions of land in between branches are low and if they have a free outlet into the sea there is a good facility of drainage. The level of the land is generally below the maximum flood level.

(vi) *Gradual flattening of the slope.* The continuous process of silt deposition into the bed of the river and the sea goes on raising the bed and thus flattening the slope. Hence the flood carrying capacity of the river decreases as it approaches the sea.

There are three important aspects of sediment (i) the quantity (concentration), (ii) quality (size, density etc.) of the sediment available at a particular cross-section in the river at a given time/different times in a river, and (iii) sediment transport capacity of a river. These aspects are applicable to a canal also.

7.3.3 Data and characteristics of Sediment in Rivers/canals

Observation done in various rivers show that sediment load in river (streams) from which canals are fed seldom exceeds 5,000 ppm with an annual average of few hundreds of ppm. Concentrations of sediment coarser than 0.075 mm diameter for various locations along the Chenab River in West Pakistan are shown in Figure 7.1. The individual lines are averages of hundreds of readings which show wide scatter in the extremes. The effect of curvature in the river is visualized by comparing values for the Trimmu for the year 1961 left bank with those for the Trimmu for the year 1961 right bank. A gradual change in curvature may account for the difference between the Trimmu left bank values for the year 1939 and 1961.

The behaviors of sediment depend on the specific gravity, size, and shape of the particles and the size distribution. A typical sample of sediment consists of a mixture of particles of various densities, sizes, and shapes. The variation of density is generally small, and for practical purpose the mean density may be used for all the sediment. The density is in most cases is so near the density of quartz (specific gravity 2.65) that this value may be used without significant error in the formulas for sediment transport.

The size distribution of a sediment mixture can be represented by soil classification and grain size distribution curve or also some times known as frequency diagram as in Figure 7.3(ii), which shows characteristic grain sizes for typical suspended sediment and bed materials. The mean or effective diameter of mixed sediment is often described by its median or 50 percent size.

Typical grain sizes of suspended and bed load, sampled in eight regime canals in West Pakistan, are shown in Figure 7.3(ii). The effect of alluvial sorting is indicated by the narrow range of grain sizes of the bed material. In most regime canals the sizes of the bed material and suspended sediments usually show a marked seasonal variation.

Finer suspended sediments (less than about 0.06 mm) have measurable effect on the performance of a canal, the impairment of canal performance usually results from an excess of coarser material (greater than about 0.06 mm). Accordingly, it is concentrations of the coarser sizes and the variation in such concentrations that are significant in canal operation. A typical annual variation, as measured near the head of the Upper Chenab Canal, West Pakistan, is show in Figure 7.3(iii).

The silt transport capability of a channel is a function of capacity, measured as the quantity of sediment which will be moved, and competence, measured by the maximum size of bed particles which will be moved. Capacity increases with decrease in particle size, and transport capability in any given size range can be for greater than the volume of sediment available.

The size of transportable grain is indicated by the tractive force, see Para 7.1.1. In a channel the bed material is usually composed of sediment which deposited during periods of decreased competency or in sufficient transporting capacity.

In the past, most valuable investigations on silt problems have been made in India. The application of the theories resulting from these investigation to similar problems in other countries can best be considered after a knowledge of the extent of silt carried by irrigation waters in India; this information is largely contained in the papers of the Punjab irrigation branch and in a book entitled "Facts, figures, and Formulas for Irrigation Engineers", compiled by R. Buckley (1905). Data of few observations carried out at the end of 18th centaury are given in table 7.2.

7.4 Classification of Sediment Laden Water

The gravitational flow of a water-sediment mixture or sediment laden water or fluid or simply often called water can be distinguished in three types of movement as under:

(i) Non Newtonian mixture

The mixture behaves Non-Newtonian, if the valumetric concentration of sediment becomes of importance, $C_s > 8\%$ (80,000 ppm). The difference between the density of the mixture and of the water is also very large, $\Delta\rho > 130$ Kg/m^3.

The flow of a Non-Newtonian fluid modifies all concepts of Newtonian hydraulics, such as the resistance to the flow, as well as the distribution of velocity and of concentration, the settling velocity is also influenced and the solid particles stay longer in suspension. The transport of sediments as hyper concentrated suspension and the debris flow, as well as hyper concentrated turbidity currents fall into this category:

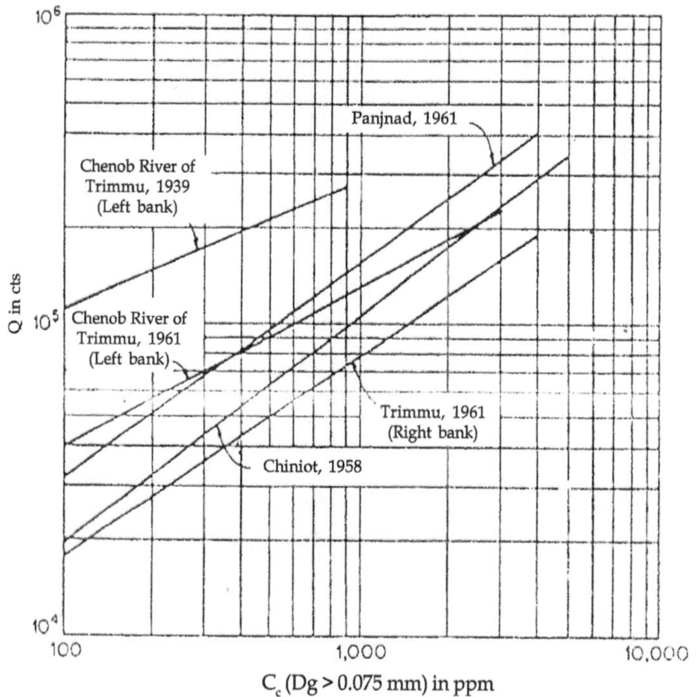

Sediment Concentration (Cc) of grain size (Dg) > 0.075 mm) in ppm

Figure 7.3 (i) Sediment Concentration v/s Discharge, Chenab River, West Pakistan

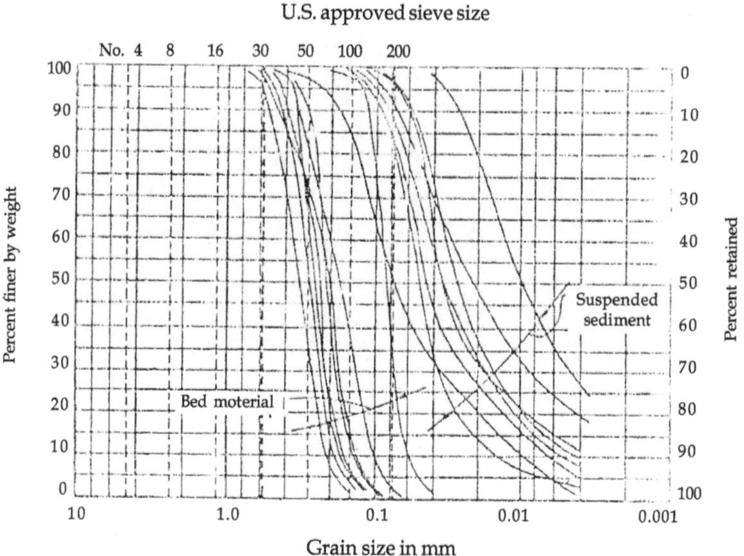

Figure 7.3 (ii) Grain Size Distribution of Suspended and Bed Sediments in regime Canals, West Pakistan

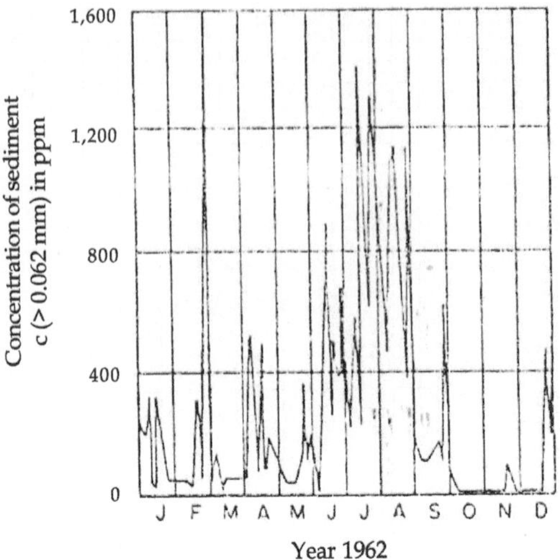

Figure 7.3 (iii) Annual Variation of Sediment Concentration-Upper Chenab Canal, West Pakistan

Table 7.2 Silt Content of Some Canal Waters

River/Canal	Period	Average per cent of silt (%)	
		Damp silt by volume	**By dry weight**
1. Water from Sutlej River entering Sirhind Canal	June to Sept. 20, 1895	0.165	0.132
do-	June to Sept. 20, 1896	0.234	0.187
do-	June to Sept. 20,1897	0.546	0.436
2. Water from Sone River entering Sone Canal	June, 1898	0.1875	0.15
do-	July, 1898	0.325	0.26
do-	Aug., 1898	0.275	0.22
do-	Sept., 1898	0.62	0.25
3. Water from Bengal Rivers near surface	Mean for whole year of 234 experiments	0.076	0.06
4. Water from Indus	Flood season	0.53	0.425
5. Water from Ganges near head of Ganges Canal (Hardwar, Uttaran Chal)	Maximum	1.00	0.80
do-	Aug and Sept.	0.16	0.13
6. Water from Nile	Flood season	0.19	0.15

- The transport of sediments as a hyper concentrated suspension is encountered in rivulets (nallah). Usually enormous quantities of sediments being of small size enter the channel due to surface erosion caused by extensive rainfalls in the catchment basin. The soil particles stay usually for long time periods in suspensions, as wash load.
- Torrential flow of debris may establish themselves at rather steep slopes, $S_0 > 15^o$. All kinds of particles, from the finest (having cohesion) to the largest (blocks of 1 m^3) participate in the movement, which is rather rare in occurrence and of short duration, and is usually caused by severe rainfall.

(ii) Quasi - Newtonian mixture

The mixture behaves quasi-Newtonian, if he volume concentration of sediment remains small, $C_s <$ 8%. The difference between the density of the mixture and of the water is important, $\Delta\rho < 130$ Kg/m^3.

The transport of sediment as concentrated suspension notably close to the bed, as well as the turbidity currents fall into this category.

(iii) Newtonian mixture

The mixture may be considered Newtonian, if the volume concentration of the particles is very small, $C_s << 1$ % (10,000 ppm). The difference between the density of the mixture and of the water, $\Delta\rho = (\rho_m - \rho) = (\rho_s - \rho)\, C_s$ remains also small, $\Delta\rho <<16$ Kg/m^3.

This type of transport of solid particles, which is most often encountered in rivers at foot hills, plains and canals (see figure 7.3).

7.5 Sediment Transport Mechanism

Sediment transport has normally been classified in two different modes of transport as shown in figure 7.4 (a) and (b) and correspond to distinctly different physical mechanisms i.e.

(i) Bed load transport and (ii) Suspended load transport

The particle move in different modes depending on the flow conditions, the ratio of densities of the fluid and the sediment, and the size of the sediment.

Sediment load or sediment discharge is the total of the sediments that move either in suspension or in contact with the bed passing through a cross-section of channel. It is the sum of suspended load and bed load. Alternatively, it is the total of bed material load and the wash load.

7.5.1 Bed Load Transport

Bed load is the sediment in almost continuous contact with the bed while carried by rolling, sliding or hopping along the bed of the stream.

Bed load is also divided into contact load and saltation load.

a) Contact load is the sediment that is rolling or sliding along the bed of the stream in substantially continuous contact with the bed.

b) Saltation load is the sediment bouncing and hopping along the bed of the stream or moved directly or indirectly by the impact of the bouncing particles.

Movement of sediment particles either by rolling or sliding along the bed is in substantially continuous, and the particles stay in close contact with the bed.

(i) Bed material is bed material, the particle sizes of which are found in appreciable quantities in the shifting portions of the bed.

(ii) Bed material load is the coarse part of the sediment load which consists of particle sizes represented in the bed (that is bed material) which is limited in its rate of movement by the transporting capacity of the channel.

For determining the rate of bed load transport various equations developed by different investigators are given in the Table 7.4. Typical example of Einstein method is given in Para 7B.5.

7.5.2 Suspended Load Transport

The stream will have only clear-water at low values of average shear stress on the bed. Further increase in the shear stress, some of the bed particles are carried into the main flow and thus lose contact with the bed. These particles travel with a velocity almost equal to the flow velocity and constitute the suspended load.

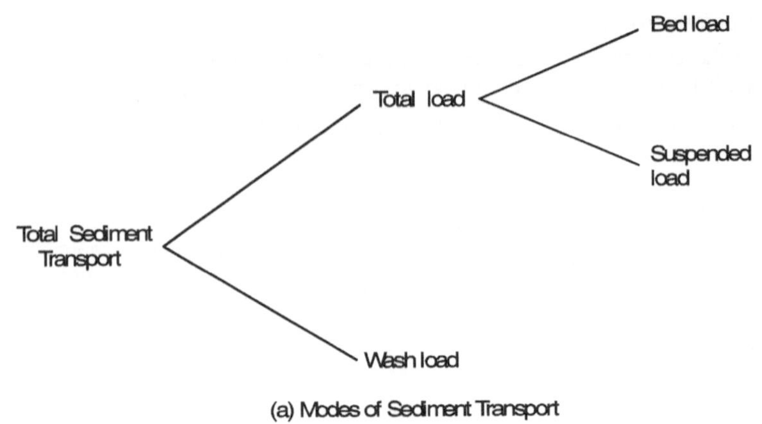

(a) Modes of Sediment Transport

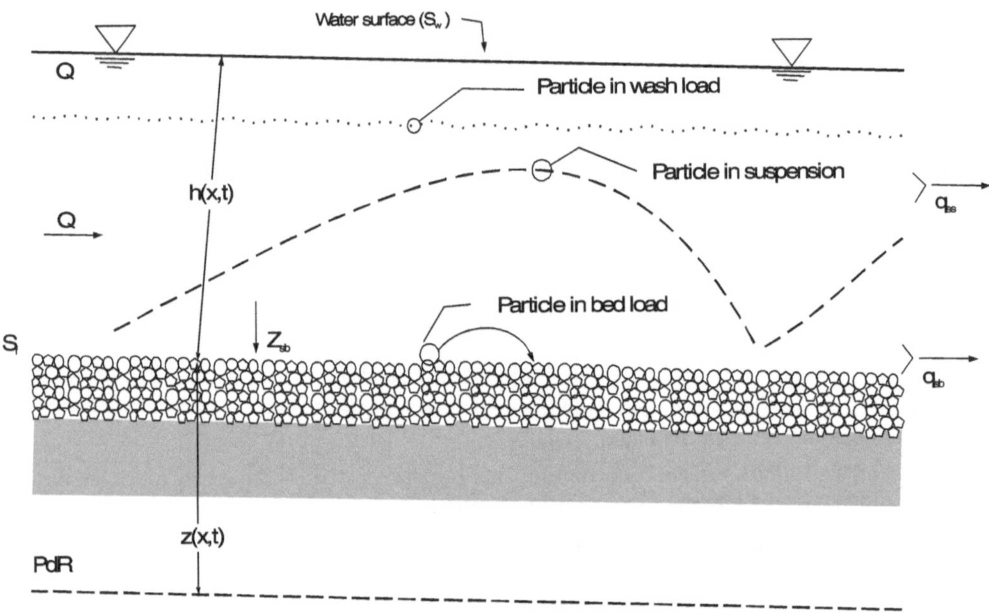

(b) Scheme of the Modes of Sediment Transport

Figure 7.4 Sediment transport Mechanism

(i) Suspended load is part of the sediment load of a stream which remains in suspension in the flowing water for considerable periods of time without contact with the stream bed, being kept up in suspension by the upward component of the turbulence or by colloidal suspension and which moves practically with the same velocity as that of flowing water.

(ii) Wash load is part of the suspended load which is composed of particle sizes smaller than those found in appreciable quantities in the shifting portions of the stream bed. It is in near permanent suspension and is transported entirely through the stream without deposition. The discharge of the wash load through a reach depends only on the rate with these particles become available in the catchments and not on the transport capacity of flow.

(iii) Sediment concentration is the ratio of dry weight of sediment in water sediment mixture to the total weight of a suspension. It is generally expressed in grams per liter or parts per million (by weight)

Contact load, saltation load, and suspended load may occur simultaneously and the border lines between these are not well defined. This difficulty is avoided in practice by dividing the total load into suspended load and bed load. The bed load moves at a lower velocity than the layer of water through which it is traveling, the traction on it being exercised through the fluid drag. The total load may also be divided into bed material load and wash load the former constituting the coarser part of the sediment load moved by the transporting capacity of the channel which may settle and the latter the fine suspended material which does not settle in the existing conditions of flow.

Suspended load transport is an advanced stage of the bed load transport. Thus in the case of uniform sediment, one would expect only bed load transport at low shear stresses, while at high shear stress both the bed transport and suspended load transport would occur. In the case of non-uniform sediment, the finer sizes of the bed material may move predominantly in suspension, while the coarse fraction of the bed material may move mostly (or totally) as bed load, if they move at all.

For determining the rate of suspended load transport various equations have been given by different investigators as summarized in the Table 7.5. A solved example of calculation is given in Para 7B.5.1.

7.5.3 Total Load Transport

The methods of computation of the total sediment transport rate can be broadly classified in two categories namely:

(i) Microscopic Method. It subdivides the total sediment load either into suspended load or load at different depth of flow and bed load.

(ii) Macroscopic Method is process of suspension based on dimensional analysis, intuition or complete empiricism.

For determining the rate of total load transport various equations given by different investigators are summarized in the Table 7.6

Two methods (a) Einstein and (b) Acker's method are explained with a solved example in Para 7B.5.2.

7.6 Factors affecting sediment transportation and deposition

The amount of material transported or deposited in the stream under a given set of conditions is the result of the interaction of two groups of variables. In the first group are those variables which influence the quantity and quality of the sediment brought down to that section of the stream. In the second group are variables which influence the capacity of the stream to transport that sediment. A list of these variables are given below.

A. Group 1 Sediment brought down to the stream.

(a) *Quality:* Size, settling velocity, specific gravity, shape, resistance to wear, state of dispersion, and cohesion.

(b) *Quantity:* Geology, and topography of watershed; magnitude, intensity, duration, distribution, and season of rainfall; soil condition, vegetal cover; cultivation and grazing; surface erosion and bank cutting.

B. Group 2 Capacity of stream to transport sediment.

(a) Geometric properties of shape of stream prism: Depth, width, form and alignment.

(b) Hydraulic properties of the stream channel: Slope, roughness, hydraulic radius, discharge, velocity, velocity distribution, turbulence, tractive force, fluid properties, and uniformity of discharge.

7.6.1 General Relationships of Sediment Transport

Kennedy and Brooks (1963) listed the variables involved in determining the behavior of alluvial channels and classified them into several sets of independent and dependent groups. Each of the dependent variables can be determined as a function of the independent variables. In some cases the functions are known and dependent variables can be determined easily. Perhaps the simplest such function is the continuity equation stating that discharge Q is equal to the product of stream width b, depth d, and mean velocity V or Q = bd V. In the first line of Table 7.3 for the case of flumes, the independent variables are fluid properties of kinematic viscosity, ν; mass density, ρ; sediment properties of density, ρ_s; geometric mean size, d_g; geometric standard deviation of sizes σ_g; fall velocity, ω; the acceleration of gravity g; and flow system characteristics Q, b, and d. The dependent variables are sediment discharge Q_s; mean velocity V; hydraulic radius of cross section, r; energy gradient, S; and Darcy-Weisbach friction factor, f.

The sediment discharge, Qs, can be expressed as

$$Q_s = f\,(Q, d, b, \nu, \rho, \rho_s, d_g, \sigma_g, \omega, g) \qquad \text{....... Eq. 7.6.1 (i)}$$

Table 7.3 Independent and dependent Variables for Sediment Flow in Alluvial streams

System	Properties of fluid, sediment, gravity, etc.	Independent Variables[a] Characteristics of flow systems, typical combinations listed)	Dependent[a] variables (not all combinations listed)
(1)	(2)	(3)	(4)
Flumes	$\nu, \rho, \rho_s, d_g\ \sigma_g,$	Q, d, b	Q_s, V, r, S, f
	ω, g	Q, Q_s, b	D, r, U, S, f
		V, d, b	Q, Q_s, r, S, f
		D, s, b	Q, Q_s, r, V, f
		R, S, b	Q, Q_s, d, V, f
		Q, S, b	G_s, d, r, V, f
Natural streams	$\nu, \rho, \rho_s, d_g\ \sigma_g,$	Q, d	G_s, b, r, V, S, f
Short term	ω, g	D, S	Q, Q_s, b, r, V,S, f
		r, S	Q, Q_s, b, d, V, f
		Q, S	Q_s, b, d, r, V, f
Long term (graded stream)	ν, ρ, ρ_s, g	Q, Q_s	b, d, r, V, S, f $d_g\ \sigma_g$, ω
Very long Term	ν, ρ, ρ_s, g Geology	Climate, man-made works	Q, Q_s, b, d, r, V, S, f, d_g, σ_g, ω

[a]Note that plan-form geometry and wash-load concentration are not considered.

In a particular flume of a given width the fluid and sediment properties can be kept constant and b and g are constant and the discharge can be replaced by V by means of the continuity equation. When the depth and sediment and fluid properties are kept constant the relation reduces to

$$Q_s = f\,(V) \qquad \text{....... Eq. 7.6.1 (ii)}$$

In canals, sediment properties changes with season only. During rainfall, more sediment enters, when off taking from weirs, such as in upper Ganga canal, off taking from Hardawar. Fluid properties change with winter/summer, day/night. Depth changes with regulation.

Kennedy (1894-95) found that for fine sand in upper Bari Doab Canal sediment transport $Q_s \propto V^{2.5}$, see Para 7.12.

7.7 Reliability of Sediment Transport Relations

Graf (1998) has commented that "none of these relations can pretend to translate the intrinsic complexity of the transport of sediments".

Most of these formulae should not be used beyond the conditions within which they were established. Table 7.6 contains a summary of the range of the parameters, d (size) and S_f, (energy slope) investigated

Table 7.4 : Empirical Equations for Bed Load Sediment Transport (Discharge)

Investigator and His Formula (1)	Units (2)	Remark (3)
7.4.1 Du Boys (1879, Brown, 1950) $q_{sb} = \psi_D \tau_o (\tau_o - \tau_c)$ $\tau_0 = \gamma_\phi R.S_0$	FPS	Value of ψ_D and τ_c obtained by Straub (1935) and reported in Brown (1950), given as functions of median size of the bed sediment (d_{50}) in Figure 7.5 (i) Includes data of Gilbert (1914) and Johnson (1943) flume experiments
7.4.2 Meyer Peter (Mayer Peter and Muller, 1948) $q_{sb}^{2/3} = 39.25q^{2/3} S_0 - 9.95 d_{50}$ $q_{sb}^{2/3} = 250q^{2/3} S_0 - 42.5d_{50}$	FPS Metric	It is valid only for beds of relatively coarse sediments for which the flow resistance due to bed forms is a small part of the total resistance; sediment size from 3.1 mm to 28.6 mm. Includes flume data of Gilbert (1914).
7.4.3 Schoklitsch (Shulits, 1935) $q_{sb} = \sum P_i \frac{25.3}{\sqrt{d_{si}}} S_0^{3/2} (q - q_{ci})$ $q_{ci} = 0.638 \frac{d_{si}}{S_0^{4/3}}$	FPS	q_{ci} is the critical value of q for initiating motion of sediment of mean size, d_{si}. For values of P_i and d_{si} a mechanical analysis of a representative sample of the bed sediment is made and a size distribution curve prepared. P_i values can be determined form the size distribution curve for corresponding to d_{si} .The formula is based on Gilbert data (1914), for graded sediment with median size from 0.3 mm to 5 mm
7.4.4 Shield (Shield, 1936) $q_s = 10qS_0 \frac{(\tau_0 - \tau_c)}{\left(\frac{\gamma_s}{\gamma_f} - 1\right)^2 d_{50}}$	Dimensionally homogeneous, FPS or metric, use consistent units both sides	τ_c is the critical bed shear stress for sediment of size d_{50} given by Shields Graph, figure 7.5(ii). Median sediment size range from 1.7 mm to 2.5mm.

Contd...

Table 7.4. Contd...

Investigator and His Formula (1)	Units (2)	Remark (3)
7.4.5. Meyer Peter and Muller Formula (1948)		
$q_s^{2/3} = \dfrac{\left[\left(\dfrac{kr}{k'r}\right)^{3/2} \gamma_f R.S - 0.097(\gamma_s - \gamma_f)dm\right]}{0.25\left(\dfrac{\gamma_f}{g}\right)^{1/3}\left(\dfrac{\gamma_s - \gamma_f}{\gamma_s}\right)^{2/3}}$	Dimensionally homogeneous, FPS or metric, use consistent units both sides	The advantage of this formula over the other Meyer Peter formula is that it can be used for graded sediments under flow conditions that give rise to dunes and other bed forms. f's is Darcy -Weisbach bed fraction factor for the sand grain roughness, shown in figure 7.5(iii). The mean diameter of sediment range from 0.4 mm to 30 mm. S^1 is portion of channel bed slope due to grain roughness, kr and k'r are determined from same value of V.
$\dfrac{kr}{k'r} = \sqrt{\dfrac{f'b}{8}}\dfrac{V}{\sqrt{g\,RS_o}}$	-do-	
$V = kr\,R^{2/3}S_o^{1/2}$; $V = k'r\,R^{2/3}\,{S^1}^{1/2}$	Metric	
$V = \sqrt{\dfrac{8}{f'b}}\sqrt{gRS^1}$	Dimensionally homogeneous	
$k'_r = \dfrac{26}{(d_{90})^{1/6}}$; $d_m = \sum_i P_i d_{si}$	Metric	
7.4.6. Modification of Einstein (1942) by Rouse, Boyer and Laursen (Rouse 1950) $q_{sb} = \phi\gamma_s F_1\sqrt{g\left(\dfrac{\gamma_s}{\gamma_f} - 1\right)d_s^3}$ $\varphi = \left(\dfrac{1}{\Psi}\right)$	FPS	$F\left(\dfrac{1}{\Psi}\right)$, function is obtained from figure 7.5(iv)through ϕ . F_1 is related to fall velocity , $\omega_0 = F_1\sqrt{\left(\dfrac{\gamma_s}{\gamma_f} - 1\right) g\,d_{50}}$

Contd...

Table 7.4. Contd...

Investigator and His Formula (1)	Units (2)	Remark (3)
$F_1 = \sqrt{\frac{2}{3} + \frac{36\nu^2}{gd_s^3\left(\frac{\gamma_s}{\gamma_f}-1\right)}} - \sqrt{\frac{36\nu^2}{gd_s^3\left(\frac{\gamma_s}{\gamma_f}-1\right)}}$		for $\frac{1}{\Psi}$ in excess of 0.09 and $\tau_o = \gamma_f$ R.So , the formula is based on Gilbert data (1994) for graded sediment with median sizes from 0.3 mm to 7 mm.
7.4.7 Einstein Bed Load Function (Einstein, 1950) $q_{sb} = \sum_i q_{si}$ $G_{sb} = bq_{sb}$ $q_{si} = q_{sbi}\left[P_r\ I_1(\eta_{oi}, z_i) + I_2(\eta_{oi}, z_i) + 1\ \right]$ $I_1(\eta_{oi} z_i) = 0.216 \frac{\eta_{oi}^{zi-1}}{(1-\eta_{oi})^{zi}} \int_{\eta_{oi}}^{1} \left(\frac{1-\eta}{\eta}\right)^{zi} d_\eta$ $I_2(\eta_{oi} z_i) = 0.216 \frac{\eta_{oi}^{zi-1}}{(1-\eta_{oi})^{zi}} \int_{\eta_{oi}}^{1} \left(\frac{1-\eta}{\eta}\right)^{zi} 1_{n\varsigma} d_\eta$ $P_r = 2.3 \log \frac{30.2\ \nu R}{d_{65}}$ $Z_i = \frac{\omega_i}{0.4V_*^1}$ $V_*^1 = \sqrt{gRS_0}$ $\omega_{*i} = \frac{q_{sbi}}{P_i\gamma_s}\sqrt{\left(\frac{\gamma}{\gamma_s-\gamma_f}\right)\frac{1}{gd_{si}^3}}$	FPS	q_{sbi} is the bed load of mean size d_{si} discharge in weight per unit width. Value of φ_* is obtained on the figure 7.5(v). φ_* as a function of Ψ_*. ξ is a function of d_{si} / x given is figure 7.5 (iv) and Y is a function of given in figure 7.5(vi). value of ηoi = 2dsi/R , η =y /R and functions, I_1 and I_2 are given in figure 7.5(viii)and ν is a dimensionless quantity in the logarithmic velocity distribution law shown in figure 7.5(ix).value of ν from annexure-1 and ω_i is given in figure 7.5(x). The formula were obtained in flume experiments with two well -sorted sediments of mean sizes 28.65 mm and 0.785 mm, respectively (for graded fine sands).

Contd...

Table 7.4. Contd...

Investigator and His Formula (1)	Units (2)	Remark (3)
$\psi_{*i} = \xi_i Y \left(\frac{\log 10.6}{\log \frac{10.64X}{d_{65}}} \right)^2 \frac{(\gamma_s - \gamma_f)\, d_{si}}{\gamma_f RS_0}$ $X = 0.77 \frac{d_{65}}{x}$ when $\frac{d_{65}}{x\delta} > 1.80$ $X = 1.398\ \delta$ when $\frac{d_{65}}{x\delta} < 1.80$; $\delta = 11.6 \frac{\nu}{V'_*}$		
7.4.8 Laursen (1958) $q_s = C_m q,$ $G_m = 0.01\gamma_f \sum_i P_i \left(\frac{d_{si}}{D}\right)^{7/6} \left(\frac{\tau_0}{\tau_{ci}} - 1\right) f\left(\frac{V_*}{\omega_i}\right)$ $\tau'_0 = \frac{\rho V^2}{58} \left(\frac{d_{50}}{D}\right)^{1/3}$ $\tau_{ci} = \tau_{*c}(\gamma_s - \gamma_f)\, d_{si}$	FPS	τ_o is Laursen's bed shear stress due to grain resistance an $f\left(\frac{V_*}{\omega_i}\right)$ is the function shown in figure7.5(xi). For value of τ_{*c} in this figure 7.5 (xi), density of fluid ρ from Table 7B.17 and τ_{*c}= 0.39 for sediment median size from 0.088 mm to 4.05 mm
7.4.9 Blench (1966) $\frac{\left(1 + 0.12x10^5 \frac{C_m}{\gamma_f}\right)^{11/12}}{1 + \frac{1}{233} x10^5 \left(\frac{C_m}{\gamma_f}\right)} = \frac{3.63gB^{1/4}q^{1/12}S_0}{k_m \nu^{1/4} \left[1.9\sqrt{(d_{50})}\right]^{11/12}}$	FPS	k_m the meander coefficient has values of 1.25 for straight reaches, 2.0 for streams with well -developed meanders and 2.75 for very sinuous streams. The formula and the constants in it were derived from regime relations and by correlating the relations mainly with data of Gilbert (1914) and Johson, (1943) obtained in small flumes with well -sorted sands ranging in size from 0.33 to 7 mm

Contd...

Table 7.4. Contd...

Investigator and His Formula (1)	Units (2)	Remark (3)
7.4.10 Colby Relations (Colby, 1964) $q_s = k_1 k_2 q_{s1}$	FPS	q_s is the sediment discharge, obtained with median size ranging from 0.2 mm to 0.3 mm in water at 60°F. k_1 and k_2 is correction factor shown in Figure 7.5(xii)
7.4.11 Engelund Hansen (Engelund , 1966 and Engelund and Hansen , 1967) $q_{sb} = 0.05\gamma_s V^2 \sqrt{\frac{d_{50}}{g\left(\frac{\gamma_s}{\gamma_f} - 1\right)}} \left[\frac{\tau_0}{(\gamma_s - \gamma_f) d_{50}}\right]^{3/2}$	FPS	The formula is based on Guy data (1966) for graded sediment with median sizes 0.19 mm, 0.27mm, 0.45 mm and 0.93 mm.
7.4.12. Inglis -Lacey (Inglis, 1968) $q_{sb} = 0.562 \frac{(vg)^{1/3}}{\omega_0} \frac{V^2}{gd} \frac{\gamma_f V^3}{g}$	FPS	ω_0 is the fall velocity of characteristic sediment particle that is arranged to be the particle having the median size of the bed material in figure7.5(x) and kinematics viscosity of fluid, vfrom annexure-1.
7.4.13Taffaleti, 1968, 1969) $q_{sbi} = M_i (2d_{si})^{1+n_v-0.758z_i}$ $M_i = 43.2 P_i C_{Li} (1+n_v) V R^{0.758z_i - n_v}$ $(C_i)_y = 2d_{si} = C_i \left(\frac{2d_{si}}{R}\right)^{-0.736z_i}$ $n_v = 0.1198 + 0.00048T$ $Z_i = \frac{\omega_i V}{C_z R S_0}$ $C_z = 260.67 - 0.667T$	FPS	z_i, n are exponents and if the concentration is in excess of 100 pcf, the concentration , C_{Li}in all equation is reduced so that gives a value of concentration equal to 100 pcf, fall velocity, ωfrom figure 7.5(x) and T is water temperature in degrees Fahrenheit. The formula is based on waterways Experiment Station of the Unified States Corps of Engineer, the bed sediment had median grain sizes ranging from 0.33mm to 0.93mm.

Contd...

Table 7.5 : Empirical Equations for Suspended Sediment Transport (Discharge)

Investigator and His Formula (1)	Units (2)	Remark (3)
7.5.1 Einstein (1950) $q_{ss} = \frac{\gamma_s}{100} 11.6 V'_* C_{2d} 2d \left\{ 2.3 \log \left(\frac{30.2 D X}{d_{65}} \right) I_1 + I_2 \right\}$ $C_{2d} = 1.12 \times 10^{-7} \left(\frac{V_*}{\omega_0} \frac{V_* d}{\nu} \right)^{5.25}$ $A = \frac{a}{D}$ $Z = \frac{\omega_0}{V_* K}$	Metric	Two integral $I_1$and I_2 numerically obtained by Einstein are shown graphically as at A and Z in figure 7.5 (viii). Fall velocity (ω_0) is obtained from figure 7.5 (x), from annexure-1 and Einstein has suggested a = 2d , k = 0.4 (Karman constant)
7.5.2 Lane and Kalinske (Grade, 2000) $q_{ss} = q C_a P \exp\left\{ 15 \left(\frac{\omega_0}{V_*} \right) A \right\},$ $\frac{C}{C_a} = \exp\left\{ \frac{-\omega_0}{V_*} \left(\frac{y}{D} - \frac{a}{D} \right) 15 \right\}$	Metric	P is a function of ω_0/V_* and V_*/KV using Manning's equation for V, V_*/KV can be expressed in term of $n/D^{1/6}$, see Figure 7.5 (xiii) and fall velocity (ω_0) is obtained from figure 7.5 (x)
7.5.3 Engelund (Grade, 2000) $q_s = 5.10 \times 10^{-5} \left(\frac{V_*}{\omega_0} \right)^4 \gamma_f q$	Metric	Fall velocity (ω_0) is obtained from figure 7.5 (x)

Contd...

Table 7.5. Contd...

Investigator and His Formula (1)	Units (2)	Remark (3)
7.5.4 Samaga et. al. (Grade, 2000) $q_{SS} = \left[\dfrac{\varphi_S \gamma_S}{\sqrt{\dfrac{\gamma_f}{\Delta\gamma_S} \times \dfrac{1}{gd}}} \right] \Big/ d$ $\varphi_s = 28.0\ \tau_*^6$ $\tau_* = \dfrac{\tau_0}{(\Delta\gamma_s d)}$	Metric	The method is by considering the individual fractions in a mixture $\varphi_{s*}\tau$, are parameters. The variation of $K_s L_s \xi_s$ with $\tau o/(\Delta\gamma s\, d_i)$ is shown in Figure 7.5 (xiv). The empirical Coefficients K_s and L_s are functions of τ_o/τ_{oc} and M respectively as shown in Tables respectively. M is Khemer's uniformity coefficient.
7.5.5 Holtorfft's (Grade, 2000) $q_{ss} = \left[\tau_0 V \times 0.055 \sum_{ib} \left(\dfrac{\tau_s}{\tau_0}\right)_i \left(\dfrac{V}{\omega_0}\right)_i \right] \Big/ \Delta\gamma_s$	Metric	$(\tau_s/\tau_o)_i$ and $(V/\omega_o)_i$ are the values corresponding to a particular size, the former being read from fig. 7.5 (xv)
7.5.6 Van Rijn's (Grade, 2000) $q_{ss} = \gamma_s FVDC_a$ $C_a = 0.15 \dfrac{d_{50} T^{1.5}}{a d_*^{0.3}}$ $T = \dfrac{V_*'^2 - V_{*c}^2}{V_{*c}^2}$; $d_* = d_{50} \left(\dfrac{\Delta\gamma_s}{\rho_f \nu^2}\right)^{1/3}$ $\dfrac{V}{V_*} = 5.75 \log \dfrac{R}{k_s} + 6.25$	Metric	C_a is the reference concentration at y = a, the reference level 'a' is taken to be equal to k_s or D/100. The correction factor F is related to a/D and Z (Suspension Parameter) as shown in fig. 7.5 (xvi)

Contd...

Table 7.6 : Empirical Equations for Total Sediment Transport (discharge)

Investigator and His Formula (1)	Units (2)	Remark (3)
7.6.1 Einstein (Grade, 2000) $G_s = Q_s + \rho_s$; $Q_s = Q_{sb} + Q_{ss}$ $Q_{sb} = q_{sb} \times b$; $Q_{ss} = q_{ss} \times b$ $q_T = q_{sb} + q_{ss}$; $q_{sb} = \varphi\sqrt{(\gamma_s - 1) g d_{35}^2}$ $\psi_* = (\gamma_s - 1)\frac{d_{35}}{R S_o}$; $\varphi = f(\psi_*)$ $q_{ss} = q_{sb}(P_e I_1 + I_2)$; $P_e = 2.303 \log(30.2)$ $\Delta = d_{65}/x$; $k_s/\delta = d_{65}/\delta$ $\delta = 11.5\, \nu/V_*$; $A_E = 2\, d_{35}/h$ $z = V_{ss}/KV_*$; $V_* = \sqrt{gRS_o}$	Metric	Intensity of transport, for Φ see figure7.5(xvii), correction term for logarithmic velocity distribution, x and apparent roughness diameter see figure 7.5(xviii) and two integral I_1 and I_2 numerical obtained by Einstein are shown in figure 7.5 (viii) and settling velocity from figure 7.5(xix)
7.6.2 Swamee and Ojha (Grade, 2000) $q_{sb} = \phi_{Bs}\ \lambda_s\left(\frac{\rho_s}{\rho_f} - 1\right)^{1/2} q^{1/2}\ d_s^{3/2}$ $q_{ss} = \varphi_{ss}\ \lambda_s\left(\frac{\rho_s}{\rho_f} - 1\right)^{1/2} q^{1/2}\ d_s^{3/2}$; $\tau_{*s}^{1} = \frac{\gamma_f R' S_o}{\Delta\gamma_{ss}}$ $\varphi_{Bs} = \left\{\left[\left(\frac{0.8}{M}\right)^{4.75} + M^{0.0004}\right]^{1.75}\left(\frac{0.0871}{{\tau_{*s}^{1}}^{M^{0.25}}}\right) + \left[\left(\frac{0.01}{M}\right)^{0.06} + M^{3.25}\right]^{1/2}\left[\frac{0.339}{{\tau_{*s}^{1}}^{9M/(8M^{7}+1)}}\right]^{1.6}\right\}^{-1}$ $\varphi_{ss} = \left\{\left[\left(\frac{0.073}{M}\right)^{4} + M^{3.8}\right]\left(\frac{0.567}{{\tau_{*s}^{1}}^{M^{0.25}}}\right)^{6} + \left[\left(\frac{0.177}{M}\right)^{2} + M^{1.455}\right]^{2}\left[\frac{0.538}{{\tau_{*s}^{1}}^{8M/(7M^{1.425}+1)}}\right]\right\}^{-1}$	Metric	M is kramer's uniformity coefficient, $\varphi_{Bs}, \varphi_{ss}$, and τ_{*s}^{1} are parameters, see figure 7.5(xxi)

Contd...

Table 7.6. Contd...

Investigator and His Formula (1)	Units (2)	Remark (3)
7.6.3 Laursen (1950) $$\frac{\bar{c}}{\left(\frac{d}{D}\right)^{7/6}\left[\left(\tau_0^1/\tau_{oc}\right)-1\right]}=f\left(\frac{V_*}{\omega_o}\right)$$	Metric	Laursen considered the following parameter to be important in the study of total sediment transport: V_*/ω_o, d/D, the total load concentration $\overline{C}$ in per cent by weight and the ratio of grain shear stress τ_0^1 to the critical shear stress τ_{0c} for the given sediment size see figure7.5(xx)
7.6.4 Garde and Dattari $$q_T = 10.39\,\tau_*^{2.02}\left(\gamma_s V_* d\right)$$	Metric	The relationship is evidently a simplification of a more complex relationship, even for uniform sediment.
7.6.5 Bagnold' Equation (1960) $$q_T=\frac{\tau_0 V}{\left(1-\frac{\rho_f}{\rho_s}\right)}\left\{\frac{eB}{\tan\alpha}+0.01\frac{V}{\omega_0}\right\},$$	Metric	es(1-eB) can be taken as 0.01 for all practical purpose and es is a constant value of 0.015. The value of tan α varies from 0.315 to 0.75 and is a function of τ_* and d as shown in table and fall velocity obtained from figure 7.5(x).
7.6.6 Bislcop, Simons and Richardson's Method $$q_T=\varphi_T\rho_s g^{3/2}d^{3/2}\left(\frac{\rho_s}{\rho_f}-1\right)^{1/2}$$ $$\psi^1=\frac{\rho_s-\rho_f}{\rho_f}\times\frac{d_{35}}{R'S_o}$$ $$\frac{V}{V_*^1}=5.75\,\log\left(\frac{12.27\,R'\,x}{d_{65}}\right)$$	Metric	The parameter ϕ and Ψ^1 are shown in figure 7.5(xxi) and d is median size of particle

Contd...

Table 7.6. Contd...

Investigator and His Formula (1)	Units (2)	Remark (3)
7.6.7 Engelud and Hansen (1967) $q_T = [\ k(\tau_*) - 0.06\sqrt{\tau_*}\sqrt{(\gamma_s - \gamma_f)d^3\rho_f}\ ]/\frac{f}{\gamma_s}\left(\frac{h}{Lf}\right),$ $\tau_*^1 - 0.06 = 0.4\tau_*^2$	Metric	$\frac{h}{Lf}$ is constant for particular value of τ_*^1 obtained from figure7.5(xxii) and f is the friction factor of the bed.
7.6.8 Ranga Raju $\tau_{*_t} = \tau_*^1 (\tau_0^1/\tau_0)^m$ $\varphi_T = 60\tau_*^{1^3}(\tau_0^1/\tau_0)^m$	Metric	Total sediment transport was analyzed using a fresh approach by Vittal that a unique relation between the parameter φ_T and τ_* should exist for a plane bed was justified by potting all available plan bed data figure7.5(xxiii). Subsequently, the effective shear stress for total load transport in case of an undulated bed τ_T was defined as the shear stress which would be required on a plane bed to cause a transport rate equal to that observed in case of the undulated bed. Obviously the effective shear stress so defined could be determined from fig. 7.5(xxiii) for the known q_T value. The variation of ϕ_T with $\tau_*^1(\tau_0^1/\tau_0)^m$ for a large amount of flume and field data based on m value is shown in fig.7.5 (xxv), m is a function of V_*/ω_0, m = 0 for $V_*/\omega_0 \le 0.5$, m = $0.2\,V_*/\omega_o$ - 0.1 for V_*/ω_o > 0.5 and ϕ_T value in the large of $0.05 \le \tau_*^1(\tau_0^1/\tau_0)^m < 1.0$.

Contd...

Table 7.6. Contd...

Investigator and His Formula (1)	Units (2)	Remark (3)
7.6.9 Ackers and white (1973)		
$Qs = Cs\, Q$ ... (i) $C_s = G_{gr} \frac{d_{35}}{h}\left(\frac{V}{V_*}\right)^n$ ... (ii) $G_{gr} = C\left[\frac{F_{gr}}{A} - 1\right]^m$ ... (iii) $V_* = \sqrt{g\, h\, S_0}$ ... (iv) $F_{gr} = \frac{V_*^n}{\sqrt{g\, d_{35}(\gamma_s - 1)}}\left[\frac{V}{\sqrt{32}\ \log\left(\frac{10h}{d_{35}}\right)}\right]^{1-n}$ ...(v) $N = 1.00 - 0.56 \log D_{gr}$ ... (vi) $A = \frac{0.23}{\sqrt{D_{gr}}} + 0.14$ ... (vii) $m = \frac{9.66}{D_{gr}} + 1.39$ ... (viii) $\log C = 2.86\ D_{gr} - (\log D_{gr})2 - 353$ ... (ix) $D_{gr} = d_{35}\left[\frac{g(\gamma_s - 1)}{v^2}\right]^{1/3}$ ... (x) $G_s = Qs \times \rho_s$ ... (xi)	Metric	Ackers and White (1973, 1980), presented a new total load sediment transport theory incorporating the advantages of dimensional analysis and physical arguments. For a direct determination of the total load transport (Q) proposed the use of some sediment logical parameters; D_{gr}, F_{gr}, C_s . hydraulic considerations and dimensional analysis. The confinement; N ,A, m and C in the above relations were determined by regression analysis, using close to 1000 experiments in the laboratory and close to 250 experiment the field. The data against which these theories were checked consisted at 840 flume experiments with sand, 180 flume experiment with high rivers. The method is more reliable as compared to other methods generally considered. Kinematics viscosity of fluid, **given in table 7B.17.**

Contd...

Table 7.6. Contd...

Investigator and His Formula (1)	Units (2)	Remark (3)
7.6.10 Yong's Equation (grade 2000) $Q_s = \overline{C}_T\, Q$, $\log \overline{C}_T = 5.435 - 0.286 \log \frac{\omega_0 d}{\nu} - 0.457 \log \frac{V_*}{\omega_o} +$ $(1.799 - 0.409 \log \frac{\omega_0 d}{\nu} - 0.314 \log \frac{V_*}{\omega_0}) \times \log \left(\frac{VS_o}{\omega_0} - \frac{V_{cr} S_o}{\omega_0} \right)$	Metric	The rate of sediment transport in an alluvial channel is primarily by the rate of expenditure of potential energy per unit weight of water i.e the unit stream power. Fall velocity, ω_o ,shown in figure7.5(x)
7.6.11 Shen and Hung's Equation (grade 2000) $Q_s = \overline{C}_T Q$ $\log \overline{C}_T = -10704.5 + 324217\, Y_* - 326309.6\, Y_*^2 + 109503.9\, Y_*^3$ $Y_* = \left[\frac{VS_o^{0.57}}{\omega_0^{0.32}} \right]^{0.0075}$	Metric	By performing regression analysis of laboratory data, obtained dimensional expression for the total load concentration, $\overline{c}_T$, in ppm by weight. Fall velocity, ω_o ,shown in figure 7.5(x) .
7.6.12 Brownlie's Equation (grade 2000) $Q_s = \overline{C}_T\, Q$ $\overline{C}_T = 7115 \quad C_F \left[\frac{V}{\sqrt{\frac{\Delta\gamma_s}{\rho_f} d}} - \frac{V_{cr}}{\sqrt{\frac{\Delta\gamma_s}{\rho_f} d}} \right]^{1.978} S_o^{0.06601} \left(\frac{R}{d} \right)^{-0.3301}$	Metric	Based on a study of merits and demerits of the parameters used by earlier investigator and by performing regression analysis of laboratory and field data, obtained equation for the concentration $\overline{c}_T$ in ppm by weight. The coefficients C_F is unity for laboratory data and 1.268 for field data.

Contd...

Table 7.6. Contd...

Investigator and His Formula (1)	Units (2)	Remark (3)
7.6.13 Karim and Kennedy's Equation (grade 2000) $\log \frac{q_T}{\gamma_s \sqrt{\left(\frac{\rho_s}{\rho_f} - 1\right) g d^3}} = -2.2786 + 2.719\, V_1 + 0.2989\, V_3 V_2 + 1.06\, V_1 V_3$ $V_1 = \log \frac{V}{\sqrt{\left(\frac{\Delta\gamma_s}{\rho_f}\right) d}}$; $V_2 = \log (D/d)$ $V_3 = \log \left[\frac{V_* - V_{*c}}{\sqrt{\left(\frac{\Delta\gamma_s}{\rho_f}\right) d}} \right]$	Metric	Carried out a regression analysis of the sediment data from laboratory flumes and natural streams.

for the establishment of few important formula by their author (s) ; other author(s) may have extended this range. Also the recommendation by the author(s) for the choice of the equivalent diameter, dx, if the granulometry is quasi or non-uniform, is given in the Table 7.7.

The formulae for the transport of sediments are often established, using laboratory data and less often using field data.

A verification of these formulae in natural channels is a very delicate task, since it is difficult to measure correctly the solid discharge in the field. Furthermore, it is often a rather subjective evaluation, since the zones of the modes of transport cannot easily be separated.

Numerous studies have been reported, comparing measurements in rivers with the different existing formulae. However this task is comparatively easy in canals. Researchers should work on canals rather in laboratories.

For a better appreciation of the validity of the above presented formulae, it will now be of interest to compare the computed results with the direct measurements of the solid discharge in the field.

Many (nineteen) of the existing formulae for the calculation of the total transport were studied by White et al. (1973) and compared with experimental results. They evaluated almost 1000 laboratory experiments with uniform and non-uniform sediments of $0.04 < d_{50}$ (mm) < 4.9, at flow depth of h < 0.4 (m), and almost 270 experiments in watercourses with sediments of 0.1 < d (mm) < 68.0 and a width/depth ratio of 9 < B/h <160.

Table 7.7 Parameters Used for Establishing the Different Sediment transport formulae.

Formula	D(mm)	S_f	D_x (mm), equivalent diameter for a non-uniform granulate
Schoklitsch	0.3 -7.0 (44.0)	1/333 – 1/10	D_{40}
Meyer-Peter	3.1 – 28.6	1/2500 – 1/50	D_m (d_{50})
Einstein (1950)	0.8 – 28.6	-	D_{35}
Graf et Acaroglu (1968)	0.3 – 1.7 (23.5)	-	D_{50}
Ackers et White (1973)	0.04 – 4.0	Fr < 0.8	D_{35}

Each formula was applied to all the data of the solid-discharge measurements. Subsequently established a ratio of the values calculated, C_{calc}, and the values observed, C_{obs}, where C = Cs is the total-load transport, expressed in concentration. Some results of this investigation are given in Fig .7.6, where one may see the success of a prediction (in percentage) for different ranges of the ratio, C_{calc}/C_{obs}. Considering only the range of $\frac{1}{2} < C_{calc}/C_{obs} < 2$, it can be seen that the percentage for the formula of

Einstein (1950), eq. 7.6.1	:	44% of success
Graf et Acaroglu (1968), eq. 6.6.3	:	40% of success
Ackers et White (1973), eq. 7.6.9	:	64% of success

This implies that with the formula of Ackers White, 64% of the experimental data can be predicted in the above-mentioned range. This is usually considered as a good (or a not-so-bad) result; more than half of the studied (nineteen) formulae give results which are less good, namely < 40%. Also noticed is that with the formula of Einstein there as slight under-estimation of the solid discharge; while the one of Graf Acaroglu gives a slight over-estimation. The comparative study of White al. (1973) is reasonably objective, but certainly not conclusive. Other studies exist (see Raudkivi, (1976, p. 227) which show clearly that an objective validation is nearly impossible.

Amongst the different existing formulae for the determination of the total-load transport, but equally for the ones of the bed-load and suspended-load transport, each one will give an answer, but none will be very precise or very true.

Result of above formulas vary from less than 50 % to 200 % i.e. 4 time variations. It is vary difficult to make a judicious choice of suitable method.

Finally, it must be said, that the results obtained with these formulae give only valuable guide-lines for the engineer. For practical purposes, it is advised to consult more than one formula; the obtained result may however render different values (see Graf, 1971).

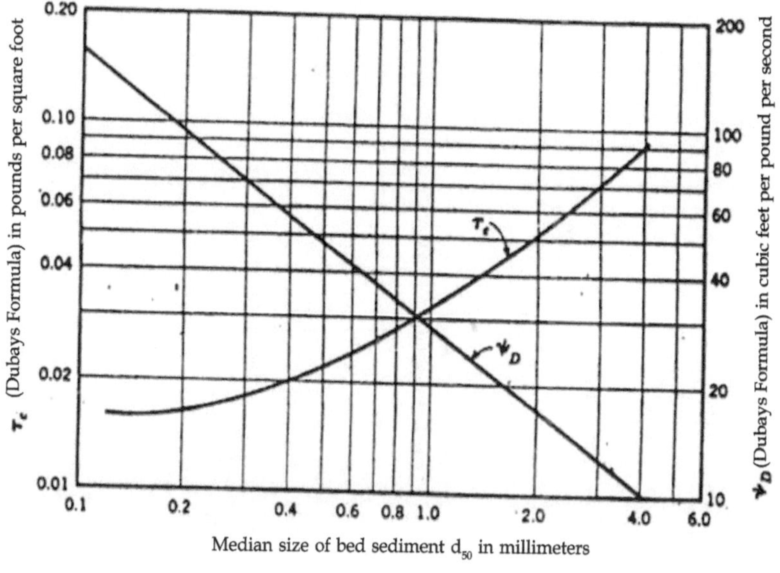

a – FPS

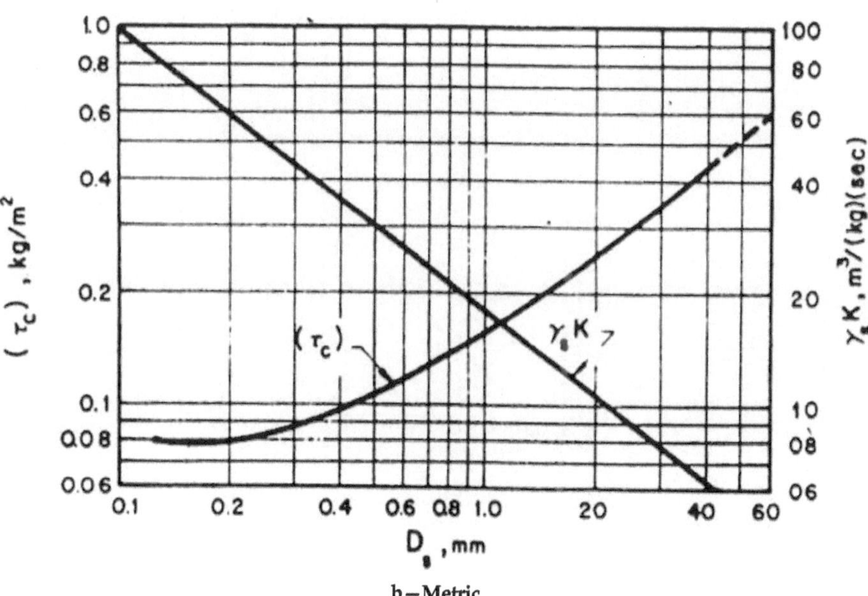

b – Metric

Figure 7.5. (i) : Sediment coefficient (Ψ_D) and critical shear stress (τ_c) for DuBoys of median size of bed sediment (Straub, 1935)

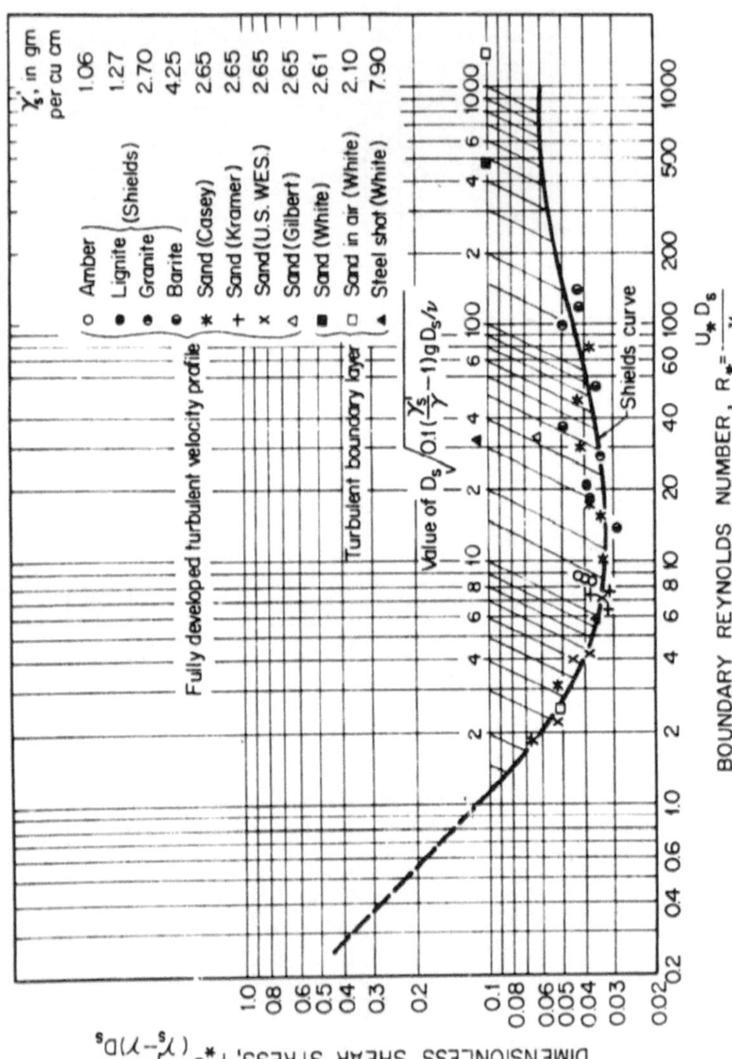

Figure 7.5 (ii) : Shield diagram with while data added (Ronse 1939) dimensionless critical shear stress.

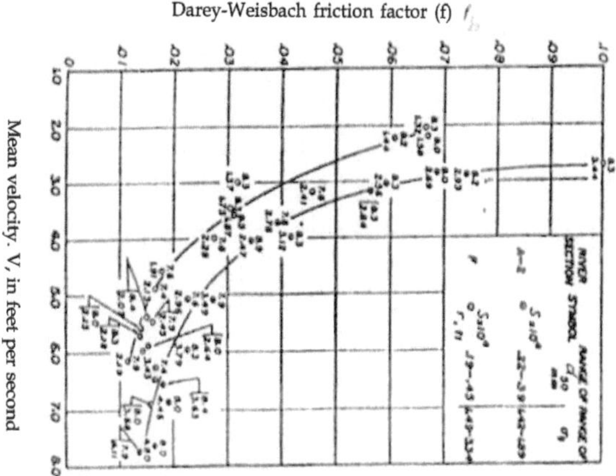

7.5 (iii) Variation of Darcy-Weisbach friction factor f_b with velocity for two section of Rio Grande in New Mexico [(data obtained by Nordin (1964), by Alam and Kennedy (1969)]

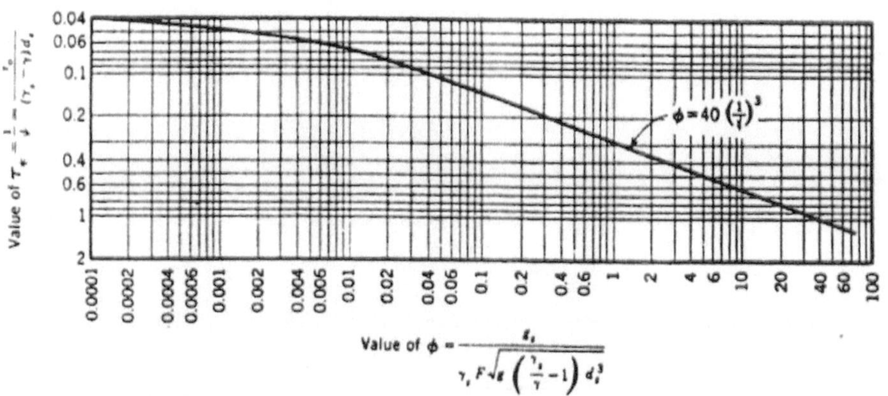

Figure 7.5 (iv) : Function $\phi = f(1/\Psi)$ for Einstein Brown (Einstein, 1942)

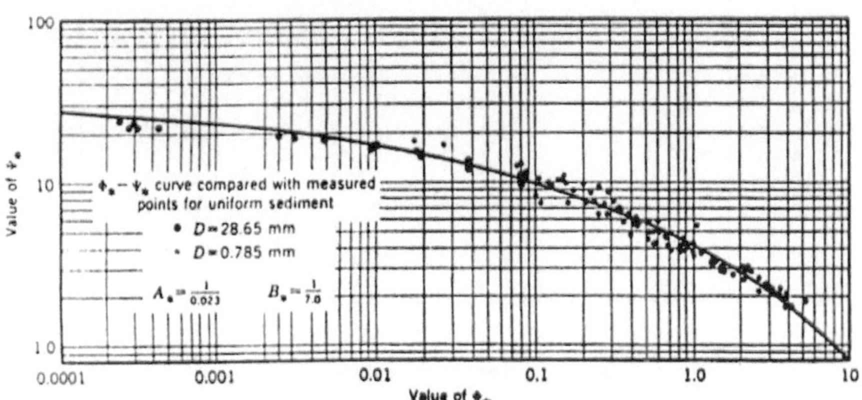

Figure 7.5 (v) : Einstein's Φ_* - Ψ_* bed load function (Einstein, 1950)

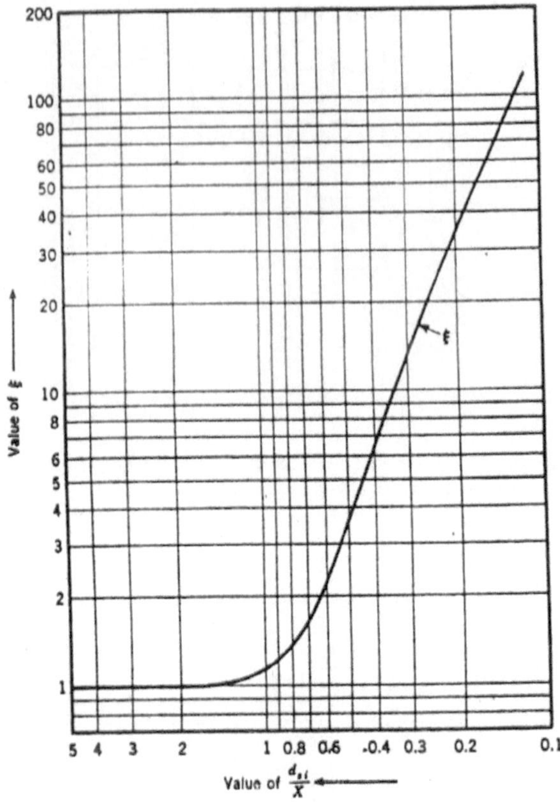

Figure 7.5 (vi) : Factor ξ in Einstein's Bed Load Function (Einstein,1950) in Terms of d_{si} / X

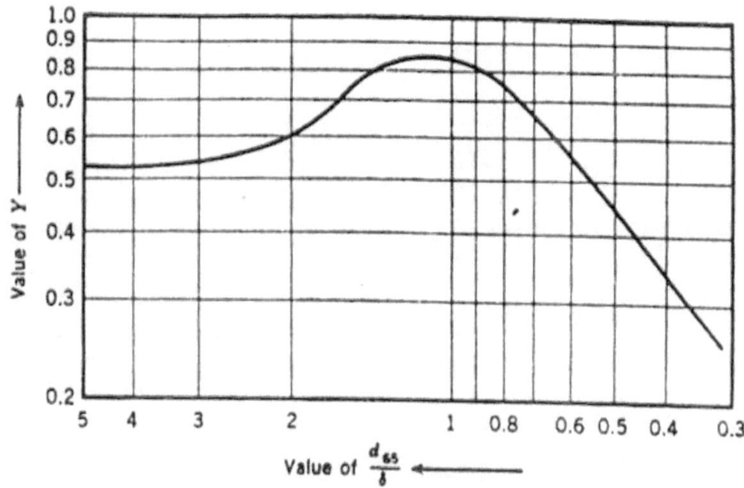

Figure 7.5 (vii) : Factor Y in Einstein's Bed Load Function (Einstein, 1950) in Terms of d_{65i}/δ

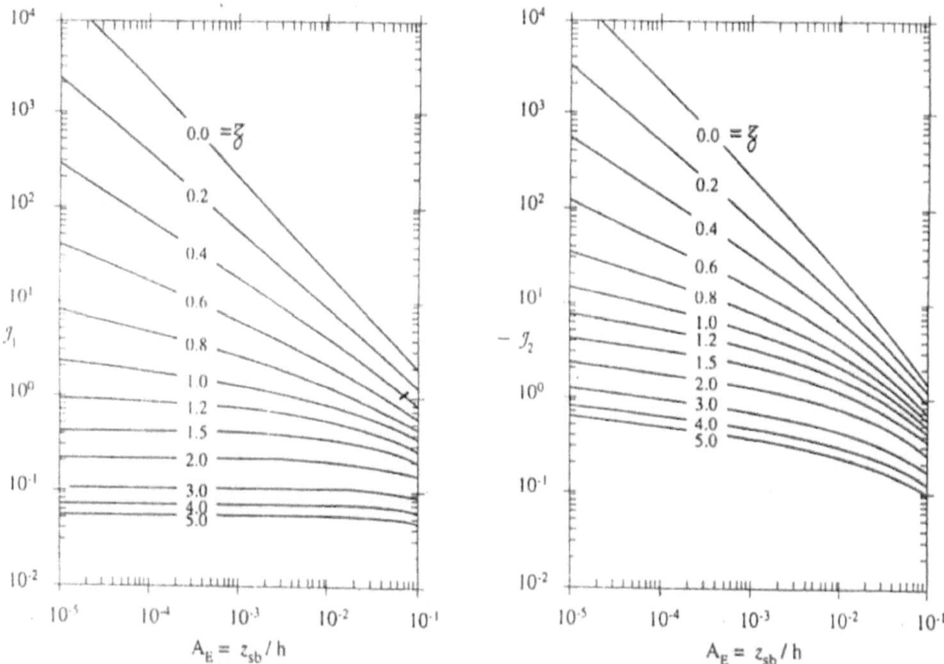

Figure 7.5 (viii) : The Integral, I_1 (A_E , z) and I_2 (A_E , z), used in the method of Einstein (1950)

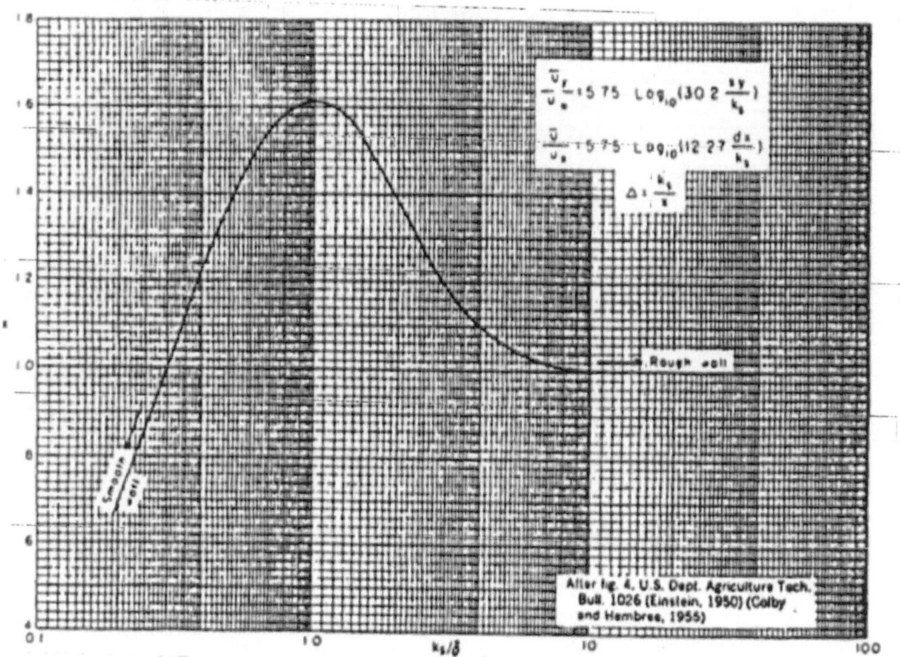

Figure 7.5 (ix) : Factor x in Velocity Distribution Equation (Einstein, 1950)

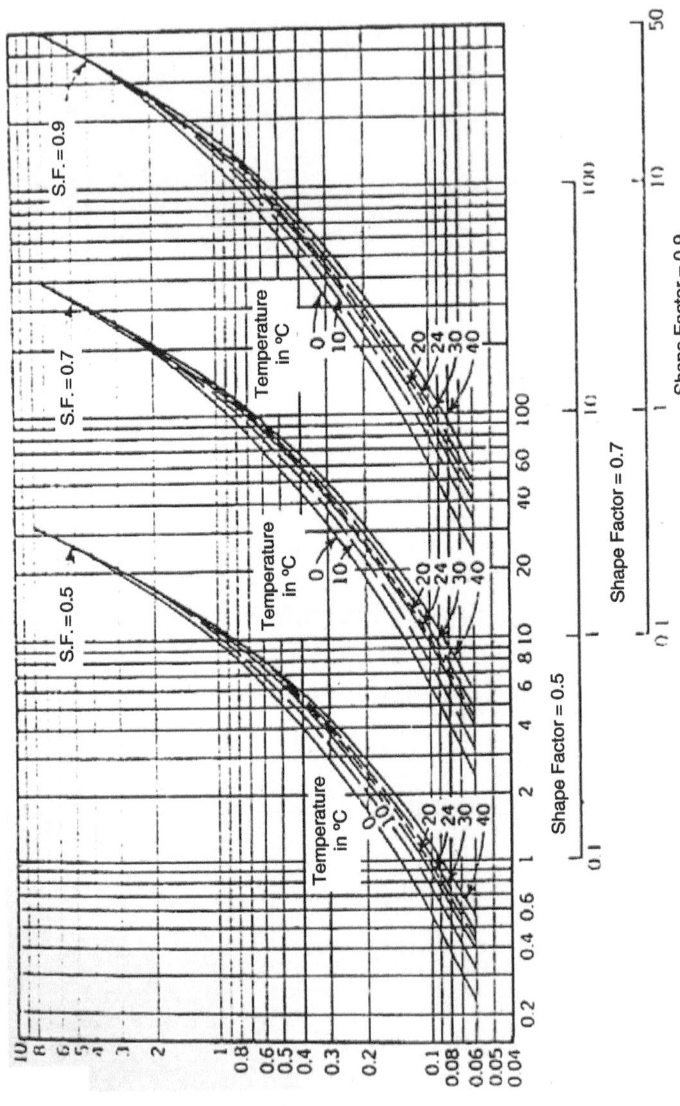

Figure 7.5 (x) : Relationship between fall velocity, sand-grain diameter, and shape factor in water. (Source: Vanoni, 1975).

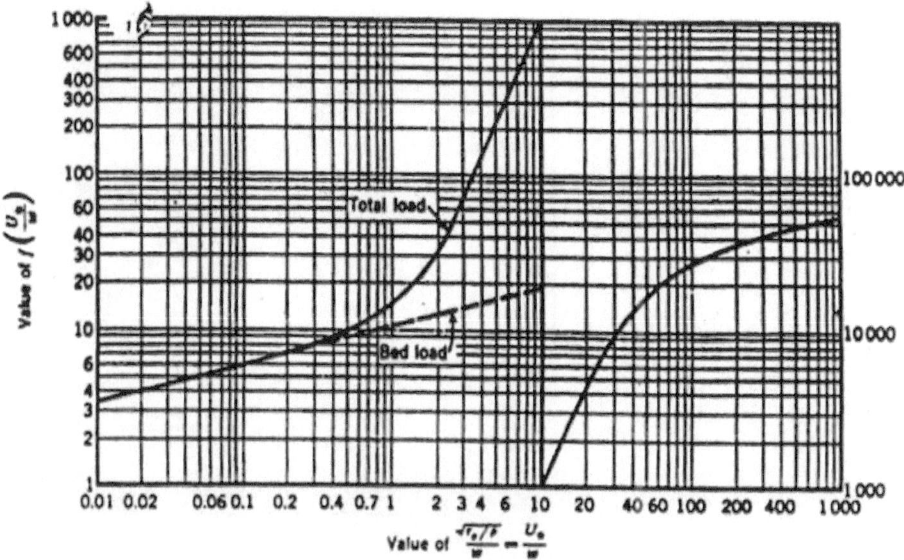

Figure 7.5 (xi) : Function f (V. /ω) for Laursen Formula (Laursen, 1958)

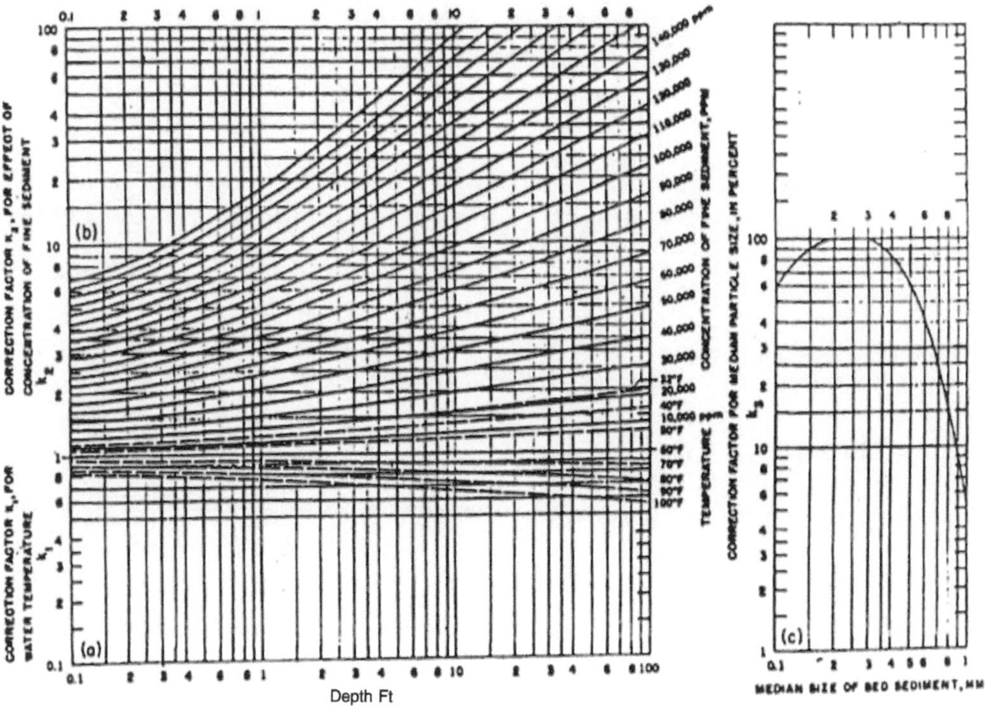

Figure 7.5 (xii) : Colby's (1964b) Correction Factors for Effect of Water Temperature, Concentration of fine Sediment, and Sediment Size to be Applied to Uncorrected Discharge of Sand (Gilbert, 1914)

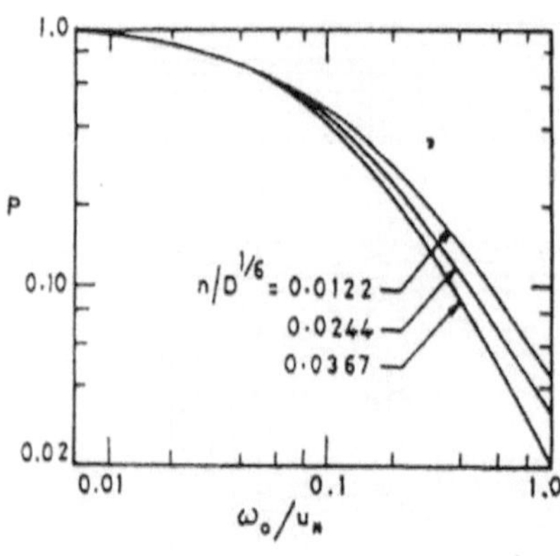

Figure 7.5 (xiii) : Variation of P with ω_o/V_* and n/D1/6 (Lane and Kalinske), 1937

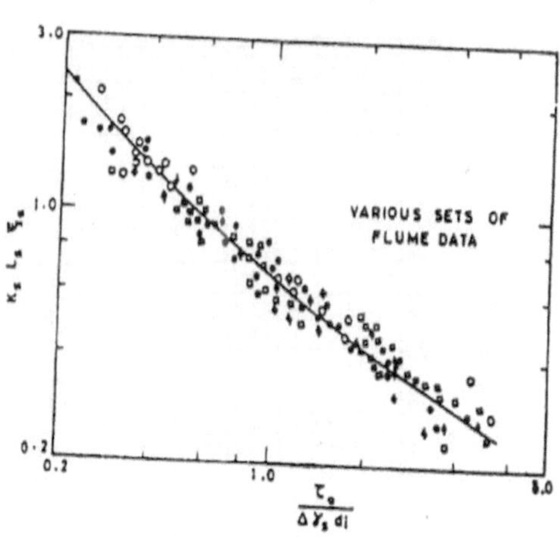

Figure 7.5 (xiv) : Variation of $K_s L_s \xi_s$ with $2D/k_s$ and K (Karman)

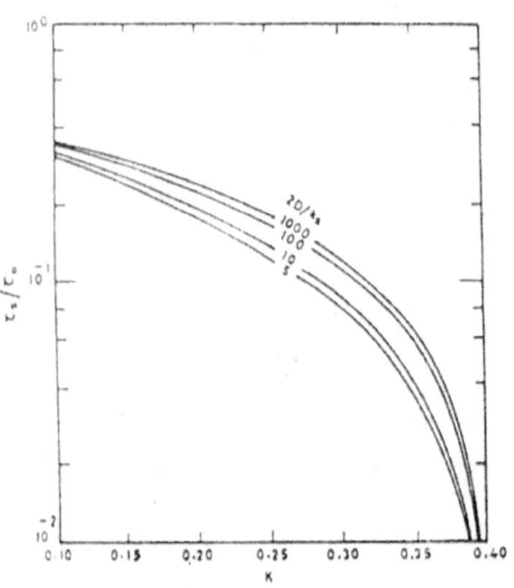

Figure 7.5 (xv) : Variation of τ_s/τ_o with $\tau_o/\Delta\gamma_s d_i$ (Samaga)

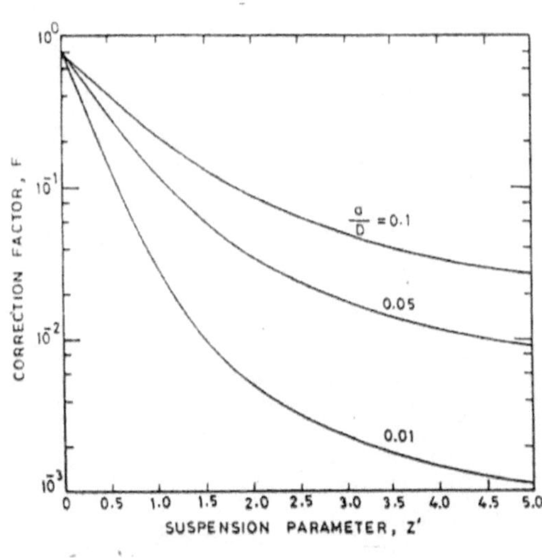

Figure 7.5 (xvi) Variation of F with Z' and a/D (Van Rijn)

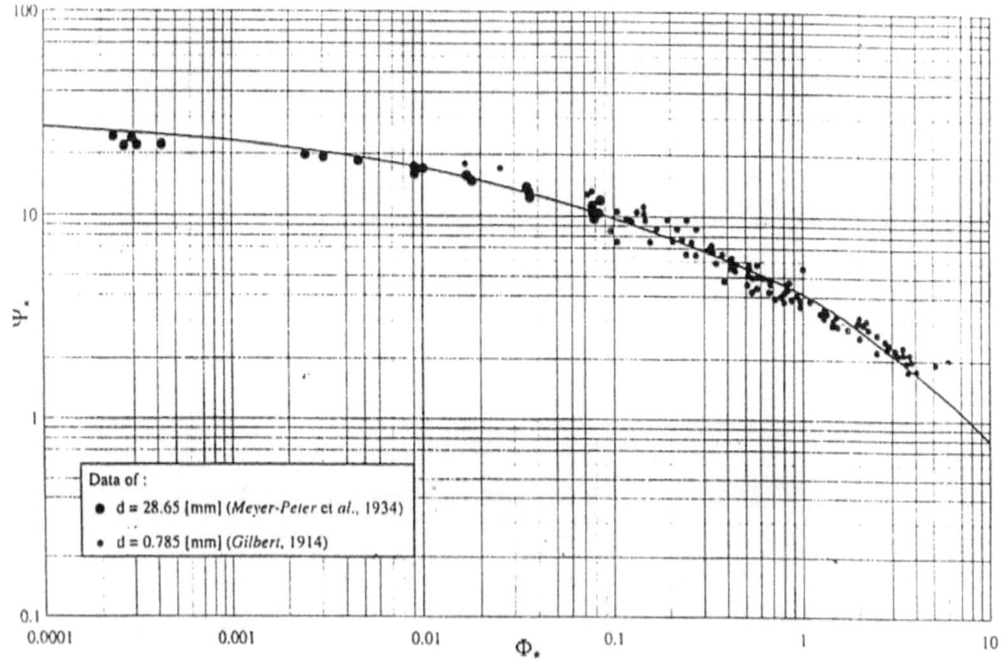

Figure 7.5 (xvii) : Equation of Bed Load, Φ_*, $f(\Psi_*)$, of Einstein (Mayer Peter and Gilbert)

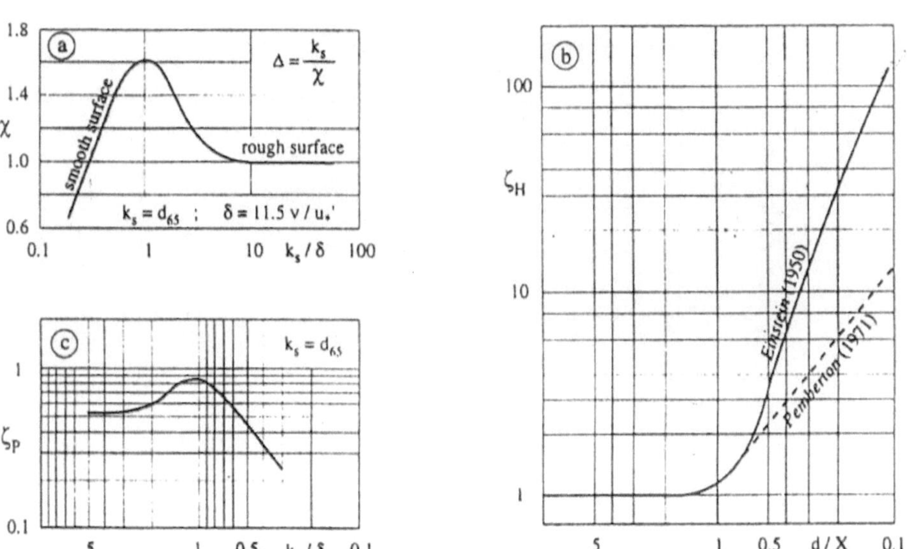

Figure 7.5 (xviii) : Correction Coefficient: (a)of velocity distribution, (b) of Hiding and (c) of Lift Force (Einstein, 1950)

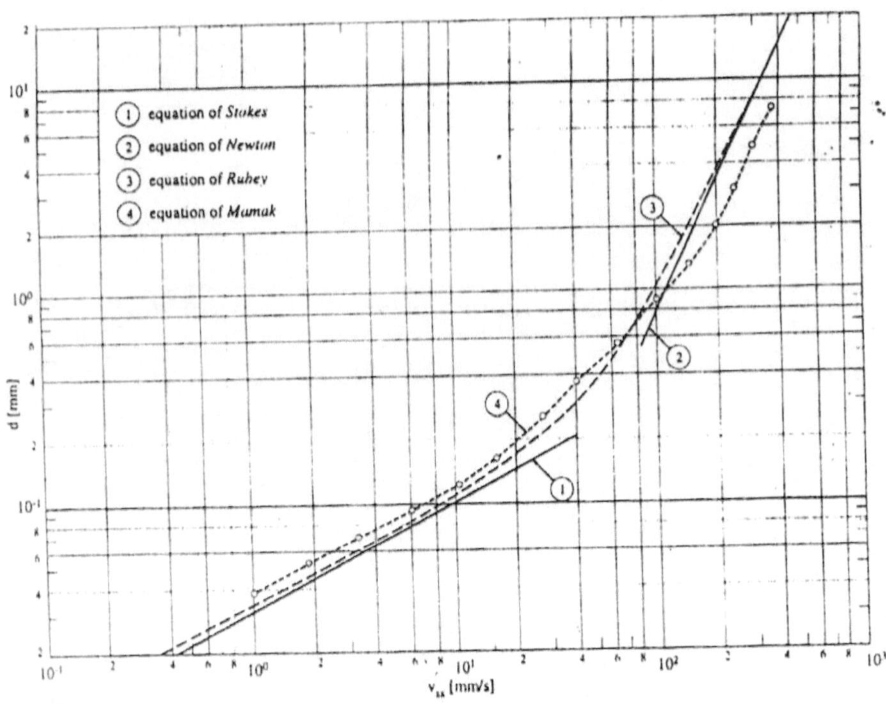

Figure 7.5 (xix) : Settling Velocity, V_{ss}, as Function of Particle Diameter, d (Rouse, 1938)

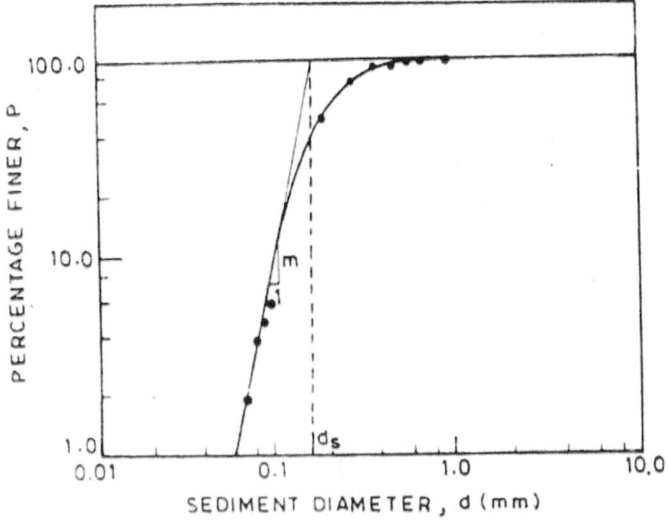

Figure 7.5 (xx) : Percentage Finer V/S sediment dia for figure (xix)

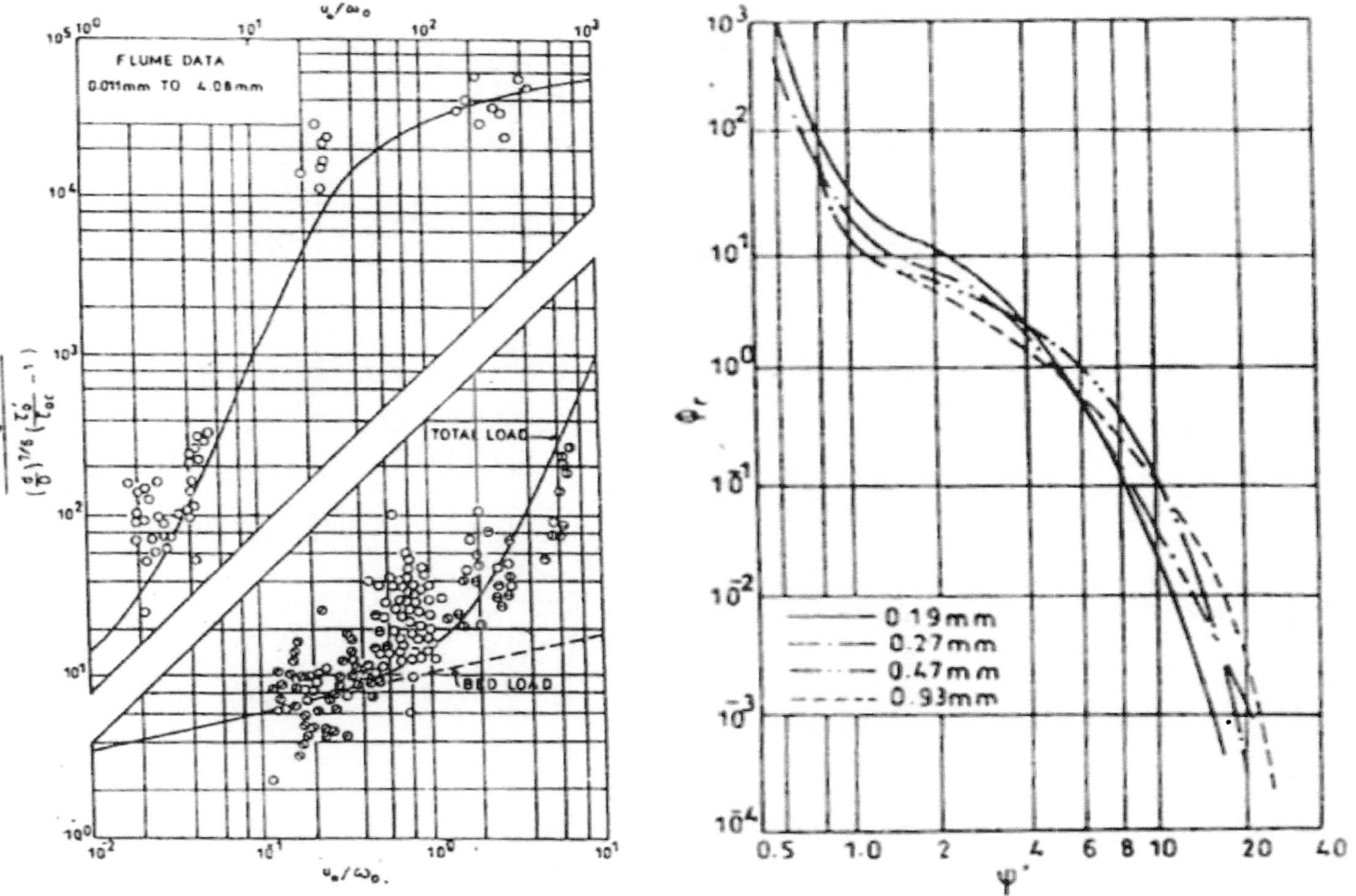

Figure 7.5 (xxi) : Laursen's Total Load Relation (Laursen)

Figure 7.5 (xxii) : ϕ_T vs Ψ' Curve of Bishhop (Bishop, Simons and Richardson)

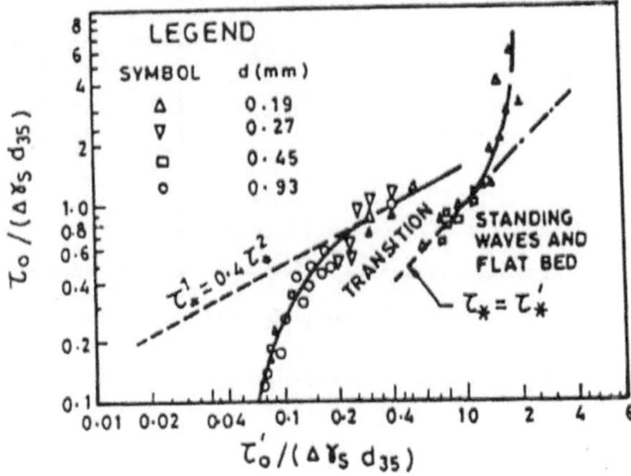

Figure 7.5 (xxiii) : Engelund's Resistance Relation Based on Flume Data (Engelund's)

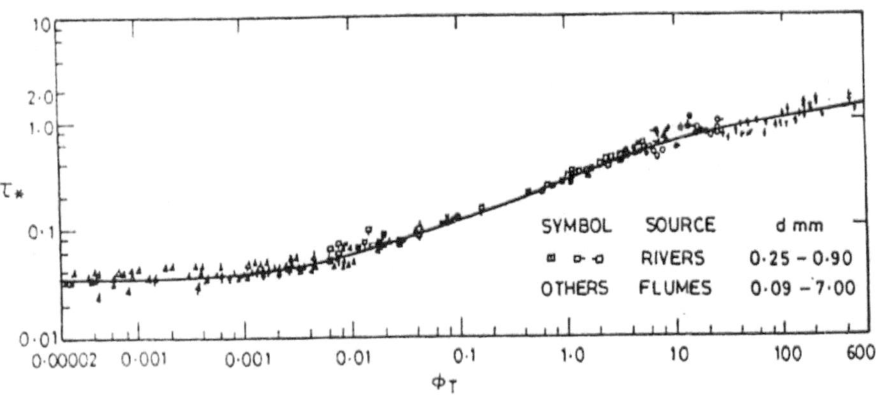

Figure 7.5 (xxiv) Variation of ϕ_T with τ_* for Plane Bed Data (Vittal)

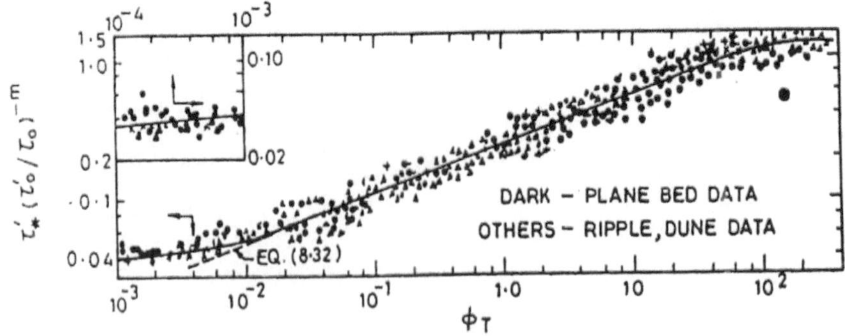

Figure 7.5 (xxv) : Sediment Transport Law for Plane, Ripple, and Dune Beds (Vittal and Ranga Raju)

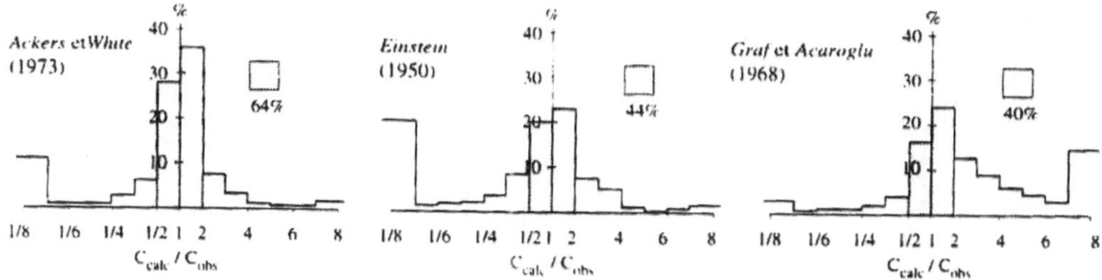

Figure 7.6 Comparison three sediment transport formula with Respect to C_{calc}/C_{obs} , of the Success of Prediction

7.8 Canal Design Theories

In general there are following three principal methods of channel design or canal design theories.

(a) Maximum permissible velocity method

(b) Tractive force theory (or Maximum permissible tractive force method) and

(c) Regime theories.

Though these methods were initially developed for unlined channels and are generally called methods of erodible channel design. But these methods have also been applied to design of lined channels either carrying silt, or some times to ensure that no silt is deposited or to ensure that there is no erosive effect on lining. Once a lined canal is silted, it starts be having as an unlined canal. Lining works as a buried membrane. Alternatively the difference may be that bed and sides can not be eroded much, but lining may be damaged and then bed and sides are also eroded. These can also be called as non alluvial canals. All the above methods are discussed in subsequent paras.

7.9 Maximum Permissible Velocity Method (Also known as critical velocity concept)

This is the simplest method, with the use of any resistance formula (table 0.1, now Manning's equation) and maximum assumed velocity. Earlier this method was most extensively used for design of unlined canals, till regime theories became popular. Later on and even now, this approach is extensively used for design of lined canal carrying either clear or silt laden water, by assuming maximum permissible velocity.

Following steps are involved in this method.

i. Knowing Q and assuming V, Compute A= Q/V

ii. Assume Z & D, and find B for trapezoidal shape.

iii. Compute R= A/P

iv. Assume N

v. Use Manning's formula $V = R^{2/3} S^{1/2} / N$, to find slope.

vi. Or alternatively, assume Z, V, N and S and compute R. From A and R, compute P, B and D. Also see Para 6.3.1 to 6.3.7 and table 6.4.

Thus out of 7 parameters, one Q is known, 4 are assumed and rest 2 is worked out from 2 equations.

Thus in this method the designer's judgment choice is more about D and S. He may design a shallow canal or a deep canal. A shallow canal will have steeper bed slope and deep canal will have flatter bed slope. Since bed slope is guided by ground slope, the choice becomes generally restricted. Further by adopting minimum B/D ratio, as discussed in chapter 3 and 4, a good unique solution is possible.

Table 7.8 : Maximum Permissible (or critical) velocity given by various investigators
* Etcheverry, 1915 *Gibson, 1952

S. No.	Material	Particle diameter in mm	Bottom velocities Du Buat (1786)* V - ft/sec			Ganguillet and Kutter (1827)*		Etcheverry, 1915 V ft/s	Fortier and Scooby, 1926 V ft/s			Parker (1932) Velocity V ft/s	U.S.S.R. (1936) Mean V ft/s
			Transportation begins	Material is kept in suspension	Silting begins	Safe bottom ft/s	Safe mean V ft/s		clear water V ft/s	water transporting colloidal silts V ft/s	Water Transporting non-colloidal silts, sands, gravels, or rock fragments		
(1)	(2)	(3)	(4)	(5)	(6)	(7)	(8)	(9)	(10)	(11)	(12)	(13)	(14)
1	**Clay**												
1.1	Stiff clay very colloidal							4.00-5.00	3.75	5.00	3.0		
1.2	Dark clay fit for pottery (sp. gravity 2.64)		0.35	0.27	--								
1.3	Fine clay											0.25	
1.4	Soft clay											0.50	
1.5	Alluvial silts colloidal	0.005							3.75	5.00	2.0		
1.6	Shales and hardpans								2.50	5.00			
2	**Silt and sand**												
2.1	Alluvial silts, non-colloidal								2.00	3.50			
2.2	Silt/loam non-colloidal								2.00	3.00	2.0		

Contd...

Table 7.8. Contd...

(1)	(2)	(3)	(4)	(5)	(6)	(7)	(8)	(9)	(10)	(11)	(12)	(13)	(14)
2.3	Silt	0.005											0.49
2.4	Firm loam, clay loam							3.00-3.75					
2.5	Ordinary firm loam								2.50	3.50	2.25		
2.6	Soft loam					0.50	0.66						
2.7	Fine sand, colloidal	0.05							1.5	2.5	1.5		0.66
2.8	Coarse yellow sand (sp. gravity 3.36)		1.07	0.71	0.62								
2.9	Finest sand											0.50	
2.10	Fine sand	0.05						0.75-1.00				0.70	0.66
2.11	Coarser sand											0.80	
2.12	Very light loose sand							1.0-1.50					
2.13	Gravel or Coarser sand											1.0	
2.14	Sand					1.00	1.32						
2.15	Medium sand	0.25											0.98
2.16	Coarse sand or light sandy soil	1.00						1.50-2.00					1.80
2.17	Average sandy soil							2.00-2.50					
2.18	Sandy loam, non-colloidal							2.50-2.75	1.75	2.5	2.0		
2.19	Soft earth											0.25	
2.20	Soft brown earth					0.25	0.33						
3	**Volcanic ash**								2.50	3.50	2.0		
4	**Gravel and Pebbles**							2.75-3.00					
4.1	Fine gravel	2.50							2.50	5.00	3.75		2.13
4.2	Gravel small size of anise seed (sp. gravity 2.55)		0.53	0.35	0.27								

Contd...

Table 7.8. Contd...

(1)	(2)	(3)	(4)	(5)	(6)	(7)	(8)	(9)	(10)	(11)	(12)	(13)	(14)
4.3	Gravel side of pea (sp. gravity 2.55)		0.71	0.62	0.53								
4.4	Gravel size of large bean (sp. gravity 2.55)		1.56	1.07	0.71								
4.5	Medium Gravel	5.0											2.62
4.6	Gravel					2.00	2.64						
4.7	Graded loam to cobbles when non-colloidal								3.75	5.00			
4.8	Grade silts to cobbles when colloidal								4.00	5.50			
4.9	Coarse gravel, non-colloidal								4.0	6.0			
4.10	Coarse gravel	10.00						5.00-6.00					3.28
4.11	Cobbles and singles								5.0	5.5			
4.12	Round pebbles 1 inch or more (sp. gravity 2.64)		3.20	2.14	1.56								
4.13	Pebbles 1 inch in diameter											2.0	
4.14	Pebbles egg size											3.0 to 3.3	
4.15	Pebbles					3.00	3.94						
4.16	Fine pebbles	15.00											3.94
4.17	Medium pebbles	25.00											4.59
4.18	Coarse pebbles	40.00											5.91

Contd...

Table 7.8. Contd...

(1)	(2)	(3)	(4)	(5)	(6)	(7)	(8)	(9)	(10)	(11)	(12)	(13)	(14)
4.19	(a) Large pebbles	75.00											7.87
	(b) Large pebbles	100.00											8.86
	(c) Large pebbles	150.00											10.83
5	Angular flint, egg size (sp. gravity 2.25)		4.00	3.20	2.14								
6	Stone 3 inches in diameter											5.0	
6.1	Broken stones or flint					4.00	5.58						
7	Boulders 6 to 8 inches in diameter											6.6	
7.1	Boulders 1 to 1.5 foot in diameter											10.0	
8	Conglomerate					5.00	6.56	6.00-8.00					
9	Concrete							15.0-20.00					
10	Stratified rock					6.00	8.20						
10.1	Hard rock					10.00	13.20	10.0-15.00					

The maximum permissible velocity is the greatest mean velocity that is possible without the occurrence of scour in earthen canals. It is primarily dependent on the boundary material and sediment load. Also, since the boundary velocity is the criteria and the velocity distribution is not uniform, it is also dependent on the depth of flow. Because of the obvious complexities, velocity in any particular case is not easy to assess and good engineering judgment based on experience is important. A number of investigators have suggested different values of velocity according to types of material, based on their experience, judgment and observations. Values given by all investigators are given in one table 7.8. These are arranged in subhead groups clays, silt, sand, gravel and pebbles for easiness and comparison in the same table. Velocity values given by the authors are indicated against the material proposed by them.

Etcheverry (1915) studied data collected from various sources viz., Flynn, Buckley, Clibborn and others and suggested the safe maximum values of mean velocities for safety against erosion. Bottom velocities were said to be 75 percent of mean velocities.

Parker (1932) gave the following very rough rules on basis of experiments made by Deacon and Lechalas and also his own observations. In this V is the velocity is feet per second and d the flow depth in feet.

i) Fine silt of mean diameter about 0.01 inch
   (a) With clear water scour would occur of $V > 0.4\sqrt{d}$
   (b) With heavily silted water, no deposition if $V > 1.05\sqrt{d}$
   (c) With heavily silted water, scour would begin if $V > 1.37\sqrt{d}$
ii) Coarse sand with mean diameter 0.04 inch
   (a) With clear water scour would occur if $V > 1.6\ d^{0.33}$
   (b) With heavily charged, no deposition if $V > 2.2\ d^{0.30}$
iii) Coarse gravel, pea size
   (a) Scour begins if $V > 2.0\ d^{0.25}$
   (b) With heavily charged water, no deposition if $V > 2.5\ d^{0.25}$
iv) Boulders would be moved if $V > 5\ d^{0.25}$

He also gave the values of bottom velocities that would just initiate movement, as given in table 7.8.

After a careful study of all available data comprising of both field and laboratory observations, the special Committee on Irrigation Hydraulics of the American Society of Civil Engineers (1926) recommended the values of maximum permissible mean velocities as reported by Fourier and Scobey (1926). These velocities were meant for straight canals and for depth of flow of 3 feet or less. For canal bends a reduction of 25% was recommended; and for depths beyond 3 feet, a mean velocity higher by 0.5 ft/sec was considered permissible.

Rubay (1937) in U.S.G.S. Professional Paper, reproduced portion of Hjulstrom's diagram showing velocities for fields of erosion, transportation and sedimentation for well sorted sediment as in figure 7.7 (a). These velocities are very low.

This chart is very similar to figure 7.7 (b) showing zone of erosion velocities for various soil grain sizes recommended by Creager and Justin (1949) on the basis of accepted engineering practice. These also do not match with velocities given by Scooby and Parker.

Data obtained from a U.S.S.R. article (1936) of permissible velocities (also see table 6.4) given in the progress Report on the studies on the design of Stable Channels by the U.S. Bureau of Reclamation, (Lane, 1953) is also included in this table. It has also proposed a correction factor for depth as in table 7.9. Besides this, various countries adopt velocity according the experience in their country, see table 6.4. The stress is more on non scouring velocity than on non silting velocity.

Kennedy's formula (1895, see Para 7.12) is also a critical velocity, concept, and can be classified under this approach. It lays stress both on silting and non scouring. It has a better concept; coefficient and exponent take care of size of sediment and depth also. This has been more largely used by many countries.

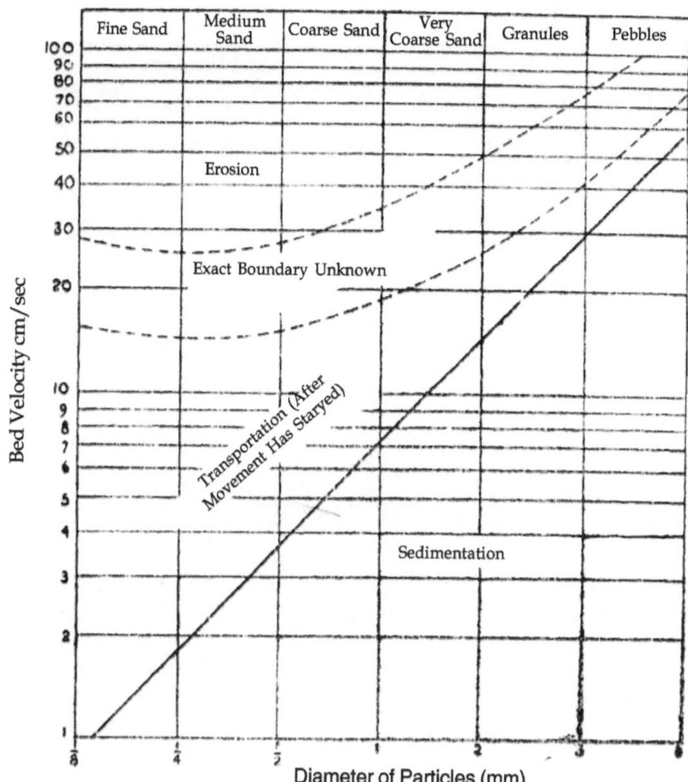

Figure 7.7 (a) Hjulstrom diagram - bed velocity versus particle diameter for the fields of erosion, transportation and sedimentation for well-sorted sediment, Rubey (1937).

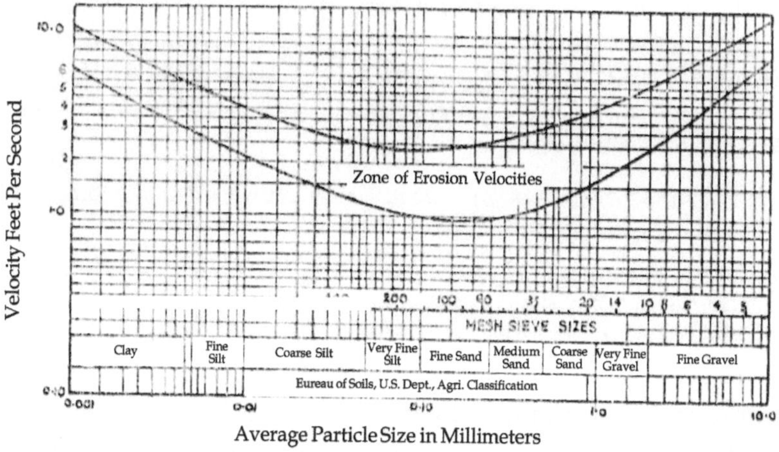

Figure 7.7 (b) Erosion velocities for various particle sizes, Creager and Justin (1949)

Table 7.9 USSR, Corelation Factor of Permissible Velocity for Depth

Depth		Corelation factor	
In feet	In meters	Non cohesive material	Cohesive material
(1)	(2)	(3)	(4)
0.98	0.30	0.80	0.80
1.97	0.60	0.90	0.90
3.28	1.00	1.00	0.95
4.92	1.50	1.10	1.00
6.56	2.00	1.15	1.10
6.20	2.50	1.20	1.10
9.84	3.00	1.25	1.20

For lined canals, maximum permissible velocity should not damage lining and is generally taken from table 2.1 and Para 2.3.

Ex. 7.8.1 illustrates a simple application of this design criterion.

Example 7.8.1

A canal is to be designed to deliver 1000 cusecs to a power house. The canal is to be constructed in natural material for which the maximum permissible velocity is estimated to be 3.25 ft/sec. Assuming a trapezoidal section with side slopes 2:1 and a depth not exceeding 9 ft, determine a suitable top width and longitudinal slope. Take N = 0.025.

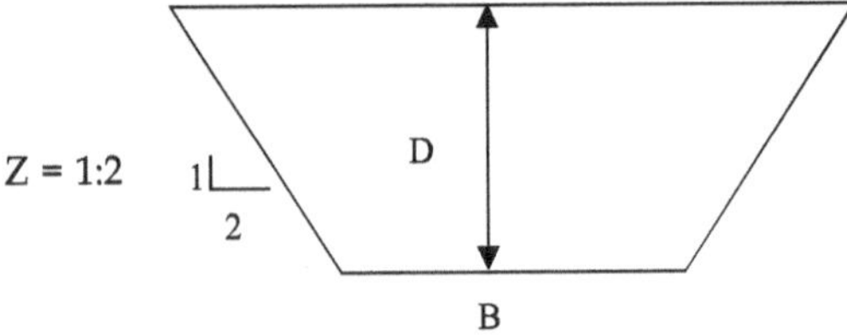

$A = BD + 2D^2$ or $D(B+2D) = A$ or $B = A/D - 2D$

In this problem, out of 7 canal parameters, following 5 parameters are given;
Q = 1000 cusecs, V = 3.25 ft/sec., Z = 2:1, D = 9 ft, N = 0.025 Now it is to find out B and S

For the given maximum velocity, A = 1000/3.25 = 307 ft². Assuming D = 9 ft, then B = A/D - 2D = 16.1 ft. Hence the top width = 52.1 ft.

On this basis R = $R = 307/(16.1+18\sqrt{5})$ = 5.45 ft. Substituting in the Manning formula, 3.25 = (1.486/0.025) x $5.45^{2/3}$ x $S^{1/2}$, from which S = 1/3200.

The problem may be extended to a determination of the unit tractive force. Thus T_0 = 62.4 x 5.45/3200 = 0.106 lb/ft², which is probably reasonable. It is interesting to note that if D is taken as 8 ft, the top width = 54.4 ft and S = 1/3080, with T_0 = 0.107 lb/ft². The section with D = 9 ft appears the most satisfactory as it approaches nearest to the hydraulic best section.

7.10 Tractive Force Theory (Maximum Permissible Tractive Force Method)

Tractive force theory also aims that no particles of soils of an earthen section are dislodged i.e. there is no soil erosion of the canal banks and bed. This is non scouring velocity.

This method is based on the concept of shear stress acting on the boundary material, (See Para 7.1.1).

Critical Tractive force is the minimum specified value of the shear stress required to be exerted on the boundary surface for commencement of movement of the soil particles. This value is designated by τ_c *or* τ_r

According to this theory, the canal section is to withstand maximum shear stress estimated and not permit erosion. The boundary material must have only critical Tractive force and no more, nor less. If Tractive force is less than critical, sediment deposition will occur.

Some important of various formulae evolved by different research workers and known after them are given below.

Du Boys equation (1879) for the unit tractive force is given in table 7.4. Maximum permissible values of τ_c are derived from laboratory and field data, see figure 7.5 (i) and (ii).

Schoklitsch (1914) gave a formula for a single grain resting freely on bed of uniform grains of the same size as below

$$WRS=\sqrt{0.201\gamma_s(\gamma_s-\gamma)\lambda D^3} \quad \ldots\ldots \quad \text{Eq. 7.10.1}$$

In which γ_s and γ are the unit weights of transported material and water respectively in Kg/m^3 and γ is a form factor varying between 1.15 and 1.35 for sand and D diameter of grain in meters. The formula was originally in terms of volume.

Schoklitsch (1937) also quoted values of critical tractive force, for different kinds of bed material, using data obtained by Government engineers in Nurnberg on rivers as in table 7.10 (i).

Dr. H.D. Krey (1921) reasoned that the resistance of sediment to motion would be proportional to particle diameter D and its specific weight in water γ (S_s - 1) where γ the specific weight of water and Ss is the specific gravity of sediment. On the basis of studies in the Prussian experimental hydraulic laboratory he found an approximate expression (1939) for the beginning of detritus transportation as

D = 8 dS; where D is the diameter of sediment particle, d the depth of flow and S the slope (use consistent units on both sides). Substituting for dS in terms of τ/γ, Krey (1953) wrote the equation as:

$$\tau_c = 0.076\,\gamma\,(S_s - 1)\,D, \text{ where } D = \text{particle diameter} \quad \ldots\ldots \quad \text{Eq. 7.10.2}$$

Table 7.10 (i) Critical Tractive Force Versus Grain Size (Schoklitsch, 1937)

S. No.	Material	τ_c in Kg/m^2
1.	Ordinary quartz sand, 0.4 to 1 mm dia.	0.25 to 0.30
2.	Ordinary quartz sand, 0.2 to 0.4 mm dia.	0.18 to 0.20
3.	Ordinary quartz sand, upto 2 mm dia.	0.40
4.	Roundish quartz gravel, 0.5 to 1.5 cm in dia	1.25
5.	Clavey soil	1.0 to 1.2
6.	Coarse quartz stones, 4 to 5 cm in dia	4.8
7.	Flat line stone particles, 1 to 2 cm thick 4 to 6 cm long	5.6

Besides Schoklitsch and Krey, investigations regarding critical tractive force have been made by several more authorities including Eisner, Schaffernak and others. Mac Dougall has presented a comprehensive summery of their work. Formulae developed by most of them contained constants which could be determined only by tests for each individual material.

Kramer (1935) developed a practical formula for critical tractive force in terms of sand size and distribution alone, not incorporating any constant to be separately determined. *Critical tractive force was defined by him as the value of bed shear which brings about general movement of the mixture, that is, it transports the largest of the particles.* It was postulated that τ_c = f (grain size, grain distribution, grain form, denseness, unit weight water) i.e. τ_c = f (D) (e) (ρs - ρ); where D = mean diameters of grain , e is voids ratio, $(\rho_s - \rho)$ effective density of material.

By grouping the variables, form of the equation was expressed as

$$\tau_c = C\frac{D(\rho_s - \rho)}{M} \quad \text{...... Eq. 7.10.3}$$

(i); Where D = mean grain (sediment) diameter and M is obtained from figure 7.7 (i, a) Hypothetical grain analysis curve.

Constant C was determined by Kramer (1935) using his own data and also of Schffernak, Schoklitsch, Krey, Engels, Gilbert etc. Value of C was found to be 100/6, that resulted the Equation as

$$\tau_c = \frac{100}{6}\frac{D(\rho_s - \rho)}{M} \quad \text{...... Eq. 7.10.3 (ii)}$$

Ranges of D and M covered by data were 0.244 mm<D<6.52 mm and 0.265 < M < 1.0. For ordinary quartz, sand average unit weight being 2.65 grams per cu. cm.

$\tau_c = 27.5\frac{D}{M}$; τ_c being in gms/sq meter and D in mm.

'O' Brien and Rindlaub (1934) examined the data of τ_c from various sources in comparison with equation of Schoklitsch and Kramer and concluded that best correlation is obtained by simple plotting of median diameter against τ_c, as given in Fig. 7.7 (i). Scatter of experimental points, was attributed to several reasons such as difference in experimental technique, errors in observation, mistakes in computation etc. Data of other investigators was further added in the same diagram by Chitale (1966).

A comprehensive study (Lane and others, 1952, 53) of available data of critical Tractive forces was undertaken by U.S.B.R. and results were reproduced as in figure 7.7 (iii) and 7.7 (iv) respectively.

U.S.B.R. made an important contribution by presentation of prototype data collected on canals in the San Luis Valley of Colorado, USA as given in figure 7.7 (v). The bed material was coarse and non cohesive and covered a large size range. The size used was such that 25 percent by weight of the material was larger than this. The Tractive force was estimated from maximum sustained flow. Fifteen reaches of the canal were considered in which discharges ranged from 45 to 1500 ft^3/s (cfs) and bed slopes 4.7 to 51 ft per mile(1 / 1123 and 1 / 103). This graph yielded the relation.

τ_c (in lbs / sft) = ½ D (in inches)...... Eq. 7.10.4

This is in close agreement with values according to quite a few formulae vide figure 7.7 (vi). To allow a certain factor of safety Lane recommended slightly lower values as τ_c = 4 D (4 times diameters in inches). Ranges of size of material for application of this relation were proposed to be between 5 and 100 mm.

Lane also attempted converting critical velocity data of Etcheverry, Fortier Scobey and U.S.S.R. into critical Tractive force values. For this conversion a bottom width of 10 ft and side slopes of 1.5: 1 (H: V) was assumed and bed slopes were estimated using Manning's formula with reasonable 'N' values. Such converted values for Fortier Scobey's velocity are given in table 7.10 (ii).

The hydraulic roughness depends on the grain size which lines (covers) the canal after some of the fines have been washed out and is important.

Permissible Tractive Force

Lane (1953) in recommending safe design τ_c values, proposed classification of bed material in three broad types vide figure 7.7 (vi) as under

(i) Coarse non-cohesive material (between 5 to 100 mm)

For coarse non cohesive more than 5 mm size, the critical tractive force (limiting stress, τ_l) is given as,

$$\tau_l = 0.736\ d_{75} \qquad \text{...... Eq. 7.10.5}$$

Where τ_l is in Kg/m² and d_{75} in mm. Due to available armoring effect the representative size used is d_{75} in place of d_{50}.

(ii) Fine non-cohesive material (between 0.1 to 5 mm)

In this fine range of sizes, τ_c is not dependent on grain diameter alone. Contents of fine sediment in water help to increase resistively and three separate curves were therefore recommended by Lane: (i) for clear water (ii) for light load of fine sediment and (iii) for heavy load of fine sediment.

Besides the armoring effect, the influence of adhesion imparted by fine clay and silt sizes suspended in the water is considered important. The roll down effect on bank particle is applicable to this case also. For both of the above materials the channel is to be so designed that the applied maximum stress on bed and banks does not exceed the resistance of either. Thus limiting stress is given in the Table 7.10 (iii).

(iii) Cohesive material

For this material the inter-particle forces are much more important than gravity forces, and hence the roll down effect may be neglected. The most important characteristic determining erosion resistances of clay soils are plasticity index and void ratio are given in Table 7.10 (iii).

If boundary material is cohesive the above data is inapplicable. U.S.S.R. data in table 6.4.1 (ii) contains some information useful for such a case. Data in tables 7.10 (ii) can in addition be used as a guide. Further work for collection of both laboratory and field data in respect of cohesive material is still badly wanting.

Simon (1957) in a review has also given curves for various types of material as shown in figure 7.7 (ix). Generally adopted tractive force given by Lane (1953) is given in table 10 (xiii).

At Bends: It may be more economical to provide local protection at bends, rather than design the channels for reduced values of tractive force.

Tractive force on sides: Tractive force is uniformly distributed only in wide channel. Normally it is greater on the bed than on the sides. The particle resting on the banks is acted upon by gravity in addition to water drag as shown in figure 7.8.

Values of τ_c recommended above for various types of materials are applicable on level bed of a channel. The force 'τ_s' necessary to cause motion on sloping sides would be less than 'τ_l' the force required for movement on a horizontal bed. Ratio of τ_s / τ_l would thus be less than 1 on sides and equal to 1 on bed. This ratio was shown by Carter (1953) to depend on the angle of sides slope and angle of repose of the material. Functional relationship derived by him was

$$K = \cos\alpha \sqrt{1 - \frac{\tan^2\alpha}{\tan^2\phi}} \qquad \text{........ Eq. 7.10.6 (i)}$$

Where α is the angle of the side slope or bank, with the horizontal and ϕ is the angle of internal friction of the bank soil.

For different values of α and ϕ, value of K can be read directly from the diagram given in figure 7.7 (vii).

Fan (1947) also independently derived K value in the form

$$K = \sqrt{1 - \frac{\sin^2\alpha}{\sin^2\phi}} \qquad \text{........ Eq. 7.10.6 (ii)}$$

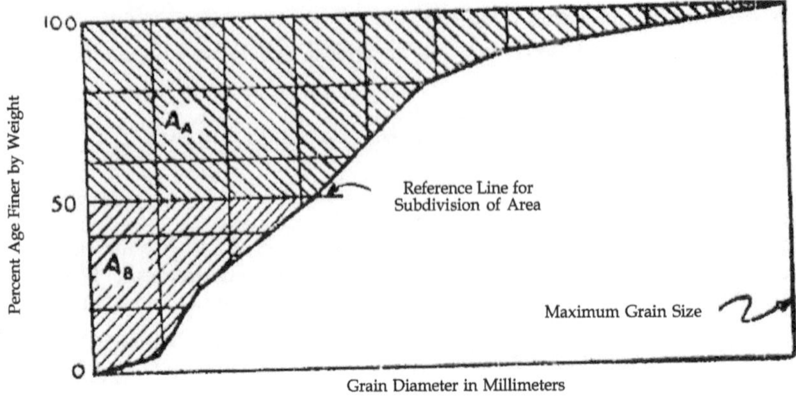

Figure 7.7 (i, a) : Hypothetical Grain Analysis Curve for Determination of M = AB / AA in Eq. 7.10.3 (i)

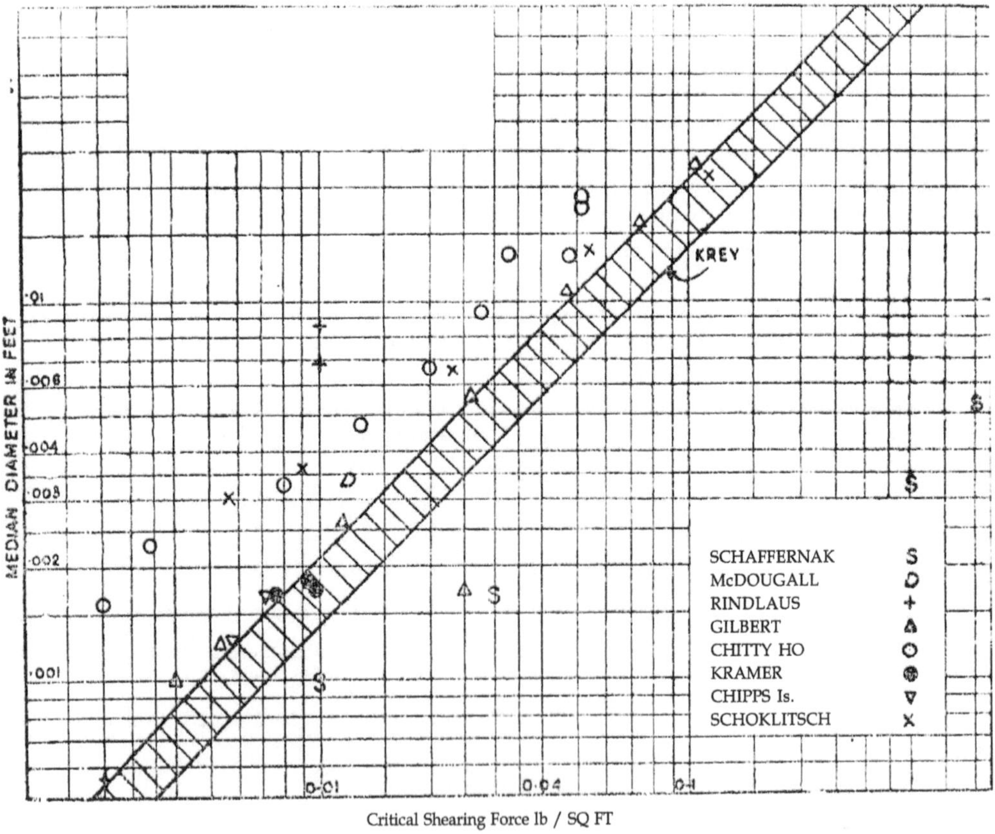

MEDIAN DIAMETER IN FEET

·01
·006
·004
·003
·002
·001

0·01 0·04 0·1

KREY

SCHAFFERNAK	S
McDOUGALL	D
RINDLAUS	+
GILBERT	Δ
CHITTY HO	O
KRAMER	●
CHIPPS Is.	▽
SCHOKLITSCH	X

Critical Shearing Force lb / SQ FT

Figure 7.7 (i, b) : Variation of Critical Tractive force with mean size of grain (O Brien and Rindlaub, 1934)

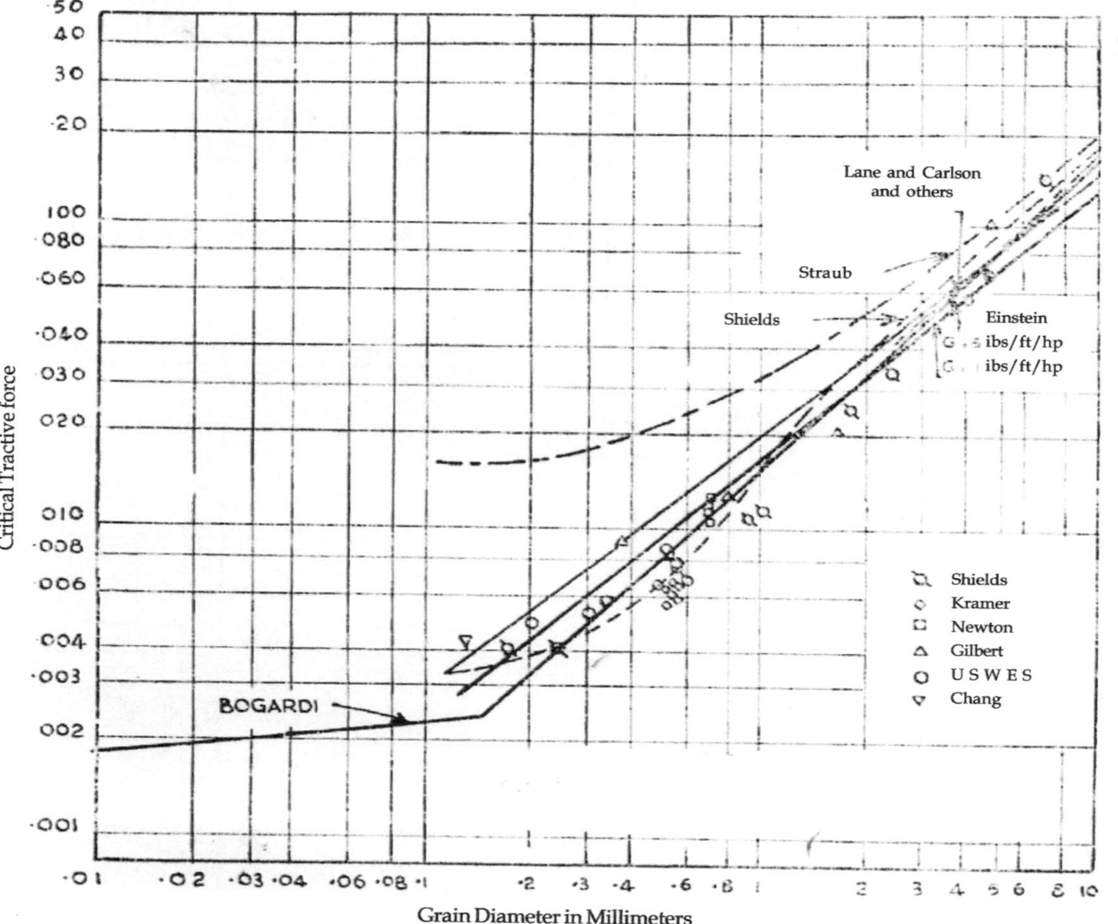

Figure 7.7 (ii) Critical Tractive Force Versus grain size (Tinney-1958)

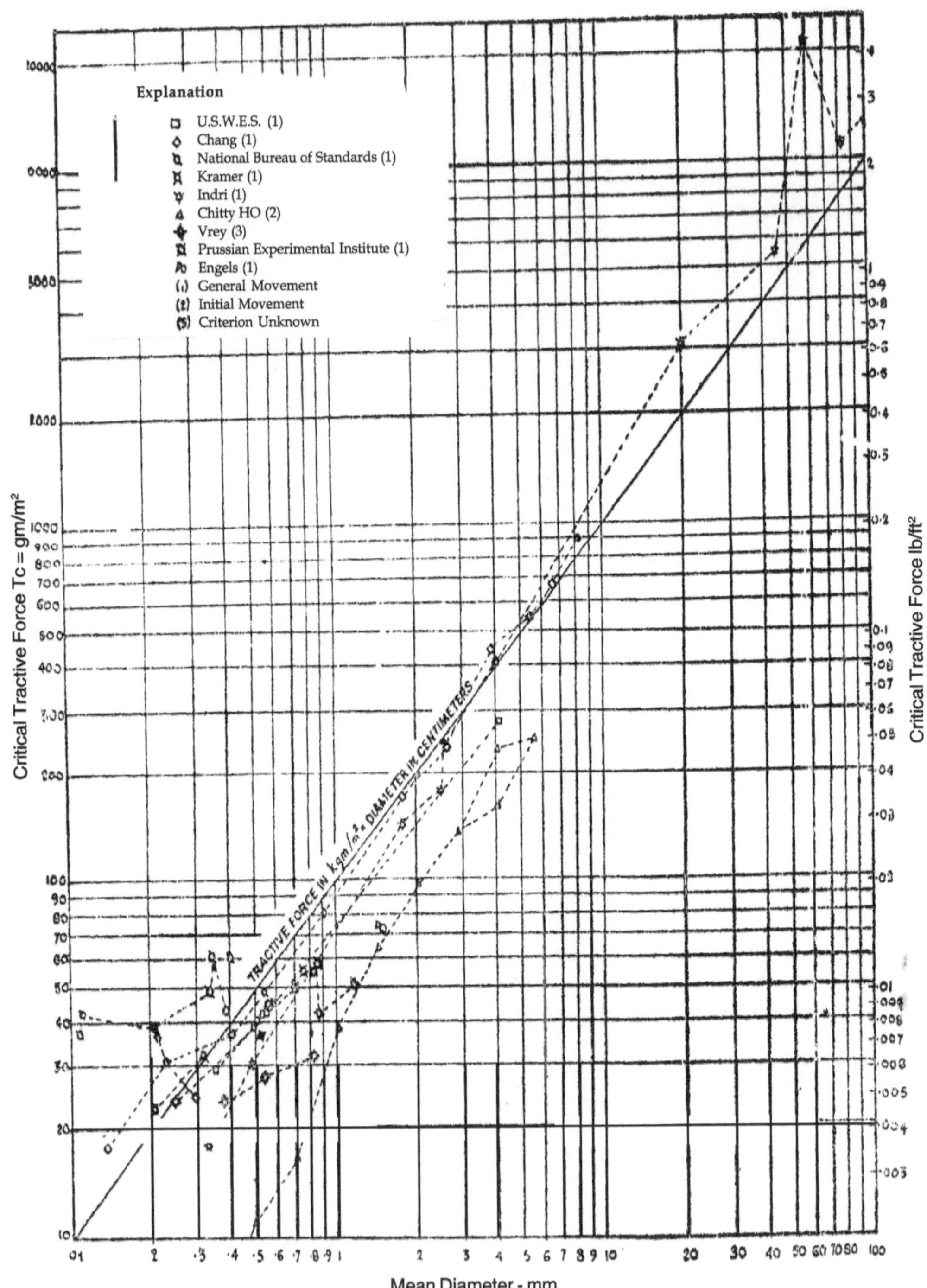

Figure 7.7 (iii) Observed Critical Tractive Forces Versus mean sediment diameter Flume Studies (Lane 1952)

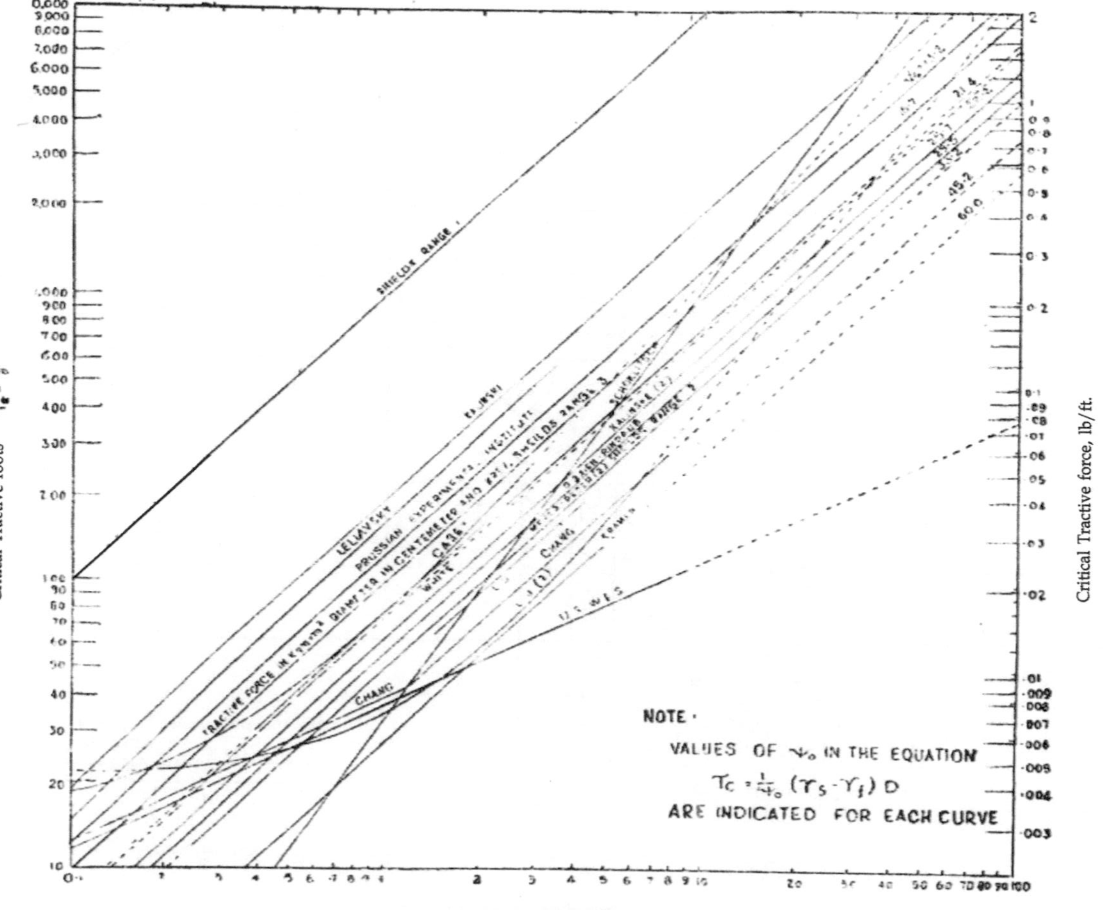

Note : Sand is assumed to have uniformity modulus (M) = 1.0, specific gravity (S_s) = 2.65

Figure 7.7 (iv) : Critical Tractive Force versus mean sediment diameter (Lane 1952)

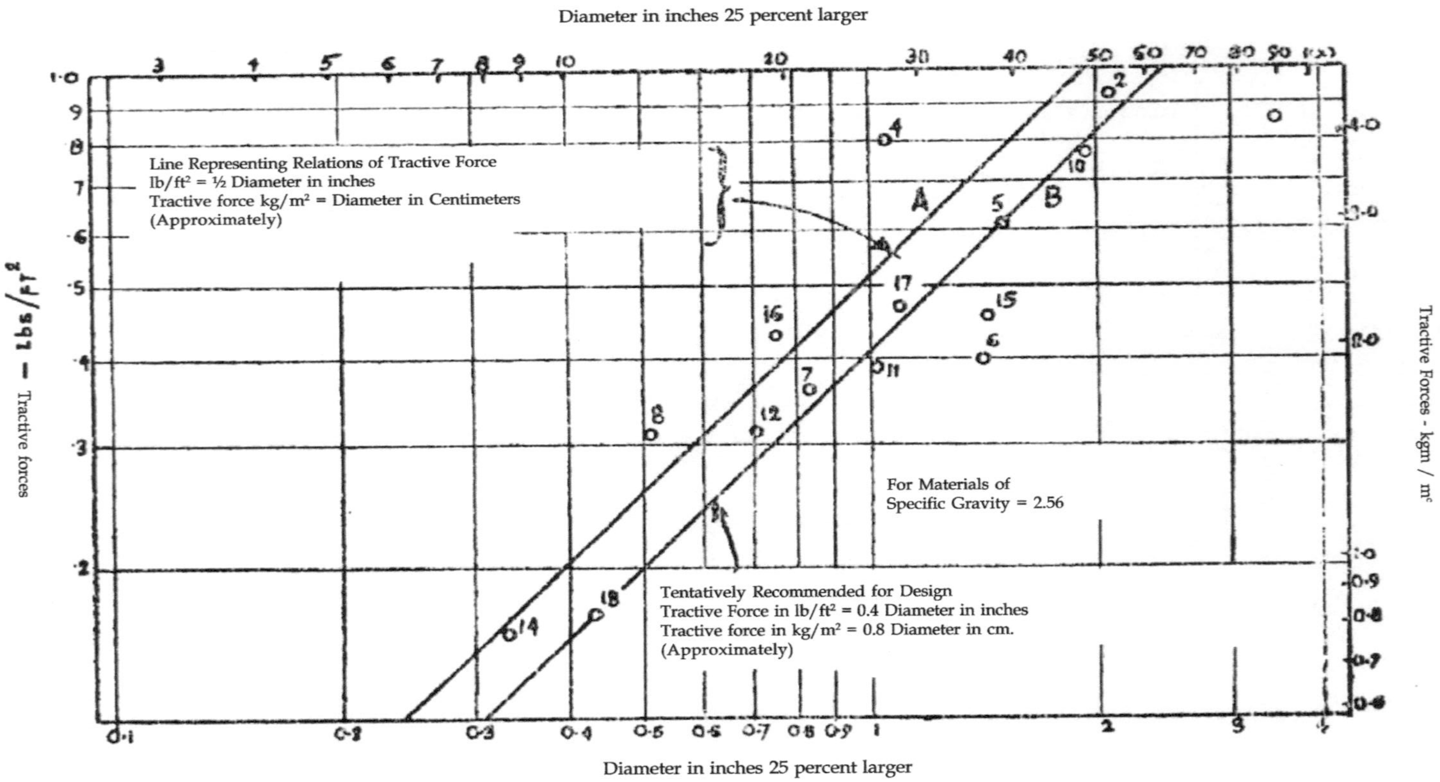

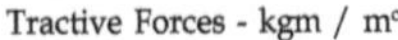

Figure 7.7 (v) : Variation of Critical Tractive Force with sediment diameter, Results of Studies on San Luis Valley Canals, Calorodo USA (USBR)

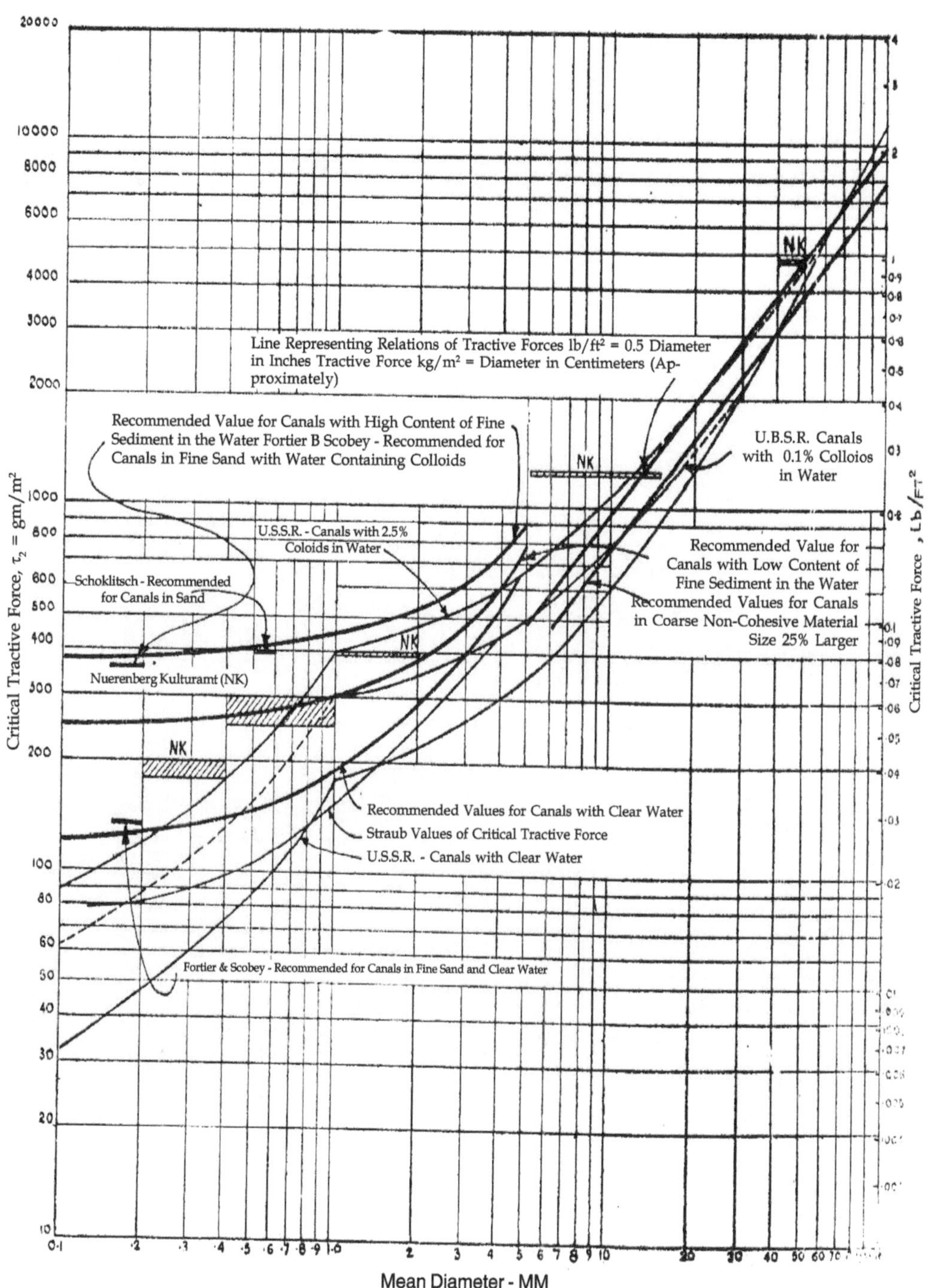

Figure 7.7 (vi) : Limiting Tractive Forces Recommended for Canals and Observed in Rivers versus Size of Bed Material (Lane 1953)

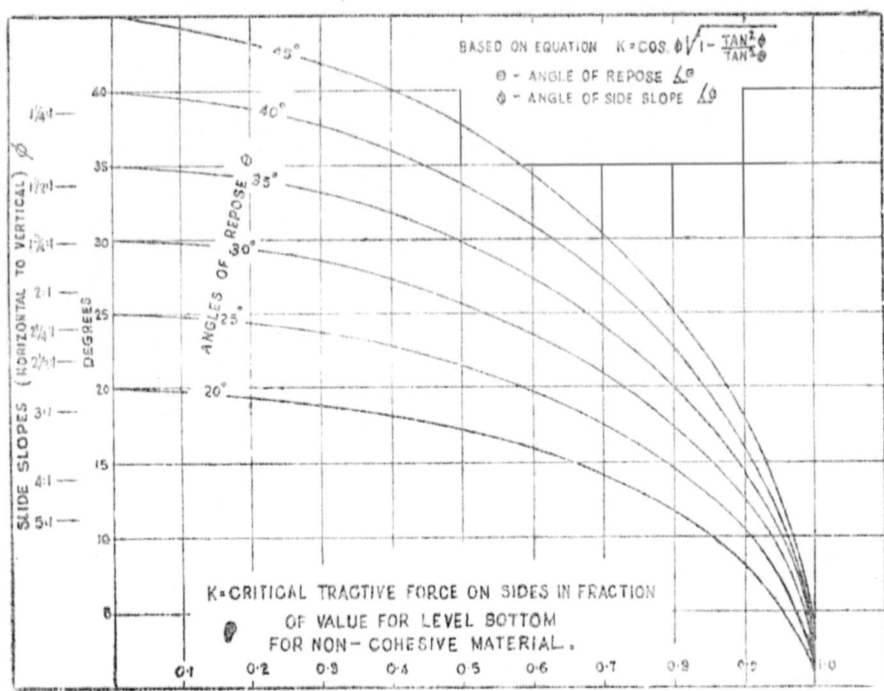

Figure 7.7 (vii) : Side Slope Versus K Values

Angle of Repose - Degrees with Horizontal

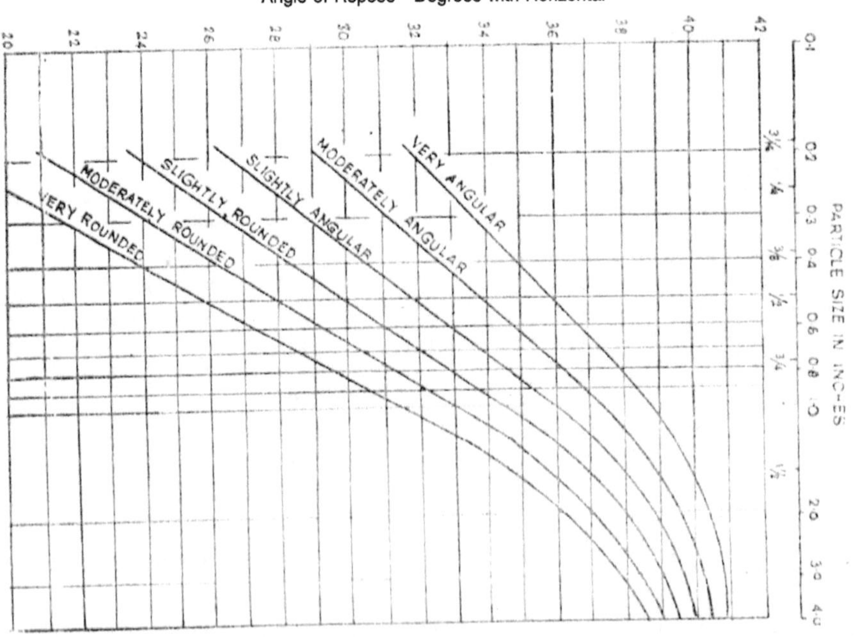

Figure 7.7 (viii) Angle of Repose of Non cohesive material versus particle size

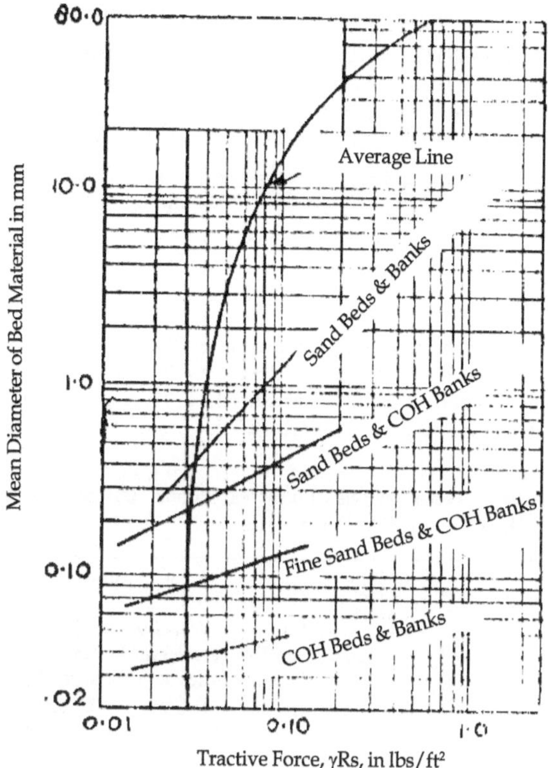

Figure 7.7 (ix) : Variation of tractive force with size of bed material, type of channel and discharge Q (Simons, 1960)

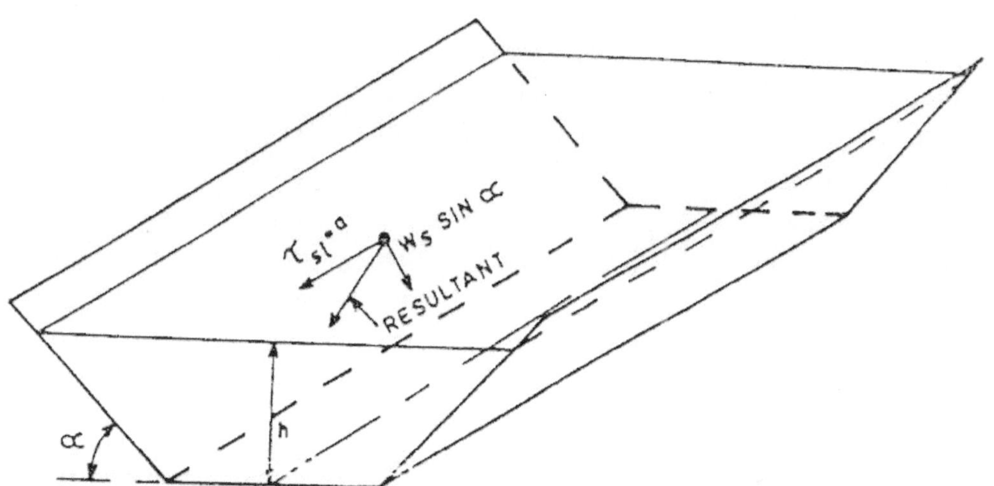

Figure 7.8 Forces Causing Movement of a Particle Resting on a Channel Bank

Although the two equations are not identical, difference in K values obtained by them is negligible.

Use of figure 7.7(vii) or equation 7.10.6 (ii) needs knowledge of angle of response of the side material. Figure 7.7 (viii) gives design values of angle of repose for various degrees of roundness. For finer material, it was observed that the cohesive forces, even with comparatively clear water, becomes so great in proportion to rolling down forces that the latter can safely be neglected.

It may be seen that there is a wide variation in the permissible tractive force and it lays stress more on non scouring velocity only, rather than non silting velocity. Therefore use of this procedure for sediment laden water is generally debatable. Also this procedure gives wider channels in comparison to the Kennedy's method for the same slopes, and may encourage silt deposition.

Design Procedure

The design Procedure is similar to velocity method as in Para 7.9. Here τ_l values are estimated from above figures and by assuming S, R is computed from Eq. $\tau = \gamma$ RS. Further by assuming N, V is computed by Manning's formula. By assuming Z and D, B is calculated. Thus again assumption of τ_l, S, N, Z and D are required. A typical example is given in Para 7B.7.5.

Table 7.10 (II) Maximum Permissible Water Velocities (Fortier and Scobey, 1901) and Corresponding Unit Tractive force (USBR) for the Design of Stable Channels (Flow Depth < 3 ft)

Material		Clear water		Water transporting colloidal silts	
	Roughness coefficient*N	Velocity (Fortier and Scobey) V(ft/s)	Unit tractive force T_0 USSBR (Lane)(lb/ft^2)	Velocity (Fortier and Scooby) V (ft/s)	Unit tractive Force T_0, USSBR (Lane) (lb/ft^2)
(1)	(2)	(3)	(4)	(5)	(6)
1. Fine sand, colloidal	0.02	1.5	0.03	2.5	0.08
2. Sandy loam, non-colloidal	0.02	1.75	0.04	2.5	0.08
3. Silt/loam non-colloidal	0.02	2.00	0.048	3.00	0.11
4. Alluvial silts, non-colloidal	0.02	2.00	0.048	3.50	0.15
5. Ordinary firm loam	0.02	2.50	0.075	3.50	0.15
6. Volcanic ash	0.02	2.50	0.075	3.50	0.15
7. Stiff clay very colloidal	0.025	3.75	0.26	5.00	0.46
8. Alluvial silts colloidal	0.025	3.75	0.26	5.00	0.46
9. Shales and hardpans	0.025	6.00	0.067	6.00	0.67
10. Fine gravel	0.02	2.50	0.075	5.00	0.32
11. Graded loam to cobbles when non-colloidal	0.03	3.75	0.38	5.00	0.66
12. Grade silts to cobbles when colloidal	0.03	4.00	0.43	5.50	0.80
13. Coarse gravel, non-colloidal	0.025	4.0	0.30	6.0	0.67
14. Cobbles and singles	0.035	5.0	0.91	5.5	1.10

*Based on a straight channel of uniform section.

Table 7.10 (iii) Limiting Stresses (Tractive Force)

For Fine Non-cohesive Material (less than 5 mm)			
Median size of material (d_{50}) mm	Limiting Tractive Force Kg/m ^{2}		
	Clear water	Light load of fine sediment	Heavy load of fine sediment
0.1	0.122	0.241	0.369
0.2	0.125	0.250	0.375
0.5	0.145	0.265	0.400
1.0	0.193	0.290	0.435
2.0	0.290	0.386	0.530
5.0	0.675	0.795	0.890
Cohesive Material			
Type of soil	**Etchevery**	**Scooby**	
		Clear water	**Silty water**
		Permissible stress Kg/m^2	
Sandy loam	3.32 to 3.97	1.74 to 2.28	3.55 to 5.20
Average loam, alluvial soil	Silty loam 3.97 to 4.78	---	---
Firm loam, clay loam	Alluvial silts	2.28	7.11
	4.78 to 7.43	3.55	7.11
Stiff clay soil	13.15 to 20.60	12.26	21.78

7.11 Regime Theories

Regime theories developed by irrigation engineers in India are wholly empirical; but have also found much favour outside the country of origin, and considerable research has been done in many countries.

The regime formulas of Kennedy, Lacey and others apply to channels with erodible boundaries in alluvial soils carrying small sediment loads. Analyzed data are mainly from irrigation channel in India and Pakistan, but also some data form Egypt, Europe and America were included. The relations were obtained by the correlation of velocity and other dimensions and slopes of apparently stable channels with discharges and sizes of bed material. These express three primary relationships, namely velocity to depth, velocity to slope, and width to discharge, while the velocity is non-silting and non-scouring. Many investigators have given different formulae; a list is given in table 7.11.

Stable and Regime Channels

A number of definition of stable or regime channels given by various investigators are under

Thomas's (1945) Classification of stable channels

He classified stable channels, broadly in three classes as under

(i) Channels, bed and banks of which can be scoured without objectionable deposits being formed. Channel carrying clear water or only small amounts of sediment are of this type. Small charge of sediment present may have been introduced along with discharge or may be due to scour of bed and banks.

(ii) Channels free from scour but in which objectionable deposits of sediment can form. A common case is lined channel carrying water heavily laden with sediment. Large dose of sediment may be due to heavy concentration in water let into the canal or may be the result of scour in an unlined portion further upstream.

Table 7.11 Regime Theories (Empirical Relationships)

Investigator and his Formula		Remark
FPS	**Metric**	
1	**2**	**3**
1 Kennedy (1895)		
$V_o = 0.84\ D^{0.64}$	$V_o = 0.546\ D^{0.64}$	- Fundamental equation
$V = 0.84\ m\ D^{0.64}$	$V = 0.546\ m\ D^{0.64}$	- Fundamental equation, where $m = V/V_o = CVR$
		- Critical velocity was affected by the grade of the silt and in order to account for the effect of the silt grade he introduced a factor 'm' called Critical Velocity Ratio (C.V.R), Where Value of m:
		For finer silt $m < 1$, ($m = 0.80 - 0.90$); For coarser silt $m > 1$, ($m = 1.10 - 1.20$) (for more detail see para 7.12 and Table 7.12).
2 Lindley (1919)		
$V = 0.95\ D^{0.57}$	$V = 0.567\ D^{0.57}$	- Fundamental equation
$V = 0.59\ B^{0.35}$	$V = 0.265\ B^{0.35}$	- Fundamental equation
$B = 9.8\ D^{1.16}$	$B = 7.86\ D^{1.16}$	- Fundamental equation
		- Bed width of a channel as a regime variable in addition to depth of flow.
		Lindley Stated "when an artificial channel is used to carry silty water, both bed and sides scour or silt, changing depth, gradient and width until a state of balance is attained at which the channel is said to be in regime".
3 Lacey (1927-1939) see table 7.14		
4 Bose (1936)		
(i) $V = 1.12\ R^{1/2}$		- Fundamental equation
(ii) $P = 2.68\ Q^{1/2}$		- Fundamental equation
(iii) $S = \frac{0.00209 d^{0.86}}{Q^{1/2}}$		- Fundamental equation
(iv) $\frac{R}{P} = \frac{S^{1/4}}{6.25d}$		- Fundamental equation
Where d is the weighted mean diameter of sediment		- Both the silt factor f of Lacey and the weighted mean diameter (d) define the size of the sediment transported in a channel, but not sediment charge or rate at which the sediment is transported.

Contd...

Table 7.11. Contd...

1	2	3
5 Malhotra(1939)		
$V = 18.18\ R^{0.63}\ S^{0.34}$		- Fundamental equation
6 White (1939)		
$V = \dfrac{0.7\ \omega^{1/4}\ g^{2/5}\ R^{1/2}}{Q^{1/20}}$		- Fundamental equation
7.Inglis (1941)		
(i) $V = \dfrac{\alpha_3\ g^{7/18}\ Q^{1/6}(C\ \omega\ d)^{1/12}}{v^{1/36}}$		(i) - Fundamental equation
(ii) $B = \dfrac{\alpha_1\ Q^{1/2}(C\ \omega)^{1/4}}{g^{1/3}\ v^{1/12}\ d^{1/4}}$		(ii) - Fundamental equation
(iii) $A = \dfrac{\alpha_2\ v^{1/36} Q^{5/6}}{g^{7/18}\ (C\ \omega\ d)^{1/12}}$		(iii)- Fundamental equation
(iv) $S = \dfrac{\alpha_5\ (C\ \omega\ d)^{5/12}}{v^{5/36}\ g^{1/18}\ Q^{1/6}}$		(iv) - Fundamental equation
(v) $D = \dfrac{\alpha_4\ v^{1/9}\ Q^{1/3} d^{1/6}}{g^{1/18}\ (C\ \omega)^{1/3}}$		(v) - Fundamental equation
(vi) $\dfrac{B}{D} = \dfrac{\alpha_5 Q^{1/6}(C\ \omega)^{7/12}}{g^{5/18} v^{7/36} d^{5/12}}$		(vi) - Fundamental equation
		- Sediment charge has a small effect on the area of a channel, relatively great effect on slope and shape and considerable effect on channel width
		- α_1, α_2, α_3, α_4 and α_5 are Coefficients, the values is about 1/400 for Uniform sand originally used in the experiment and probably about 1/233 for natural river bed sand.
(vii) $V = (Fb\ Fs\ Q)^{1/6}$		(vii) - Fundamental equation
(viii) $S = \dfrac{Fb^{5/6}\ Fs^{1/12}\ v^{1/4}}{3.63\ (1+aC)\ g\ Q^{1/6}}$		(viii)- Fundamental equation
(ix) $B = \left(\dfrac{Fb\ Q}{Fs}\right)^{1/2}$		(ix) - Fundamental equation

Contd...

Table 7.11. Contd...

1	2	3
(x) $D = \left(\frac{FsQ}{Fb^2}\right)^{1/3}$		(x) - Fundamental equation - Fb defining bed sediment factor or bed factor, is related to erosive action on bed
(xi) $Fb = \frac{V^2}{D}$		- Fs defining side factor, is related to erosive action on sides. Values for side factor in design are as follow:
(xii) $Fs = \frac{V^3}{B}$		- Fs = 0.1 for bank materials of slight cohesiveness - Fs = 0.2 for bank materials of medium cohesiveness - Fs = 0.3 for bank materials of high cohesiveness - $Fs = Fb^2/8$ for rounded gravel
8. Manning (1889 ,1936)		
$V = \frac{1.49}{N} R^{2/3} S^{1/2}$	$V = \frac{1}{N} R^{2/3} S^{1/2}$	- Fundamental equation - N is roughness coefficient and depends on many factors shown in para 2.1 and Table 2.14 - This formula was developed from seven different formula viz. Chezy, Ganguillet and Kutter, Bazin, Powell and others. Based on Bazin's experimental data, and further verified by 170 observations. Owing to its simplicity of form and to the satisfactory results it lends to practical applications. The Manning formula has become the most widely used of all uniform flow formulas for open channel computation.

(iii) Channels, scouring and silting can both occur under suitable conditions. Canals dug through soil formed of sediment same or similar to that transported by the canal waters are of the third type. Alluvial canals in the Indo-Gangetic plains are typical illustrations.

Kennedy (1895) defined a channel in regime as one in which the bed material and silty side berms are self derived and long continued, and do not change. Channel is stable and requires no maintenance.

Lane (1952), presented, an excellent definition of stable or regime channels as under-

'A stable channel is unlined earth canal for carrying water, the banks and bed of which are not scoured by the moving water and in which objectionable deposits of sediment do not occur.'

Minor seasonal adjustments by scour and deposits are not ruled out but these should compensate each other on a yearly basis.

The word regime is often used to be synonymous with word stable as has been done by Lane. This may confuse a reader, since expressions such as high flood regime, low river regime or even changing regime of a river are also common. The word regime in these expressions means merely a set of prevalent conditions. Thus Lindley (explained) that when an artificial channel is used to convey sediment laden water, both bed and banks scour or fill, changing depth, gradient and width until a state of balance is attained at which a channel is said to be in regime. That means canal in regime means a stable canal but canal regime only refers to canal conditions which may or may not be stable.

According to Lindley (1919), dimensions of a canal in regime carrying a given discharge and sediment load are uniquely fixed by nature. Lacey commented that this concept of regime postulates too delicate and precise a dynamic balance in all the dimensions which can very seldom be achieved. For such regime to be established, the fundamental requirements are that discharge should be constant, the channel flows uniformly in unlimited incoherent alluvium of the same character as that transported and the sediment grade and charge are also constant. In practice such conditions are rarely possible.

The term regime was used by Lacey (1930 - 58, also see Para 7.13) to mean stability as was done by Lane. He however made a fine distinction between these two. According to him-

(i) A stable channel is one that is active (meaning having mobile bed) and neither silting nor scouring. Stability may be instantaneous or the average of a complete cycle of events.

(ii) A regime channel is a stable channel transporting a regime sediment charge.

(iii) A regime sediment charge is the minimum transported sediment load consistent with a fully active bed. Full activity is such that any reduction would lead to effective partial rigidity and at the limit to complete rigidity and immobility of bed. Also see Para 7.14.1.

Inglis (1941) observed that regime sediment charge had really little significance and it was in fact merely a yard-stick. Real meaning and significance of Lacey equations was lucidly explained by him. Under uniform flow conditions (which imply constant discharge, temperature and silt charge) there is a natural regime slope and section towards which a channel tends to develop and to which it will eventually attain, either by accretion, if the slope be too flat or the section too large, or by scouring, if the channel be too small or the slope too steep-assuming the channel to be flowing in material which can scour freely.

The theory also implies that for any particular discharge and mean diameter of bed sand, the sediment charge is normal; that is to say, the quantity of bed material and its size distribution is normal.

In practice, water temperatures vary throughout the year, the variation being much greater in some parts of the country than in others and also greater in small than large channels. Similarly, silt and sand charge, entering the canals and hence the bed-material size distribution varies widely at different times of year. Hence it must be clearly understood that average values for his data, were subject to variation from regime at different times of the year in spite of which equilibrium conditions were closely approached on an annual and long term basis. Also see Para 7.17

Channels which do not alter appreciably from year to year-though they may vary during the year-were therefore said to be in regime.

It would appear that-

(i) The expression 'channel in regime' has been used to mean's stable channel' which does not change appreciably from year to year though seasonal variations may occur.

(ii) Lacey regime represents a special case of equilibrium or stable conditions, in which, charge in movement is 'the regime silt charge' as defined by Lacey.

It will be evident as to why it is often said that Lacey channels are regime or stable channels but not all stable channels are Lacey channels.

7.12 Kennedy's Silt Theory (1894-95): (Also Known As Critical Velocity Theory or Concept)

Kennedy collected data in the year 1890 in the Upper Bari Doab Canal systems, one of the oldest in the Punjab constructed around the year 1850. The reaches were from 90 miles length of lower parts of the Upper Bari Doab canal system. These reaches were selected because it had achieved stability and required no maintenance.

Canal dimensions and discharge in Kennedy's observations varied between the following ranges.

Q in cusecs	Bed width (ft)	Depth of flow at full supply (ft)	Mean velocity (ft/sec)
26 to 1250	8 to 70	2.2 to 7.0	1.30 to 2.90

The form of cross-section was nearly rectangular, the sides being near vertical and of fine sediment and bed horizontal and of coarse sand.

Observations were made at thirty sites on the channels of the canal system in about the middle reach. He selected channels which had not been silt-cleared for the past 30 years and had become stable. These channels were neither silting nor scouring. Mean velocity was derived from known full supply discharges and measured cross-sectional area of the channel. Regulation was such that each distributory as far as possible always discharged with its full supply.

This data when plotted showed that width had no effect on velocity but the mean velocity V_0 was a function of depth D alone, see figure 7.11(i), the equation being.

$V_0 = 0.84\, D^{0.64}$ (FPS Units) $= CD^n$ or $V_0 = 0.55\, D^{0.64}$ (Metric Units) Eq. 7.12.1

Where,

V_0 = Critical velocity in m/sec (non-silting and Non-scouring)

D = Depth of flow in meters; m = Critical velocity ratio; C and n = Constants

The above formula is without any reference to the silt which the water carries in the earthen irrigation channel sections.

Sediment in the flowing canal is kept in suspension solely by the vertical component of the constant eddies, which can always be observed over the full width in any stream boiling up gently to the surface. These eddies rise on account of the roughness of the bed and work up against the depth of the channel. From the sides also, some such eddies may occur to a much smaller degree but any such eddies must be for the greater part horizontal and of no silt supporting power. The silt supporting power in a stream is, therefore proportional to the width of the stream and not the total wetted perimeter. *According to Kennedy, a Regime Channel is that which neither silts up nor scours.*

This equation is applicable to channels flowing in sandy silt of the same grade as that which exists on the Upper Bari Doab canal system on which the observations were made. Kennedy later recognized that the grade of silt played a part in this relationship and introduced another factor in the equation, called by him as "the critical velocity ratio" and given the symbol `m'. The equation was then written as

$V = 0.84\ m\ D^{0.64}$ (FPS Units) ...... Eq. 7.12.2

Kennedy made no correlation between water surface slope of regime channels and the mean velocity or the vertical depth. He relied on Kutter's equation to give slopes to the channels.

Soon after Kennedy's work, research was carried out in many countries and various investigators gave different values to the coefficient C and exponent in to suit the condition in their country, as given in table 7.12.

Mean or the critical velocity

Kennedy defined critical velocity as the mean velocity which will just keep the channel free from either silting or scouring. This is denoted by the letter V_0.

This expression gave the non-silting velocities for channels, which had silt of the same character as in the observed channels of Kennedy on the Upper Bari Doab Canal. But if the channels had silt of different size and grade, they run non-silting with a velocity different from V_0 (mean velocity).

C.V.R. (critical velocity ratio)- Critical velocity ratio is the ratio of he actual velocity in a channel to the critical velocity calculated according to the Kennedy's V_0 formula:

$V_0 = m.C.D^n$ .... Eq. 7.12.3

C.V.R. represents a factor, which does not reflect departure from regime, but a factor, which is a measure of the variation in the silt conditions from the standard silt conditions on Upper Bari Doab Canal in the Kennedy's observations. It is also a factor, which takes into account the scale effect of the dimensions of the channels. Kennedy considered the question of bank erosion, quite distinct from the silt question, though of undoubted importance in fixing the limiting depth and the safe limiting velocity of channel. He considered the safe velocity in the Punjab soil to guard against side erosion as 3.5 feet per second. This would mean a depth not more than 10.0 feet.

Silt Transportive power of channel

Kennedy found expression for silt transportive power in a channel as below:

Let it vary as V_0^n then Total silt carried = $k.B.V_0^n$

k = a factor; B = width; V_0 = non silting velocity

Let, p = a percentage of silt by weight in water and Q = Discharge. In stable channel all silt is carried away,

$\therefore k.B.V_0^n = pQ$; and since in rectangular channels, $Q = B.D.V_0$

$k.B.V_0^n = p.B.D.V_0$ or $V_0^{n-1} = (D)p/k$

$$\text{or } V_0 = \left(\frac{p}{K}\right)^{\frac{1}{(n-1)}}.D^{\frac{1}{(n-1)}}$$

This is of the same form as Kennedy's $V_0 = C.D^n$

Equating the values of index of D in the Kennedy's Equation 7.12.1; $0.64 = 1/(n-1)$ or $0.64(n-1) = 1$ or $n = 1.64/0.64 = 2.56$

Therefore n = 2.56 or Say 2.5

$\therefore$ The silt transportive power varies as $V_0^{2.5}$ .... Eq. 7.12.4

All canals lined/unlined carrying silt laden water higher than transporting capacity have adjustable shape by silt deposit. Sides, berms and bed get adjusted.

This relation is similar to Eq 7.2 (ii) Para 7.6.1

Table 7.12: Values of C and n in Kennedy Equation for non silting. Non-Scouring Velocities given by various investigators

S. No.	Factors Applying to Kennedy Eq. In English Units		Locality	Authority	Limit	Reference
	Coefficient C	Exponent n				
(1)	(2)	(3)	(3)	(4)	(5)	(6)
1.	0.381	0.64	Egypt	Buckley	Lower	Irrig. Dept. of Egypt.
2.	0.46	0.64	Egypt	Buckley	Upper	"Irrigation Practice," P. 207
3.	0.39	2/3	Egypt	Molesworth &	Lower	Minutes of Proceedings, Inst. C.E.
4.	0.475	2/3	Egypt	Yenidunia	Upper	P. 260 Vol. 229, Also P. 285.
5.	0.391	0.727	Egypt	K.D. Ghaleb	...	Vol. 223.
6.	0.56	0.64	Egypt	...	...	U.S. Dept. of Agri. Tech. Bulletin No. 67 P. 44
7.	0.38	0.64	Mozaffargorh	G.W. Duthy	Lower	Proceeding, Punjab Eng. Congress
			D. Punjab, India	G.W. Duthy	Upper	P.P. 44 and 48, 1919
8.	0.63	0.64	Sind...	F.W. Woods	...	Engineer P. 648. Vol 40 Parker "control of Water"
	0.63	0.64				P. 678
9.	0.67	0.55	Godavari Western delta Madras	Kennedy	...	Minutes of Proceedings Inst. C.E.P. 260 Vol. 229 Madras Publi9c Works Dept. Oct. 9, 1 912 Dist. 1872
10.	1.01	0.44	Rio Hegro Argentina	R.E. Ballester	...	Muniutes of Proceedings Inst. C.E.P. 280 Vol. 223
11.	0.52	0.66	Siam.......		...	U.S. Dept. of Agri. Technical Bulletin No. 67 P. 44.
12.	0.93	0.52	Madras (Kistna)	Kennedy	...	Minutes of Proceed ings, Inst. C.E.P. 260 Vol. 229
13.	0.67	0.64	Sutlej India	F.W. Woods	...	Engineer, June 17, 1927 P. 648
14	0.91	0.57	Burma (Shwebo)	Kennedy	...	Minutes of Proceedings Inst. C.E.P. 260 Vol. 229 (1929 -1930) Parker "Control of Weter"
15.	0.95	0.57	Chenab, Punjab	Lindley	...	Minutes of Proceedings Inst. C.E.P 260 Vol. 229 Punjab Eng. Congress Proceedings. 1919
16.	0.756	0.64	Sirhind Punjab	W.B. Harwey	...	Proceedings Punjab Eng. Congress 1919 P. 58.
17.	0.84	0.64	Bari Doab	Kennedy	...	Hydraulic Diagrams Kennedy, P.W.D India 1917.
18.	0.924	0.64	Penner R iver	J.M. Lacey	Lower	Minutes of Proceedings Inst. C.E.P. 333. Vol. 229
	0.924	064	Cauvery Delta	J.M. Lacey	Lower	
19.	1.09	0.64	Penner River	J.M. Lacey	Upper	
20.	0.966	0.64	Cauvery Delta	J.M. Lacey	Upper	
21.	0.98	0.64	Imperial Valley	Rothery	...	Minutes of Proceedings Inst. C.E.P. 179 Vol. CCXVI
22.	1.26	0.64	Cauvery Delta	J.M. Lacey	Extreme	Minutes of Proceedings Inst. C.E.P. 333. Vol. 229
23.	1.33	0.61	Imperial Valley	Collings	Lower	Transactions AM. Soc. C.E. Vol. 99.
24.	1.83	0.61	Imperial Valley	Collings	Upper	(1934) P. 549
25.	0.42	0.64	Behera Delta	R.G. Kinder	...	Proceedings Punjab Eng. Congress 1919 P. 74 J.
26.	**	**	Jamrao Sind	W.L.C. Trench	...	Minutes of Proceedings Inst. C.E.P. 307. Vol.223, 1926-27

Kennedy's Diagrams (Revised 1907)

He took simply the Kutter's formula and gave the value of N (coefficient of roughness) equal to .0225 as the average value for all irrigation channels. He designed the hydraulic diagrams known as Kennedy's Diagrams, printed at the Thompson College (IIT in the year, 2002) Press at Roorkee, India. He gave diagrams for different slopes showing discharge, depth, velocity and C.V.R.

He also proposed for stable channels, minimum longitudinal bed slopes, velocity of flow for various discharges and full supply depth as given in table 7.13.

Table 7.13 Kennedy's minimum longitudinal slopes and velocities of flow in channels, required to prevent silting for given discharges and for various full supply depths

Discharge in Cubic feet/sec	For Maximum Probable Depths			For Moderate Depths			For Minimum Probable Depths		
	Depth in feet D	Minimum mean Velocity V_0 ft/s	Minimum fall per 1,000 S	Depth in feet D	Minimum mean Velocity V_0 ft/s	Minimum fall per 1,000 S	Depth in feet D	Minimum mean Velocity V_0 ft/s	Minimum fall per 1,000 S
(1)	(2)	(3)	(4)	(5)	(6)	(7)	(8)	(9)	(10)
10	2.1	1.3	0.5	2	1.3	0.48	1.8	1.21	0.43
25	2.5	1.51	0.43	2.4	1.46	0.37	2.2	1.39	0.34
35	2.8	1.63	0.39	2.6	1.55	0.34	2.4	1.46	0.31
50	3.1	1.74	0.34	2.9	1.66	0.31	2.6	1.55	0.29
75	3.4	1.85	0.32	3.1	1.74	092	2.9	1.66	0.27
100	3.7	1.94	0.31	3.3	1.82	0.28	3	1.7	0.26
125	4	2.04	0.3	3.6	1.91	0.27	3.2	1.77	0.26
150	4.2	2.1	0.29	3.7	1.95	0.26	3.4	1.85	0.25
175	4.3	2.13	0.28	3.9	2.01	0.25	3.5	1.88	0.24
200	4.5	2.2	0.27	4	2.04	0.24	3.6	1.91	0.23
250	4.8	2.26	0.26	4.2	2.1	0.24	3.8	1.98	0.23
300	5	2.35	0.26	4.5	2.2	0.23	4	2.04	0.22
350	5.3	2.41	0.25	4.6	2.23	0.23	4.2	2.1	0.22
400	5.5	2.5	0.25	4.8	2.29	0.22	4.3	2.13	0.22
450	5.7	2.56	0.25	5	2.35	0.22	4.4	2.17	0.21
500	5.8	2.62	0.24	5.1	2.38	0.22	4.6	2.23	0.21
600	6.1	2.66	0.24	5.3	2.44	0.21	4.8	2.29	0.21
700	6.3	2.73	0.23	5.5	2.5	0.21	5	2.35	0.2
800	6.6	2.8	0.23	5.8	2.62	0.21	5.2	2.41	0.2
900	6.8	2.86	0.22	5.9	2.61	0.2	5.3	2.44	0.19
1000	7	2.92	0.22	6	2.64	0.2	5.4	2.47	0.19
1500	7.8	3.12	0.21	6.8	2.85	0.2	6.1	2.66	0.19
2000	8.5	3.31	0.20	7.3	3.00	0.19	6.6	2.8	0.18

It was soon found out that the value of N in Kutter's formula was not constant. Kennedy himself suggested the value of .02 for large canals. Similarly in the case of small irrigation channels, the value of N has been found to be .025.

Sharma (1936) worked out the value of N in the Manning's formula for the earthen irrigation channels and found to vary from .018 to .028.

Garret's Diagrams

In practice, the design can be carried out with the help of Garret's diagrams. It provides a graphical solution of Kennedy's and Kutter's equations. The procedure for their use is as follows:

(i) Follow the given discharge curve till it meets the given horizontal slope line and mark off their point of inter-sections.

(ii) Draw a vertical line through the point of intersection. This will intersect (cross) several bed width curves.

(iii) Choose the point of intersection of the vertical line on any bed width and read the corresponding value of water depth (D), and critical velocity V_0, on the right hand ordinate.

(iv) Calculate the actual velocity of flow (V) for the chosen bed width and the corresponding value of the water depth. Determine the critical velocity ratio `m' (V/V_0) taking V as calculated and V0 as read.

(v) If the value of V/V_0 is not the same as given, repeat the procedure with other values of the bed width till the C.V.R. approximates to the given value.

These diagrams were drawn for side slopes of ½:1 on the assumption that irrigation channels, after some time acquire this slope due to silt deposition on the berms, irrespective of the initial side slope provided. Please note that this side slope of 0.5: 1(V: H) is very near to the optimum side slope of 0.577: 1 (V: H) by Manning's formula. These curves are for N = 0.025. These can also be utilized for any other value of N with the help of a monogram provided at the top of the curves. When using for any other value of N, the vertical line drawn at the intersection of the discharge and slope line has to be shifted, either to the left or to the right side to the extent given in the monogram. Further procedure for determining bed width and depth remains unaltered.

7.13 Lacey's Silt Theory (1930-1958)

The so called Regime theory was progressively developed by Lacey in a number of papers written over a period from 1930 to 1958. Earlier formulae were changed in the last paper of 1958. Summary of all the formulae in various papers is given in table 7.14.

(i) First Paper (1930)

In his first paper of 1930 Lacey introduced a set of simple formulae which uniquely fixed the size, shape and slope of a stable alluvial channel.

He started with the examination of Kennedy equation $V_0 = 0.84\ d^{0.64}$ (FPS units). He replotted Kennedy's, Lindley's and Madras data in terms of V to R (R being hydraulic mean radius).

Critical mean velocity was known to be dependent also on characteristics of bed material. Bed sand was classified by critical velocity ratio V/Vo that is, the ratio of regime velocity actually obtaining with a given depth to the velocity for the same depth that Kennedy formula would indicate. The difference in actual bed material from the one for which Kennedy's formula was applicable resulted in variable coefficient in Kennedy's type formulae. To get over this difficulty, Lacey introduced 'f' a silt factor defined as $V/V_0 = f^{0.5}$ in the V to R formula, and obtained $V_0 = C\sqrt{fR}$. Plot of regime canal data as in figure 7.9 (iii) resulted equation.

$V = 1.17\ \sqrt{fR}$ ..... Eq. 7.13.1 Fundamental Eq. FPS units. In which both the coefficient 1.17 and exponent 0.5 were constant.

For design of channel dimension, second formula would be necessary giving either sectional area or wetted perimeter. Lacey choose to evolve a relationship between velocity and area. For this it was assumed that any two cross- sections in a regime canal system would be completely geometrically similar as shown in figure 7.9 (i) and 7.9 (ii).

abc represents one section while abc represents another stable channel with the same velocity but a different silt factor 'f '. By virtue of equation 7.13.1, f'R'=fR. Law governing wetted perimeters, P was considered to be the same as one governing R.

fP = f'P'

When mean velocities were constant these two identities gave
$Af^2 = A' f'^2 = A'' f''^2$ etc.

In other words Af^2 was dependent only on the value of mean velocity and therefore $Af^2 = f(V)$. Actual relationship found by Lacey vide figure 7.9 (iv) is his second regime equation as under:

$Af^2 = 3.8\ V_0^5$ ..... Eq. 7.13.2 Fundamental Eq., FPS units

The wetted perimeter P being equal to A/R, it could be obtained directly from equations 7.13.1 and 7.13.2. Thus

$P = \frac{A}{R} = \frac{3.8V^5}{f^2} x \frac{1.17^2 f}{V^2} = 5.2\frac{V^3}{f}$ ..... Eq. 7.13.3 derived from 7.13.1 and 7.13.2, FPS units

This was also transformed to introduce Q by using equation 7.13.2

$Af^2 = 3.8\ V^5$. Therefore $Q\ f^2 = 3.8\ V^6$ or $Q = 3.8\ \frac{V^6}{f^2}$

or $Q^{1/2} = 1.95\ \frac{V^3}{f}$ ..... Eq. 7.13.4, FPS Units

Substituting from Eq. 7.13.3 in Eq. 7.13.4

$P = 2.668\ Q^{0.5}$ ..... Eq. 7.13.5 FPS units

Important deduction from equation 7.13.5 was 'that for a given discharge, the wetted perimeter of a stable channel was constant and independent of fineness of the silt transported, the silt merely controlling the shape figure 7.9 would serve to clarify implication of this statement. Different sections in figure 7.9 have same P and pertain to same Q but have different sediment sizes. Field observations showed that artificial channels flowing on and within banks composed of sediment of the same kind as the bed tend to assume elliptical cross- section. Coarser the material flatter the semi ellipse.

As a proof of validity of equation 7.13.5, plot of P against Q was prepared by Lacey as in figure 7.9 (v) in which similar curves due to other investigators were also drawn for comparison.

Lacey contended that equation 7.13.5 was applicable also to rivers and showed that the relation W = 2.67 Q0.5 fitted observed data fairly well.

After fixing the dimensions of an alluvial canal by formulae 7.13.1, and 7.13.5, a third formula to determine requisite slope was introduced. For this purpose Lacey originally adopted Manning's equation.

$V = \frac{1.486}{N}\ R^{2/3}\ S^{1/2}$ FPS units.

A direct relationship between N and f was envisaged. An approximate equation given by him was;
$N = 0.22\ f^{0.2}$ ..... Eq. 7.13.6

Substitution for N in terms of f and for V and R yielded.

$S = \frac{f^{1.5}}{2587Q^{1/9}}$ Or $S = \frac{0.000387f^{3/2}}{Q^{1/9}}$ ..... Eq. 7.13..7 FPS units.

The silt factor f was also correlated with size of material D. A rough qualitative formula was given as $D = f^{2/64}$; Eq. 7.13.8 FPS units;

Where D = diameter in inches of the predominant type of material transported.

Table 7.14 Lacey's Formulae, See Para 7.13

S. No.	Formula	In I paper (1930)	In II paper (1935)	In C.B.I. Publication No. 20 (1939)	In III paper (1946)	In IV paper (1958)
(1)	(2)	(3)	(4)	(5)	(6)	(7)
1.	V, R, S relation Fundamental Eq.	$V = 16 R^{2/3} S^{1/2}$ - FPS $V = 10.8 R^{2/3} S^{1/2}$ Metric				$V = 16.046(d_m RS)^{1/3}$ - FPS d_m = mean depth
2.	V, f, R relation Fundamental Eq.	$V = 1.17\sqrt{fR}$ - FPS	$V = 1.15\sqrt{fR}$ or $V = 2\sqrt{fR}/\sqrt{3}$ - FPS $V = 0.639\sqrt{fR}$ - Metric			$V = 1.15\sqrt{fd_m}$ - FPS d_m = mean depth
3.	A, f, V relation Fundamental Eq.	$Af^2 = 3.8 V^5$-FPS	$Af^2 = 4V^5$ - FPS $Af^2 = 140.0V^5$ - Metric			
4.	V, R, S, N relation Fundamental Eq.	$V = \frac{1.3458R^{3/4}S^{1/2}}{N}$ - FPS $V = \frac{1}{N_a} R^{3/4} S^{1/2}$ - Metric	$V = \frac{1.3458ER^{3/4}S^{1/2}}{N}$ (FPS) E = a shock factor	$V = \frac{1.3458}{N} R^{3/4} (S-S')^{1/2}$ (FPS) S′ = fraction of S lost in shock.		$V = \frac{1.346}{N} d_m^{1/4} R^{1/4} S^{1/2}$ E = 1.10 was assumed d_m = mean depth
5.	N, f relation, Fundamental Eq.	$N = 0.0225 f^{0.25}$ – FPS $N = 0.22 f^{0.20}$ – FPS				
6.	f, m relation Fundamental Eq.	$D = f^2/64$ or $f = 8\sqrt{D}$ - FPS D= diameter of particle in inches	$fr = 1.76m_r^{1/2}$ m = diameter of particle in mm			$fr = 1.60m^{1/2}$ subscript 'r' denotes regime conditions. m = diameter of particle in mm
7.	S, f, R, Relation	$S = 0.000383 \frac{f^{3/2}}{R^{1/2}}$ - FPS	$S = 0.000376 \frac{f^{3/2}}{R^{1/2}}$ - FPS			

Contd...

(1)	(2)	(3)	(4)	(5)	(6)	(7)
	Derived by raising the power of Eq. (1) & (2) by (3) and then eliminating V^3		or $f = 192R^{1/3}S^{1/2}$ - FPS or $f = 287R^{1/3}S^{1/2}$ - Metric			
8.	P, Q relation Derived by raising the power of Eq. 2 by 4 and then eliminating f^2 between this and Eq. 3. Then multiplying both sides by A and replacing V by Q/A, & A/R by P.	$P = 2.67\sqrt{Q}$ - FPS $P = 4.75\sqrt{Q}$ - Metric				$W_s = 2.67\sqrt{Q}$ -FPS
9.	(i) S, f, Q relation, Derived from Eq. 2 (ii) S, f, q relation With approximation of P = B and q = Q/B for very wide and low depth channels.	$S = 0.00056\dfrac{f^{5/3}}{Q^{1/6}}$ FPS $S = 0.000178\dfrac{f^{5/3}}{q^{1/3}}$	$S=0.000547\dfrac{f^{5/3}}{Q^{1/6}}$ or $S = \dfrac{f^{5/3}}{1830Q^{1/6}}$ -FPS $S = \dfrac{f^{1/3}}{3340Q^{1/6}}$ Metric Derived from Eq. 7 and 8			
10.	(i) V, f, Q relation (ii) A, f, Q, relation Derived from Eq. 3	$V = \left(\dfrac{Qf^2}{140}\right)^{1/6}$ - Metric $A = 1.199\dfrac{Q^{5/6}}{f^{1/3}}$ - FPS	$A = 1.26\dfrac{Q^{5/6}}{f^{1/3}}$ - FPS $A = \dfrac{2.279Q^{5/6}}{f^{1/3}}$ - Metric			$V = 0.7937Q^{1/6}f^{1/3}$ - FPS $V = 0.4389Q^{1/6}f^{1/3}$ - Metric

Contd...

(1)	(2)	(3)	(4)	(5)	(6)	(7)
11.	Regime scour	$R = 0.4725\left(\frac{Q}{f}\right)^{1/3}$ - FPS $R = 0.47\left(\frac{Q}{f}\right)^{1/3}$ - Metric R = maximum scour below water surface				Holds good for wide and approx. rectangular channels only. Derived from equation 2 and 9. This is to be used when looseness factor* is more than one (reference : IS.6966)
12.	Regime scour	$R = 1.35\left(\frac{Q}{f}\right)^{1/3}$ - Metric $R = 0.91\left(\frac{q^2}{f}\right)^{1/3}$ R = maximum scour below water surface				Derived from equation 2 and 9. This is to be used When looseness factor is less than one.

Notes: Lacey equations are empirical; Range over which they are applicable is important; Lacey himself gave this range as Q from 5 to 500 cfs; m from 0.15 to 0.4; Bed load small.

Ws to Q relation gives minimum stable width of a sandy bedded river during high flood valid up to discharges not exceeding 2.5 million cusecs. In addition, bed material ought to be in cohesive and bed form to comprise of ripples and not waves.

Before adopting his formulae for general application, Lacey cautioned that it would be necessary to determine values for constants by making observations on existing stable channels.

**Looseness factor:* The ratio of the overall length of weir/barrage provided to be theoretically computed minimum stable width of river, at the design flood obtained by using Lacey's equation.

Silt factor 'f': This can be calculated by the average particle size m_r in the soil using the relationship $f = 1.76\sqrt{m_r}$. It may be checked from Lacey's relationship $f = 290\,(R^{1/2})^{2/3}]$.

Types of soil	Value of f	Type of soil	Value of f
Fine silt	0.5 to 0.7	Medium sand	1.25
Medium silt	0.85	Coarse sand	1.50
Standard silt	1.0		

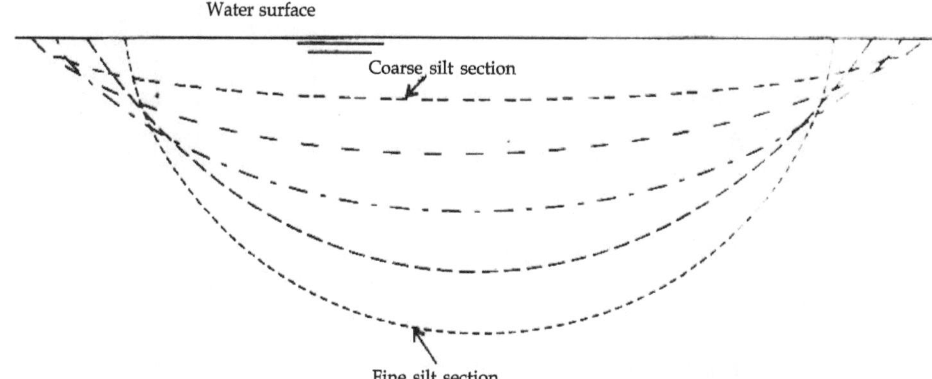

Figure 7.9 (i) Lacey's concept - Effect of sediment size on shape of section

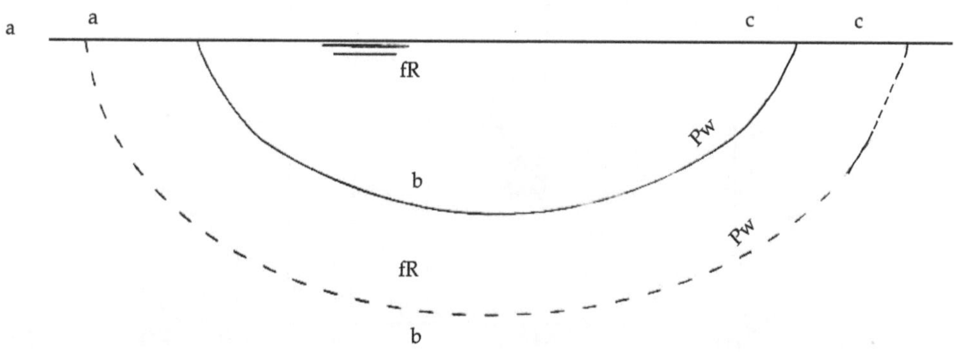

Note :

abc - Stable channel section

abc - Another channel with same mean velocity but with smaller silt factor

Figure 7.9 (ii) Lacey's stable sections - Lacey's concept of shape due to effect of silt factor on stable channel section

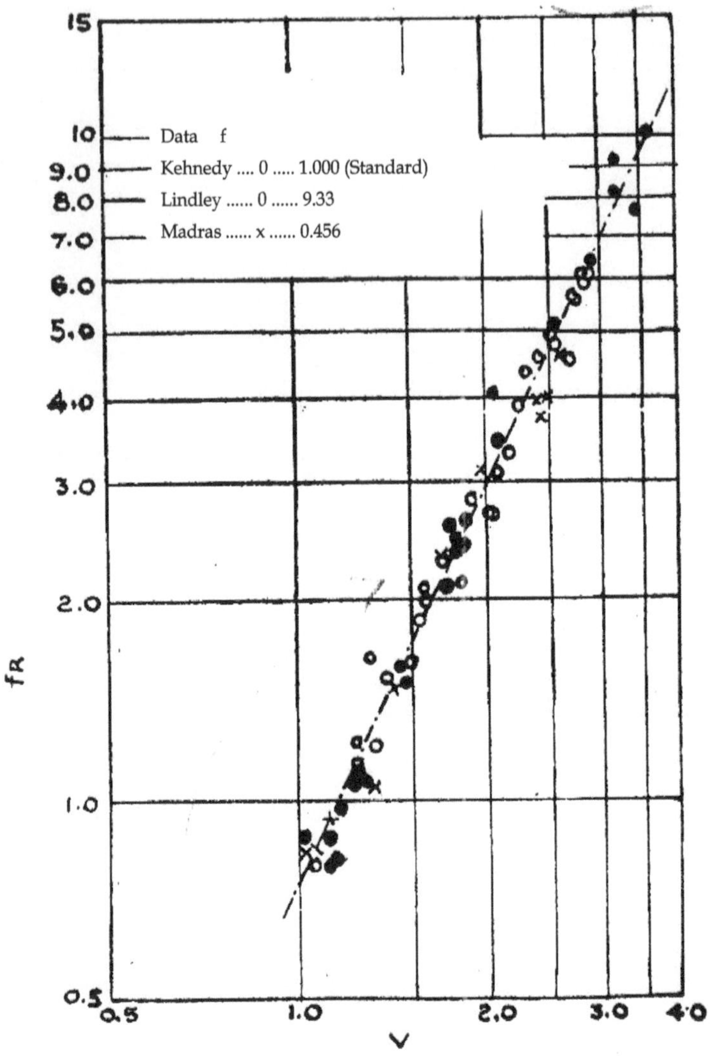

Data f

Kehnedy .... 0 ..... 1.000 (Standard)

Lindley ...... 0 ...... 9.33

Madras ...... x 0.456

fR

V

Figure 7.9 (iii) Relation between Velocity, silt factor f, and R (V= 1.17 √fR ; Lacey, 1930)

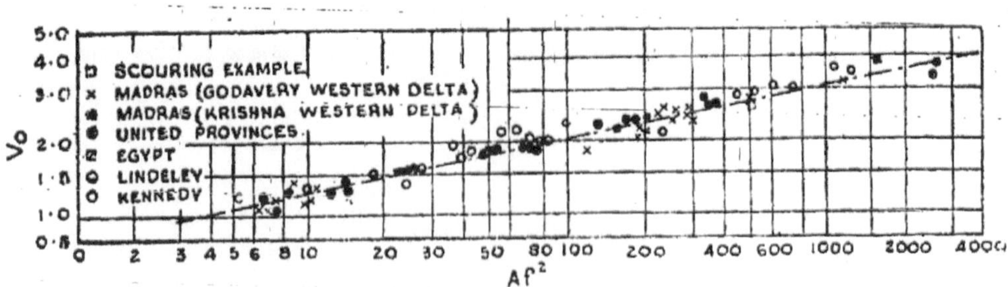

Figure 7.9 (iv) Relation between V0, A, and f (Af2=3.8V05 ; Lacey, 1930)

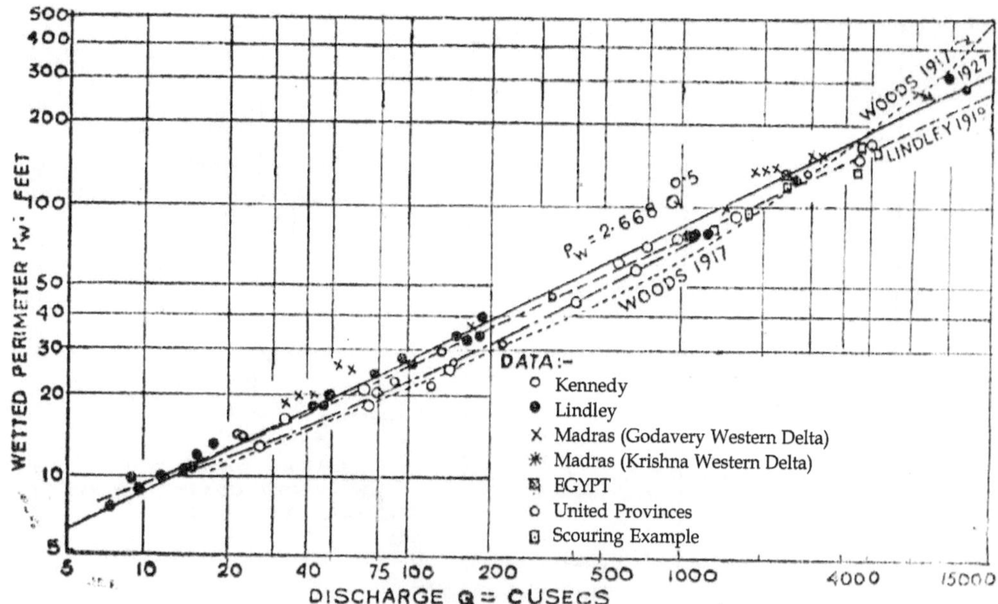

Figure 7.9 (v) Correlation between P and Q (Lacey 1930)

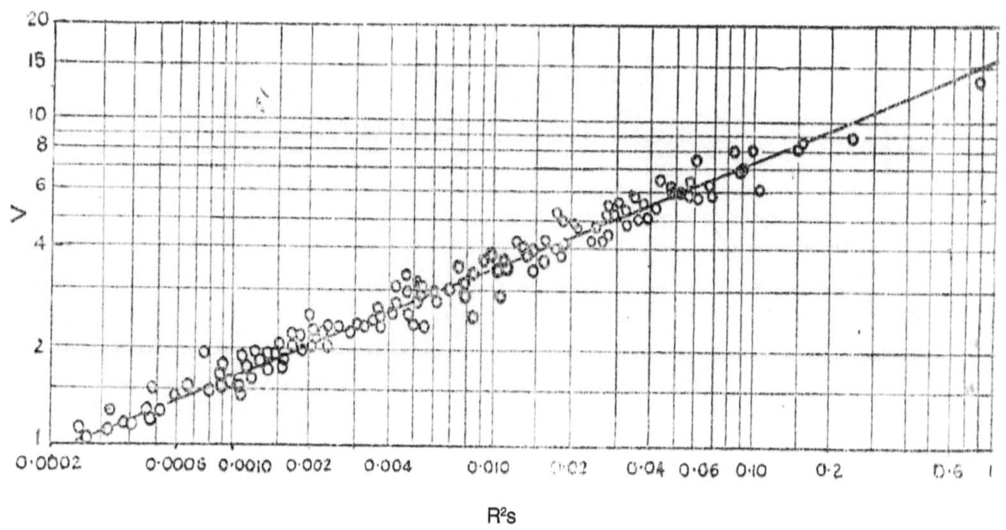

Figure 7.9 (vi) Correlation between V and R2S (Lacey's, 1935)

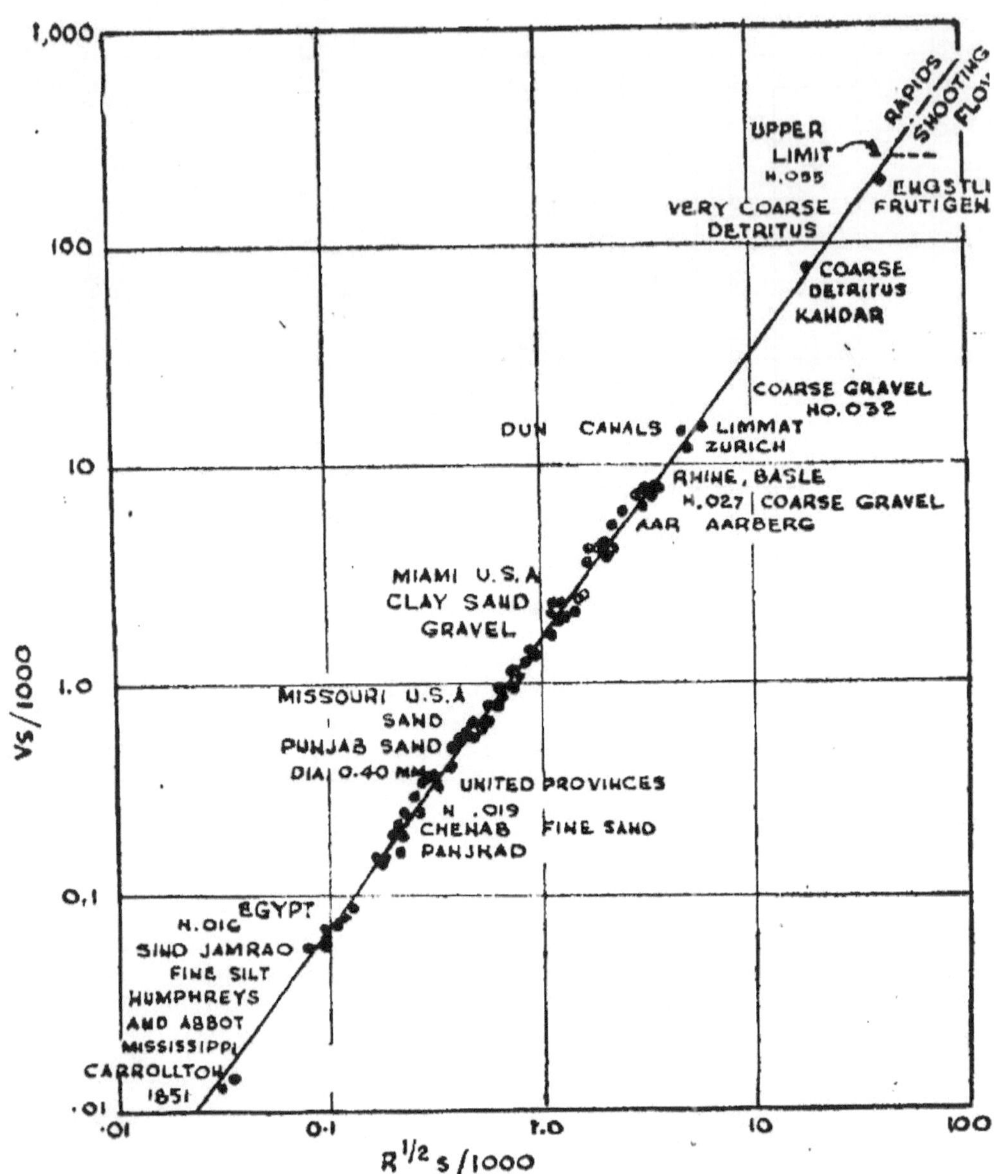

Figure 7.9 (vii) Correlation between V, S, and R (Lacey, 1946)

Second Paper (1935)

Object of the second paper was to develop a general theory of flow applicable to channels, both rivers and canals, flowing uniformly in incoherent alluvium.

To start with, using Kennedy, Madras and Lindley data of stable canal, Lacey showed that with constant bed material size (i) mean velocity varies directly as the square root of hydraulic mean depth i.e. $V \alpha \sqrt{R}$ Or $V = C' R^{0.5}$ and

(ii) The slope varies inversely as the square root of the hydraulic mean depth i.e. $S \alpha \frac{1}{\sqrt{R}}$ or $SR^{0.5}$ = constant. From these two relations it was deduced that, the general expression for flow with constant silt grade should be

$V \alpha R^{1/2} (R^{1/2}S)^n$ Assuming n = 0.5

Or, $V = KR^{3/4} S^{1/2}$ ..... Eq. 7.13.9

Coefficients C′ and K depend on the bed material size and were obviously interrelated. In order to determine their relation, data of characteristic channels including regime canals as well as rivers and covering a wide range of values of C′ and K were choosen. This indicated that K varied inversely as the square root of C′ or $KC'^{1/2}$ = constant. From the three identities viz. $V = C'R^{1/2}$, $V = KR^{3/4} S^{1/2}$ and $KC'^{1/2}$ = constant, it was possible to eliminate both C' and K and obtain

$V = K'R^{2/3} S^{1/3}$ Eq. 7.13.10; FPS units. K′ another constant

Lacey observed that effectively size of material was implicit and introduction of separate rugosity coefficient as in Kutter and Manning, formulae or of silt factor as in his V to R relation was really not necessary when the general formula $V = K'R^{2/3} S^{1/3}$ was used.

Third Paper (1946)

Third Paper provide physical basis for the general regime flow formula

$V = 16 R^{2/3} S^{1/3}$ .... Eq. 7.13.11; FPS units, previously developed by him. It was also shown that the equation applied to regime and non regime channels alike and that discordant data could be reconciled when the variables silt charge and 'shock' were taken into account.

Fourth Paper (1958)

On re-examining his choice of P and R an objection became evident. Lacey equations in terms of P and R involved a limiting semi-circular cross-section as shown in figure 3.9 and a limiting velocity of 0.88 ft/sec for any type of material. Originally it was believed that below this limiting velocity, turbulent flow in alluvial channel could not be maintained. Subsequent evidence on models, however, showed this contention to be incorrect. Lacey, therefore, now considered that as far as cross-section was concerned, the correct parameters to employ were not P and R but the horizontal width and vertical depth, appropriately measured. In all equation involving the gradient S, however, without exception the hydraulic mean depth R played an important part.

7.13.1 Lacey's concept of regime

According to Lacey, the regime conditions shall be established when the following fundamental requirements are fulfilled.

(i) Discharge should be constant

(ii) Channel flowing uniformly in unlimited incoherent alluvium of the same character as that transported. Incoherent alluvium is supposed to be the loose granular material, which can be scoured out as readily as it can be deposited. Evidently all of these requirements are difficult or rather impossible to be fulfilled. Lacey, therefore, qualified the regime conditions as below:

True Regime : The channel should flow in an unlimited alluvium plain of the same grade as the material transported. There should in theory be complete freedom for lateral movement. Sandy rivers in alluvium plains achieve to some extent this freedom and by meandering adjust there length and slope.

A constant discharge transporting silt of a given grade and flowing in a self transported alluvium plain of the same grade tends eventually to assume a gradient solely determined by the discharge and silt grade, the mean velocity, hydraulic mean depth and wetted perimeter will also tend to unique determination. Such a constant discharge will also tend also to transport a fixed regime silt grade.

Artificial channels such as irrigation channels can never be in their true regime unless they are neglected like rivers in plains. They may acquire initial regime generally or final regime in some cases.

Initial Regime: Channels excavated in the first instance with defective slopes and to somewhat narrow dimensions are free by immediately throwing down the incoherent silt on the bed to increase their slopes and by generation of the increased velocity to achieve a non-silting "Initial Regime'. Such channels if berms are grassed which is usually the case of irrigation channel, will be subject to a considerable lateral restraint. The channels of this type in the Initial Regime attain a working stability and do neither silt nor scour, but there are no in final regime.

Final Regime: It represents the conditions set up eventually in theory when all the variables are equally free to vary. In the final regime, the wetted perimeter, far from being a variable, is constant.

Lacey's Diagrams

Lacey published his diagrams in Technical Paper No. 1, P.W.D., United Provinces, India. These diagrams were based on his original equations.

Example : 7.2

Design a regime channel by Lacey formula for the following given data:

(i) Full supply discharge, Q = 300 cusec (85.02 cumecs)

(ii) Silt factor of standard grade silt = 1.0

(iii) Inner side slope of channel, Z = 0.5 : 1

From Lacey's theory

(i) $V = \left(\frac{Qf}{140}\right)^{1/6} = \left(\frac{85.02X^1}{140}\right)^{1/6} = 0.607^{1/6} = 0.92\text{m/sec.}$

(ii) $\text{Area } A = \frac{85.02}{0.92}\left(\frac{Q}{V}\right) = 92.413\text{m}^2$

(iii) $P = 4.75\sqrt{Q} = 4.75\sqrt{85.02} = 43.798\text{m}$

(iv) b = P-2.236d = 43.798-2.236X2.32 = 38.608m

$$\text{F or } Z = 0.5:1, d = \frac{P - \sqrt{P2 - 6.944A}}{3.472}$$

$$= \frac{43.798 - \sqrt{43.798^2 - 6.944x92.41}}{3.472}$$

$$= \frac{43.798 - 35.729}{3.47} = 2.32m$$

(v) b = P-2.236d = 43.798-2.236X2.32 = 38.608m

(vi) Hydraulic mean radius

$$R=\frac{5}{2}\frac{V^2}{f}=\frac{5}{2}X\frac{0.92X0.92}{1}=2.116$$

$$\text{Also } R=\frac{bd+d^2/2}{b+d\sqrt{5}}=\frac{38.61X2.32+\left(\frac{2.32}{2}\right)^2}{38.61+2.32X2.236}=\frac{92.27}{43.8}=2.11 \quad \text{Hence checked.}$$

(vii) $$\text{Slope } S=\frac{F}{33400^{1/6}}=\frac{1}{3340(85.02)^{1/6}}=\frac{1}{7004}$$

Hence channel bed width b = 38.61m & a depth = 2.32m.

The longitudinal slope $S=\frac{1}{7004}\approx 7000$

7.14 Simons Formulae (1960)

A special research project was conducted by Simons (1957, 1960) with the following objectives:

(i) To investigate the validity of the regime theories developed in India.

(ii) To investigate, expand and possibly improve the tractive force method of stable channel design.

(iii) To relate the regime theories so far as possible to the tractive force.

For this purpose, canal data was first collected from the following sources:

(i) Data collected by Simons and Bender during summers of 1953 and 1954 of straight reaches of stable irrigation canals in Wyoming, Colorado and Nebraska, U.S.A. These canals were of diverse types but could be divided mainly in two classes, one having sandy bed and banks and the other having cohesive bed and banks.

(ii) Data of 15 straight uniform reaches of canals in the San Luis Valley, Colorado USA (collected by U.S.B.R). The canals were located on an alluvial fan and the boundary material was coarse and non-cohesive.

(iii) Data of 42 stable sites on Punjab canals, India.

(iv) Data of 28 stable sites on Sind canals. Both, Punjab and Sind canals had cohesive banks formed of silty berms while active sand movement occurred on bed.

(v) Data of 4 irriagtion canals in the Imperial valley U.S.A. These carried very high sediment load ranging from 2500 to 8000 ppm and their bed and banks were similar to those of Punjab and Sindh canals.

Range of important variables in these data are given in table 7.15

All the above data could be classified in the following four distinct groups:

(i) Canals formed in coarse non-cohesive material of the type studied by the U.S.B.R.

(ii) Canals formed in sandy material with sandy bed and banks.

(iii) Canals having sandy beds but cohesive banks as in Punjab and Sind canals.

(iv) Canals having cohesive bed and banks.

After separation and grouping the data as above, regime type equations correlating various parameters were evolved separately for the four groups. More important of these are abstracted in Table 7.15 and figures 7.10 (i) to (vi).

Design procedure becomes much simplified by use of above figures and equations. General approach for a canal design suggested was.

(a) Estimate area of section from figure 7.1 (i)
(b) Estimate P or W from figure 7.10(ii) and (iii)
(c) Workout R or D by A/P or A/W
(d) Check against values R or D obtained from figure 7.10 (iv) or (v)
(e) From Q and A, find V
(f) Estimate S either from figure 7.10 (vi)

Table 7.15 Compilation of Stable Canal data by Simons (1957)

Size of bed material (mm)	Source of data	Range of variation in respect of				Sediment Concentration or transport
		Q (cfs)	$S \times 10^3$	Av. Width (ft.)	Depth (ft.)	
(1)	(2)	(3)	(4)	(5)	(6)	(7)
.029 to 7.60	Simons & Bender	43 to 1031	0.058 to 0.388	10.47 to 72.80	2.61 to 8.50	Max. 516 Tons/ day
20.1 to 82.0	U.S.B.R.	16.6 to 1500	0.80 to 9.65	7 to 65	0.6 to 4.87	
0.15 to 0.43	Punjab Canals	5.79 to 9005.39	0.12 to 0.34	2.0 to 264	2.05 to 10.90	Max. Suspended 318 ppm. Mean suspended 238 ppm
0.0211 to .0196	Sind Canals	305 to 9057	0.0592 to 0.0831	41.01 to 259.4	4.67 to 13.14	Max. suspended 4550 ppm
	Imperial valley Canals	600.7 to 10021	0.255 to 0.440	26.4 to 117.0	4.28 to 6.49	2500 to 8000 ppm suspended.
0.02 to 82.0	All Canals together	5.79 to 10021	0.058 to 0.440	2.0 to 264	2.61 to 13.14	Max. suspended 8000 ppm

Main contribution of Simons lay in extension of the regime theory to cover almost all types of canals. In addition, his regime type formulae for W and d provides a guide for determining W/d (average width/depth) ratio in case of canals in coarse non-cohesive material for which the tractive Force method was basically sound but failed to indicate an appropriate W/d ratio.

Table 7.16 Simmon's Regime Formulae

	Correlation of	In canals with sandy beds and cohesive banks	In canals with coarse non-cohesive bed and bank material	In canals with sandy beds and banks	In canals with cohesive bed bank material	Plot of curves shown in figure number	Remarks
(1)	(2)	(3)	(4)	(5)	(6)	(7)	(8)
1	A to Q	$A=1.076\ Q^{0.873}$	$A=0.45\ Q^{0.873}$	..	..	7.10 (i)	Similar to Lacey
2	P to Q	$P=1.076\ Q^{0.51}$	..	..	..	7.10 (ii)	Do.
3	P to W	..	..	..	..	7.10 (iii)	Do.
4	R to Q	$R=0.43\ Q^{0.361}$	$R=0.247\ Q^{0.361}$	..	..	7.10 (iv)	Do.
5	R to D	..	..	..	..	7.10 (v)	Do.
6	V to R^2S	$V=16.0\ (R^2S)^{1/3}$	$V=17.9\ (R^2S)^{0.286}$	$V=13.86\ (R^2S)^{1/3}$	..	7.10 (vi)	Do.
7	$\frac{V^2}{gDS}$ to $\frac{VW}{\nu}$	$\frac{V^2}{gDS}=0.525\left(\frac{VW}{\nu}\right)^{0.37}$	..	..	..	7.10 (vii)	Similar to Blench King.

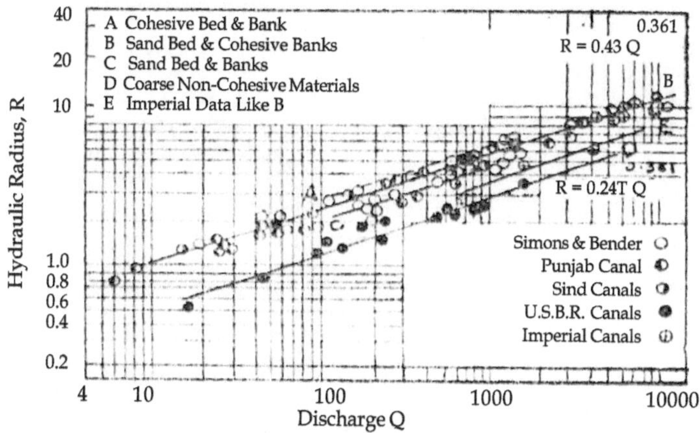

Figure 7.10 (i) Variation of Hydraulic radius R with discharge Q and Type of Channel (Simons, 1960)

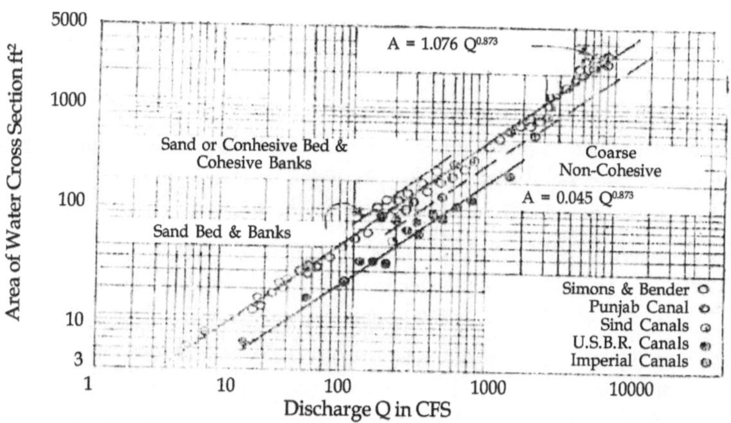

Figure 7.10 (ii) Variation of Area of Water Cross Section A with Discharge Q and Type of Channel (Simons 1960)

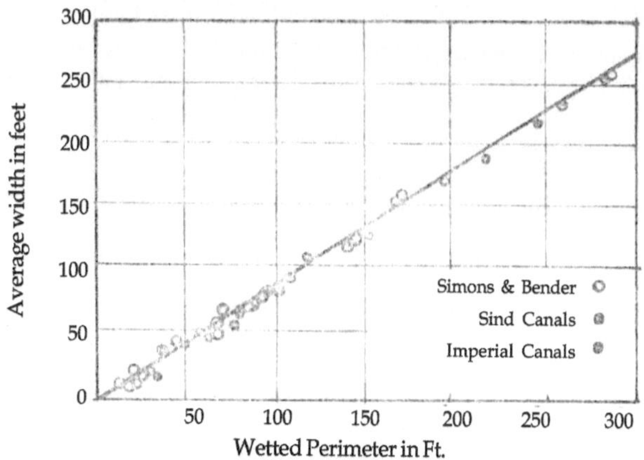

Figure 7.10 (iii) Variation of average width W with wetted perimeter P (Simons, 1960)

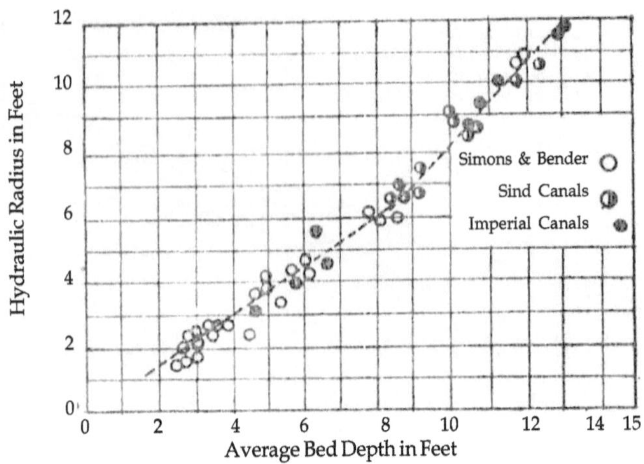

Figure 7.10 (iv) Variation of Hydraulic radius R with depth D (Simons, 1960)

7.15 Comparison of Manning's, Kennedy's and Lindley's Formula (May be called as velocity formula)

All these Equations, give velocity depth relationship only.

The equations by Kennedy's and Lindley's formula can be easily compared with the Manning's Master Eq. 1.3 (also see para 4.3) for best section, as under:

(i) $V = 0.546\ D^{0.64}$ Kennedy's (1895 initial equation, metric units)

(ii) $V = 0.567\ D^{0.57}$ Lindley (1919 initial equations, metric units,) and

(ii) $V = 0.63\ D^{2/3}\ S^{1/2}\ /\ N$, Manning's (Master Eq.1.3 metric units Para 1.6.1)
$= 0.63\ D^{0.667}$ (Treating $S^{1/2}/N = 1$, metric units)

By treating slope (S) and rugosity coefficient N as constants and $S^{1/2}\ /N$ as unity in Manning's formula (generally specified in any canal design), it reduces to a simple form of $V = 0.63\ D^{0.66}$ (For optimum side slope, area is minimum, see Para 1.8).

This is very similar to Kennedy's or Lindley's Equation or in other words these empirical relationships are near to the Manning's Equation of most efficient trapezoidal section.

From these equations, velocities for different depth of flow are as given in table 7.17: These are plotted in Figure 7.11 (ii). The Kennedy's observed velocities are also indicated in the figure 7.11 (i)

For a bed slope of 1/10000 and N = 0.01, $S^{1/2}\ /\ N = \frac{1}{0.01}\left[\frac{1}{10000}\right]^{1/2}$

$= \frac{0.01}{0.01} = 1$, for N = 0.02, and to attain same velocity, slope is to be raised to 1/2500 (i.e. 4 times) then

$$S^{1/2}\ /\ N = \frac{1}{0.02}\left[\frac{1}{2500}\right]^{1/2} = 1$$

Alternatively if slope can not be raised, the only solution is to increase depth. (Higher depth will increase R and V, see figure 1.3 section will no longer be best, and will be narrower. Also see Para 4.

The above three relationships plotted in Figure 7.11 (ii), for depth up to 8.00 m, normally maximum so far adopted show that Lindley and Kennedy's velocities are close to each other and Manning's velocity for best section are little higher.

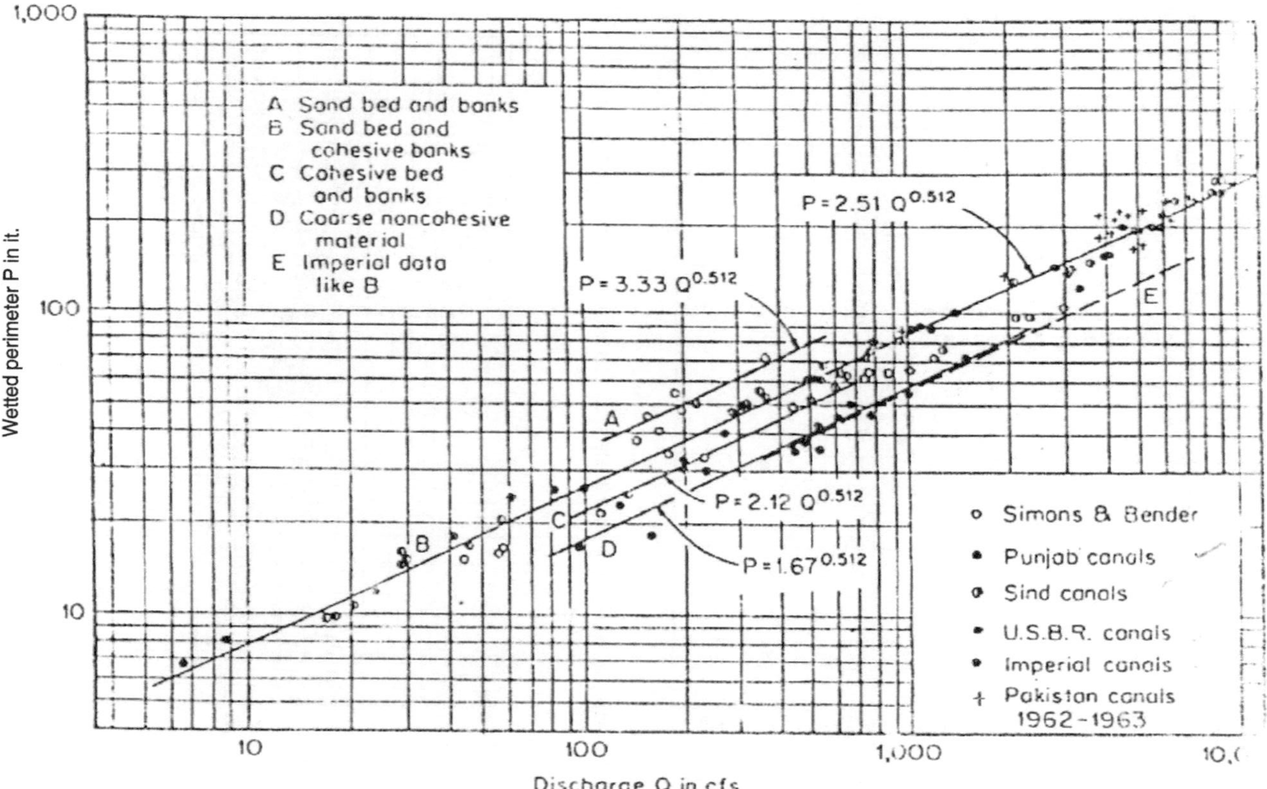

Figure 7.10 (v) Variation of wetted perimeter P with Discharge Q and Type of Channel (Simons, 1960)

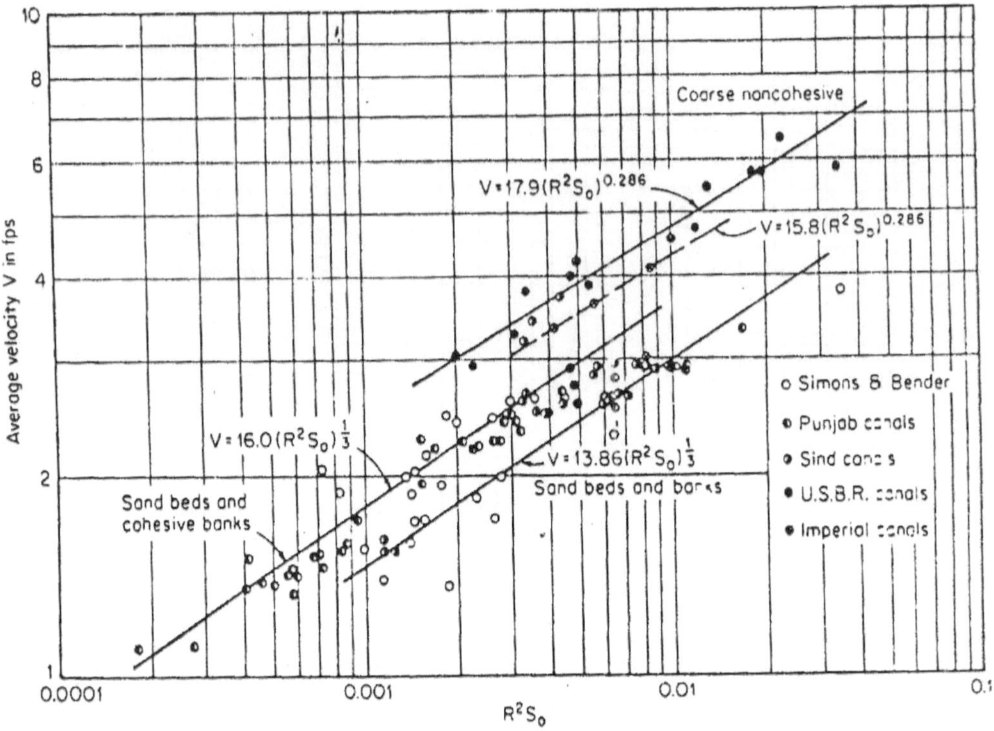

Coarse noncohesive
V=17.9(R²S₀)^0.286
V=15.8(R²S₀)^0.286
V=16.0(R²S₀)^1/3
V=13.86(R²S₀)^1/3
Sand beds and cohesive banks
Sand beds and banks
o Simons & Bender
Punjab canals
Sind canals
U.S.B.R. canals
Imperial canals
Average velocity V in fps
R²S₀
10
8
6
4
2
1
0.0001
0.001
0.01
0.1

Figure 7.10 (vi) Variation of Average Velocity with Hydraulic Radius, Slope and Type of Channel. (Simons, 1960)

c²/g = V²/gds = 0.525(VW/ν)^0.37
COHESIVE BED & BANKS
SAND BED AND COHESIVE BANKS
SAND BED & BANKS
SIMONS & BENDER
PUNJAB CANALS
SIND CANALS
V²/gds = C²/g
VW/ν
1000
100
10
10⁵
10⁶
10⁷
10⁸

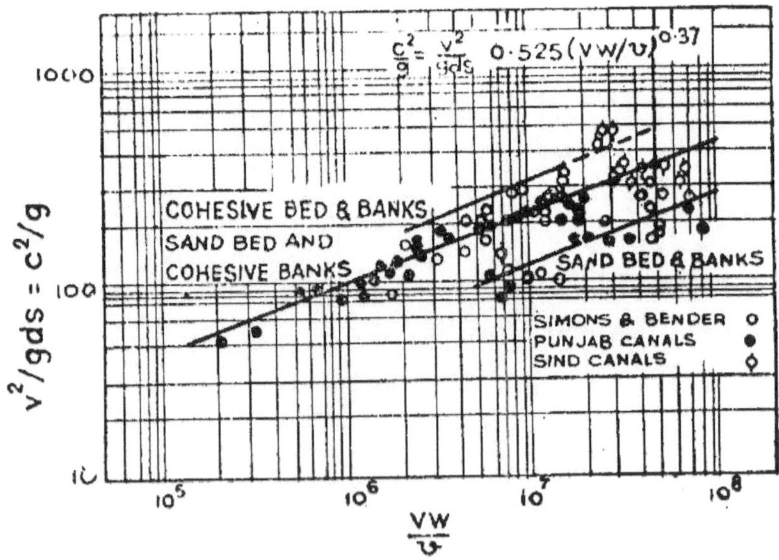

Figure 7.10 (vii) Variation of V^2/gds with VW/v Type of Channel (Simons, 1960)

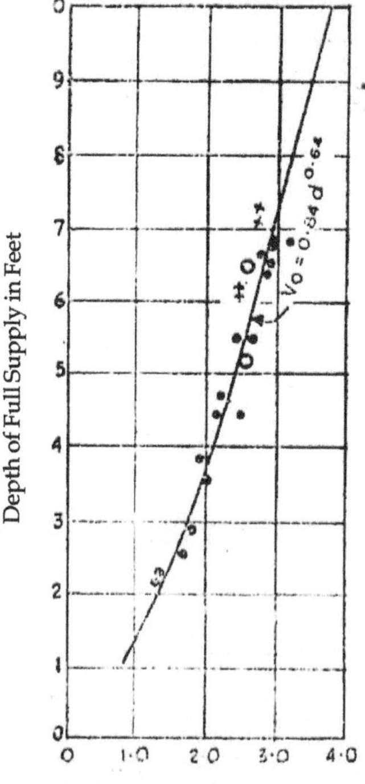

Note : The points market X are for reaches where the water level has lately been raised and the silting has not yet reached. The points marked are for the upper Sutlej canal system

Figure 7.11 (i) Relation of critical mean velocity (V) to depth (D) for full supply, plotted for sites where a permanent regime has be

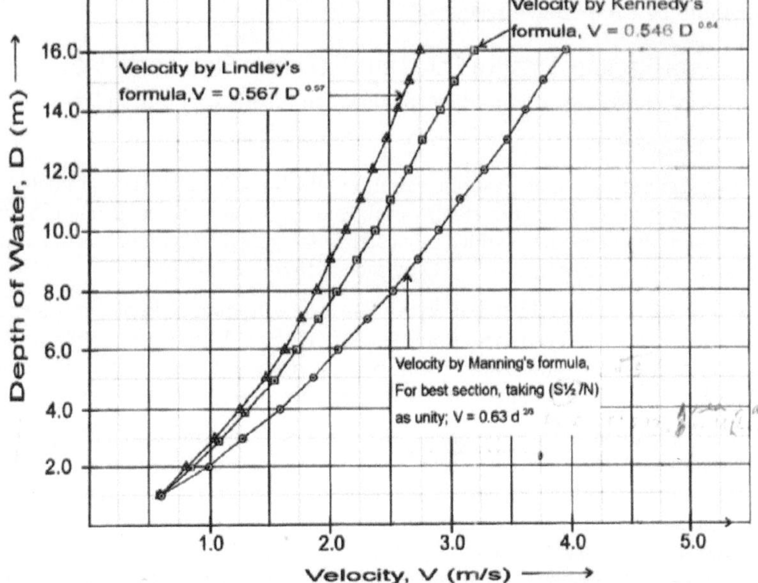

Figure 7.11 (ii) Comparison of Kennedy's Lindley's and Manning's formula Relation of Depth to Allowable Velocity

Table 7.17 Relation of Depth to Velocity by Kennedy's, Lindley's and Manning's (Best Section) Formula

Depth (m)	Velocity by Kennedy's formula (m/s)	Velocity by Lindley's formula (m/s)	* Velocity by Manning's formula (m/s)
(1)	(2)	(3)	(4)
1.	0.546	0.567	0.7
2.	0.851	0.842	1.0
3.	1.103	1.061	1.30
4.	1.326	1.250	1.65
5.	1.529	1.419	1.85
6.	1.719	1.574	2.05
7.	1.897	1.719	2.30
8.	2.066	1.855	2.5
9.	2.228	1.984	2.74
10	2.383	2.107	2.90

* Velocity is for Optimum Best Section

Inferences: For a given velocity, Kennedy's depth of flow is more, i.e. section is deeper than the optimum best trapezoidal section.

For a given depth of flow (D), Kennedy's velocity is less than Manning's velocity, i.e. section is wider than optimum best trapezoidal section.

7.16 Comparison of Regime, Tractive Force and Resistant Flow Formulae

A qualitative comparison of all important canal design theories is given in table 7.18.

There are 7 Parameters in a channel, i.e. - Q, V, N, S, Z, B and D.

Any flow formula must have at least one equation which contains all the parameters and 6 more equations which may have 2 to 6 parameters/variables to find their values. If one is assumed say Q, then still 5 more relations are required which must satisfy the general 7 parameter equation. In running canal, the problem is just reverse, i.e. to find Q by measuring, B, D and Z.

If sediment transport is to be included in the flow formula than at least 2 more variable that is sediment concentration (CS) and sediment grade/quality (f) must be included in the general equation of 7 variables. That means there should be at least one equation of 9 parameters and 7 more equations are required besides the continuity equation Q = A.V, to solve the problem.

None of the regime theories have all the nine parameters. Therefore channel Regime theories such as Kennedy's, Lacey's and others, only explains the regime condition of flow through velocity i.e. non - silting and non scouring, to a limited extent. Generally these equations give limiting velocity of flow which may not cause erosion of soil particles of canal bed and sides or velocity sufficient to carry limited suspended sediment load and does not permit it to settle. Also there is lot of controversy about non silting and non scouring velocity to be same. However this is resolved by treating stability over a year.

Table 7.18 : Comparison of Important Canal Design Theories

Maximum permissible Velocity method	Tractive force theory	Regime Theories		
		Kennedy	Lacey	Simons
1	2	3	4	5
Aims at maximum velocity (V) which will not dislodge soil particles, i.e. Non scouring velocity. V is taken from tables. V in empirical tables is low. Generally gives wide channel. Uses Manning's formula. N, S and Z are assumed from experience. Judgment about D is required. For high Q, V may result in silting. **More suited to clear water and small channels.** Method is simple	Aims at shear force that will not dislodge the soil particles. V is practically same as in column 1. τ values generally converted from permissible velocity or empirical relations are low. τ is taken from empirical relation. R is computed from T= VRS. N, S and Z are assumed or adopted from experience. Judgment about D is not required. Generally gives wide channels. **For high Q, V may result in silting.** **More suited to clear water and very small channels.**	Aims at transport of some minimum sediment concentrations of given grade. But does not indicate both. Also does not permit erosion. V is taken from empirical relation with D. V increases with depth. Uses Manning's formula. No Z assumption is required. It is near vertical or Z = 0.5:1 (H:V) very near to optimum trapezoidal section. √S/N =1. That mean slope has to be increased for increase in N. V is very near to V for optimum best section (semi circular) **More suited to sediment laden water for all discharges.**	Aims at transport of some minimum sediment concentration of grade defined by 'f. But does not indicate concentration. Also does not permit erosion. There are four relations between V-R-S, V-f-R, A-f-V, and V-N-R-S and five derived relations between S-f-R, P-Q, S,f,Q,V-R f and S-f-Q, No Z assumptions required, Z=0.5:1(H: V) or nearly vertical. Gives many solutions on the assumed values of N and S. **Generally gives wide channels.** **More suited to small channels / discharges**	It is modification of Lacey's and other empirical relations to include various types of silt grade and boundary material. Equations are similar to Lacey's formula with modified Coefficients/exponents. Better than Lacey's Theory.

Regime equations also do not indicate sediment-transporting capacity of the channel. Velocity and discharge also becomes questionable in the case of change in sediment concentration or size. In an actual practice, when silt concentration changes, it changes the bed gradient, surface and energy gradient, size (B, D and Z) ,shapes from trapezoidal to parabolic, steeper or vertical side slopes from flatter slopes. All changes can be seen in any canal lined or unlined, carrying silt laden water. Then flow characteristics of such channels can only be analyzed by Manning's formula (most common) and not by the Regime flow formula.

In fact regime theories also use Manning's equation in their development.

All the regime theories indirectly have a concept of maximum permissible velocity along with other empirical relations. A comparison of methods for a high discharge canal-IGMC is done in chapter 7B.

Presently complete hydraulics of flow is explained only by Manning's Master Equation, Para 4.3 which include all the seven parameters (Q, B, D, Z, S, N, V,).

Out of 7, only one is known i.e. Q and two equations (i) Q=AV and (ii) Manning formula are available for design. Therefore determination of 6 variables from two equations involves lot of assumptions trail and error, and there are many answers. The effect of all the variables on each other is well explained only by the Manning's Master Equation (Para 4.3) and from it is easy to choose appropriate answer to suit the practical conditions/limitations. Therefore above two equations and some velocity equations from regime theories are more helpful in solving the problem.

In spite of considerable research, none of the above theories or even more recent advances can take care of all instabilities in a channel (see Para 7.17), which are uncontrollable and occur as a natural process, typically sediment concentration and its grade, availability of water i.e. discharge etc.

So many factors causing instability in canals discussed in Para 7.17, are not included in these regime equations. Because of high variations in these factors with in a year, it may also be not possible to include all of them.

Still some reasonable stable and regime channels can be designed. For that purpose velocity criteria is more easy and logical, and a simple unified procedure is developed in Para 7.20.

Perhaps by some suitable modification of equations of regime theories, in combination and also to include Manning's Master equation can develop a series of equations to give a better solution. Researchers may work on it.

7.17 Instability in Canals-Factors Influencing Instability

A list of the major factors involved is as follows:

1. *Hydraulic Conditions:* Discharge variation
   a. Steady discharge; very rare, canals are generally for peak discharge.
   b. Variation of discharge with water availability or crop needs; capacity factor varies from 1 to 0.6.
   c. Periodic operation;
   d. Design discharge exceeded;
   e. Surface drainage water inflow/outflow at cross drainage works (typically at level crossing).
2. Water temperature: changes with season/weather and during day/night
3. Vegetal growth
4. Effect of wind action on stability of channels:
   a. Estimate of Tractive shear exerted on the bed of a channel dune waves;
   b. Effect of percolation caused by waves on the stability of a canal bed;

c. Effect of waves on bank stability; and

d. Wind effect during the non-operating season.

5. Bed Slope/Velocity;
6. Sediment inflow Magnitude and range in size of total sediment load; (Quantity and quality)
7. Boundary materials: Soil conditions;

   Alluvial canals

   a. Cohesive materials;

   b. Non-cohesive materials, sand range;

   c. Non-cohesive materials, larger then sand;

   Non alluvial canals
8. Channel dimension and its Shape
9. Berms and Bank stabilization:

   a. Deposition of wash load;

   b. Shape of channels possessing berms;

   c. The influence of berm formation on bank vegetation;
10. Secondary circulation of transverse flow;
11. Seepage inflow or outflow;

All the above factors are highly variable and cause instability/change regime condition of the channel as discussed below:

1. Hydraulic Conditions-Discharge Variations

There are different levels of a canal as shown in figure 7.12 depending upon the source, such as Minor, Medium and Major or large project, on a minor project there may be one or two direct canals with out lets on it, where as on a medium and major project, canal net work may be of two, three or four levels with branches, distributaries minors and then out lets. These canals may be operated in accordance to water availably and crop needs (according to climate). On a minor project a canal may be operated once or twice a year only that two for a short duration of 1 to 4 weeks, for one or two watering during the critical stages of crop.

On a medium project, canal running period may be for one to 4 months with intermittent closures, where as on a major project main canal may be perennial, branches and its distributaries may also run for whole year with intermittent closures

Discharge in the canal system is not necessarily conveyed at uniform flow conditions. In practice, regulators are built across the canals at suitable intervals and the heads of the distributaries through which water is distributed to the fields are constructed to take off just above the regulators, so that by heading up water at regulators, the desired degree of 'command' can be achieved. This heading up of water at regulators causes a marked flattening of the water surface slope upstream, see figure 7B.6 and Para 7B 4.1. It so accelerates deposition of silt until a dominant regime conditions becomes established, see figure 7 B.10, and 7B.6.2.

Typical discharge curves at the head of Upper Ganga Canal and that of its Mat Branch are shown in figure 9.16

In India canal distributaries are sometimes either run full or closed down, the total discharge of the canal being only sufficient to provide at one time for say nine out of the twelve minor channels; so that usually, neglecting losses, the sum of the discharges of the distributaries open at any time is equal to the supply in the canal. Short supplies are distributed and adjusted.

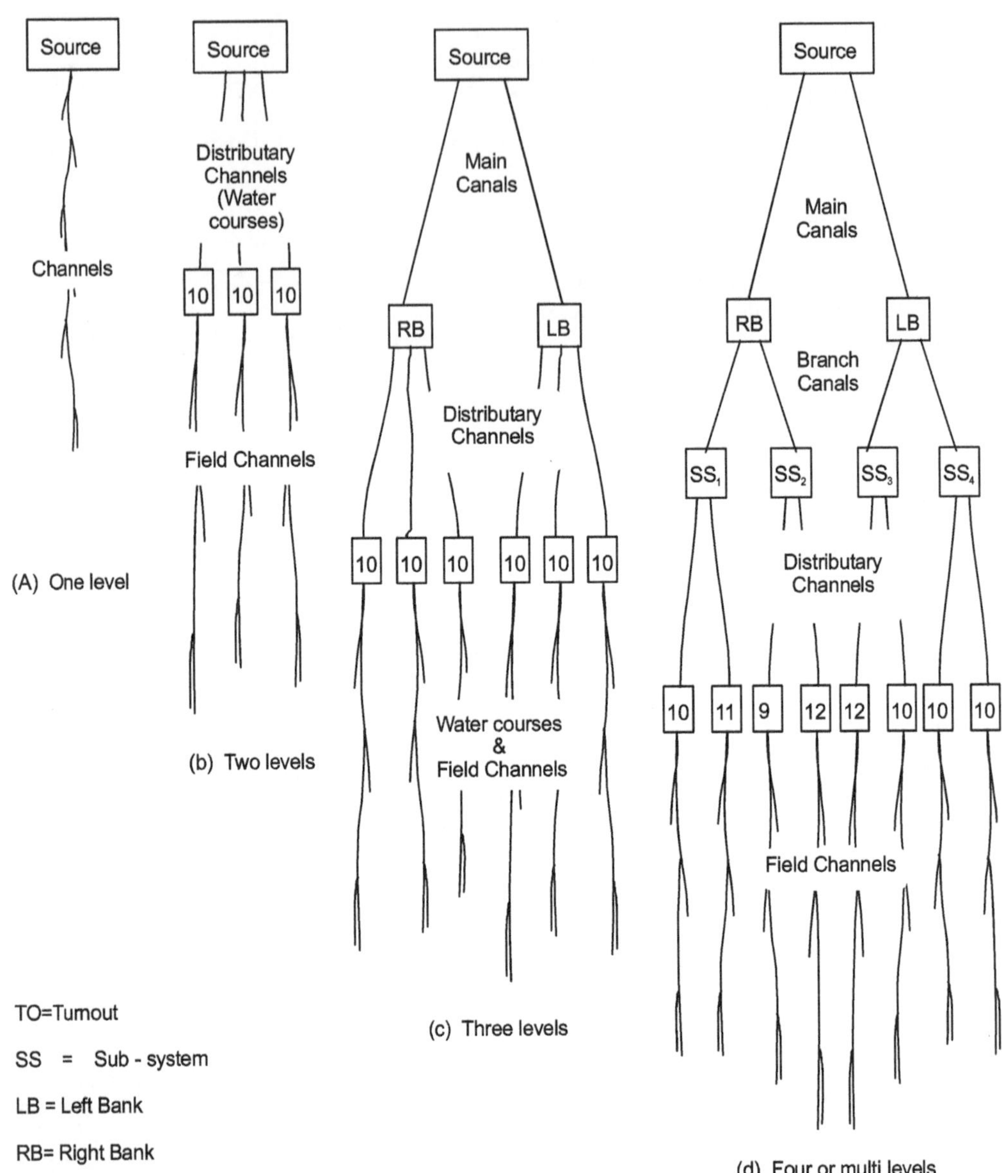

Figure 7.12 : Configuration of An Irrigation System With Different Numbers of Levels of Main, Branch, Distributory, Minor And Water Courses.

Thus hydraulically discharge regulation in canals may be classified as follows:

(i) Canals operated at full supply through out the irrigation season.

(ii) Canals operating in accordance with water availability and crop needs which can vary widely over the irrigation season. Most usual case, capacity factor varies from 1 to 0.6. One is only in one month in a year. Usually it is 0.7 and 0.8.

In this method, different parts of the cross-section come into operation with increased supplies. Velocity changes see Para 7 B.4.1 and figure 7B.6. Various forms of vegetation have then a chance to grow. Bank stability is thereby increased, berm growth is encouraged and overall roughness is increased resulting in loss of carrying capacity and some times deposition on bed. Periodical maintenance is essential in these canals.

(iii) Canals operating periodically during the irrigation season such that they are either run at full supply or else closed down.

Closure of canal though may be slow, yet brings about forces on canal side slopes similar to those on slope of an earthen dam under sudden draw down conditions. This causes sloughing of banks unless banks are composed of cohesive material, or damage to lining of sides.

(iv) Design discharge is sometimes exceeded due to additional demand of irrigation water or due to additions from cross drainage during storms if these are allowed to empty into the canal.

Typical example is upper Ganga canal. This situation is normally more of an exception. The most common damage due to excess discharge may be either scour or silting. The damage may or may not be of much significance depending on duration of excess discharge and the degree of protection afforded by berm silt and vegetation.

Effect of a particular type of regulation on canal regime would evidently depend on its period of operation. All these factors viz. mode of operation, duration and effect of control structures have to be considered in studying stability conditions.

In new canals, specially in the initial running period also known as initial silting period see figure 6.25 to 6.27, width of the channel is excessive and water is headed up to feed the off taking channels and command the area, the flow is non uniform and encourages silting as above may be for many years.

2. Temperature

In the north Indian plains temperature variation is quite considerable. Summer temperature may rise as high as 1200 F while winters may be cold enough to produce frost at night. These conditions impose a wide range of variation in water temperatures varying from about 500 to 850 F during the year.

Predominant effect is on sediment transport due to changes is viscosity of water. It has been verified by many investigators that increase in suspended load transport occurred due to lowering in temperature, see table 7.3, Para 7 B 5.1, 7 B 5.2 and figure 7 B. 8.

According to Blench equations, the stable slope of a canal varies directly with kinematic viscosity to one-fourth power. Along with viscosity, temperature variation also causes a change in density. Boundary shear should consequently be affected by temperature.

3. Vegetal Growth (weeds)

Vegetal growth (weeds) can occur both on bed as well as on banks. It increases with the formation of berms which consist of fine silt. The root system of vegetation provides additional reinforcement of bank besides cohesive strength of silt. From hydraulic view, the vegetation or weed growth may be of two types.

(i) A high dense weed growth which causes reduction in velocity and encourages berm formation.

(ii) Short grass growth which does not appreciably increase resistance to flow but lends more strength against erosion due to roots.

Also see Para 9.5 for various types of weeds.

4. Wind Action

Wind may affect stability of channel due to two effects (i) action of wind waves and (ii) direct wind erosion and deposition.

Regarding magnitude of shear transferred to be due to wave action, Simons (1957, 1960) has given theoretical computation using velocity distribution created by an oscillating flat plate. These indicate that wave action might increase the resultant shear on the bed by as much as 50 to 60 per cent for shallow canals where depth is less than 2 feet down to a negligible quantity for canals deeper than 6 ft.

Besides bed shear, waves may also cause seepage forces that reduce stability. This is due to water flowing into part of the sandy bed beneath the crest of the wave and emerging in portions of the bed under the troughs adjacent to the wave crests. The resulting flow pattern can be visualized by drawing a flow net. This flow in the permeable bed set up seepage forces that reduce the stability of the bed under wave troughs.

Magnitude of the effect of wind waves can be gauged from the fact that bed erosion to n extent of 2 to 3 ft. in a length of 300 ft. was observed to have occurred in Bijon canal due to waves 2 ft. high. The waves thus set up forces of a magnitude which might play an important role in channel stability.

Beside bed erosion, cutting of banks is also common as a result of waves. Wave action causes a rapid rise and fall of water surface at the water line. In this process water spills over the banks and then drains back. Erosion is encouraged both by flow of water along the canal banks and by seepage forces. Lined free board and vertical masonry walls reduce this effect; however wind blown sand in the free board portion is washed downwards.

Another important process by which wind may affect channel stability is by either depositing or by scouring and removing sand from canal beds when they are completely dry. This may happen in small canals of minor and medium irrigation projects, which run for a short period of 2 or 3 months in a year.

In arid conditions as in Rajasthan, a large amount of top sandy soil may be transported by wind. As the air laden with sediment sweeps over canal sections some of the sediment is dropped out due to reduction in wind velocity over the canal. Typical example is Indira Gandhi canal projects see chapter 7B.

When canals in such tracts are dry typically distributaries and wind flows along the canal direction it may pick up and remove fine bed material from canal bottom and carry it to the first bend where it may be deposited.

5. Bed Slope and Velocity

If designed slope is too steep, canal starts degrading at upper end and vice-versa. Maintenance of capacity and stability of a canal is a function of its slope in conjunction with channel dimensions, sediment charge, type and condition of bed and banks etc.

Velocity

Velocity in a channel is obviously not an independent variable but depends primarily on parameters like depth, slope, boundary roughness, shape of channel etc. Bed sand moves in the form of ripples and dunes which increases roughness. Suspended sediment on the contrary reduces rugosity coefficient or friction factor and increases velocity due to damping of turbulence. Quantitative assessment of these effects and methods for their evaluation at the present state of knowledge is far from satisfactory.

Besides head loss, other problems related with velocity are of canal stability. It is an observed fact that there is a limiting permissible velocity for a particular soil of which the canal banks are made. If the velocity increases beyond this, erosion may set in.

Such limiting velocities have been recommended by many authors as explained in Para 7.9. A curious fact has however been observed that the same velocity in greater depths may not cause equal erosion. Also see Para 7. 12.

6. Boundary Material- Alluvial and Non-Alluvial Channels

Alluvial channels or canals

Soils formed through the action of running water are known as alluvial or water-laid soils. Channels flowing in soils of their own making are termed as alluvial channels. Sediment transported by them is similar to that of which the channel boundaries are formed. The sediment may generally be either sand, fine or coarse, or silt and clay, or a mixture of all the grades. Distribution of various sizes of the material along the wetted perimeter may also be non uniform. The banks may consist of silty clay, while bed may carry coarser sand grades in ripple movement. These are the normal characteristics of a typical north Indian canal. Conditions may, however, vary in different tracts and the size; gradation etc of boundary material may be very dissimilar resulting in different shapes and dimensions of channels. This is true of both alluvial canals and rivers. One characteristic is however true of all alluvial channels, that they are free to form their shapes, dimensions and bed gradient, either by scour or deposition. There is no restraint on this process except that banks and bed may locally offer less or more resistance to these natural processes depending upon their physical characteristics.

Different types of material which may form the channel boundary may be classified as below:

(i) Cohesive materials

These are more resistant to erosion and scour because of cohesive strength and tendency to promotion of easy growth of vegetation. Since this type of material presents a smoother boundary to flow a smaller B/D ratio i.e. ratio of width to depth of channel suffices than in canals with non-cohesive material.

(ii) Non-cohesive material, sand range

Due to lack of cohesion and little reduced ability to support vegetal growth, these materials rely primarily on their weight, shape and surface texture to resist erosive action. B/D ratio in sandy material may be little larger.

(iii) Non-cohesive material, coarse range

Due to greaters weight of individual particles, and steeper angle of repose, coarser material is much more stable than sand. Greater latitude in selection of shape of canal and permissible velocities is therefore possible. Such materials are often used for bank stabilization.

In alluvial canals, only the above three types of materials are met with.

A typical alluvial canal in India has a sandy bed with ripple or dune form and banks made up of silty berms. Description of the development of such a canal section has been given by Blench (1955). If the canal is 'in cut' the berms are excavated in natural soil; if it is 'in fills' berms are formed by deposition due to heavy charge of suspended sediment obtained during flood season. The original bank is purposely retired to accommodate a larger channel section. The channel section then builds up the bed from sand and side berms from fine sand, silt and clay.

To accelerate deposition and to prevent tendency for meandering, the berm formation requires some assistance. This is provided by constructing short brush wood spurs along both banks, their spacing varying from 2 to 5 times the length.

A naturally deposited berm consists almost entirely of material finer than 0.08 mm size, highly breach resistant and effective in reducing seepage losses.

That the width of a channel is dependent inter alia on nature of bank material was realized by many authors. Blench has introduced side factor into channel equations and recommended specific values, for different types of materials met with in practice. Side factor Fs was defined to be equal to Y^3/B. This was shown by him as a measure of the square of the shear stress on sides.

Canal beds in North India consist normally of sand which is of a grade more than 0.08 mm mean diameter, except when temporarily blanketed by fine material. Gradation of mean size changes from coarser in the middle of the channel to relatively finer towards the sides. The material of the plains in which these canals flow consists of surface soil say of silty clay loam under which are found great depths of clean river sand, grades varying from 0.1 to 0.6 mm with a mean of 0.25 mm. Density of the sand is remarkably constant and averages 2.65.

In case of canal data on which Lacey formulae were based, the size of bed material decreased as $Q^{1/10}$. Kennedy and other engineers considered it was due to upper channels drawing off more than their share of coarse material; still others to attrition and chemical action. Inglis was of the view that the size depended on the ratio of side slope area (bounded by smooth turbulent flow) to the bottom are (bounded by rough turbulent flow). This ratio increased when the discharge decreased. The degree of turbulence therefore decreased and the bed material size became finer as the discharge decreased.

The sand on canal beds normally moves in the form of ripples. Lacey explained that his formulae were, applicable to canals with rippled beds. It has, however been observed that, ripples and waves form in the major canal system in the Punjab, particularly those taking off the rivers in proximity to the hills. The bed wave is marked in the upper reaches of main canals, while ripples persisted in the lower parts of the system.

7. Sediment inflow Concentration and grade

Considerable changes may occur in the beds of alluvial canals when quantity of suspended material changes from season to season, see Para 7.3.3 and figure 7.3 (iii). During the flood season, when heavy charge of suspended load enters the head of a canal system from the river, part of the silt falls to the bed and mixes with the bed material. Weighted mean diameter of the material exposed on the bed changes. This results in corresponding increase/decrease in the rugosity coefficient, in Manning's and Lacey's formulae.

When floods subside, water contains much less silt, and, on entering the canals gradually removes the finer particles from the bed mix. The bed material becomes progressively coarser and builds up into ripples and the rugosity factor increases in some case to N=.025 or higher near the head. This change progressively reduces downstream, see typical example of upper Ganga canal in Para 7.19.

Non-Alluvial Channels

All other types of materials which may occur in non-alluvial canals are as below:

(i) Lensed materials having alternate layers of sand and clay
(ii) Cemental materials
(iii) Shales
(iv) Different types of rocks.

In a non-alluvial channel, sediment transported is dissimilar in character to that of boundary material. The source of material in transport is solely from the watershed and is the result of soil erosion. Sediment load of such streams is not governed by bed conditions and hence rate of transportation cannot be determined as a function of bed characteristics. Channel bed and banks are often resistant to erosion and offer restraint to free adjustment of channel section and gradient by the process of scour and deposit.

Artificial and Natural Channels

Similarity between rivers and alluvial canals would appear to be conspicuous. Channel dimensions and slope in both are governed by the discharge and sediment supply and there is a continuous process of adjustment to achieve stability. Along the length of the channel, width, depth, slope, bed material size., velocity etc. all vary with discharge. Correlation between discharge and other variables mentioned is markedly similar in artificial and natural channels. In case of width, rate of its change with discharge has been proved to be almost same in both. Lacey has shown that his relation $V=16R^{2/3}S^{1/3}$ is valid both for rivers and canals.

It is therefore natural that analogy between these two types of channels be considered and inferences drawn for river data be applied in developing laws of canal hydraulics. It is for this reason that Lacey, and others have widened the scope of some aspects of their work to cover both. This has enabled the theory of flow for alluvial channels to be generalized.

Still limitations of dealing with river and canal hydraulics jointly need, however, to be fully appreciated and well differentiated. In a river, discharge increases in the downstream direction while in a canal it reduces. This results in the river bed profile being concave to the sky while it may be convex in case of a canal. Discharge in a canal being artificially regulated to meet demand of irrigation or power, the hydrographs of river and canal are different. Another important difference is that a river in flood plain may either be a braided or a meandering channel. Canal is devoid of these features. Pattern of velocity distribution may also be appreciably dissimilar in the two. Besides, velocities in a canal may be invariably smaller, than in a river.

Divergence in the behavior of rivers and canals in respect of some of the above aspects is basic and caution has therefore to be exercised in stretching the analogy and inferring laws of one from the other.

Natural Channel

Rocky rivers/channels are subject to the fundamental laws of "fixed boundary" hydraulics. In practice the irregularity of the boundaries and the consequent difficulty of determining dimensions and roughness coefficients make accurate theoretical analysis impossible. The most frequent problem is that of ascertaining the water profile and other flow conditions for various discharges, and is best solved by site observation and gauging.

8. Channel Dimensions and Shape

Sectional Dimensions

In an alluvial canal, B, D and S width, depth and slope are free to adjust. Whether a unique shape is possible under these conditions is a moot question. According to Lindley hypothesis, for a given discharge and sediment charge, dimensions of a canal viz. B, D and S are automatically fixed by nature. This is the basic concept of regime theory and Lacey equations provide these dimensions in terms of discharge and silt factor.

According to tractive force concept it is possible to design alternative canal sections which carry a particular discharge and resist bank erosion. If a condition is laid down that tractive force should be critical at all points along the canal perimeter a unique section can be fixed.

Dimensions of a canal be conceived in terms of B and D and Z. Kennedy and Lindley equation were in terms of these parameters and Z was near vertical.

It was Lacey who first deviated from this practice and adopted wetted perimeter and hydraulic mean radius instead. This change over was not accepted by many like Blench who maintained that for alluvial channels, B and D were the correct parameters to be used in flow equations. Lacey's formulae were in terms of B_m, the mean width and dm, the depth from water surface to mean sand bed level. Lacey later reconciled to this view partially, in his 1958 paper, he observed that as far as cross-section was concerned, the correct parameters to employ were not P and R but the horizontal water surface width Ws and vertical depth d_m (d_m =A/ W_s). In all equations involving slope S, however, without exception the hydraulic mean depth R played an important role.

Thus some authors prefer B and D while others use Ws and dm as parameters, but in any case Ws/ dm=BD=A. Which pair are best representative remains to be decided and their exact determination is awaited.

Shape

Apart from canal dimensions, shape of section is also important as distribution of velocity and boundary shear is governed by it. In designing a canal, the standard section is taken to be trapezoidal with side slopes of about 0.5 to 1 (H:V). Actually when berms are formed of fine cohesive silt, the shape of big canal section consists of a fairly flat bottom with sides conforming to parabolic, elliptic or semicircular curves which are tangential to the channel bed. Small channels appear only parabolic.

Lacey observed that if channels are visualized as having semi-elliptical cross section then shape would be determined by the ratio of the major axis to semi-minor axis. This would depend on the type of sediment. The coarser the sediment, the flatter is the semi-ellipse and vice versa. All stable channels of the same mean velocity have geometrically similar shapes, the silt factor determining the only difference, which is one of scale. He further argued that for a given cross-sectional area and given water surface width, a segment of a circle involved the minimum welted perimeter. If the channels were wide this would be equivalent to a parabolic section. If it was stipulated that all the isotachs without exception terminated normal to the water surface, the minimum wetted perimeter probably conformed to a semi-elliptical cross-section. Since the finer particles were responsible for banks, the semi-ellipse was somewhat flattened.

An engineer engaged in the design of irrigation channels may mean by the term shape, the ratio of the horizontal B to D i.e. B/D. If the channel is irregular in shape he may employ the ratio of water surface width Ws to the mean vertical depth d_m i.e. W/d_m.

A convenient way of comparing channel cross-sections may be by means of the ratio of the area of the channel below water surface to the area of rectangle enclosing the section. This ratio was proposed by Lane and termed as form factor. For a rectangle its value would be 1.0, for an ellipse 0.79, for a parabola 0.67 and for a triangle 0.5. Lane found that form factors for various channels investigated by him ranged from 0.56 to 0.92.

Shape of a channel section depends on the type of boundary material. The channel in coarse material may form a semi elliptical cross-section as assumed by Lacey. It is not necessary that for obtaining such a section grade of boundary material must be uniform. Lacey regime theory merely assumed that in similar cross-sections the size of material along the wetted perimeter would present at all points a constant difference in scale and that the variation in size would be continuous from center towards the edges. If, however, the boundary material changed suddenly, the shape of the section would be different. Canals with berms of fine silty material and bed composed of coarse sand, present a nearly horizontal bed as observed by Kennedy on Bari Doab system. There are thus two types of cross-sections, the first with a continuous boundary as assumed by Lacey and the second with shape differing from a semi-ellipse due to discontinuity in boundary material.

9. Berms and bank stabilization (Also see Para 8.)

Typical Berm formation in silt laden water in shown in figure 7.13 (i), with this canal section are established.

10. Secondary Circulation

Even in a straight channel, secondary circulation has been noticed that results in affecting distribution of velocity and sediment. The extent to which secondary circulation affects stability of a channel is however unknown, so far.

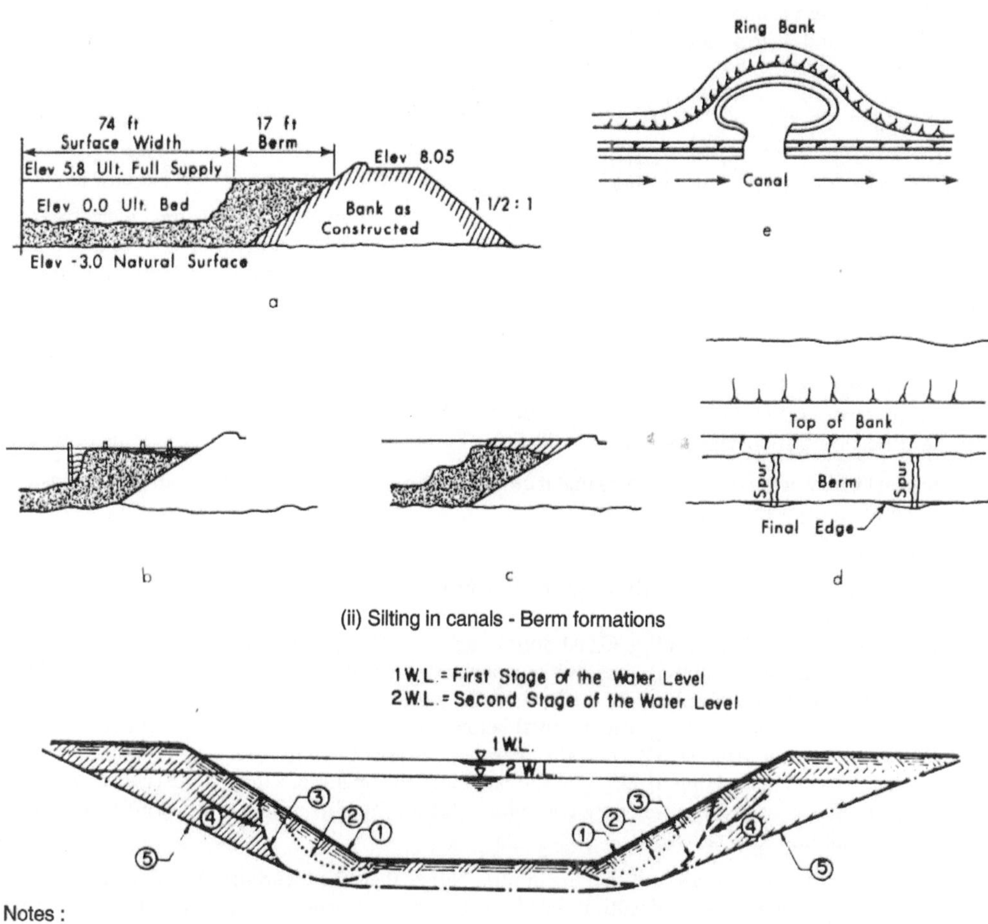

Notes :

Stage 1 : Main-made trapezoidal section.

Stage 2 : Intial erosion at the toe of the side slopes.

Stage 3 : Advanced stage of bank erosion.

Stage 4 : Sides sliding into the channel, may be stable depending upon type of soil and veolcity or

Stage 5 : May be final shape of the section depending upon stability of slopes under various hydraulic conditions, see chapter 8.

As per experience, generally alluvial soils are stable at 1:0.5 (v:h) for FSD upto 2.50 m and v = 1.0 m/s

(ii) Erosion of canal Banks

Figure 7.13 Berm Formation in Channels

Inference

To date, no theory has been conceived capable of adequately considering all the above factors causing instability and their influence on channel behavior.

General behavior of canals carrying silt laden water

Almost all investigators/researchers and authors have indicated that when the sediment concentration is more than the transporting capacity of the channel, it settles in the head reaches, and showily decreases in the middle/tail reaches, where stable sections can be seen. In this process it changes the bed gradient and energy slope and do not remain uniform in a long length, However sides remain near vertical.

Such canals can be improved by raising the top bank in the head reaches, providing more energy slope in head reaches if possible.

7.18 Silting Behavior of Upper Ganga Canal, India

Some brief details of this canal are given in Para 6.5.2. As the canal off takes from a weir/barrage at the foot hills, lot of silt, mostly in suspension enters into the canal, in spite of silt excluders. Silt concentration in river water changes considerably within the monsoon period and there after. So silt concentration entering the canal also changes. Canal discharge duration curve for the year 2002-2003 is shown in figure 9.16 (i). Discharge characteristics are more or less similar in most of the years. Since the year 1951 the canal runs almost at a constant discharge of 10,000 cusecs (285 cumecs) from May to September except for 1 or 2 or 3 days closures taken intermittently during the monsoon period from June to September, to avoid entry of heavy silt concentration in the canal. The closures are taken when ever the silt concentration exceeds 3000 to 5000 ppm. There is an operation manual for this purpose.

The past history of the Canal indicates that it often gets silted during monsoon and after the monsoon is over, the canal capacity again increases.

Tragedy of 1970 and lessons

During last week of July and early August, 1970, heavy silting of Ganga Canal took place. Later it was known that it was due to land slides in Alakhnanda Valley, resulting in very heavy sediment concentrations, presumably exceeding 30,000 ppm in the river Ganga. As there is little opportunity for exclusion of sandy material, moving mostly in suspension, these high concentrations also entered the canal through its Mayapur Headworks. It silted up the canal from its head to Pathari Power House, located at M-7-3 (Miles and Furlong-British units) below the head. It was so badly silted that the canal had to be closed down in August 1970, and silt clearance carried out by manual and mechanical means at a huge cost. The events leading to silting of the canal and the operations during the crucial period of silting have been analyzed by an enquiry committee headed by Shri S.G.K. Murti (1970). After that operation Manual has been revised.

The slit clearance was carried out to a line a little higher than the theoretical bed line approved in 1951. The canal was opened after desilting on 14-10-70 and water was run with full gauge from mid night of 14-15 October, 1970. With in a fortnight of its running, the canal bed attained the non-monsoon regime conditions and the bed scoured below theoretical levels. There was no problem upto May, 1971.

Typical bed profiles observed in the years 1950, 1980, 1990 are shown in the figure 7.14.

Though this canal has not been designed by any regime theory, still its behavior can be said as stable channel over a year. Some silt deposits in head reach during monsoon season due to heavy silt concentration and erosion of same occurs during non monsoon season of low silt concentration.

Most of the silt in the canal passes through the branches and distributaries, their bed or crest at the off taking point being very near to the bed levels of the main canal. Discharge in the off taking channels is regulated through ungated constricted type low crest weirs.

7.19 Proposed Unified Procedure for Design of Lined and Unlined Canals

Different theories give different canal sections for the same discharge, N and bed slope. There is lot of controversies in Researchers and designers, as to which theory is better and also there is much confusion. Perhaps theories may be incomplete. It has not been possible to include instability of flow as given in Para 7.17 in any theory. Therefore with lot of discussion in chapter 1 to 6, this chapter, also 8th and 9th, a unified simple procedure of design of either lined or unlined channel carrying clear or sediment laden water can be developed with the help of Kennedy's and Manning's formula as under :

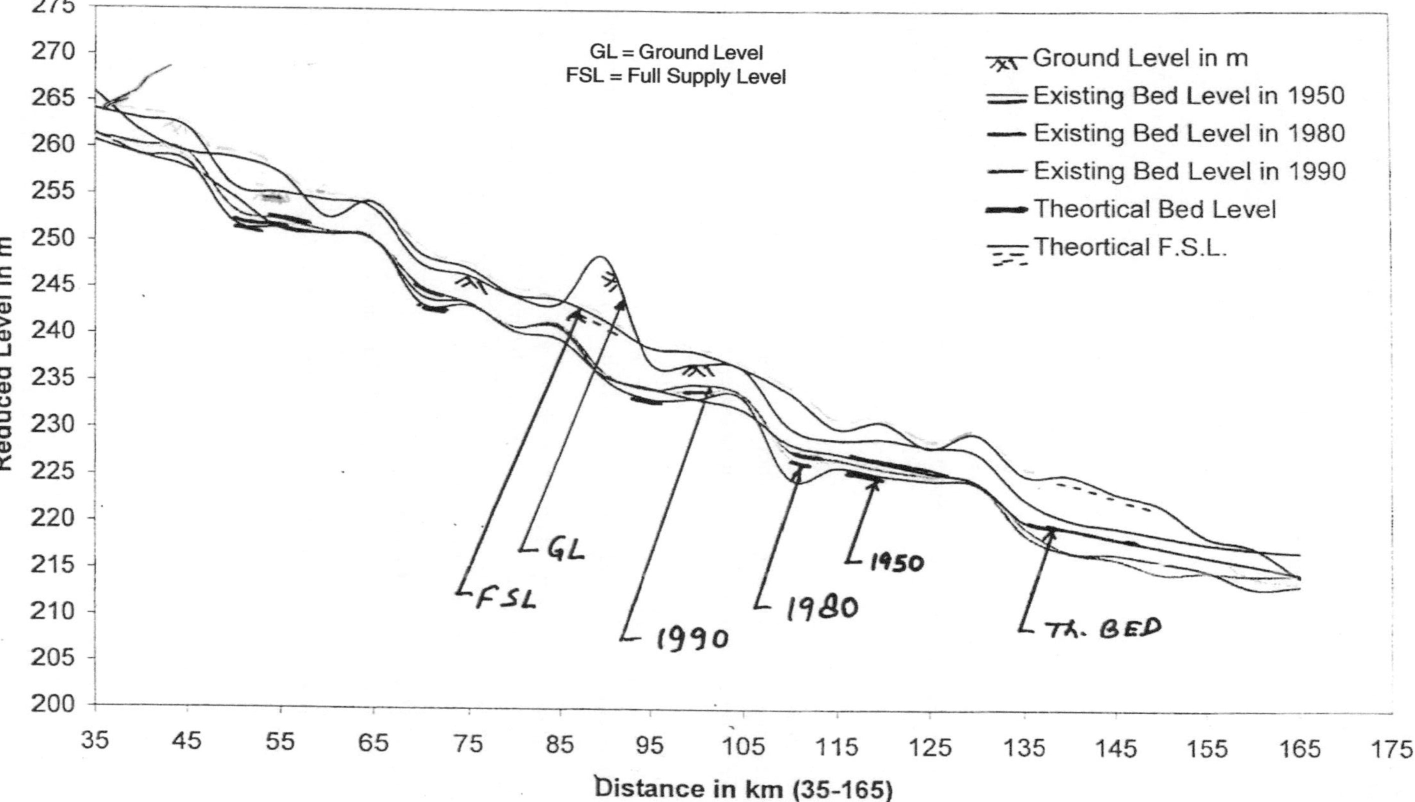

Figure 7.14 L Section of UGC showing bed profile during year 1950, 1980 & 1990 and comparison with theoretical bed

1. Consideration of lining cost

Because of heavy cost of lining, little advantage over unlined canal, a good lining policy (Para 9.18), canals may be partly lined (in some portion i.e. in patches or in bed or one/two sides.

Section and bed slope may change from reach to reach with suitable transitions at structures, or otherwise. Sections may also be designed in such way that future lining is easy, and economical.

2. Critical velocity

For most of the soils, or even lined canals, Kennedy's critical velocity given in figure 7.15 can be adopted, which increases with depth. This velocity is good for clear or sediment laden water and is non - scouring and non silting. High velocity in unlined canals can also be permitted by a protective cover of gravel or stones etc. USSR Practice (Para 7.9) can be followed.

3. Bed Width to Depth Ratio (B/D = X)

Higher B/D ratio reduces velocity. D is to be manipulated according to stability of slopes.

For Q upto 50 cumecs, a fixed B/D ratio for best section (chapter 3) should be adopted.

For Q > 50 cumecs, a rigid requirement is not warranted. B/D ratio of 1 to 4 (preferably between 1 to 2) are beneficial for economy less land, earth work, perimeter and lining and structures, see figures in chapter 6), reduction in transit losses (Para 9.3), good velocity, low slit deposit and low weed growth (Para 9.5).

4. Longitudinal Bed Slopes

High capacity irrigation canals are long, flat slope is essential /unavoidable.

Low capacity irrigation canals requires little steep slope, (unavoidable). See table 9.5 for general guidance.

In practice, the general country slope is followed for ridge canals and for sidelong-ground, suitable slope to give total desired command but giving higher than minimum permissible velocity needs to be adopted. A certain slope is adopted and design proceeded with to check the above criteria for sidelong canals.

For economy in construction, the slope should be such as to yield maximum permissible velocity V. For providing maximum possible irrigation, flatter slope guided by minimum permissible velocity would be desirable. The slope is to be within these limiting values. This is energy slope and may be provided either in the bed or in the water surface in zero bed slope canal (refer Para 9.14.1).

5. Inner Side Slopes (Z =1 Vertical to Horizontal)

Role of flatter Z is to reduce velocity, Z > 0.5, decreases velocity (Para 3.1) and encourages silt deposition on sides.

The following side slopes are recommended, for design in all types of soils for discharge upto 50 cumecs and depth of flow upto 2 m, Z = 1: 0.5 to 1

For D>2.00 m and Q > 50 cumecs, Z = 1 to 1.5.

As far as possible Z may be kept near 1 to 1.25 and a little wider channel may be preferred. For rocks etc. Z may be kept according to geology, see Chapter 8 and 9.

In very deep cuts, although above side slopes may generally be satisfactory for depth which rarely exceeds 5 or 6 meters, suitable berms must be kept at intervals above the flow section, so that the stability of slope is ensured, (Para 8). The spoil bank should also be thrown sufficiently away from the edge of cutting and suitable drainage provided. It is advisable to design deep cut or filling reaches of more than 6m depth by stability analysis based on soil tests, of embankment and foundations, (Chapter 8).

6. Coefficient of Rugosity

It should not be merely decided on the basis of type of surface, but should be decided on the basis of maintenance programme (silt and weed clearance). See Table 6.4 and Para 9.15.2, irrespective of lined or unlined canal.

If the canal is not maintained in good condition by removing silt, weeds, etc, the rugosity coefficient will be substantially higher as much as 100 to 200% than the values given in table 2.1 to 2.3. It is desirable to make sufficient allowance of 20 to 30% for this purpose.

7 Design Procedure or Steps

(i) From considerations of water requirements or availability of water, Q is determined for a particular reach, with flexibility of sitting, peak demand, etc.

(ii) Area of cross-section of the canal is worked out by continuity equation Q = A.V.

(iii) Value of R (Hydraulic mean radius) is worked out from Manning's formula.

$$V = \frac{1.486}{N} R^{2/3} \; S^{1/2} \text{ (BritishUnits)}$$

$$\text{or } V = \frac{R^{2/3} \; S^{1/2}}{N} \text{ (MetricUnits)}$$

(iv) Compute perimeter, from P = A/R

(v) Knowing P and A, values of D (depth) and B (width) of the canal section are found out. The design is finalized adopting the criteria of B/D ratio (X) and V/V_0). Help of Table 9.5 and nomogram in figure 4.3 is very useful.

(vi) Following equation is useful in the above procedure for finding values of D.

$$D = \frac{P \pm \sqrt{P^2 - 4A\left(2\sqrt{Z^2+1} - Z\right)}}{2\left(2\sqrt{Z^2+1} - Z\right)}$$

There will be two values of D except when

$$P^2 = 4A\left(2\sqrt{Z^2+1} - Z\right)$$

In this case a single value resulting in section of minimum perimeter for given velocity is obtained. Some times the expression below the root sign in the above equation is negative, giving imaginary values of D. In such a case, lower value of V is arbitrarily assumed for obtaining a solvable equation. In certain cases, the B/D ratio with obtained values would be too high. In such instance, higher than maximum permissible velocity cannot possibly be fixed and the procedure becomes more trial some. A suitable flatter slope need be tried and procedure repeated. As is obvious, the procedure may require successive trials to get the desired ranges of X value.

The procedure becomes simple with the help of Nomogram in figure 4.1.

8. Typical Example

An unlined canal is to be designed to carry a discharge of 30 m^3/s with following data:

(i) V chosen as 1 m/sec for murum and hard soil and 5 percent excess also proposed to be allowed.

(ii) N = 0.0225, chosen - on the type of surface, slit and weed growth.

(iii) Z = 1.5 chosen since long embankment reaches are also involved in the alignment of the canal.

(iv) X = 2.5 to 5 proposed to be adopted.

(v) The general ground slopes are given below and it is decided that steeper slope upto 5 percent more than the ground slope could be allowed without much loss of command of the canal.
(a) 1 in 4200, (b) 1 in 5000, (c) 1 in 4400,
(d) 1 in 3500, (e) 1 in 3000

The typical answers of the above cases are noted for reference in Table below:

Case No.	Finally chosen values				Remarks
	X	V (m/sec)	D (m)	B (m)	
(a) 1 in 4200	3.0 (obtained)	1.0	2.58	7.75	(1) X-V relation used for (b), (c) and (d)
(b) 1 in 5000	Chosen 2.5	0.945	2.81	7.03	(2) For V = 1 m/sec the design is not possible in case (b) indicating that slope is flatter than minimum slope for that velocity
(c) 1 in 4400	Chosen 4	0.963	2.38	9.52	(3) In case (c) X values of 2.5 could be provided to give more efficient section.
(d) 1 in 3500	Chosen 4.5	1.04	2.19	9.87	(4) X-V and V-S relations used in case (e); Also checked by X-S relation.
(e) 1 in 3000 changed to 1 in 3750*	Chosen 5	1.0	2.15	10.75	

* Falls of 0.5 m per 7500 m are proposed so as to follow the general slope of 1 in 3000

A canal may be designed by any formula, but in day to-day operation, hydraulics of flow can be computed only by Manning's formula. Therefore it is better to design any canal by the above procedure.

In canals off taking from weirs/ Barrages, having silt concentration, silt excluders are essential, silt entry from slopes by rain water/drains or cross drainage works should be avoided.

Silt ejectors or silt chambers are essential only in power canals.

River Inter Linking Canals

A big programme of interlinking of rivers in India is under examination of the National Water Development Agency (NWDA). Such canals may generally off take from weirs. These canals may have lot of silt problems and therefore little higher velocity/non silting velocity then the non scouring velocity may be better criteria. Also these canals will cause afflux in the meeting river/channel and needs a separate analysis. This afflux will cause a reduction in carrying capacity in the up stream of the main meeting channel and rise in water level and flooding of banks.

7.20 Summary and Conclusions

Empirical Equation for Sediment Discharge

Many formulas have been given by various investigators, after Du Boys (1879) presented his tractive force. The results of different formulas differ drastically. So far it has not been possible to determine positively which method gives most realistic or reliable result. Therefore selection is not straight forward. Still Acker's (1973) and Einstein's (1950) are more wide accepted and can be used canals.

Some of the important formulas, which are used by many Engineers, are presented in Table 7.4 to 7.6 in a systematic way as under,

a. Table 7.4 bed load transport equation
b. Table 7.5 suspended load transport equation
c. Table 7.6 total load transport equation.

Non-Scouring Velocity

The velocity at which there will be no erosion of bed. At a velocity greater than this flow of water starts eroding the bed/side. This velocity depends upon the type of soil.

Non-Silting Velocity

The velocity at which the sediment in suspension, remain in motion and does not settle. This velocity depends upon sediment con centratian and size.

The above two velocities can not be identical. Non Scouring Velocity is generally lower than Non-Silting Velocity. Also velocity increases with depth, see Para 4.6.

Stable velocity

Another concept may be that velocity may be little higher than non-scouring that erodes little bed in the head reaches during little/ low sediment concentration and permits some sediment deposition during little higher concentration. Canal may be stable over the year. This behaviour has been noticed in upper Ganga canal (Para 7, 18). Also this definition has been accepted by some authors, see Para 7.11.

The flow of silt in the canal/river depends upon the flow pattern i.e. discharge, depth and velocity rating and duration curves of the canal.

The silts carried by the canals either get deposited in it or goes out according to the change in discharge, velocity and sediment characteristics. Silt is deposited in head to middle reaches only. Middle to tail reaches are generally stable. Typical discharges, velocity versus depth curves (rating curves) are shown in figure 4.5 to 4.6, see Para 4.6 . These curves are valid up to uniform flow in the canal. Discharge duration curve at head for upper Ganga Canal, U.P. (India) is shown in figure 9.16 (i)

The flow pattern changes according to the availability of water from the source and demand i.e. season and crop water requirements. Flow characteristics are further changed by cross-regulators.

A canal is generally designed for a maximum discharge at full supply level. This maximum discharge is only for a short period in a year. For rest of the time the discharge is generally low/ less. However full supply level (FSL) is maintained at the cross-regulators, generally only in main canal. The flow in few kilometers upstream of cross-regulator is not uniform flow, but is gradually varied flow, with higher depth but low velocity. There (near cross-regulators) silting is inevitable.

However branches and distributaries are run on rotation intermittently for maximum discharge (full supply discharge) and at full supply level (FSL).

During the phase of intermittent running/closing the canal passes through a transitional phase of rising discharge and water level, (during opening) and falling discharge and water level, (during closing). This phase also helps in reduced velocity and silting. There are many other factors causing instability in canal as discussed in Para 7.17. No theory can take care of all above variable factors.

The interrelationship of discharge (Q), velocity (V), channel geometry (B/D ratio X and side slope Z), energy slope (S), and channel roughness (rugosity N) can be more easily understood by the Manning's Master equations, (see Para 4.3).

Q in a river/canal is an independent variable as surface runoff or release in a canal. Still Q in river gets effected at junctions by back water flow. However Q in canal gets effected by increased head loss, rise in FSL in canal, reducing the discharging head. The effect of other parameters as discussed in Para 4.3 can be summed up as under.

(a) $X \alpha Q$ (see 1 (ii), Para 4.3) i.e. X can be adjusted to carry the same discharge.

(b) $N \alpha Q^{1/3}$ (see 2 (vii), Para 4.3) i.e. N slightly increases with increase in Q.

(c) $S \alpha 1/Q^{2/3}$ (see 3 (viii), Para 4.3) i.e. energy slope (or bed slope for uniform flow will adjust by aggradations/degradation (sediment deposit or erosion). With increase in discharge S will slightly increase.

(d) $V \alpha Q^{1/4}$ (see 4 (x), Para 4.3) i.e. velocity increases with increase in Q.
Again velocity is a function of X, Z, N, S, see Para 4.3.

Therefore Sediment transport can be easily related to Q and V, or Froude number Fr and size of sediment D sediment transport Qs = f (Q, V, or Fr, D) and researchers should attempt upon it.

Silt clearance is utilized in strengthening the canal banks and service road in filling, though it is costly and cannot be favoured in design. It is avoidable only through silt excluders, ejectors, silting tanks etc. and should invariably be provided and operated.

Study and practical design for stable channel has progressed along two different lines. In one approach, imposed and resistive forces on channel boundary are considered to result in stability or other wise. Primary force involved is boundary shear, also termed as tractive force. Theoretical approach along this line is known as 'tractive force method'. Principal contributors of this school are Lane and others of the study team of the U.S.B.R. (Para 7.10).

The other school is older and empirical given by Lacey, Kennedy, Inglis and others (Table 7.11). Their attempt was to determine simple emperical relationship between various canal dimensions and flow parameters, on the basis of observed data from stable canals. The method provided a practical and simple tool to design engineer, where none existed before. This achieved immense popularity for the method and firmly established itself, especially in India and commonly known as regime theory.

Critical velocity concept

Perhaps the simplest of all bed traction theories, is the one based on critical velocity concept. Most common observation that sand particles are moved by smaller velocities while boulders require much higher velocities would suggest that some correlation may exist between the type and size of material and corresponding competent velocity necessary to either just move or keep it in motion or just permit deposition. Kennedy's concept seems much better in this regard.

A comparison of all the methods of canal design is discussed in Para 7.16 and Table 7.18 with their limitation. Each method has its own limitation and may not serve all the qualities of a good canal as discussed in Para 9.1. Comparison of Kennedy's and Manning's formula is made in Para 7.15 and it is seen that Kennedy's velocity are very reasonable and near to most hydraulic efficient section, and follows laws of hydraulics.

None of theories take care of all instability in canal flow and as well explain complete hydraulic of flow. Therefore hydraulics of flow in day to-day operation is computed by Manning's formula. Also see para 7B.3. With discussion of all the theories complexities, a unified procedure of design for both lined and unlined canals, with the help of Kennedy's and Manning's formula is proposed in Para 7.19.

7B

Review of Indira Gandhi Main Canal (IGMC)

Formerly Known as Rajasthan Main Canal, India and Lessons for the Future

A brief description of IGMC is given in Para 6.6 with details of the canal sections. Here an attempt is made to review the hydraulic performance of the Main Canal.

7B.1 Study of Flow Characteristic of the Main Canal

The main canal is designed for a maximum discharge of 524 m3/s at head i.e. at Km. 0. Large of number of branches, distributaries and minors off take from the main canal at different location all along the canal. These are shown in the L- section, Figure 7B.1 and a statement off taking canals is given Table 7B.1. Many distributaries such as lift canals have been planned subsequently. Also discharges of few off taking channels have been changed. A clear and updated data are not readily available. So data available are indicated in the Table and plotted in the L - section. A half bottom plan of the canal is also shown in Figure 7B.2

The L-section shows following important features

(i) The bed slope is very mild, 1 in 12000 and uniform through out the entire length.

(ii) Canal has a maximum depth of 6.4 m at head to 5.06 m at tail (Km 444). Decrease in depth is done by an up step in the canal.

(iii) Designed discharge varies from 524 m^3/s at head to 151 m^3/s in tail.

(iv) B/D = X ratio varies from 11 near head to 2 in tail.

(v) Maximum velocity of flow at designed discharge is 1.524 m/s at head and gradually decreases to 1.18 m/s at the tail.

(vi) There are 9 cross-regulators in the main canal and flow is regulated through them.

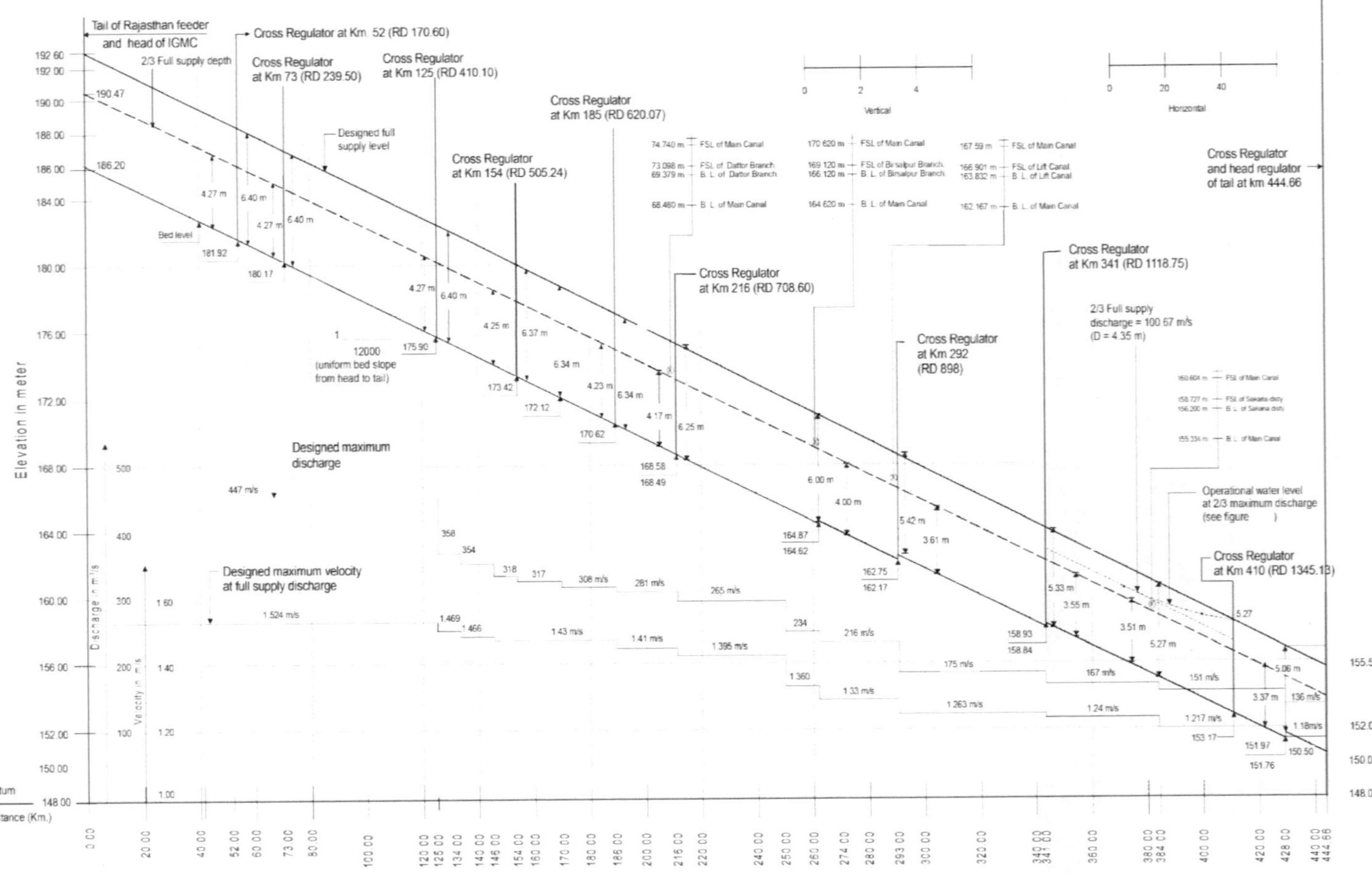

Figure 7B.1 : L - Section of Indira Gandhi Main Canal (IGMC) Showing designed behaviour operational behaviour near Km 410 (see chapter 6 for details)

Note: All off taking channels not shown. For details see table 7B.1 *Source* IGNB - Revised Project Estimates (1996)

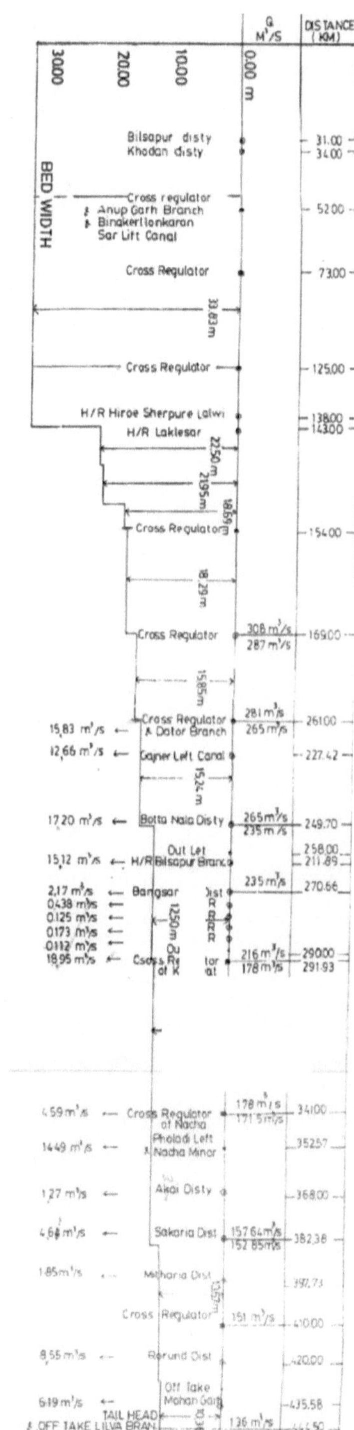

Figure 7B.2 : Half Plan of Indira Gandhi main Canal showing off taking channel and bed width.

Source : IGMN Revised Project Estimates

Table 7B.1 : Details of off Taking Channels of Indira Gandhi Main Canal, (a) Stage I from Km 0 to Km189.

		MAIN CANAL				OFF TAKING CHANNELS *		
S. No.	Km	H/R of Off Taking Channels and Cross Regulator's	Discharge, Q (m^3/s)	Full Supply Level, FSL (m)	Bed level (m)	Discharge, Q (m^3/s)	Full Supply Level, FSL (m)	Bed level (m)
1	2	3	4	5	6	7	8	9
1	31	*Birsalur disty*	447	190.057	183.657	N.A	N.A	N.A
2	34	Khodan disty	447	189.84	183.44	N.A	N.A	N.A
3	52	Anupga rh Branch	447	188.32	181.92	N.A	N.A	N.A
4	52	Bikaner Loonkaran Sar (Kanwar Scn Lift Canal)	447	188.32	181.92	N.A	N.A	N.A
5	52	Cross -Regulator	447/447	188.32	181.92	N.A	N.A	N.A
6	73	Cross -Regulator	447/447	186.57	180.17	N.A	N.A	N.A
7	125	Cross Regulator	447/358	182.30	175.9 0	N.A	N.A	N.A
8	138.8	H/R Hiroe sherpura & Lalwali disty.	358	181.09	174.69	N.A	N.A	N.A
9	143	H/R Laklesar disty	318	180.79	174.33	N.A	N.A	N.A
10	154	Cross Regulator	318/317	179.49	173.12	N.A	N.A	N.A
11	189	Cross Regulator	308/281	176.96	170.62	N.A	N.A	N.A

* Incomplete due to unavailability of data. N.A = not available.

Table 7B.1 : Details of Off Taking Channels from Indira Gandhi Main Canal, (b) Stage II from Km 189 to Km 444 (Tail).

			MAIN CANAL			OFF TAKING CHANNELS		
S. No.	Km	H/R of Off Taking Channels and Cross Regulator's	Discharge, Q (m^3/s)	Full Supply Level, FSL (m)	Bed level (m)	Discharge Q (m^3/s)	Full Supply Level, FSL (m)	Bed level (m)
1	2	3	4	5	6	7	8	9
1	216	**Cross regulator**	281/265	174.740	168.480			
2	216.43	H/R of Dattor branch, Cross Regulator	281	174.634	168.294	15.83	173.098	169.379
3	227.42	H/R of Gajner lift	265	173.644	167.395	12.664	173.644	170.73
4	249.70	H/R of Botta Nala Disty	252.506	171.861	165.609	17.021	170.154	167.814
5	258	Direct outlets	234.927	171.17	164.917	0.485	170.35	169.55
6	261.89	H/R of Bilsalpur Branch	235	170.852	164.604	15.115	169.120	166.120
7	270.66	H/R of Bangsar lift disty	219.885	170.360	164.335	2.166	169.120	167.620
8	275.85	Minor	217.719	169.58	163.58	0.113	167.78	166.28
9	280.42	Minor	217.606	169.17	163.17	0.154	166.67	166.47
10	284.38	Minor	217.452	168.84	162.84	0.127	167.34	166.14
11	286.52	Minor	217.325	168.66	162.66	0.112	166.96	165.96
12	290	Direct out lets	217.213	168.37	162.37	1.719	168.33	167.33

Contd...

Table 7.B.1. Contd...

S. No.	Km	H/R of Off Taking Channels and Cross Regulator's	MAIN CANAL			OFF TAKING CHANNELS		
			Discharge, Q (m^3/s)	Full Supply Level, FSL (m)	Bed level (m)	Discharge Q (m^3/s)	Full Supply Level, FSL (m)	Bed level (m)
13	291.93	H/R Kolayat lift canal	215.00	167.59	162.170	18.95	166.901	163.832
14	291.93	Cross Regulator	215.00/175	167.59	162.170			
15	293	Charanwala Branch	175	167.34	161.92	15.251	166.560	164.060
16	301	Direct Outlet	159.749	166.67	161.25	0.830	162.478	161.478
20	352.57	Nachina Minor	158.919	162.37	156.95	0.190	158.049	157.449
21	368	Awai distributory	158.729	161.08	155.66	1.265	158.498	157.449
22	384.05	H/R of Sarkria disty	157.464	160.604	155.334	4.615	158.727	156.200
23	397.73	H/R of Mitharia disty.	152.849	159.553	154.279	1.789	157.553	156.653
24	410	Cross regulator	151	158.44	153.17			
25	416.04	H/R of Rorund disty	149.849	158.029	152.755	8.549	154.414	153.621
26	428.25	Mohan garh disty.	141.300	157.012	151.738	5.048	155.216	53.216
27		H/R of Digha branch , Tail IGMC	136.252	159.450	150.570	40.006	155.450	153.450
28	444	H/R of Lilva branch,	136.252	159.450	150.570	99.246	155.450	152.570

The maximum discharge is to pass only for a short duration in year. For the rest of period discharge varies according to capacity factor. The capacity factor in lean month is around 0.55 and generally around 0.6 to 0.7, except few months when it may be 0.8 or 0.9. It is 1 only in peak demand of 1 month.

With decrease in discharge velocity further decreases. At the cross-regulators the velocity is not uniform. It varies gradually according to discharge, see figure 7 B.6.

The canal alignment passes through undulating topography of high sand dunes and valleys. Two typical cross-sections of heavy cutting and filling are shown in Figure 7B.3 and 7B.4

It is reported that the canal get silted due to heavy dust storms. There is no silt ejector in the canal.

In this study an attempt is made to know the flow characteristics of main canal, sediment transporting capacity and comparison with the designed practices.

7B.2 Effect of Cross Regulators

From consideration of regime of channels it will be better if there are no structures across any channel but to carry on regulation without cross regulators is very difficult. The absence of cross regulators envisages constant supply to all off taking channels. But the demand on any channel is not constant with respect to time. Also a particular channel may have to be closed due to some repairs or on account of breach in it. The supply in the parent channel is not constant. If the supply is reduced the level will fall and other off taking channels will not be able to draw full supply discharge. Thus cross regulators are a necessary evil. Still these are unavoidable and have fallowing advantages.

Advantages of cross regulators

(i) When there is no head regulator on the main canal the cross regulators to a certain extent minimize the disadvantages for want of head regulator.

(ii) When the water level in the main canal is low, they help in raising the water level and feed the off taking channel.

(iii) They help in raising water level above the designed and thus give full supply to lands slightly above the command level of the canal.

(iv) They enable parent channel to be divided into sections for easy regulation.

(v) They help in absorbing fluctuations in various sections of the canal and thus reduce possibility of flooding tail reach and causing breaches there. The excess discharge is therefore retained all along the canal.

(vi) They help in closing of breaches in lower sections.

(vii) They facilitate working of the whole system by rotation in days of low supply and thus reduce silting in the branch canals.

It is a well-known principle in irrigation engineering that a canal should either run full or dry. When designed for high level and if it runs at low level, considerable silting occurs as the ratio of mean to critical velocity falls.

(viii) They help in increasing revenue by ensuring full supply discharge to most of the tracts.

(ix) They facilitate construction of road bridges with little additional cost.

(x) If the cross regulators were not there all the branches could not be designed with a high full supply level so that much of the area will be thrown out of command.

Disadvantages of cross regulators

(i) While cross regulators on the parent channel prevent silting of off taking channels by making system of rotation possible, but they by heading up water, cause the parent channel to silt. Part of this silt may be washed away when the cross regulator is opened. Still it must be admitted that too frequent heading up in case of a canal provided with many cross regulators will affect its working.

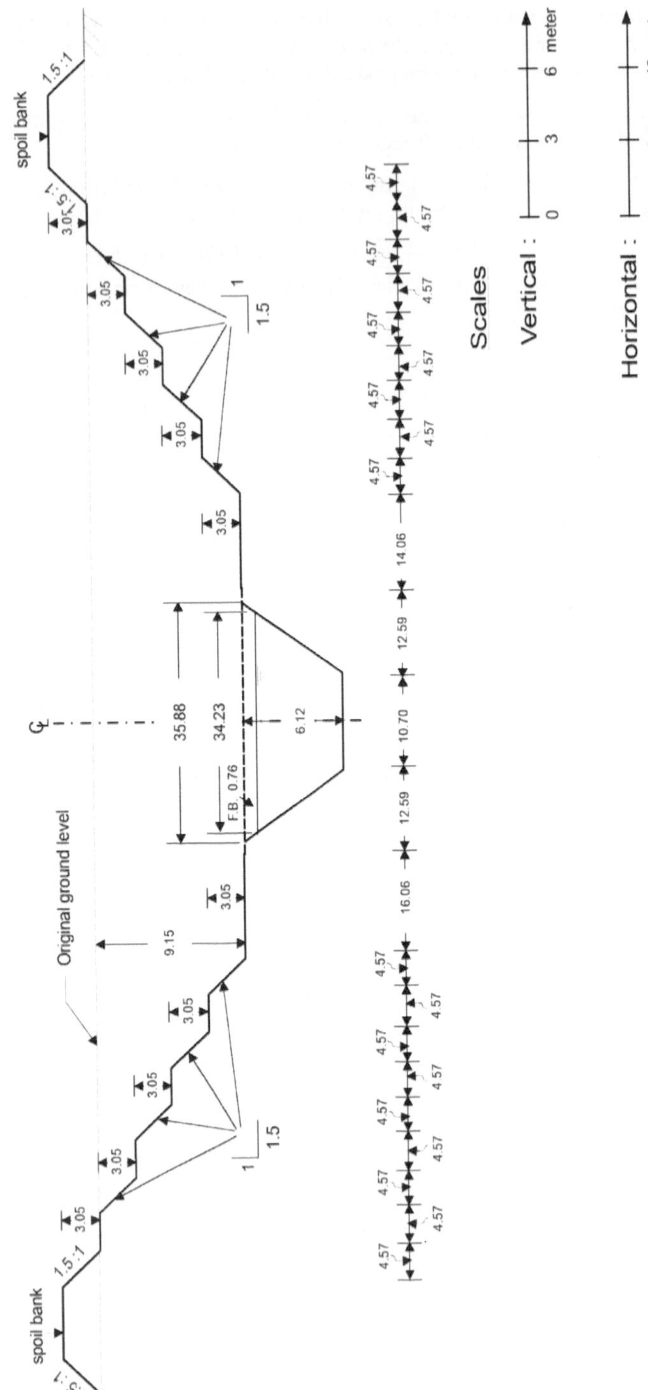

Figure 7B.3 : General Typical Cross Section of Hevey Cutting Above Bed up to 15.25 m of Indira Gandhi Main Canal, Internal Sections between Km. 384 to 410 (RD 1260 - 1345)

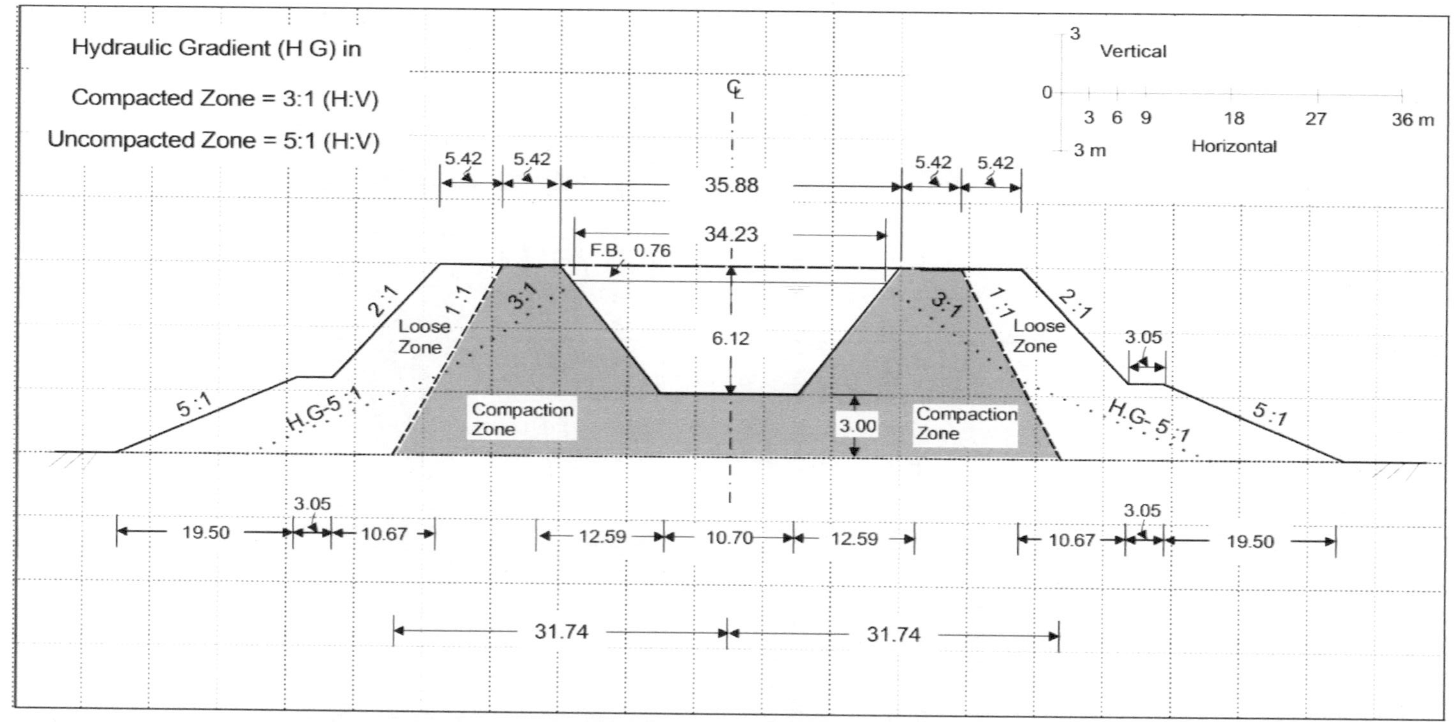

Figure 7B.4 : General Typical Cross Section of Hevey Bank Filling up to 9.00 m of Indira Gandhi Main Canal between Km. 380 to 410 (RD 1260 - 1345)

A relieving factor however is that in the lower reaches where the cross regulators are more frequent, water is usually clearer.

(ii) Cross regulators put in an undue large power in the hands of low paid establishment, who for personal gain are apt to misregulate. A strict watch, control by rotation and surprise visits, though are the remedies for this, but cost much.

On the whole advantages and convenience from regulators outweigh their few disadvantages.

7B.3 Methodology of Computation of Flow Profile Upstream of A Cross Regulator

By heading up water during low supplies to maintain FSL, a backwater curve is formed for some distance upstream of the cross regulator. The computation of flow profiles involves basically the solution of the dynamic equation of gradually varied flow (Chow 1973). Broadly there are three methods of computation; namely.

(i) The graphical integration method.

(ii) The direct integration method

(iii) The direct step method

The direct step method is a simple step method applicable to prismatic channels. It is characterized by dividing the channel into short reaches, and computing step by step form one end of the reach to the other.

According to the principle of conservation of energy, the total energy head at the upstream section (i) should be equal to the total energy head of the downstream section (ii) plus the loss of energy between the two sections, see figure below.

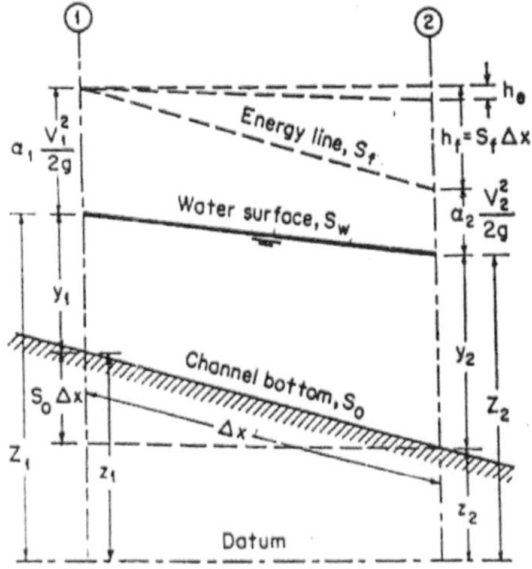

A Channel Reach for The Derivation of Step Method

$$S_0\ \Delta x + y_1 + \alpha_1 \frac{V_1^2}{2g} = S_f\ \Delta x + y_2 + \alpha_1 \frac{V_1^2}{2g} + h_f$$

Solving for Δx, we get

$$\Delta x = \frac{E_2 - E_1}{S_0 - S_f} = \frac{\Delta E}{S_0 - S_f}$$

and for low velocity uniform flow $\alpha_1 = \alpha_2 = 1$ and for short length, hf may be taken equal to 0.

$S_0 \Delta x + E_1 = S_f \Delta x + E_2$ Also, $y = V^2/2g = E$, where E is specific energy

That is specific energy is equal to the sum of the depth of water and the velocity head measured with respect to the channel bottom the total energy in the flow of the section with reference to a datum line is the sum of the elevation z, the piezometric height y, and the velocity head $V^2/2g$. Where y the depth of flow, V the mean velocity, α is the energy coefficient = 1, S_0 = the bottom slope, S_f = the friction slope. S_f= the average value of S_f. In Manning's formula, the friction slope is expressed by

$S_f = \frac{N^2V^2}{R^{4/3}}$ Where, N = roughness; V= mean velocity; R = hydraulic radius

This requires a discharges rating curve of the canal. Computations of this are done in Para 7B.4 And Table 7B.2 It is shown in Figure 7B.5 The computation of backwater are done with the with the help of this Figure and by the above step by step procedure as explained in Table 7B.3 The explainatory notes for simplicity are also given below the Table.

The back water curve is plotted in Figure 7B.6. This curve is computed for 2/3 full supply discharge, as the same is considered most predominant.

7B.4 Development of Flow Rating Curve of Indira Gandhi Main Canal Between Km 384 to Km 410 (RD 1260 to RD 1345) for Uniform Flow

Project Data

Channel Parameters,

Discharge, Q = 151,375 m3/s ; Bed width, B = 10.67 m ; Depth, h = 5.27 m

Bed slope, S0 = 1/12000 ; Roughness, n = 0.017 ; Side slope, H: V = 2:1

Cross section of canal, trapezoidal section with curved ends as below:

Cross Section of I.G.M.C between Km 380 to Km 410 (RD1260 to RD 1345) Computation of discharge versus gauge

(i) Up to ht. 0.557 m - curved portion

$A = B \times h + 2\ (\pi h^2\phi/360 - ½\ 4.713 \times 2.357)$

$= 10.67 \times 0.557 + 2[\pi \times (5.27)^2 \times 26.57/360 - ½\ 4.713 \times 2.357]$

$= 5.943 + 1.771 = 7.714\ m^2$

$P = B + 2\ (2\pi h \times \phi/ 360)$

$= 10.67 + 2(2\pi \times 5.27 \times 26.57/360) = 15.558$ m

$R = A/P = 7.714/15.558 = 0.496$ m

$V = R^{2/3} \times s_0^{1/2} / n$

$= (0.496)^{2/3} \times (1/12000)^{1/2} / 0.017 = 0.336$ m/sec

$Q = A \times V = 7.714 \times 0.336 = 2.596$ cumecs

(ii) Similar calculations are done for every 0.5 meter depth up to F.S.L. i. e. up to height 1.00 m, 1.5 m,2.0 m,2.5 m,3.0 m, 3.5 m,4.0 m, 4.35 m, 4.5 m, 4.6 m,4.8 m,5.0 m, and lastly up to height 5.27 m (F.S.L.).

$A = 115.529 + [(33.156 + 34.236)/2] \times 0.27 = 124.627\ m^2$

$P = 35.427 + 2\sqrt{(0.27)^2 + 0.54^2} = 36.634$ m

$R = 124.627/36.634 = 3.402$ m

$V = (3.402)^{2/3} \times (1/12000)^{1/2} / 0.017 = 1.215$ m/sec

$Q = 124.627 \times 1.215 = 151.37$ cumecs

Table 7B.2 : Flow Characteristics of Indira Gandhi Main Canal from Km 384 to 410 Km (RD 1260 to1345) for Uniform Flow, Project Design

S.No.	Depth of Flow Above Bed Y (m)	Top Width T (m)	Cross-Sectional Area of Flow A (m^2)	Wetted Perimeter P (m)	Hydraulic Radius R = A/P (m)	Velocity V (m/s)	Discharge Q = A.V (cumecs)	Froude Number $F_r = V/\sqrt{gA/T}$ No.
1	2	3	4	5	6	7	8	9
1.	0.557	15.384	7.714	15.558	0.496	0.336	2.596	0.1515
2.	1.000	17.156	14.905	17.539	0.850	0.482	7.182	0.1651
3.	1.500	19.156	23.983	19.775	1.213	0.611	14.698	0.1743
4.	2.000	21.156	34.061	22.011	1.547	0.718	24.470	0.1807
5.	2.500	23.156	45.139	24.247	1.862	0.813	36.686	0.1860
6.	3.000	25.156	57.217	26.483	2.161	0.897	51.348	0.1899
7.	3.500	27.156	70.295	28.719	2.448	0.975	68.557	0.1935
8.	4.000	29.156	84.373	30.955	2.726	1.048	88.405	0.1967
9.	4.500	31.156	99.451	33.191	2.996	1.116	110.993	0.1994
10.	5.000	33.156	115.529	35.427	3.261	1.181	136.429	0.2020
11.	5.270 (Full supply depth)	39.230	124.627	36.634	3.402	1.215	151.375 (Full supply discharge)	0.2033

Note : For detailed calculation see para 7B.4, development of flow rating curve of Indira Gandi Main Canal at Km 410.

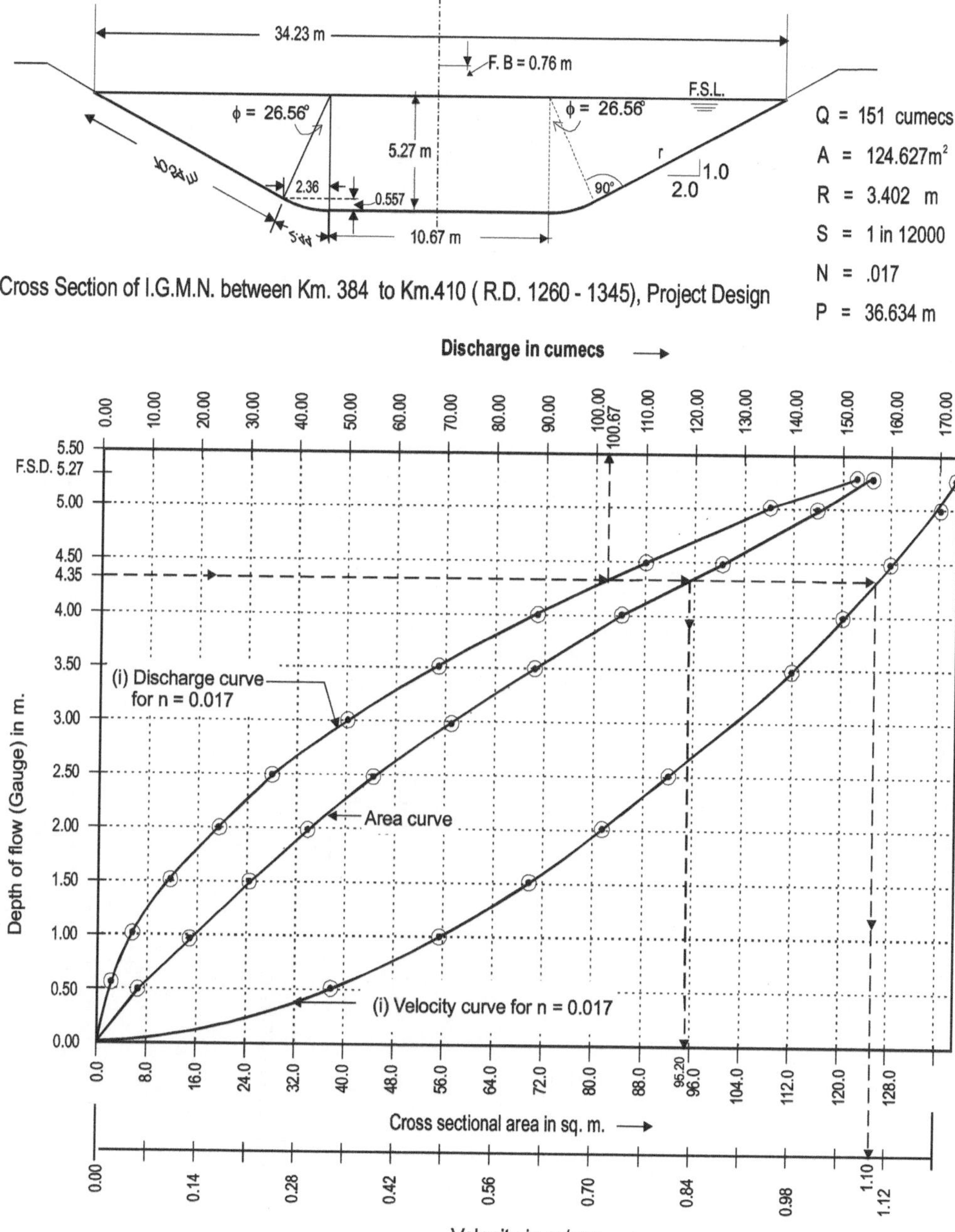

Figure 7B.5 : Canal Rating Curve or Flow Characteristics in Trapezoidal Channel With Curved ends, Indra Gandhi Main Canal between Km. 384 to Km. 410 (RD 1260 to RD 1345), Project Design

Notes : To know discharge for a given gauge move arrow horizontally to meet discharge curve and then vertically upwards and read discharge. Similarly for area move upto area curve and down wards to read area scale. Similarly for velocity , move upto velocity curve and than down wards to read velocity scale.

Table 7B.3 : Indira Gandhi Main Canal, Computation of the Flow Profile Upstream of Cross Regulator at Km 410 (RD 1345), [Back Water Profile] for Gradually Varied Flow, for Q = 2/3 Full Supply Discharge = 100.67 cumecs , N = 0.017, S_0 = 1/12000, α = 1. 00, Project Design

S. No.	Depth Above Bed, h	Cross-Sectional Area of Flow, A	Hydraulic Radius, R=A/P	$R^{4/3}$	Velocity V=Q/A (100.67) / Col. 3	$\alpha v^2/2g$	Specific Energy E	ΔE	S_f $\left(\frac{N^2V^2}{R^{4/3}}\right)$	$\overline{S_f}$	$S_0 - \overline{S_f}$	Δx, up Stream of Regulator	X, up Stream of Regulator (Distance)
	(m)	(m^2)	(m)	($m^{4/3}$)	(m /s)	(m)	(m)	(m)				(m)	(m)
(1)	(2)	(3)	(4)	(5)	(6)	(7)	(8)	(9)	(10)	(11)	(12)	(13)	(14)
1.	5.27	124.627	3.402	5.117	0.808	0.033	5.303	-	0.0000369	-	-	-	-
2.	5.00	115.529	3.261	4.836	0.871	0.039	5.039	0.264	0.0000453	0.0000411	0.0000422	6,256	6,256
3.	4.80	108.888	3.153	4.623	0.925	0.044	4.844	0.195	0.0000535	0.0000494	0.0000339	5,752	12,008
4.	4.60	102.587	3.050	4.423	0.981	0.049	4.649	0.195	0.0000629	0.0000582	0.0000251	7,769	19,777
5.	4.35	94.823	2.916	4.166	1.062	0.061	4.407	0.242	0.0000782	0.0000706	0.0000127	19,055	38,832

Explainatory Notes:

1. X, h and V are plotted in figure 7B.6
2. Explainatory notes are given at the end of Para 7B.4.

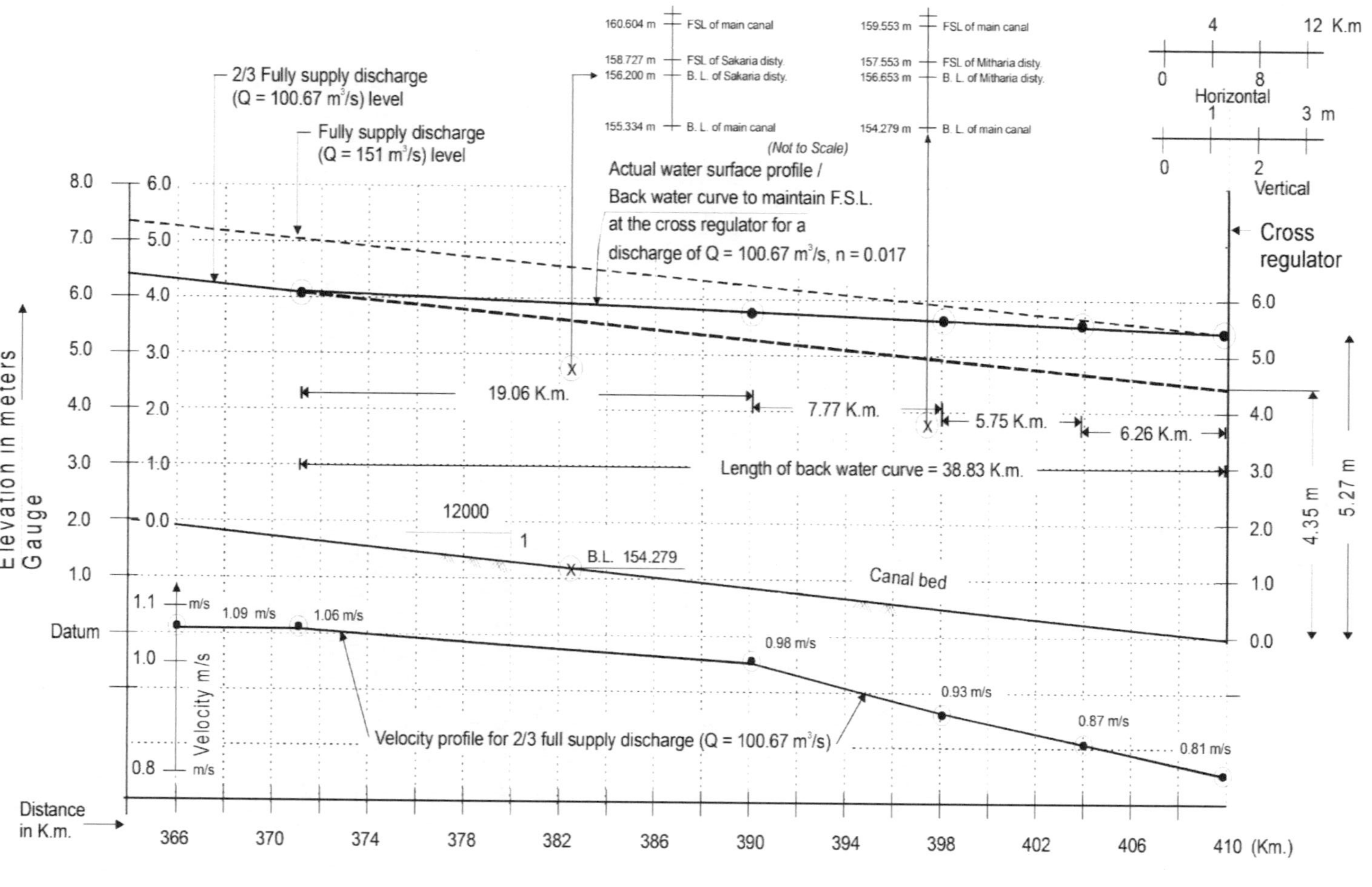

Figure 7B.6 : Velocity and Back Water Surface Profile at Cross Regulator at Km. 410 (RD 1345.13) To Maintain FSL For 2/3 Discharge in Indra Gandhi Main Canal (IGMC) For Existing / Project Design

All these data are tabulated in Table 7B.2 and plotted in Figure 7B.5 which gives discharge rating curves. From this table, discharge can be read for any gauge (stage/depth of flow) or vice versa. This table is further used in development of backwater curve.

Explanatory notes to Table 7B.3

Col.2 Depth of flow above bed is m, arbitrarily assigned from 5.27 o 4.35 m.

Col.3 Water area in m^2 corresponding to the depth y in col.2, where A=94.667 taken from canal rating at discharge curve, Figure 7B.5.

Col.4 Hydraulic radius from Table 7B.2 in m corresponding to y in col.2.

Col.5 Four thirds power of the hydraulic radius (R), i.e. $R^{4/3}$; row (1), $R^{4/3} = (3.402)^{4/3} = 5.117\ m^{4/3}$

Col.6 Mean velocity in m/s obtained by dividing 100.67 cumecs by the water area in col.3 = Q/A; for row (1), V = 100.67/124.627 = 0.808 m/s

Col.7 Velocity head in m, $\alpha V^2/2g$; for row (1), $\alpha V^2 / 2g = (1 \times 0.808^2)/ (2 \times 9.81) = 0.033$ m

Col.8 Specific energy in m obtained by adding the velocity head in col.7 to the depth of flow in col.2 i.e. col. 2 + col. 7; row (1), E = 5.27 + 0.033 = 5.303 m

Col.9 Change of specific energy in m, equal to the difference between the E value in col.8 and that at the previous step. i.e. row (2)-(1) and so, row (2): ΔE = 5.3035.039 = 0.264 m

Col.10 Friction slope computed by $n^2V^2 / R^{4/3}$ with n = 0.017 and with V as given in col.5 and $R^{4/3}$ in col.5. row (1); In row (1) $S_f = (0.017^2 \times 0.808^2) / 5.117 = 0.0000369$

Col.11 Average friction slope between the steps, equal to the arithmetic mean of the friction slope just computed in col.10. And that of the previous step i.e. row (1); row (2):
$S_f = (0.0000369 + 0.0000453) / 2 = 0.0000411$

Col.12 Difference between the bottom slope 0.00008 and the average friction slope; row (2):
$S_0 - S_f = 0.0000833 \quad 0.0000411 = 0.0000422$

Col.13 Length of the reach in m between the consecutive steps, by dividing the value of ΔE in col.9. by the value in col.12 ; row (2) : $\Delta X = 0.264 / 0.0000422 = 6.256$ m

Col.14 Distance from the section under consideration to the gate site (upstream from regulator); row (2): X = 6,256 + 0 = 6.256 = 6,256 m.

7B.4.1 Inferences and lessons

With 2/3 discharge i.e. 151x2/3 = 100.67m^3/s. the depth for uniform flow is 4.35 m. In order to maintain FSL at the cross regulator, it is partially closed. The water level at upstream of the regulator is maintained at FSL (corresponding to 151 m^3/s) and there is a gradually varied flow in a length of 38.8 Km (known as back water curve). The velocity at uniform flow for 100.67m^3/s is 1.09 m/s and it goes on reducing in the zone of back water till the regulator to 0.808 m/s. See Figure 7B.6 therefore this zone shows a higher rate of silting, also see Para 7B.5.1 and 7B.5.2. With silting cross sectional area goes on decreasing, there by further reducing discharge. With reducing carrying capacity, water level starts heading upstream, endangering safety of canal.

Though the purpose of cross-regulator is to provide FSL to all off taking channels in its zone in the up stream, yet it may not meet the full requirement. In this example the Sakaria distributory does not get the same design head (H) for 2/3 Q in the main canal and H is much less than that at Q. It is, because that the back water profile for 2/3 Q does not have the designed FSL at this point of off take. Now if the off taking channel is silt free, FSL may not be visible in it at head of full gate opening. Therefore excess gate opening is normally done to pass the required discharge (some time appear excess Q), which may or may not happen depending upon the flexibility of gate design. But normally there is some flexibility.

When the off taking channel is silted or has weed growth, the FSL in the channel will rise due to excessive head loss showing apparently F.S.Q. actually that may not be there.

When the main canal is silted, the effect can be seen in figure 7B.10. The length of back water profile is slightly more and there may be more head available, still less than designed H. Also see Para 7B.6.2.

To avoid the above problem, solution is to provide more free board at the cross regulator and raise FSL at it so that the back water profile for 2/3 Q provides the design head to each distributory with in the zone of this cross regulator. That means this extra free board may be as much as the difference between the FSL and the uniform flow level for 2/3 FSQ. In this example this is equal to 5.27 - 4.35 = 0.92 m. This will increase the cost of lining in the back water length of 38.8 km., corresponding to depth equivalent to half of the difference i.e. 0.46m depth. It can also be a vertical wall section.

Another alternative is to have a top level bank in this reach, which may be called as a top level bank canal. Also see Para 7B.5.4, 7B.6.2 and 9.14.1 and 9.14.2.

Since the cross regulator is provided only in the cutting reach, (from canal safety point) the cost of earthwork will not increase, rather decrease due to higher top Bank level, see figure 7B.3. Also service road need not be parallel to FSL or canal bed level. On some canals, service road is not parallel to the FSL, but follows up and down grade, according to topography, i.e. up grade is provided in cutting reaches and brought to a down grade in filling reaches and then parallel to free board or FSL. Typical example is Sunder Nagar Hydel channel. Earthwork is considerably economized in such canals.

Similar curves are required at every regulator.

7B.4.2 Modification of Existing Canals

By suitable raising the top bank levels in the upstream of cross-regulators.

The existing unlined/lined canals can also be improved/modified to resolve the problem of low head feeding of distributaries/minors, without constructing more cross-regulators/across walls in the canals. The earthwork for raising may be from silt clearance or from the heavy cuttings itself or from spoils

Even otherwise raising the top bank levels is more beneficial for easy regulation, better feeding of tails, and canal safety. However proper drawings of L-section and cross-section should be prepared before such raising.

7B.4.3 Discharge through cross-Regulator

(i) The discharge through the cross-regulator depends upon (i) gate opening and (ii) the difference of water level at the upstream of regulator and down stream of regulator, not immediately, but at a sufficient distance after the hydraulic jump, where stable levels are available. The water levels in the hydraulic jump are of no use, as it shows the levels corresponding to pre jump depth (d1).

Generally cross-regulator discharge curves are prepared only on the gate opening, treating that the difference of water level remains constant. But this is not so. The water level upstream of cross regulator depends upon heading of water and in down stream depends upon the down stream channel conditions. Both are independent.

7B.5 Sediment in Flow and Its Transport

Lot of sediment (fine sand for classification, see Figure 7B.7) enters in the entire length of the canal, through dust storms, a peculiar phenomena of the desert area. General typical cross-sections shown in Figure 7B.3 and Figure 7B.4, show that the canal passes through heavy cutting and filling. A review of L- Section indicates that the entire alignment is in alternate heavy cutting and heavy filling, again a peculiar phenomenon of desert dune areas. More sediment enters in heavy cutting and adjoining filling reaches.

Grain Size Distribution (Sieve analysis)

Diameter (mm)	% finer
710	100
500	100
300	99
212	95
150	46
075	1
063	0

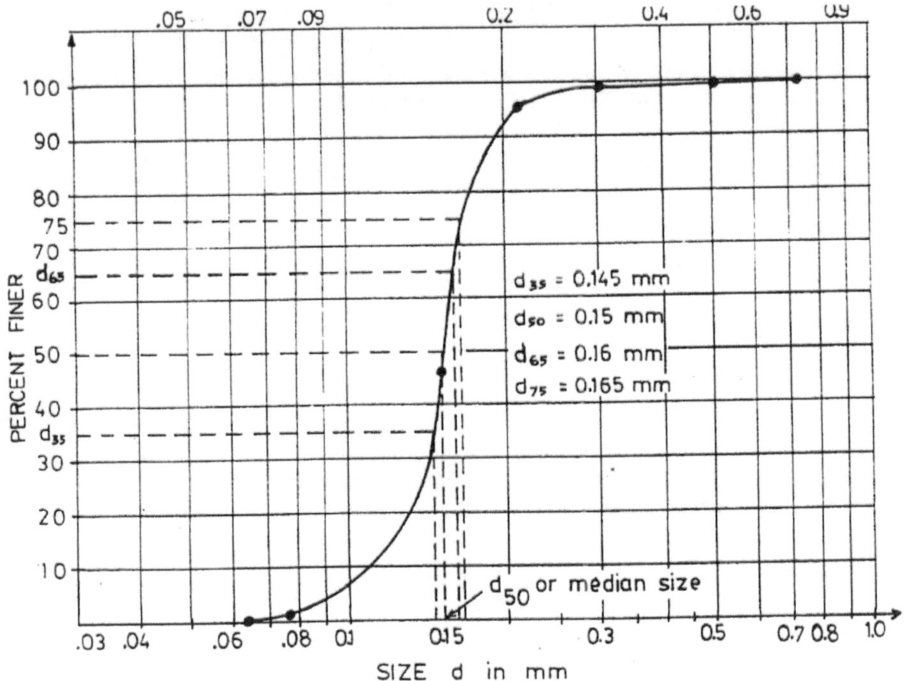

Figure 7B.7 : Cumulative Soil Classification Curve, IGMC

Though plantations along the canal have been done in the area, yet the magnitude of reduction in sediment is not certain. Silt clearance is done almost every year, but detailed analysis of observations of sediment is not available. The aspect of sediment transport is not available in the revised project report.

Also because of high level of off taking channels, as compared to bed level of main canal very little silt go out of main canal.

Here an effort is made to know the sediment transporting capacity of channel.

Temperature

Indira Gandhi canal suffers extreme variation of temperature. Winter is quite cold and at places the mean minimum temperature is normally recorded in the month of January and varies from 4.7 to 7.9°C. The hottest months are from April to September with the peak temperature being mostly in the month of May when the mean maximum temperatures vary from 41.5 to 42.0°C. Even in a season day and night temperatures vary considerably.

Wind

The general wind direction in the region is southwest. The wind speed remains highest in Bikaner and Phalodi and Jaisalmer through out the year and gradually decreases as one move towards North-East. Dust storms are very common during summer when hot winds prevail. Maximum numbers of dust storms occur in April to June. Due to poor rainfall, humidity, in this tract is extremely low, sand storms are of frequent occurrence during major part of the year. Though good vegetation, have been developed all along the canal side, yet sand storms are frequent. Though with increase is cultivated area, sand storms will reduce, yet will continue considerably, due to a high ratio of gross uncommand to command area.

7B.5.1 Total Sediment Transport by Einstein Method, for Uniform Flow between Km 384 to Km 410 of IGMC At 20°C of Water, Medium Season - March, July to November, for Project Designed Section.

Project Data

(i) Channel parameters

Discharge Q = 151.375 m/s ; Bed width, B = 10.67 m; Top width , B_T = 34.23 m

Full supply depth, D = 5.27 m : Bed slope, $S_0 = 1/12000$

Average bed width, Ba = $\frac{B+B_T}{2} = \frac{10.67+34.23}{2}$ = 22.45 m

Flow characteristic of uniform flow are given in Table 7B.2 and Figure 7B.5

(ii) Water properties

Temperature of water, T = 20° C ; Density at 20° C, = 998.2 Kg/m^3

(See Annexure -1)

Specific weight of fluid, γf = $1 Ton/m^3$; Viscosity at 20° C, $\nu = 1.003 \times 10^{-6}$ m/s (see table 7B.17)

(iii) Sediment properties

Diameter of sediment, d_{35} = 0.145 mm, d_{50} = 0.15 mm, d_{65} = 0.16 mm

(From Figure 7B.7)

Density of solid particle, ρ_s = 2650 Kg/m^3 ; Specific weight of sediment,

γ_s = 2.65 Ton/m^3 ; Settling velocity, v_{ss} (d35) = 0.017 m/s (from Figure 7.28)

(iv) Computations of sediment transport

Sediment transport is calculated by Einstein's method given at 7.6.1 in Table 7.6, for different depths of flow in the main canal. The depth of flow will change with the change in discharge. The computations are done in a tabular manner as given in Table 7B.4. Explanatory notes for calculations are also given below the Table. The total sediment transport (discharge) is plotted in Figure 7B.8 versus gauge and also shown against corresponding discharge.

Explanatory notes to Table 7B.4

Col.2.	h	Flow depth (from Table 7B.2)
Col.3.	R	Hydraulic radius (from Table 7B.2)
Col.4.	V_*	Friction velocity, $V_* = \sqrt{gRS_0}$
Col.5.	δ	Thickness of viscous sub layer, $\delta = 11.5\ \nu/V_*$; row (1); $\delta = 11.5\ (1.003 \times 10^{-6}/0.02)$ $= 5.767 \times 10^{-4}$ m

Table 7B.4 : Indira Gandhi Main Canal, Computation of Stage – Solid – Discharge Curve at 20 °C, by Einstein Method.

S. No	Flow Depth h (m)	Hydraulic Radius R (m)	Shear Velocity V_* (m/s)	Thickness of Viscous sub layer δ (m)	Relative Roughness k_s / δ	Correction Term X	Apparent Roughness Diameter Δ (m)	Transport Parameter Pe	Intensity of Shear ψ_*	Intensity of Transport Φ_*	Bed Load Transport: Volume by unit width Q_{sb} ($m^3/s/m$)	Bed Load Transport: Volume Q_{sb} (m^3/s)	Bed Load Transport: Volume Q_{sb} (m^3/day)
1	(2)	(3)	(4)	(5)	(6)	(7)	(8)	(9)	(10)	(11)	(12)	(13)	(14)
1	0.557	0.496	0.020	$5.767x10^{-4}$	0.277	0.600	$2.667x10^{-4}$	11.054	6.029	0.40	$2.810x10^{-6}$	$6.308x10^{-5}$	5.450
2	1.000	0.850	0.026	$4.436x10^{-4}$	0.361	0.673	$2.377x10^{-4}$	11.754	3.518	1.50	$1.054x10^{-5}$	$2.366x10^{-4}$	20.442
3	1.500	1.213	0.031	$3.721x10^{-4}$	0.430	0.764	$2.094x10^{-4}$	12.287	2.465	2.50	$1.756x10^{-5}$	$3.942x10^{-4}$	34.059
4	2.000	1.547	0.035	$3.296x10^{-4}$	0.485	0.891	$1.796x10^{-4}$	12.728	1.933	3.50	$2.459x10^{-5}$	$5.521x10^{-4}$	47.701
5	2.500	1.862	0.038	$3.035x10^{-4}$	0.527	0.964	$1.660x10^{-4}$	13.030	1.606	4.71	$3.309x10^{-5}$	$7.429x10^{-4}$	64.187
6	3.000	2.161	0.041	$2.813x10^{-4}$	0.569	1.018	$1.572x10^{-4}$	13.267	1.384	6.00	$4.215x10^{-5}$	$9.463x10^{-4}$	81.760
7	3.500	2.448	0.044	$2.621x10^{-4}$	0.610	1.055	$1.517x10^{-4}$	13.457	1.222	6.50	$4.566x10^{-5}$	$1.025x10^{-3}$	88.560
8	4.000	2.726	0.046	$2.508x10^{-4}$	0.638	1.127	$1.420x10^{-4}$	13.656	1.097	7.50	$5.269x10^{-5}$	$1.183x10^{-3}$	102.211
9	4.500	2.996	0.048	$2.403x10^{-4}$	0.666	1.182	$1.354 x10^{-4}$	13.822	0.998	7.75	$5.444x10^{-5}$	$1.222x10^{-3}$	105.581
10	5.000	3.261	0.051	$2.262x10^{-4}$	0.707	1.236	$1.295x10^{-4}$	13.972	0.917	8.75	$6.147x10^{-5}$	$1.380x10^{-3}$	119.232
11	5.270	3.402	0.052	$2.218x10^{-4}$	0.721	1.255	$1.275x10^{-4}$	14.040	0.879	9.00	$6.322x10^{-5}$	$1.419x10^{-3}$	122.631

Contd...

Table 7.B.4. Contd...

S. No.	Dimension less Height A_E	Rouse Exponent	Integral I_1	Integral I_2	Suspended Load Transport: Volume by Unit Weight q_{ss} ($m^3/s/m$)	Volume Q_{ss} (m^3/s)	Volume Q_{ss} (m^3/s)	Total Sediment Transport: Volume Q_s (m^3/s)	Volume Q_s (m^3/s)	By Mass G_s (Kg/s)	By Mass G_s (Ton/day)
(15)	(16)	(17)	(18)	(19)	(20)	(21)	(22)	(23)	(24)	(25)	(26)
1	5.207×10^{-4}	2.125	0.182	-1.667	9.700×10^{-6}	2.175×10^{-5}	1.879	8.483×10^{-5}	7.329	0	0
2	2.900×10^{-4}	1.635	0.380	-3.000	1.546×10^{-5}	3.471×10^{-4}	29.989	5.837×10^{-4}	50.432	2	173
3	1.933×10^{-4}	1.371	0.600	-3.500	6.800×10^{-5}	1.527×10^{-3}	131.933	1.921×10^{-3}	165.974	5	432
4	1.450×10^{-4}	1.214	0.900	-5.000	1.587×10^{-4}	3.563×10^{-3}	307.843	4.115×10^{-3}	355.536	11	950
5	1.160×10^{-4}	1.118	1.333	-7.250	3.348×10^{-4}	7.516×10^{-3}	649.382	8.258×10^{-3}	713.491	22	1901
6	9.667×10^{-5}	1.037	1.750	-9.000	5.993×10^{-4}	1.345×10^{-2}	1162.080	1.440×10^{-2}	1244.160	38	3283
7	8.286×10^{-5}	0.966	2.500	-12.000	9.882×10^{-4}	2.219×10^{-2}	1917.216	2.322×10^{-2}	2006.208	62	5357
8	7.250×10^{-5}	0.924	3.250	-15.000	1.548×10^{-3}	3.475×10^{-2}	3002.400	3.593×10^{-2}	3104.352	95	8208
9	6.444×10^{-5}	0.885	4.150	-16.250	2.238×10^{-3}	5.025×10^{-2}	4341.600	5.147×10^{-2}	4447.008	136	11,783
10	5.800×10^{-5}	0.833	5.100	-20.000	3.151×10^{-3}	7.074×10^{-2}	6111.936	7.212×10^{-2}	6231.168	191	16,513
11	5.503×10^{-5}	0.817	5.500	-22.000	3.491×10^{-3}	7.837×10^{-2}	6771.168	7.979×10^{-2}	6893.856	211	18,269

Conclusion: Maximum Sediment discharge capacity at full supply discharge of 151.375 m^3/s = 6893.856 m^3/ day Explainatory notes for computations are given in paragraph 7B.5.1.

Table 7B.5 : Indira Gandhi Main Canal, Computation of Stage – Sediment Discharge Curve at 32ºC for Uniform Flow between Km 384 to Km 410 for Project Section by ACKERS METHOD

S. No.	Depth Above Bed h (m)	Hydraulic Radius R (m)	Shear Velocity V (m/s)	Velocity V (m/s)	Discharge Q (m^3/s)	Parameter of Mobility F_{gr}	Transport Parameter G_{gr}	Concentration by Volume C_s	Total Sediment Transport: Concentration by Volume $C_s \times 10^6$ (ppm)	Total Sediment Transport: Volume Q_s (m^3/s)	Total Sediment Transport: Volume Q_s (m^3/day)	Total Sediment Transport: By Mass. Gs (Kg/s)	Total Sediment Transport: By Mass G_s Tons /day
(1)	(2)	(3)	(4)	(5)	(6)	(7)	(8)	(9)	(10)	(11)	(12)	(13)	(14)
1.	0.557	0.456	0.020	0.336	2.596	0.353	$3.353 \times (10^{-4})$	$5.300 \times (10^{-7})$	0.530	$1.38 \times (10^{-6})$	0.119	0	0
2.	1.000	0.850	0.026	0.482	7.182	0.467	$4.720 \times (10^{-3})$	$4.450 \times (10^{-6})$	4.450	$3.196 \times (10^{-5})$	2.761	0	0
3.	1.500	1.213	0.031	0.611	14.648	0.561	$1.693 \times (10^{-2})$	$1.107 \times (10^{-5})$	11.070	$1.622 \times (10^{-4})$	14.014	0	0
4.	2.000	1.547	0.035	0.718	24.470	0.637	$3.677 \times (10^{-2})$	$1.849 \times (10^{-5})$	18.490	$4.525 \times (10^{-4})$	39.096	1	10.4
5.	2.500	1.862	0.038	0.813	36.686	0.698	$6.182 \times (10^{-2})$	$2.555 \times (10^{-5})$	25.550	$9.373 \times (10^{-4})$	80.983	2	215
6.	3.000	2.261	0.041	0.897	51.348	0.755	$9.456 \times (10^{-2})$	$3.304 \times (10^{-5})$	33.040	$1.697 \times (10^{-3})$	146.621	4	389
7.	3.500	2.448	0.044	0.975	68.557	0.810	$1.365 \times (10^{-1})$	$4.122 \times (10^{-5})$	41.220	$2.826 \times (10^{-3})$	244.166	7	647
8.	4.000	2.726	0.046	1.048	88.405	0.852	$1.764 \times (10^{-1})$	$4.744 \times (10^{-5})$	47.440	$4.194 \times (10^{-3})$	362.362	11	960
9.	4.500	2.996	0.048	1.116	110.995	0.893	$2.229 \times (10^{-1})$	$5.399 \times (10^{-5})$	53.990	$5.993 \times (10^{-3})$	517.795	16	1372
10.	5.000	3.261	0.051	1.181	136.424	0.945	$2.937 \times (10^{-1})$	$6.386 \times (10^{-5})$	63.860	$8.712 \times (10^{-3})$	752.717	23	1995
11.	5.270	3.402	0.052	1.215	151.375	0.965	$3.248 \times (10^{-1})$	$6.739 \times (10^{-5})$	67.390	$1.020 \times (10^{-3})$	881.280	27	2335

Conclusion: Sediment Discharge capacity at full supply discharge at 151.375 m^3/s at F.S.L. = 67.39 ppm = **881.28 m^3/ day**

Table 7B.6 : Indira Gandhi Main Canal, Computation of Stage – Sediment Discharge Curve at 20 °C for Uniform Flow between Km 384 to Km 410 for Project Section by ACKERS METHOD

S. No.	Depth Above Bed h (m)	Hydraulic Radius R (m)	Shear Velocity $_*V$ (m/s)	Velocity V (m/s)	Discharge Q (m^3/s)	Parameter of Mobility F_{gr}	Transport Parameter G_{gr}	Concentration by Volume C_s	Total Sediment Transport: Concentration by Volume $C_s \times 10$ (ppm)	Volume Q_s (m^3/s)	Volume Q_s (m^3/day)	By Mass. Gs (Kg/s)	By Mass G_s Tons /day
(1)	(2)	(3)	(4)	(5)	(6)	(7)	(8)	(9)	(10)	(11)	(12)	(13)	(14)
1.	0.557	0.496	0.020	0.336	2.596	0.369	$1.806 \times (10^{-4})$	$3.200 \times (10^{-7})$	0.324	$8.300 \times (10^{-7})$	0.072	0	0
2.	1.000	0.850	0.026	0.482	7.182	0.494	$3.789 \times (10^{-3})$	$4.050 \times (10^{-6})$	4.050	$2.909 \times (10^{-5})$	2.513	0	0
3.	1.500	1.213	0.031	0.611	14.648	0.601	$1.697 \times (10^{-2})$	$1.262 \times (10^{-5})$	12.620	$1.849 \times (10^{-4})$	15.975	0	0
4.	2.000	1.547	0.035	0.718	24.470	0.687	$4.155 \times (10^{-2})$	$2.382 \times (10^{-5})$	23.820	$5.829 \times (10^{-4})$	50.363	2	173
5.	2.500	1.862	0.038	0.813	36.686	0.756	$7.543 \times (10^{-2})$	$3.560 \times (10^{-5})$	35.800	$1.306 \times (10^{-3})$	112.838	3	259
6.	3.000	2.161	0.041	0.897	51.348	0.821	$1.231 \times (10^{-1})$	$4.916 \times (10^{-5})$	49.160	$2.524 \times (10^{-3})$	218.074	7	605
7.	3.500	2.448	0.044	0.975	68.557	0.885	$1.893 \times (10^{-1})$	$6.537 \times (10^{-5})$	65.370	$4.482 \times (10^{-3})$	387.245	12	1037
8.	4.000	2.726	0.046	1.048	88.405	0.933	$2.541 \times (10^{-1})$	$7.824 \times (10^{-5})$	78.240	$6.917 \times (10^{-3})$	597.629	18	1555
9.	4.500	2.996	0.048	1.116	110.995	0.980	$3.324 \times (10^{-1})$	$9.226 \times (10^{-5})$	92.260	$1.024 \times (10^{-2})$	884.736	27	2333
10.	5.000	3.261	0.051	1.181	136.424	1.040	$4.572 \times (10^{-1})$	$1.139 \times (10^{-4})$	113.900	$1.554 \times (10^{-2})$	1342.656	41	3542
11.	5.270	3.402	0.052	1.215	151.375	1.063	$5.132 \times (10^{-1})$	$1.220 \times (10^{-4})$	122.000	$1.847 \times (10^{-2})$	1595.614	49	4234

Conclusion: Sediment Discharge capacity at full supply discharge at 151.375 m^3/s at F.S.L. = 122.0 ppm = 1,595.6 m^3/ day .All additional sand due to dust storms will be deposited in the canal. Explainatory Notes are given in para 7B.5.2.

Col.6. k_s/δ Relative roughness, $k_s/\delta = d_{65}/\delta$ row (1) ;
$k_s/\delta = 1.45 \times 10^{-4} / 5.767 \times 10^{-4} = 0.277$

Col.7. x Correction term for logarithmic velocity distribution from Fig 7.5 (ix)

Col.8. Δ Apparent roughness diameter,
$\Delta = d_{65}/X$, row (1) ;
$\Delta = 1.6 \times 10^{-4}/0.6 = 2.667 \times 10^{-4}$ m

Col.9. Pe Transport parameter,
$P_e = 2.203 \log (30.2\ h/\Delta)$, row (1) ;
$P_e = 2.203 \log (30.2 \times 0.557/2.667 \times 10^{-4}) = 11.054$

Col.10. Ψ_* Intensity of shear, $\Psi_* = (\gamma_s - 1)d_{35}/RS_0$, row (1);
$[2.65-1) \times 1.45 \times 10_{-4}]/(0.496 \times 0.00008) = 6.029$

Col.11. ϕ_* Intensity of transport, $\phi_* = f(\Psi_*)$, (from Figure 7.5 (iv))

Col.12 q_{sb} Solid discharge, as bed load, by volume and per unit width,

$q_{sb} = \phi_* \sqrt{(\gamma_s - 1)\ g d_{35}^3}$, row (1)

$= 0.4 \times \sqrt{(2.65-1) \times 9.81 \times (1.45 \times 10^{-4})^3} = 2.810 \times 10^{-6}$ m³/s/m

Col.13. Q_{sb} Solid discharge, as bed load, by volume, row (1);
$Q_{sb} = 2.810 \times 10^{-6} \times 22.45 = 6.308 \times 10^{-5}$ m³/s

Col.14. Q_{sb} $Q_{sb} \times 12 \times 3600 \times 24$ in tons/day, row (1) ;
$Q_{sb} = 6.308 \times 3.600 \times 24 = 5.450$ m³/day

Col.15. S.No

Col.16. A_E Dimensionless height, $A_E = 2d_{35}/h$, row (1) ;
$A_E = (2 \times 1.45 \times 10^{-4})/0.557 = 5.207 \times 10^{-4}$

Col.17. z Rouse exponent, $z = V_{ss}/KV_*$,
V_{ss} Settling velocity from Figure 7.5 (x), row (1) ;
$z = 0.017/(0.4 \times 0.02) = 2.125$

Col.18. I_1 Einstein's first integral, from Figure 7.5 (viii)

Col.19. I_2 Einsten's second integral, from Figure 7.5 (viii)

Col.20. q_{ss} Solid discharge, as suspended load, by volume and by unit width,
$q_{ss} = q_{sb}(P_e I_2 + I_1)$, row (1) ;
$q_{ss} = 2.810 \times 10^{-6} (11.054 \times 0.182)\ 1.667 = 9.7 \times 10^{-6}$ m³/s/m

Col.21. Q_{ss} Solid discharge, as suspended load, by volume,
$Q_{ss} = q_{ss} \times B_m$, row (1) ;
$Q_{ss} = 9.7 \times 10^{-6} \times 22.45 = 2.175 \times 10^{-5}$ m³/s

Col.22. Q_{ss} Solid discharge, as total load, by volume, row (1);
$Q_{ss} = 2.175 \times 10^{-5} \times 3600 \times 24 = 1.879$ m³/day

Col.23. Q_{ss} Solid discharge, as total load, by volume,
$Q_s = Q_{sb} + Q_{ss}$, row (1) ;
$Qs = 6.308 \times 10^{-5} + 2.175 \times 10^{-5} = 8.483 \times 10^{-5}$ m³/s

Col.24. Q_s Solid discharge, as total load, by volume, col. 23 x 3600 x 24, row (1) ;
Q_s = 8.483 x 3600 x 24 = 7.329 m³/day

Col.25. Gs Solid discharge, as total load, by mass, Gs = Qs x s , row (1) ;
$G_s = 8.483 \times 10^{-5} \times 2650 = 0.00008 = 0$ Kg/s

Col.26. G_s G_s x 3600 x24 in tons/day, row (1) ;
$G_s = 0 \times 3600 \times 24 \times 10^{-3} = 0$ tons/day

7B.5.2 Total Sediment Transport by Ackers Method, At 32°C of Water, Very Hot Season, April, May and June, for Project Designed Section for Uniform Flow between Km 384 to Km 410 of IGMC

Project Data

(i) Channel parameters

Discharge Q = 151.375 m^3/s ; Bed width, B = 10.67 m ; Top width , B_T = 34.23m

Full supply depth, D = 5.27 m; Bed slope, S_0 = 1/12000

Average bed width $B_a = \frac{B+B_T}{2} = \frac{10.67+34.23}{2} = 22.45$ m

Flow characteristic of uniform flow are given in Table 7B.2 and Figure 7B.5

(ii) Water properties

Temperature of water, T = 32° C; Density at 32° C, ρ = 994.923 Kg/m^3 (See Table 7B.17)

Specific weight of fluid,

γ_f = 1 Ton/m^3; Viscosity at 32° C, $\nu = 0.768 \times 10^{-6}\ m^2/s$ (see table 7B.17)

(iii) Sediment properties

Diameter of sediment, d_{35} = 0.145 mm, d_{50} = 0.15 mm, d_{65} = 0.16 mm (from Figure 7B.7); Density of solid particle, ρ_s = 2650 Kg/m^3; Specific weight of sediment, γ_s = 2.65 Ton/m^3; Settling velocity, V_{ss} (d_{35}) = 0.017 m/s (Figure 7.5(xix))

(iv) Computations: (from Eq. 7.6.9 in Table 7.6 Ackers and White)

Dimensionless diameter of grain D_{gr}, Eq. (x) = $d_{35}\left[\frac{g(\gamma_s-1)}{\nu^2}\right]^{1/3}$

$= 1.45 \times 10^{-4}\left[\frac{9.81(2.65-1)}{(0.768\times10^{-6})^2}\right]^{1/3}$

= 4.3737

Parameter A, E_q. (vii) = $\frac{0.23}{\sqrt{D_{gr}}}+0.14 = \frac{0.23}{\sqrt{4.3737}}+0.14 = 0.2500$

Parameter N, E_q (vi) = 1.00-0.56 log D_{gr} = 1.00-0.56 log 4.3737 = 0.6411

Parameter m, E_q. (viii) = 9.66/D_{gr} + 1.34 = 9.66/4.3737 + 1.34 = 3.5487

Parameter C, E_q. (ix) log C = 2.86 log D_{gr} (log Dgr)² -3.53

Log C = 2.86 log 4.3737 (log 4.3737)² - 3.53 = -2.1079

or C = 0.0078

Sediment transport computation for different discharges, according to flow rating curve given in Figure are done in a tabular manner given in Table 7B.5.

Explanatory notes to Table 7B.5

Col.2 h Flow depth. (From Table 7B.2)

Col.3 R Hydraulic radius. (from Table 7B.2)

Col.4 V_* Total shear velocity, $V_* = \sqrt{g\ R S_0}$, row (1);

$V_* = \sqrt{9.81 \times 0.496 \times 0.00008} = 0.02 m/s$

Col.5 V Average velocity (from Table 7B.3)

Col.6 Q Water flow (Liquid) discharge (from Table 7B.3)

Col.7 F_{gr} Parameter of mobility,

$$F_{gr} = \frac{V_*}{\sqrt{(\gamma_s - 1)\ g\, d_{35}}} \left[\frac{V}{\sqrt{32}\log(10h/d_{35})}\right]^{(1-n)}$$ row (1) ; Fgr =

$$\frac{(0.02)^{0.6411}}{\sqrt{(2.65-1) \times 9.81 \times 1.45 \times 10^{-4}}} \left[\frac{0.336}{\sqrt{32} \times \log\ (10 \times 0.557) / 1.45 \times 10^{-4}}\right]^{(1-0.6411)} = 0.353$$

Col.8 G_{gr} Transport parameters, $G_{gr} = C\left[\frac{Fgr}{A} - 1\right]^m$, row (1) ;

$Ggr = 0.0078 = \left[\frac{0.353}{0.25} - 1\right]^{3.5487}$ 3.353×10^{-4}

Col. 9 C_s Concentration by Volume, (this is gross total sediment transport concentration and should not be confused with point concentration)

$C_s = Ggr\frac{d_{35}}{h}\left(\frac{V}{V_*}\right)^n$, row (1); $C_s = 3.353 \times 10^{-4} \times \frac{1.45x10^{-4}}{0.557}$

$\left[\frac{0.336}{0.02}\right]^{0.6411} = 5.30 \times 10^{-7}$

Col.10 $C_s \times 10^6$, row (1) ; $C_s = 5.30 \times 10^{-7} \times 10^{-6} = 0.530$ ppm

Col.11 Q_s Solid discharge, as total load, by volume, $Q_s = C_s \times Q$,

row (1) ; $Q_s = 5.30 \times 10^{-7} \times 2.596 = 1.38 \times 10^{-6}\ m^3/s$

Col.12 $Q_s \times 60 \times 60 \times 24$, row (1) ; $Q_s = 1.38 \times 10^{-6} \times 3600 \times 24$

$= 0.119\ m^3/day$

Col.13 Gs Solid discharge, is total load, by mass, $G_s = Q_s\ \rho_s$, row (1) ; $G_s = 1.38 \times 10^{-6} \times 2650 = 0.0037 \approx 0$ Kg/s

7B.5.3 Total Sediment Transport by Ackers Method, at 20°C of Water, Medium Season, March and July to November, for Uniform Flow between Km 384 to Km 410 of IGMC

Project Data

(i) Channel parameter

Discharge $Q = 151.375\ m^3/s$; Bed width, B = 10.67m ;

Top width , $B_T = 34.23$ m

Full supply depth, D = 5.27 m : Bed slope, $S_0 = 1/12000$

Average bed width, $B_a = \frac{B+B_T}{2} = \frac{10.67+34.23}{2} = 22.45$ m

Flow characteristic of uniform flow are given in Table 7B.3 and Figure 7B.6

(ii) Water properties

Temperature of water, T = 20°C; Density at 20° C, ρ = 998.2 Kg/m^3 (Table 7B.17)

Specific weight of fluid, γ_f = 1.0 Ton/m^3 ;

Viscosity at 20° C, ν = 1.003 x 10^{-6} m/s (see Table 7B.17)

(iii) Sediment properties

Diameter of sediment, d_{35} = 0.145 mm, d_{50} = 0.15 mm, d_{65} = 0.16 mm (Figure 7B.8)

Density of solid particle, ρ_s = 2650 Kg/m^3; Specific weight of sediment, γ_s = 2.65 Ton/m^3 ; Settling velocity, V_{ss} (d_{35}) = 0.017 m/s (from Figure 7.5 (xix))

(iv) Computations: (from equations at 7.6.9 in Table 7.6 Ackers and White)

Dimensionless diameter of grain D_{gr}, $E_{q.}$ (x) = $d_{35}\left[\frac{g(\gamma_s-1)}{\nu^2}\right]^{1/3}$

$$= 1.45 \times 10^{-4}\left[\frac{9.81(2.65-1)}{(1.003x10^{-6})^2}\right]^{1/3} = 3.6606$$

Parameter A, $E_{q.}$ (vii) = $\frac{0.23}{\sqrt{D_{gr}}}+0.14 = \frac{0.23}{\sqrt{3.6606}}+0.14 = 0.2602$

Parameter N, E_q (vi) = 1.00-0.56 log D_{gr} = 1.00-0.56 log 3.6606 = 0.6844

Parameter m, $E_{q.}$ (viii) = 9.66/D_{gr} + 1.34 = 9.66/3.6606 +1.34 = 3.9789

Parameter C, $E_{q.}$ (ix) log C = 2.86 log D_{gr} - (log D_{gr})^2 -3.53

Log C = 2.86 log 3.6606 - (log 3.6606)^2 - 3.53 = - 2.2358 or C = 0.0058.

Sediment transport computation for different discharges, according to flow rating curve given in Figure are done in a tabular manner given in Table 7B.7.

7B.5.4 Comparison of Sediment Transport Methods

The sediment transporting capacity of the channel according to discharge for uniform flow is plotted in Figure 7B.8 against gauge. Since the day and night temperature and also month to month temperature varies considerably (see Para 7B.5) sediment transporting capacity should be worked out for a range of different temperatures. Here it is worked out for two temperatures of water at 20° C and 32° C by Acker's method. These are plotted in Figure 7B.8. With increase in temperature, sediment transport decreases. This is due to decrease in viscosity of water with rise in temperature.

Also sediment transport by Einstein Method is shown in the graph. This method gives more sediment transport by 332.04 % at peak discharge than by Acker's Method.

Acker's method is more preferable in canals, being on conservative side.

Sediment transport and deposits upstream of cross regulator

Sediment transporting capacity of the channel in the zone of gradually varied flow computed by Acker's Method is shown in Table 7B.7 for 2/3 discharge. 2/3 full supply discharge is considered more prominent, next to the full supply discharge. For any other discharge it can be worked by above procedure.

This Table shows additional silt deposits in the up stream of cross-regulator as under:

Table 7B.7 : Indira Gandhi Main Canal, Sediment Flow at 20°C for Gradually Varied Flow between Km 384 to Km 410 for Project Section (existing) by ACKERS METHOD, for 2/3 Full Supply Discharge Q = 100.67 m³/s , with FSL at Cross Regulator

Km Of Canal	Depth Above Bed h (m)	Hydraulic Radius R (m)	Shear Velocity V_* (m/s)	Velocity V (m/s)	Discharge Q (m^3/s)	Parameter of Mobility F_{gr}	Parameter Transport C_{gr}	Concentration by Volume C_s	Concentration by Volume, $C_s x10^6$ (ppm)	Total Sediment Transport: Volume Q_s (m^3/s)	Volume Q_s (m^3/day)	Mass C_s (Kg/s)	Mass C_s (T/day)
(1)	(2)	(3)	(4)	(5)	(6)	(7)	(8)	(9)	(10)	(11)	(12)	(13)	(14)
390.22	4.60	2.916	0.0489	0.981	100.67	0.879	$1.822 x 10^{-1}$	$4.472 x 10^{-5}$	44.72	$4.502 x 10^{-3}$	388.873	12	1037
398.00	4.80	3.153	0.0497	0.925	100.67	0.871	$1.730 x 10^{-1}$	$3.866 x 10^{-5}$	38.66	$3.892 x 10^{-3}$	336.260	10	864
403.74	5.00	3.262	0.0506	0.871	100.67	0.865	$1.663 x 10^{-1}$	$3.382 x 10^{-5}$	33.82	$3.405 x 10^{-3}$	294.163	9	780
410	5.27	3.502	0.0520	0.808	100.67	0.859	$1.598 x 10^{-1}$	$2.874 x 10^{-5}$	28.74	$2.894 x 10^{-3}$	250.004	8	663
(X-Regulator													

Conslusion : In the zone of gradually varied flow upstream of cross –regulator, sediment transport is gradually decreasing from 389 m³/day at Km 390.2 to 250 m³/day at cross-regulator at Km 410. (in 20 Km). This reach of canal will silt at an additional rate of 139 m³/day. During heavy dust storms when ever there is more sediment in the canal, only above rate will be transported and all balance will be deposited. Also see para 7B.5.4.

Explainatory notes : same as in Table 7B.6

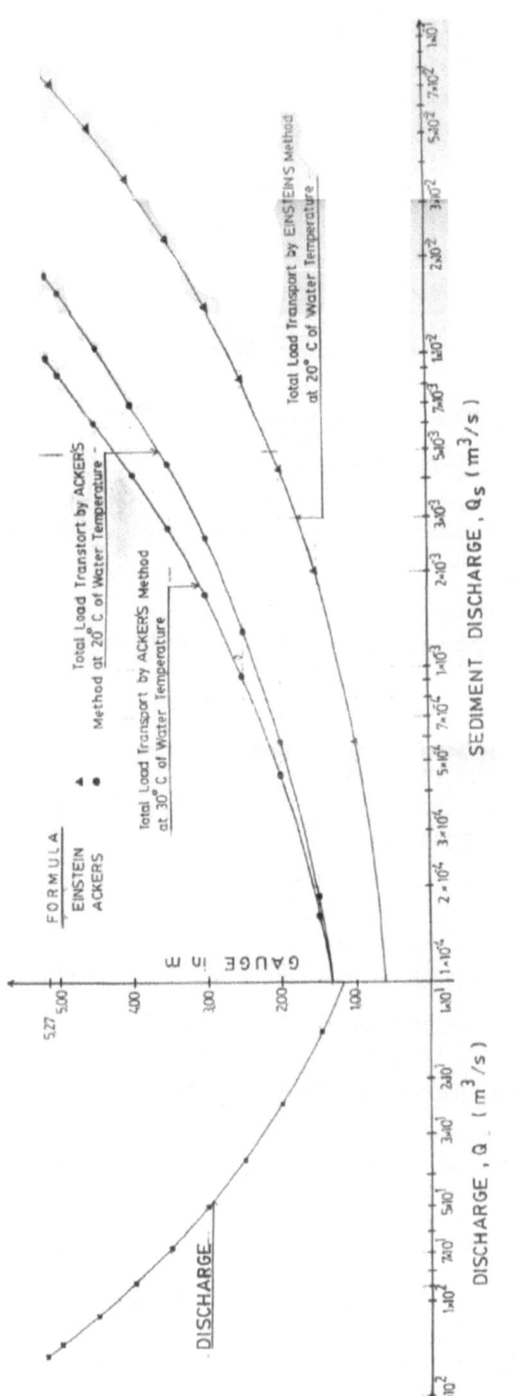

"Difference in sediment transport at full supply discharge between Einsteints Method and Ackers Method = 6.893.86 - 1.595.64 = 5298 22 m³/day

Figure 7B.8 : Guage-Discharge and Sediment Discharge Curve for Indira Gandhi main Canal between Km 384 to Km 410 (RD 1260 to RD 1345)

Volume of channel between 20.0 Km of back water length of curve (taking half approx).

Area of cross Section at 390 Km, corresponding to 2/3 full supply,

Discharge i.e. at gauge 4.35 m = 94.0 m^2

Area of cross section at 410 Km, corresponding to FSL i.e. at gauge 5.27 m = 124.6 m^2

Average area = 109.3 m^2

Volume of canal between back water curve = 109.3 x 20,000 = 2,186,000 m^3

(i) % loss of volume by Acker's Method for water at 20° C for gradually varied flow = [(388, 87 250) /2186, 000] x 100 = 0.0064% per day

(ii) If same % is considered through out the year then 0.0064 x 365 = 2.34% additional capacity, more than normal silt deposit is lost or that much of additional sediment will have to be cleared every year. Otherwise this will reduce discharging capacity. Also see Para 7B.4.1.

(iii) Any additional free board in the zone of back water curves will ensure desired capacity/flow profile with higher safety of canal from over topping, even with rise in water level.

7B.6 Review of Project Design (For Canal Performance)

It has been reported that canal get silted by wind blown sand more on sides. Also weed grows in the canal. Then canal starts behaving as unlined canal. Therefore same is reviewed by taking N = 0.02, 0.025 and 0.03 computations are done for the reach.

7B.6.1 For Uniform Flow, N = 0.02

For cross section of canal, trapezoidal section with curved ends and computation of A, P, and R see Para 7B.4

Computation of discharge versus gauge

i) Up to Height 0.557 m curved position

$A = 7.714\ m^2$; $P = 15.558$ m ; $R = 0.496$ m

$V = R^{2/3} \times S_0^{1/2} / n = (0.496)^{2/3} \times (13/12000)^{1/2} / 0.02 = 0.286$ m/sec

$Q = A \times V = 7.714 \times 0.286 = 2.206$ cumecs

(ii) Up to height 1.00 m

$A = 14.905\ m^2$; $P = 17.539$ m ; $R = 0.850$ m

$V = (0.850)^{2/3} (1/12000)^{1/2} / 0.02 = 0.410$ m/s

$Q = 14.905 \times 0.410 = 6.105$ cumecs

Similar calculation are done for every interval of 0.5 m up to F.S.L. i. e. up to height 1.50 m, 2.0 m, 2.5 m, 3.0 m, 3.5 m, 4.0 m, 4.5 m, 5.0 m, and lastly up to 5.27 m (F.S.L.)

Also computation of discharge versus gauge For Uniform Flow, for N = 0.025, N = 0.03 are done.

All these data are tabulated in Table 7B.8 and plotted in Figure 7B.9 which gives discharge rating curves. From this Table, discharge can be read for any gauge (stage/depth of flow) or vice versa. This Table is further used in development of backwater curve for N = 0.025 in Table 7B.9. The back water profile for N = 0.025 is plotted in figure 7B.10.m

7B.6.2 Inference

This Figure shows that for N=0.02 discharge reduces from 151.375 m^3/s to 128.00 m^3/s and further increase in N to 0.025, discharge reduces to 102 m^3/s i.e 67% of original discharge. These computations do not take into account the reduction in cross-sectional area due to silting, as no data could be collected. The discharge will further reduce. With reduction in discharge velocity reduces and so more silting.

Table 7B.8 : Comparison of flow Characteristics of Indira Gandhi Main Canal for N = 0.02, N = 0.025 and N = 0.03, from Km 389 to 410 Km (RD 1260 to1345) for Uniform Flow, Project Design Section

		N = 0.02			*N = 0.025*			*N = 0.03*		
S. No.	**Depth of flow above bed Y** (m)	**Velocity V** (m/s)	**Discharge** Q=A.V (cumecs)	**Froude Number** $F_r = V/\sqrt{gA/T}$ No.	**Velocity V** (m/s)	**Discharge** Q=A.V (cumecs)	**Froude Number** $F_r = V/\sqrt{gA/T}$ No.	**Velocity V** (m/s)	**Discharge** Q=A.V (cumecs)	**Froude Number** $F_r = V/\sqrt{gA/T}$ No.
1	2	3	4	5	6	7	8	9	10	11
1.	0.557	0.286	2.206	0.1290	0.229	1.765	0.1033	0.191	1.471	0.0861
2.	1.000	0.410	6.105	0.1404	0.328	4.884	0.1124	0.273	4.070	0.0935
3.	1.500	0.519	12.451	0.1481	0.415	9.960	0.1184	0.346	8.300	0.0987
4.	2.000	0.611	20.795	0.1537	0.488	16.636	0.1228	0.407	13.863	0.1024
5.	2.500	0.691	31.183	0.1580	0.553	24.946	0.1265	0.461	20.789	0.1054
6.	3.000	0.763	43.652	0.1615	0.610	34.922	0.1291	0.509	29.102	0.1078
7.	3.500	0.829	58.279	0.1645	0.663	46.623	0.1316	0.553	38.853	0.1097
8.	4.000	0.891	75.151	0.1672	0.713	60.121	0.1338	0.594	50.101	0.1115
9.	4.500	0.949	94.337	0.1696	0.759	75.470	0.1356	0.632	62.891	0.1129
10.	5.000	1.004	115.969	0.1717	0.803	92.767	0.1373	0.669	77.306	0.1144
11.	5.270 (Full supply depth)	1.032	128.671 (Full supply discharge)	0.1727	0.826	102.937 (Full supply discharge)	0.1382	0.688	85.781 (Full supply discharge)	0.1151

Note : For detailed calculation see para 7B.4, development of flow rating curve of Indira Gandhi Main Canal at Km 410.

Table 7B.9 : Indira Gandhi Main Canal, Computation of the Flow Profile Upstream of Cross Regulator at Km 410 (RD 1345), [Back Water Profile] for Gradually Varied Flow, for Q = 2/3 Full Supply Discharge = 100.67 cumecs , N = 0.017, S_0 = 1/12000, α = 1, 00, Project Design

S. No.	Depth Above Bed H (m)	Cross-Sectional Area of Flow A (m^2)	Hydraulic Radius R=A/P (m)	$R^{4/3}$	Velocity V=Q/A (100.63) / Col. 3 (m/s)	$\alpha v^2/2g$	Specific Energy E (m)	ΔE (m)	S_f $\left(\frac{n^2V^2}{R^{4/3}}\right)$	$\overline{S_f}$	$S_0 - \overline{S_f}$	Δx, Up stream of Regul-ator (m)	X, Up stream of Regulator (Distance)
(1)	(2)	(3)	(4)	(5)	(6)	(7)	(8)	(9)	(10)	(11)	(12)	(13)	(14)
1.	5.27	124.627	3.402	5.117	0.551	0.015	5.285	-	0.0000371	-	-	-	-
2.	5.00	115.529	3.261	4.836	0.594	0.018	5.018	0.267	0.0000456	0.0000414	0.0000419	6, 372	6, 372
3.	4.80	108.888	3.153	4.623	0.630	0.020	4.820	0.198	0.0000537	0.0000497	0.0000336	5, 893	12, 265
4.	4.60	102.587	3.050	4.423	0.669	0.023	4.623	0.197	0.0000632	0.0000585	0.0000248	7, 944	20, 209
5.	4.35	94.823	2.916	4.166	0.724	0.027	4.377	0.246	0.0000786	0.0000709	0.0000124	19,838	40, 047

Explainatory Notes : same as in Table 7B.3

34.23 m
F. B = 0.76 m
F.S.L.
ϕ = 26.56
ϕ = 26.56
5.27 m
2.36
0.557
90°
r
1.0
2.0
10.67 m

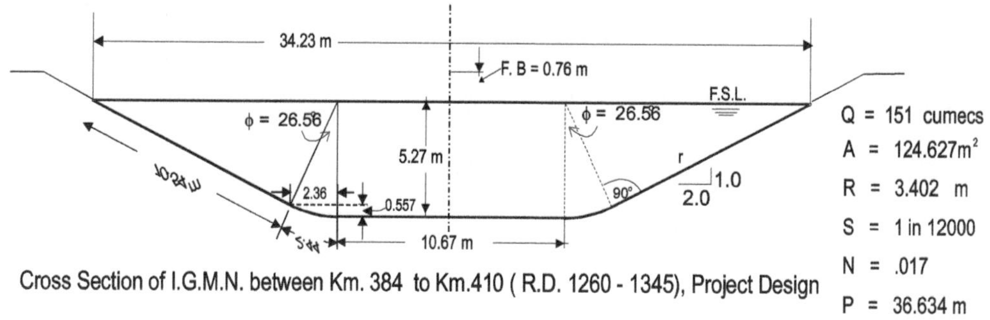

Cross Section of I.G.M.N. between Km. 384 to Km.410 (R.D. 1260 - 1345), Project Design

Q = 151 cumecs
A = 124.627m^2
R = 3.402 m
S = 1 in 12000
N = .017
P = 36.634 m

Discharge in cumecs
0.00
10.00
20.00
30.00
40.00
50.00
60.00
68.63
70.00
80.00
90.00
100.00
100.67
110.00
120.00
130.00
140.00
150.00
160.00
170.00
5.50
F.S.D. 5.27
5.00
4.50
4.35
4.00
3.50
3.00
2.50
2.00
1.50
1.00
0.50
0.00
Depth of flow (Gauge) in m.
(iv) Discharge curve for n = 0.030
(iii) Discharge curve for n = 0.025
(ii) Discharge curve for n = 0.020
(i) Discharge curve for n = 0.017
Area curve
(iv) Velocity curve for n = 0.030
(iii) Velocity curve for n = 0.025
(ii) Velocity curve for n = 0.020
(i) Velocity curve for n = 0.017
0.0
8.0
16.0
24.0
32.0
40.0
48.0
56.0
64.0
72.0
80.0
88.0
95.20
96.0
104.0
112.0
120.0
128.0
Cross sectional area in sq. m.
0.00
0.14
0.28
0.42
0.56
0.70
0.74
0.84
0.98
1.10
1.12
Velocity in m/sec.

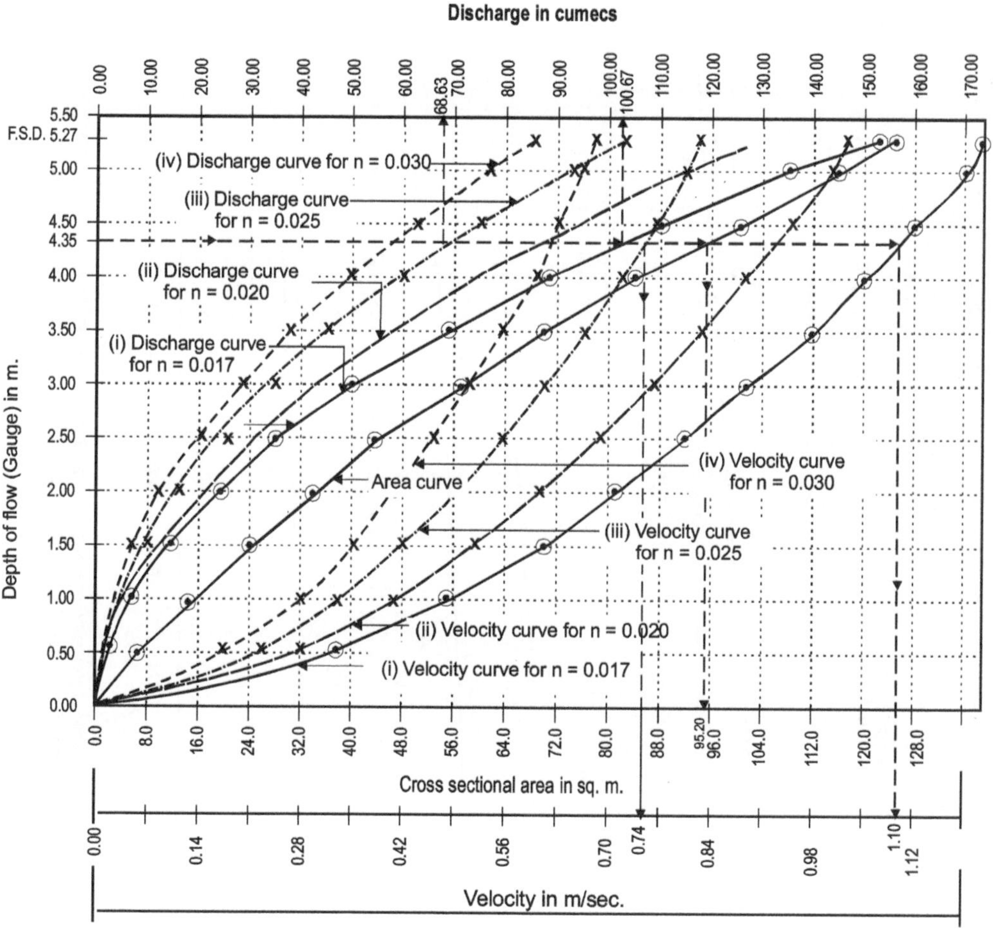

Figure 7B.9 : Canal Rating Curve or Flow Characteristics in Trapezoidal Channel With Curved ends, Indra Gandhi Main Canal between Km. 384 to Km. 410 (RD 1260 to RD 1345), Project Design

Notes : To know discharge for a given gauge move arrow horizontally to meet discharge curve and then vertically upwards and read discharge. Similarly for area move upto area curve and down wards to read area scale. Similarly for velocity , move upto velocity curve and than down wards to read velocity scale.

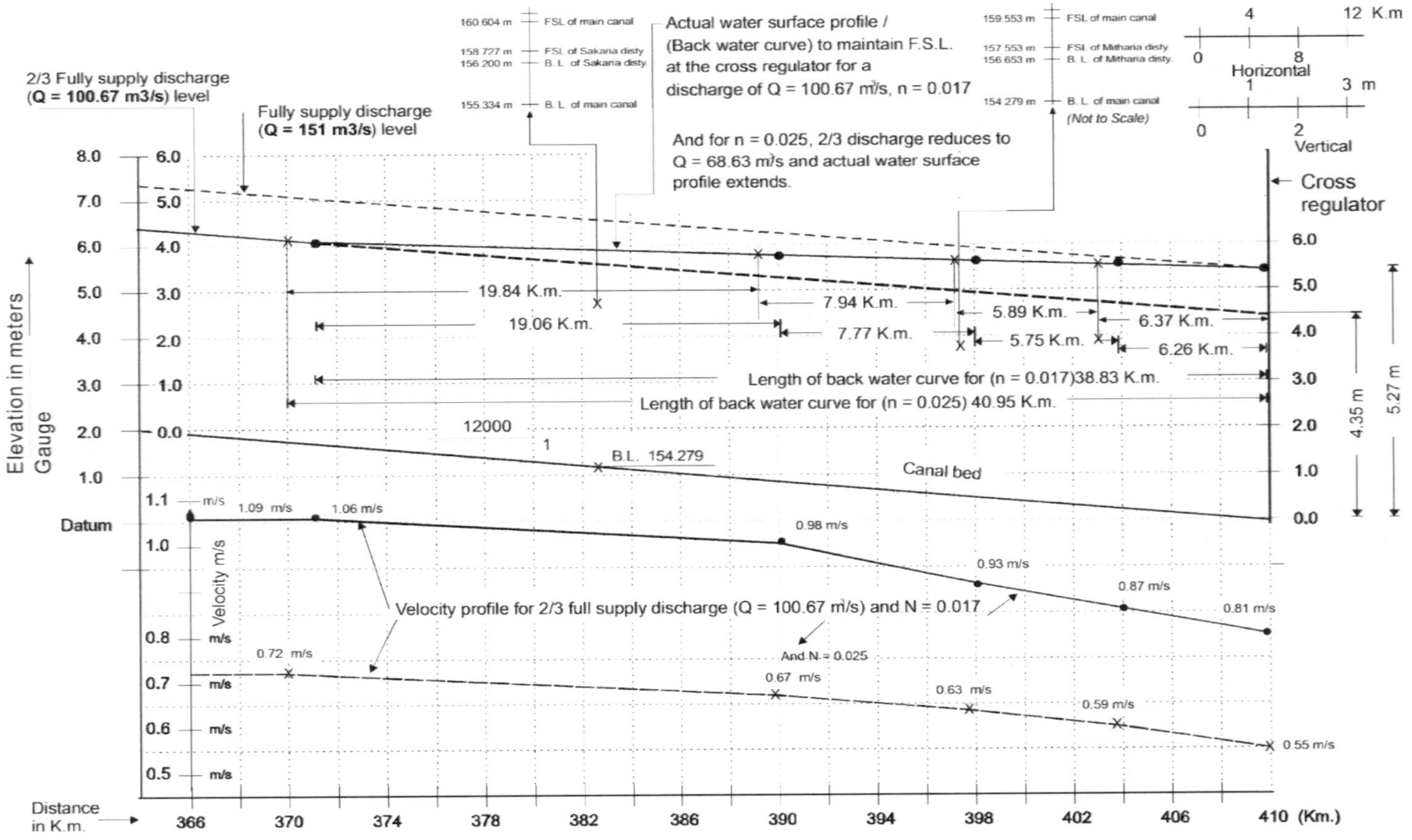

Figure 7B.10 : Comparative analysis of Velocity and Back Water Surface Profile for N = 0.017 and 0.025 at Cross Regulator at Km. 410 (RD 1345.13) To Maintain FSL For 2/3 Discharge in Indra Gandhi Main Canal (IGMC) For Existing / Project Design

Actual data on reduction of cross section should be observed and more realistic rating curve prepared. In many situations discharge may be less than 50%. Also see Para 7B.4.1.

The problem can be resolved by provision of extra free board or higher value of N or both according to maintenance programme. Also see table 6.5.

Another alternative is to raise the bank, (see Para 7B.4.1) or provide a top level bank canal, (Para 9.14.2) or canals of zero bed slope (Para 9.14.1). The flow is provided through sufficient energy gradient (S_f). It may be more preferable to provide a top level bank canal. The existing canals can be also easily improved by above modifications, see Para 7 B 4.1. This ensures better feeding of tail areas.

The back water profile, (without including any silting effect) show a slight increase in length by 0.78 Km, with a slight increase in head at sakaria distributory, still less than the design head at FSL.

7B.7 Alternative Hydraulic Design

Indira Gandhi Main Canal, between Km 384 to Km 410 (RD 1260 to 1345).

7B.7.1 Review For Various B/D Ratio, but same side slope Z and check for Kennedy's critical velocity ratio

With view to understand and explain the mechanics of flow for different B/D ratio, a revised alternate design is made. The purpose is not to reconstruct the canal, but only to explain the effect of B/D ratio in canal design and draw lessons for the future. Also it is compared with the critical velocity ratio given by Kennedy.

A comparison of canals section for the same bed slope of 1 in 12000 and side slope Z = 2:1 (H:V) (as per project design) for various B/D (bed width : depth) ratios along with Kennedy's critical velocity is made in Table 7B.10, 7B.11 and shown in Figure 7B.11 and 7B.12 From this Table it is observed as under:

Explanatory notes to table 7B.10

Col.2 B/D is ratio, where B is bed width and D is depth of flow

Col.3 Depth of flow in (m) is computed by equation,

$$D = \left[\frac{Q \cdot N}{S^{1/2}}\right]^{3/8} \text{ in (m)}$$

Col.4 Bed width (m) corresponding to the depth of flow in col. 3, B = XD

Col.5 Top width (m) corresponding to D in col. 3, $B_T = D(X + 4.47)$

Col.6 Water area (m²) corresponding to D in col. 3, $A = D^2 (X + 2.463)$

Col.7 Wetted parameter (m) corresponding to D in col. 3, $P = D(X + 4.926)$

Col.8 Hydraulic radius (m) corresponding to D in col. 3, R = A/P

Col.9 Manning's velocity (m/s), $V = R^{2/3} S^{1/2} / N$

Col.10 Kennedy's critical velocity (m/s), $V_0 = 0.55\, D^{0.64}$, corresponding to D in col. 3.

Col.11 Critical velocity ratio obtained by dividing col. 9 by col. 10.

Col.12 Discharge (m³/s), Q = A x V

Col.13 Froud number, $F_r = V/\sqrt{g\ A/B_T}$

At X = 1.6 or say 1.7 the depth of flow is 5.519 m, and 5.458 m, and velocities are 1.223 m/s and 1.222 m/s and F_r is also near to 0.20. This is optimum solution.

Table 7B.10 : Indira Gandhi Main Canal, Computations of Sections for Q = 151. 375 m³/s, S_o = 1/12000 N = 0.017, for Uniform Flow, Trapezoidal Sections with Curved Ends. Z = 2:1 (H: V) For Different B: D Ratios and Comparison with Kennedy's Critical Velocity, from Km 384 to Km 410 (RD 1260 to RD1345)

S.No.	X=B/D	Depth of Flow D (m)	Bed Width B (m)	Top Width $T(B_T)$ (m)	Area of flow A (m^2)	Wetted Perimeter P (m)	Hydraulic Radius R (m)	Manning's Velocity V (m/s)	Kennedy's Critical Velocity V_0 (m/s)	Critical Velocity Ratio $m=V/V_0$	Discharge Q=AV (m^3/s)	Fraud Number Fr
(1)	(2)	(3)	(4)	(5)	(6)	(7)	(8)	(9)	(10)	(11)	(12)	(13)
1	0.5	6.421	3.211	31.913	122.168	34.841	3.506	1.239	1.808	0.685	151.397	0.2022
2	0.6	6.318	3.791	32.034	122.281	34.915	3.502	1.239	1.790	0.692	151.487	0.2025
3	0.7	6.220	4.354	32.159	122.384	34.996	3.497	1.237	1.772	0.698	151.375	0.2025
4	0.8	6.127	4.901	32.287	122.477	35.081	3.491	1.235	1.755	0.704	151.286	0.2025
5	0.9	6.039	5.435	32.428	122.639	35.182	3.486	1.234	1.738	0.710	151.346	0.2026
6	1.0	5.955	5.955	32.571	122.787	35.287	3.480	1.233	1.723	0.716	151.415	0.2028
7	1.1	5.874	6.462	32.721	122.956	35.399	3.473	1.231	1.708	0.721	151.336	0.2028
8	1.2	5.796	6.956	32.866	123.074	35.509	3.466	1.230	1.694	0.726	151.418	0.2029
9	1.3	5.724	7.441	33.026	123.282	35.636	3.459	1.229	1.680	0.731	151.462	0.2031
10	1.4	5.651	7.912	33.173	123.372	35.750	3.451	1.227	1.666	0.736	151.340	0.2031
11	1.5	5.584	8.376	33.338	123.579	35.884	3.444	1.225	1.654	0.741	151.437	0.2031
12	**1.6**	**5.519**	**8.830**	**33.498**	**123.767**	**36.014**	**3.436**	**1.223**	**1.641**	**0.745**	**151.269**	**0.2031**
13	1.7	5.458	9.278	33.674	124.000	36.163	3.429	1.222	1.629	0.750	151.506	0.2033

Contd...

Table 7.B.10. Contd...

S.No.	X=B/D	Depth of Flow D (m)	Bed Width B (m)	Top Width $T(B_T)$ (m)	Area of flow A (m^2)	Wetted Perimeter P (m)	Hydraulic Radius R (m)	Manning's Velocity V (m/s)	Kennedy's Critical Velocity V_0 (m/s)	Critical Velocity Ratio $m=V/V_0$	Discharge Q=AV (m^3/s)	Fraud Number Fr
(1)	(2)	(3)	(4)	(5)	(6)	(7)	(8)	(9)	(10)	(11)	(12)	(13)
14	1.8	5.397	9.714	33.837	124.153	36.298	3.420	1.219	1.618	0.754	151.380	0.2032
15	2.0	5.284	10.568	34.187	124.609	36.597	3.405	1.215	1.596	0.761	151.439	0.2032
16	2.5	5.031	12.578	35.067	125.622	37.361	3.362	1.205	1.547	0.779	151.324	0.2033
17	3.0	4.816	14.448	35.975	126.701	38.171	3.319	1.194	1.504	0.794	151.34	0.2031
18	3.5	4.629	16.201	36.891	127.761	39.002	3.276	1.184	1.466	0.808	151.299	0.2031
19	4.0	4.466	17.865	37.828	128.915	39.865	3.234	1.174	1.433	0.819	151.322	0.2030
20	6.0	3.969	23.812	41.552	133.294	43.362	3.074	1.136	1.329	0.855	151.434	0.2025
21	8.0	3.626	29.008	45.216	137.565	46.869	2.935	1.100	1.254	0.877	151.340	0.2014
22	10.0	3.369	33.690	48.750	141.459	50.286	2.813	1.070	1.197	0.895	151.425	0.2006
23	12.0	3.168	38.018	52.180	145.173	53.625	2.707	1.043	1.150	0.906	151.348	0.1996
24	15.0	2.933	43.999	57.111	150.256	58.449	2.571	1.008	1.095	0.920	151.452	0.1984
25	20.0	2.650	53.006	64.853	157.784	66.062	2.388	0.960	1.026	0.935	151.397	0.1964
26	25.0	2.446	61.162	72.098	164.373	73.213	2.245	0.921	0.975	0.945	151.448	0.1948
27	30.0	2.291	68.748	78.957	170.330	80.002	2.129	0.888	0.935	0.950	151.310	0.1931

Note : Explanatory notes are given on next page

Table 7B.11 Comparison of Sections for B/D = 1.6 and 2.0 for Z = 2:1 (H: V)

Parameter	B/D Ratio		% Saving /Excess
	B/D = 2.05 (As executed)	B/D = 1.6	
Q	151 m^3/s	151 m^3/s	Nil
D	**5.27 m**	**5.52 m**	+ 4.74
B	10.67 m	8.83 m	(-17.24)
A	124.627 m^2	123.76 m^2	(- 0.69)
R	3.402 m	3.436 m	-
S	1 in 12000	1 in 12000	-
N	0.017	0.017	-
P	36.634 m	36.014 m	(- 1.69)
V	**1.215 m/s**	**1.223 m/s**	+ 0.65
T	**34.23 m**	**33.50 m**	- 2.13

This section with B/D = 1.6, besides being more efficient, also has a saving of 1.69 % in the perimeter i.e. approximate cost of lining and 2 % saving in land and earth work which is considerable.

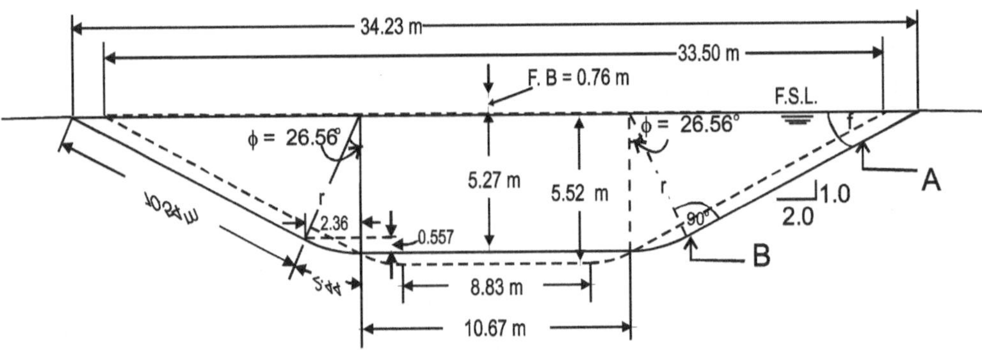

Fig. 7B.11 Comparison of cross sections IGMC for B/D = 1.6 and 2.0, Project Design between Km 384 to Km 410 (RD 1260 to RD 1345)
Notes : A = Project design for B/D = 2.0 B = project design for B/D = 1.6

1. Kennedy's velocity of 1.808 m/s at X = 0.5 decreases to 0.935 m/s at X = 30.0
2. Kennedy's critical velocity ratio (m) increases very slowly from 0.685 at X = 0.5 to 0.950 at X = 30 where the velocity are considerably reduced.
3. Manning's velocity decreases from 1.239 m/s at X = 0.5 to 0.888 m/s at X = 30.0
4. Froud number is gradually increasing (very little) from 0.2022 at X = 0.5 to 0.2033 at X = 1.7 and then gradually decreases to 0.1931.
5. *Thus the concept of critical velocity ratio of m $\geq$ 1 gives very large bed width and uneconomical section, for the very flat bed slope and flat side slope of 2:1 (H:V). It may not be practical to achieve m as unity*
6. For the project design, critical velocity ratio (CVR or m) is 0.7625 < 0.8 much less than 1

   The back water profile, (without including any silting effect) shows a slight increase in length by 0.78 Km, with a slight increase in head at Sakaria distributory, still less than the designed at FSL.

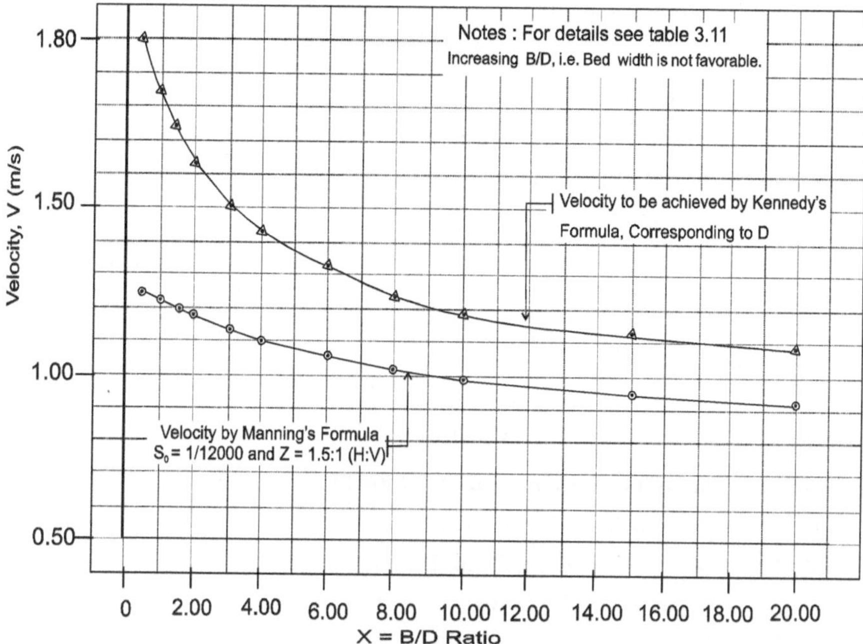

Figure 7B.12 : Indira Gandhi Main Canal, From Km 384 to Km 410 (RD 1260 to RD 1345), Computations of Sections for Q = 151.375 m3/s, S_0 = 1/12000, N = 0.017 and Z = 2:1 (H:V) For Different B/D Ratio and Comparison with Kennedy's Critical Velocity, For Uniform Flow

7. *At X = 1.6 or say 1.7 the depth of flow is 5.519 m, and 5.458 m, and velocities are 1.223 m/s and 1.222 m/s and Fr is also near to 0.20. This is optimum solution, see figure 7B.11 and table 7B.12*

8. In fact there appears little justification for increasing X greater 1.5 or say 1.6.There after neither depth decreases rapidly, nor critical velocity ratio increases rapidly.

9. *Comparison with Indian Standard Codes:*

   IS 10430 - 1982 lays down as under in Para 6.8, "Critical velocity ratio is not applicable in lined canals but the possibility of silting can not be neglected. Hence the critical velocity ratio should be aimed at higher than unity or by any other method; it should be ensured that silting would not take place in the lined canal".

   IS 10430 - 2000 also lays down that the critical velocity ratio should be aimed at higher than unity or by any other method, it should be ensured that silting will not take place in the lined canal.

   It is not practical to achieve above criteria in very flat Bed slopes. Also it may not be economical and practical. For this either bed slope or side slope should be increased.

7B.7.2 Design by Kennedy's Method

Channel parameters through Manning's formula

Depth of flow, h = 5.27 m; Bed slope, S_0 = 1/12000; Discharge, Q = 151.375 m^3/s

Velocity, V = 1.215 m/s; Trapezoidal section with curved ends side slope Z = 2:1 (H:V).

Velocity by Kennedy's formula for the adopted depth,

$V_0 = 0.55\ h^{0.64} = 0.55\ (5.27)^{0.64} = 1.5934$ m/s

Critical velocity ratio,

$m = V/V_0 = 1.215/1.5934 = 0.7625 < 0.8$

Much less than 1, designed section do not satisfy m=1, for that V should be1.593 m/s

Corresponding to this velocity, required area of flow,

$A = Q/V = 151.375/1.5934 = 95.001\ m^2$

Depth of flow By Manning's formula for best trapezoidal section (most optimum) of Z = 1: 0.577 (V:H), for the given slope of 1/12000,

$D = [\ (Q \times N)/\ (1.091\ S^{1/2})\ ]^{3/8} = [151.375 \times 0.017/1.091 \times \sqrt{(1/12000)}\ ]^{3/8} = 8.028$ m

Maximum Velocity,

$V=(D/2)^{2/3} \times S^{1/2} \times 1/N = (8.028/2)^{2/3} \times \sqrt{(1/12000)} \times 1/0.017 = 1.356$ m/s.

This is still less than 1.593 m/s. (from Kennedy's formula)

But V_0 by Kennedy's formula for D = 8.028 m

$V_0 = 0.55\ (8.028)0.64 = 2.086$ m/s This is furthermore.

As depth goes on increasing, V_0 also increases and Manning's velocity will be lower than V_0 for the given slope of 1/12000.

Alternative may be to compute slope by Manning's formula for the adopted Kennedy's Velocity.

Slope by Manning's formula for Kennedy velocity

a) For V = 1.593 m/s
   $1.593 = (5.27/2)^{2/3} \times S^{1/2} \times 1/0.017$ (R = D/2 for best section)
   or $1.593 = 112.22\ S^{1/2}$
   or S = 0.0002 = 2/10000 = 1/5000, (1: 5000) say 1 in 5000

b) For V = 2.086 m/s
   $2.086 = (8.028/2)^{2/3} \times S^{1/2} \times 1/0.017 = 148.572\ S^{1/2}$
   or $S = 0.000197 = \frac{197}{1000000} = \frac{1}{5073}$ (1 : 5073) say 1 in 5000
   (Slightly steeper and not flatter).

7B.7.3 Design by Lindley Method

Velocity by Lindley's formula for the adopted depth of h = 5.27 m,

$V = 0.567\ D^{0.57} = 0.567\ (5.27)^{0.57} = 1.462$ m/s

Corresponding to this velocity, required area of flow,

$A = Q/V = 151.375/1.462 = 81.428\ m^2$

Depth of flow by Manning's formula for best trapezoidal section (most optimum), for the given slope of 1/12000 and Z = 1:0.57 (V : H) is 8.028 m and maximum velocity of 1.356 m/s. Corresponding to this depth and velocity, required bed width is 5.873 m and top width is 41.762 m and area of flow is 205.933 m^2.

But V by Lindley's formula for D = 8.028 m

$V = 0.567\ (8.028)\ 0.57 = 1.859$ m/s

Table 7B.12 : Comparison of Sections for Q = 151.375 m³/s, S_o = 1/5000, N = 0.017, Trapezoidal Section With Curved Ends, Z = 2:1 (H:V), For Various B/D Ratio

S.No.	X=B/D	Depth of Flow D (m)	Bed Width B (m)	Top Width $T(B_T)$ (m)	Area of Flow A (m^2)	Wetted Perimeter P (m)	Hydraulic Radius R (m)	Manning's Velocity V (m/s)	Kennedy's Critical Velocity V_0 (m/s)	Critical Velocity Ratio $m = V/V_0$	Discharge Q=AV (m^3/s)	Froud Number Fr
(1)	(2)	(3)	(4)	(5)	(6)	(7)	(8)	(9)	(10)	(11)	(12)	(13)
1	0.50	5.449	2.724	27.081	87.972	29.566	2.975	1.721	1.628	1.057	151.397	0.3049
2	0.75	5.239	3.929	27.345	88.174	29.734	2.965	1.717	1.587	1.081	151.354	0.3052
3	1.00	5.053	5.053	27.640	88.418	29.944	2.953	1.712	1.551	1.104	151.415	0.3056
4	1.25	4.888	6.110	27.959	88.711	30.188	2.939	1.706	1.518	1.124	151.355	0.3058
5	1.50	4.739	7.108	28.290	88.988	30.451	2.922	1.702	1.489	1.143	151.437	0.3063
6	1.75	4.605	8.059	28.645	89.356	30.746	2.906	1.695	1.462	1.160	151.461	0.3064
7	2.00	4.484	8.968	29.011	89.730	31.055	2.889	1.688	1.437	1.175	151.439	0.3064
8	2.50	4.269	10.673	29.757	90.460	31.704	2.853	1.673	1.392	1.201	151.324	0.3063
9	3.00	4.087	12.260	30.527	91.237	32.391	2.817	1.659	1.354	1.225	151.344	0.3064
10	3.50	3.928	13.748	31.305	91.999	33.096	2.780	1.645	1.320	1.246	151.299	0.3063
11	4.00	3.790	15.160	32.101	92.831	33.829	2.744	1.630	1.290	1.263	151.322	0.3061
12	6.00	3.368	20.206	35.260	95.983	36.796	2.609	1.578	1.196	1.319	151.434	0.3053
13	8.00	3.077	24.616	38.370	99.060	39.773	2.491	1.528	1.129	1.353	151.340	0.3036
14	10.00	2.859	28.589	41.368	101.864	42.672	2.387	1.487	1.077	1.380	151.425	0.3025
15	15.00	2.489	37.337	48.464	108.196	49.599	2.181	1.400	0.986	1.420	151.452	0.2991
16	20.00	2.249	44.980	55.033	113.619	56.059	2.027	1.332	0.924	1.442	151.397	0.2961

Explainatory notes: Same as in Table 7B.10

Table 7B.13 : Comparison of Section for Q = 151.375 m³/s, S_o = 1/6300, N = 0.017 Trapezoidal Section With Curved Ends, Z = 2:1 (H:V) for Various B/D Ratio

S.No.	X=B/D	Depth of Flow	Bed Width	Top Width	Area of Flow	Wetted Perimeter	Hydraulic Radius	Manning's Velocity	Lindley's Velocity	Velocity Ratio	Discharge	Froud Number
		D	B	$T(B_T)$	A	P	R	V	V	V_m/V_c	Q=AV	Fr
		(m)	(m)	(m)	(m^2)	(m)	(m)	(m/s)	(m/s)		(m^3/s)	
(1)	(2)	(3)	(4)	(5)	(6)	(7)	(8)	(9)	(10)	(11)	(12)	(13)
1	0.50	5.691	2.845	28.283	95.956	30.878	3.108	1.578	1.528	1.033	151.397	0.2735
2	0.75	5.471	4.103	28.559	96.176	31.054	3.097	1.574	1.494	1.054	151.354	0.2738
3	1.00	5.277	5.217	28.867	96.442	31.273	3.084	1.570	1.463	1.073	151.415	0.2742
4	1.25	5.105	6.381	29.200	96762	31.528	3.069	1.564	1.436	1.089	151.355	0.2743
5	1.50	4.949	7.424	29.546	97.065	31.802	3.052	1.560	1.411	1.106	151.437	0.2748
6	1.75	4.810	8.417	29.917	97.466	32.110	3.035	1.554	1.388	1.120	151.461	0.2749
7	2.00	4.683	9.366	30.299	97.874	32.434	3.018	1.547	1.367	1.132	151.439	0.2749
8	2.50	4.459	11.147	31.078	98.670	33.111	2.980	1.534	1.329	1.154	151.324	0.2748
9	3.00	4.268	12.804	31.883	99.517	33.829	2.942	1.521	1.300	1.173	151.344	0.2748
10	3.50	4.102	14.358	32.695	100.349	34.566	2.903	1.508	1.268	1.189	151.299	0.2748
11	4.00	3.958	15.833	33.526	101.256	35.331	2.866	1.494	1.242	1.203	151.322	0.2746
12	6.00	3.517	21.103	36.825	104.695	38.429	2.724	1.446	1.161	1.246	151.434	0.2740
13	8.00	3.214	25.708	40.073	108.050	41.538	2.601	1.401	1.103	1.270	151.340	0.2723
14	10.00	2.986	29.858	43.205	111.109	44.566	2.493	1.363	1.058	1.288	151.425	0.2713
15	15.00	2.600	38.995	50.615	118.019	51.801	2.278	1.283	0.977	1.313	151.452	0.2683
16	20.00	2.349	46.977	57.476	123.931	58.547	2.117	1.222	0.923	1.324	151.397	0.2656

Explainatory notes:

Col. 2 Col. 2 to col. 9 same as in Table 7B.10

Col. 10 Lindley's velocity in m/s, $V = 0.567\ D^{0.57}$

Col. 11 Col. 11 to col. 13 same as in Table 7B.10

Table 7B.14 : Comparison of Section for Q = 151.375 m³/s, S_o = 1/12000, N = 0.017, Trapezoidal Section With Curved Ends, Z = 1.5:1 (H:V)

S.No.	X=B/D	Depth of Flow D (m)	Bed Width B (m)	Top Width $T(B_T)$ (m)	Area of Flow A (m^2)	Wetted Perimeter P (m)	Hydraulic Radius R (m)	Manning's Velocity V (m/s)	Kennedy's Critical Velocity V_0 (m/s)	Critical Velocity Ratio $m = V/V_0$	Discharge Q=AV (m^3/s)	Froud Number Fr
(1)	(2)	(3)	(4)	(5)	(6)	(7)	(8)	(9)	(10)	(11)	(12)	(13)
1	0.50	6.733	3.366	27.605	117.318	31.483	3.726	1.291	1.864	0.692	151.402	0.1999
2	0.75	6.438	4.829	28.006	117.639	31.715	3.709	1.286	1.811	0.710	151.290	0.2003
3	1.00	6.185	6.185	28.451	118.127	32.013	3.690	1.281	1.765	0.726	151.327	0.2007
4	1.25	5.960	7.450	28.906	118.575	32.339	3.667	1.276	1.724	0.740	151.336	0.2012
5	1.50	5.762	8.643	29.385	119.114	32.704	3.642	1.270	1.687	0.753	151.304	0.2014
6	1.75	5.585	9.774	29.879	119.713	33.096	3.617	1.264	1.654	0.765	151.366	0.2017
7	2.00	5.423	10.846	30.370	120.233	33.494	3.590	1.260	1.623	0.776	151.467	0.2021
8	2.25	5.278	11.876	30.877	120.847	33.917	3.563	1.252	1.595	0.785	151.281	0.2020
9	2.50	5.146	12.864	31.388	121.479	34.352	3.536	1.247	1.569	0.795	151.462	0.2024
10	2.75	5.023	13.814	31.897	122.072	34.790	3.509	1.241	1.545	0.803	151.508	0.2026
11	3.00	4.911	14.732	32.411	122.698	35.239	3.482	1.233	1.523	0.810	151.336	0.2024
12	3.50	4.710	16.485	33.442	123.971	36.155	3.429	1.221	1.483	0.823	151.331	0.2024
13	4.00	4.536	18.145	34.476	125.277	37.089	3.378	1.208	1.448	0.835	151.352	0.2023
14	6.00	4.011	24.066	38.505	130.117	40.815	3.188	1.163	1.338	0.869	151.342	0.2020
15	8.00	3.654	29.229	42.382	134.663	44.486	3.027	1.124	1.260	0.891	151.300	0.2012
16	10.00	3.390	33.897	46.100	138.894	48.053	2.890	1.089	1.201	0.907	151.307	0.004
17	15.00	2.945	44.171	54.772	148.178	56.468	2.624	1.022	1.098	0.931	151.380	0.1983
18	20.00	2.658	53.151	62.718	155.999	64.249	2.428	0.971	1.028	0.944	151.403	0.1965

Explainatory notes:

Col. 2 Col. 2 to col. 4 same as in Table 7B.10

Col. 5 Top width in m corresponding to D in col. 3, B_T = D (x+3.60)

Col. 6 water area in m² corresponding to D in col. 3, A = D² (x +2.088)

Col. 7 watted parameter in the corresponding to D in col. 3, P = D (x+4.176)

Col. 8 Col. 8 to col. 13 same as in Table 7B.10

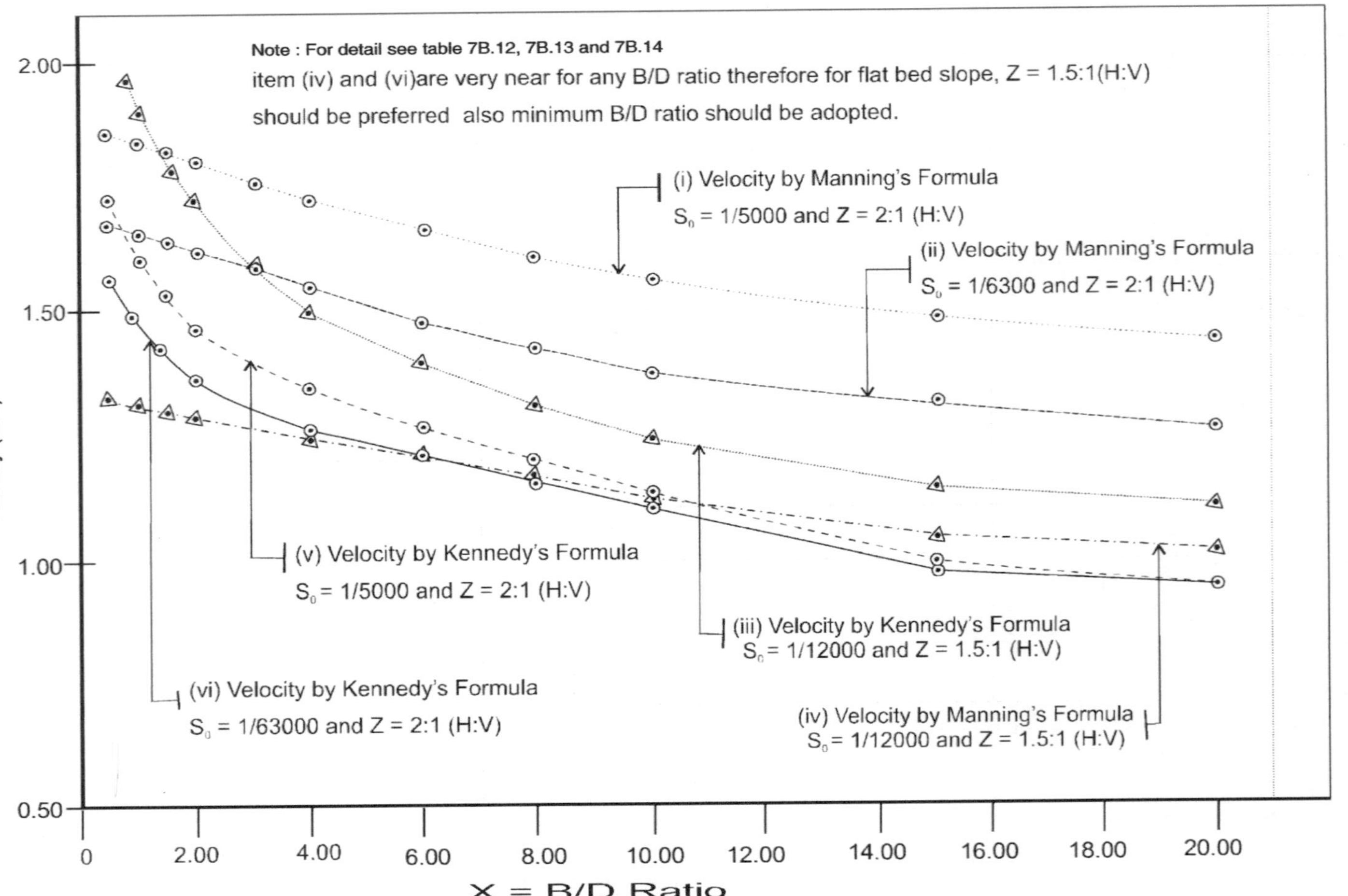

Figure 7B.13 : Comparison of Cross Sections for Q = 151.375 m3/s, N = 0.017, Project Design and Different B/D Ratio with Kennedy's Critical Velocity and Lindley's Velocity For Uniform Flow, IGMC

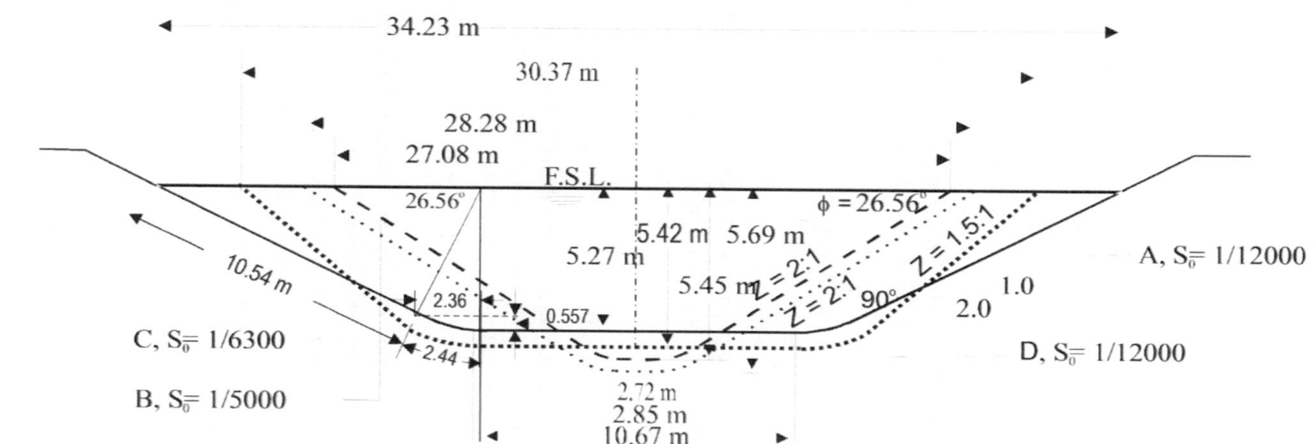

Figure 7B.14 : Comparison of Cross Section For Different Bed Slopes and side Slopes with Project Design Between Km 384 to Km 410 (RD 1260 to RD 1345)

Note : A = Project design for Z = 2:1 (H:V), S_0 = 1/12000 and X = B/D = 2.0

B = (i) Alternative design forZ = 2:1 (H:V), S_0 = 1/5000 and X = B/D = 0.5

C = (ii) Alternative design for Z = 2:1 (H:V), S_0 = 1/6300 and X = B/D = 0.5

D = (iii) Alternative design for Z = 1.5:1 (H:V), S_0 = 1/12000 and X = B/D = 2.0

For the Same Bed Slope Alternative (iii) i.e. D is better. Also see table 7B.15

This is furthermore. As depth goes on increasing, V also increases and Manning's velocity will be lower than V of Lindley for the given slope of 1/12000.

Alternative may be to compute slope by Manning's formulae for the adopted Lindley's Velocity.

Slope by Manning's formula for Lindley velocity

a) For V = 1.462 m/s
$1.462 = (5.27/2)^{2/3} \times S^{1/2} \times 1/0.017 = 112.22\ S^{1/2}$
or S = 0.00017 = 1/0.00017 = 5891 m , (1: 5800) say 1 in 5800

b) V = 1.859 m/s
$1.859 = (8.028/2)^{2/3} \times S^{1/2} \times 1/0.017 = 148.572\ S^{1/2}$
or S = 0.000157 = 157/1000000 = 1/6387, (1 : 6387) say 1 in 6300
(Slightly flatter and not Steeper then in (a) above)

7B.7.4 Comparison of Cross Sections for Various B/D Ratios and Bed Slopes

7B.7.4.1 Computation for comparison of canal sections for the bed slope of 1 in 5000 for Kennedy's velocity and side slope Z = 2:1 (H:V) for various B/D (bed width : depth) ratio along - with Kennedy's critical velocity is made in Table 7B.12 and shown in Figure 7B.14 From this Table it is observed as under:

1. Kennedy's critical velocity (m) is slightly greater than 1, even at X = 0.50
2. Kennedy's critical velocity ratio (m) increases very slowly from 1.057 at X = 0.5 to 1.442 at X = 20 where the velocity are considerably reduced.
3. Kennedy's velocity decreases slowly from 1.628 m/s at X = 0.5 to 0.924 m/s at X = 20.0
4. Manning's velocity decreases from 1.721 m/s at X = 0.5 to1.332 m/s at X = 20.0
5. Froud number is gradually increasing (very little) from 0.3049 at X = 0.5 to 0.3064 at X = 1.75 and then decreases to 0.2691.
6. At X = 0.5 the depth of flow is 5.45 m and velocity is 1.721 m/s and Fr is also near to 0.30.

In fact there appears little justification for increasing X greater then 0.5. There after neither depth decreases rapidly, nor critical velocity ratio increases rapidly. But may not be practical to achieve the bed slope of 1 in 5000. But this canal has more silt transporting capacity.

7B.7.4.2 A comparison of canal sections for the bed slope of 1 in 6300 for Lindley's velocity and side slope

Z = 2:1 (H:V) for various B/D (bed width : depth) ratio along - with Lindley's critical velocity is made in Table 7B.12 and shown in Figure 7B.14. From this Table it is observed as under:

1. Lindley's velocity decreases from 1.528 m/s at X = 0.5 to 0.923 m/s at X = 20.0
2. Manning's velocity decreases from 1.239 m/s at X = 0.5 to 0.888 m/s at X = 20.0
3. Froud number is gradually increasing (very little) from 0.2735 at X = 0.5 to 0.2749 at X = 1.75 and then decreases to 0.2656.
4. At X = 0.5 the depth of flow is 5.691 m and velocity is 1.578 m/s and and F_r is also near to 0.27

Again there appears little justification for increasing X greater 0.5. There after depth decreases slowly and velocity also decrease. Again it may not be practical to achieve this bed slope.

7B.7.4.3 Therefore a third alternative is considered to steeper (increase) the side slope

A comparison of canal sections for the same bed slope of 1 in 12000 (as per project design), but side slope Z = 1.5:1 (H:V) for various B/D (bed width: depth) ratio along - with Kennedy's critical velocity is made in Table 7B.13 and shown in Figure 7B.14 From this table it is observed as under:

1. Kennedy's velocity decreases from 1.864 m/s at X = 2.0 to 1.028 m/s at X = 20.0
2. Kennedy's critical velocity ratio (m) increases very slowly from 0.692 at X = 0.5 to 0.944 at X = 20 where the velocity are considerably reduced.
3. Kennedy's velocity for side slope, Z = 1.5:1 (H:V) are greater than for side slope, Z = 2:1 (H:V)
4. Manning's velocity decreases from 1.291 m/s at X = 2.0 to 0.971 m/s at X = 20.0
5. Froud number is gradually increasing (very little) from 0.1999 at X = 0.5 to 0.2026 at X = 2.75 and then decreases to 0.1965.
6. At X = 2.0 the depth of flow is 5.42 m and velocity is 1.260 m/s and F_r is also near to 0.20.

In fact there appears little justification for increasing X greater 2.0 there after neither depth decreases rapidly, nor critical velocity ratio increases rapidly.

Thus for flat slope of 1 in 12000, a side slope of 1:1.5 in more desirable. It gives D = 5.423 m say 5.43 m, at X = 2.0, B=10.85 m, Manning's V=1.26 m/s and Kennedy's required V = 1.623 m/s, and Kennedy's critical velocity ratio of 0.776, though still less than 1.

In general it may be concluded that Kennedy's formula $V = 0.546\ m\ D^{0.64}$ gives velocity little higher velocity than that by Manning's formula for best section $V = 0.63\ d^{2/3}\ S^{1/2}/N$ and are more suitable to carry or (transport) the sediment entering into canal. Kennedy's formula also gives bed slopes with the help of Manning's formula.

A comparison of all the sections is shown in table 7B.15 and figure 7B.14. For the project slope of 1 in 12000, a side slope of 1:1.5 (V: H) is much better, otherwise change in slope is desirable.

Table 7B.15 Comparison of Cross Sections of Project Design with Various Bed Slopes and Side Slopes

Parameter	Sections with Z = 2:1 (H:V) and S_o = 1/12000	S_o = 1/5000	S_o = 1/6300	Section with Z= 1.5:1(H:V) and S_o = 1/12000	Comparison of Section D with section
	A	B	C	D	A
Q	151 m³/s	151 m³/s	151 m³/s	151 m³/s	0.0
D	5.27 m	5.45 m	5.69 m	5.42 m	+ 2.85 %
B	10.67 m	2.72 m	2.85 m	10.85 m	+1.69 %
B/D	2.0	0.5	0.5	2.0	0.0
A	124.627 m²	87.97 m²	95.96 m²	120.23 m²	- 3.5 %
R	3.402 m	2.98 m	3.11 m	3.59 m	-
N	0.017	0.017	0.017	0.017	0.0
P	36.634 m	29.57 m	30.88	33.49 m	- 9.3 %
V	1.215 m/s	1.721 m/s	1.578 m/s	1.260 m/s	+ 4.1%
T	34.23 m	27.08 m	28.28 m	30.37 m	- 11.2 %

For the same bed slope alternative (iii) i.e. section D is better

7B.7.5 Design by Lacey's Method

Sediment properties: Mean particle size, $m_r = d_{50}$ = 0.15 mm.

Step by step procedure is as fallows:

Silt factor, $f = 1.76\ \sqrt{m_r} = 1.76 = 0.6816 = 0.7$

Velocity of flow, $V = [Qf^2/140]^{1/6} = [151.375\ (0.7)^2/140]^{1/6} = 0.8995$ m/s

Area of channel, $A = Q/V = 151.375/0.8995 = 168.2879\ m^2$

Computation of bed width, B:

$P = 4.75\ \sqrt{Q} = 4.75 \times \sqrt{151.375} = 58.441$ m

Equating the area of flow with the general relation of B and D for side slope Z = 1:2

A = B x D + 2 (πD2 x 26.57/360 + 0.5 D x 2D) = 168.288 .... eq. (1)

or BD +2.464 D^2 = 168.288

P = B + 2 (πD x 26.57/180 + 2D) = 58.441

or B + 2 D (πx 26.57/360 + 2) = 58.441or B + 4.927 D = 58.441

or B = 58.441 - 4.927D .... eq. (2)

Substituting eq. (2) to eq. (1) :

BD +2.464 D^2 = 168.288 or (58.441 - 4.927D)D + 2.464 D^2 = 168.288

or 58.441D - 4.927 D^2 + 2.464D^2 = 168.288

or -2.464 D^2 + 58.441D - 168.288 = 0

Using, $D = \frac{-b+\sqrt{b^2-4ac}}{2a}$

$= \frac{-58.441+\sqrt{58.441^2-4x(-2.464\ \ x-168.288)}}{2x(-2.464)} = 3.354\ \text{m}$

or B x 3.354 + 2.464 x 3.354^2 = 168.288 or B = 41.911 m

Check for correctness of computations

Hydraulic radius,

$V=\sqrt{\frac{2}{5}f.R}$ or R = 5/2 x V^2/f = 5/2 x $(0.8995)^2$/0.7 = 2.890 m

also, $R = A/P = \frac{BD+2\ (\pi D^2 26.57/360+D^2)}{B+2\ (\pi D\ 26.57/180+2D)}$

$= \frac{(41.911x\ 3.354) + 2\ (\pi x3.354^2\ x\ 26.57/360 + 3.354^2)}{41.911 + 2\ (\pi x\ 3.354\ x\ 26.57/180 + 2\ x\ 3.354)}$

= 2.880 m (Hence checked)

Bed slope, $S = \frac{f^{5/3}}{3340\ x\ Q^{1/6}} = \frac{(0.7)^{5/3}}{3340\ x\ (151.375)^{1/6}}$

= 0.000072 (1 : 13889) say 1 in 13800.

Discharge, Q = A x V = 168.2879 x 0.8995 = 151.375 m^3/s

Thus Lacey's theory gives S = 1/13800, V = 0.90 m/s,

B = 41.91 m, D =3.35 m.

7B.7.6 Design by Tractive Force Method

Water properties: Specific weight of fluid, γ_f = 1 ton/m^3.

Sediment properties: Diameters of sediment, d_{75} = 0.165 mm (the diameter of 75% (by weight) is finer sediment particles from figure 7B.7.

Step by step procedure is as follows:

Permissible shear stress,

τ_L (from table 7.10 (iii) by interpolation) = $\frac{(0.15-0.1)}{(0.2-0.1)} \times (0.250-0.241)+0.241$

= 0.247 kg/m^2

Limiting Tractive force,

$\tau_L = \gamma_f$ D.S or $D = \tau_L/\gamma_f S$ = (0.247 x 12000)/1000 = 2.964 m, (say 3.00 m)

Using Manning equation and continuity equation,

$Q = \frac{1}{N} A R^{2/3} S^{1/2} = \frac{1}{0.017} A R^{2/3} (1/12000)^{1/2}$ .... (i)

Area of cross-section, A = B x D + 2 (π D$^2\phi$/ 360 + 0.5 D x 2D)

= 2.964 B + 2 (π x 2.964^2 x 26.57/360 + 2.964^2) = 2.964 B + 21.645

Wetted perimeter of cross section,

P = B + 2 (π D ϕ/ 180 + 2D)

= B + 2 (π x 2.964 x 26.57 / 180 + 2 x 2.964) = B + 14.605

Hydraulic radius, $R = \frac{A}{P} = \frac{2.964\ B + 21.645}{B + 14.605}$

Substituting in equation (i)

151.375 = $\frac{1}{0.017}$ (2.964 B + 21.645) x [(2.964 B + 21.645) /

(B + 14.605)]$^{2/3}$ x (1/ 12000)$^{1/2}$

B is the only unknown, solving by trial we get, B = 43.14 m (this trial is little lengthy process and not given here)

A = 2.964 B + 21.645 = 2.964 x 43.14 + 21.645 = 149.512 m^2

P = B + 14.605 = 43.14 + 14.605 = 57.745

R = $\frac{A}{P}$ = 149.512 / 57.745 = 2.589 m

V = $\frac{1}{0.017} (2.589)^{2/3} (1/12000)^{1/2}$ = 1.013 m/s

Q = A x V= 149.512 x 1.013 = 151.381 m^3/s

Tractive force theory gives, for S = 1/12000, B = 43.14 m and D = 3.0 m

7B.7.7 Comparison of project canal sections with sections by Tractive Force and Lacey's Method

A comparison of project canal sections with these methods is given in table 7B.16 and figure 7B.15 and 7B.16

1. Tractive force's velocity is 1.013 m/s which is greater than Lacey's velocity i.e 0.899 m/s
2. Tractive force's area of cross section is 149.512 m2 which less than Lacey's area of cross section i.e. 168.288 m2. But the bed width by tractive force is more than that by Lacey's Method. Also depth of flow is less in tractive force method than that in Lacey's method.
3. Though cross section by tractive force and by Lacey's methods are slightly different, yet are comparable with each other.
4. These sections are much more costlier than project design also have low velocity i.e. less silt carrying capacity. Even unlined canals may result in silt deposition.

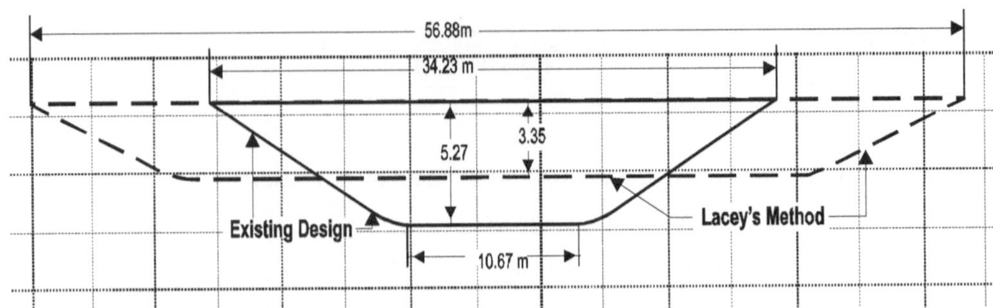

Figure 7.15 : Comparison of Cross Section of IGMC For Exiting Design and Lacey's Method Between Km 384 to Km 410 (RD 1260 to RD 1345

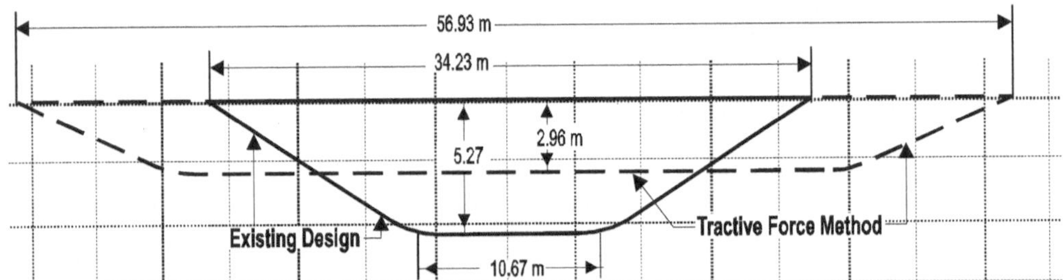

Figure 7.16 : Comparison of Cross Section of IGMC For Exiting Design and Tractive Force Method Between Km 384 to Km 410 (RD 1260 to RD 1345

Table 7B.16 Comparison of Project Canal Sections of with Lacey and Tractive Force Method

Parameter	Canal Sections		
	Existing Project Design	By Lacey is Method	Tractive Force Method
Q	151 m3/s	151 m3/s	151 m3/s
D	5.27 m	3.37 m	2.964 m
B	10.67 m	41.63 m	43.14 m
A	124.627 m2	168.288 m2	149.512 m2
R	3.402 m	2.890 m	2.589 m
S	1 in 12000	1/13800	1/12000
N	0.017	0.017	0.017
P	36.634 m	58.441 m	57.745 m
V	1.215 m/s	0.899 m/s	1.013 m/s
T	34.23 m	56.70 m	56.93 m

Table 7B.17 : Physical Properties of Water in SI Units

Temperature, ^0C	Specific Weight γ, kN/m^3	Density ρ, kg/m^3	Viscosity $\mu \times 10^3$, $N.s/m^2$	Kinematic. Viscosity $v \times 10^6$, m^2/s	Surface Tension σ, N/m	Vapor Pressure P_v, kN/m^2,abs	Vapor Pressure Head P_v/γ, m	Bulk Modulus of Elasticity $E_v \times 10^{-6}$, kN/m^2
0	9.805	999.8	1.781	1.785	0.0756	0.61	0.06	2.02
5	9.807	1000.0	1.518	1.519	0.0749	0.87	0.09	2.06
10	9.804	999.7	1.307	1.306	0.0742	1.23	0.12	2.10
15	9.798	999.1	1.139	1.139	0.0735	1.70	0.17	2.14
20	9.789	998.2	1.002	1.003	0.0728	2.34	0.25	2.18
25	9.777	997.0	0.890	0.893	0.0720	3.17	0.33	2.22
30	9.764	995.7	0.798	0.800	0.0712	4.24	0.44	2.25
40	9.730	992.2	0.653	0.658	0.0696	7.38	0.76	2.28
50	9.689	988.0	0.547	0.553	0.0679	12.33	1.26	2.29
60	9.642	983.2	0.466	0.474	0.0662	19.92	2.03	2.28
70	9.589	977.8	0.404	0.413	0.0644	31.16	3.20	2.25
80	9.530	971.8	0.354	0.364	0.0626	47.34	4.96	2.20
90	9.466	965.3	0.315	0.326	0.0608	70.10	7.18	2.14
100	9.399	958.4	0.282	0.294	0.0589	101.33	10.33	2.07

Source : Change (1985). Fluid Mechanics with Engineering Applications, 8th ed., McGraw Hill, New York.

These theories may be good to the extent that no soil particales of the canal move forward. But raises a very a good question that how much of silt/sediment coming into the canal will settle or get transported (move in suspension).

Therefore these theories are good for clear water or very little sediment laden water entering in the canal, so that no soil particles of earthen canal sections are eroded. But when some silt enters at any other point such as through natural drains or side slope slides, etc. these theories will not be able to transport the sediment.

Therefore a general approach of limiting velocity with use of Manning's formula may be more practical both for lined/unlined canals. The limiting velocity should be based on depth, see table 9.5, and figure 7.11.Also see para 7.19 for a unified approach.

References

Abdul Ghani, Safi, A Review on Design of Stable Channel by Empirical Formula, M.E. Dissertation, WRDTC, University of Roorkee, 1975.

Albertson, M.L., Effect of Shape on the Fall Velocity of Gravel Particles, Proceedings, 5th Hydraulics Conference, University of Iowa Studies in Engineering, Iowa City, Iowa, 1953.

Ambadas, G. Mirajgaoker, Design of Stable Channels in Alluvial Soils, M.E. Dissertation, WRDTC, University of Roorkee, 1957.

ASCE, M&R No. 54, Sedimentation Engineering, Report of America Society of Civil Engineers Task Committee for the Manual on Sedimentation of the Sedimentation Committee of the Hydraulics Division, Headquarters of the Society 345 East 47 TH ST, New York, 10017, 1975.

ASCE, Sediment Transportation Mechanics: Initiation of Motion, by the Task Committee on Preparation of Sedimentation Manual, Vito A. Vanoni, Chmn., Journal of The Hydraulics Division, ASCE, Vol. 93, No. HY5, Proc. Paper 5408, Pp. 297-302, Sept., 1967.

Bagnold, R.A., The Movement of A Cohesionless Granular Bed by Fluid Flow Over It, British Journal of Applied Physics, Vol. 2, Pp. 29-34, 1951.

Ballester, R.E., Muniutes of Proceedings Inst. C.E.P. 280 Vol. 223.

Barr, D.I.I., and Herbertson, J.G., Discussion of Sediment Transportation Mechanics: Initiation of Motion, By The Task Committee on Preparation of Sedimentation Manul, Committee on Sedimentation of the Hydraulics Division, Vito A. Vanoni, Chmn., Journal of The Hydraulics Division, ASCE, Vol. 92, No. HY6, Proc. Paper 4959, Pp. 248-251, Nov., 1966.

Bhakara, Canal Manual Guidelines, General Instructions for the Design of Lined and Unlined Channels of Bhakra Canal System, 1954.

BIS (Bureau of Indian Standards), Code of Practice for Design of Cross-Section of Lined Canals, IS No 4745 - 1968.

Blench, T., Discussion of Sediment Transportation Mechanics: Initiation of Motion, by The Task Committee on Preparation of Sedimentation Manual, Committee on Sedimentation of The Hydraulics Division, Vito A. Vanoni, Chmn., Journal of The Hydraulics Division, ASCE, Vol. 92, No. HY2, Proc. Paper 4891, pp. 287-288, Sept., 1966.

Bretting, A.E., Stable Channels, Technical Paper No. 144 Presented at the Sixth International Conference on Coastal Engineering, University of Florida, USA, December, 1957.

Brien, O' M.P., And Rindlaub, B.D., The Transportation of Bed Load by, Streams, Trans. AUG Vol. 15, P. 597, 1934.

Brown, C.B., Sediment Transportation, Chapter XII Engineering Hydraulics, by H. Rouse, Ed., John Wiley And Sons, Inc., New York, N.Y., 1950.

Buckley, Irrigation Department of Egypt, Irrigation Practice, P. 207.

Buckley, Irrigation Pocket Book, 1893, 2nd Ed. 1905.

Butt, M. A., Some Recent Developments In Fluvial Hydraulics, M.E. Thesis University of Roorkee, 1973.

Cartier, A.C., Critical Tractive Forces on Channel Side Slopes, U.S.B.R. Hydraulic Laboratory Report No. Hyd.- 366 February, 1953.

Cartier, A.C., Critical Tractive Forces Which Start Movement of Sediment in A Channel, U.S. Bureau of Reclamation, Hyd. Lab.Report, HYD 296, 1950.

CBIP, (Central Board of Irrigation & Power), Manual on Current Practices in Canal Design in India, Malcha Marg, Chanakyapuri New Delhi, June 1968.

CBIP, Design Practice for Unlined Incised Canals, Technical Report -7 May 1970.

CBIP, Pub. NO. 144: Symposium on Operation and Maintenance of Canal Systems.

CBIP, Sediment Control in Rivers and Canals, 1966.

CBIP, Symposium on Canal Lining, 1967.

Chang, Y.L., Laboratory Investigation of Flume Traction and Transportation, Trans. ASCE, Paper 2037, 1939.

Clement, M., Ottmarsheim Development, Travaur (Paris), Jan. 1951.

Clibbron, 1901

Colby, B.R., Practical Computation of Bed-Material Discharge, Journal of The Hydraulics Division, ASCE, Vol. 90, No. HY2, Proc. Paper 3843, Pp. 217-246, Mar., 1964.

Collings, Transactions ASCE, Vol. 99, P. 549, 1934.

Creager, W.P. & Justin, J.D., Hydroelectric Handbook, Second Edition, Page 623, 1949.

CBIP/CWC, Draft Manual on Irrigation Channels, Guidelines, 1960, 1968 and 1984.

Davis, C.V., Handbook of Applied Hydraulics, Mc. Graw-Hill Book Company Ltd., 1952.

Duboys, P., Le Rohne Et Les Riveres A Lit Affouillable, Annales Des Ponts Et Chaussees, Series 5, Vol. 18, Pp. 141-195, 1879.

Duthy, G.W., Proceeding, Punjab Eng. Congress, P.P. 44 and 48, 1919.

Dutta, R.R. A New Method of Finding The Channel Section to Satisfy The Kennedy's Equation; Journal of Central Board of Irrigation and Power, May 1962.

Einstein, H.A., Formulas for the Transportation of Bed Load, Transactions, ASCE, Vol. 107, Paper No. 2140, Pp. 561-573, 1942.

Einstein, H.A., The Bed Load Function For Sediment Transportation in Open Channels, Technical Bulletin 1026, United States Department of Agriculture, Soil Conservation Service, Washington, D.C., 1950.

Engelund, F., and Hansen, E., A Monograph on Sediment Transport in Alluvial Streams, Teknisk Vorlag, Copenhagen, Denmark, 1967.

Engelund, F., Hydraulic Resistance of Alluvial Streams, Journal of The Hydraulics Division, ASCE, Vol. 92, No. HY2, Proc. Paper 4739, Pp. 315-327, Mar., 1966.

Fan, C.H., A Study of Stable Channel Cross-Sections, Hydraulic Engineering (Published by The Chinese Society of Hydraulic Engineers) Vol. 15, 1947, P.71. Also See U.S.B.R. Hydraulic Laboratory Report No. Hyd. 366 of February 1953.

Flaxmn, E.M., Discussion of, Sediment Transportation Mechanics: Initiation of Motion, by The Task Committee on Preparation of Sedimentation Manual, Committee on Sedimentation of The Hydraulics Division, Vito A. Vanoni, Chmn., Journal of The Hydraulics Division, ASCE, Vol. 92, No. HY6, Proc. Paper 4959, Pp. 245-248, November, 1966.

Fortier, S. and Scobey, F.C. "Permissible Canal Velocities", Trans. ASCE, Pp. 940-984, 1926.

Garde, R.J. and Ranga Raju, K.G., Mechanics of Sediment Transportation and Alluvial Stream Problems, Third Edition, Published by K.K. Gupta for New Age International (P) Ltd., 4835/24, Ansari Road, Daryaganj, New Delhi - 110002.

George, W. Pickels: Run-off Investigations in Central Illinois, University of Illinois, Engineering Experiment Station, Bulletin 232, Vol. 29, No. 3, September, 1931.

Ghaleb, K.D., Minutes of Proceedings, Inst. C.E., Vol. 223.

Gilbert, G.K., Transportation of Debris by Running Water, Professional Paper No. 86, United Sates Geological Survey, 1914.

Glover, R.E., and Florey, Q.L., Stable Channel Profiles, U.S.B.R. Hydraulic Laboratory Report No. Hyd. 325, September 19.

Graf, W.H., Hydraulics of Sediment Transport, Mcgraw-Hill Book Co., Inc., New York, 1971.

Guilford, L. Molesworth, Pocket-Book of Engineering Formulae (Useful Formulae and Memoranda) for Civil and Mechanical Engineers, E & F.N. Spon, London, 7th Ed., P. 176, 1871.

Harwey, W.B., Proceedings Punjab Eng. Congress 1919 P. 58, 1986.

Houck, Irrigation Engineering, Vol II., 1951.

Inglis, C.C., Discussion of, Systematic Evaluation of River Regime, by Charles R. Neill and Victor J. Galey, Journal of the Waterways and Harbors Division, ASCE, Vol. 94, No. WW1, Proc. Paper 5774, Pp. 109-114, Feb., 1968.

Kandap, D.M., and Garde, R.J., Design of Stable Alluvial Channels; Journal of Central Board of Irrigation and Power, Oct. 1979.

Karmar H., Sand Mixtures and Sand Movement in Fluvial Models, Trans. ASCE, Paper No. 1909, 1935.

Kazmann, R.G., Safe Yield in Ground Water Development Reality or Illusion, J. Irrg. Drain. D.V. Proc. A.S.C.E. - 1982, 1956.

Kennedy, Minutes of Proceedings Inst. C.E.P. 260 Vol. 229, 1929 - 30.

Kennedy, Minutes of Proceedings Inst. C.E.P. 260 Vol. 229, Madras Public Works Dept., Dist. 1872, Oct. 9, 1912.

Kennedy, R.G., Hydraulic Diagrams for Channels In Earth, First Edition 1895, Second Edition 1907, P.W.D India, 1917.

Kennedy, R.G., The Prevention of Silting in Irrigation Canals, Paper No. 2826, Minutes of Proceedings of the Institution of Civil Engineers Vol. CXIX, Pp. 281-290, 1895.

Kinder, R.G., Proceedings Punjab Engineer Congress P. 74 J, 1919.

Lacey, G., In Discussion on Lane's Paper, Stable Channels in Erodible Material, Trans. A.S.C.E., 1937.

Lacey, G., In Reply to Discussions on his Paper No. 6274, Flow in Alluvial Channels with Sandy Mobile Beds, Proc. I. of Civil Engineers, October, 1958.

Lacey, G., Regime Flow in Incoherent Alluvium, Central Board of Irrigation Publication No. 20, 18th July, 1939.

Lacey, G., Shock in Regime Channels, Punjab Engineering Congress, February 23, 1940.

Lacey, G., Stable Channels in Alluvium, Paper No. 4736, Proc. of the Last of Civil Engineers, Vol. 229, Section 1929-930, 28th January 1930.

Lacey, Gerald, A General Theory of Flow in Alluvium, Paper No. 5518 Journal of the Institution of Civil Engineers London, November, 1946.

Lacey, Gerald, Flow in Alluvial Channels with Sandy Mobile Beds, Papers No. 6274 Proc. The Inst. of Civil Engineers. Feb., 1958.

Lacey, Gerald, Uniform Flow in Alluvial Rivers and Canals" Paper No. 4893 Minutes of Proc, Institution of Civil Engineers London, Vol. 237, Pp. 421-453, 1935.

Lane, E.W., and Kalinske, A.A., The Relation of Suspended to Bed Material in Rivers, Transactions, American Geophysical Union, Part IV, Washington, D.C., Pp. 637-641, 1939.

Lane, E.W., Progress Report on Results of Studies on Design of Stable Channels U.S.B.R. Hy. Lab Report No. Hyd. 352, P-2, June, 1952.

Lane, E.W., Progress Report on Studies on The Design of Stable Channels by the Bureau of Reclamation, Project A.S.C.E., Vol. 79, Separate No. 280, of Irr. and Drainage Division, September, 1953.

Laursen, E.M., The Total Sediment Load of Streams, Journal of The Hydraulics Division, ASCE, Vol. 54, No. HY1, Proc. Paper 1530, Pp. 1-36, Feb., 1958.

Leliavsky, Serge, An Introduction of Fluvial Hydraulics, P. 45, 1954.

Lindley, E.S., Regime Channels, Punjab Engineering Congress, Lahore, P 63 to 74, 1919.

Lindley, Minutes of Proceedings Inst. C.E.P 260 Vol. 229, Punjab Eng. Congress Proceedings, 1919.

Mac, Dougall C.H., An Experimental Investigation of Bed Sediment Transportation.

Mahapatra, J., Design of Stable Channel in Alluvium, M.E. Thesis University of Roorkee, 1972.

Malhotra, S.L., and Ahuja P.R., A Reveiw of The Progress on Theory and Design of Stable Channels in Alluvium, Processing of The Regional Technical Conference on Flood Control in Asia and The Far East, Flood Control Series, No. 3 U.N. Bangkok, P. 133, 1952.

Mavis, F.T., and Laushey, L.M., Discussion of, Sediment Transportation Mechanics, Initiation of Motion, by The Task Committee for Preparation of Sedimentation Manual, Committee on Sedimentation of The Hydraulics Division, Vito A. Vanoni, Chmn., Journal of The Hydraulics Division, ASCE, Vol. 92, No. HY5, Proc. Paper 4891, Pp. 288-291, Sept., 1966.

Meyer Peter, E., and Muller, R., Formulas for Bed-Load Transport, Report on Second Meeting of International Association for Hydraulic Research, Stockholm, Sweden, Pp. 39-64, 1948.

Mirajgaoker, A.G., Design of Stable Channels in Alluvial Soils, M.E. Dissertation, WRDTC, University of Roorkee, 1957.

Molesworth, & Yenidunia, Minutes of Proceedings, Inst. C.E., P. 260 – 285 Vol. 229.

Murti, N. G. K., and Others, Alakhnanda Enquiry Committee Report, 1970.

Punjab Irrig. Research Institute Report for The Year Ending, Pp. 55-81, April, 1938.

Punmia, B.C., and Pande, B.B.L., Irrigation and Water Power Engineering, Standard Publishers Distributors, New Delhi, 1977.

Puttaswamy, V. H., Design of Unlined Sediment Transporting Channel; M.E. Thesis University of Roorkee, 1972.

Rahman, M. A., Nagarjuna Sagar Right Canal Vol. XIV, No.9, P. 383-386, Sep. 1969.

Rajasthan Canal Board, Canal Lining, Sanchai Bhawan, Bhawani Singh Road, Jaipur.

Ranga Raju, K. G. And Misri, R. L. Simplified Canal Design Procedure Using Kennedy's Theory; Journal of Central Board of Irrigation and Power, April 1979.

Ranga, Raju P., Resistance Studies on Rough open Channels Having Rigid Boundary; M.E. Thesis University of Roorkee, 1973.

Rathbun, R.E., and Goswami, A., Discussion of Sediment Transportation Mechanics, Initiation of Motion, by The Task Committee on Preparation of Sedimentation Manual, Committee on Sedimentation of the Hydraulics Division, Vito A. Vanoni, Chmn., Journal of The Hydraulics Division, ASCE, Vol. 92, No. HY6, Proc. Paper 4959, Pp. 251-253, November, 1966.

Raudkivi, A.J., Discussion of "Sediment Transportation Mechanics: Initiation of Motion," by The Task Committee on Preparation of Sedimentation Manual, Committee on Sedimentation of The Hydraulics Division, ASCE, Vol. 92, No. HY6, Proc. Paper 4959, Pp. 253 -2 55, November, 1966.

Rothery, Minutes of Proceedings Inst. C.E.P. 179 Vol. CCXVI

Rouse, H., An Analysis of Sediment Transportation in The Light of Fluid Turbulence, Soil Conservation Service Report No. SCS-TP- 25, United States Department of Agriculture, Washington, D.C., 1939.

Rouse, H., Critical Analysis of Open-Channel Resistance, Journal of The Hydraulics Division ASCE, Vol. 91, No. HY4, Proc. Paper 4387, Pp. 1-25, July, 1965.

Rouse, H., Engineering Hydraulics, John Wiley and Sons, Inc., New York, P. 769-857, 1950.

Rubey, W.W., The Force Required to Move Particles on A Stream Bed, U.S.G.S. Professional Paper 189 – E, 1937.

Sahgal, P. P., Direct Method of Alluvial Channel Design By Using Kennedy's Equation; Journal of Central Board of Irrigation and Power, July 1972.

Saifi, A. G., A Review on Design of Stable Channels by Empirical Formula M. E. Thesis WRDTC University of Roorkee, 1975.

Schoklitsh, A., Hydraulic Structures, Vol. I, Translated by Schlits, Pp. 133, 1937.

Scobey F.C.: Flow of Water in Irrigation and Similar Canals, U.S. Department of Agriculture, Technical Bulletin No. 652, February, 1939.

Shields, A., Anwendung Der Aenlichkeitsmechanik Und Der Turbulenzforschung Auf Die Geschiebebewegung, Mitteilungen Der Preussischen Versuchsanstalt Fur Wasserbau Und Schiffbau, Berlin, Germany, Translated to English By W.P. Ott and J.C. Van Uchelen, California Institute of Technology, Pasadena, Calif., 1936.

Shulits, S., and Hill, R.D., Jr., Bedload Formulas, Bulletin, Department of Civil Engineering, Hydraulics Laboratory, Pennsylvania State University, University Park, Pa., Dec., 1968.

Shulits, S., The Schoklitsch Bed-Load Formula, Engineering, London, England, Pp. 644-646, P. 687, June 21, 1935, and June 28, 1935.

Sikka, D. R., Distribution & Management of Canal Supplies in Rabi Tracts in Madhya Pradesh Vol. XV, No.6, P.29-30, June 1965.

Simon, D.B., and Albertson M.L., Uniform Water Conveyance Channels in Alluvial Materials, ASCE Jour Hyd Div May, 1960.

Simon, D.B., Theory and Design of Stable Channels in Alluvial Materials, Colorado State University, Fort Collins, May 1957.

Singam, S. R., Analysis of Some Data of Stable Unlined Channels, M.E. Thesis Colorado, University of Roorkee, 1974.

Singh, Bharat, Fundamental of Irrigation Engineering, Printed in India, By N.C. Jani at The Roorkee Press, Roorkee (UP), 1982.

Standards for Permissible Non-Eroding Velocities, Bureau of Methodology of The Hydroenergo Plan, Gidrotenkhni Eheskoye Stroitel, Slovo, Obedinennoe Nauchono-Tekhnicheskoe Izatel'stvo, Moscow, USSR, May, 1936.

Straub, L.G., A Practical Theory of Detritus Trasportation and Design of Stable Channels, IAHSR (Second Proposed Meeting) P. 37, 1939.

Straub, L.G., Missouri River Report, House Document 238, Appendix XV, Corps of Engineers, United States Department of The Army to 73rd United States Congress, 2nd Session, P. 1156, 1935.

Sutherland, A.J., Discussion of Sediment Transportation Mechanics: Initiation of Motion, by The Task Committee on Preparation of Sedimentation Manual, Committee on Sedimentation of The Hydraulics Division, Vito A. Vanoni, Chmn., Journal of The Hydraulics Division, ASCE, Vol. 92, No. HY6, Proc. Paper 4959, Pp. 255-257, Nov., 1966.

Thomas, A.R., "Annual Report" of The Central Board of Irrig. of India, 1945, 1972.

Tinney E.R., In Discussions Published in ASCE Jour. of Hyd. Div., P. 1558-23 on Paper of, Mechanics of Sediment-Ripple Formation, February 1958.

U.S. Waterways Experiment Station, Vicksburg, Studies of River Bed Materials and Their Movement with Special Reference to The Lower Missippi River, Paper 17, 1935.

United Nations Economic Commission for Asia and The Far East, The Sediment Problem, Flood Control Series No. 5, P. 21, 1953.

Vanoni, V.A., A Summary of Sediment Transportation Mechanics, Proceedings, Third Midwestern Conference on Fluid Mechanics, University of Minnesota, Minneapolis, Pp. 129-160, 1953.

Varshney, Theory and Design of Irrigation Structures, Nem Chand & Bros Civil Lines, Publishing Company Ltd. , 2000.

White, C.M., Equilibrium of Grains on Bed of Streams, Proc. Royal Society (London) Vol. 174A, Pp. 322-334, 1940.

White, C.M., The Influence of Transported Solids Upon Rivers, British Research, Report of The International Union of Geology and Geophysics General Assembly, Washington Sept., 1939.

White, W.R., H. Milli Et A. Crabbe, Sediment Transport An Appraisal of Available Methods. Hydr.Res. Stat., Rept. INT 119, Wallingford, G.B, 1973.

Willis, J.C., Discussion of, Sediment Transportation Mechanics: Initiation of Motion, By The Task Committee on Preparation of Sedimentation Manual, Committee on Sedimentation of The Hydraulics Division, Vito A. Vanoni, Chmn., Journal of The Hydraulics Division, ASCE, Vol. 93, No. HY1, Proc. Paper 5059, Pp. 101-107, Jan., 1967.

□□□

Chapter 8

Embankment and Other Details

8.1 Design of Canal Embankments

There are two aspects of canal embankment design.

(i) The general requirement, such as free board, dowel, service/inspection road, drainage of banks etc.

The aspect of free board is discussed in Para 8.2 to 8.3 and that of service/ inspection road in Para 8.4 and drainage of banks in Para 8.5 in this chapter.

(ii) Safe design of embankment i.e. Hydraulic and structural requirement.

It includes stability of inner and outer slopes, Structural safe design of inner cut slopes in deep cuttings in different types of soils/rocks of varying nature and ground water conditions. Stability of slopes may include drainage, so as to prevent rock falls/debris in canal. For outer slopes in embankment (filling sections) hydraulic gradient/pharetic line, toe drains etc, similar to the earthen dams, on different types of soil/rock foundation with respect to varying ground water level conditions are important.

The canal embankments in filling may even fail through its foundations by sloughing or liquefactions.

All these aspects of canal embankments are discussed in this chapter in Para 8.4 to 8.9.

8.2 Free Board in Canals

Free board in a canal is usually defined as the vertical distance from the water surface to the top of the banks while water is being conveyed at full capacity, or at full supply level.

Sufficient freeboard should be specified in the design of canals to prevent overtopping of the banks during sudden rises in water surface.

Sudden rises in water surface may result from obstructions, chocking / checking of flows in improper operation of gates, accidents in operation, wave action, landslides, inflow during heavy rainfalls, or other causes. Gradual rises above design levels also result from excessive growth of vegetation or accumulation of silt deposits.

Limited encroachments in freeboards are sometimes permitted during checking operations or during a few days of maximum irrigation demand. However, full freeboard usually should be maintained, in order to insure safety against damage from unanticipated occurrences.

Freeboards provided in some unlined/lined canals in the world are discussed in Para 8.3. Normal freeboards may have to be increased around the outside edges of curves, to account for rise in water level i. e. super elevation.

Tops of compacted cores in embankments along canals usually are placed about 0.30 m (1 foot) above maximum water surface.

Where bank materials are likely to slough, freeboards should not be greater than heights needed to prevent overtopping. Excesses of canal excavation should be used in widening the banks rather than in increasing the heights.

Freeboards to tops of banks along lined canals are generally about the same as in unlined canals of similar sizes. However, tops of linings can be placed at lower elevations. Suitable distances from water surfaces to tops of concrete linings depend principally on the dimensions of the flow section. In small farm ditches, 3 inches may be sufficient. It may vary from about 0.15 m (6 inches) in small laterals to about 0.6 m (2 feet) in large canals.

8.2.1 Criteria for fixing free board

Adequate freeboard in canals depends on dimensions of flow sections, conditions of flow, bank materials and methods of bank construction, proximity of waste ways, and damage that may result if the banks are overtopped.

Free board is generally governed by consideration of the following aspects:

(i) Size of canal and its locations. top water surface width, depth, Side slopes wind velocity, silting, weed growth and general maintenance programme.

(ii) Hydraulic gradient.

(iii) Water surface fluctuations.

(iv) Soil Characteristics.

(v) Percolation.

(vi) Service road requirements.

8.2.2 General Practices of Free Board

Free boards in unlined canal sections generally vary from about 0.15 m (0.5 foot) in small water courses (ditches) to about 6 feet in large canals. For sections of intermediate size, free boards are sometimes determined by adding 0.30 m (1 foot) to 25 percent of the flow depth.

Generally U.S. Bureau of Reclamation practices, (1952), CWC (India) or IS Codes are followed in many countries and are discussed in subsequent Para.

Generally minimum free boards for various canal discharges are as below:

Canal Discharge	Free Board
More than 10 cumecs	0.75 m
Between 3 to 10 cumecs	0.60 m
1 to 3 cumecs	0.50 m
Less than 1 cumec	0.30 m
Less than 0.1 cumec (Water course)	0.15 m

A horizontal cement concrete coping 100 mm to 150 mm thick, depending upon size of canal is usually provided at the top of lining to check the ingress of rain water behind the lining of the side slope of the canals. This also protects the top of lining. The width of coping at the top is normally not less than 225 mm for discharge up to 3 cumecs, 350 mm for discharge more than 3 cumecs and 550 mm for discharge more than 10 cumecs.

Typical cross-sections of unlined/lined canal are shown in Chapter 6 and 9.

8.2.3 Dowla (Also Known Dowel or Dwarf Bund)

Dowel is a small raised embankment or masonry wall on the banks of the canal, see figure 8.1

Suitable dowels are generally essentially provided on the canal side of the service road and preferably both the banks, either lined or unlined canal.

It serves the following purposes.

(i) Protects canal banks from the formation of rain cuts and soil entering into the canal

(ii) Protects the inspection vehicles, other traffic and stray animals from falling into the canal. Also provides safety to the foot walkers from falling into canal

(iii) It is also an extra safety of free board particularly against wave action. By this dowel, free board can be restricted.

The size i.e. height and top width of the dowel depends upon the type and size of the lined or unlined canal. From economic or aesthetics or both, additional safety and low maintenance considerations, dowels may be replaced by parapet walls, particularly in hilly regions or in case of high embankments. However, the dowel or parapet should not be considered additional free board. The height of dowel portion should not be used for free board purposes.

Dowel is generally constructed on the inner side of the bank, while in those places, where rains are heavy, dowels are provided on both sides of the bank to prevent soil erosion by rains and formation of rain cut on outer slope. Water is then led out at intervals through artificial gutters or drains see Para 8.6.

The height of the earthen dowel is generally kept 0.5 m, with side slopes 1.5:1 and top width as 0.5m. It can be more with vehicular traffic. For vehicular traffic dowel of 1 to 1.5 m height are required for safety of the vehicle to avoid falling in the canal.

8.3 Free Board Practices in Various Countries

1. Australia, Victoria-State Rivers and Water Supply Commission

In fixing the Free Board, canal gradient and regulation procedure are taken into account. In earthen channels minimum value adopted is 0.38 m, (1.25 ft) though during construction suitable allowance is made for 10 per cent subsidence and 10 per cent wear. The minimum free boards adopted for different discharge ranges, both in unlined as well as lined canals, are as in Table below:

Free board adopted in Australia, Victoria

Capacity of canal		**Minimum Free Board values**			
		Unlined (earthen)		**Lined (concrete)**	
ft³/s	**(m³/s)**	**ft**	**(m)**	**Ft**	**(m)**
Up to 30	(up to 0.85)	1.25	(0.38)	0.75	(0.23)
31 too 100	(0.85 to 2.83)	1.50	(0.46)	0.75	(0.23)
101 to 300	(2.83 to 8.50)	2.00	(0.61)	1.00	(0.30)
301 to 1,000	(8.50 to 28.32)	2.50	(0.76)	1.25	(0.38)
Above 1,000	(Above 28.32)	3.00	(0.91)	1.25	(0.38)

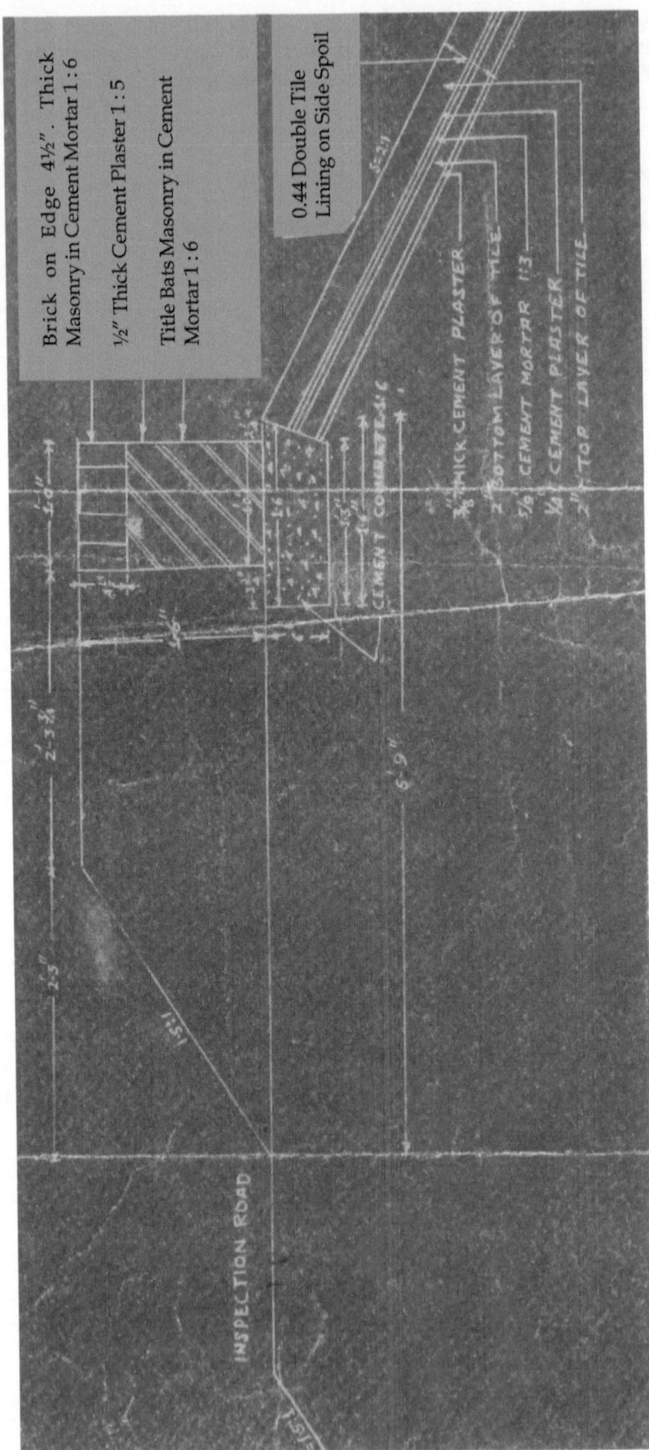

Figure 8.1 Details of Masonry Dowel in Sandy Reaches Along Indira Gandhi Main Canal

The general practices for unlined earthen canals are:

1.0 ft (0.30 m) + 15 per cent to 20 per cent of water depth, subject to minimum 1.25 ft (0.38 m) and maximum 4.0 ft (1.22 m);

And for concrete-lined channels- 20 percent to 25 per cent of water depth, subject to minimum 0.50 ft (0.15 m) and maximum 2.0 ft (0.61 m).

2. Bulgaria

Generally Free Board is adopted as under

(i) For small unlined canals = 0.25 m (0.8 ft);

(ii) For medium canals = 0.40 m (1.25 ft),

(iii) For large canals (mains) = up to 0.60 m (2.0 ft).

(iv) For small lined canals in water-resistant and loess soils = 25 cm (10 inches).

(v) For medium size lined canals in water-resistant heavy clay and loess soils, = 30 cm (12 inches).

(vi) For large lined canals in water resistant and loess soils, = 40 cm (16 inches). The berm width for manually-constructed small canals is at least 0.40 m (1.25 ft) and for large canals more than 1.0 m (3.3 ft).

3. Canada

Free Board generally provided is 0.4 d (water depth) with maximum of (3.0 ft) 0.91 m for larger canals and minimum of 0.30 m (1.0 ft) for smaller canals.

4. Colombia

According to the USBR standards (see item 18 relating to the 'United States of America').

It is observed that the free-board given by the formula of U.S.B.R. is vary much higher than those recommended by Indian Standards.

5. Cyprus

Free Board of 3 inches (7.6 cm) is provided to both rectangular as well as circular concrete channels.

6. France

(i) For small earthen canals, = 0.30 m (0.98 ft) and

(ii) For very large canals = 1.00 m (3.28 ft)

Allowance is made to account for bulking of soil and ultimate settlement.

For lined canals, the lining ends little below the top of the banks, the Free Board between the upper edge of the lining and the water surface level varies between 0.15 m (0.50 ft) and 0.60 m (2.0 ft) depending on the importance of the canal. In execution, it is convenient to take into account final swellings and settlement of soils. For embankments, not particularly eroded, one can allow final settlements 15 to 30 percent of the height of the banks depending on the nature of materials.

7. Greece

Standards of United States Bureau of Reclamation (USBR) are followed.

8. India

(i) Andhra Pradesh

For unlined and lined canals these are generally adopted as given in para 8.2.2

(ii) Bihar

River Valley Project Department-Gandak Project; Kosi Project; Damodar Valley Corporation and Irrigation Department; Sate Electricity Board.

Various Standards for Free Board, are adopted as given in para 8.2.2 above or as given by CWC.

(iii) Gujarat

Design data for Free Board, are adopted according to the channel capacity as given para 8.2.2 above or as given by CWC.

(iv) Jammu and Kashmir

Maximum Free Board provided is 0.60 m (2 ft) for channels with discharges of 400 ft^3/s (13.3 m^3/s) and minimum 0.37 m (1.2 ft) for minor discharges.

(v) Kerala

A Free Board of 1.5 ft (0.46 m) to 3 ft (0.91 m) is generally allowed in all canals depending upon their capacities. Free Board and Berm width normally used in design are given in para 8.5.1.

(vi) Madhya Pradesh

Free Board of value of 3 ft (0.91 m) is provided for canals having discharges between 2.83 m^3/s (100 ft^3/s) and 9.91 m^3/s (350 ft^3/s).

(vii) Maharashtra

The Free Board provided depends upon discharging capacity of the canals. Values, given in para 8.5.1 are generally adopted.

(viii) Mysore

According to the specifications given in Mysore P.W.D. Hand Book. Values, given in para 8.5.1, are generally adopted.

(ix) Punjab

The minimum values of Free Board provided are given in Table below.

Free Board

Type of canal	**Free Board**	
	Ft	**m**
1. Unlined distributaries [Q up to 350 ft^3/s (9.91 m^3/s)]	1.5	0.46
2. Unlined main canals and branches	2.5	0.76
3. Lined main canals and branches	2.5	0.76

(x) Rajasthan

Free Board, adopted are as below:

Name of project	**Discharge** Ft^3/s (m^3/s)		**Free Board** ft	**(m)**
Bhakra canal system	Less than 25 (0.71)		1.0	(0.3 m)
Ganga (Bikaner) canal system	-		2.0	(0.61 m)
Chambal system	Less than 500 (14.16)		2.0	(0.61 m)
	500 to 1,000 (14.16 to 28.32)		2.5	(0.76 m)
Minor irrigation projects	-	in cutting	1.5	(0.46 m)
		in filling	2.5	(0.76 m)

(xi) Tamil Nadu

Free Board for lined and unlined canals is adopted as below:

Discharge Ft³/s (m³/s)	Free Board Unlined canals Ft	(m)	Lined canals Ft	(m)
Less than 2 (less than 0.06)	1.5	(0.46)	0.25	(0.08)
25 to 2 (0.71 to 0.06)	1.5	(0.46)	0.50	(0.15)
100 to 25 (2.83 to 0.71)	2.5	(0.76)	0.75	(0.23)
300 to 100 (8.50 to 2.83)	3.0	(0.91)	1.0	(0.30)
Above 300 (8.50)	3.0	(0.91)	1.25	(0.38)

(xii) Uttar Pradesh

Free Board is adopted according to capacity of channel as below:

Discharge (ft³/s)	(m³/s)	Free Board Ft	(m)
2 to 100	(0.06 to 2.83)	1.5	(0.46)
100 to 1,000	(2.83 to 28.32)	2.0	(0.61)
1,000 to 2,500	(28.32 to 70.80)	2.5	(0.76)
2,500 and above	(70.80 and above)	3.0	(0.91)

(xiii) West Bengal

Free boards are provided according to discharging capacity of the channel as given below:

Discharge ft³/s (m³/s)	Free Board adopted in the Kangsabati Project ft. (m)	Free Board adopted in the Mayurakshi Project ft. (m)	Berm width
Less than 5 (0.14)	1.0 (0.30)	1.5 (0.46)	
5 to 25 (0.14 to (0.71)	1.5 (0.46)	1.5 (4.46)	5 ft or $d_{(cut)}$ 1.52 m or $d_{(cut)}$
25 to 50 (0.71 to 1.42)	1.5 (0.46)	2.0 (0.61)	10 ft (3.05 m, on one side and 5 ft (1.52 m) or $d_{(cut)}$ on other side whichever is greater.
50 to 100 (1.42 to 2.83)	2.0 (0.61)	2.0 (0.61)	
100 to 200 (2.83 to 5.66)	2.5 (0.76)	2.5 (0.76)	
200 to 500 (5.66 to 14.16)	2.5 (0.76)	3.0 (0.91)	
500 to 1,000 (14.16 to 28.32)	3.0 (0.91)	3.0 (0.91)	10 ft or $d_{(cut)}$+5 (3.05 m or $d_{(cut)}$ + 1.52m)
1,000 to 2,000 (28.32 to 56.64)	3.5 (1.07)	3.5 (1.07)	15 ft or $d_{(cut)}$+5 (4.57 m or $d_{(cut)}$ + 1.5m)
2,000 and above (56.64 and above)	3.5 (1.07)	4.0 (1.22)	20 ft or $d_{(cut)}$+5 whichever is greater (6.10 m or $d_{(cut)}$ + 1.52 m)

(xiv) Central Water and Power commission

The Free Board in a channel is governed by consideration of the size of the canal, its location, water surface fluctuation, percolation, grades, etc. Free Board required under ordinary conditions is provided according to U.S. Bureau of Reclamation practice which is given by the formula

$$FB = \sqrt{Cy}$$

where FB = Free Board in ft

C = coefficient

y = depth of water in canal in ft.

The coefficient 'C' varies from 1.5 where the canal capacity is 2.5 ft^3/s (0.07 m^5/s) to 2.5 for canal of 3,000 ft^3/s (84.96m^3/s) or more according to the requirements. However, greater Free Boards than those given above are used under exceptional circumstances.

9. Japan

The amount of Free Board required for both unlined and lined canals is calculated by the following formula derived on basis of observations made on the canals:

$FB = 0.05\ y + h_v$ + allowance and

$= 0.07\ y + h_v$ + (allowance) - For general canal sections of reinforced concrete or concrete flume.

Where FB = Free Board (m); y = water depth (m); h_v = velocity head of the flow (m)

Allowance for the fluctuation in water level is 0.05 to 0.15 m (0.16 to 0.49 ft)

Beyond the Free Board obtained by the above equation, embankments are raised to 0.2 to 0.5 m (0.66 to 1.64 ft) from the top of lining according to the canal size.

It is also the practice to adopt one-third of the designed flow depth as standard value of the Free Board.

10. Malaysia

Standards, given below, are adopted.

Type of canal	Free Board
(i) Small canals	1.5 ft (0.46 m)
(ii) Large canals	3.0 ft (0.91 m)
(iii) Drainage canals	1.0 ft (0.30 m)

11. Pakistan

Standards, as given below, are adopted.

Type of canal		Free Board in ft (m)
Unlined	Main canal	3.0 (0.9) (min)
	Branch canal	2.5 (0.76) (min)
	Distributory	1.5 (0.46) (min)
Lined canal		2.5 to 4.0 (0.76 to 1.22)

12. Poland

Free Board is recommended on discharge ranges both for unlined as well as lined canals as given below. Effect of wind is generally not taken into account, up to $Q = 100\ m^3/s$ (3,531 ft^3/s), and considered for more discharges.

Capacity of canal		Free Board			
		Unlined earth canals		Lined canals	
m^3/s	ft^3/s	m	ft	m	Ft
Up to 1.0	Up to 35.31	0.20	0.66	0.10 to 0.15	0.33 to 0.49
1.0 to 10.0	35.31 to 353.10	0.30	0.98	0.20	0.66
10.0 to 20.0	353.10 to 706.20	0.40	1.31	0.30	0.98
20.0 to 50.0	706.20 to 1,765.50	0.50	1.64	0.35	1.15
50.0 to 100.0	1,765.5 to 3,531	0.60	1.97	0.40	1.31

13. Portugal

The Free Board adopted varies between 0.30 m and 0.40 m (0.98 ft and 1.31 ft). Minimum being 0.20 m (0.66 ft). These values are generally fixed by experience.

14. Rhodesia

Adopted according to the USBR practice.

15. Spain

Adopted according the estimated future requirement of canal for serving larger area. If such an increase is not anticipated, free Boards, as given below are provided depending upon depth of flow.

In case of unlined canals, generally higher values of Free Board are provided with a maximum of 1.5 m (4.9 ft) for large canals.

Depths m (ft)		Free Board cm (inches)
Up to 0.5	(1.64)	
0.5 to 1.00	(1.64 to 3.28)	10 (3.94)
	15 (5.31)	
1.0 to 1.5	(3.28 to 4.92)	20 (7.87)
1.50 to 2.0	(4.92 to 6.56)	25 to 30 (9.84 to 11.81)
2.0 to 4.0	(6.56 to 13.12)	30 to 50 (11.81 to 19.69)

16. Turkey

Standards evolved by the USBR are followed while fixing the Free Board.

17. U.A.R.

The standards are based on the area served by a canal. As given below:

Area served by canals in Acres (hectares)		Free board m	(ft)
Very small canals		0.5	(1.6)
Up to 4,000	(1,619)	0.75	(2.5)
4,000 to 10,000	(1,619 to 4,047)	1.00	(3.3)
Above 10,000	(4,047)	1.25 to 1.50	(4.1 to 4.9)

18. U.S.A. (United States of America) USBR Practice

Free Board in a canal is normally governed by the considerations of its size, location, velocity, storm-water inflow, water surface fluctuations caused by cross bunds/checks, wind action, soil characteristics, percolation gradients, operating road requirements, and availability of excavated material. The recommended minimum Free Boards i.e. the height of bank above the water surface is shown in Figure 8.2, which varies from 1 for small canals to 6 for large canals. These illustrations are based upon average USBR practice, and it is emphasized that they will not necessarily serve all conditions.

19. U.S.S.R.

Free Board over maximum water level is provided according to Table 8.

In canals carrying discharges in excess of 100 m^3/s (3,530 ft^3/s) and also in case of canals of smaller capacity between 50 and 100 m^3/s (1,765 and 3,530 ft^3/s) but located in regions exposed to heavy wind, the free board is provided with due regard to height of wind waves.

Canal discharge in m^3/s (ft^3/s)	Free Board in m (ft)	
	Unlined canals	Lined canals
Up to 1.0 (35.3)	0.2 (0.7)	0.10 to 0.15 (0.3 to 0.5)
1 to 10 (35.3 to 353)	0.3 (1.0)	0.20 (0.7)
10 to 30 (353 t 1,059)	0.4 (1.3)	0.30 (1.0)
30 to 50 (1,059 to 1,765)	0.5 (1.5)	0.35 (1.15)
50 to 100 (1,765 to 3,530)	0.6 (2.0)	0.40 (1.30)

Free board may be determined by the formula

$FB = \sqrt{C.D}$ (as in U.S.A.)

Where,

F = free board in meters.

C = Coefficient.

D = Depth of water in canal in meters.

The coefficient `C' varies linearly from 1.5 where the canal capacity is 0.7 cumecs to 2.5 for canal of 85 cumecs or more. However, greater free boards and bank heights than those within these ranges may be used in exceptional circumstances.

20. Venezuela

Under normal conditions, Free Board = 1/3 rd full supply depth is provided. Under abnormal conditions, such as with high wind speed or considerable fluctuations due to regulation of supplies, the following formula is used:

$FB = \sqrt{C.y}$ (MKS units)

Where FB = Free Board; C = coefficient varying from 1.5 to 2.5; and y = water depth.

For unlined earthen canals values given below are generally adopted which are based upon the bed width.

Bed width	Free Board
0.8 m (2.62 ft)	0.4 m (1.31 ft)
0.8 m to 1.5 m (2.62 ft to 4.92 ft)	0.5 m (1.64 ft)
1.5 m to 3.0 m (4.92 ft to 9.84 ft)	0.6 m (1.97 ft)
3.0 m to 20.0 m (9.84 ft to 65.62 ft)	1.0 m (3.28 ft)

For lined canals USSR standards are followed.

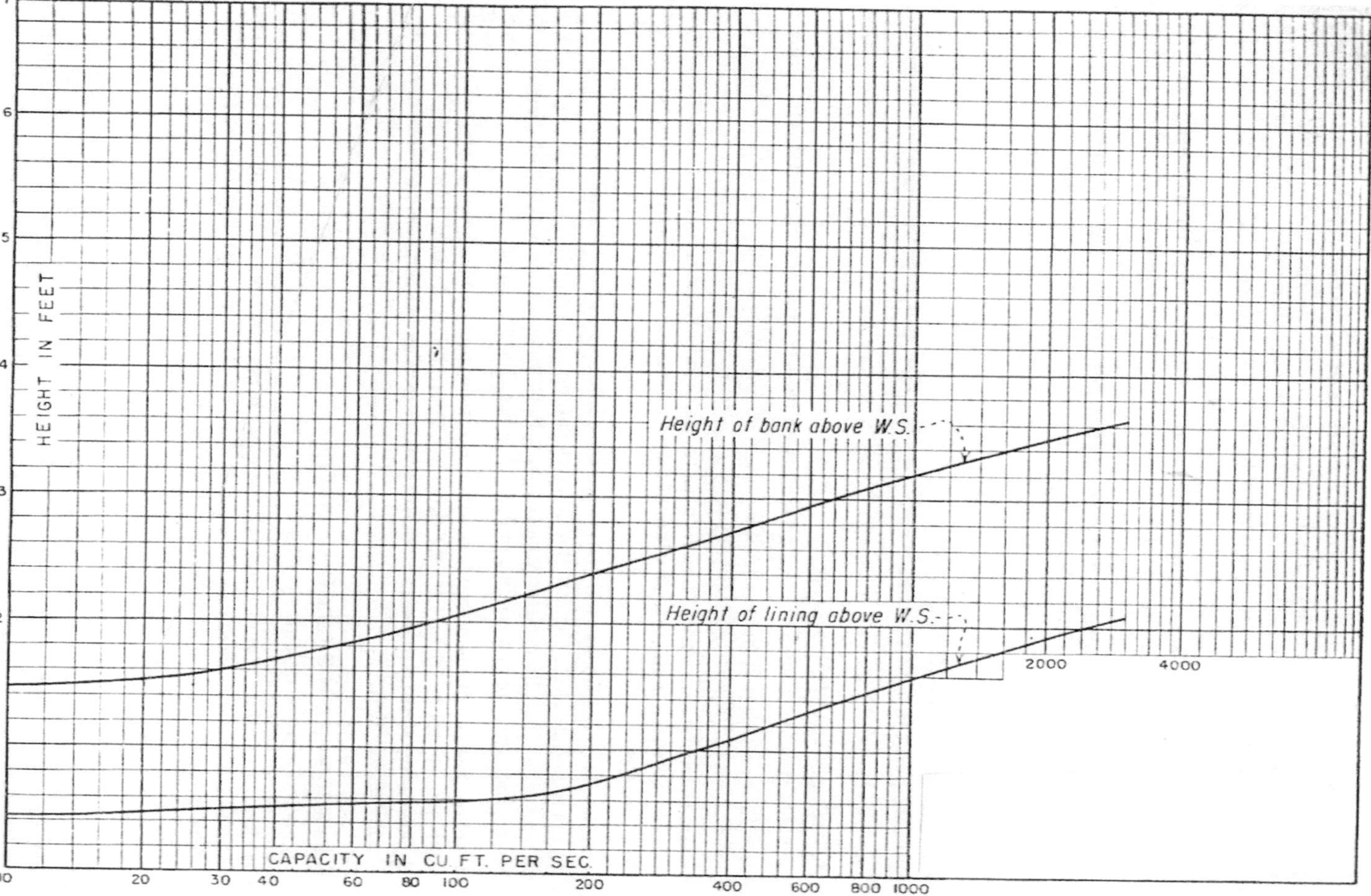

Figure 8.2 Freeboard and Height of Banks for Lined Canals, USBR

Inferences

The free Board practice varies from country to country. Even within the country there is no uniform practice. It varies from project to project. The reason being any slight change effects cost considerably in heavy filling and heavy cutting reaches, the most usual or more often cases in a long canal. For this reason Free Board provision has been revised many times in the 18000 Km length of IGNP, during the years 1960 to 2000.

To reduce the cost of free board without endangering the safety of canal, a dowel preferably of masonry see figure-8.1 is generally preferred. However sufficient free board should include provision for silting of canal, rise in F.S.L., wind /wave action, surface width of canal water, maintenance programmes etc and should be invariably provided. This ensures more safety to traffic. For traffic safety high dowels, preferably 1 to 1.5 m high are required, and should be provided.

There is a general acceptance of USSR practice and followed by many countries, and appears reasonably sound.

8.4 Design of canal banks (embankments)

The primary purpose of banks is to retain water. But they also serve the purpose of inspection road/traffic road (also generally known as service road).

8.4.1 Service Road/ Top Width of Bank

It is usual to provide roadway except on minors on the left bank of a channel, for inspection of works. But this practice can be deviated depending upon topography and alignment typically along hill sides or in contour channels (see typical section in chapter 6 and 9).

The width of driving bank should generally not be less than 5 m. Central water commission (CWC), India has recommended minimum road width of 6.1 m. (also see Para 8.5-16) However, this may not be feasible in all cases.

The road level is provided 0.5 m to 1.00 m above the full supply level, generally at free board level, depending upon the size of the channel. But in heavy cutting reaches, it can have up / down suitable gradients, not steeper than 1:6 preferably flatter, say 1:10 or more. On some roads there may be local traffic than a flatter gradient is desirable, preferably as per road standards.

As a normal measure of safety in driving, dowels usually minimum 0.5 m high and 0.5 m wide at the top with side slopes of 1:1 or 1.5:1 are provided along the bank. Some time masonry dowel is provided. Dowels incidentally prevents erosion of soil due to rains. With increasing public traffic higher dowels should be provided.

8.4.2 Hydraulic gradient (H.G.) and Cover on it

H.G. Line is the top most saturation line.

For a channel running in embankment, the top most percolation line through the bank though actually curved is assumed straight and sloping at 1:3 to 1:6 (V:H) gradient depending upon the bank material. This is known as hydraulic gradient line.

Now there is no more concept of H. G. line, but it is phreatic line, i. e. top saturation line. It depends upon the drainage of the outer toe. It is parabolic in shape.

When the height of bank is such that percolation line intersects the outer slope above ground level an extra layer of earth is put to provide a minimum cover of 0.30m. The surface of this cover is either kept parallel to the hydraulic gradient or is in horizontal steps. It should be wide enough so that a minimum cover on the H. G. (Hydraulic Gradient) line is available.

The purpose of cover on the saturation line is that soil particles do not move under hydraulic pressure, (known as soil boiling).

The designs of canal bank require the knowledge of geophysical engineering, rock and soil mechanics, stability of slopes, design of earthen dams etc.

Hydraulic gradient in different soils has been generally taken as under:

Type of soil		H.G.
1. Loamy, clay, loam and sandy soils (compacted zone)	=	1: 3 to 1: 4
2. Light soil (un compacted or loose zone)	=	1: 5 to 1: 6

To fulfill these criteria, counter berms may some time become necessary, and are provided accordingly.

A soil having hydraulic gradient flatter than 1:6 is not suitable for canal banks.

For embankments more than 5m height, true position of saturation line shall be worked out and stability of slope should be checked as in earth dams.

8.4.3 Berms

Berm is a horizontal space on slope. It brakes continuous high slopes and is used to increase stability of slopes.

Berms at ground level

It is a horizontal space initially left at ground level between the inner or outer toe of bank and excavation. In lined canals inner berm are avoided up to free board to reduce cost of lining. In unlined canals, silt is deposited on berms and strengthens the banks, and hence generally provided. The principal objectives for providing inner berms are:

a. To provide wider waterway above FSL to limit the height of rise of water at the time of positive fluctuations.

b. To afford less pervious silt lining on the sides of the silt carrying channels to decrease absorption losses.

c. To recede the banks away from the direct action of water to prevent breaches.

d. To provide a scope for future widening of canal.

e. To provide space for borrowing earth during construction at times of necessity and later on for maintenance when the berms are formed.

f. To break the flow of rain water down the bank slope

g. It also helps in providing space for silt deposit, its clearance and strengthening of banks/ increasing top width.

The following practice has generally been followed:

i. When the F.S.L. is above G.L. but the bed is below G.L., that is, the canal is partly in cutting and partly in filling, berm may be kept at N.S.L. equal to 2D in width where D is F.S.D.

ii. When the F.S.L. and the C.B.L. are both above the G.L. that is, the canal is in filling, the berm may be kept at the F.S.L. equal to 3D in width.

iii. When the F.S.L. is below G.L., that is the canal is completely in cutting, the berm may be kept at he F.S.L. equal to 2D in width. In embankments, adequate berm may be provided so as to retain the minimum cover over the hydraulic grade line.

Berms are also provided in all cuttings, when the depth of cutting is more than 3m. Where a canal is constructed in a deep through cut requiring waste banks, berms should be provided between the canal section cut and the waste bank.

Berms between water section and inner bank slopes are needed in canal sections, where bank materials are susceptible to sloughing. In deep cuts, berms are needed between tops of excavated slopes and adjoining waste banks that rise to higher levels, as shown in Figure 9.4. Berms along one or both banks may also be needed at lower elevations in deep cuts, so that operating roads can be maintained at convenient levels a few feet above maximum water surface.

Where berms are used as operating roads, widths should be great enough to meet road requirements. Berms along the edges of large earth canals can be made wider if desired. Berms along the sand-dune section of the All-American canal are 20 feet wide, see figure 6.18 (iv) (c).

Berms are commonly provided along both banks of lined canals and ditches, to facilitate the use of trimming and lining equipment as well as to catch sloughing bank materials and provide access for maintenance. Berm widths may vary from about 2 feet along small ditches to more than 6 feet along the larger canals.

Berms along both lined and unlined channels should be graded toward the water sections, to facilitate drainage. Along lined sections, berms should be constructed, as far as possible, in such a manner as to keep drainage water away from percolating along the back of the lining.

In deep cut reaches of canals with discharge capacity exceeding 10 cumecs, it is desirable to provide berms of 3 m to 5 m width on each side for stability, facility of maintenance, silt clearance, etc (see Fig. 9.4). In such sections, the inner sides above the berms may be provided with turfing.

Toe Berms

Toe Berms are a horizontal space initially left at ground level between the toe of bank and excavation. The principal objective for providing berms is to provide sufficient cover on H.G. line, and safety of outer slope.

Few practices of berm width in India are as given below.

Name of project	Discharge ft³/s (m³/s)	Berm width
Bhakra canal System-Punjab, Haryana, Rajasthan.	5,000 to 1,000 (141.50 to 28.32)	Equal to depth of cutting with a minimum 3 ft (0.91 m)
Chambal Project -Madhya Pradesh, Rajasthan.	1,000 to 300 (28.32 to 8.50)	4 ft (1.22 m) or $d_{(cut)}$ whichever is more
	Less than 300 (8.50)	10 ft to 15 ft (3.05 to 4.57 m) at natural surface level
Minor irrigation projects	-	5 ft (1.52 m) berm at interval of 10 ft (3.05 m) height when canal is in deep cutting

8.5 Top Width of Bank- Practices in the World

1. Australia

(i) Victoria-State Rivers and Water Supply Commission

Minimum crest width depends upon the construction equipment available and also to meet the requirement of a maximum hydraulic gradient of 1 in 5 from full supply level to outer toe of bank. The minimum crest widths used are:

i. For $Q < 0.85$ m³/s (30 ft³/s), crest width of 1.22 m or 1.83 m (4 ft or 6 ft) and
ii. For $Q > 0.85$ m³/s (30 ft³/s), crest width = 2.44 m (8 ft).

(ii) New South Wales-Water Conservation and Irrigation Commission.

Minimum crest width depends upon whether the canal is in cutting or in embankment. When it is in fully or partially in fill, the minimum crest width adopted varies from 0.91 m (3 ft) to 3.05 m (10 ft) for discharge range from 0.71 m^3/s to 70.79 m^3/s (25 ft^3/s to 2,500 ft^3/s) respectively. When vehicular traffic is expected, this width is increased to 3.66 m (12 ft). In fixing the kerb width, due consideration is also given to the normal hydraulic gradient (phreatic line) which is adopted as 1 in 4.25 and crest usually 0.61 m (2 ft) wide is placed outside the kerb.

When the channels are wholly in cutting, one of the banks is provided with 1.83-m (6-ft) wide crest while the other, with 4.57-m (15-ft) where spoil is disposed of.

In case of lined canals, the crest width usually equals the free Board.

(iii) Queensland-Irrigation and Water Supply Commission

On small canals, the minimum crest width used (except in some special circumstances) is 3.66 m (12 ft) since the banks are used for traffic, and for purposes of operation and maintenance.

Where intermediate berms are provided in sections fully in cut, the minimum width of berms is 4.88 m (16 ft) to provide for traffic.

In the case of inclined ground, channels in fill, the hydraulic gradient is also taken into consideration in fixing the bank crest widths.

2. Canada

Minimum bank width is taken as 0.6 B (i.e. 0.6 times bed width) with minimum of 0.91 m (3.0 ft) and maximum of 3.66 m (12.0 ft) except where bank carries maintenance roadway which needs bigger width. Bank widths for canals in partial or full filling are kept the same. Bank widths are ignored when canal is wholly in cutting unless excessive cut is encountered when further cut backs are used to provide berms to prevent sliding.

3. Colombia

Generally top width of banks is used for maintenance roads and hence is kept more than 3 m (9.84 ft) wide.

4. France

The width of bank depends upon several considerations, such as the nature of the soil, side slope and whether the banks are to be used for traffic or not. Generally, top width of banks adopted is equal to the water depth in the canal with a minimum of about 0.80 m (2.62 ft). If one of the spoil banks has to be used as central track, its minimum width is kept as 3.50 m (11.48 ft). If it is to be used as normal track, its width is increased to 6 m (19.68 ft) and if the bank is to receive metal road, its width is kept at 8 m (26.24 ft) minimum.

The safety of the banks is ensured by proper sectioning against danger of piping.

5. Greece

Following bank widths are normally provided;

Type of canal	Top width
Canals in filling	1.5 m to 2.0 m (4.92 ft to 6.56 ft) (minimum)
Canals in cutting	0.8 m to 1.0 m (2.62 ft to 3.28 ft) (minimum)
If operation road is provided on bank	4.0 to 6.0 m (13.12 ft to 19.68 ft)

6. Malaysia

The standards for bank width are as follows:

1.22 m (4.0 ft) minimum.

1.83 m to 3.05 m (6 ft to 10 ft) for light access road.

3.66 m to 19.50 m (12 ft to 64 ft) for roads.

7. Poland

The minimum top width of bank is fixed according to the height of embankment as given below:

Height of embankment	Top width (minimum)
Up to 2.0 m (6.56 ft)	1.0 to 1.5 m (3.28 ft to 4.92 ft)
2.0 to 3.0 m (6.56 ft to 9.84 ft)	2.0 m (6.56 ft)
Above 3.0 m (9.84 ft)	3.0 m (9.84 ft)

If the traffic road is required to be carried over the top, the top width is increased accordingly.

8. Rhodesia

Generally 1.22 m (4 ft) to 1.82 m (6 ft) widths is provided depending upon whether the canal is in embankment or in cutting.

9. Spain

The width depends upon the requirement of whether the top of the banks are to be used for service road and whether mechanical equipment is to be used for construction. For the later, a minimum of 2.50 m (8.20 ft) is provided. If a service road is to run over, the top width of 4.50 m (14.76 ft) for pavement and 0.75 m (2.46 ft) for path on each side is provided. Additional width of 1 to 1.50 m (3.28 ft to 4.92 ft) for berm is generally left.

10. Turkey

For both lined and unlined canals top widths are fixed according to USBR standards.

11. U.A.R.

General practice is to provide traffic road on one of the banks. The bank width in such cases vary from 4 to 4.5 m (13 to 14 feet) as shown below where as width of other bank is between 2 m to 5 m (6.5 ft to 16.0 ft).

Type of canal	Bank width Where traffic road is provided	Where traffic road is not provided
(i) Main Canal	8 m to 12 m (26.2 ft to 39.3 ft)	4 m to 5 m (13.1 ft to 16.4 ft)
(ii) Branch Canal	4 m to 6 m (13.1 ft to 19.7 ft)	2 m to 3 m (6.6 ft to 9.8 ft)

12. U.S.A.

The top bank width for both lined as well as unlined canals generally varies from 3.66 m to 6.10 m (12 ft to 20 ft) or more for the canals with discharge of 2.83 m^3/s (100 ft^3/s) to 70.80 m^3/s (2,500 ft^3/s) and more respectively to provide for operation and maintenance roads. A minimum of 0.91 m (3 ft) wide bank is adopted for small laterals when the banks are not proposed to carry any road.

13. Japan

Following standard top widths are adopted both for unlined as well as lined canals and for canals in cutting as well as in filling embankment.

Canal banks having vehicular traffic, 3.5 m (11.48 ft) and above
Canal banks having no vehicular traffic, 2.0 m (6.56 ft)
Small canals 1.0 m (3.28 ft)
If the canal is in filling the width of first embankment is as follows.
Q < 3.0 m3/s (105.94 ft3/s) = 0.5-1.5 m (1.64 to 4.92 ft)
Q > 3.0 m3/s (105.94 ft3/s) = 1.5-2.5 m (4.92-8.20 ft)

14. U.S.S.R.

Top width of canal banks are provided according to the design discharge as given below.

Discharge		**Top width of canal banks**	
m3/s	**(ft3/s)**	**M**	**(ft)**
<0.5	(<17.6)	0.5 to 0.8	(1.5 to 2.5)
0.5 to 1.0	(17.6 to 35.3)	0.8 to 1.0	(2.5 to 3.0)
1.0 to 5.0	(35.3 to 176.5)	1.0 to 1.25	(3.0 to 4.0)
5.0 to 10.0	(176.5 to 353)	1.25 to 1.50	(4.0 to 5.0)
10.0 to 30.0	(353 to 1059)	1.50 to 2.0	(5.0 to 6.5)
30.0 to 50.0	(1,059 to 1765)	2.0 to 2.50	(6.5 to 8.0)
50. to 100.0	(1,765 to 3,530)	2.5 to 3.0	(8.0 to 10.0)

Top widths are increased where necessary to suit size of excavators working on the job and also when the bank top is used as a road.

Berms are provided in deep cuts exceeding depth of 5 m (16.4 ft). They are made at every 5 m (16.4 ft) of the height and width is kept not less than 1.0 m (3.3 ft). If a berm is to be used as a road or for canal clearing machinery, its width could be larger as required.

15. Venezuela

Width of bank at top varies according to the capacity the canal. Normally adopted are as below.

Discharge		**Top width**	
m^3/s	**(ft^3/s)**	**m**	**(ft)**
0.90	(31.78)	2.0 to 3.0	(6.56 to 9.84)
20.00	(7,06.29)	3.0	(9.84)
30.00	(1,059.44)	4.6	(14.76)
40.00 or more	(1,412.59 or more)	5.0	(16.40)

16. India

(i) Central Water and Power Commission

The top widths are provided according to the channel capacity, as given below.

Discharge	**Top width**		
ft^3/s	**(m^3/s)**	**Ft**	**(m)**
Below 10	(0,28)	3.0	(0.91)
10-50	(0.28-1.42)	4.0	(1.52)
50-150	(1.42-4.25)	5.0	(1.52)
150-350	(4.25-9.91)	6.0	(1.83)
350-500	(9.91-14.16)	8.0	(2.44)
500- 1,000	(14.16-28.22)	12.0	(3.66)
1,000- 5,000	(28.32-141.60)	15.0	(4.57)

(ii) Jammu and Kashmir

The bank width is provided according to capacity of the channel as given below:

Discharge ft³/s (m³/s)		Bank width			
		Inspection road		Non-inspection road	
		m	(ft)	m	(ft)
400 to 150	(11.33 to 4.25)	8.84	(29)	1.83	(6)
150 to 20	(4.25 to 0.57)	5.18	(17)	1.83	(6)

(iii) Kerala

The top width of embankments of canals vary from a minimum of 1.06 m (3.3 ft) to 4.50 m (14.75 ft) where motorable road ways are provided parallel to the canal.

(iv) Madhya Pradesh

Generally one of the banks is made 3.66 m (12 ft) wide to accommodate inspection path and on the side of the non-inspection path the bank is made 1.52 m (5 ft) wide.

(v) Maharashtra and Mysore

Practice followed in fixing the top width is given below:

Type of canal	Minimum top width			
	Service road		Non-Inspection path	
	m	(ft)	m	(ft)
1. Main canal and branches	4.27	(14.0)	2.44	(8)
2. Major distributaries Q up to 0.28 m³/s (10 ft³/s)	3.66	(12.0)	1.22	(4)
3. Minors Q = 0.28 to 0.09 m³/s (10 to 3 ft³/s)	1.52	(5.0)	0.91	(3)
4. Watercourses Q = 0.09 to 0.03 m³/s (3 to 1 ft³/s)	0.91	(3.0)	0.91	(3)

(vi) Punjab

Guidelines, given below, are adopted while fixing top width of banks for lined and unlined canals.

Channel capacity		Bank width				Boundary road			
		Left		Right		Left		Right	
ft³/s	m³/s	ft	m	ft	m	Ft	m	ft	m
Above 500	Above 14.16	35.0	10.67	12.5	3.81	15.0	4.57	15.0	4.57
500-350	14.16-9.91	35.0	10.67	8.0	2.44	15.0	4.57	15.0	4.57
350-200	9.91-5.66	8.0	2.44	8.0	2.44				

(vii) Tamil Nadu

These are adopted according to the design discharge of the channel as given below.

	Capacity of canal m³/s (ft³/s)					
	141.6 to 28.32 (5,000 to 1,000)	28.32 to 8.50 (1,000 to 300)	8.50 to 4.25 (300 to 150)	4.25 to 0.71 (150 to 25)	0.71 to 0.28 (25 to 10)	Less than 0.28 (10)
Maximum crest width of widest bank in m (ft)	4.57 (15.0)	1.83 (6.0)	1.83 (6.0)	1.83 (6.0)	1.22 (4.0)	0.91(3.0)
Berm Width	4+Q/1,000+y/2	4+y/2	2+y/2	2+y/2	2+y/2	2+y/2

(viii) Uttar Pradesh

Minimum bank widths, as given below, are adopted for channels of different capacities.

Service roads are provided on left bank on all channels more than 3 miles (4.83 km) long. The width of service road is also given below.

Discharge ft^3/s	(m^3/s)	Minimum top width Ft	(m)	Width of service roads ft	(m)
2 to 20	(0.06 to 0.57)	4	(1.22)		
20 to 500	(0.57 to 14.16)	7	(2.14)	12	(3.66)
500 to 1,000	(14.16 to 28.32)	9	(2.75)	14	(4.27)
1,000 to 2,500	(28.32 to 70.80)	10	(3.05)	16	(4.88)
2,500 to 5,000	(70.80 to 141.60)	12	(3.66)	- do	
5,000 and above	(141.60 and abov e)	16	(4.88)	- do	

(ix) West Bengal

When the canal is in embankment, the banks are made 10 ft (3.05 m) to 15 ft (4.57 m) wide on the inspection side and 0.30 m (1 ft) to 1.83 m (6 ft) on the non-inspection side. When the canal is in cutting, its banks are made 20 ft (6.10 m) wide or depth of cutting whichever is less subject to a minimum of 12 ft (3.66 m).

17. Pakistan

The top width of banks depends upon whether the bank is to be used for patrolling (inspection) or not. The bank width for unlined main canals and branches is fixed as 6 m (20 ft) with 1.5 m (5 ft) additional width for avenues and 1.5 m (5 ft) for dowels, etc. In case of lined canals this width varies from 6 m to 9 m (20 ft to 30 ft) with additional width provided for dowels, etc. Where the bank is not used for locating a road, its width is adopted according to the capacity of unlined canal given below.

Canal discharge ft^3/s (m^3/s)	Main canals and branches 500 (14) and above	499 to 350 (14 to 10)	Distributaries 350 to 200 (10 to 5.7)	199 to 50 (5.7 to 1.4)	49 (1.4 and below
Width of non patrol Inspection bank ft (m)	12 (3.7)	8 (2.4)	6 (1.8)	5 (1.5)	4 (1.2)

8.5.1 Proposed Canal Standards Including Top Width of Bank

The various practices of free board, dowel, top width of bank etc., also known as canal standards, given above varies slightly from country to country. The width of the banks may vary according to the importance, capacity of the canal, and the construction equipment.

In case of distributaries, service road is generally provided on one bank for inspection and maintenance purpose. However, in case of main and branch canal or other important canals, service road or driving banks are some times generally provided on both the banks.

The minimum values generally recommended for top width of the bank, free board and cover on H. G. line is as given below in table 8.1.

Table 8.1 : Proposed Canal Standards

S No.	Discharge m³/s	Minimum Top Width of Bank		Minimum Width of Road Way (m)	Minimum Free Board (m)	Depth of cover over saturation gradient (m)
		Inspection (Service) (m)	Non Inspection (m)			
(1)	(2)	(3)	(4)	(5)	(6)	(7)
1	0.15 to 1.5	3.0	1.5	generally No road way is provided	0.3 to 0.4	0.5
2	1.5 to 3.0	3+ dowel+ drain size	2.0	--do--	0.50	0.5
3	3.0 to 10.0	4.0 + dowel+ drain size	2.5 +Dowel	3.5	0.60	0.5
4	10.0 to 30.0	5.0 + dowel	4.0 +Dowel	3.5	0.75	0.5
5	30.0 and above	6.0 + dowel + drain size	5.0 + dowel+ drain size	6.0	1.0	0.75 to 1.0

Notes

1. Bank widths given above may be altered when justified by specific conditions.
2. For distributaries canals carrying less than 3.0 cumecs and minor canals, it is generally not economical to construct a service road on top of bank as this usually requires more materials than the excavation provides. In such cases, service road suitably lowered below top of lining may be provided on natural ground surface and adjacent to the bank; however, the importance of providing adequate service roads where they are needed should always be kept in view. The service road should be above normally encountered high flood level (HFL) with some free board.
3. Where the stability of the embankment is required, wider bank widths can be provided. Turfing should be provided on the outer slopes.
4. In hilly terrain where it is not possible to provide above bank widths, the bank widths may be suitably reduced.
5. When the bank widths are reduced on exceptional ground, refuges cover taking space, after every 100 m should be provided for passing and sheltering of opposing traffic.
6. For drains, see Para 8.6.

Width of berms

In lined canals, inner berms are generally avoided. In unlined canals berms may be provided as per Para 8.4.2.

(i) Berm width in cutting.

(ii) Berm width in unlined canals in partly filling.

Maximum D + width due to difference in slope in cutting and embankment, limited to 2 D at F.S.L.

Minimum width due difference in cutting and slope embankment at ground level, minimum D at F.S.L.

(iii) Berm width is entire filling Minimum Nil ; Maximum 3 D.

Berms are provided above F.S.L. depending upon depth of cutting.

Land width

(i) Widths of land to be acquired clear of bank when canal is in less than balancing depth of cutting. (Extra land is required for borrow pits)	Width due to full height of bank + 5 m	Width due to full height of bank above ground + 1.5 m
(ii) Widths of land to be acquired clear of banks when canal cutting is more than balancing depth.	As per actual drawing +5 m	As per actual requirements + 1.5 m
(iii) Extra land may be acquired for plantation/ forests.		

8.6 Drainage of Canal Banks

Proper disposal of rain water on canal banks is essential so as to avoid formation of rain cuts either in filling reaches or in cutting reaches. In cutting reaches rain water causes erosion of inner slopes and this soil/silt flows in the canal, causing silt deposition in the canal known as silting. Heavy land erosion or even a land slide may enter the canal. Many times this silt inflow is so heavy that it completely blocks the canal and water way. Because of this, water may over top somewhere in upstream in filling reaches, causing a breach in the canal and consequently washing away of the bank. Extent of damage depends upon the height of embankment on the ground and the discharge. Restoration is very costly, besides loss of crops and to the people of the area. In some sandy areas, soils and heavy wind storms or in soft rocks of sliding nature blocking of canal is very common and a serious problem. In filling reaches rain water erodes the outer slopes of the canal banks and damages them considerably. Many times sufficient bank width is not available, canal become unsafe for driving vehicles and accidents of driving vehicles occur at these rain cuts. Many times canal banks also become unsafe hydraulically to retain water.

Also drainage of surface water is essential to avoid seepage into the banks, which creates back water pressure on lining. Due care of this problem can be taken by suitable design of drainage of banks.

In sandy soils, no drainage of banks is required as long as rain can be absorbed by soil. The experience on the left Bank of IGMC, where good earth was laid on sandy soils for good service road, showed a bad experience. The rain cuts were more on this bank, than the left bank, which was left sandy. All the rain water was absorbed in the sandy soil, where as on good soil, the rain water collected and on accumulation found its way towards outer edge and damaged the banks creating big rain cuts. Therefore for good soil, adequate drainage of banks is essential.

All these problems can be resolved by some design modifications and by providing suitable surface drainage arrangements. Here drainage of the banks is discussed for safe disposal of rain water and to safe guard canal blockage and formation of rain cuts. This reduces maintenance cost besides ensuring regular and reliable water supply in the channels.

There may be several procedures for safe disposal of rain water. It may be decided with respect to the following field situations/conditions:

(i) Lined canals
   (a) Filling/partial filling Sections
   (b) Cutting reaches/Section

(ii) Unlined canals
   (a) Filling/partial filling Sections
   (b) Cutting reaches/Section.

Further there are two aspects of drainage

1. Drainage into the canal
2. Drainage to the outer toe of the canal, (i.e. on outer slope).
   The choice of any one of above two depends upon the unlined/lined, along with type of lining and economics.

(i) Filling/partial filling Sections (Drainage into the canal)

In filling or in partial filling outer toes of the banks should be 3" to 6" (7.5 cm to 15 cm) higher than the normal level of the earth banks near the dowel. This means that there should be some cross slope from top outer toe, of the bank towards dowel at the inner edge, so that the water on the bank width flows towards dowel. This will avoid flow of water towards the outer slope and the natural ground and there will be no formation of rain cuts. This water near the dowel is further collected in a very small drain and discharged into the canal itself, through 4" diameter cement pipes of reasonable length say 3". This water is further discharged in the canal through a pucca (masonry or concrete) open drain in unlined canals or through a drain on the lining. These pipes may be provided at an interval of 30 m or so depending upon the rain fall in the area. A close interval in high rain fall area and large interval in low rain fall area is preferred. These pipes can be constructed simultaneously with the construction of the dowel so that there is no damage to the dowel or dismantling/reconstruction is not required. At the outlet point a plate form of the size 2 x 2' as shown in the drawing No. 8.3 (i) and (ii) is provided. This platform is 6" to 9" (15 to 25 cm) depressed than the bank level so that the silt coming with the rain water is deposited in the pit. This soil can be further utilized for restoration of the banks. Thus the rain water is very smoothly disposed off with a very low velocity without causing any erosion of the canal banks and soil. Also restoration is economical.

Drainage to Outer Slope of the Canal

Alternatively water is collected and drained towards the outer bank. A longitudinal drain is provided on the outer side of the bank and masonry (pucca) drains are provided on the outer slope. This may be costly if height of bank is more.

Cutting Reaches/Section

In cutting reaches earth drain of size 3 to 5' (1m to 1.5m) width at top and 1 to 2' at bottom is provided depending upon the soil/rock condition. Drain may be 1 to 2' (30 to 60 cm) deep with side slope of 1:1 or steeper/flatter. This drain will have steep longitudinal slope towards the lower NSL that is towards the filling side. The last part of the drain connecting general ground level can be made Pucca(masonry or concrete) as shown in the drawing. Some details of drains are given in figures in chapter 6 and 9.

Drainage of Common Bank of Twin Canals

Drainage of rain water on the surface of common bank of twin canals of IGMC and Sirhind feeder is shown in figure 8.3 (i) and (ii). Also see figure 6.7 (i) for drainage arrangement on the common bank of Upper Ganga canal and Parallel Upper Ganga Canal,(Lined).

Wherever additional spoil banks are to be provided on the land side of the embankment, adequate drainage channels shall be provided with suitable slope on the roadway sloping away from the canal side. No rain water shall be allowed to flow or percolate towards the canal slope behind the linings.

8.7 Radius of Curvature

Bends in canals are inevitable in its alignment. These are negotiated through circular curves from 10 to 20 times bed width of lined/unlined canals. Practices of minimum Radius of curvature, varies from country to country and are given below.

1. Australia

(i) New South Wales Water Conservation and Irrigation Commission
Unlined canals half the toe to toe width of earth work section
Lined canals - kerb to kerb width of lining

(ii) Queensland Irrigation and Water Supply Commission
Unlined canals - 20 times water depth, with minimum of 50 ft
Lined canals - 10 times water depth, with minimum of 50 ft

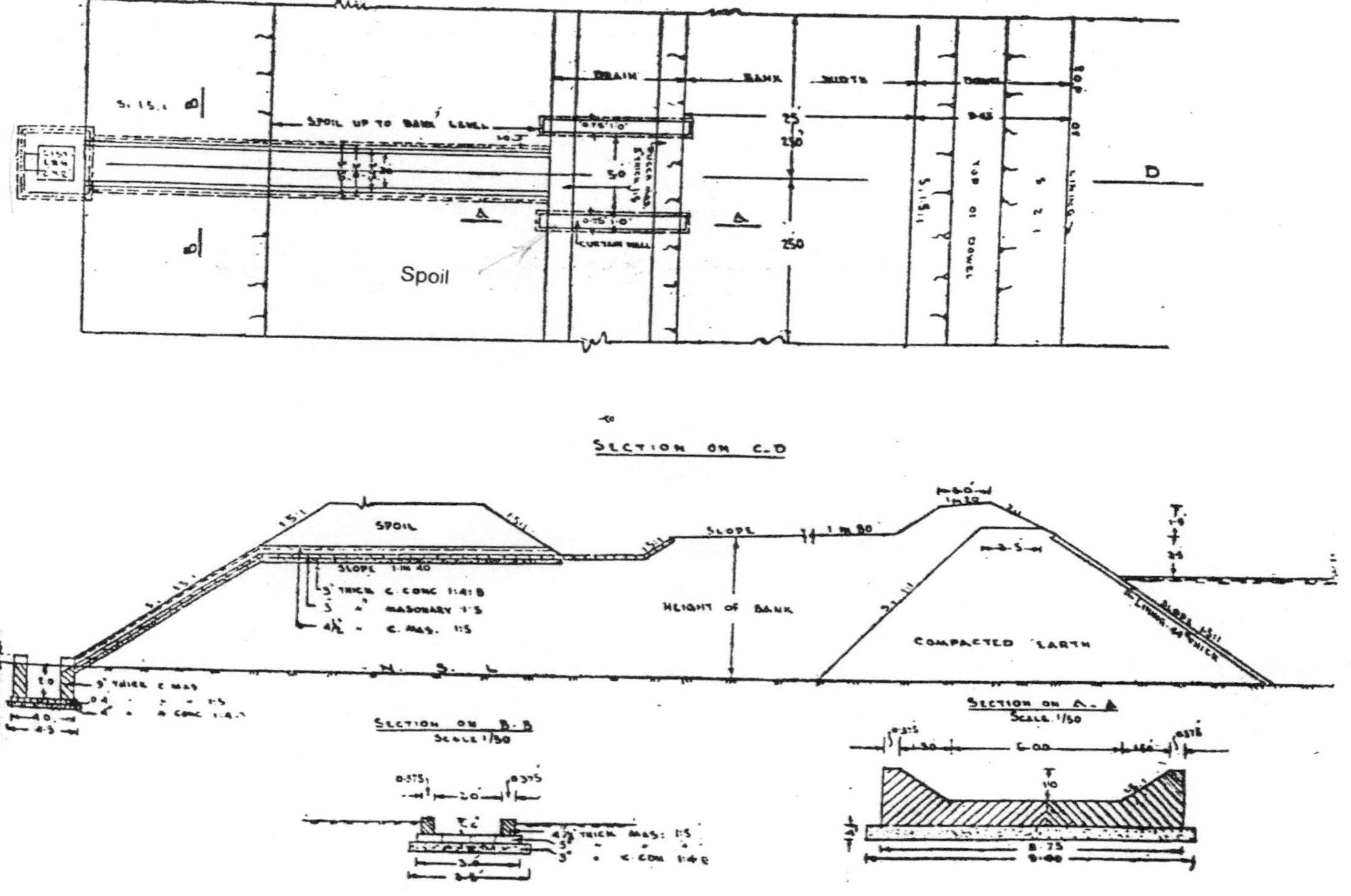

Figure 8.3 (i) Surface Drainage Outfall on Left Bank of Rajasthan Feeder

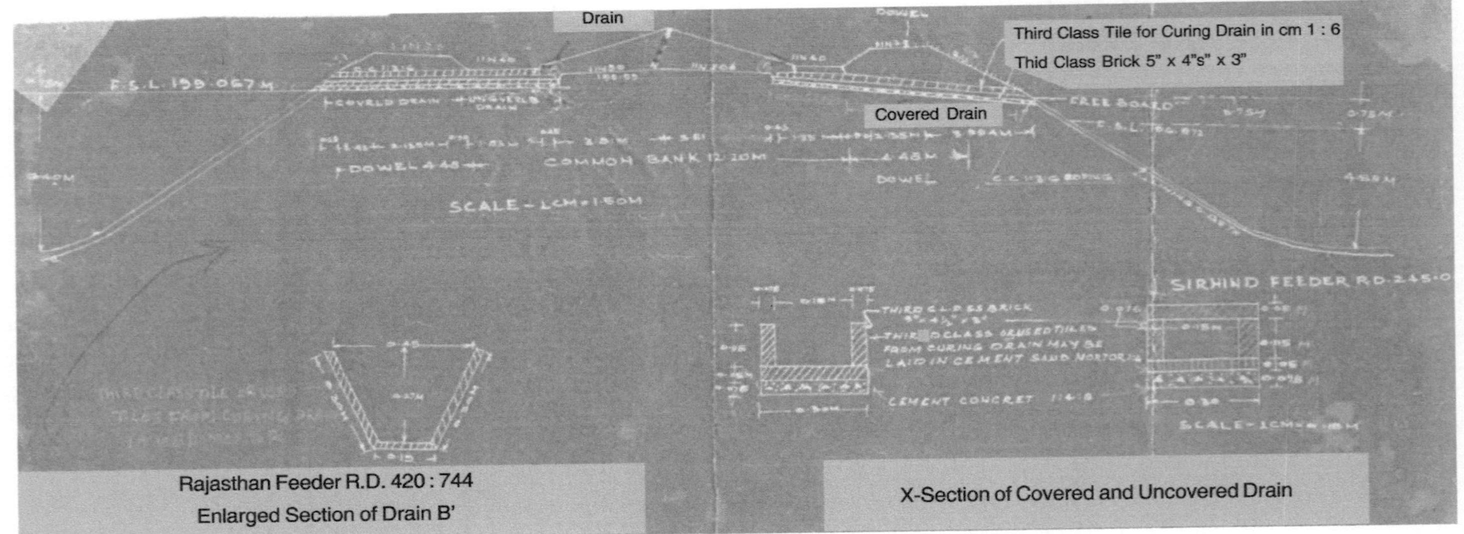

Note : Rain water in drained into the canals on both sides

Figure 8.3 (ii) Surface Rainfall Drainage Outfall on Common Bank of Sir hind Feeder

2. Canada

Unlined canals-10 times canal bed width, except in reaches of riprap or stone pitching.

3. Colombia

According to USBR Practices.

4. France

Unlined canals 20 times the bed width of canal.

5. Greece

Unlined canals - 15 times depth or 7 times full supply level width.
Lined canals - 15 times depth or 5 times full supply level width.

6. India

(i) *Andhra Pradesh*
Unlined / lined canals 10 to 20 times of bed width.

(ii) *Bihar*
It is based on discharge as under:

Discharge ft³/s (m³/s)	Minimum radius ft (m)	
< 10 (0.28)	100	(30.48)
10 to 100 (0.28 to 2.83)	500	(152.40)
100 to 500 (2.83 to 14.16)	1,000	(304.80)
500 to 1,000 (14.16 to 28.32)	2,000	(609.60)

(iii) *Gujarat*
Unlined canals - 20 times bed width or smaller as per the judgment of the Designer.

(iv) *Kerala*
Lined/unlined - 10 times bed width of canal

(v) *Madhya Pradesh*

Unlined channels 20 to 40 times bed width depending upon the discharge as under.

		Minimum radius	
Discharge ft³/s	(m³/s)	ft	(m)
Above 3,000	(84.96)	5,000	(152.4)
3,000 to 1,000	(84.96 to 28.32)	3,000	(914.4)
1,000 to 500	(28.32 to 14.16)	2,000	(609.6)
500 to 100	(14.16 to 2.83)	1,000	(304.8)
100 to 10	(2.83 to 0.28)	500	(152.4)
Below to	(0.28)	300	(91.4)

(vi) *Maharashtra*
Unlined/lined canals 20 times bed width, curves are generally located in cutting reaches.

(vii) *Mysore*
Unlined/ lined canals 20 times bed width.

(viii) Punjab according to discharge and lining as under

Channel capacity		Minimum radius			
		Lined		Unlined	
ft³/s	m³/s	ft	m	ft	m
Less than 10	Less than 0.28	..	..	300	91.43
10-100	0.28-2.83	..	..	500	152.40
100-500	2.83-14.16	..	..	1,000	304.80
500-1,000	14.16-28.32	..	..	2,000	6.9.60
1,000-1,500	28.32-42.48	..	...	3,000	914.40
1,500-2,500	42.48-70.80	1,000	304.80	3,000	914.40
2,500-3,000	70.80-84.96	1,500	457.20	3,000	914.40
3,000-5,000	84.96-141.60	1,500	457.20	5,000	1524.00
5,000-7,500	141.60-212.40	2,000	609.60	5,000	1524.00
7,500-10,000	212.40-283.20	2,000	609.60	5,000	1524.00
Greater than 10,000	Greater than 283.20	3,000	914.40	5,000	1524.00

(ix) *Rajasthan*
As in Punjab

(x) *Temil Nadu*
Unlined canals - For discharge < 25 cusecs, - 10 times bed width. For discharge of > 25 ft³/s, (0.71 m³/s) 15 times bed width.
Lined canals - 10 times bed width

(xi) *Uttar Pradesh*
Unlined canals - It is adopted according to discharge as under.

Discharge ft³/s (m³/s)	Minimum radius of curvature ft	(m)
Up to 10 (up to 0.28)	200	(60.96)
11 to 100 (0.28 to 2.83)	500	(152.40)
101 to 500 (2.83 to 14.16)	1,000	(304.80)
501 to 1,000 (14.16 to 28.32)	2,000	(609.60)

(xii) Central Water Commission, (New Delhi)

Several factors influence the choice of the radii of curvature for unlined channels. The primary factors are the size or capacity of the channel, the velocity of flow, the characteristics of the soil and dimensional elements of the canal section. Velocities in unlined channels very from 1.5 to 3.5 ft/s (0.46 to 1.07 m/s) and have an appreciable influence on canal curvature. The character of soil will also influence because the soil may range from extremely firm to extremely shifting and its stability may be quite sensitive to the curvilinear flow of the water.

One suggested rule is to make the minimum radius 15 times the water depth. Such a radius may be sufficient for conditions in limited capacity ranges. It may be too small particularly for extremely small shallow distributaries or for canals with large capacities. Another rule is that the radius of curvature should be 3 to 7 times the water surface width. These rules have very limited application and should be considered in the nature of minimum requirements.

7. Japan

Unlined/lined canals - 10 times water surface width.

8. Malaysia

Unlined canals - 5 times bed width, also at bends bed width is progressively increased from entry and exit ends to 1.5 times the normal bed width, at the centre.

9. Pakistan

According to discharge as under

Capacity of canal		**Minimum radius**	
ft³/s	**m³/s**	**ft**	**m**
Above 3,000	85	5,000	1,524
3,000 to 1,000	85 to 28	3,000	914
1,000 to 500	28 to 14	2,000	610
500 to 100	14 to 2.8	1,000	305
100 to 10	2.8 to 0.3	500	152
10 and below	0.3 and below	300	91

10. Poland

Unlined canals

The radius of curvature is worked out from the following formula for both lined and unlined canals not required to carry floating ice or slush:

$$R_{(curv)} = 11\, v^2 \sqrt{A_{(w)}} + 12$$

Where $R_{(curv)}$ = radius of curvature in m; V = mean velocity of flow m/s, $A_{(w)}$ = wetted area, m^2

Minimum radius of curvature is restricted to the one given by the one following equation:

$$R_{(curv)} = 5\,B$$

If the canal is required to carry floating ice or slush the minimum radius of curvature is limited to the following values:

$$R_{(curv)} = 10\,B \text{ when } \alpha < 35^\circ$$
$$= 14\,B \text{ when } 35^\circ < \alpha < 55^\circ$$
$$= 20\,B \text{ when } \alpha > 55^\circ$$

Where,

B = bed width of canal; α = angle of bend.

11. Portugal

Lined canals - No fixed value but generally it is 15 m.

12. Rhodesia

USBR - practice is adopted

13. Spain

Unlined / lined canals - > 5 to 6 times water surface width.

14. Turkey

USBR Practice

Unlined canals - 10 to 15 times bed width.

15. U.A.R.

Usually minimum radius of 10 to 15 times the bed width is provided in earthen canals.

16. U.S.A. (United States of America)

Minimum radius to the canal centre line is generally kept between 3 and 7 times the water surface width for unlined earth canals and erodible linings (the larger rations for the larger capacities) depending upon the size or capacity of canal, the soil characteristics and the velocity.

For lined canals (hard surface) minimum radius of 3 times the water surface width is adopted.

17. Venezuela

Minimum radius adopted is 15 times the depth of water or approximately 6 times the water surface width.

Inferences

There are generally following practices:

i) Minimum radius of 15 times the water depth.

ii) Minimum radius of 5 to 40 times the bed width

iii) Minimum radius is generally between 3 and 7 times the water surface width for unlined earth canals (the larger ratios for the larger capacities) depending upon the size or capacity of canal, the soil characteristics and the velocity.

iv) For lined canals (hard surface) minimum radius is 3 times the water surface width.

v) Fixed radius according to discharge.

Adoption of appropriate value depends upon the experience of the designer, based on the topography, angle of bend, discharge, lined/unlined and velocity of flow.

8.8 Failures of Slopes

According to Terzaghi and Peck (1969) "Slides may occur in almost every conceivable manner, slowly or suddenly, and with or without any apparent provocation". These slope failures are usually due to either a sudden or gradual loss of strength by the soil or a change in geometric condition, e.g. steepening of an existing slope. Typical slides that can be expected to occur in soil slopes are illustrated in figure 8.4 and 8.5. These usually take the form of either (1) translational, (2) plane or wedge surface, (3) circular, or (4) noncircular, or a combination of these types.

The Geologic conditions that influence the shape and development of the potential failure surface are summarized in Table 8.2. The planar failures surfaces are usually expected in slopes where a soil layer, or relict jointing. A relatively low strength and strongly influences the shape of the failure surface. The translational type of failure occurs in shallow soils overlying relatively stronger materials, and circular failure surfaces usually occur in slopes consisting of homogenous materials.

As most soils are generally heterogeneous, noncircular surfaces, consisting of a combination of planar and curved sections, are most likely. Often, retrogressive failures consisting of multiple curved surfaces can occur in layered soils, as shown in Figure 8.2. Such failures are typical where the first slip tends to over steep the slope, which then leads to additional failures.

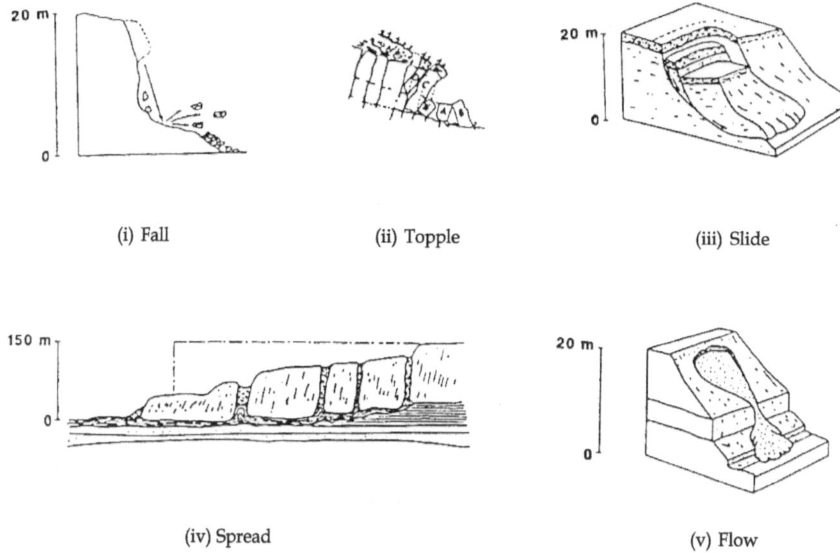

(i) Fall (ii) Topple (iii) Slide

(iv) Spread (v) Flow

Figure 8.4 Types of Land Slides

Note : Broken lines indicates the original ground surfaces. Arrows show portion of the trajectories of individual particles of the displaced mass.

Table 8.2 Geologic factors controlling shape of potential failure surface

Geologic Conditions	Potential Failure Surface
Cohesion less soils Residual or colluvial soils over sallow rock stiff fissured clays and marine shales within the upper, highly weathered zone	Translational with small depth/ length ratio
Sliding block Inter bedded dipping rock or soil Faulted or slackened material. Intact stiff to hard cohesive soils on steep slopes	Single planar surface
Sliding blocks in rocky masses Weathered inter bedded sedimentary rocks clay shale and stiff fissured clays stratified soils side hill fills over colluviums	Multiple planar surfaces
Thick residual and colluvial soil layers Soft marine clays and shale's soft to firm cohesive soils	Circular or cylindrical shape

8.8.1 Shear Strength Factors Contributing to Slope Failures

Slope failures are often caused by two main processes

(i) Processes that increase shear stresses on the soil mass or

(ii) Processes that decrease shear strengths of the soil mass.

8.8.2 Factors that Cause Increase in Shear Stresses in Slopes

Processes that most commonly increase in the shear stresses acting on slopes are listed below:

1. Removal of support (i.e. under cutting, see figure 8.11).
   A. Erosion
      1. By streams and rivers
      2. By glaciers

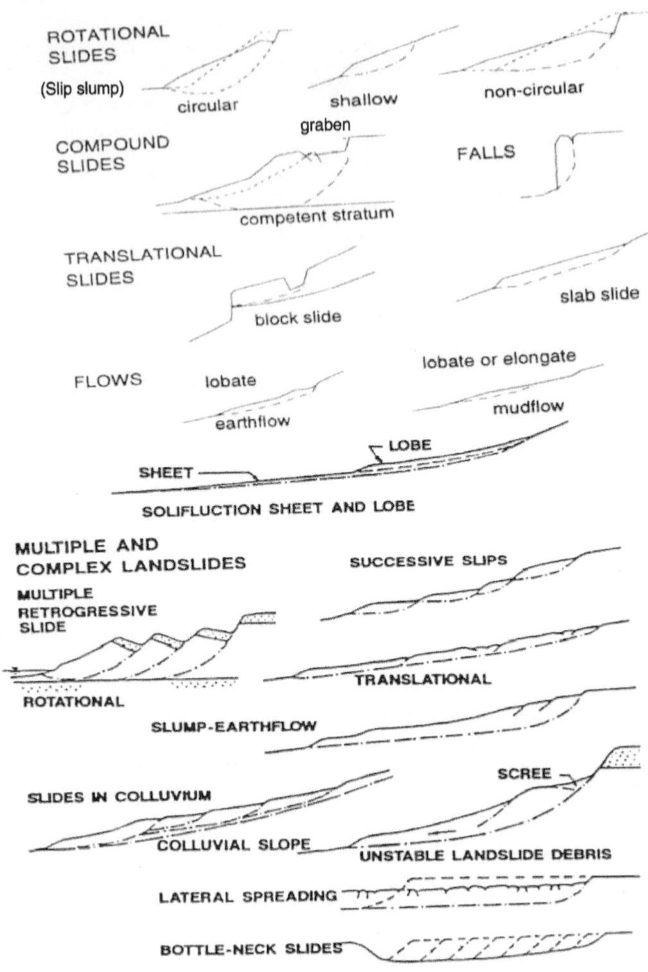

Figure 8.5 (i) : Types of movement in clay slopes

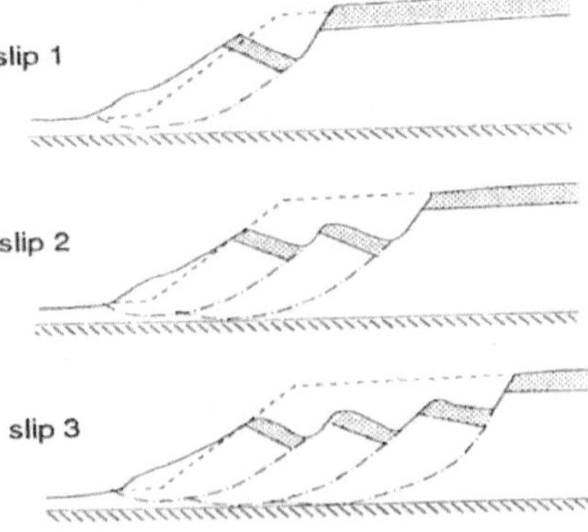

Figure 8.5 (ii) : Typical retrogressive slides

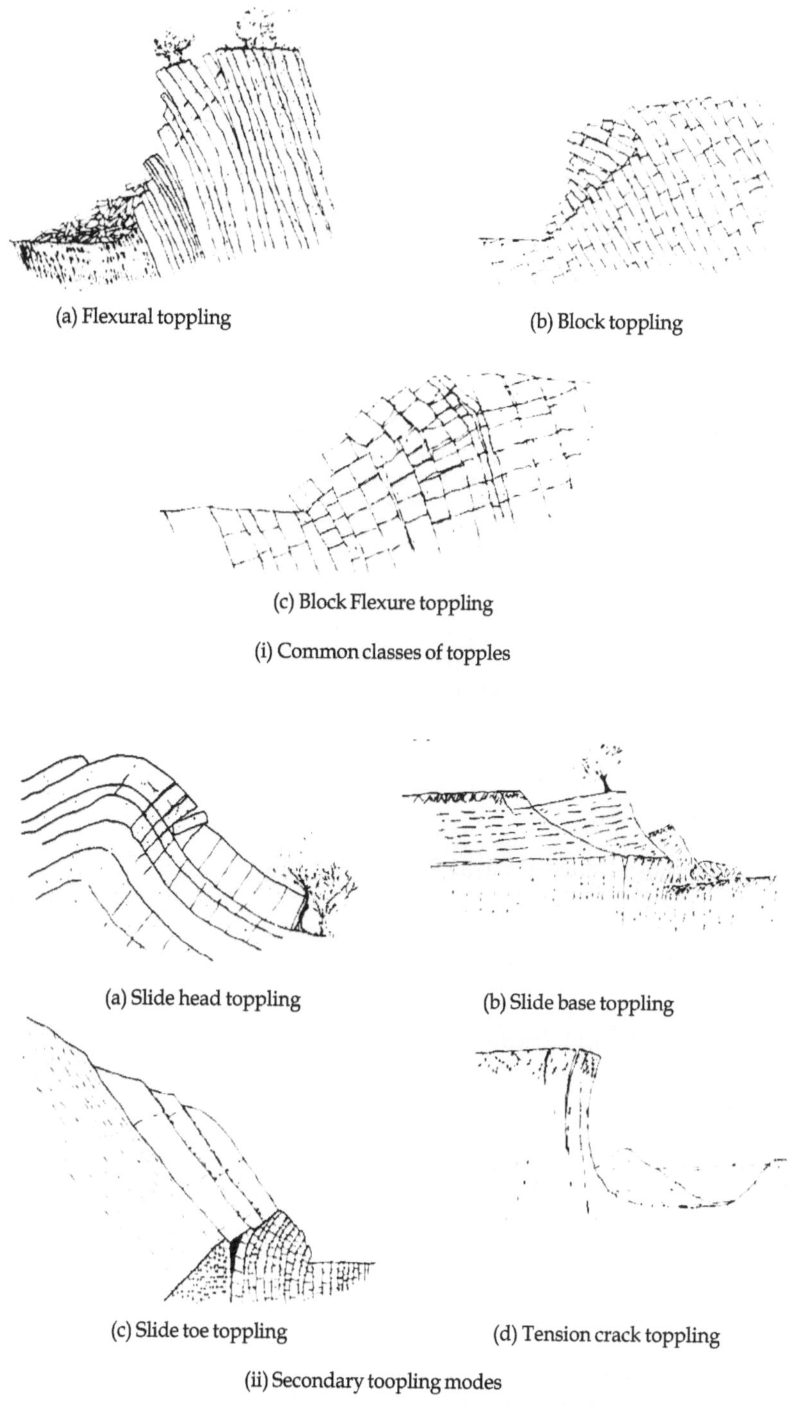

(a) Flexural toppling

(b) Block toppling

(c) Block Flexure toppling

(i) Common classes of topples

(a) Slide head toppling

(b) Slide base toppling

(c) Slide toe toppling

(d) Tension crack toppling

(ii) Secondary toopling modes

Figure 8.6 Failures of Rock Slopes by Toppling

3. By action of waves or marine currents
4. By successive wetting and drying (e.g. winds, freezing)

B. Natural slope movements (e.g., falls, slides, settlements)

C. Human activity
1. Cuts and excavations
2. Removal of retaining walls or sheet piles
3. Drawdown of bodies of water (e.g., lakes, lagoons)

2. Overloading
   A. By natural causes
      1. weight of precipitation (e.g., rain, snow)
      2. Accumulation of materials because of past landslides
   B. By human activity
      1. Construction of fill
      2. Buildings and other overloads at the crest
      3. Water leakage in culverts, water pipes, and sewers
3. Transitory effects (e.g., earthquakes)
4. Removal of underlying materials that provided support
   A. By rivers or seas
   B. By weathering
   C. By underground erosion due to seepage (piping), solvent agents, etc.
   D. By human activity (excavation or mining)
   E. By loss of strength of the underlying material
5. Increase in lateral pressure
   A. By water in cracks and fissures
   B. By freezing of the water in the cracks
   C. By expansion of clays.

8.8.3 Factors that Cause Reduced Shear Strength in Slope

Processes that most commonly cause a decrease in the shear strength of slope materials are listed below:

1. Factors inherent in the nature of the materials
   A. Composition
   B. Structure
   C. Secondary or inherited structures
   D. Stratification
2. Changes caused by weathering and physiochemical activity
   A. Wetting and drying processes
   B. Hydration
   C. Removal of cementing agents
3. Effect of pore pressures
   A. Changes in structure
   B. Stress release
   C. Structural degradation.

Zones, relict joints, and soil dikes: Relict joints and structures in residual soils often lose strength when saturated. Slickened seams or weak dikes may also pre-exist in residual soil and weathered rock slopes. Faults, bedding surfaces, cleavages, and foliation's have more influence on rock slope stability than soil slope stability.

The main items required to evaluate the stability of a slope are: (1) shear strength of the soils. (2) Slope geometry, (3) pore pressure or seepage forces, and (4) loading and environmental conditions. These are discussed in Para 8.11.

Typically, the engineer will develop a subsoil model of the slope and foundation soils through a site exploration study and subsequently select appropriate shear strengths for the soils. Ideally, the shear strengths should be provided as undrained strength, S_u, or the more typical Mohr-Coulomb parameters, c and Φ. The slope geometry may be known for existing, natural slopes or may be a design parameter for embankments and cut slopes. A major contributor in too many slope failures is the change in effective stress caused by pore water pressures. These tend to alter the shear strength of the soil along the shear zone and may even cause destabilizing forces to develop if surface runoff infiltrates cracks at the head of the failure surface. Later sections discuss the methods used to account for pore water pressure effects within the analytical procedures.

8.9 Consequences of Uncontrolled Seepage

Most people know the story of little Dutch boy, who held his finger in a leak in a dike until help arrived. He knew that a small trickle could develop into a torrent that would flood his homeland. Failures by piping or erosion as shown in figure 8.7 and 8.8 are the most easily understood. They represent one of the two basic kinds of failures caused by uncontrolled seepage. The second kind, which may be less obvious to the layman, is caused by seepage forces and pressures within soil and rock as shown in figure 8.12 (a) and (b). In brief, most failures caused by groundwater and seepage can be placed in one of following two categories.

1. Those which take place when soil particles migrate to an escape exit, causing piping or erosional failures.

2. Those which are caused by uncontrolled seepage patterns and lead to saturation, internal flooding, excessive uplift, excessive seepage forces, leading to land slides and liquefaction.

Some common type of failures of Category 1 and 2 are listed in Table 8.3 First category of failures are very easily seen by most people and well known. 2nd category of failures are little complex and not very easily understood and needs more knowledge/experience on behavior of internal water pressures in saturated mass. It includes failures of dams, dry docks, retaining walls, etc., caused by excessive saturation seepage forces, and uplift pressures. In this category are listed the deterioration and failure of pavements from internal flooding, the uplifting of canal linings after drawdown, failures of fills and foundations caused by seepage forces and uplift pressures in trapped water, landslides, and similar cases.

Seeping water flows through the interconnected pores (characterized by porosity n) between the solid particles [which have a volume (1-n) in unit volume]. In the normal state, total pressure head of water is dissipated as viscous friction producing a frictional drag, acting in the direction of flow, on the solid particles. Seepage forces act on the particles of a soil in addition to gravitational forces.

Filtration deformations are caused by unfavorable field conditions characterized by poor bank/ subsoil and river bed subsoil which give rise to a large water height, increased velocities, flow rate and pressures within a short distance.

The high porosity, the unfavorable grain size distribution and other factors mentioned above allow water inflow into the subsoil and under flowing first stage. In the second stage, the soil particles are washed away and transported from the upper side to the downstream toe. In the third stage, a continuous channel on the contact surface of loam and sands (or gravel) is formed. After that the channel grows in size and picks up the soil of the sub base and of the bank body (fourth stage). In this way, the disaster stage becomes obvious. It begins in the sinking and its final stage is when the soil mass is carried away by water, figure 8.7 to 8.10.

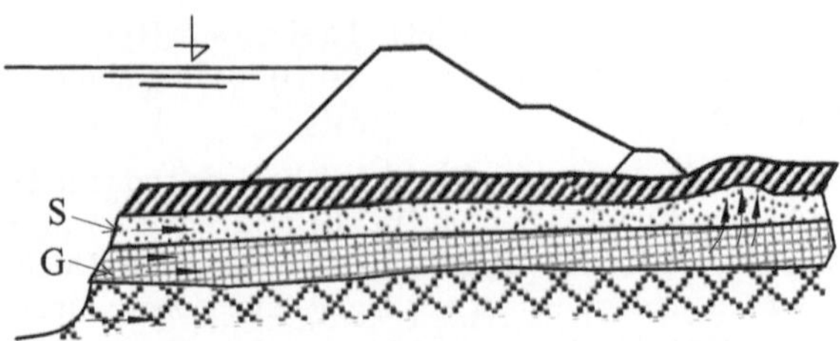

(a) Hydraulic structure and seepage acting on sand (S) - and Gravel, (G) - Layer in its subsoil.

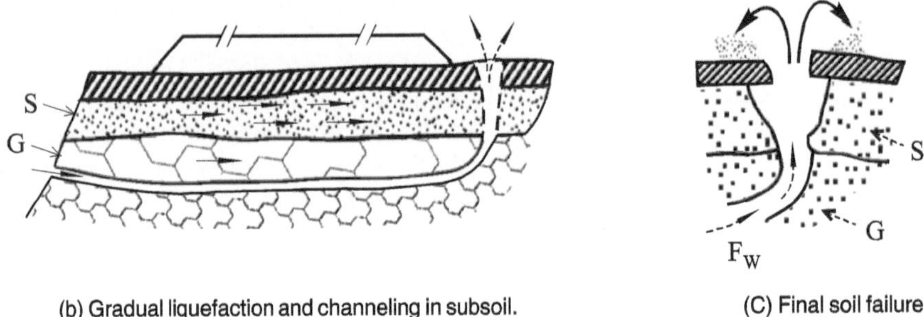

(b) Gradual liquefaction and channeling in subsoil.

(C) Final soil failure.

Figure 8.7 : Typical Example of the Embankment and Soil failures.

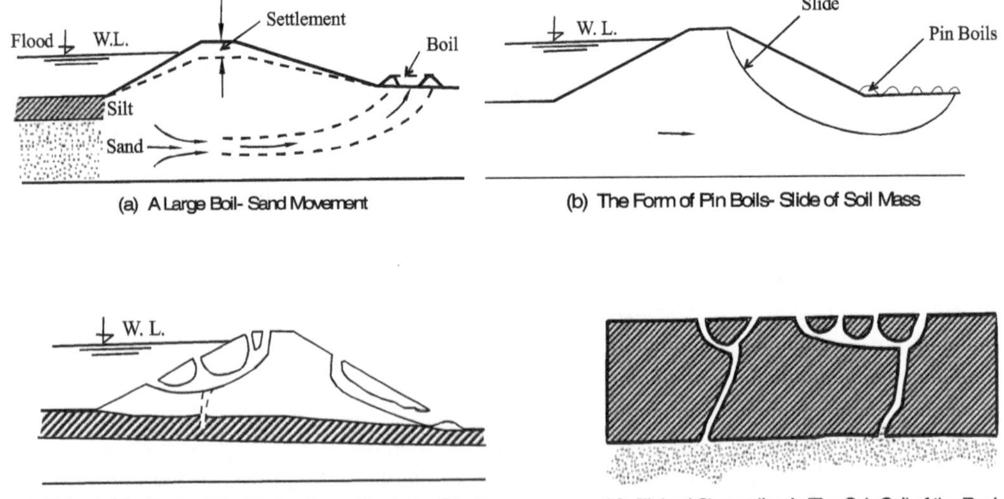

(a) A Large Boil- Sand Movement

(b) The Form of Pin Boils- Slide of Soil Mass

(c) Piping in The Body of The Bank- Channeling in the Structure

(d) Piping/ Channeling in The Sub Soil of the Bank

Figure 8.8 Piping phenomena in the body of embankment and subsoil (foundation)

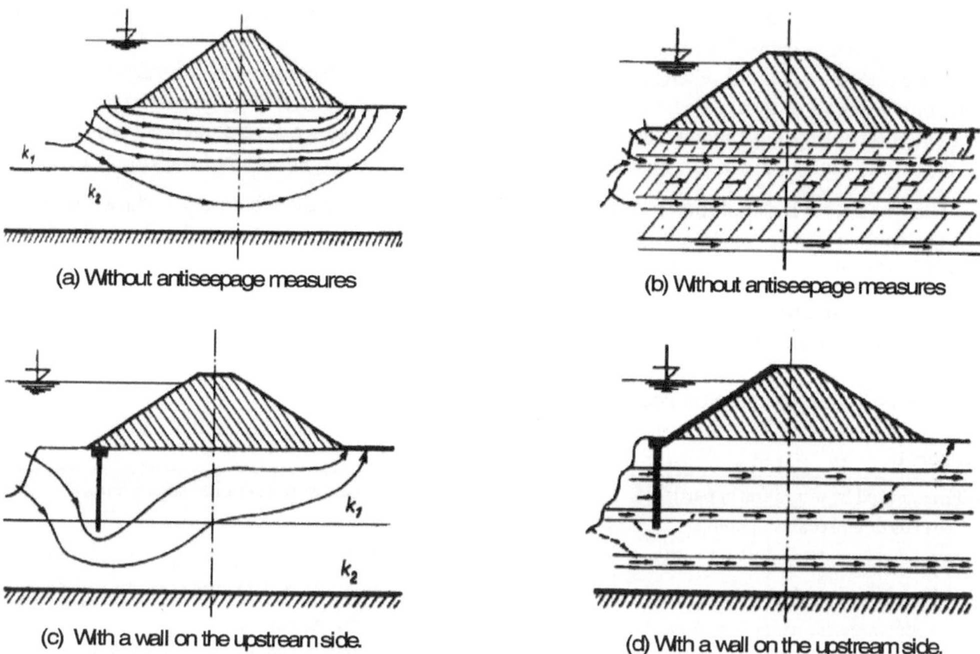

(a) Without antiseepage measures

(b) Without antiseepage measures

(c) With a wall on the upstream side.

(d) With a wall on the upstream side.

Figure 8.9 : Leakage through a layered subsoil before and after a short cut off wall erection.

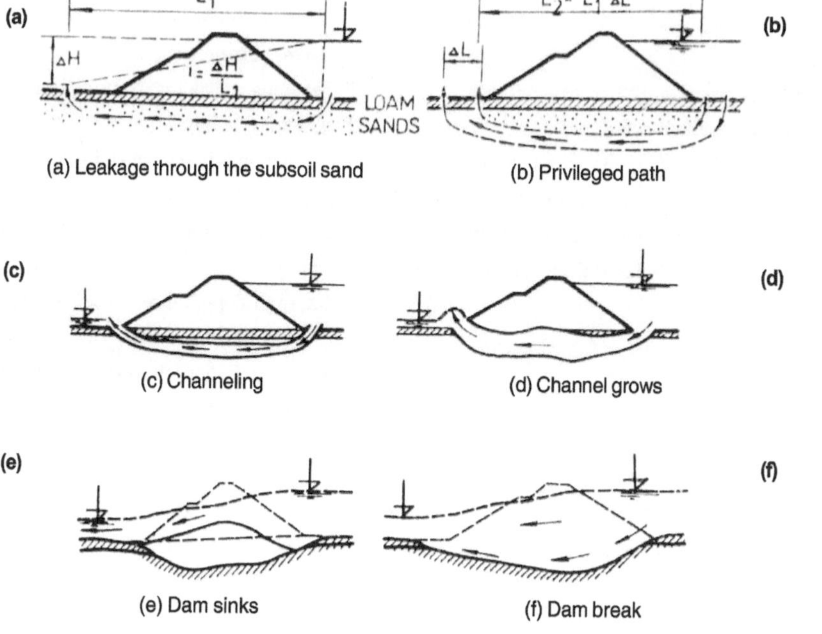

(a) Leakage through the subsoil sand

(b) Privileged path

(c) Channeling

(d) Channel grows

(e) Dam sinks

(f) Dam break

Figure 8.10 Stages of failure through seepage channel

In a permeable double layer (Fig. 8.9) or multilayer strata, the seepage is concentrated in the upper layers. Being in the vicinity of the ground level, great uplift forces result in overburden damage. Piping or similar phenomena occur when seepage water issues from the embankment or ground surface under sufficient pressure and at a sufficient velocity that the particles comprising the material are swept away. It is obviously desirable to stop this phenomenon. Sheet pile or a several meters deep and long cut-off wall is preferred with an upstream impermeable blanket several hundred meters long.

This process of the evolution of filtration damage as observed in the laboratory is shown in figure 8.10.The contact failure was investigated in the field in the case of the earth cut in the Danube valley is shown in figure 8.11.

A typical liquefication failure through foundation is shown in figure 8.12 (a) and (b). Common aspects of hydraulics and soil mechanics should be respected.

Table 8.3 Typical failures caused by uncontrolled seepage water

Category 1	Category 2
Failures caused by migration of particles to free exits or into coarse openings	Failures caused by uncontrolled saturation and seepage forces.
1. Piping failures of dams, levees, and reservoirs, caused by:	1. Most landslides, liquefaction failures, including those in highway or other cut slopes, reservoir slopes, etc. caused by saturation.
a. Lack of filter protection	2. Deterioration and failure of road bed caused by insufficient structural drainage.
b. Poor compaction along conduits, in foundation trenches, etc.	3. Fill foundation failures caused by trooped . groundwater
c. Gopher holes, rotted roots, rotted wood, etc.	4. Earth embankment and foundation failures caused by excess pore pressures.
d. Filters or drains with pores so large soil can wash through.	5. Retaining wall failures caused by unrelieved hydrostatic pressures.
e. Open seams or joints in rocks in dam foundations or abutments.	6. Canal linings basement and spillway slabs uplifted by unrelieved pressures.
f. Open-work gravel and other coarse strata in foundations or abutments.	7. Dry dock failures caused by unrelieved uplifted pressures.
g. Cracks in rigid drains, reservoir linings, dam cores, etc. caused by earth movements or other causes.	8. Dam and slope failures caused by excessive seepage forces or uplift pressures.
h. Miscellaneous man-made or natural imperfections.	9. Most liquefaction failures of dams and slopes caused by earthquake shocks.
2. Clogging of coarse drains, including French drains.	10. Loss of Bearing capacity, soil becoming marshy.

8.9.1 Damage by Animals

A variety of animals such as hamsters are also responsible for channeling found in the body of an embankment. Such many failures occurred in Mahi canals, Banswara, India and Chambal canals, Kota, India.

This kind of damage is different than hydraulic laws.

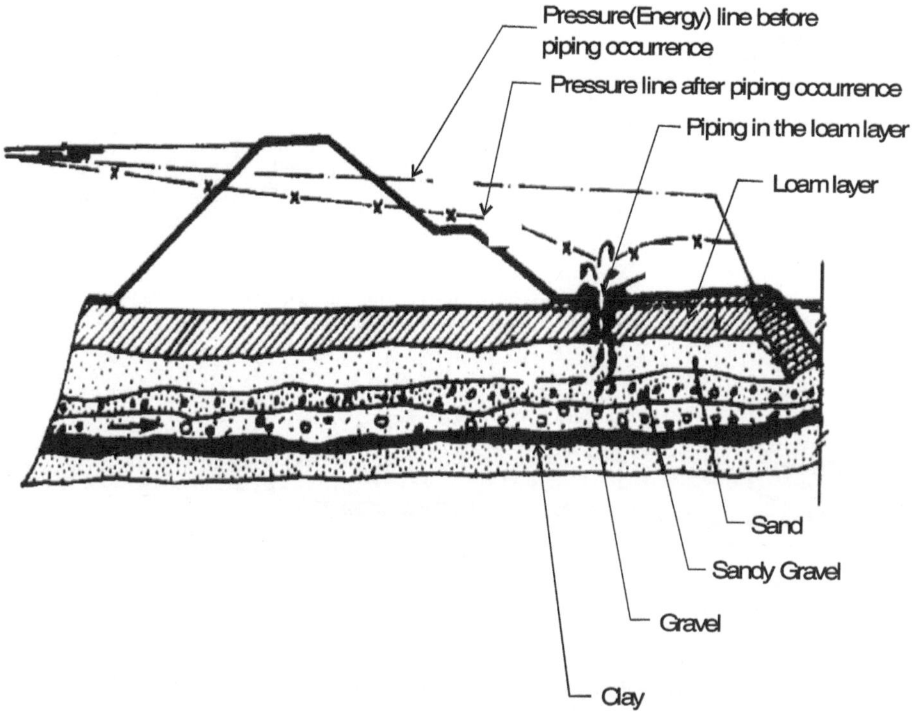

Pressure(Energy) line before piping occurrence
Pressure line after piping occurrence
Piping in the loam layer
Loam layer
Sand
Sandy Gravel
Gravel
Clay

(a) Piping and sand boiling

0.0
- 2.5
- 6.5
- 10.0
- 13.0
- 21.5
Gravel + Sand
Loam
Sand
Sandy Gravel
Gravel
Clay

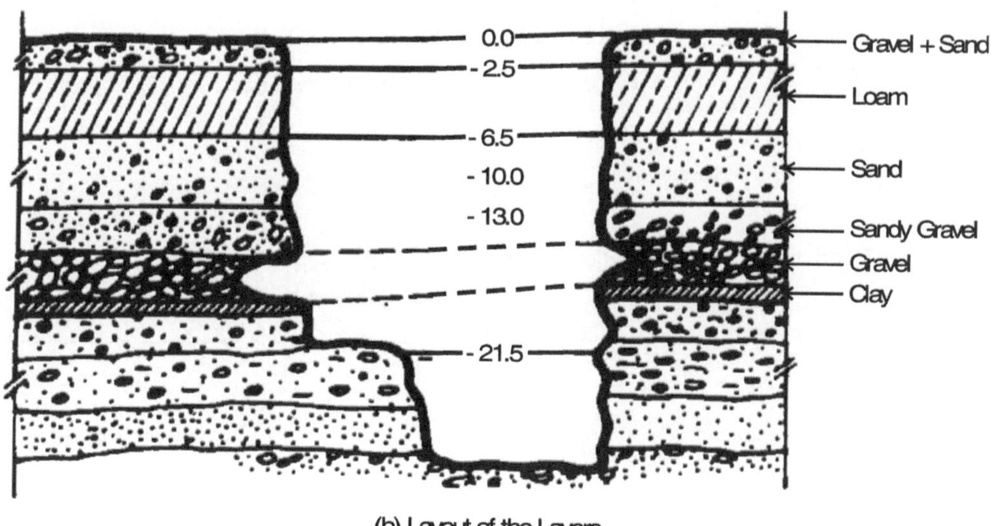

(b) Layout of the Layers

Figure 8.11 Composition of soils influencing the occurrence of piping near the Danube

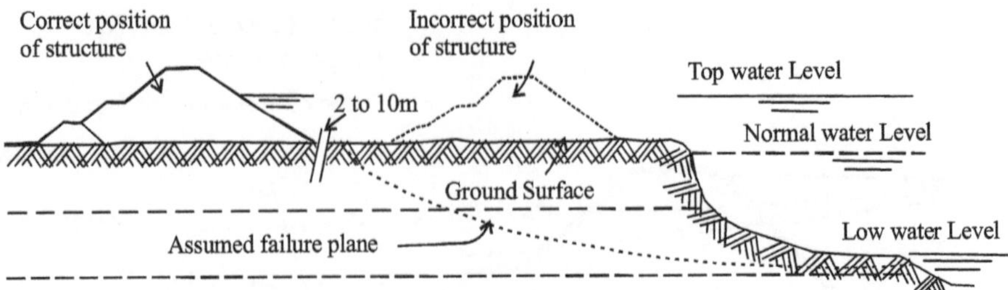

(a) **Land slide Danger**- Typical in clay soils in partial cutting/ filling or in heavy cutting where spoil bank is very close. This type of failure may also occur due-to traffic in saturated condition.

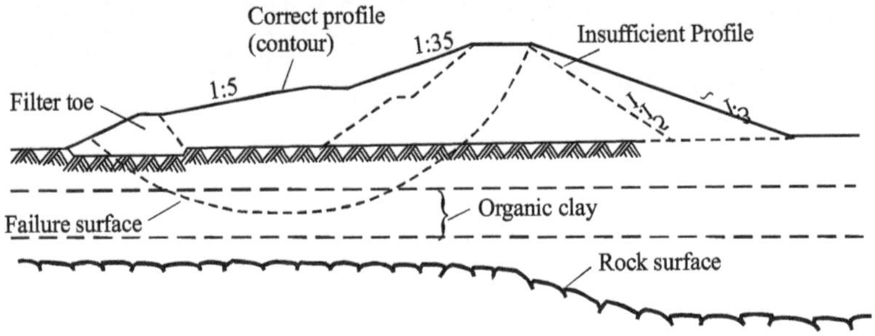

(b) **Land slide danger**, failure of embankment through foundations

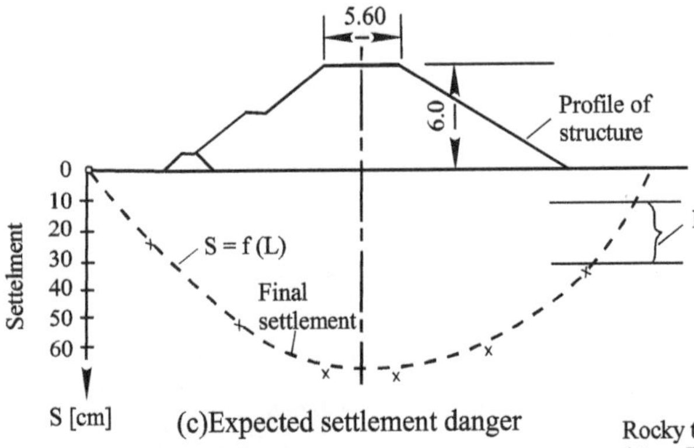

(c)Expected settlement danger

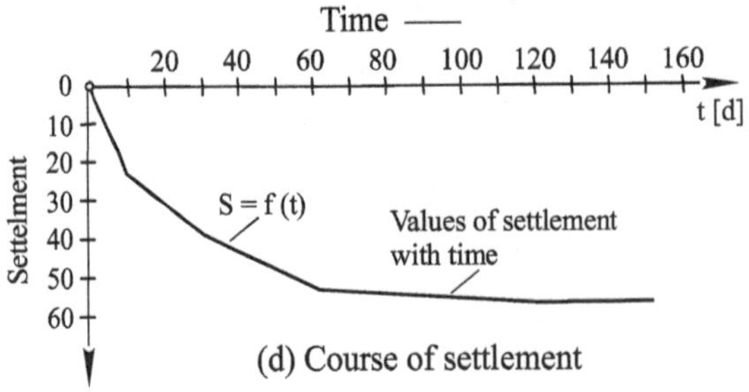

(d) Course of settlement

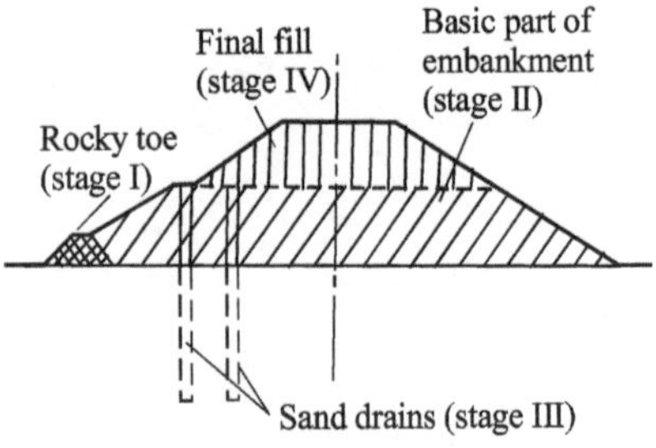

(e) Stages of material placement

Figure 8.12 : Typical example of failure and settlement of embankments

8.9.2 Collection and Removal of Seepage Water

Canal embankments act as boundaries between two different potentials, given by the different levels of surface and underground water. The water travels through the boundary and under it in the direction of the hydraulic gradient (and the gradient arising from capillary action and occasionally other forces affecting the potential). This water has to be controlled both from the seepage point of view (collection and drainage), and also from the point of view of distribution of potential, weakening of uplift force, stabilization of surface, etc. From this standpoint, linear drainage and relief wells comprise an important element of the so-called active protection of the land behind embankment, and the subsoil under it.

To this end, part seepage ducts as the simplest arrangement for collecting and disposing of water, and in part pebble drains, various sorts of pipe drain and wells are used as relatively demanding and expensive structures with which to control the water even at great depth figure 8.13.

Seepage quantity and the effect on the phreatic curve can be calculated by means of the Zhukovskil function that is by a transformation to a plane.

Seepage on hill slopes can be controlled through wells as shown in figure 8.14 (a). Stability of hill slopes is increased by benching/berms as in figure 8.14 (b).

8.10 Experiences of Bank Failures of IGMC (Indira Gandhi Main Canal), Rajasthan, India

A brief description of this project is given in Para 6.1 and chapter 7B. Typical cross sections of canal in heavy filling and cutting are shown in figure 7B.4 and 7B.5. The embankment is a homogenous section of sandy soil, laid over natural ground. The canal had double tile lining on sides and single tile lining in bed.

The right bank of main canal failed at different times at locations as given below, CWC Team Report (1984) and many others (references given at the end)

Bank Failure/Slides on Rajasthan Main Canal (Now IGNP)

S. No. (1)	Date (2)	Location (3)	Type of Failure (4)	Length (5)
1	1978	RD 774.300	Right Bank Failure	300 ft.
2.	1978	RD 762.500	Right Bank Failure	700 ft.
3.	1978	RD 768.800	Right Bank Failure	200 ft.
4.	1978	RD 772.500	Right Bank Failure	500 ft.
5.	1978	RD 773.500	Right Bank Failure	200 ft.
6.	1979	RD 764.500	Right Bank Failure	600 ft.
7.	1979	RD 782.500	Right Bank Failure	800 ft.
8.	1979	RD 776.400	Right Bank Failure	1100 ft.
9.	1979	RD 775.500	Right Bank Failure	150 ft.
10.	15.04.79	RD 766.000	Right Bank Failure	450 ft.
11.	17.01.80	RD 781.000	Right Bank Failure	150 ft.
12.	12.02.80	RD 769.300	Right Bank Failure	30 ft.
13.	13.02.80	RD 762.500	Right Bank Failure	40 ft.
14.	1982	RD 834.820	Right Bank Failure	20 ft.
15.	1984	RD 1036.37	Right Bank Failure a few ft. above lining bed	150 ft.
16.	1984	RD 1121.000	Failure due to Water action	100 ft.

In almost all cases the bank was in heavy filling and slipped towards the down stream side.

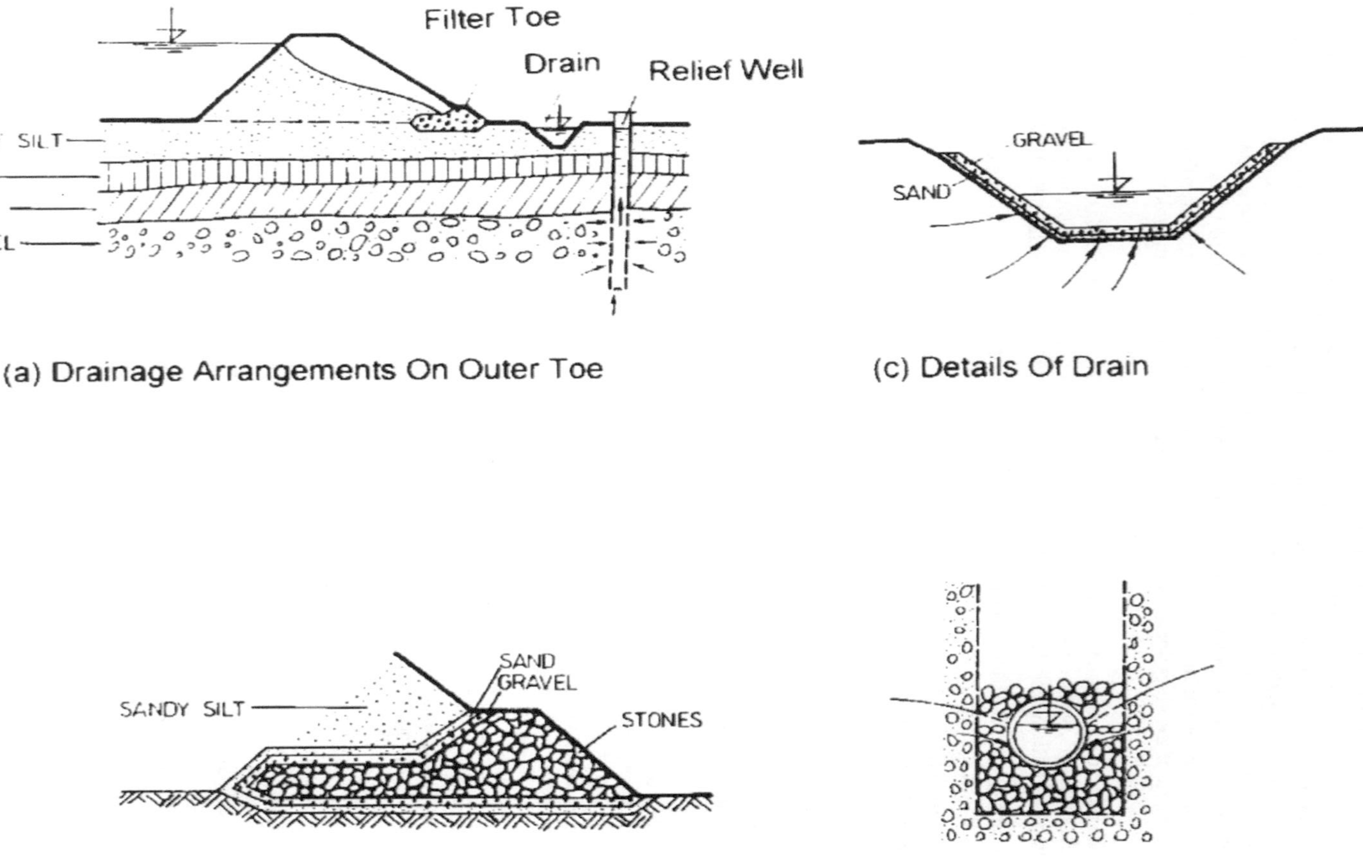

Figure 8.13 Different types of drainage collectors and relief structures

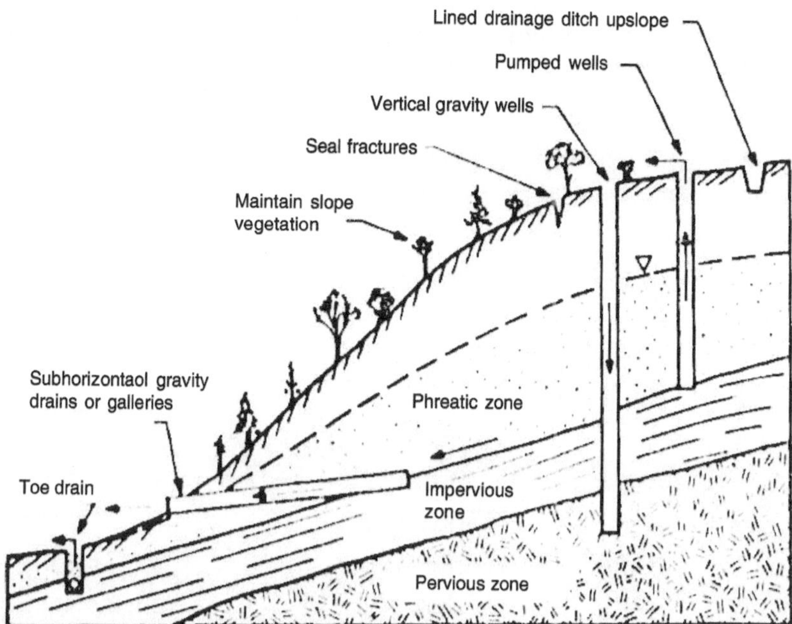

(a) Control of seepage forces through wells. Well may be to charge deeper ground water and drain upper surface, or it may pump well.

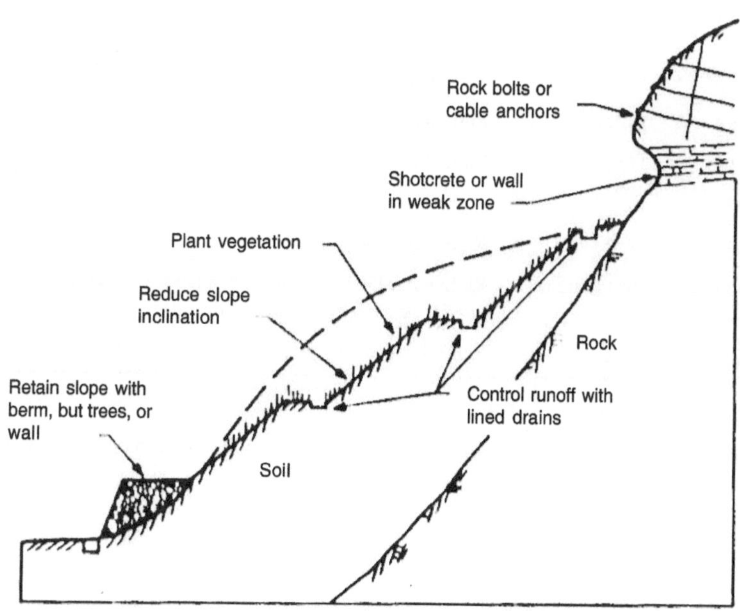

(b) Reducing the driving forces and increasing the resisting forces

Figure 8.14 The general methods of slope stabilization in hilly areas

Generally all failures occurred during fully saturated conditions of down stream ground and foundations. Most of the failures occurred when the canal was carrying much less discharge than the full supply and water level was much lower. The failures can be grouped as under:

1978 5 failures little scattered.

1979 5 failures all close, within 1 to 0.5 Km.

1980 3 failures, within 6 Km. length

1982 One failure

1984 Two failures occurred, nearly 25 Km. apart.

The above failures can be regrouped RD wise as under:

S. No.	Location	Date	Remarks
1	RD 762.5	1978	Right Bank 700 ft., again in 1980.
2	RD 762.5	13.02.80	Right Bank 40 ft.
3	RD 764.5	1979	Right Bank 600 ft.
4	RD 766	15.04.79	Right Bank 450 ft.
5	RD 768.8	1978	Right Bank 200 ft.
6	RD 769.3	12.02.80	Right Bank 30 ft.
7	RD 772.5	1978	Right Bank 500 ft.
8	RD 773.5	1978	Right Bank 200 ft.
9	RD 774.3	1978	Right Bank 300 ft.
10	RD 775.5	1979	Right Bank 150 ft.
11	RD 776.4	1979	Right Bank 1100 ft.
12	RD 781.0	17.01.80	Right Bank 150 ft.
13	RD 782.5	1979	Right Bank 800 ft.
14	RD 834.8	1982	Right Bank 20 ft.
15	RD 1036-37	1984 (January)	Right Bank 150 ft., At the time of failure Q = 60 cumecs and D = 7 ft. Section in complete filling.
16	RD 1121	1984	Right Bank 100 ft.

8.10.1 Soil Characteristics

All locations of failures have practically same type of soil and foundations. The type of soil and soil tests carried on it are given below in table.

Details of soil tests of IGMC at failure locations.

S. No.	Particulars	Sample No. 1	Sample No. 2
1.	Specific gravity	2.688	2.694
2.	Dry density (gm/cc)		
	(a) In loose State	1.455	1.455
	(b) After four minutes of vibration.	1.655	1.690
	(c) At optimum moisture content.	1.640	1.645
3.	Optimum moisture content	12.5%	12.5%
4.	Coefficient of permeability (Cm/Sec.)		
	(a) In loosest State	$4.808\text{x}10^{-3}$	$5.816\text{x}10^{-3}$
	(b) After four minutes of vibration	$2.018\text{x}10^{-3}$	$2.296\text{x}10^{-3}$
	(c) At optimum moisture content	$1.797\text{x}10^{-3}$	$2.105\text{x}10^{-3}$
5.	Shear parameters		
	(A) Cohesion	Nil	Nil
	(B) Angle of internal friction		
	(a) Loose State (dry)	34	36.5°
	(b) Loose state (under water with drainage)	-	35°
	(c) Loose State (under without drainage)	-	33°
	(d) Loose State (under water with drainage) fast shearing	-	33.5°
	(e) Loose State (under water with out drainage) fast shearing	-	33°
	(f) Compacted at CMC (shear under water)	33°	35.5°
	(g) Compacted with low energy (dry)	39°	38°
	(h) Compacted with high energy (dry)	41.5°	40°
	(i) compacted with high energy (under water)	36°	35.5°

6. Grain size distribution, dry sieve analysis, Sample No. 1

Diameter (mm)	% finer Field sample	Afternoon paction (Wet)	After compact (dry)
0.71	100	100	100
0.50	100	100	100
0.30	99	99	99.3
0.212	95	90.5	94.3
0.150	46	35	44.6
0.075	1	1	9
0.063	--	1	2
D_{10}	0.09	0.095	0.09
D_{30}	0.125	0.14	0.13
D_{60}	0.165	0.18	0.17

Test results of sample 2 are almost identical (practically same)

The area was also investigated by trail pits. It was found that in the reach RD 815 to RD 819 there was a kankar mixed compact sand layer which is less permeable than over-lying strata. In the reach RD 770 to RD 780 a thin layer varying in thickness from 1/8 to 3/4 of cemented material is found at the depth of 13 from the N.S.L. The samples of this material were collected and tested and found that it contains 36% of $CaCO_3$ hence it is classified as calcarious tufa. The material is practically impermeable for the passage of water and probably this is the main reason why the water is standing since it does not get the passage for its down-ward movement.

In addition to the above, some unconventional tests regarding the occurrence of quick conditions were also performed. In general, it has been found that if soil is in very loose state, the bearing capacity of the soil reduces to almost nil even at a very small hydraulic gradient. A complete quick condition take place early in loose soil as compared to that compacted under dry condition.

From all the above tests, the following was inferred.

1. The soil can be compacted even in dry state to achieve good density.
2. Increase in density, increases the co-efficient of internal friction considerably and hence the strength of the soil is also improved.
3. Increase in density reduces the permeability considerably.
4. The effect of water on the co-efficient of internal friction is to reduce the angle of internal friction by about 2° and hence the strength of the soil.

Besides the above, many other failures have also occurred on water supply channel and many branches and distributaries.

8.10.2 Seepage and Its Effect

There was a water supply channel on the right sides of the main canal at a distance varying from 250 to 300 m from the main canal, to carry a discharge of about 350 cusecs for construction purposes of the main canal. The water supply canal is unlined passing through the same sand dunes as the main canal. Lot of seepage from this canal created saturated conditions on its both sides, extending upto the main canal and even crossing it to the left side.

The seepage of water from water supply channel was noticed in the year 1978, in the reaches between RD 750 to RD 780, RD 815 to 819 and near RD 860 of main canal. It is found that water is seeping in heavy quantity and especially near RD 780 and RD 816 a pool of water has been formed.

At some locations there was ample seepage from the lined main canal. This aggravated the situation.

The seepage attained a menacing stage in the area between the water supply channel and main canal in the above reaches soil became fully saturated and small gullies of water gazing out could be noticed at many places. In some reaches the strata attained the floatation gradient and unable to with stand any super imposed load. The direct effect of this seepage is manifold.

1. The safety of the embankments of water supply Channel and Main canal was endangered.
2. The heavy seepage resulted in loss of water which could be other-wise utilized for useful purposes.
3. It became impossible to increase the discharge in the water supply channel, since any increase in water depth may result in breach.
4. The seepage advanced to the extent that it crossed the right bank of Rajasthan Main canal, the bed and left bank, water is visible as pond near RD 780.
5. The bed of Rajasthan main Canal which was not yet lined also became saturated, there by rendering it unfit for laying lining.

8.10.3 Type of Failure

In the opinion of this author, seepage caused practically over saturated conditions and bank failed by sloughing through the foundation, above the hard Kankar surface, in all the reaches from RD 762.5 to RD 1121. The failure circle may be of the form as in figure 8.12 (a) and (b).

The clay layer may loose all the strength / bearing capacity in marshy saturated condition. Researchers must dig the old files, compile data in a better way and analyze the failures more scientifically.

Typical Example of Seepage

A few typical cross section of water supply channel, Main canal and extent of seepage near RD 1370 of Main Canal is shown in figure 8.15. The seepage was observed in the year 1986, in the depression near road on the left side. It travelled more horizontally due to an impervious barrier of hard Kankar mixed clay at 10 to 15' below main canal bed. The hydraulic gradient ranged from 1 in 30 to 1 in 100 or even more in sand dunes.

The seepage of the water supply channel near RD 1370 of main canal was stopped by LDPE film, before any breach. Lining of main canal could be done after that. Also LDPE film was used in the lining of bed of main canal. This has worked satisfactorily and no problem has been reported till the year 2005.

8.10.4 Suggested Remedial Measures and Comments

The Remedial measures suggested by various authorities are:

(i) Loading berms and filling of depressions

At the places where the extent of seepage is menacing, the stability of the embankments may be critical. Therefore immediate action to backfill the low lying areas and to construct properly compacted loading berms should be taken.

In the opinion of this author this is not much fruitful. It only helps in delaying the eminent danger. Soon the hydraulic gradient becomes flat, as seen in the cutting reaches in figure 8.15. Again section can fail by sloughing. More over this is very costly.

(ii) Check seepage by densification of the embankment material

The techniques of compaction by vibro-floatation and small explosive charges can not be used in already completed reaches, because it may induce flow slides and damage the canal lining.

(iii) Bentonite grouting and L.D.P.E. film lining : Details are given below

(iv) Positive cut off by sheet pile driving

Pile driving may not be successful because in absence of confining pressures around canal bank, the increase in density may be marginal only. More over this is very costly.

(v) Vertical drainage

Accelerated consolidation by well point drainage appeared to be the most appropriate and safe remedial measure for the completed reaches.

(vi) Plantation of trees

Plantation of trees and creation of green belt in the depressions on both sides of the canal. Trees suck the seepage water for its plant growth and trans-pirates the seepage water. It has proved to be very helpful and effective. But it is time consuming and slow process till trees grow for 4 to 6 years.

Details of polyethylene lining

a) LDPE film of 175 microns (700 gauge) in a minimum width of 12 ft.
b) The over lapping of films to the extent of 6" and joined together by plastic Cement (emulsion)
c) Covering material local soil which has been excavated.

Steps in existing Lined Canal

a) Excavation of the existing bed and sides upto F.S.L. to a depth of 1.5.
b) Remove the excavated soil in the first panel of 12.

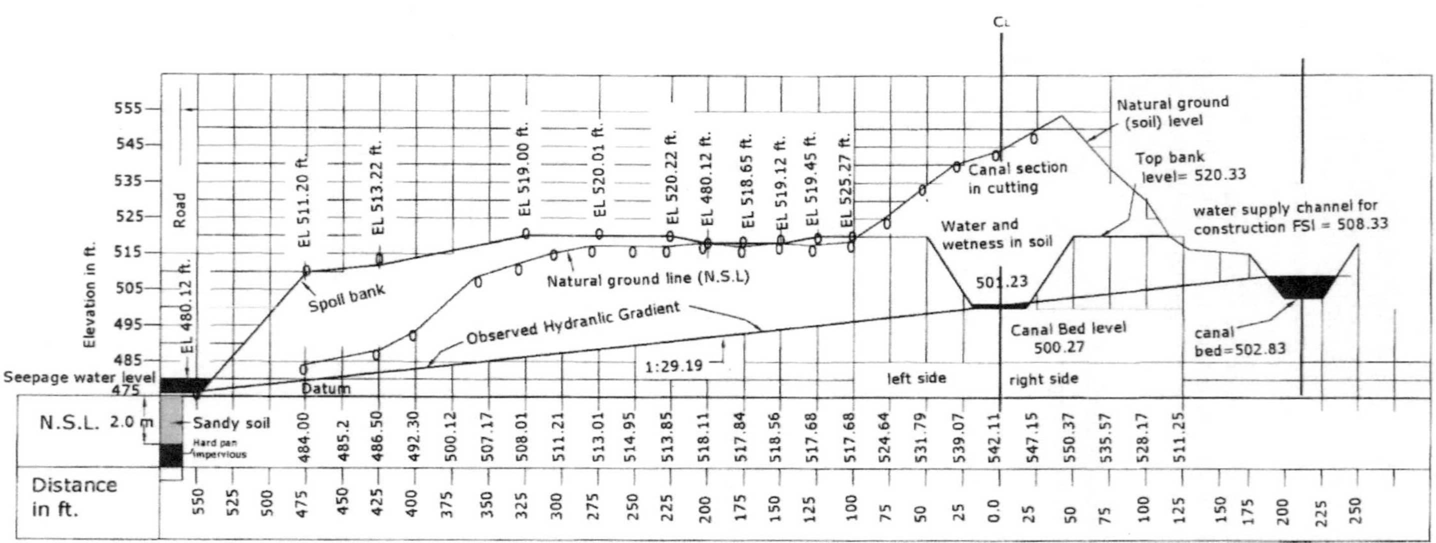

Figure 8.15 (a) Typical cross-section of IGMC at RD 1371.7 showing the water supply channel and seepage from it (Observed in the year 1986).

Notes to figure 8.15 (a) : There is a hard pan (Kankar and clay mixed), practically impermeable at 2 m below El. 480 ft in the left side depression. This hard pan is practically horizontal at about 10 below canal bed and extends beyond water supply channel. This along with the high permeable sand dunes appears the causes of flatter hydraulic gradient.

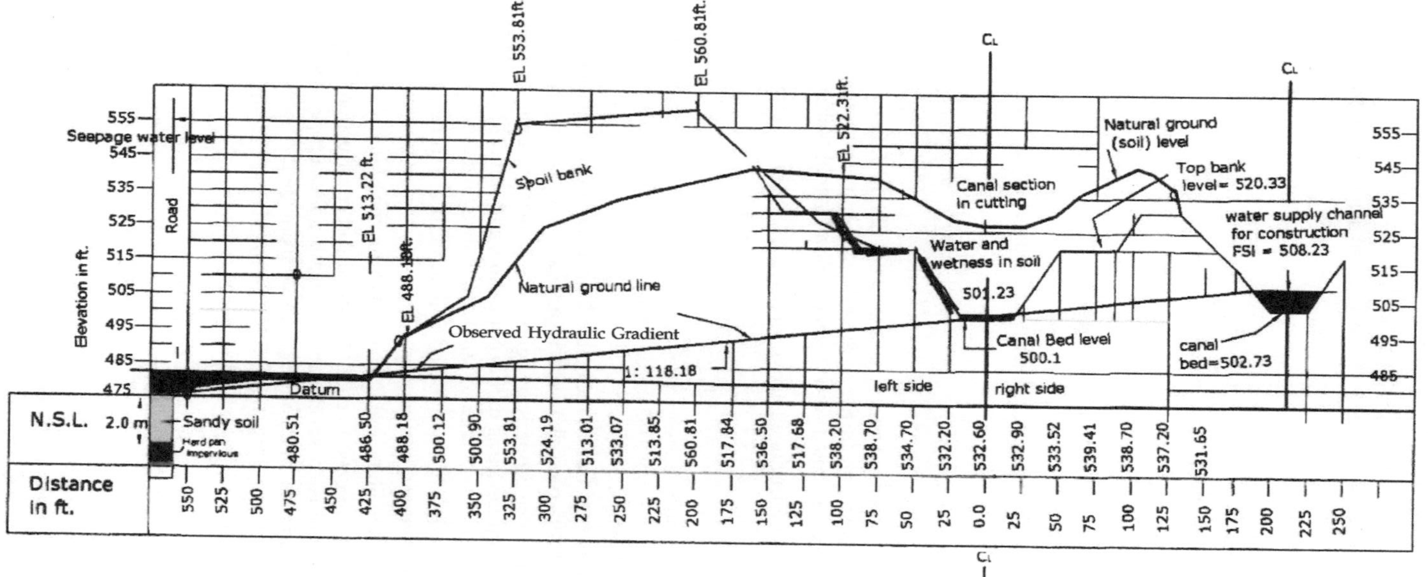

Figure 8.15 (b) Typical cross-section of IGMC at RD 1372.65 showing the water supply channel and seepage from it. (observed in the year 1986).

Notes to figure 8.15 (b) : There is a hard pan (Kankar and clay mixed), practically impermeable at 2 m below El. 480 ft in the left side depression. This hard pan is practically horizontal at about 10 below canal bed and extends beyond water supply channel. This along with the high permeable sand dunes appears the causes of flatter hydraulic gradient.

c) Lay the polythene film by unrolling the roll.
d) Ends may be keyed in a cut off trench is slopes in a depth of 1.
e) Excavate adjacent panel upto one foot and throw the excavated soil on the Polythene film already laid upto 11-6".
f) Prepare the next panel as before.
g) Join the film to last 6".
h) Lay the film as in the first panel.
i) Repeat the cycle until the desired length is achieved.

Details of Vertical Sand Drain

1. Position the rig at toe of the bank.
2. Bore the hole of 6" upto 25" minimum from bed of R.M.C.
3. Pour graded sand i.e. Bikaner Bajari.
4. Lift the casing gradually by pouring the graded sand until hole is full to the top.
5. Remove the casings.
6. Shift to the next hole.

8.10.5 Remedial measures taken

Seepage of water supply channel was stopped by laying LDPE film in it. Later on around the year 1987, a filter toe in filling reaches as shown in the figure 8.16 and another arrangement of laying a porous drain was approved for the constructed canal by the Indira Gandhi Nahar Board.

Some of the remedial measures as mentioned above were also adopted to some extent in different reaches.

Preventive Measures for future

The following measures and precautions were proposed for the construction of the main canal in the remaining portion, as well as for the construction of the branch canal and distributaries to prevent the seepage and ensure stability of banks.

(a) Un compacted loose fill may not be used in the embankment. The quality of earthwork should be controlled in accordance with the prescribed specifications.
(b) The embankment sections are to be properly designed after considering all the parameters of the soil used for the embankment and existing in the foundation, and carrying out necessary stability analysis. Suitable drainage is to be provided to safely drain away the seepage water.
(c) Adequate drainage and protection of the exposed soil surfaces are to be provided, by surface drainage system drainage system and turfing.
(d) The bricks and mortar used in the lining are to be checked for their quality of imperviousness.
(e) Strict quality control should be enforced to check the quality of construction of the embankment and the lining to satisfy the stipulated specifications. This includes proper compaction of earth work and testing of the soil samples taken out from the embankment.
(f) The overall imperviousness of the canal lining is to be improved after carrying out in-situ tests to find the quality of the lining laid with regard to the seepage losses.

When the in-situ tests show that seepage losses are substantial due to inherent pervious nature of soil in spite of the precaution taken and the tiles do not prove to be effective in preventing seepage losses it would be advisable to provide polyethylene lining and the section of the canal suitably modified. In the opinion of this author, a Zoned Section as adopted for many canal, see chapter 6 would be better and economical. Positive cut off of clay-Kankar from deeper strata or near by depression could be used. Alternatively a positive cut off of brick masonry could be used. Also a treatment adopted by Col. Cautley in upper Ganga Canal, as in figure 6.7 (vi) could be considered. Of course plantation is a ruminative solution.

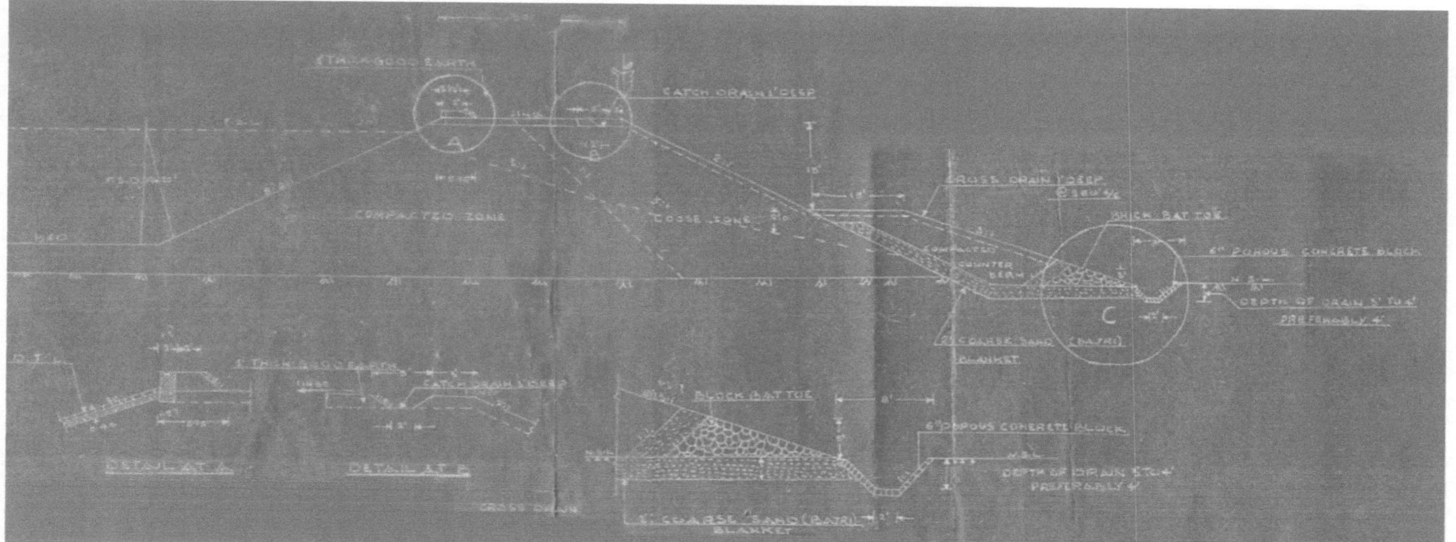

Figure 8.16 (i) Details of subsequent modification of filter blanket toe and counter berms for filling reaches of Indira Gandhi Main Canal India.

Notes

1. Brick bats toe should consist of well burnt brick tiles of size 4 to 8.
2. Porous concrete blocks shall be precast concrete blocks of mix 1: 6 i.e. one part cement and six parts of stone grit. The maximum size of stone grit shall be 12 to 20 mm.
3. Filter material shall consist of coarse sand (Bajari), finer material less than 0 : 3 mm shall be removed from the filter material as recommended by director (E.R.D.D.), C.W.C. New Delhi.
4. Thickness of the filter material has been kept as 2 feet as recommended by the I.G.N.P Board, Jaipur. Filter is to be laid in two layers of one feet each.
5. The cross drains connecting the longitudinal catch drains shall be provided at an interval of 250 feet. The size of cross drains and longitudinal catch drain are same. Catch drain should be lined.
6. The seepage water collected in the toe drain shall be disposed off by means of suitable cross drain to the lower depressions according to the topographical suitability and site conditions.

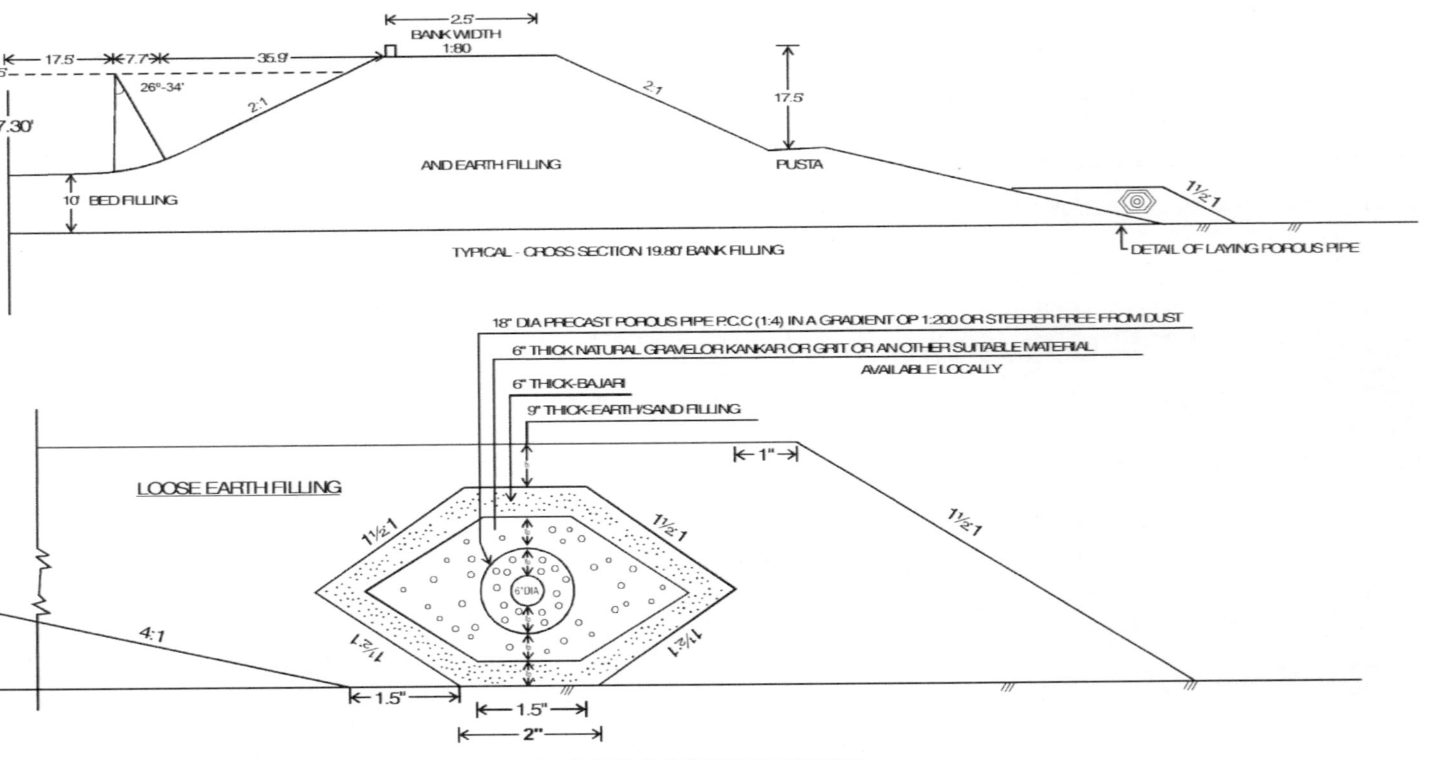

Figure 8.16 (ii) Details of alternative subsequent modification of filter toe in heavy filling reaches of Indira Gandhi Main Canal.

8.10.6 Importance of Foundation investigations in canals

Above breaches of the canal indicates that the foundation of filling sections, and even heavy cutting reaches should be adequately investigated. In filling reaches bore hole/trail pits may be laid below ground level to at least for a depth equal to 0.25 to 0.5 times the height of embankment (H) and not the full supply depth. These need not be at the centre but may be about H to 2H times away or in the near by depressions.

In cutting reaches, deep bore holes may not be easily feasible, then these can be laid in the near by depressions with in 1 Km from the alignment, and under ground geology may be inferred from it.

Alternatively underground investigation can also be carried out by geophysical methods such as electrical resistive, electromagnetic etc.

At every stage, it is very important to take into account not only the height of the embankment (H) but also the water elevation-head above a chosen datum.

There are two things:

(i) The changes of water level, head energy as a first principle, and

(ii) The influence of water as a "hidden" energy and not in pure static form-neglecting its dynamic aspect.

8.11 Slope Stability Analysis

Slope stability analysis is performed to assess the potential for failure of the slope by rupture. Slopes that can be analyzed fall into limited categories, as illustrated in figure 8.17. Mathematical/ analytical analysis may not be possible in many of the cases shown in figure 8.4 to 8.14.

Slope failure forms that cannot be analyzed in the present state of the art include avalanches, flows, failure by lateral spreading, and progressive failure. All of these forms can be initiated by a slide failure at the toe of the mass.

The primary objective of a stability analysis is to determine the factor of safety (FOS) of a particular slope to likely imminence of failure and suggest economical treatment when necessary or design an economical section.

Analytical techniques can be applied where peak strength occurs essentially simultaneously at every point along single or multiple failure surfaces and the mass moves as a unit or group of units. The sliding surface may be circular, planar, or irregular, also see Para 8.7.

Two basic method of evaluating the structural stability in common uses are:

(i) Limit Equilibrium Method and

(ii) Unit stress method.

Both the above methods are also called as Analytical methods. Both also use the total and the effective stress concept, see Para 8.11.1.

In the limit equilibrium approach, the margin of safety between the available shear strength and the developed shear stress along a defined sliding surface (Circular or planer) of a soil mass is evaluated.

In this category are the methods of slices (see Para 8.11.2), or sliding wedge methods in which the general considerations are:

(i) The sections are considered two-dimensional,

(ii) Failure takes place along a surface which is assumed to be potential depending upon type of embankment and foundation,

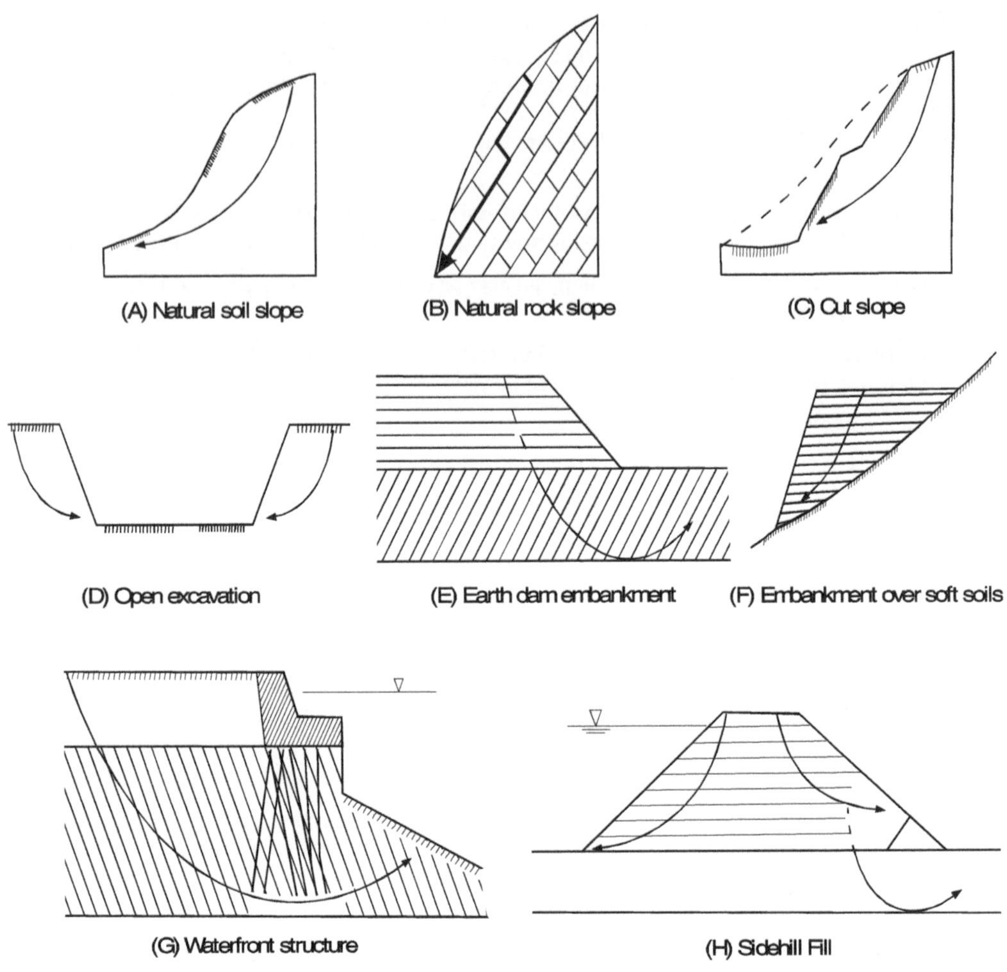

Figure 8.17 Categories of slope stability Analysis

(iii) The developed shear stresses and available shear strength are computed along the failure surface, and

(iv) The safety margin is obtained by comparing magnitudes of the components in (iii) above.

It is an approximate solution that considers a state of equilibrium between the forces acting to cause failure (driving forces) and the forces resisting failure (mobilized shear stresses). General cases illustrating various definition of factor of safety are given in Figure 8.18.

Limit equilibrium analysis further assume the following:

1. The failure surface is of simple geometric shape (planar, circular, log-spiral).
2. The distribution of stresses acting along the failure surface causing failure is determinate.
3. The same percentage of mobilized shear strength acts simultaneously along the entire failure surface.

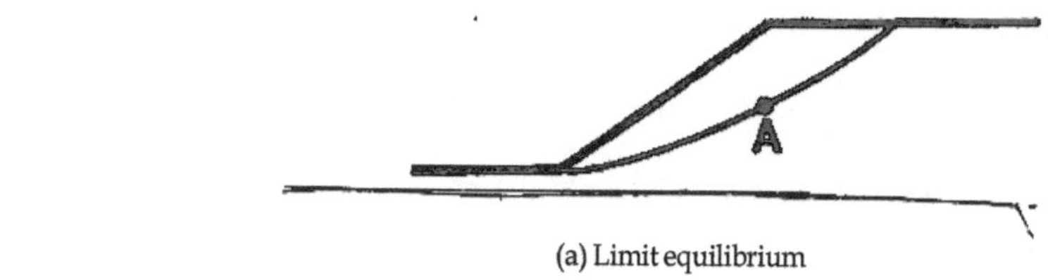

(a) Limit equilibrium

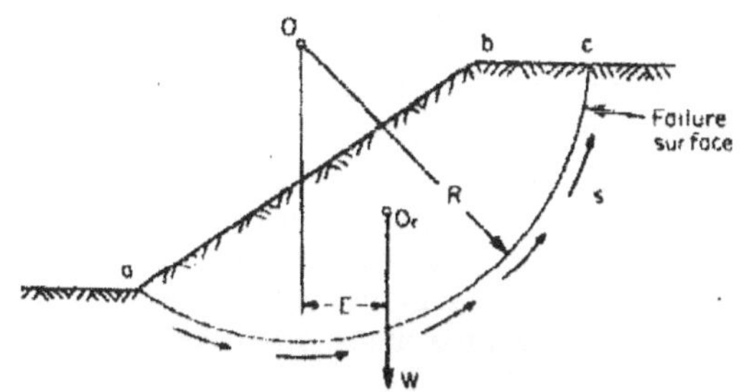

(b) Rotational Cylindrical Failure Surface with Length L - Moments

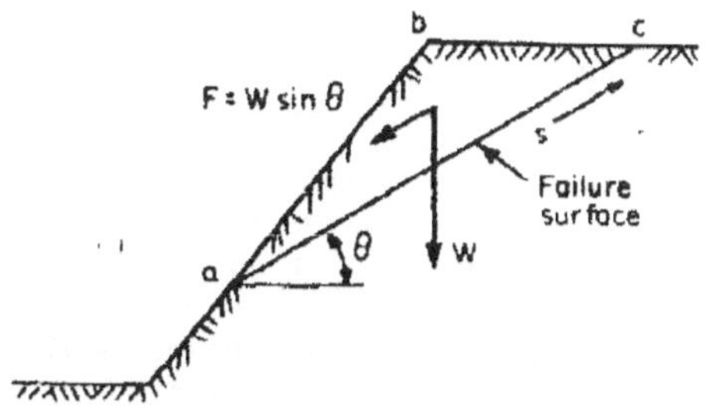

(c) Simple Wedge Failure on Planner Surface with Length L

$$FOS = \frac{Resisting\ movement}{Overturning\ moment} = \frac{R\int S_u ds}{Wx} = \frac{sLR}{EW} \qquad FOS = \frac{Summation\ of\ resisting\ force}{Summation\ of\ mobilized\ force} = \frac{sL}{W \sin\theta}$$

Figure 8.18 Various definitions of factor of safety (FOS)

Limit equilibrium methods of analysis typically use the Mohr-Coulomb failure criterion, in which

$$s = c + \sigma_n \tan\varphi \qquad \text{..... Eq. 8.11.1}$$

Where s is the shearing resistance, c is the cohesion, n is the normal stress, and is the angle of internal friction. Various definations of factor of safety (FOS) are explained in figure 8.18.

FOS can also be considered as the ratio of the total shearing strength available along the sliding surface to the total shearing stresses required to maintain limiting equilibrium, given as

$$FOS = \frac{cL + N\tan\varphi}{c_m L + N\tan\varphi_m} \qquad \text{..... Eq. 8.11.2}$$

Where L is the length of the failure surface, N is the normal force on the assumed failure surface, c_m is the mobilized cohesion at equilibrium, ϕ_m is the mobilized friction angle at equilibrium, and c/Cm = ϕ / ϕ_m.

8.11.1 Total and Effective Stress Concepts

A soil mass subjected to compressive stress exhibits two kinds of stresses the stress at the contact point of any adjacent grains, and the pressure in the pore fluid between grains. The former is known as 'effective stress' and later, as 'pore pressure'. The combined stress acting on any surface is the total stress. Since the shear strength of fluid is zero, strength of a compacted embankment is essentially the inter grainular or effective stress.

Two different approaches to the stability analysis are in common use:

(i) The total and

(ii) The effective stress methods.

In the 'effective stress method', pore-pressure along the sliding surface is approximated; and Eq. 8.11.1 is used to express the shear strength.

Pore pressures that exist on the potential failure surface (within the embankment and the foundation) are estimated and allowed for separately from normal stress. The shear strength of soils is also determined in terms of effective stress.

In the total stress method, because of the difficulty in approximating the pore-pressure along the sliding surrounding face are approximated; and Eq. (8.11.1) is used to express the shear strength.

In the total stress method, total stresses, without allowance for pore water pressures, are taken into account. The shear strength of the soils adopted is also determined in terms of total stresses without pore pressure measurements. It is assumed that the pore pressures which will develop in the sample during the laboratory test will be equal to those which would develop in the embankment/slope at the time of failure. In other words the shear strength is assumed to include the influence of the pore pressure during loading as well as shearing.

Because of the difficulty in approximating the pore-pressure, actual pore-pressure values as obtained theoretically or from measurements of real field conditions are applied to the specimen to be tested in laboratory.

Appropriate total stress or effective stress strength parameters for specific loading conditions are given in Table 8.4.

Table 8.4 Total versus effective stress Analysis

Condition	Preferred Method	Comment
Stability at intermediate times	$\bar{c},\bar{\phi}$ analysis with estimated pore pressures	Actual pore pressures must be field checked.
End of construction; partially saturated soil; construction period short compared to soil consolidation time	Either method: $\bar{c}_u$, $\bar{\phi}_u$ from CU tests, or $\bar{c},\bar{\phi}$ plus estimated pore pressures	$\bar{c}$, $\bar{\phi}$ Analysis permits check during construction using actual pore pressures.
End of construction; saturated soil; construction period short compared to consolidation time	Total stress or S_u analysis with $\phi = 0$ and $c = S_u$	$\bar{c},\bar{\phi}$ Analysis permits check during construction using actual pore pressures.
Long-tern stability	$\bar{c},\bar{\phi}$ Analysis with pore pressures given by equilibrium ground water conditions.	Stability depends on amount of water-table rise and pore-pressure increase.

8.11.2 Methods of slices

Two simplified methods of slice in common use are:

(i) Fellenius (1936) also known as standard or ordinary method of slices or Swedish Method and

(ii) The Bishop's (1955) slice method and Modified (Janbu et. al., 1956) also called as Simplified Method.

Difference in both the methods is shown in figure 8.19.

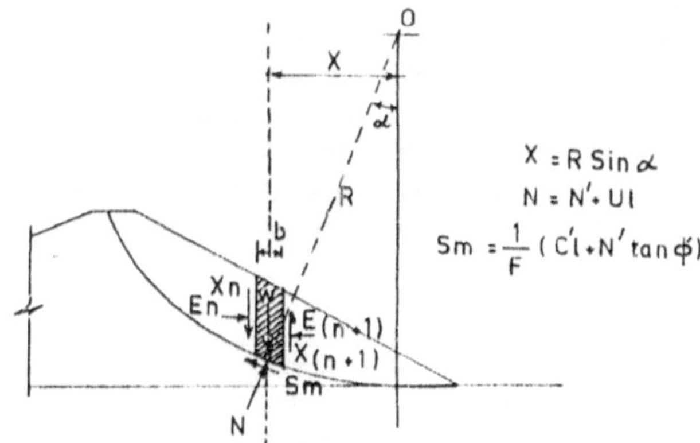

(a) Forces Acting on a Typical Slice of a Sliding Mass

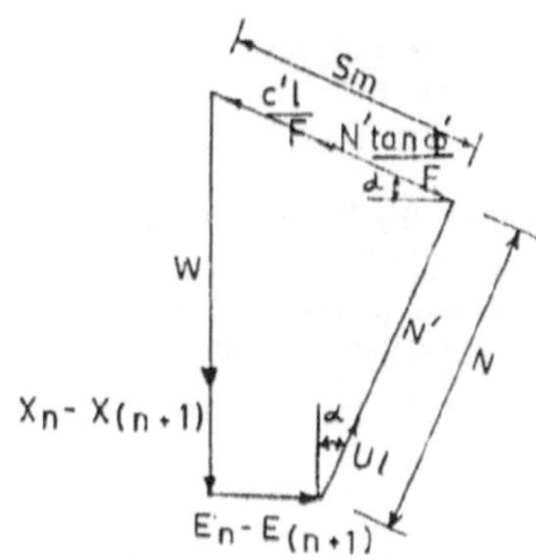

(b) Force Polygon

(a) Feelenius Methods

(i) Forces are Resolved Normal to sheding surface

(ii) $N' = W \cos \alpha - ul$

(b) Bishop's Simplified Method

(i) Forces are Resolved Vetrically

(ii) $$N' = \frac{W - u\cos\alpha - \frac{c'l}{E}\sin\alpha}{\cos\alpha + \frac{\tan\phi}{F}\sin\alpha}$$

Figure 8.19 Stability Analysis by method of slices

The slice methods can be further divided into two groups:

(i) *Non-rigorous:* It satisfies either force or moment equilibrium, and

(ii) *Rigorous:* It satisfies both force and moment equilibrium. The factor of safety estimated is relatively insensitive to the assumptions made to obtain determinacy (Duncan, 1992; Espinoza et al., 1992, 1994).

However, non-rigorous solutions can produce significantly different estimates of safety depending on the assumptions made. In general, a non-rigorous solution satisfying only moment equilibrium is superior to one satisfying only force equilibrium and will provide solutions close to a rigorous method.

The difference in two is explained in figure 8.18.

Both methods of slices assume a circular slip surface. The generalized procedure of slices and the Morgenstern Price (1965) method are commonly used for research problems, and for comparison of results obtained by the other methods.

The location of the critical failure surface is found by determining the lowest value of safety factor obtained from a large number of assumed failure surface positions.

Common to all slice methods is the assumption that the assumed soil mass and failure surface can be divided into a finite number of slices. Equilibrium conditions are considered for all slices. The problem is strongly indeterminate, requiring several basic assumptions regarding the location of application or resultant directions of applied forces.

Swedish method is a trial-and-error technique that locates the critical failure surface, or that circle with the lowest value for factor of safety (FOS). It is not a rigorous solution, because the shear and normal stresses and pore-water pressures acting on the sides of the slice are not considered. In general the results are conservative. In slopes with low angles and moderate inclinations, FOS may be 10 to 15% below the range of the more exact solutions; with high and slope inclination, FOS can be underestimated by as much as 60%.

For the ϕ=0 case, normal stresses do not influence strength, and the ordinary method provides results similar to rigorous methods (Johnson, 1974).

8.11.3 Sliding Wedge Methods

In the wedge or sliding block method, the failure is assumed to occur along a simple single or a number of planes; in which the sliding mass is divided into active and passive wedges (sometimes with a neutral wedge in between), see figure 8.18 (c) and 8.20 (a) and (b). The factor of safety is obtained after a number of trial computations, until a static equilibrium of forces is reached.

This method is generally applicable under the following two circumstances:

a) Where one or more horizontal layers of weak soil exist in the upper part of the foundation, and

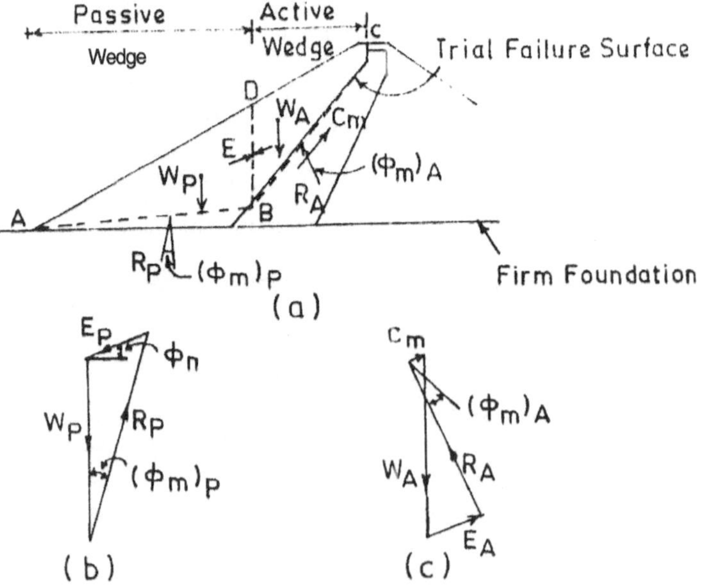

Figure 8.20 (a) Method of wedge analysis (two wedges)

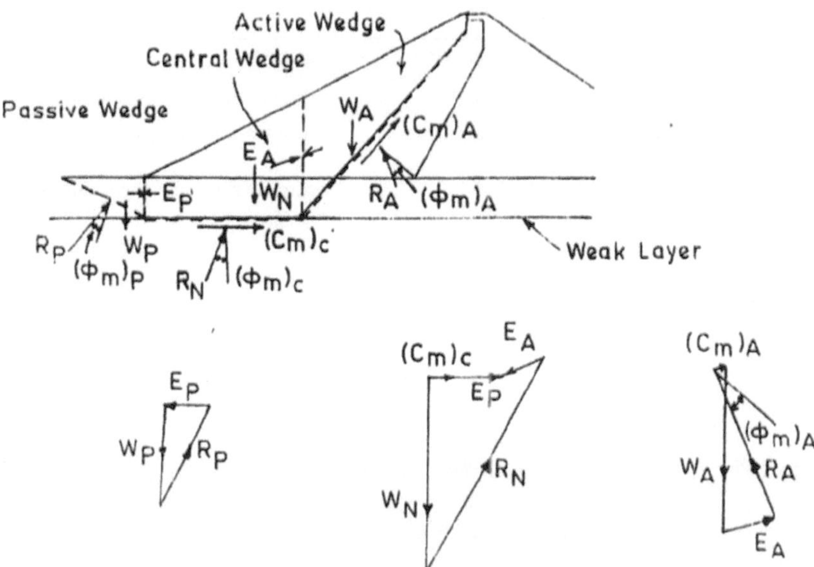

Figure 8.20 (b) Method of wedge analysis (three wedges)

b) Where the foundation consists of hard stratum through which failure is not anticipated and dam resting on it has a core of fine-grained soil with relatively large shells of dense granular material.

In this method, the trail sliding mass is divided into two or three segments, the top segment is called active wedge and the bottom segment is called passive wedge. The middle wedge in case of a three wedge system is called the central block. The resultant of the forces acting on the active wedge and the passive wedge are first determined. These resultants acting on the central block along with other forces on the block or otherwise shall give a closed polygon for stability. For procedure for this analysis readers are requested to refer text books on soil mechanics.

Unit Stress Method: The unit stresses and strains are computed for the entire embankment and foundation sections using the elastic theory. Development of computer facilities and the finite element technique have created a new dimension and interest for applying this method in the recent years.

It is necessary to closely define material properties, slope geometry, and the initial state of stress, which often are difficult to assess accurately.

8.11.4 Concepts of Factor of Safety (FOS)

One recognized function of the FOS is to account for uncertainty, and thus to safe guard against ignorance about the reliability of strength parameters, pore pressure distribution, and straitigraphy. In general, the lower is the quality of the site investigation, the higher the desired FOS. Another role is that it constitutes an empirical tool whereby deformation stability performances are limited to tolerable amounts within economic restraints. In this way, the choice of the FOS is greatly influenced by the accumulated experience with a particular soil mass. Since the degree of risk that can be taken is also greatly influenced by experience, the actual magnitude of the FOS used in design will vary with material type and performance requirements.

The minimum values of FOS given by Wilson (1979) are given in table 8.5 with notes for use.

Table 8.5 Minimum Values for FOS for Earth and Rock-fill dams, Wilson (1979)b

Case	Design Conditions	FS_{min}	Shear strength[c]	Remarks
I	End of construction	1.3[d]	Q or S[c]	Upstream and downstream slopes
II	Sudden drawdown from maximum pool	1.0[f]	R, S	Upstream slope only, use composite envelope
III	Sudden drawdown from spillway crest	1.2[f]	R, S	Upstream slope only, use composite envelope
IV	Partial pool with steady seepage	1.5	(R + S)/2 for R < S	Upstream slope only, use intermediate envelope
V	Steady seepage with maximum pool	1.5	(R + S)/2 for R < S	Upstream slope only, use intermediate envelope
VI	Earthquake (Cases I, IV, and V with seismic leading)	1.0	.........g	Upstream and downstream slopes

Notes

[b]*Not applicable to embankments on clay shale foundations; higher FS values should be used for these conditions.*

[c]*Q = quick (unconsolidated undrained test), S = slow (consolidated drained test), and R = intermediate (consolidated undrained test).*

[d]*For embankments more than 50 ft. high over relatively weak foundation use FSmin = 1.4.*

[e]*In zones where no excess pore water pressures are anticipated use S strength.*

[f]*FOS should not be less than 1.5 when drawdown rate and pore water pressures developed from flow nets are used in stability analysis.*

[g]*Use shear strength for case analyzed without earthquake (Values for FS are based on pseudostatic approach).*

8.11.5 Design loads and Conditions

In Normal cases, the following forces are acting on the embankment and foundations

(1) The weight of the structure.
(2) The water pressure static and dynamic loading.
(3) Seepage forces and pore pressures.
(4) Wind forces: waves from wind, vessels, etc.
(5) Earth pressures.
(6) Frost effects: expansion of water and soils.
(7) Forces induced by other volume changes swelling, shrinkage.
(8) Forces leading to internal instability of soils piping, hydraulic fracturing liquefaction.
(9) Earthquake and other dynamic loading (Vibration).
(10) Man made forces; prestressed construction.

From the point of view of mechanics, these forces can be divided into surface and volume forces. Most have a variable character and change with water head. This is a leading factor as a transfer medium working on the structure.

(i) Construction Condition with or without Partial Pool (for Upstream and Downstream Slopes)

This situation is when the structure is just constructed. In this condition the pore pressures developed as a result of material compression due to the overlying fill are not dissipated or are only partly dissipated.

If the rate of rising of embankment is less than about 15 meters per year, this condition may not become critical as the residual pore pressure in the embankment and foundation are expected to be negligible, except in highly clayey foundation with high water table, for example, marshy areas.

Construction pore pressures may some time exceed the pore pressures likely to be developed due to the seepage from the reservoir and consequently may control the design. The magnitude and distribution of these pore pressures depend primarily on the construction water content in embankment and natural water content in the foundation, the properties of the soil, rate of raising, the height of embankment and the rate at which dissipation may occur due to internal drainage.

(ii) Reservoir Partial Pool (for Upstream Slope)

This condition corresponds to the initial canal filling in which it is assumed that a condition of steady seepage has developed at the intermediate stages. The stability of upstream slope shall be investigated for various water levels on upstream, usually levels corresponding to one-third to two-thirds height of head of water to be stored at full supply level and minimum value of factor of safety worked out. The analysis should account for reduction, if any, in the effective normal stresses where the pore pressures developed during construction are not dissipated before a partial pool condition can develop.

All the zones above phreatic line drawn for upstream water level under consideration be considered as moist for working out resisting and driving forces and zones below it be taken with their submerged weights for working out both resisting and driving forces.

Partial pool condition may not prove to be critical for all embankments and hence analysis for this condition needs to be carried out only in cases where it is considered necessary. This condition is likely to be critical in cases of high dams where the range of drawdown is small as compared to the height of dam.

(iii) Sudden Drawdown (for Upstream Slope)

Canal embankment remains saturated due to prolonged full supply levels. Sudden drawdown condition may occur due to the subsequent lowering of canal gates, so that canal water level falls faster than pore water can dissipate. All possible efforts are made to avoid this situation by proper canal regulation orders. Yet it may occur in emergency cases of canal breaches, or in intermittent running canals. This induces unbalanced seepage forces and excess pore water pressures. This condition becomes critical if the materials of the upstream portion of the embankment are not freely draining.

(iv) Steady Seepage (for Downstream Slope)

This is most common condition as canals generally run at full supply level or little more. Steady seepage is developed and the seepage lines are established in the embankment. This is most critical condition for the downstream slope.

(v) Steady Seepage with Sustained Rainfall (for Downstream Slope)

Generally in rainfall areas or where there is a possibility of sustained rainfall slope is analyzed on a arbitrary assumption that a partial saturation of shell material due to rainfall takes place.

8.11.6 Choice of Method

Selection of the proper method depends, upon a number of factors, as listed below:

(i) Type of slope to be analyzed, such as natural or cut slope in soil, or rock (Bjerrum, 1973; Patton and Hendron, 1974; Brand, 1982; Leonards, 1979, 1982) or rock (Deere, 1976), earth-dam embankments (Lowe, 1967).

Specific details can be found in referenced works.

(ii) Embankments over soft ground or side hill fill (Chirapuntu and Duncan, 1976; Ladd, 1991).

(iii) Location, orientation, and shape of a potential or existing failure surface which is controlled by material type and structural features. The shape of the rupture zone can have single or multiple surfaces, and can be composed of single or multiple wedges. Failure surfaces can be planer, cylindrical or log-spiral, or irregular, See Para 8.7.

(iv) Material distribution (stratigraphy) within and beneath the slope, which can be divided generally into homogeneous zones wherein the properties are more or less similar in all directions, and non homogeneous zones wherein soils are stratified and rock masses contain major discontinuities.

(v) Material types and representative shear strength parameters; angle of internal friction , cohesion c, residual strength r, and undrained strength su (See table 8.4)

(vi) Drainage conditions; appropriateness for either drained or undrained analysis, which depends on the relative rates of construction and pore pressure dissipation. Often considered as short-term (during construction) or long-term (post construction or natural slope) conditions; that is, total versus effective stress analysis (See Table 8.2).

(vii) Distribution of piezometric levels (pore-or cleft-water pressures) along the potential failure surface and an estimate of the maximum value that may prevail.

(viii) Potential earthquake loadings, which are a transient factor.

A detailed discussion of these necessary considerations for the various slope types are given by Hunt (1984, 1986).

8.11.7 Chart solutions for Homogenous Slopes

Various chart solutions have been developed for simple homogenous slopes, such as Taylor (1937, 1948), Janbu (1968), Hunter and Schuster (1968), and Cousins (1978).

8.11.8 Computer Programs

Most slope stability analyses are computationally intensive. Many computer programs have been developed to perform stability analyses on personal computers. Commonly used packages include PCSTABL (Purdue University) and UTEXAS3 (University of Texas). Data is input for the problem geometry, material stratigraphy, and the phreatic surface based on a coordinate system, and material properties are then entered. The programs search for the potential failure surface that produces the lowest value for FOS. Many programs have also been developed in India by many Engineering colleges / universities.

8.12 Characteristics and Behavior of Soils

Coarse Grained Soils (Gravel and Sand)

Gravel and sand have essentially the same engineering properties, differing mainly in degree. Well graded compacted gravels or sands are stable materials. The coarse grained soils when devoid of fines are pervious, easy to compact, little affected by moisture and not subjected to frost action. Although grain shape and gradation, size affect these properties, gravels are generally more pervious, more stable and less affected by water or frost than are sands, for the same amount of fines.

As sand becomes finer and more uniform, it approaches the characteristics of silt with corresponding decrease in permeability and reduction in stability in the presence of water.

The soil classification based on particle size distribution and Atterberg limits are given in most soil mechanics text books and relevant codes of Bureau of Indian Standard.

Silts

Silts are the non plastic fines. They are inherently unstable in the presence of water and have a tendency to become quick when saturated. Quick silts are often called bull's liver by construction men.

Silts are fairly impervious, difficult to compact and are highly susceptible to frost heaving. Silt masses undergo change of volume with change of shape (the property of dilatancy). The dilatancy property together with the "quick" reaction to vibration affords a means of identifying typical silt in the loose, wet state. When dry, silt can be pulverized easily under finger pressure (i.e. very slight dry strength)

Silts differ among themselves in size and shape of grains which are reflected mainly in the ability to compress. Higher the liquid limit of a silt, more compressible it is. The liquid limit of a typical bulky-grained, inorganic silt is about 30 percent, while highly micaceous or diatomaceous silts (so called elastic silts) consisting mainly of flaky grains may have liquid limit as high as 100 percent.

Soils containing large quantities of silt and clay are the most troublesome to the engineer. These materials exhibit marked changes in physical properties with change of water content. Dry clay is hard and suitable as a foundation for heavy loads, but may turn into marsh when wet. Many of the fine soils shrink on drying and expand on wetting, which may adversely affect structures founded upon them or constructed of them.

Clays

Clays are plastic fines (particle size less than 0 .002 mm). The behavior of all soils and particularly clays varies considerably with water content. Clay may be almost like a liquid or it may be very stiff depending upon its water content. They have low resistance of deformation when wet, but when dry they are hard, cohesive masses. Clays are virtually impervious, difficult to compact when wet and impossible to drain by ordinary means. Large expansion and contraction with changes in water content are characteristics of Clays. The small size, flat shape and type of mineral composition of clay particles combine to produce a material that is both compressible and plastic. The clays having higher liquid limit are more compressible. At the same liquid limit, the higher the plasticity index, the more cohesive is the clay.

Small amount of organic matter in colloidal form in clay will result in appreciable increase in liquid limit without increasing plasticity index. Clays with high organic matter create voids through decay.

Warning

1. The optimum moisture content (OMC) should always be lesser than shrinkage limit; otherwise on exposure to sun, cracks will develop in such soil. If such soil has to be used in embankments, then it should be covered with good suitable soil, so that moisture reduction in such soils could be avoided.
2. Soils having shrinkage limit (S.L.) greater than 10 should not be used in embankment, as the dam section is likely to develop extensive cracks.
3. If S.L. is in between 10 to 15, than it is likely to develop cracks in the embankment.
4. Specific gravity of most inorganic soils is in between 2.60 to 2.80.values less than 2.60 indicate possible presence of organic matter in appreciable quantity.
5. Permeability of downstream portion of embankment should be higher than upstream portion; otherwise dam may fail on this account only.
6. The soils containing appreciable quantities of colloidal particles (less than 0.001 mm in diameter), sodium, clay mineral Montmorillonite & Illite are prone to dispersion. Higher values of percent sodium in soil make it depressive.
7. The depressive soil may be used safely in dam embankment except near masonry / concrete works and near foundation rock, if the permeability of embankment may be ensured as 1x10Ï cm/sec or less
8. Soils of high swelling potential are not suitable for embankment.

9. The soils containing appreciable quantities of sulphates should not be used adjoining concrete / masonry works. If it is essential to use, suitable admixture should be added in concrete /mortar to make them safe.

8.13 Soil Test

Generally following soil tests are essential before designing an earthen embankment. These tests should be conducted for embankment, foundation and borrow area.

(i) Mechanical Analysis
(ii) Atterberg Limits
(iii) *In situ* moisture content and density test
(iv) Proctor's maximum dry density & Optimum moisture content
(v) Total soluble solids with EC, pH, carbonates, bicarbonates, and sulphates
(vi) Specific gravity
(vii) Permeability of (a) disturbed samples (b) Field permeability
(viii) Triaxial shear test for (a) Cohesion (in Kg/cm^2) and (b) Φ (in degrees)

Table 8.6 Summary of Common Shear Strength Tests

Test Type	Applicability	Advantages	Disadvantages
Direct shear	a. Effective strength parameters for coarse-grained and fine-grained soils.	a. Simple and inexpensive b. Thin sample allows for rapid drainage of fine-grained soils	a. Only for drained conditions b. Failure plane forced to occur at joint in box c. Non-uniform distribution of stress and strain d. No stress-strain data
Triaxial	a. Effective and total strength parameters for coarse-grained and fine-grained b. Compared to direct shear tests, triaxial tests are preferred for fine-grained soils	a. Easy to control drainage b. Useful stress-strain data c. Can consolidate sample hydrostatically or to insitu K_0 state of stress d. Can simulate various loading conditions	a. Apparatus more complicated than other types of tests b. Drained tests on fine-grained soils must be sheared very slowly
Direct simple shear	a. Most common application is undrained shear strength of fine-grained soils	a. K_0 consolidation b. Gives reasonable values of undrained shear strength for design use	a. Non-uniform distribution of stress and strain
Unconfined	a. Undrained shear strength of 100% saturated samples of homogenous, un fissured clay b. Not suitable as the only basis of design on critical projects	a. Very rapid and inexpensive	a. Not applicable to soils with fissures, silt seams, varves, other defects, or less than 100% saturation b. Sample disturbance not systematically accounted for
Lab vane	Same as for unconfined test	Same as for unconfined test	Same as for unconfined test

(ix) Dispersivity by (a) Pin hole test and (b) SCS Double Hydrometer test
(c) Chemical analysis of pore water extract (d) Crumb test

(x) Free swell index

(xi) Swelling pressure

(xii) Compressibility

(xiii) Ionic concentration of river water

(xiv) Organic matter

Important Terminology of Soil Mechanics

The relationships between void space and the volume occupied by the particles are fundamental and are characterized by the following definitions.

$$\text{Porosity}\quad n = \frac{\text{volume of voids}}{\text{total volume of soil}}$$

$$\text{Voids ratio}\quad e = \frac{\text{volume of voids}}{\text{volume of soil particles}}$$

$$\text{Hence}\quad e = \frac{n}{1-n} \quad \text{and} \quad n = \frac{e}{1+e} \qquad \text{.........(1)}$$

$$\text{Degree of saturation}\quad S_r = \frac{\text{volume of water}}{\text{volume of voids}}$$

$$\text{Water content}\quad w = \frac{\text{weight of water}}{\text{weight of soil particles}}$$

$$\text{Hence if } G_s = \text{specific gravity of soil particles,} \qquad w = S_r \frac{e}{G_s}$$

$$\text{For fully saturated soils } S_r = 1 \text{ and } w = \frac{e}{G_s}$$

$$\text{Bulk density}\quad \gamma = \frac{\text{total weight of soil}}{\text{total volume of soil}}$$

$$\text{Hence}\quad \gamma = \frac{G_s + S_r e}{1+e}\gamma_\omega \qquad \text{..............(2)}$$

where γ_ω = density of water = 62.4 lb / cu ft (1 g / cc)

$$\text{For fully saturated soils } \gamma = \frac{G_s + e}{1+e}\gamma_\omega \qquad \text{..............(3)}$$

$$\text{Dry density}\quad \gamma_d = \frac{\text{weight of soil particles}}{\text{total volume of soil}}$$

$$\text{Hence}\quad \gamma_d = \frac{G_s}{1+e}\gamma_\omega \quad \text{.......(4)} \qquad \text{and} \quad \gamma_d = \frac{\gamma}{1+w} \quad \text{.........(5)}$$

$$\text{also } n = 1 - \frac{1}{G_s}\frac{\gamma_a}{\gamma_\omega} \quad \text{.........(6);} \qquad \text{and} \quad S_r = w \Big/ \left(\frac{\gamma_\omega}{\gamma_d} - \frac{1}{G_s}\right) \quad \text{..........(7)}$$

Submerged density, of soils below the water table

$$\gamma_s = \frac{G_s - 1}{1+e}\gamma_\omega \quad \text{.......(8);} \qquad \text{or} \quad \gamma_s = \frac{\gamma(G_s - 1)}{G_s(1+w)} \quad \text{.......(9)}$$

Alternatively,

$$\text{Percentage air voids } A = \frac{\text{volume of air}}{\text{total volume of soil}} \times 100$$

$$A = 1 - \frac{\gamma_d}{\gamma_\omega}\left(\frac{1}{G_s} + w\right) \qquad \text{................(10)}$$

References

CBIP, (Central Board of Irrigation & Power), Manual on Current Practices in Canal Design in India, Malcha Marg, Chanakyapuri New Delhi, June 1968.

Framji, K.K., Design Practices of Irrigation Canals in the World, Report of International Commission of Irrigation and Drainage, Central Office, 48 Nyaya Marg, Chanakyapuri, New Delhi - 21, 1972.

Hunt, R.E., Geotechnical Engineering Techniques and Practices, McGraw Hill, New York, 1986.

Lambe, T.W. and Whitman, R.V. Soil Mechanics, John Wiley & Sons, New York, 1969.

Nagarkar, P. K., Sohoni, S. B. Rokade, S. N. and Ankaikar, B. N., Mula Right Bank Canal Slips Probable Causes of Failure and Remedial Measures, CBIP Proceeding 45th Annual Research Session, Hyderabad, Andhra Pradesh, Vol. I (Hydraulics) pp. 79-85, June 1976.

Prasad, J., Balaji Shear Strength In Stability Analysis of Slopes; M.E. Thesis, University of Roorkee, 1970

Sahay, B. K., Stability of Slopes of Earth Dams, M.E. Thesis University of Roorkee, 1971.

Vallabha Rao, G. A Study of Slope Protection for Embankment Dams; M.E. Thesis University of Roorkee, 1972.

Wali, Quadri S. A., J. Lakshmana Murthy, Bashir Mohammed and A Venkataramana. A Study of the Slope Failure in Nagarjunasagar Left Canal, CBIP Proceeding, 45th Annual Research Session, Hyderabad, Andhra Pradesh, Vol. I (Hydraulics) pp. 97-105, June 1976.

Wilson, S.D. and Marsal; R.J., Current Trends in Design and Construction of Embankment Dams, ASCE, New York, 1979.

References on Bank Failures of Indira Gandhi Main Canal

Notes by

Basu, J. K., A report on quaternary Geology, Geomorphology and Environmental Appraisal in a part of Rajasthan Desert area, Jaisalmer District (Unpublished) G.S.I. report, (1980-81).

Bithu, B.D., Superintending Engineer, IInd Stage circle, R.C.P. Bikaner (1983)

Chaturvedi, D.C., Superintending Engineer, S&I Circle IGNP, Bikaner.

CWC Team Report, Problem of Seepage and Bank failure in Rajasthan Main Canal Stage and suggested remedial measures (1984)

Harkauli, A.N, Member, C.W.C. New Delhi (1979).

Kapoor, A.S., Chairman, Rajasthan Canal Board, Jaipur. (1983 & 84).

Kashyap, S., Chief Engineer, R.C.P. Bikaner.

Kulkarni, Research Officer, R.C.P. Bikaner (1978).

Ranthidevan, R.V., Director, C.S.M.R.S. New Delhi (1980).

Research Laboratory, R.C.P. Bikaner, Test Report.

Sen, H.P., Director (B & CD-II), C.W.C. New Delhi (1979).

Sethi, S.N., Chief Engineer (Mec-III), C.W.C. New Delhi. (1982).

Sharda, S.C., Prof. and Head of Civil Engineering Department, MRE College Jaipur (1984).

Singh, Alam & Purnia, B.C., Engineering College, Jodhpur (1980).

□□□

Chapter 9

Channels of Most Suitable and Economical Cross Section

So far much have been talked about the best hydraulic sections, as these appeared to be most economical sections. But as it appears may not be always true and there are exceptions to each individual site conditions.

9.1 Suitable and Economical Cross Section

Best hydraulic sections are most economical in lined and unlined canals. In lined canal this section is economical as long as the cost of earth work (excavation and embankment) is less than the cost of lining per unit length of channel and this is the most usual case. These sections, besides having least area and perimeter also have minimum top width. Therefore land width required is also minimum, in comparison to other wide sections. *But if the cost of earth work / excavation is more, such as in rocks, or in very heavy cutting/filling, best hydraulic section do not remain most economical, and recourse to alternative section must be worked out.*

Even in unlined channels in deep excavation or in deep filling recourse must be taken to narrower and deeper sections as far as possible.

Also the type of lining in the bed may be cheaper than the lining on sides and in such cases, lining of best section is no more most economical and cost economics of various alternatives including increase in bed width has to be worked out, see para 4.1 and 6.3.

Also for lining of existing canals of low depth, a rectangular sections seems better choice from durability, ease of construction, low operation and maintenance cost, minimum modification of existing structures etc,. so may be over all cost economic.

Thus economic suitability is site specific and that is aimed in para 9.6. Besides cost economics a good section should also fulfil following impotant creteria.

1. Minimum cost and practical economical section, see para 9.2 and 9.6
2. Minimum operation and maintenance cost i.e.
   (i) Minimum seepage loss section see Para 9.3

(ii) Minimum silting and minimum abrasion on lining (for lined canals only) i.e. transport the sediment entering into the canal without damaging lining, see chapter 7.

(iii) Above (ii) is also applicable to unlined canals to the extent of non silting and non scouring. When clear water is entering, it should be non scouring, see chapter 7.

3. Minimum weed growth section, see Para 9.5
4. Maximum permissible velocity section, see Para 2.3 and chapter 7.
5. Minimum and Maximum permissible bed slope section, see para 2.3.and 9.13.
6. The embankment section should also be safe (structurally stable), see chapter 8 and have maximum hydraulic efficiency, see chapter-1 and para 9.9 to 9.10 in most of the normal working conditions.

9.2 Cost of Lined Canal Section

There are 3 principle items of cost as under :

(i) Cost of lining
(ii) Cost of earth work i.e. excavation/filling
(iii) Cost of land

(i) Cost of lining

The cost of lining is proportional to the perimeter of canal, including freeboard. This lining may or may not be uniform (same) for the entire perimeter. The lining of bed or bottom curve may be different then the lining of sides, dowel etc., see Para 4.1, 6.3 and 6.4. In addition to it, there is a third cost of relief valves and under drainage system. So the combined cost should be minimum, i.e.

$C_p = C_b + C_s + C_d$ should be minimum ... eq. 9(i)

Where

C_p = total cost of lining of perimeter
C_b = cost of lining of bed, including its dressing and
Cs = cost of lining of side slopes including lip cutting, dressing of sides etc.
C_d = cost of relief valves and - drainage system.

(ii) Cost of Earthwork (excavation and banks)

The cost of earthwork i.e. excavation and filling of banks depends upon the terrain of the canal through which the alignment is passing. In heavy cuttings and particularly in rocks, the cost is high. Similarly, cost is high in heavy filling. Cost also depends upon the requirement of service roads, dowels etc. The cost is minimum in balanced cutting/filling. Though, every effort is made in laying the alignment for the balanced cutting filling yet the above situations are unavoidable. The cost of filling further increases to include extra earthwork for lip cutting, cost of compaction and the arrangement of water and watering for compaction. However, cost of filling may be reduced to the extent excavation (earth from cuttings) is utilized by economical leads. There are detailed instructions in IGNP, for double use of earth work. The cost of earthwork can be represented as

$C_{ew} = Ce + C_f.$ ... Eq. 9(ii)

Where,

C_e = cost of excavation including leads and lift
C_f = cost of filling, including leads, lifts watering and compaction

(iii) Cost of Land

Land width required varies according to the type of section, i.e. top water width, heavy cutting, heavy filling, requirement of service road etc. In general, smaller top width requires less land width, see example 3.2, 4.1, 4.2, chapter 6 and figure 6.2 and 6.3. With increase in top width, because of side slopes and berms, land width increases in heavy cutting and filling sections, see para 9.6. Cost of land varies along the length, depending upon the type of land through which the alignment is passing.

Even if the land is cheap in uncultivated areas, forest or otherwise minimum land should be acquired from environmental consideration. However, extra land acquired for forests etc. may not be added to this cost.

Besides above three costs, cost of following more items also requires consideration for a true comparative analysis.

(iv) Cost of maintenance and repairs (M&R)

In addition to above 3 costs, there is a 4th cost of maintenance and repairs of lining. This cost depends upon silt inflow, sediment deposit, rock falls, and land slides, weed growth etc. Many times this cost may be more than that for unlined canals, see table 9.6 and Para 9.15.2. Many times a closed section is the only solution for small discharges, see table 9.2. It is difficult to estimate these costs, and therefore is generally avoided, though not desirable.

(v) Lining of existing canals

In it, many times the cost of other items associated with it and land is not included. Only the cost of lining and earth work is added. But this may not be always correct, see Para 6.3, 9. 15.2, table 9.6., Aspects of parallel canals, diversions, modification of structures etc. mentioned in the para 6.3 and their cost should be considered.

(vi) Other Costs

It includes cost of incidental structures such as water arrangements, storages, pumps etc. Also cost of above (iv) and (v) may be included in it.

(vii) Total cost of lining

The total cost of lining is thus the sum total of all above 5 costs including (iv), (v) and (vi) where applicable. This can be represented as:

$$C_T = C_p + C_{ew} + C_l + C_o \qquad \text{... Eq. 9 (iii)}$$

Where,

C_T = Total cost of lined section

and C_l = Cost of land

C_o = other costs; other notations are same as above.

Thus total cost of lining will change from reach to reach due to change in topography and geology. Also see para 9.2.1 for a detailed list of items.

For economic section of a lined canal, this total cost should be minimum.

This cost is generally minimum, when the top width of the water surface is minimum. However, besides minimum cost, the section should satisfy the important aspect of the channel command and hydraulics, see table 9.3. Therefore, a compromise is often/frequently made. Still some analysis is always required. A typical list of items of lining is given in para 9.2.1 and a methodology of cost, comparatively analysis and economics including saving in water are given in para 9.15.

It may not be possible to economies the section for a long reach. Therefore, it may be done in suitable part lengths (generally called reach).

9.2.1 List of Items for Lining

A typical list of items to be considered for comparative analysis of cost for different types of sections and linings in *new canals* is as under:

(a) Earth Work

1. Earthwork in excavation and filling
   (i) From cuttings with extra leads,
   (ii) From borrow area with extra leads
2. Compaction, including watering etc.
3. Lip cutting and its adjustment
4. Slope protection in (i) upstream in heavy cutting and (ii) downstream in heavy filling
5. Downstream filter toe arrangements.

(b) Lining

1. Dressing of side slopes including removal of irregularities up to 7.5 cm, and extra for height
2. Providing and laying water proof paper on sub-grade
3. Preparing and laying (1:3:6 concrete or double tile etc.) lining on side slopes
   (i) Up to 3.5 m
   (ii) Above 3.5 m
4. Scaffolding and template allowance
5. Dressing of bed including removal of irregularities up to 7.5 cm depth.
6. Preparing and laying (1 : 3 : 6 cement concrete or tile etc.) lining in bed.

The above are common items and change according to type of lining and specifications, for example in rock cutting there is nothing like lip cutting.

(c) Drainage Arrangements (only where ever required)

Cost of drainage arrangements including laying (depending upon the type of soil, water table etc).

(d) Cost of land

(i) Cultivated land
(ii) Uncultivable land

(e) Dowel on top of lining

It may be earthen or masonry or combination of both.

For masonry dowel the items are:

1. Earthwork excavation in foundation of dowel
2. C.C 1 : 3 : 6 in foundation of dowel
3. Brick/Tile/stone masonry in cement mortar 1: 6 Cement plaster on inner or outer surface or bats surface of dowel (as the case may be only if both are used)
4. C.C. 1 : 3: 6 for coping, including shuttering.

(f) Other costs

1. Cost of air entraining agents.
2. Water Pumping charges, including cost of arrangement for water and its storage.
3. Curing charges for 28 days.
4. Filling expansion joints.

(g) Ancillary or incidental structures for lining of existing canal

1. Cross bunds
2. Diversion and connections
3. Removal of above
4. Modification of structures.

Note: Specifications of items and unit cost change according to availability of material, as per site conditions and reach to reach.

Reach may be from 1 Km to 5 Km in which earth is adjusted.

Typical lead statement for working out cost of material in a canal reach is as under (Change according to reach).

S.No.	Items	Source	Cost of material	Lead*	Cost of Carriage
				Kaccha & Pucka*	Total
1	Cement				
2	Sand				
3	Stone grit				
4	Brick /stone				
5	Other materials (specify)				

*Kaccha = earthen cart road or fair weather road

*Pucka = Black topped or at least WBM (water bound macadam, not bitumen) road, still good for vehicles in all seasons

For working out the cost comparison of lined/unlined canals, see Para 9.15 table 9.4 and 9.5.

9.3 Canal Section for Minimum Seepage Loss and its Suitability

The seepage of water from irrigation canals is a complex process and depends on many factors, some listed below.

I. Soil Factors

1. Soil characteristics of the canal bed and side slopes, (cutting/ embankment) permeability
2. Entrained air in the soil, compaction.
3. Position of impermeable layer in relation to canal
4. Other geological factors - rocks, joints, crevices
5. Soil moisture tension.

II. Topographical Factors

1. Ground slope at right angle to canal cutting/filling/one bank canal
2. Drainage levels of nearby drain and its location.

III. Ground Water

1. Position of ground water table in relation to canal, bed.
2. Velocity of flow of ground water, permeability.

IV. Canal Design and Operational Factors

1. Wetted area or shape of the canal; lining and its effectiveness (life, cracks, joints, damages)
2. Age of the canal
3. Amount of sediment and its grade
4. Salt concentration in the canal water and soil
5. Frequency of canal uses, i.e. regulation, perennial/non perennial, (intermittent) discharge and water levels in the canal, along with its duration.
6. Surrounding vegetation of canal, weed growth.
7. Temperature of water and soil.

Above large number of factors, more than 15, affect the conveyance losses considerably. Two typical seepage flow nets, shown in figure 9.1 indicate wide variation in losses.

All the above factors are not stationary all along the entire length of the canal and in time also. All the factors change considerably due to heterogeneity of soil, as the canal pass through different types of topography and geology. In time scale weather is not constant in a year so canal regulation is not constant. Also ground water level changes in a year. All parameters change year after year. More discussions on the details of influence are dealt separately under seepage losses.

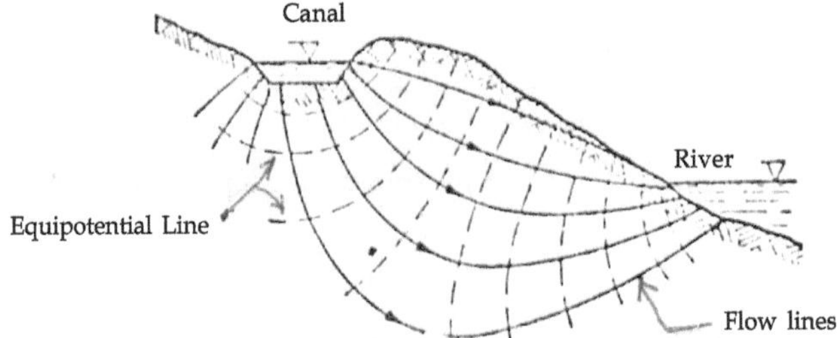

(i) Seepage from elevated canal towards the river

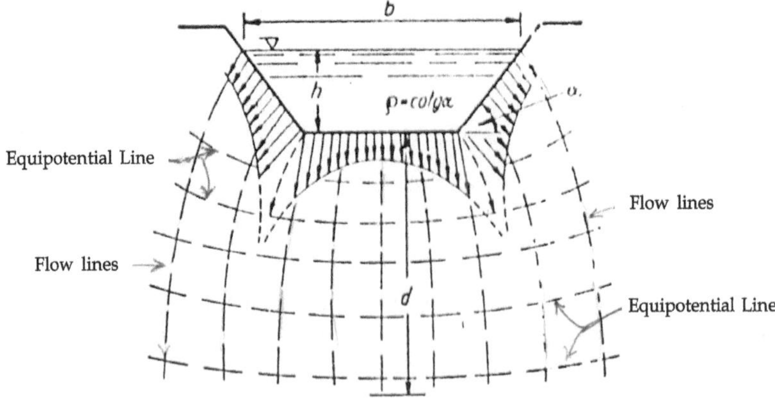

(ii) Flow pattern for seepage from the canal in case of deep ground-water table (see *Vedernikot's* method)

Figure 9.1 Seepage loss patterns.

Still it has been generally accepted that the seepage loss varies mainly with the texture of the canal soil and subsoil and the depth to the water table. It is taken as proportional to the wetted perimeter and is affected by the depth of flow. Though there is probably no simple relation between the seepage loss and the depth, yet Etcheverry (1915) made an assumption that the intensity of seepage loss (S per unit of cross-sectional wetted area varies with the square root of the depth

i.e. $S \propto \sqrt{d}$ or $S = C\sqrt{d}$

Where

C = a constant depending on texture of soil depth to water - table, etc.

d = depth of water in the canal at any point

He further assumed that the average intensity for the sides is 2/3 of the intensity on the base. Thus the total seepage loss for a given canal cross section at full water depth was deduced as under :

$S_P = C\sqrt{d}\,(b) + 2/3\, C\sqrt{d}\,(2d/\sin\theta)$

$S_P = C\sqrt{d}\,(b + 4\,d/3\sin\theta)$

Where, S_P = seepage loss per unit length of canal

For trapezoidal section, $A = (X\,d^2 + d^2 \cot)$

or $d = \sqrt{A\,/\,(X + \cot\,\theta)}$ and $b = Xd = X\sqrt{A\,/\,(X + \cot\,\theta)}$

Substitute for d and b in terms A, X (b / d ratio) and θ, then

$$S_P = C\left(\frac{A}{X + \cot\theta}\right)^{3/4}\left(X + \frac{4}{3\,\sin\theta}\right)$$

To obtain minimum value of S_p, the first derivative in terms of θ is placed equal to zero, which gives

$X = (1-\cos\,\theta)\,/\,\sin\,\theta\ = 4\,\tan\theta/2$

or $d = \sqrt{\dfrac{A\,\sin\theta}{4-\,3\,\cos\theta}}$ Eq. 9 (iv) and $b = 4d\,\tan\dfrac{\theta}{2}$ Eq. 9 (v)

This value of B: D ratio (X of bottom width to depth), for minimum seepage loss, is twice the value required for maximum hydraulic radius or maximum velocity, or most economical section.

Khushalani (1954) has also shown that when seepage $S \propto d^{3/4}$

$d = \sqrt{\dfrac{A\sin\theta}{8-7\cos\theta}}$ - Eq. 9 (vi) and $b = 8d\,\tan\theta/2$ Eq. 9 (vii)

This width is twice that when $S \propto d^{1/2}$ and four times that of most efficient hydraulic section.

Since the perimeter of such section is much more, the above formulae did not become very popular/ common in practice for lined canals.

Also seepage loss should be taken as function of what power of depth is highly debatable and is no more acceptable. A much more acceptable is that it is proportional to perimeter.

Considerable research has further been recently attempted by many research scholars to find a optimal canal section of minimum cost and minimum seepage and evaporation by making many simplified assumptions. Vedernikov (Harr 1962) gave seepage functions from trapezoidal canals in isotropic, porous media of infinite depth and water table lying at infinity. Muskat (1946) gave analytical solution for finite depth but laid over drainage layers. In actual practice, both above situations do not exist in the fields. Ground geology is highly heterogeneous with varying layers of permeable and impermeable layers (horizontal / non horizontal or inclined) of changing porosity and permeability of ground above water table.

Chahar (2000) further extended these theories with unit cost of earthwork and lining and gave section coefficients.

But in actual practice, the factors effecting seepage (9.3), and units cost (Para 9.2) are highly variable all along the canal. Therefore the use of such equations is highly debatable. The work needs further verification on the existing lined and unlined canals in the fields.

In fact, the best hydraulic section is the section of minimum perimeter and therefore appears to be a section for minimum seepage loss also.

More over there appears little justification to have enhanced bed width (Eq. 9(v)) for minimum seepage and then to line it to reduce seepage. This increases cost unnecessarily. The concept of minimum seepage loss section is more important in unlined canals, and the effort of earlier investigators, was for unlined canal section.

Still the minimum seepage loss section is not the only criteria for design of lined or unlined canals. The section should also satisfy the other important criteria's given in Para 9.1

More over the design of unlined or lined canals is based on hydraulic laws which over governs the section for minimum seepage loss. In fact all the problems and solution of a section suitable for minimum seepage, or cost, or non silting or non weed growth, or non erosive bed slopes, etc. can be reduced to the acceptable permissible velocity.

In actual practice, best section can/has been adopted upto 55 cumecs (2000 cusecs). After that discharge i.e. for high capacity canals, best section is deep and may not be practical. Therefore, B/D ratio of more than 2 to 9 has generally been adopted (See Para 6.2 and Chapter 6) Based on several discussion, in this text, a B/D ratio of 1.5 to 5 may be reasonable.

In fact the best criteria would be a reasonable velocity, depth of flow and a minimum B/D ratio for that section. Thus no arbitrary selection of any B/D ratio is required.

9.4 Partly Lined canal section

Either bed or sides lined (some times called partially lined section)

Analog studies of partially or partly lined canal section using resistance net work have been reported by Bouwer (1965). Subramanya, et al (1973) aimed at obtaining solutions to problems of partially lined trapezoidal canals with only sides lined, and only bed lined, see figure 9.2(i) and (ii). This study for deep water table, when only the sides of the canal are lined, indicate that the slope of the sides has negligible influence on seepage loss. As B/D ratio increases, seepage loss increases.

In the case, when only canal bottom is lined and seepage occurs through the sides only, the inner side slope of the canal banks has considerable influence on the quantity of seepage, see figure 9.2 and 9.3. In figure 9.2 (i), q is the seepage loss from sides per unit length of the canal. Figure 9.3 shows the variation of the percentage reduction of seepage due to lining of either bed or side slopes with B/D for various side slopes Z, ($Z = \cot \beta \Pi$). This figure indicates that for a canal having side slope 2.5:1(H: V) and B/D = 9.0, if only the bed is lined, a 50% reduction in the quantity of seepage is expected and if only the sides are lined, a 25% reduction in the seepage quantity would be achieved.

This figure shows following important aspects

(i) For $\beta = 1/4$, and at B/D = 2, seepage loss is identical from bed and sides

(ii) For $\beta = 1/6$, and at B/D = 3.4, -------- do--------

(iii) For $\beta = 1/8$, and at B/D = 4.8, ----------do--------

That means for above sides slopes and B/D ratio, the section is the minimum seepage loss section.

Experiments conducted at U.P. Irrigation Research Institute, Roorkee on electrical analogy model for shallow water-table condition indicate that even a very small open area results in seepage losses nearly equal to those taking place from unlined canal. Part lining of a canal with shallow water-table condition is, therefore, not considered economical from seepage considerations.

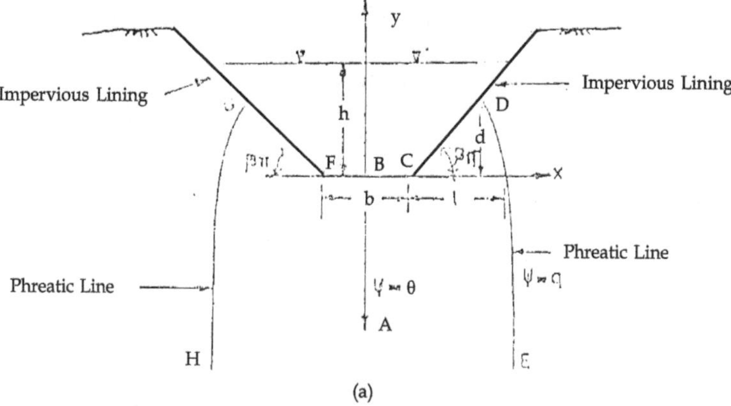

(i) Canal with sides lined and bottom unlined

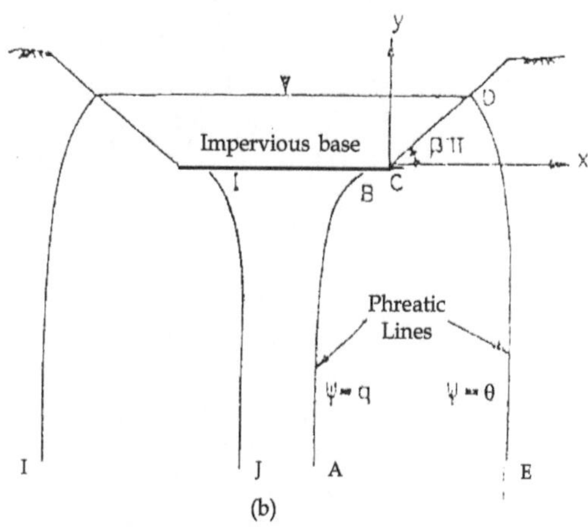

(ii) Canal with bottom lined and sides unlined

Figure 9.2 Seepage loss patterns for partly lined canals

9.5 Canal Section for Minimum Weed Growth

(Minimum velocities and side slopes required to prevent weed growth)

All canals whether lined or unlined are subject to some silt deposition (see Para 9.4) and weed growth. Weeds either enter the canal from the source or may get germinated & generated in the canal itself.

Magnitude differs with the source and the canal. From deep reservoirs, only clear water enters the canal. But in shallow reservoirs a weed get germinated and enters the canal in varying quantities depending upon the design of diversion arrangement.

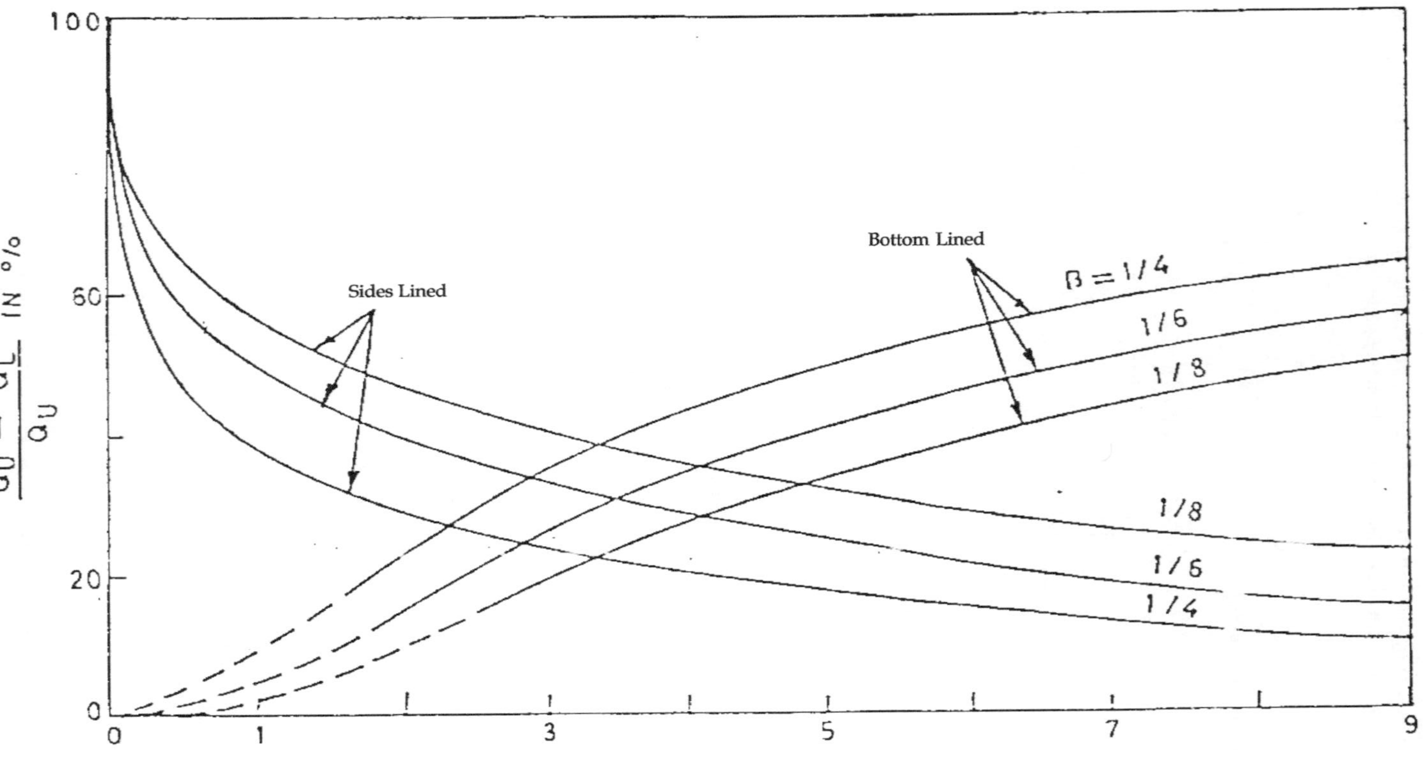

Figure 9.3. Variaton of percentage reduction in seepage due to linning with B/D for various values of (angles of side slopes)

In shallow reservoirs weeds grow near the fringe of water only when water remains standing for few weeks and then flow towards the canal outlets. Otherwise with depletion of reservoir, weeds at the fringe die. Situation of weed growth in small and medium reservoir also often occur when the canals run intermittently. Also such situations can be seen in reservoirs when a particular water level is maintained either by release from a upper reservoir or by regeneration (flow of ground water or excess irrigation water).

Typical examples are Survaniya and Harsora dam, medium irrigation project Banswara, Rajasthan India. With the incoming of Mahi canals in their commands the storage level has been restricted by the local political pressure. Water is spread over large area with low depth and remains constant for most of the time. This germinates huge weed and transmits to canal off taking from them.

Weirs/barrages, direct river water carry some weeds and seeds in the canal. Seeds are also brought in the canal, by wind and birds.

Besides, that both urban and rural people add some wastes/pollutants, such as cow dung, waste water, agricultural, organic wastes etc, in the canal, enhance the weed growth. Though additions of all such wastes is not desirable, still total ban (100% stoppage) may be difficult to achieve.

The type of weeds, and their magnitude, vary according to climate and locality i.e. weeds are season and site specific.

The weeds encountered in the channels can be broadly divided into :

(i) Ditch Bank Weeds.

(ii) Floating Weeds.

(iii) Submerged Weeds, and

(iv) Emerging (or Immessed) Weeds.

By their biological properties they are called monocotyledons or di-cotyledons depending upon their narrow parallel varied leaves or netted broad leaves.

Ditch Bank Weeds : They grow along the banks of canal just above water level, over hanging the flow and when they fall into water they catch other floating weeds and reduce the velocity of flow. Their leaves often fall in the canal.

Floating Weeds : These weeds float along the water surface and are troublesome on account of their clogging nature. It has been found that dense masses of these weeds can greatly obstruct the flow at bridge piers. Failure of two bridges on Kasur Nallah had to be attributed to this phenomenon. Floating weeds have been found to double the friction factor. Typical examaple in water Hyacinth.

A single plant of water Hyacinth can give rise to as many as thousand in one season; the active growing season being September to October. It is called "Million Dollar" weed in U.S.A. because it affects water control, wild life and navigation. The experience of the author is, it grows luxuriously wherever there is stagnant water, sewage contamination, etc.

Submerged Weeds : they generally take root at the bottom of the canal and produce long strands of dense growth which may be as long as 7-8 m and these also cause clogging of the bridge piers. Presence of submerged weeds may result in increase of the frictional resistance twenty-fold.

Villisnaria, Hydrilla Verticellata, Potamogeton, Pectinatus are some of the typical submerged weeds.

Emergent Weeds : they emerge above the water surface with their roots in the bed and sides. They reduce the channel capacity and raise the water-level resulting in additional seepage losses. They catch debris, entangle other aquatic growth thus further reducing the flow. A common variety is 'Typha angustata'.

The growth of aquatic plants and moss is usually most when the following favorable conditions are obtained:

(i) Clear, slightly warm, shallow water exposed to sun light.

(ii) Small velocity.

Shallow and stagnant water are the biggest source of pollution and weed growth.

Algae and moss do not begin to develop to any extent unless the temperature of the water is 65° to 70° F. Turbid waters or deep waters interfere with the action of sunlight and are unfavorable to plant growth. In fact water when travels at a reasonably good velocity it gets self-cleaned. (nature's way of cleaning, and when at low velocity or stagnant becomes dirty and foul).

Weed growth reduces carrying capacity to ¼ or less, depending upon their magnitude and helps in silt deposition.

In large, deep canals the growth of aquatic plants is largely confined to the sides of the canal in the shallow water along the edges. Additional obstruction of the flow is frequently produced by the weeds. Grass or alfalfa growing along the water edges, drops and trails in the water.

Therefore flatter inner side slopes should be avoided. Also flatter side slope reduces velocity, see para 3.3. Steeper the slopes, more is the velocity and lesser silt deposition and weed growth. Therefore for small depth of 1 to 1.5 m vertical side walls (preferably with batter of 0.4: 1 of (H:V) on both sides) may be economical and most suitable for easy operation and maintenance.

Also see para 4.6 figure 4.5 which indicates that velocity is function of $d^{2/3}$ and increases slowly with depth in all types of sections.

The growth of aquatic plants and the formation of moss are directly affected by the velocity of flow.

The velocity required to prevent the growth of weeds and aquatic plants and the occurrence of moss is dependent on the depth of flow, character of the area, vegetation and on the extent to which the canal is operated.

The growth of grass and weeds on the beds and sides of small channels (ditches) carrying water intermittently for short periods may be directly affected by the use of a velocity sufficiently high to carry the seeds of plants through the ditches to the fields.

Small distributaries, minors, and water courses operated intermittently only for a short time, not more than a few days at a time, and the period of submergence may not be long enough to drain out or kill the ordinary weeds/grasses, and these may grow rapidly, especially during the period of no water in them. Still higher velocity, narrower section retards the growth of aquatic plants.

Flynn (1892) states that in Spain velocities of 2 to 2½ feet per second are considered necessary. Fortier (1901) concluded from a large number of observations on canal systems, mostly in Utah, USA that a mean velocity not less than 0.75 m/s to 0.85 m/s (2 ½ fee per second and preferably 2 ¾ feet per second) is desirable to prevent aquatic plants, moss and silt deposits.

English and Gokhale (1938) have discussed in detail weed problems in canals in Southern India and concluded following important aspects.

(i) *Infection* : Under South Indian (Deccan) conditions, water weeds grow profusely in all nalas and ponds i.e., infection is universal and this enquiry showed that there is no relation between weed growth in canals and intensity of infection.

(ii) *Temperature of water* : In the Deccan water temperatue varies between 20° and 30°C, and hence it is highly favourable for weed growth.

(iii) *Chemical Compositon of water* : There is no traceable relation between intensity of growth in various canals and composition of water, but alkalinity by causing despersion of caly has a secondry effect ; because it increases turbidity and reduces the amount of light at the bed.

(iv) *Velocity* has little effect except where it exceeds 2 feet per second; but it has an indirect effect, in that, when the velocity is in excess of regime velocity, deposition of seeds and cuttings does not occur.

(v) *Depth of water* : Water weeds can grow in clear water at depths down to 18 feet,

(vi) *Deposition of silt* : It is undoubtedly favourable to weed growth, but deposition only occurs where the velocity is considerably below regime velocity and such low velocities do not occur in properly designed channels. When canals are newly opened, they are generally run with low discharges and water has to be headed up at regulators. This reduces the velocity and causes deposition of silt especially along the banks. Rendering the water clear; so that all conditions are then favourable for weed growth. Thereafter weeds tend to persist.

(vii) *Reduction of light due to trubidity* : There is reason to anticipate that exclustion of light will reduce weed growth, in that, light is required for carbon assimilation and tissue building (photo-synthesis) and there are no cases in which weeds grow where the water is turbid throughout the year, whereas weeds grow profusely when water is clear. Weeds generally grow luxuriantly when several factors are favourable. Weeds do not grow where silt is depositing, unless the water is also relatively clear.

In small canals under 2 feet in depth and with a mean velocity of flow less than 2 feet per second, the entire cross section, if not properly cleaned at intervals, may become entirely filled with aquatic plants and moss.

Very little research/observations has been reported in the journals on this aspects. Field engineers should carry out more careful observations and report on this aspect in journals.

Fortunately, the semi circular or best trapezoidal with steepest side slopes (as much as possible) is also the best section for minimum weed growth, and may be attempted as far as possible.

It is a matter of common sense that wider the channels, more is the problmes of weeds. Deeper and norrow section will have lesser problems of weeds.

Weeds can be seen in all lined canals ranging from few cusecs to 18500 cusecs canals of Indira Gandhi canal project, Mahi project, and many other small or medium projects in Rajsthan, India.

Many of the weeds are independent of the lined or unlined surface.

Weeds of different types, grass can be easily seen on the roofs of concrete or lime of all buildings. They need very little soil and moisture to grow. Their hair roots penetrate into the joints of masonry/ concrete.

Grass, bushes/shurbs and even trees (particularly Pepal, north Indian local name) can be seen in free board portion of lined canal above FSL. There hair roots penetrate into the joints of cement mortor, and travel down towards water and soil below lining. Even in hot seasons little water of wave action is sufficient to sustain them.

As regards weeds all lined canals have shown behaviour as unlined canals.

On Narwana Branch of the Bhakra Canal system, the efficency of plantations on the canal embankment to weed off weed growth has been established. There had been a very thick tree plantation on both banks of the canals and this had been instrumental in preventing direct sunlight falling upon the water surface. Some reaches of the canal had, however, been recently denuded of plantation and in those reaches, floating weeds called 'jala' had emerged. It caused considerable reduction in the capacity of the canal. Where thick plantation still exists, there is hardly any trace of jala. This phenomenon apparently points to significant role of plantation in the maintenance of canal systems. On canals fed from reservoirs, the water is silt-free and this leads to emergence of algae (jala). The algae is a food for vegetarian type of fish. So the development of pisciculture with vegetarian fish which feeds on algae on channels with silt-free water can deduce the emergence of algae.

Weed Control and its Removal

The types of weed control measures may be divided into :

(i) *Design aspects* : see above para; higher velocity, steeper side slopes, deeper canals or increasing the velocity of flow by re-grading i.e. (changing the bed slope) of the channel. This remedy was adopted on the Bhakra canals, according to Superintending engineer, Bhakra Main Line Circle, and channels were re-graded so that the minimum velocity was not lower than 0.61 m (2 ft)/sec. Bhakra Canals carry silt free water, yet there is acute weed problem in such channels.

(ii) Regulation measures, alternate wetting and drying,

(iii) Biological measures- fishiculture and plantation,

(iv) Chemical measures,

(v) Mechanical, methods of weed removal/cleaning (ii) to (v) are parts of an O of M programme and again site specific, i.e. size of canal, magnitude and frequency of weeds etc.

This is beyond the present text.

9.6 Practical Economical Sections

Different types of economical sections adopted in different topographical situations are given in figure 9.4 and 9.5 and further discussed as under :

(a) Partly cut and fill section, figure 9.4 (i) (also generally known as partial cutting /filling section) : The earth from excavation is utilized in making banks. This is the most economical and common section and always an attempt is made to keep the alignment as far as possible in balance cutting/ filling (i.e. all earth from excavation is utilized on banks in filling and no extra earth is brought from outside, no borrow pits required.). This is the most ideal layout of a canal and normally very difficult to achieve throughout the length of a canal. In such cases cost of earth work is less than cost of lining. Almost all best sections - semi circle, rectangle, trapezoidal / Mehboob sections with steep side slopes as feasible, (discussed in chapter 3) or wider channel sections depending upon discharge (discussed in chapter 4) are generally adopted, also see Table 6.1.

(b) Cut into ground or cutting section, figure 9.4 (ii)
Normally section is just fully in cutting . All earth from cutting is either disposed on spoils or utilized in adjoining filling reaches longitudinally to the extent of economical leads. Cost of such section though higher than (a), still generally less than cost of lining, therefore all best sections as in (a) above are generally adopted. Cost of excavation is compensated to some extent, the earth is utilized (i.e. double use).

(c) Canal section in very deep cutting, figure 9.4 (iii) (also vagouly known as deep cutting section): All efforts are made to avoid such sections in laying the alignment of canal, still some such section are unavoidable, in hilly or undulating topography see figure 9.8. Such Open section besides being costly, also poses serious problems of stability of slopes in soils or rocks, drainage, of upland, falling of debris, land slides and consequently choking of canal. These canals are also hazards for grazing animal on the upland. When deep cuts are involved, the slopes are usually subdivided by berms generally about 5 ft. wide. Similar berms are desirable at ground level, if spoil banks are formed alongside the canal, and at rock level if the excavation is partly in rock. Storm water should not be allowed to flow at random into the canal because it will erode the inner slopes and carry material into the canal. Suitable drainage arrangements should be provided to lead storm water into properly constructed intakes / escapes or drainage ditch.

Typical examples of such deep cutting can be seen almost for all discharge ranging from 524 cumecs to 0.05 cusecs (18500 cusecs to 3 cusecs) in Indira Gandhi canal Project (formerly known as Rajasthan Canal Project) in the sandy soils of great Indian Desert known as Thar Desert. Subsequently a serious thought led to a closed canal section, rectangular or circular for discharge say upto 1.5 cumecs (50 cusecs).

Other examples of deep cutting in rocks are Right and Left canals of Mahi Bajaj Sagar Project, Banswara, Rajasthan, India, see para 6.8 and figure 6.10. These canals have shown problems of roots of shrubs and trees, penetrating behind lining and damaging the same seriously.

More complex problems of slidding/slippages of rocks can be seen in the Baroda Branch canals of Som Kamala Amba Project, in Dungarpur district Rajasthan, India.

Many times a good choice may be tunnels or a cut and cover section. Some typical cross-sections are shown in figure 9.12 also see para 9.11.

(d) Canal in filling, figure 9.4 (iv), (v) and 9.5

Sections in figure 9.4 (iv) though costlier than section shown in (i) are still very common almost in all types of canal and in all types of soils. The embakment is composed of good soil. Section in figure 9.4 (vii) on high fills are many times unavoidable, though all possible efforts should be made to avoid the same. Many times the cost of earthwork is more than the cost of lining.

For small discharges, flume sections can be used as shown in figure 9.5 But for large discharges flumes are costly and heavy filling is done. Typical examples of section in heavy filling can be seen almost in all sizes of canals, right from a small discharge watercourse to medium discharge distributory, branch or a large discharge main canal of Indira Gandhi Nahar Project (IGNP), [Main canal carries a discharge from 524 cumecs (18000 cusecs) to 230 cumecs (8000 cusecs)] . Design of large / higher discharge canal banks is to be done like an earthen dam and lining section are choosen by cost economics of various alternatives.

The embakment may be provided with a core wall to ensure water tightnes and should be designed with the same care as any other embankment dam.

The bank must either be wide enough to prevent excessive leakage, or must have a impervious core or lining. The top width is sometimes made equal to the depth of water, with a minimum of 1.2 m (4 ft.). Also see para (e).

(e) Canal section in side cutting, figure 9.4 (vii) to (xxvi)

Such sections are along hills or sloping ground or contour canals in all type of soils/rocks.

Many engineers prefer to have them wholly in excavation figure 9.4 (vii). Some times such excavation may be costly. Therefore this practice by no means is universal. Canals are often built with one side formed by artificial embankment, see figure 9.4 (viii, ix, xii, h, i) so that excavation and filling are balanced. The lined section need not be uniform on both sides of the centre line such as rectangular or trapezoidal, but can have varied shape on either side of the centre line, see figure 9.4 (x).

The downhill side of the canal can not be constructed as an embankment formed from the rock material excavated from the canal. This side of the canal would then become a rock-fill dam. A lining is essential in this case.

Concrete Walls

As an alternative to rock filling, the downhill side may be formed by a masonry concrete wall This construction lends itself to the provision of spillways. The choice between rock filling and concrete walling is a matter of economics.

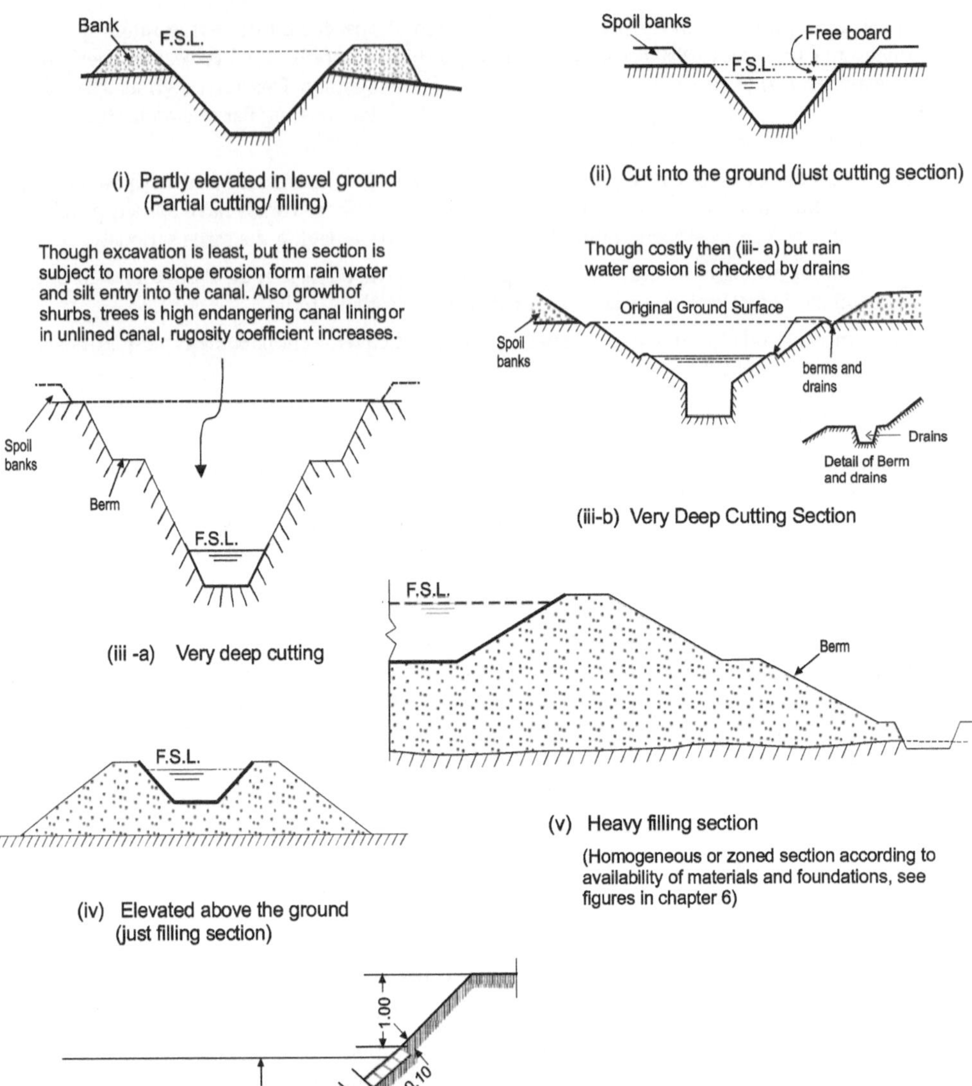

(i) Partly elevated in level ground (Partial cutting/ filling)

(ii) Cut into the ground (just cutting section)

(iii -a) Very deep cutting

(iii-b) Very Deep Cutting Section

(iv) Elevated above the ground (just filling section)

(v) Heavy filling section

(Homogeneous or zoned section according to availability of materials and foundations, see figures in chapter 6)

(vi) Different side slopes in deep canals.
Canal lining.
Peggau development, Mur river
(Steirische Elektrizitatsgesellschart, Graz)

(vii) Cut-and-fill cross section in gently sloping ground

Figure 9.4 : Types of canal sections according to topographical conditions

contd...

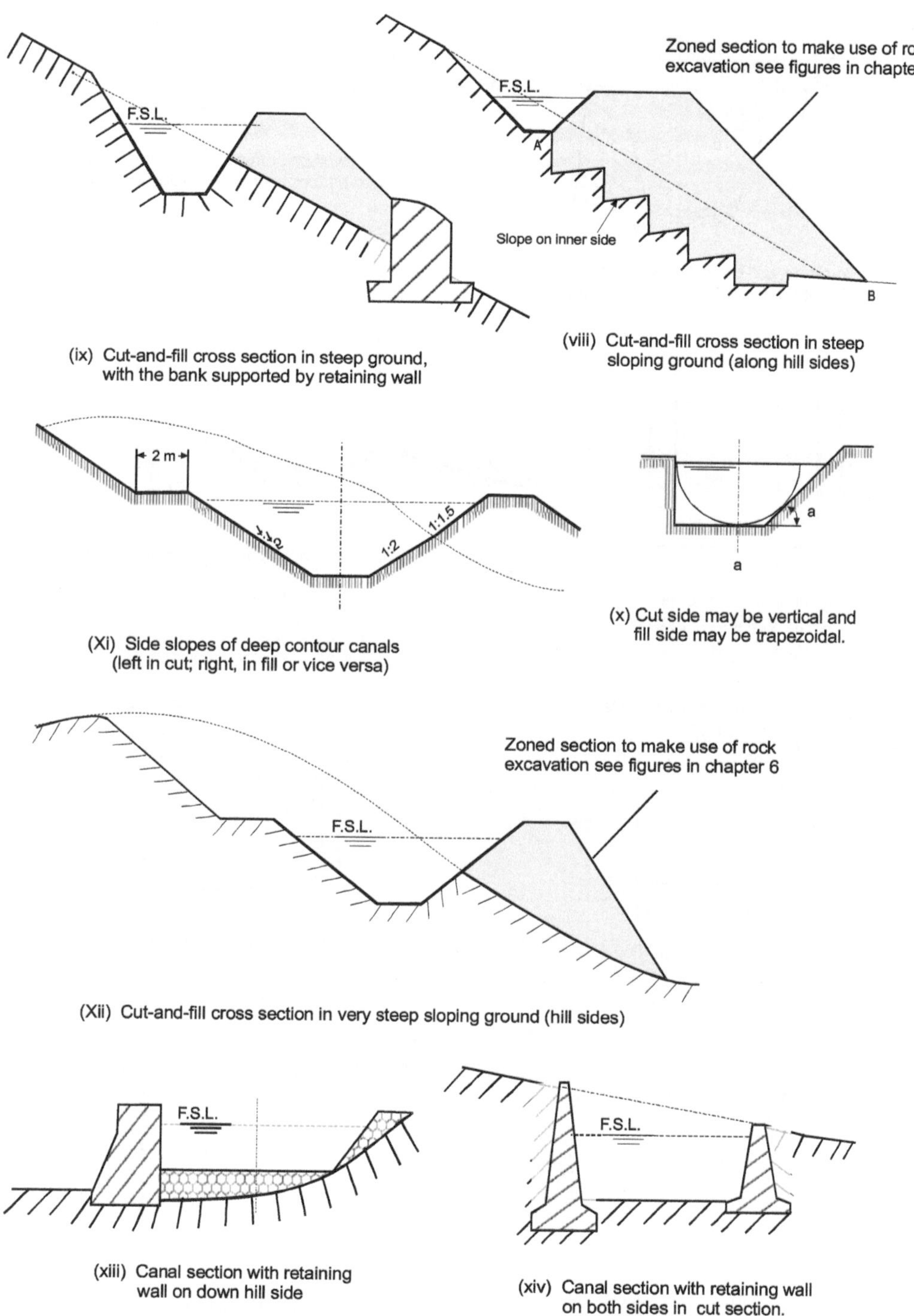

(ix) Cut-and-fill cross section in steep ground, with the bank supported by retaining wall

(viii) Cut-and-fill cross section in steep sloping ground (along hill sides)

(Xi) Side slopes of deep contour canals (left in cut; right, in fill or vice versa)

(x) Cut side may be vertical and fill side may be trapezoidal.

(Xii) Cut-and-fill cross section in very steep sloping ground (hill sides)

(xiii) Canal section with retaining wall on down hill side

(xiv) Canal section with retaining wall on both sides in cut section.

Figure 9.4 : Types of canal sections according to topographical conditions contd...

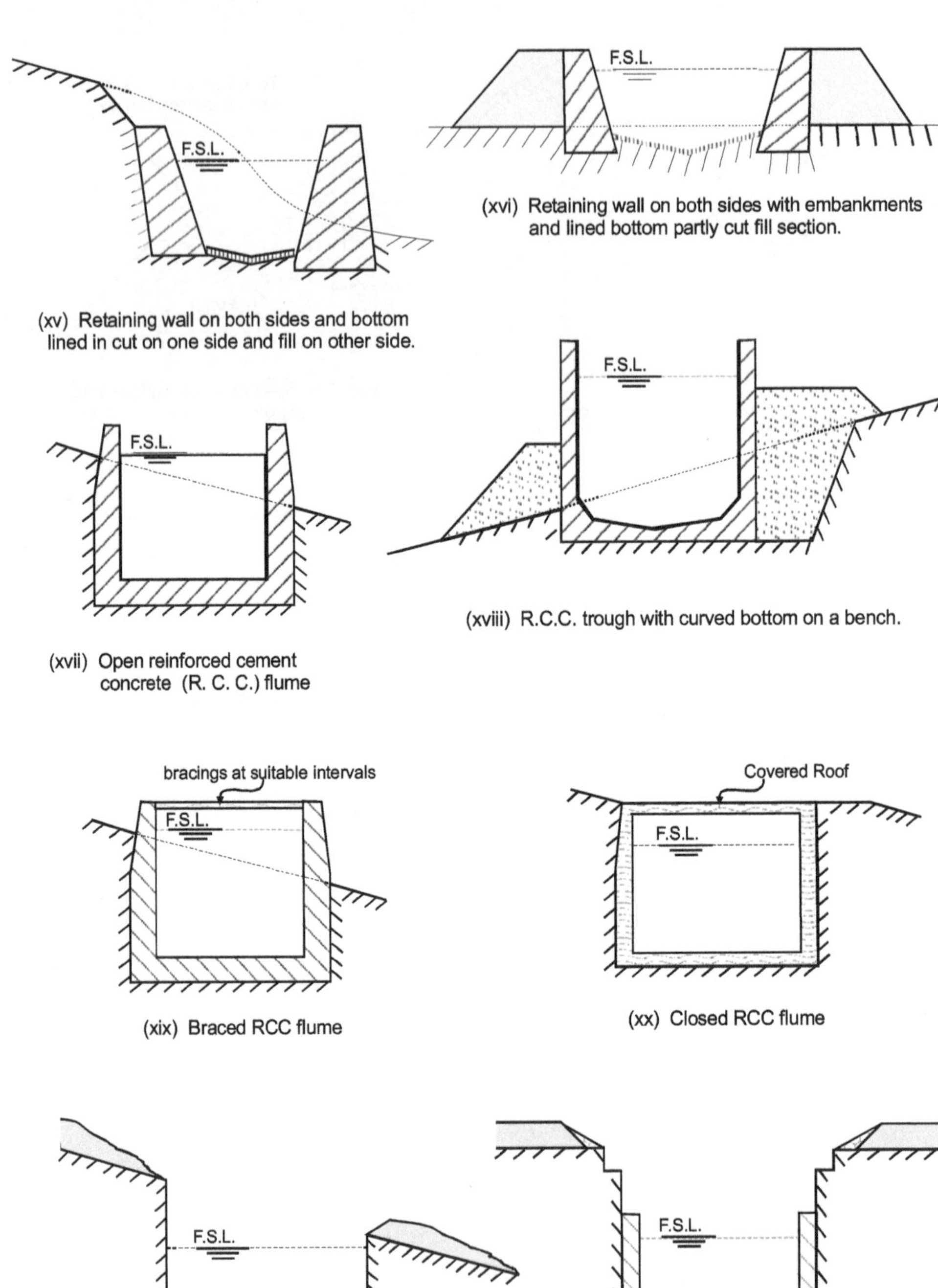

(xv) Retaining wall on both sides and bottom lined in cut on one side and fill on other side.

(xvi) Retaining wall on both sides with embankments and lined bottom partly cut fill section.

(xvii) Open reinforced cement concrete (R. C. C.) flume

(xviii) R.C.C. trough with curved bottom on a bench.

(xix) Braced RCC flume

(xx) Closed RCC flume

(xxi) Unlined cut in rock

(xxii) Lined in rock

Figure 9.4 : Types of canal sections according to topographical conditions contd...

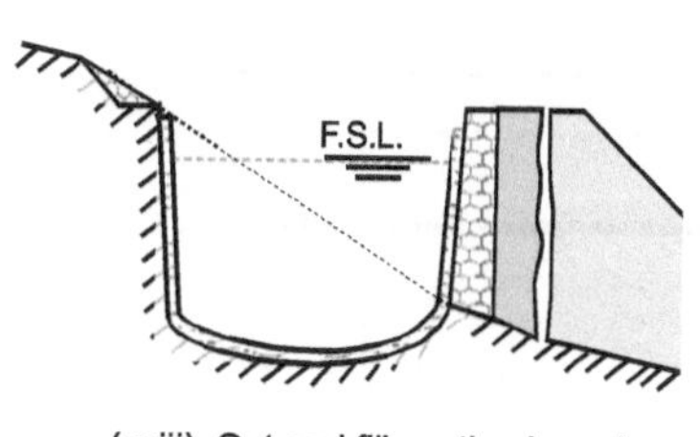

(xxiii) Cut and fill section in rock

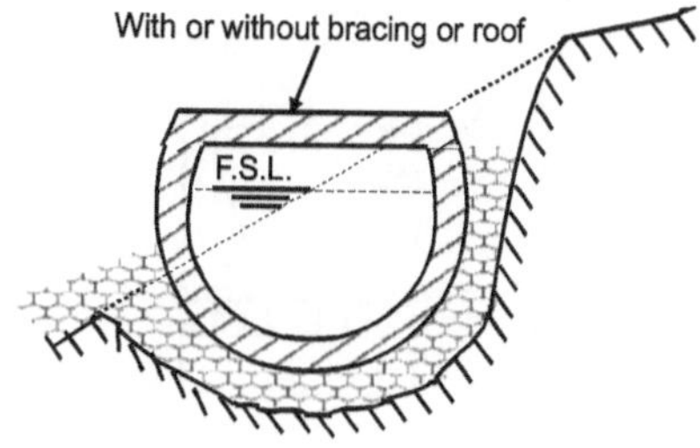

(xxiv) Semi circular RCC flume in rock

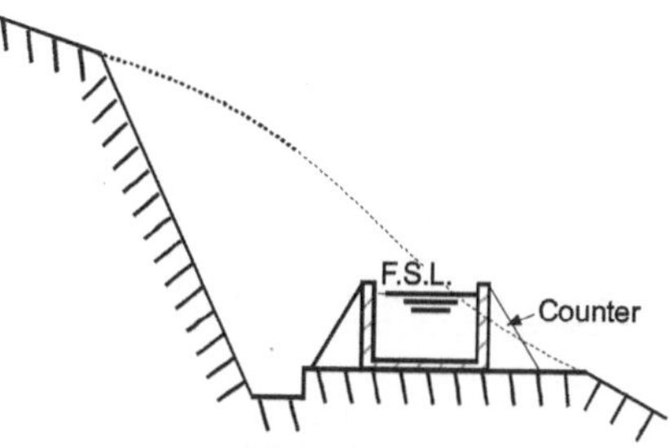

(xxv) Counter foot type RCC flume on rock bench

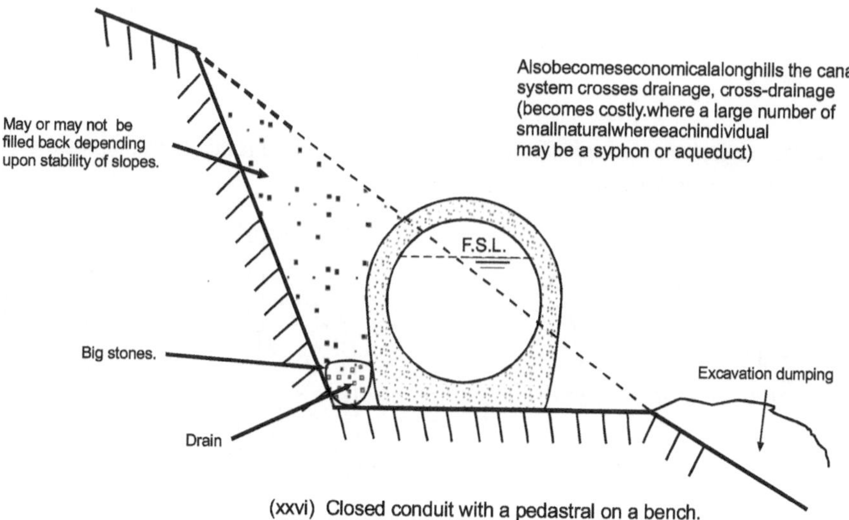

(xxvi) Closed conduit with a pedastral on a bench.

Figure 9.4 : Types of canal sections according to topographical conditions

Notes :

1. These sections shows only lined portion of cross section of canal and do not show complete section of canal, such as drains, dowel, service roads etc. A complete section depends upon discharge and site specific conditions, type of soil available, materials, a few more detailed section are shown in figure.
2. There can be several combination depending upon discharge, construction materials available, topography, type of soils and stability of slopes etc.
3. For detailed cross-section, see chapter 6 and Para 9.11

Brick / stone
masonry / RCC trough
Stone
slabs
or RCC
slab
G.L.
G.L.

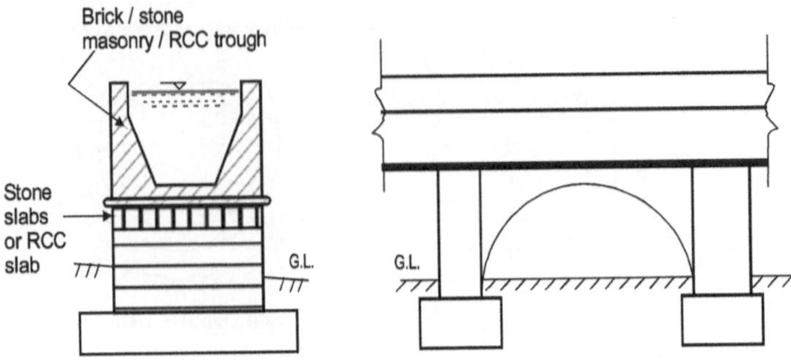

(i) Open flume supported on piers or arches.

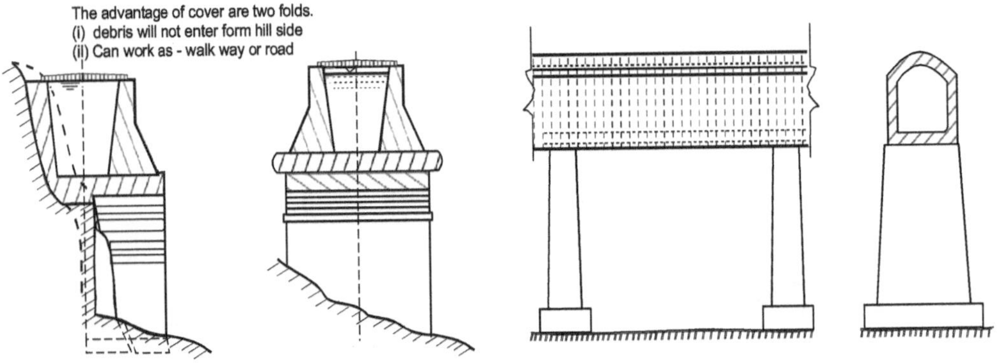

The advantage of cover are two folds.
(i) debris will not enter form hill side
(ii) Can work as - walk way or road

(ii) Covered concrete flume supported on piers and arches, along hill sides.

Closed concrete flume on piers

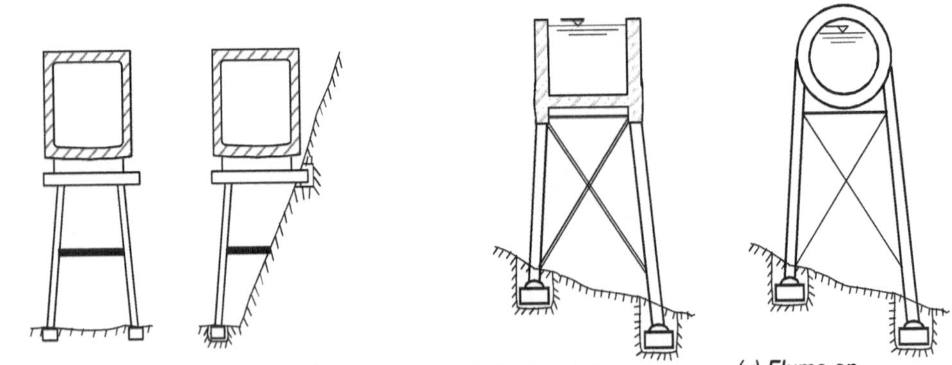

(iii) Closed reinforced-concrete flume on trestles / coloumns. (framed structures)

(iv) Open reinforced concrete flume on trestles / coloumns

(v) Flume on trestles / columns

Figure 9.5 Canal Flumes (or aqueducts) on supports.

Drainage

Efficient drainage of the sub-grade is just as essential as in lined alluvial canals. A common system is to provide a longidutinal drain at the base of one or both side slopes and to connect them at suitable intervals by transverse drains, which can discharge freely on the down hill side of the canal.

If the natural ground slope is steep the foundation should be benched as shown in figure 9.4 (viii).

The topographic conditions mainly the steepness of the proposed hillside greatly influence the type of cross-section to be adopted. On steep mountain sides the construction of fills or fill parts of cut-and-fill sections may be uneconomical or sometimes impossible. In such cases the valley side embankment need to secured to foundation by rock fills or retaining walls, and a wall may even be necessary to bound the wetted section. Further more it may sometimes be economical to apply flumes of reinforced concrete, the slenderness of which can be increased by tie beams connecting the upper edges of the structure. For typical sections see figure 9.4 (viii) to (xxvi).For embankment details see above para (e).

The bottom width, A-B in figure 9.4 (viii), is usually between 5 and 10 times the depth of water, depending on the nature of the ground.

On sloping terrain embankments should be constructed with staggered foundations having horizontal or a gentle counterslope contacting surfaces. The foundation should extend sufficiently deep through weathered soil and contact with sound rock.

Sometimes the sides of the canal are formed by permeable rock, while the bottom extends into the underlying impervious layer. Over such sections the lining of the slopes with proper key wall foundations maybe neccessery and sufficient. However, if the impermeable bottom rock is likely to loosen and to introduce therby sediment into the water, the bottom should be protected by a layer of gravel or crushed stone.

The lower part of cross sections extending into sound rock may be constructed with vertical side walls.

An embankment forming a side of a canal must be designed with the same care as any other embankment dam . The stability of the bank must be analysed by the more exact methods outlined as for an earth dam. For complete details of cross sections see chapter 6 and 8.

Sometimes a canal is placed wholly in filling to avoid a lengthy detour; for example parts of the Shannon headrace canal are in filling about 60 ft. deep, (see para 6.13).

Access road

It is highly desirable to provide an access road along any canal, so that points of incipient leakage or failulre are readily accessible and may be rapidly dealt with. The provision of a road will govern the top and bottom width in many cases. Many times a road may not be feasible on top. In such cases, a road along the bottom toe of filling reaches duly acessible is highly desirable.

Seriousness of bank failures (breaches)

The failure of a canal embankment is likely to be quite disastrous as that of an embankment dam particularly in high capacity canals. Though the discharge is limited to the capacity of the canal intake and the length of canal above the breach. Nevertheless such a failure can /will cause loss of life and serious damage to crops/land. Generally canal breaches near filling and heavy retrogation occurs in bed and banks causing repairs costly and time consuming. In a long canal it may be considered desirable to provide auto control spillways (like ungated weir) at suitable points to prevent the canal banks from being overtopped by abnormal flow, particularly if there is likelihood of storm-water entering the canal. Gates should be provided at the canal intake, and should be capable of rapid closure against full discharge. However rapid closures are not desirable. Also see para 9.7 and 9.8 for canals in rocks and along hill sides.

9.7 Canals and Tunnels in Rocks

The nature of the rock and its formation have a vital bearing on canal/tunnel in hills. The engineer should study the geology of the area through which the canal/tunnel is to pass. Expert geological opinon should be obtained at the project stage of any canal /tunnel in hilly areas.

When canals are excavated in rock, the sides are usually much rougher than in soft ground, even when they are carefully trimmed. These characteristics are, however, influenced by the nature of the rock; some types of sandstone, for example, can be accurately excavated to a relatively smooth surface without much trimming. On the other hand such rocks are sometimes very porous and may deteriorate rapidly under the action of weather and water. The risk of appreciable leakage may be more or less depending upon fracturing/joints/cleavages in rocks. Limestone is notoriously leaky, and canals excavated in this rock can be very treacherous.

The cross-section of a canal in rock is usually trapezoidal. In sound horizontally beded rock the sides may be vertical or nearly so. When the natural beds are inclined, as in figure 9.6, there is considerable risk of slips on one side of the canal, particularly if the beds contain seams of clay which become lubricated by the canal/rain water. The rock loads are so heavy that any kind of lining fails and the entire rock slips in to the canal. Such typical examples are canals of Som Kamla Amba Project, Rajasthan, India, where rock slopes blocked the entire canal section. In these conditions it may probably be better to adopt a circular or closed section. Rock can be cut and the closed section can be covered back with rocks. Sufficient drainage prefarbly within the canal should be provided to avoid any excessive water pressures and corosion of steel reinforcement, see figure 9.12.

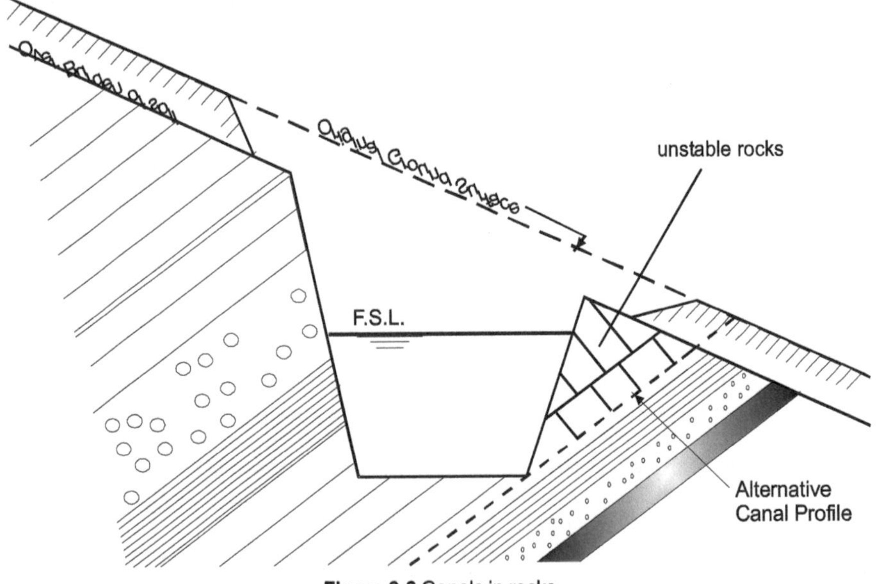

Figure 9.6 Canals in rocks

The hillock strata can be classified as :

(i) hard stable rocks

(ii) hard but unstable rocks due to dips/faults or shear zone or clay seems between the rocks.

(iii) Soft stable rocks

(iv) Soft unstable rocks

(v) Murrum

(vi) Soil mixed with murram, kankar, pebbles and boulders/stone.

(vii) Besides above hillocks may have loose soil cover of varying nature on different types of rocks (i) to (iv) above.

Therefore detailed geological investgations are very essential in laying canal alignment in/along hills and choosing a suitable cross-section.

The area and gradient are normally determined by economic considerations, which are sometimes greatly affected by the value of the frictional head-loss. The velocity must not, however, be excessive and Etcheverry (1915) suggests the following limiting velocities :

Conglomerates, Cemented gravel, Soft slate, tough hardpan and Soft sedimentary rock	2 to 2.5 m/s (6.0 to 8.0. ft/sec)
Hard rock	2.5 to 4.5 m/s (10.0 to 15. ft./ sec)

The value given for hard rock is possibly somewhat conservative. Also see Chapter 2 and table 7.8.

9.7.1 Tunnels

Tunnels become essential or are imperative when the alignment comes acoross long and wide mountaineous ridges and detouring is costly, see figure 9.8.

When the characteristics of the rock formation are fairly well known, it may be possible to evolve a section which is likely to fit the shape into which the rock will break naturally. Thus while a horse shoe or D-section with its semi-circular roof is fairly easy to obtain in some formations, there are others where the tunnel crown tends to break into a form more nearly square, and if there is no risk of heavy external pressure upon the lining, or if the tunnel is to be unlined, there is no reason why the designed cross-section of the tunnel should not be made to suit the characteristic of the rock. Recent Swedish tunnels, such as the Hjalta tunnel are of quasi-rectangular section.

A pear - shaped section, or "lazy D" section is sometimes adopted for steel - lined tunnels in order to give adequate working space for welding, lining and for filling behind it with concrete,see figure 9.7. e. These are generally for power tunnels.

Massive igneous rocks have given the least trouble, as exemplified by the tunnels of very large diameter which have been excavated, and left permanently unlined, in Sweden and United states. The stability of bedded rocks depends upon the thickness of the beds, their inclination to the tunnel axis, and upon the frequency of joints.

When the rock is of a sedimentary or metamorphosed type, its behaviour is greatly influenced by the thickness and inclination of beds to tunnel axis, as well as by the number and disposition of the joints. Horizontally bedded rock is liable to heavy falls if the joints run parallel to the tunnel axis, but can be surprisingly stable if they are normal to it, even in comparatively thinly bedded formations. Steeply dipping rocks generally give far less trouble when the tunnel axis is normal to the line of strike than when the two are parallel or nearly so, see figure 9.7.

Rock formations are not always constant in character for any considerable distance. They may for instance be heavily faulted or folded, and not only may the bedding planes vary widely in their inclination to the tunnel axis, but the tunnel may pass through a number of entirely different formations, each with its own particular risks.

Folding and faulting are themselves sources of weakness. As seen in Figure 9.7 (d) the fractures on the outside of folds may not only be sources of structural weakness, but may also lead to the inflow of water. Faults with their crush-zones can give rise to very serious difficulty, and considerable deviation of the tunnel may be justified in order to cross them at as steep an angle as possible and to restrict the trouble to the shortest possible length. Faults may also continue to be active and may cause fracture of the tunnel.

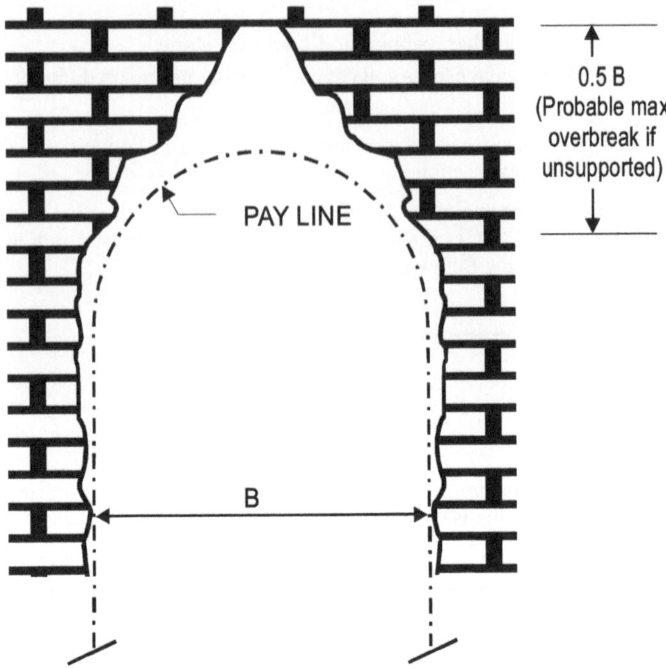

(a) Over break in Horizontally Stratified Rock

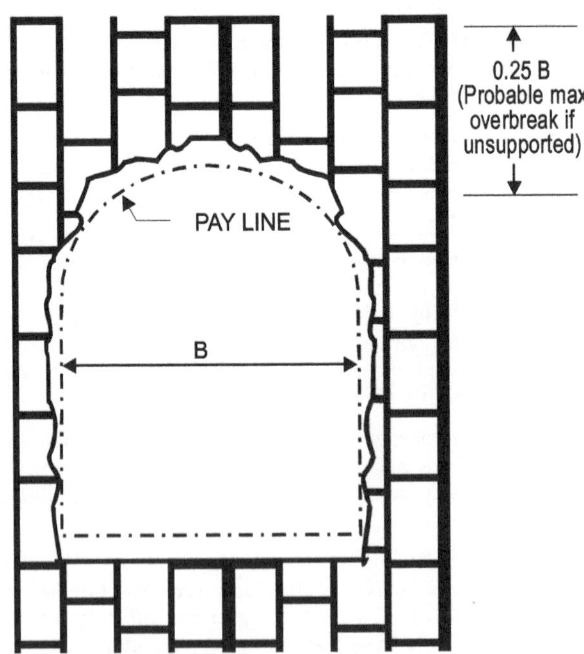

(b) Over break in Vertically Stratified Rock

Figure 9.7 Load Transference on Tunnels

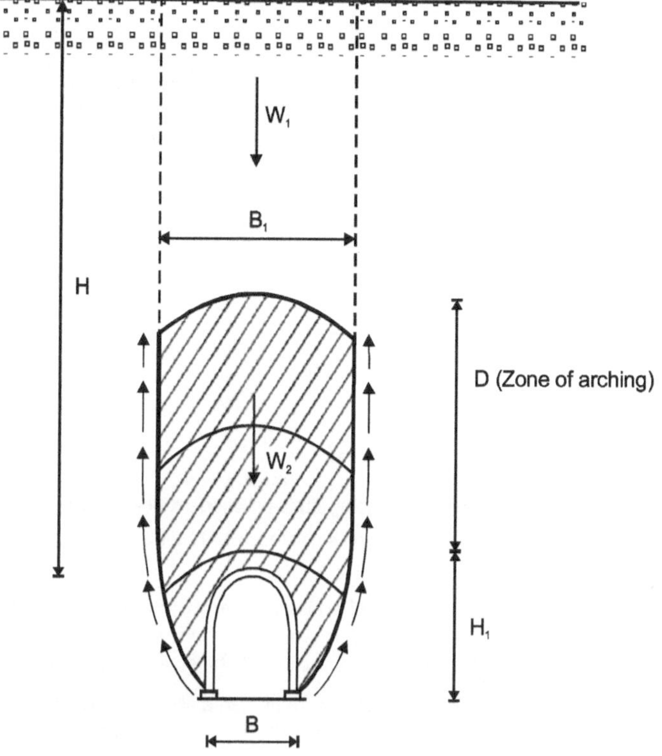

(c) Load transference tunnels in stratified/loose Fractured Rock

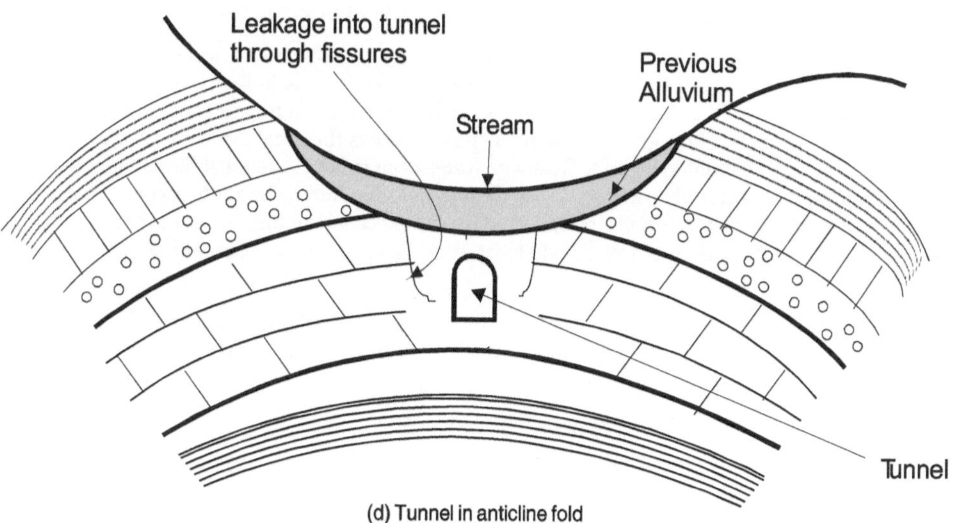

(d) Tunnel in anticline fold

Figure 9.7 Load Transference on Tunnels

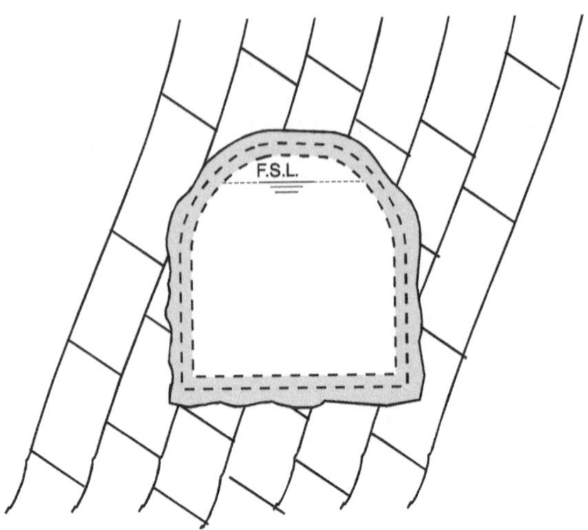

(e) Tunnel in steeply sloping strata (inclined bedding plane).

Figure 9.7 Load Transference on Tunnels

Quartzitic instrusions sometimes exist in igneous or metamorphosed rocks, and in cases of extreme toughness may slow up the drilling. Veins of pure quartz, when inclined to the axis of the drill hole, may cause deflection of the drill tip resulting in binding of the steel on further penetration.

In deep tunnels rock may be under high pressure and in a state of high internal stress; excavation of the tunnel bore upsets the stress equilibrium, with the result that fairly large pieces of rock may fly off.

9.8 Canals along Hill Sides

Over sufficiently uniform terrain the canal may be designed with an open cross section through cuts, over fills and in cut - and- fills as shown in figure 9.4. (i) to (iv).

But on hill slopes, and specially on steep mountain sides, the alignment closely follow the contour lines of the terrain. On rugged mountain slopes it may not be possible to follow the irregular contour lines : deep valleys are to be bridged by aqueducts (such as elevated flumes or canal bridges), and high ridges crossed by tunnels. Through moderately deep side valleys the canal can be located in a fill, while smaller ridges can be negotiated by cuts. Drainage water from side valleys must be conveyed under the canal through specially carefully designed cross-drainage works. Along deep cuts of considerable length in hill sides, the run - off water should be collected by special canals or flumes constructed at the uphill side of the canal. A typical layout of canals located on steep, rugged mountain slopes is shown in figure 9.8.

A canal located according to the above principles may involve the construction of relatively more expensive structures, such as bridges and tunnels, yet the resulting alignment may still be more economical than that strictly following the reliefs of the rugged terrain, because

1. The length of the canal may be significantly reduced.
2. The head loss will also be smaller, not only on account of the reduced length, but also by eliminating losses due to a great number of sharp, closely spaced bends.
3. Also operation and maintenance of such canals are less costly and they run trouble free. Otherwise they are often subject to choking by silt / over fall of rock burden.

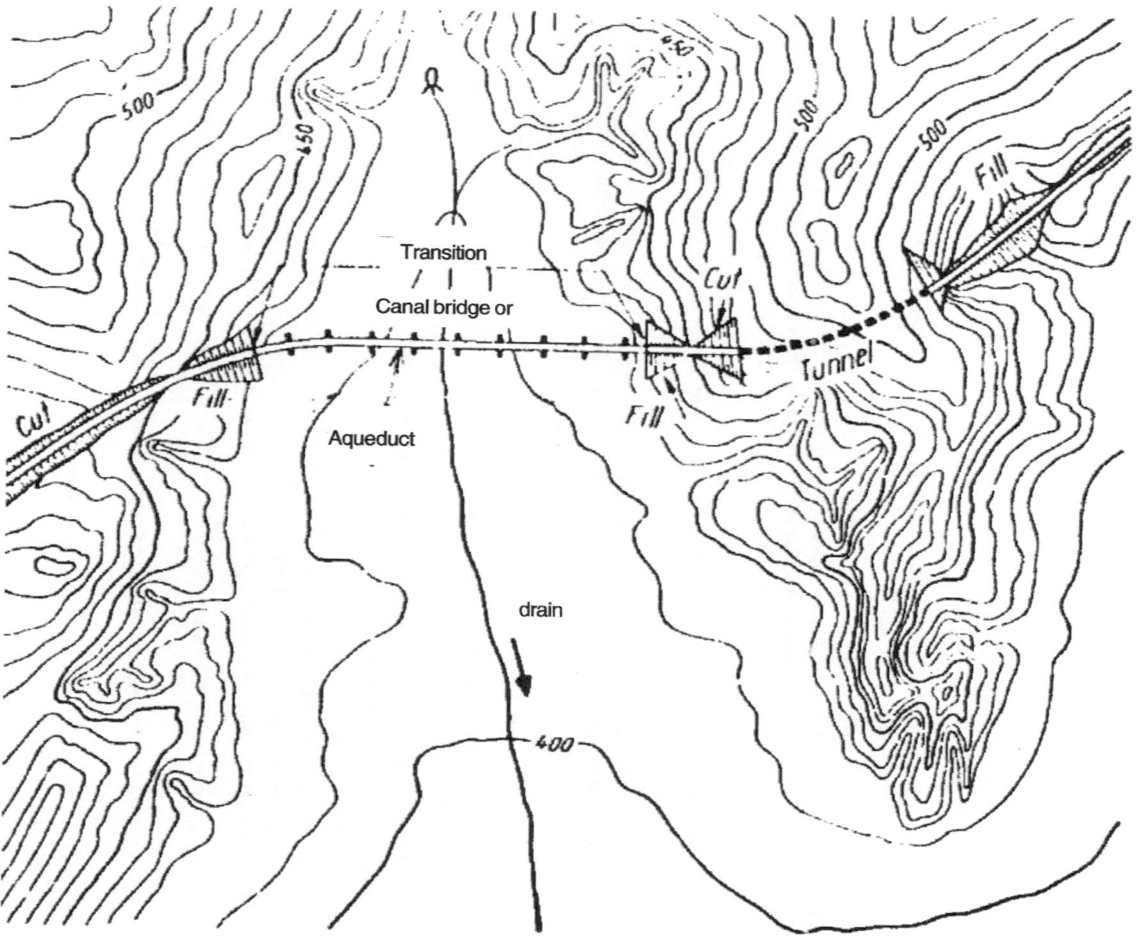

Figure 9.8 Alignment of a canal in rugged terrain, - tunnels, aqueducts, heavy cuts and fills.

The correct solution between free location and full adherence to the terrain should, of course, be determined by comparing the relative economics of several possible alternative solutions.

It should be kept in mind, that geological conditions of the terrain influence the location of the canal. In order to establish reliable bases for tracing the alignment and determining the cross sections, the geologic formation, the dip of layers, the quality of rock (degree of fissuring, permeability, strength, frost resistance, tendency to weathering, etc.) should be explored very thoroughly over sections in cut and in fill. The proper solution will be governed by

a. the permissible slope of banks and embankments,
b. the depth of the cut, respectively the height of the fill,
c. the dimensions and foundations of canal walls, backfills if required,
d. the extent and quality of the lining.

Tunnels provide protection against the formation of frazil ice, and in forest regions against great quantities of leaves, snow and stone avalanches.

On steep mountain sides, on terrain susceptible to sliding and even on gentle slopes the structural stability of the slope may be jeopardized by intensive earthworks. Therefore on such sites a very careful geological exploration is indispensable with special regard to the action of water leaking from the canals, which may further create sliding surfaces (in clay, shale rock and flysch).

Thickness and quality of lining in free - flow tunnels are governed by rock pressure, hydraulic requirements of water conveyance and watertightness and sometimes by the aggressivity of the waters conveyed . Internal water pressure generally need not be considered.

Adequate protection against freezing only, can be provided by a closed canal, or even by a simple cover over a canal of conventional type. These kinds of canals are also called bench flumes. Examples of closed canal types are shown in figure 9.1. Canal sections crossing urban settlements are frequently constructed with a closed cross section in order to occupy less ground area, to provide protection against pollution and to meet safety requirements.

If the cover is designed as a monolithic structure with the side walls and the foundation, the canal is of the closed type.

The construction of open canals generally meet difficulties when

1. the mountain slope is not stable,
2. the mountain slope is too steep,
3. the mountain slope above the canal is likely to produce much rubble,
4. snow avalanches are to be expected, or
5. in extremely severe and long winter periods the water freezes in the canal.

Problems 1 to 3 can be seen on many canals in unstable hill areas and 4 to 5 are only related to cold climates

Above conditions may occur separately or simultaneously. In such reaches following solutions can be applied.

a) Types of open or closed flumes as shown in figure 9.5 may be constructed of steel, stone, concrete or reinforced concrete.

b) The water may be conveyed in tunnels , A few types of cross section of tunnel are shown in figure 9.4 and 9.7. Also see chapter 5. In order to reduce head losses, such tunnels should be designed for free - surface flow conditions. Free - flow tunnels can be protected against over pressure caused by occasional surges in the power canal by installing spillways (waste ways) in the open channel at suitable locations. Excess water may be most conveniently disposed at the canal crossing a natural rivulet, (nallah).

c) Alternatively a cut and cover solution can be adopted, i.e. a closed canal section is built adjoining cutting and next section is excavated. Excavations are dumped on the first section and canal section built, and so on. Sufficient drainage is provided either to fall within canal or led away to adjoining drain. Drainage is essential to avoid heavy water pressure/loads and avoid corrosion of steel.

Dealing with Ice

In very cold climates, allowance must be made for the effect of ice, which may take the form of surface ice, anchor ice or *frazil ice.*

Anchor ice consists of a mass of coarse ice crystals clinging to the channel bed. It occurs only with a clear sky, and when the surface ice is transparent or non-existent. It is often held to be caused by the loss of heat resulting from radiation, though more recent investigation tends to disprove this theory.

9.9 Selection of Inner Side Slopes

In addition to the points discussed in para 3.3, with the flatter side slopes, the lining area of free board portion is more than that with steeper side slopes and this also increases the cost of lining, top width (i.e. cost of earth work) and land width (cost of land). However in some cases free board may be provided in the form of vertical walls/dowel, see chapter 6, and 8.

One other disadvantage with flatter slopes is that the weight of lining and the weight of water impounded on it, acts more on the side slopes causing more stresses and lining act as a slab/beam on a flexible earthen strata. Any weakness in the earthen embankment may result in hair cracks in lining, as the lining may not be that strong enough to resist bending moments. These stresses are much less in case of steeper side slopes, On the contrary, earth and back water pressure may be little more on steeper slopes than an flatter slopes. Some times, back water pressure causes severe cracks and slipping of linning in steeper slopes, as the counter weight of lining on soil is less, Thirdly flatter side slopes encourages silt deposition and weed growth below and above free board. Weeds are highly deterimental to lining, roots of which penetrate into lining, see para 9.5, 9.15.2 and experiences of lined canals in chapter 6.

Thin hard surface lining should rest on stable side slopes of natural soil. Slopes should be such that no earth pressure is exerted on the back of lining. Therefroe a very judicious judgement is needed.

General Practices

The general practices adopted by various design organisations and given by many authors in their text books, are given in table 9.1 for different types of soils, rocks and depth of flow. In some exceptional case slopes may be up to 1V :3H.

Table 9.1 General practices of inner side slopes

S. No. (1)	Type of soil (2)	Side slopes (H: V) (3)
1.	Very light loose sand to average sandy soil.	2:1 to 3:1
2.	Sandy loam	1.5:1 to 2:1 (in cutting) 2:1 (in embankment)
3.	Sandy gravel / murrum	1.5:1 (in cutting) 1.5:1 to 2:1 (in embankment)
4.	Black cotton	1.5:1 to 2.5:1 (in cutting) 2.1 to 3.5:1 (in embankment)
5.	Clayey soil	1.5:1 to 2.5:1 (in cutting) 1.5:1 to 2.5:1 (in embankment)
6.	Rock	0.25:1 to 0.5:1

Note: The above slopes are generally recommended for depth of cuting/height of embankment upto 6 m. For depth/height in excess of the above, special studies for the stability of slopes are required.

The above gudelines are very general and should not be followed blindly, see chapter 8 and para 8.1.6 and para 9.6.

Bhakra Canal Manual (1954) recommends the side slopes, for channels in its command area as under:

"The side slopes may be kept as 1:1 for radii less then 12 ft. (approx 4.0 m) and 1.25:1 for radii over 12 ft. (> 4 m), provided, the angle of repose of the soil is not flatter than 1.5:1. Here radii means depth of flow as the recommendations are for Mehboob section, see figure 3.1.

These guide lines gives more efficient hydraulic and over all economic section then the above general guide lines. Also see para 6.3.

Stable side-slopes depend mainly on the natural stability of the saturated material in relation to the surrounding water table, including its variation and the extent and rate of variation in the canal water-level. Where chances of sudden draw down in the canal are considerable, proper slip circle analysis should be made for stable side slopes, see chapter 8.

The various modes of failures of slopes, and stability analysis are also discussed in chapter 8.

Some weak fractured and jointed rocks which appear very stable, become unstable in saturated conditions. In most such cases, no stable slopes may be feasible. Rock loads are very heavy and fail many strong linings. In such cases a cut and cover solution may be better, see figure 9.12.

For Rocks, geological mapping should be carried out and dips and likely slopes of weak rocks should be established, before adopting above side slopes, see figures in chapter 8. – and figure 9.6.

It is also proposed that for black cotton and clayey soils, c and f values for saturated conditions should be determined and stability of slopes analysed for sudden draw down conditions. The various modes of failures of slopes, stablity analysis are discussed in chapter 8.

Details of sides slopes adopted in practice on a number of projects is given in table 6.1 to 6.3.

This table shows that it has so far been a matter of individual experience and judgment.

Therefore a judicious judgement is always required in the design of inner side slopes of lined canals, giving due care to the stability of soils, rock for various depths, and back water pressure, and low silting on sides.

The inner side slopes of lined canals should be designed/chosen from structural, hydraulics and economical point of view so as to have maximum stability of slopes, minimum silting, weed growth and minimum perimeter. A little flatter slope may be more desirable from stability point of view of earthen embankments, but is not desirable from silting and economic point of view. A big controversy in lined canals. With flatter slopes lined canals carrying silt water starts behaving as unlined canals with parabolic shape with steep side slopes as observed by Kennedy and Lacey. (para 7.12 and 7.13).

Generally similar side slopes are adopted on both sides, but it is not nesessary and many times it may not be economical/possible in hilly regions, see figure 9.10.

Generally inner side slopes may be preferred as steepest as possible, preferably 1:1 (V:H) for depth upto 2.0 m, 1:1.25 (V:H) to 1:1.5 (V:H) for depth between 2 to 4.0 m. for soils and 1: 0.5 or 1: 0.6 for rocks, with due regard to discussions in para 9.7. Also see chapter 8.

9.10 Proposed Internal or Hydraulic Canal Sections

With lot of theoretical discussions regarding the best hydraulic section for minimum perimeter in Chapter-1 and practical discussion as above regarding (i) practical limitations of velocities (permissible velocity), (ii) limitations of bed slopes, (iii) limitations of depth of flow and consequently smaller velocities, (iv) feasibility of side slopes, according to type of soil and depth, stability of banks, (v) effect of side slopes on hydraulic efficiency (vi) Surface roughness N, (vii) topography such as plain, alluvial soils, undulating topography of sand dunes (typical Indian Thar desert), rocky strata, uneven topography, sloping ground, along hill sides etc. through which a canal alignment may be passing, cost economics restricts the adoption of a few suitable economical sections.

Suitable economical canal sections for different site conditions, different discharges, based upon the topography and many other conditions as discussed above are proposed in table - 9.2.

In the proposed sections due care has been incorporated for efficient operation and maintenance.

Table 9.2 : Proposed lined canal sections according to discharge, type of soil and topography

S.N 1	Type of soil and topography 2	Suitability/economical lined canal section 3
I. Discharge from 1 to 15 cusecs (.03 to 0.5 cumecs approx.) These are generally watercourses and minors.		
(A)	Coarse/fine sands/gravel etc. (GP, GM, GW, GC, SC, SP, SW, ML, MH, etc)	
	(i) Partial cutting/filling	Trapezoidal with inner side slope of 1:1 or Rectangular with or without curved bottom saves cost of land, or Semi circular precast cement concrete of high quality preferably 5 to 8 cm thick, saves land and are economical.
	(ii) Just cutting / filling	-do-
	(iii) Heavy cutting	Circular or closed rectangular with curved bottom [open channel get choked, besides heavy cost of earth work.
	(iv) Heavy filling	Circular, semicircular or rectangular, or trapezoidal with steep inner side slope of 1:1 (V:H) or aqueducts/flumes on Pedestals or prefer Hume pipe siphons with manhole (vertical wells) as these are more better and requires less maintenance.
	(v) Sloping ground (Contour Canal along Hills) (vi) Undulating topography (sand dunes and ravines)	Circular or closed rectangular with curved bottom [open channel get choked, besides heavy embankment on one side] Closed section under pressure
(B)	Clay, (CL, CH)	
	(i) Partial cutting/filling	Only circular sections (Hume pipe preferable), can with stand pressures of clay.
	(ii) Just cutting / filling	-do-
	(iii) Heavy cutting	-do-
	(iv) Heavy filling	-do-
	(v) Sloping ground	-do-

Contd...

Table 9.2. Contd...

(C)	Organic silts and silt clays with low/high plasticity.	
	(i) Partial cutting/filling	No lining is proposed as these are subject to heavy settlement and cracks. However pipes may be used with flexible joints only
	(ii) Just cutting / filling	**-do-**
	(iii) Heavy cutting	-do-
	(iv) Heavy filling	-do-
	(v) Sloping ground	-do-
(D)	Hillocks/rock cutting only little soil, or soil mixed with kankar, pebbles etc. available for embankment	
	(i) Partial cutting filling	Trapezoidal with inner side slope of >1:1 (V:H)or Rectangular with or without curved bottom
	(ii) Just cutting	-do-
	(iii) Heavy cutting	Circular or closed rectan gular, with or without curved bottom. Excavation filled back on roof, i.e. cut and cover section. Suitable drainage of back fill is essential to avoid any heavy pressures and corrosion of steel reinforcement. Vertical wells are provided at suitable interva l for cleaning. Section is free from day to day choking besides saving in heavy cost of excavation.
	(iv) Heavy filling	Canal aqueduct/flume on pedestals or siphons or a combination of above.
	(v) Sloping ground (around hills)	Rectangular with or without curved bottom or circular closed canal preferable depending upon the steepness of slopes and their stability, as in (iii) above.
	(vi) undulating topography (sand dunes and ravines)	Closed section under pressure only if possble or combinations of all above, see figure 9.4 (vii)-(xxvi) and – 9.6, to 9.11.
Various types of linings, shapes such as rectangular, circular, semi circular or trapezoidal, details of embankments / cuttings has been discussed by Bairathi (1982)		

Contd...

Table 9.2. Contd...

II.	Discharge from 15 to 100 cusecs (0.5 to 3 cumecs approx.) These are generally minors and distributaries	
(A)	Coarse/ fine sands/gravel etc. (GP, GM, GW, GC,SC, SP, SW, ML, MH,etc.)	
	(i) Partial cutting/filling	Mehboob or best trapezoidal sections with Z upto 1:1 (V:H) or Rectangular with curved bottom in exceptional cases.
	(ii) Just cutting	-do-
	(iii) Heavy cutting	-do- or circular or rectangular with curved bottom as a closed canal
	(iv) Heavy filling	-do- or aqueducts or siphons.
	(v) Sloping ground	One bank canals, preferably rectangular on one side and trapezoidal on other or closed canal or circular section on a bench. The later reduces maintenance cost.
	(vi) undulating topography (sand dunes and ravines)	Closed section under pressure
(B)	Clay (CL, CH.)	
	(i) Partial cutting/filling	Only circular sections, (Hume pipe preferable) can with stand pressures of clay.
	(ii) Just cutting	-do-
	(iii) Heavy cutting	-do-
	(iv) Heavy filling	-do-
	(v) Sloping ground	-do-
(C)	Organic silts and silt clays with low/high plasticity.	
	(i) Partial cutting/filling	No lining proposed as these are subject to heavy settlement and cracks. However only pipes may be used with flexible joints. However flexible lining or stone /brick pitching may be considered, if felt very necessary
	(ii) Just cutting / filling	-do-

Contd...

Table 9.2. Contd...

	(iii) Heavy cutting	-do-
	(iv) Heavy filling	-do-
	(v) Sloping ground	-do-
(D)	Hillocks/rock cutting only little soil, or soil mixed with kankar, pebbles etc. available for embankment	
	(i) Partial cutting filling	Trapezoidal or Rectangular with or without curved bottom
	(ii) Just cutting	-do-
	(iii) Heavy cutting	Circular or closed rectangular with curved bottom. Excavation filled back on roof. Suitable drainage of back fill is essential to avoid any heavy pressures and corrosion of steel reinforcement. Vertical wells are provided at suitable interval for cleaning. Section is free from day to day choking besides saving in heavy cost of excavation.
	(iv) Heavy filling	Canal aqueduct / flume on pedestals or siphons or a combination of above.
	(v) Sloping ground (around hills) (vi) Undulating topography (sand dunes and ravines)	Rectangular with or without curved bottom or circular / closed canal preferable depending upon the steepness of slopes and their stability, as in (iii) above Closed section under pressure
III. Discharge from 200 to 3000 cusecs (3.0 to 100 cumecs approx.) These are generally distributaries and branches.		
(A)	Coarse/fine sands/gravel etc (GP, GM, GW, GC, SC, SP, SW, ML, MH, etc.)	
	(i) Partial cutting/filling	Mehboob or Best trapezoidal section with inner side slopes of 1:1 to 1.25 depending upon depth and type of soil.
	(ii) Just cutting (iii) Heavy cutting	-do- Circular or closed canal section with cut and cover or tunnel, may be under pressure with some head loss.
	(iii) Heavy filling	Mehboob or Best trapezoidal section with inner side slopes of 1:1 to 1.25 depending upon depth and type of soil. Alter natively siphons or flumes

Contd...

Table 9.2. Contd...

	(v) Sloping ground	One bank canals with different side slopes on either side, preferably rectangular on one side and trapezoidal on other or circular section or closed canal on a bench with cut and cover.
	(vi) undulating topography (sand dunes and ravines)	Closed section under pressure
(B)	**Clay (CL, CH.)**	
	(i) Partial cutting/filling	Only circular sections can with stand pressures of clay. Preferably no lining proposed or flexible lining or stone pitching if felt very necessary.
	(ii) Just cutting	-do-
	(iii) Heavy cutting	-do-
	(iv) Heavy filling	-do-
	(v) Sloping ground	-do-
(C)	Organic silts and silt clays with low/high plasticity.	
	(i) Partial cutting/filling	No lining proposed as these are subject to heavy settlement and cracks. However for smaller discharges only pipes may be used with flexible joints
	(ii) Just cutting / filling	-do-
	(iii) Heavy cutting	-do-
	(iv) Heavy filling	-do-
	(v) Sloping ground	-do-
(D)	Hillocks/rock cutting only little soil, or soil mixed with kankar, pebbles etc. available for embankment	
	(i) Partial cutting filling	Mehboob or Best trapezoidal section with inner side slopes of 1:1 to 1.25 depending upon depth and type of soil.
	(ii) Just cutting	-do-

Contd...

Table 9.2. Contd...

	(iii) Heavy cutting	Tunnels - circular or horse shoe or closed rectangular with curved bottom. Excavation filled back on roof. Suitable drainage of back fill is essential to avoid any heavy pressures and corrosion of steel reinforcement. Vertical wells are provided at suitable interval for cleaning. This section is free from day to day choking besides saving in heavy cost of excavation.
	(iv) Heavy filling	Canal aqueduct/flume on pedestals or siphons or a combination of above.
	(v) Sloping ground (around hills)	One bank canals with different side slopes on either side, preferably rectangular on one side and trapezoidal on other or circular section or closed canal on a bench with cut and cover.
	(vi) Undulating topography (sand dunes and ravines)	Closed section under pressure
IV. Discharge above 3000 cusecs (100 cumecs approx.) These are generally major branches/ main canals		
(A)	Coarse/fine sands/gravel etc.(GP, GM, GW, GC, SC, SP, SW, ML, MH, etc.)	
	(i) Partial cutting/filling	Trapezoidal with curved ends, bed width may be more than depth, inner side slopes according to stability of slopes.
	(ii) Just cutting	-do-
	(iii) Heavy cutting	-do-
	(v) Sloping ground	One bank canals with different side slopes on either side, preferably rectangular on one side and trapezoidal on other or circular section or closed canal on a bench with cut and cover. Also see table
(B)	Clay (CL, CH.)	
	(i) Partial cutting/filling	Preferably no lining proposed or flexible lining or stone pitching if felt very necessary.
	(ii) Just cutting	-do-

Contd...

Table 9.2. Contd...

	(iii) Heavy cutting	-do-
	(iv) Heavy filling	-do-
	(v) Sloping ground	-do- One bank canals with different side slopes on either side.
(C)	Organic silts and silt clays with low/high plasticity.	
	(i) Partial cutting/filling	No lining proposed as these are subject to heavy settlement and cracks.
	(ii) Just cutting / filling	-do-
	(iii) Heavy cutting	-do-
	(iv) Heavy filling	-do-
	(v) Sloping ground	-do-
(D)	Hillocks/rock cutting only little soil, or soil mixed with kankar, pebbles etc. available for embankment	
	(i) Partial cutting filling	Trapezoidal with curved ends, bed width may be more than depth, inner side slopes according to stability of slopes.
	(ii) Just cutting	-do-
	(iii) Heavy cutting	-do-
	(iv) Heavy filling	-do- If suitable soils are not available canal banks must be designed as a rock fill dam with core wall.
	(v) Sloping ground (around hills) (vi) Undulating topography (sand dunes and ravines)	One bank canals with different side slopes on either side, preferably rectangular on one side and trapezoidal on other. Closed section under pressure Combination of all above see figure – 96, para 9.11

9.11 Proposed Typical Over all Cross Sections Including Service Road, Drainage etc.

The typical cross sections for lined canals given by CWC Manual (1984, CBIP, Publication No. 191) are given in figure 9.9 and those given by IS – 104300 – 2000 are given figure 9.10 These are of general nature and more suited to normal alluvial soils and flat topography. In actual practice there will be considerable difference, depending upon topography, strata, availability of soils etc. as per para 9.10. Also a zoned section may have to be preferred in deep fillings and soft foundations, as adopted in Narmada canals, as shown in figure 6.32.

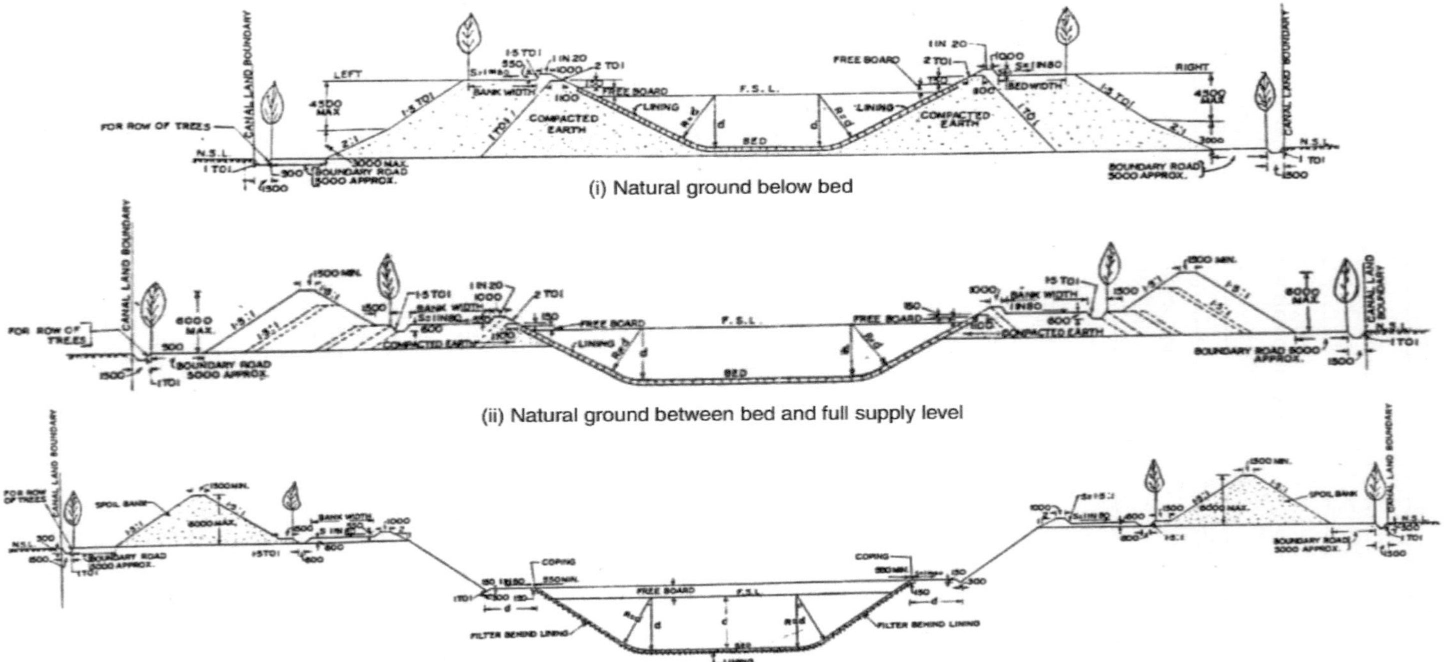

(i) Natural ground below bed

(ii) Natural ground between bed and full supply level

(iii) Natural ground is above top of lining

Figure 9.9 Typical cross sections for lined canals (CWC/CBIP-1984)

Notes

1. These sections are similar to the section adopted by IS 10430-1982, see figure-9.10

   Comments – proposed cautions for use:
2. These sections are suitable for
   (i) Large discharge (ii) ordinary soils, such as SC/SM/SW
3. These sections with modifications, for masonry dowel have been adopted in IGMN and its many branches and for lining in existing canals. IGMN has failed at many places, see chapter 8
4. For heavy cutting and filling reaches due care of stability of slopes is required.
5. There is no mention about type of soil and foundations/ (natural soil on which embankments are placed)
6. Zoned section with adequate foundation should be adopted for high capacity or high embankment canals, as shown in figures in chapter 6.
7. As regards trees, see the notes to figure 9.11.

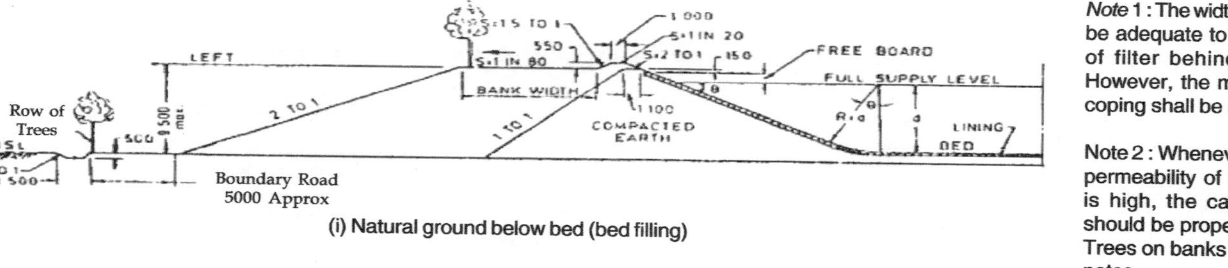

(i) Natural ground below bed (bed filling)

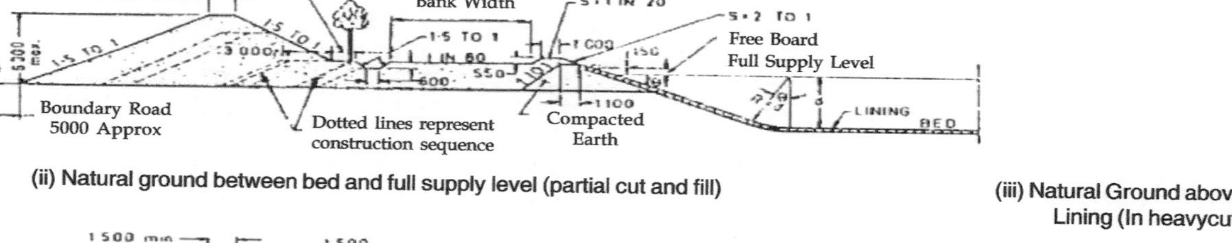

(ii) Natural ground between bed and full supply level (partial cut and fill)

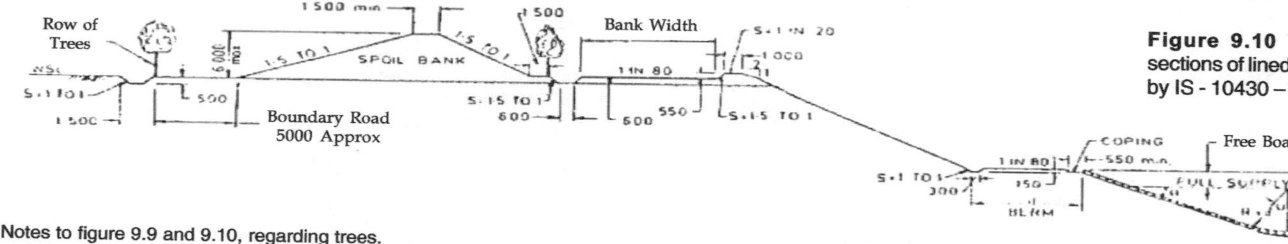

(iii) Natural Ground above top of Lining (In heavycut)

Note 1 : The width of coping should be adequate to cover the layers of filter behind lining, if any. However, the minimum width of coping shall be 550 mm.

Note 2 : Whenever the permeability of back fill material is high, the catch water drain should be properly pitched. 1/3 : Trees on banks are harmful, see notes.

Figure 9.10 Typical cross-sections of lined canals (adopted by IS - 10430 – 1982)

Notes to figure 9.9 and 9.10, regarding trees.

(i) That there is a row of trees on the banks of lined canal in figure 9.9 and 9.10. But trees removed from IS 10430-2000. Trees on banks of lined canal are undesirable as roots penetrate behind lining and damage the lining. They should preferably be out side the Toe of bank of lined canal.

(ii) After realizing the great damage by the roots of the trees on the banks and near the structures, on Indira Gandhi Canal project, it was decided in July 2006, that no plantation on canal banks and near the structures should be done. Plantation at a distance of 30 m from toe of bank of canal may be done. This may be still partly complete as there is no decision about trees in the cutting reaches of the canal. Of course on the bank, there cannot be any trees. In the opinion of this author, trees may be outside the spoil bank. However, bushes, and grass may be permitted on the inner slopes of cutting reaches, as well as on the outer slopes of the filling embankments. This will reduce the soil erosion by rain water.

However trees may be permitted on the banks of unlined canal.

Above sections with judicious modifications as per the discussion in chapter 6 and 8, and Para 9.6, and table 9.2 may be adopted.

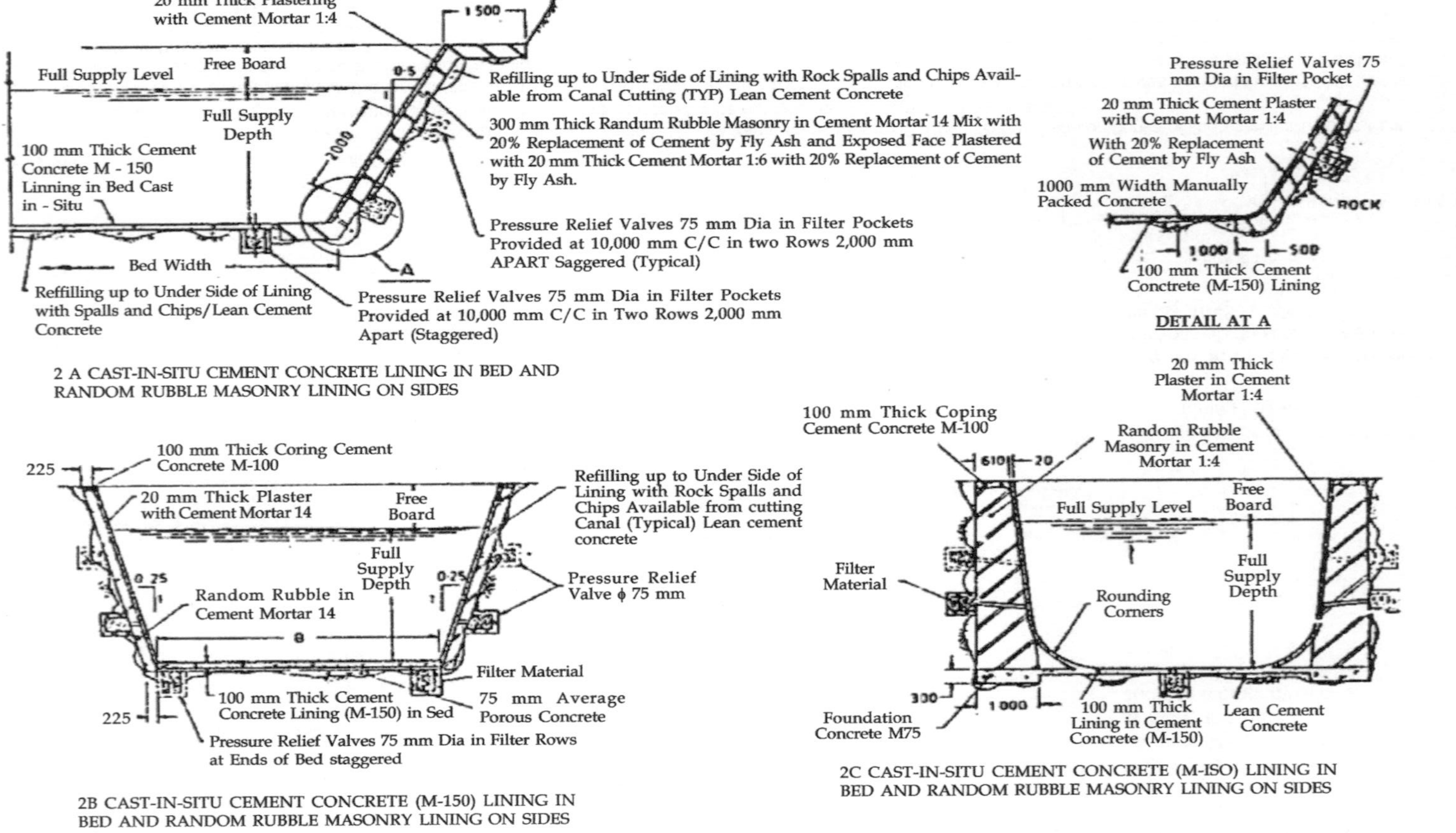

Note 1 : In case of sound rock, cement concrete lining may be provided in bed and sides. On sides, the lining may be anchored sitably to the rock behind

Note 2 : All the dimensions are in millimeters unless otherwise specified.

Figure 9.11 Typical cross-sections of lined canals in rock cutting (adopted in IS 10430 -1982)

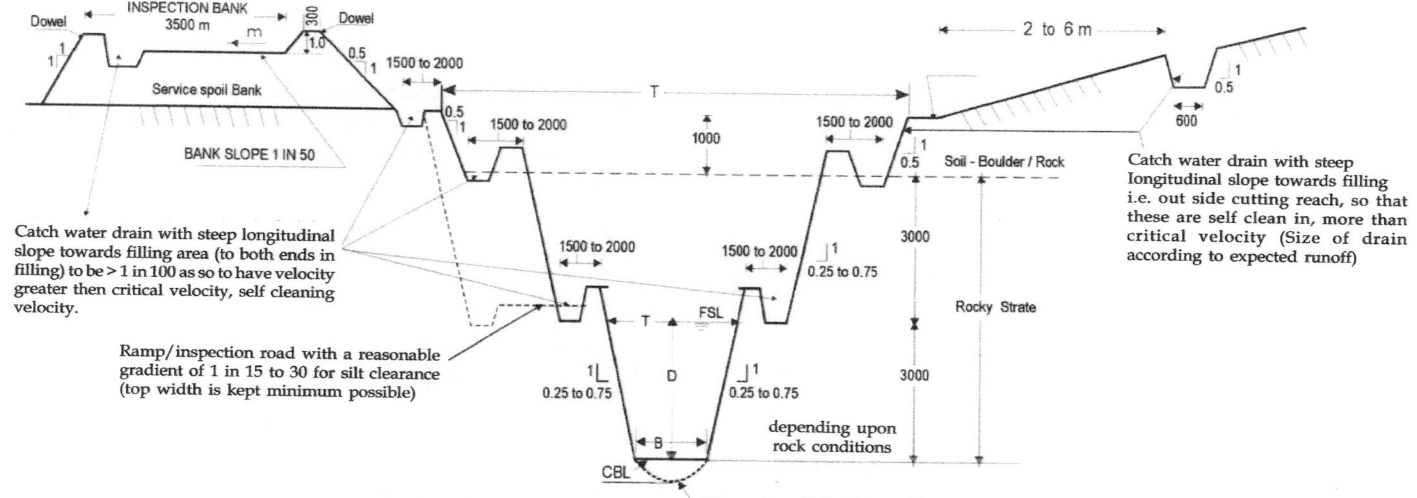

Notes : 1. Service road can be on any side of canal depending upn topography

2. All dimensions are in mm unless otherwise specified.

3. Discharge range : Medium and high capacity canals, see table 9.3

Figure 9.12 (i) Proposed Typical Cross Section for canals in Heavy cutting in stable Rocks / Bedding Planes, Joints, Faults etc.

Notes to figure 9.12 (i)

1. All earthen cuttings are to be utilized in filling reaches to the extent of economical leads and also to increase the bank width.
2. Stones to be utilized in masonry works as for as possible.
3. Maximum cutting up to 7.0 m, Beyond 7.0 m, alternatives of a tunnel or cut and cover section should be worked out, see figure 9.15(vi).
4. In L – section, left, centre and right levels are to be shown, so as to indicate sloping nature of ground.
5. Typical cross section of maximum cutting to be prepared and got approved, from the next higher authority, then approving the L-section.
6. Suitable ramps with gradients of 1 in 10 to in 20 to be provided in cuttings at intervals of 1km. or so. At least one ramp in each cutting to a minimum road width of 3.5 m, is essential or any one side, for ease of construction and future cleaning and be shown in L - section and plan of canal. This will also work as a inspection road to canal, as it is difficult inspect canal from top service road.
7. For such canals NSL should be taken at every 15 m interval longitudinally and at 2 m interval on either side of centre line in transverse direction. Before finalization of L – section. Bore hole investigation should be done at least at 1 Km. interval, or at closer intervals as per the advice of geologist. Geological mapping of soil / rock should be done, and stabilities of slopes be established.
8. Spoil banks should not be continuous. There should be a gap of at least 10 m after every 300 - 500 m interval.
9. This section is costly for discharges up to 3.0 cumecs and lined cut and cover section may be economical, see figure 9.15 (v) for alter native section.

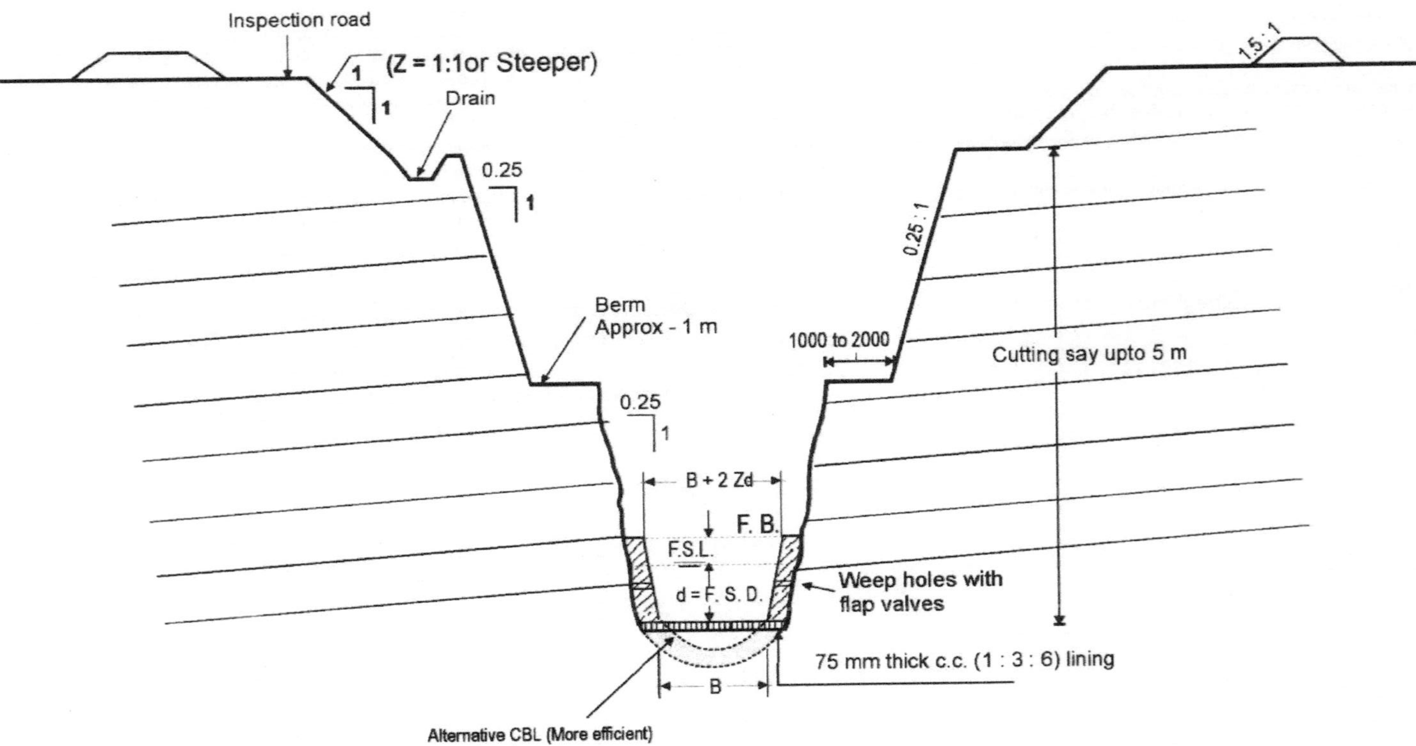

Notes : 1. All dimensions are in mm unless otherwise specified.

2. Discharge range : Medium, see table 9.3

Figure 9.12 (ii) Typical cross-sections fro canals – stone masonry side walls for lining in deep cuttings, in stable rocks.

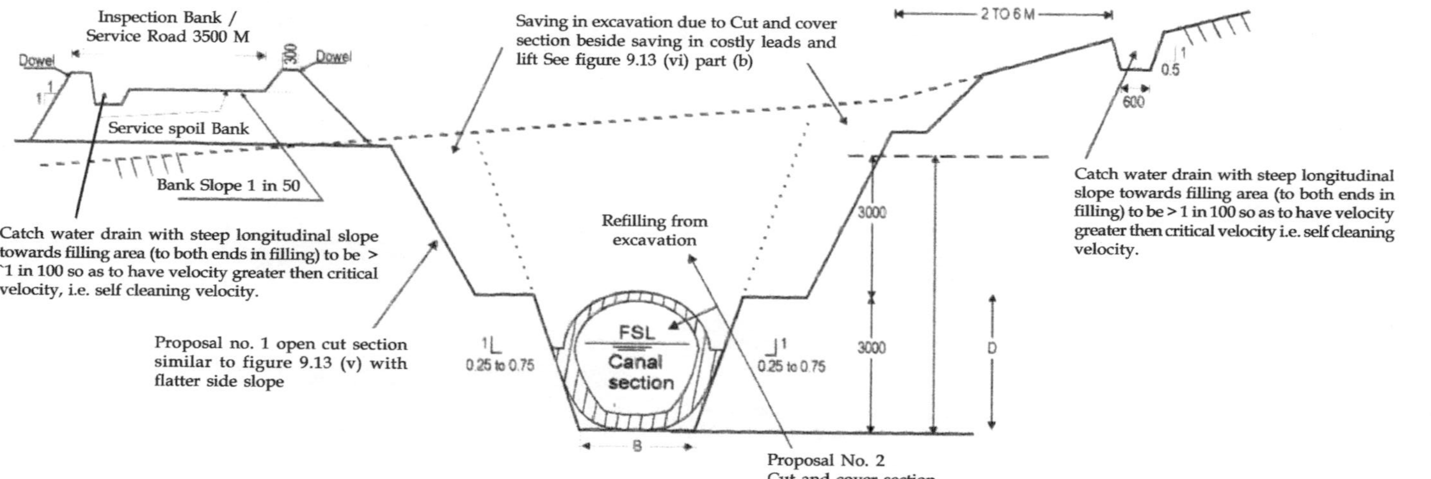

Notes : 1. Service road can be on any side of canal depending upn topography

2. All dimensions are in mm unless otherwise specified.

3. Discharge range : Low, Medium and high capacity canals, see table 9.3

Figure 9.12 (iii) Part (a) Proposed typical cross- section for cut and cover, lined canal in unstable rocks/ bedding plan in deep cutting- circular or arch roof section to save extra, excavations.

Notes to figure 9.12 (iii)

1. Cutting quantity i.e. 'A' and 'B' are saved due to timely planning the covering of the canal.
2. All cutting to be utilized in filling reaches.
3. Stones to be utilized in masonry works as for as possible/otherwise used in lining or slope protection.
4. Maximum cutting up to 7.0 m, Beyond 7.0 m depth tunnels may be considered.
5. In L- Section of levels centres left, right are to be shown.
6. Typical cross section of maximum cutting to be prepared and got approved, from a higher authority.
7. Suitable ramps with gradients of 1 in 10 to 1 in 20 to be provided at inlets and outlets of deep cutting and at every intervals of 1 km for a minimum road width of 3.5 m. This is essential for ease of construction and future cleaning it should be shown in L-Section and plan of canal.
8. Bottom curved sections are more economical hydraulically and structurally and may be preferred.
9. Even Hume pipe can be laid, preferably of 1 to 2 M. dia. For more dia special Hume pipes can be got manufactured.
10. Suitable main holes and vent holes of 1.5 to 2 m dia should be provided at every interval of 0.5 km Iron ladder should be provided in them. Also sump pit should be provided in these main holes.
11. Wear type structure should be provided at inlet to avoid entry of debris.
12. These conduits are can also be designed as pressure flow.
13. In free flow, steeper velocities are proposed through steeper gradient.

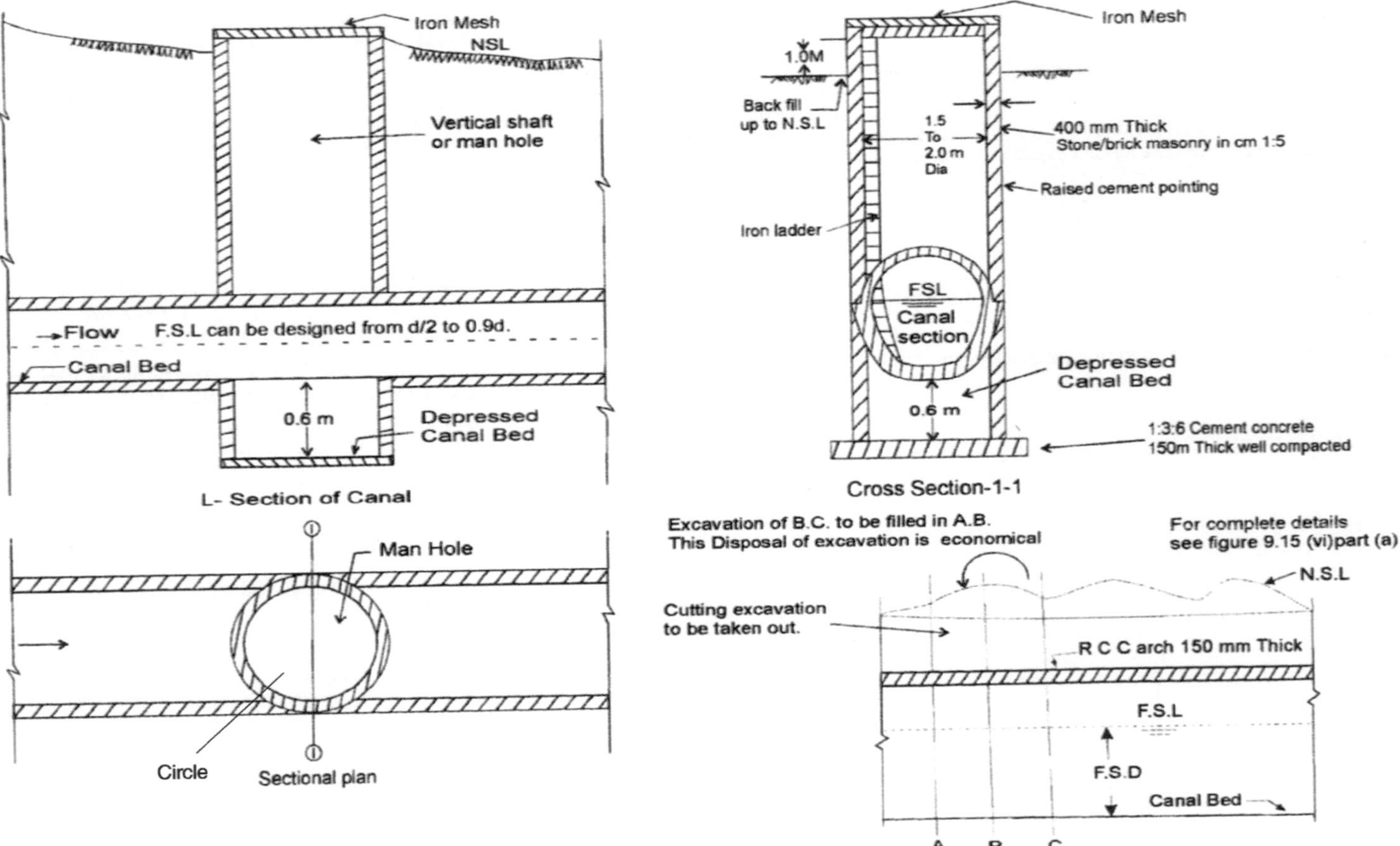

Figure 9.12 (iii) Part (b) Typical cross-section for cut and cover, lined cnal in unstable rocks in deep cutting circular setion concrete arch to save extra excavations.

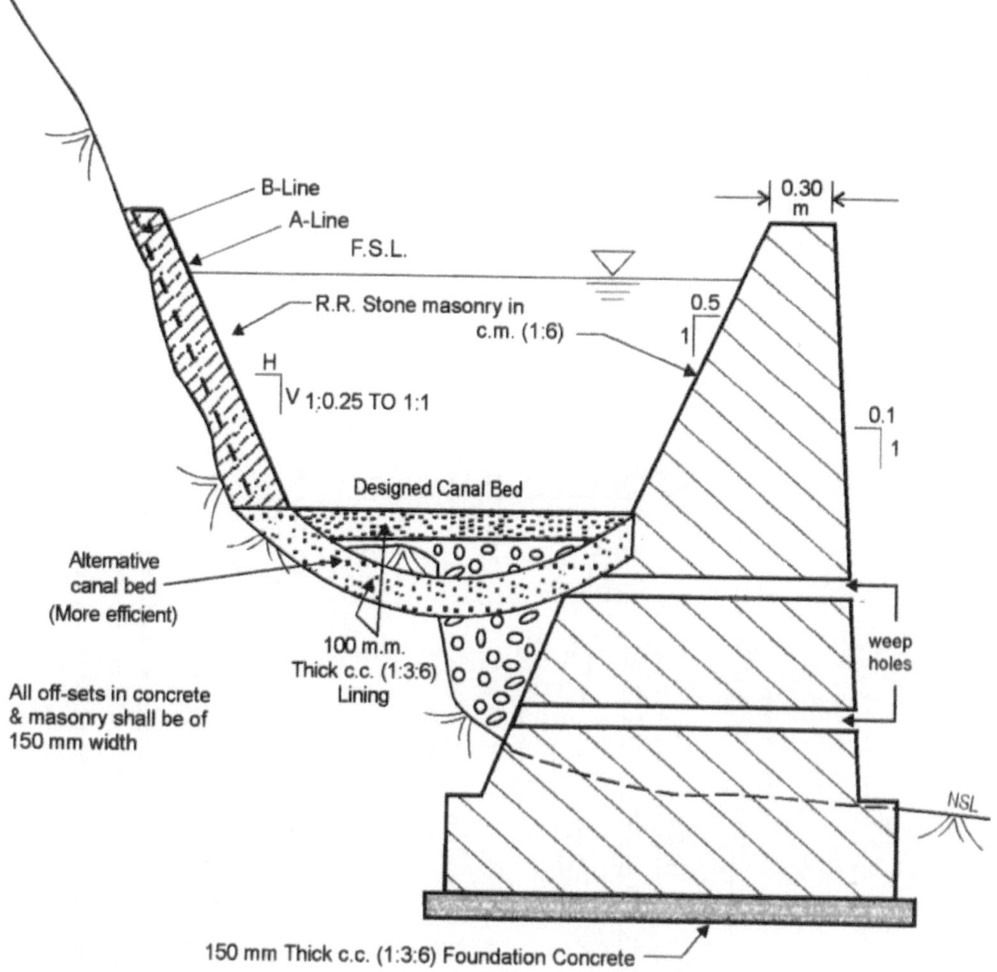

Notes

1. Canal Excavation & lining work shall have to be executed in the same agreement so as to make use of stones.
2. Left side innder side slope may as stee as 1:0.25 for sound stable rocks to 1:1 for soft disintegrated rocks, depending upon geological conditions & stability criteria.
3. For unstable rocks (see figure ...), special treatment required.
4. A. Line - Finished line surface ; B. Line - Line of clear excavation (payment line).
5. Canal bed lining shall consisit of 100 mm. thick c.c. (1:3:6) M-20 grade.
6. Over cuts/over excavation shall be made good by the contractor and no extra payment shall be allowed for this.
7. The thickness of slope lining shall be 200 mm or 250 mm depending upon lining height.
8. On R/S stable masonary wall with provision of weep hoes as depicted shall be provided.
9. Stability of retaining wall be ensured & checked.
10. Depression below bed shall be filled with cohesion-less material, quarry spalls or natural filter material (well compacted).
11. For economy service road can be avoided near top. Alternatively it can be provided at toe of hillock on retaining wall side, at ground level.

Figure 9.12 (iv) Typical cross-section of canal lining when canal is aligned through toe of hill rock, partial cutting partial filling.

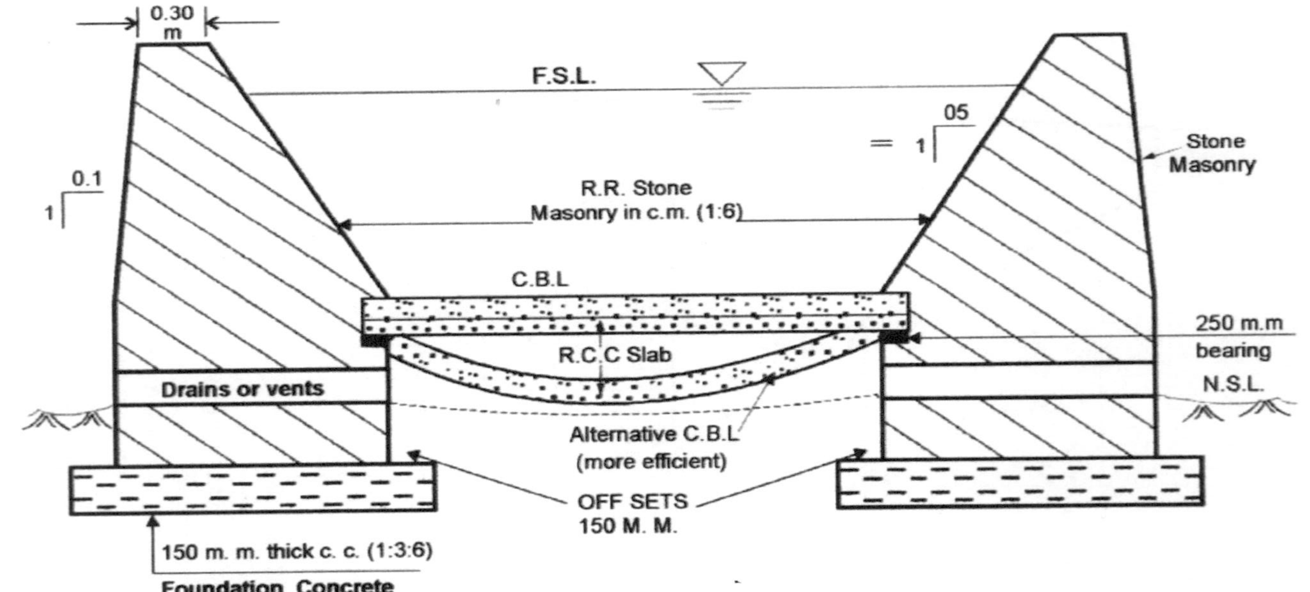

Notes

1. R. R. Stone masonry retaining section with inner slope 1:0.5, rear slope of 1:0.1 and top width 300 m.m. shall be provided on either bank alter.
2. Masonry wall shall continue at least 300 m.m. below N.S.L. and shall rest on 150 m.m. thick c.c. (1:3:6) A-20, foundation concrete.
3. For canal bed R.C.C. Slab spanning between masonry walls shall be laid in c.c. (1:2:4)
4. Thickness of R.C.C. slab may be 150 m.m. to 200 m.m. for a water column of < 2.5 m.
5. Bearing beneath R.C.C. slab shall be 300 m.m. wide & 150 m.m. thick with nominal reinfor cement.
6. No foreign material shal l be carted for filling along side masonry walls.
7. Trimming of natural surface shall be done to provide smooth road surface for service bank, on any one side.
8. Suitable vents / Pipes shall be provided below C.B.L. to drain out storm water also to economise masonry cost.

Figure 9.1.12 (v) Typical cross-section of canal lining in filling reaches (filling<2.0m below c.b.l.) in rocky terrain, where no suitable earth is available with in economic haulage for small canal sections having bed width < 2.5 m.

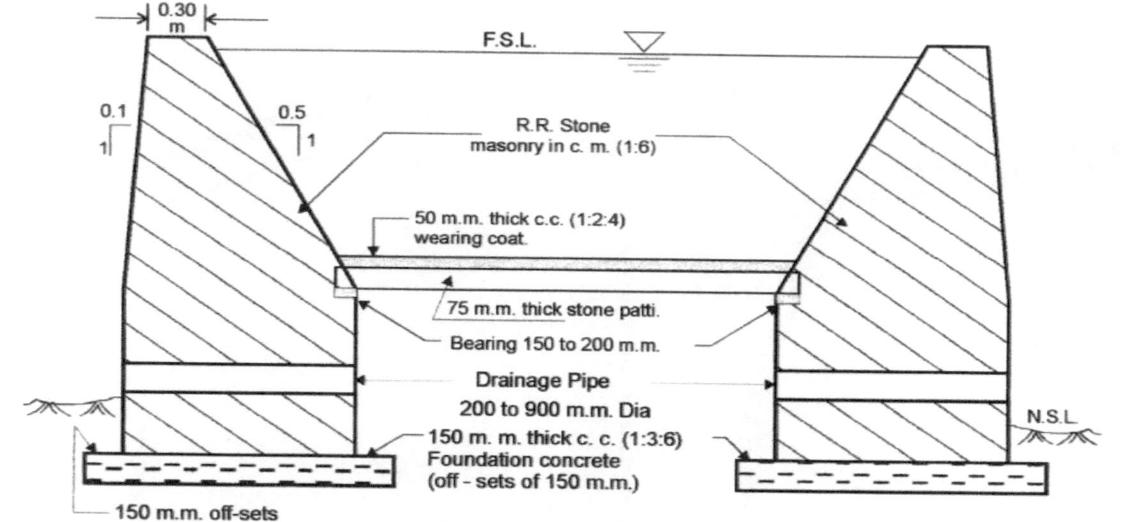

Notes

1. R. R. Stone masonry retaining section with inner slope 1 :0. 5 rearslope 1:0.1 and top width 300 m.m. shall be provided on either bank.
2. Masonry wall shall continue atleast 300 m.m. below N.S.L. and shall rest on 150 m.m. thick c.c. (1:3:6) A-20, foundation concrete.
3. For canal bed 75 m.m. Thick stone patties shall be laid spanning between retaining walls.
4. 50 m.m. Thick C.C (1 : 2: 4) resembling to flooring shall be laid over stone patties.
5. 150 m.m bearing shall be on either end of patties.
6. No foreign material shall be carted for filling along side masonry walls.
7. Trimming of natural surface shall be done to provide smooth reading surface on service bank on N.S.L.
8. Suitable vents / Pipes shall be provided below C.B.L. to drain out storm water.

Figure 9.12 (vi) Typical cross-section of canal lining in filling reaches (filling < 2.0 m below bed) where no suitable earth is available with in economic haulage, and stones are available in vicinity or stone masonry section and stone slabs is cheaper (for small canal sections having bed width < 2.5 m.).

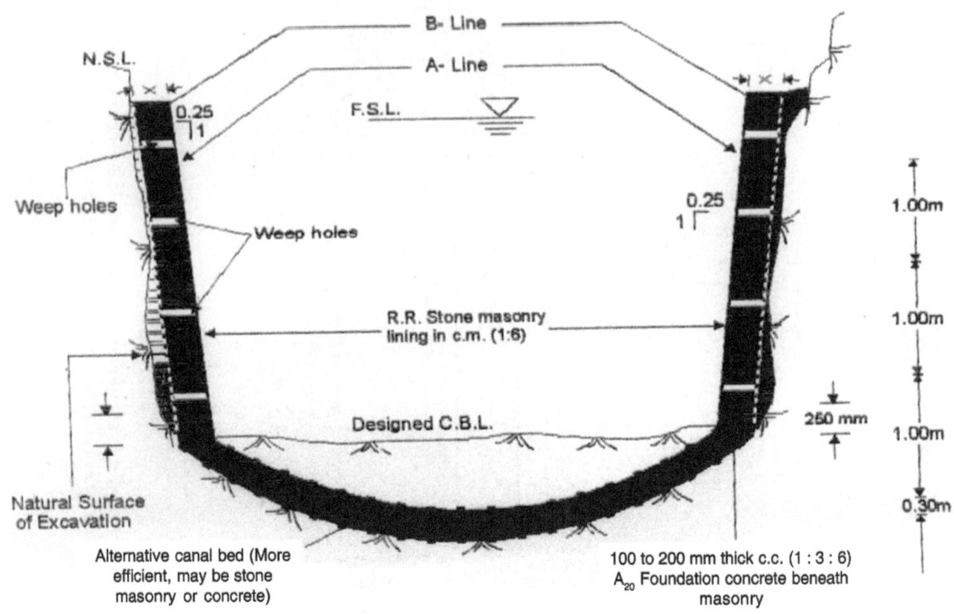

(a) Lining in hard stable rock cutting reaches

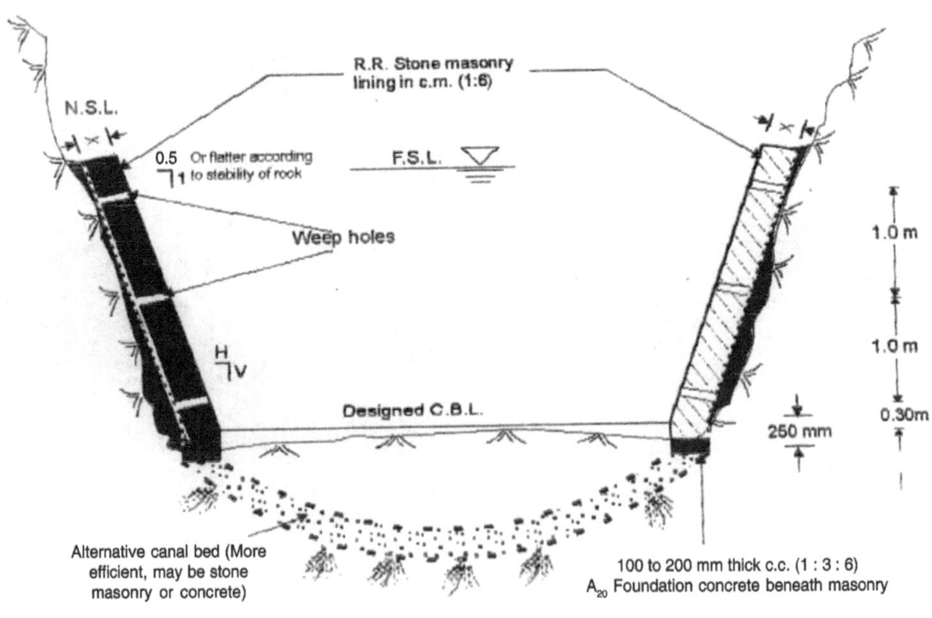

(b) Lining in soft rock cutting reaches

Figure 9.12 (vii) Proposed typical cross-section of canal lining in rock cutting reaches

Notes to figure 9.12 (vii)

1. Normally no lining is required in hard rock to reduce seepage or increase N. No lining may be done if the length of reach is 100 m. or more. These canals generally being in cutting, bring lot of seepage water during rains and also otherwise. However some seepage occurs in dry season, depending upon water table. High value of N is taken care by excess excavation as per B line.
2. No lining in canal bed may be done.
3. However excavation for canal lining i.e. excavation to make room for lining thickness shall be done, at the time of excavation. Thus section may be little more.
4. In hard rock reaches, the inner side slope of canal section shall be 1: 0.25 (V:H)
5. In soft rock reaches the inner side slopes shall depend upon Geological conditions, see para 9.7 and chapter 8. for generally may be as under

   (a) For depth of cutting up to 2.00 m, Z=1: 1 (V: H)

   (b) For depth of cutting > 2.00 m and up to 4.00 m, Z= 1: 0.75 (V: H)

   (c) For depth of cutting > 4.00 m, Z=1: 0.5 (V:H)

   Flatter slopes increases excavation cost very much.
6. Where lining is done, the thickness of masonry lining may be as below

   (a) 200 mm thick for canals having lined height up to 1.5 m (b) 250 mm thick for canals having lining height more than 1.5 m

   Some favour lining to make use of stones, as the same may be taken away by local people or contractors working for other works.
7. A -Line - Line of finished lined surface

   B -Line - Payment line - Line of clear excavation.

   No projection over hangs shall be allowed beyond B-line inside canal section
8. Weep holes shall invariably be provided for stability of lining and to reduce backwater pressure.

   Horizontal spacing - 2.0 m c/c; Vertical spacing - 1.0 m c/c and be staggered.

   The first hole shall be placed 0.30 m above canal Bed Level (CBL)

   Size of weep holes:- 50 to 75 mm Square/circular holes as deemed appropriate by the site engineer.
9. Over excavation /over cuts, if any shall be make good by the contractor and no payment for this shall be made. However, the clear horizontal thickness of lining shall not be less than as specified in note no. 5.
10. For over all section, refer to drawing No. (iv) and (v).

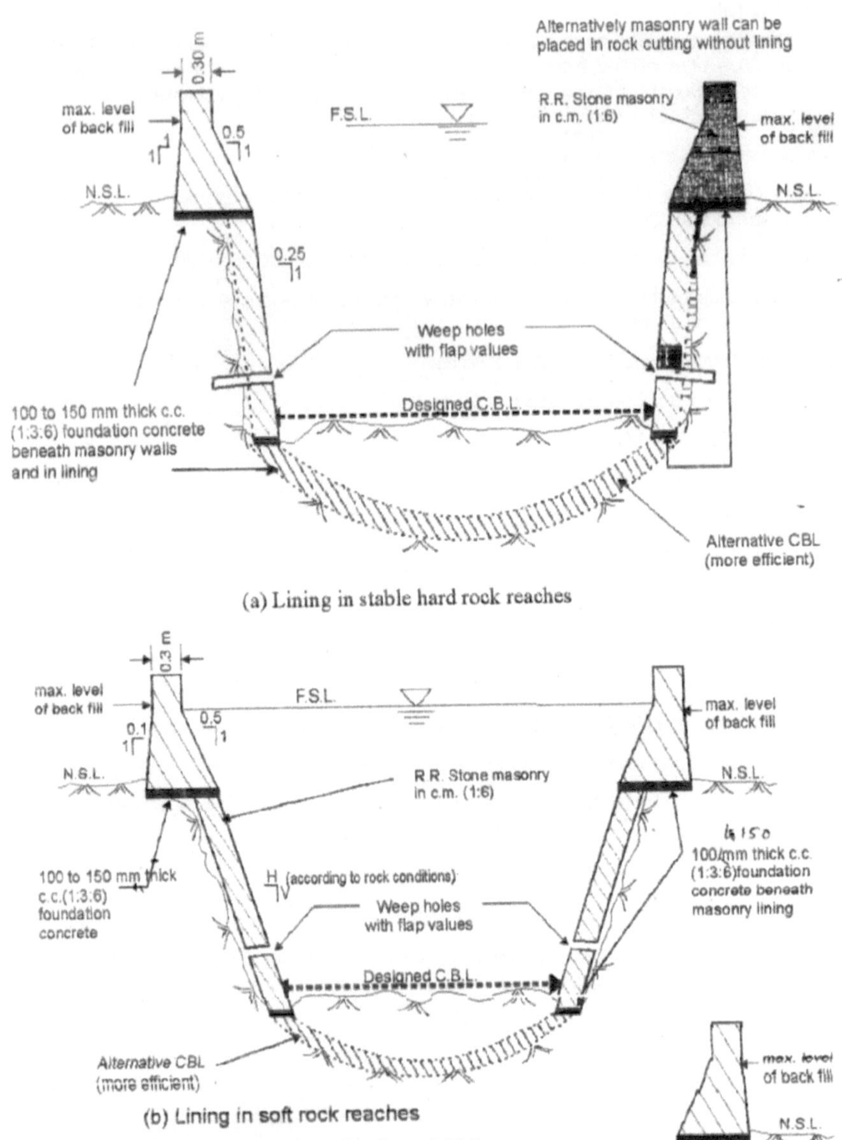

Note : for details of bank, service road etc. See figure 9.12 (i)

Notes to figure 9.12 (viii)

1. For lining in cutting portion, refer to drawing. no. (viii) and their notes.
2. Back fill behind masonry wall shall consist of excavation material available from canal excavation. No foreign material be carted for back filling.
3. In filling zone R. R. stone masonry retaining wall in cement mortar (1:4 to 1:6 with inner slope as 1: 0.5 may be provided. Top width of retaining wall shall be 300 mm. The top 80 to 50mm masonry shall be Kharanja type.
4. No plastering or pointing may be done on exposed surface of stone masonry to ensure good quality work.
5. The joints be flushed finished during construction, and ensure filling of all crevices, and no seepage. The maximum level of back filling shall be at par with F.S.L. of canal.

Figure 9.12 (viii) Proposed typical cross-section of canal lining in partial rock cutting & partial filling reaches

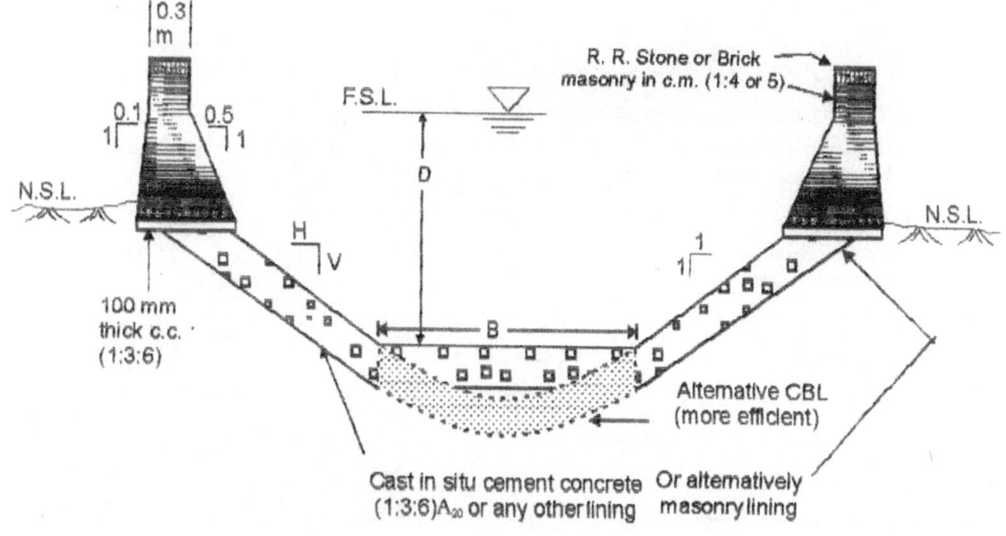

(a) Partial cutting / filling section

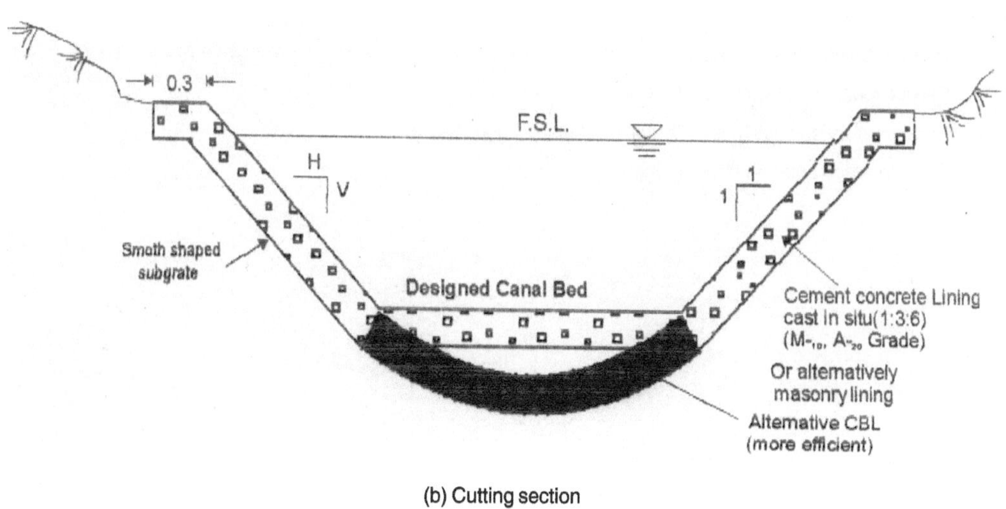

(b) Cutting section

Figure 9.12 (ix) Proposed typical cross-section of canal lining where no soil is available within economical leads such as in murrams etc. and Stones/ aggregates are available (other than rock cutting & expansive soil).

Notes to figure 9.12 (ix)

1. The Inner side slopes in cutting shall not be steeper than 1:1. (V:H). It may be little flatter say 1:1.25 (V:H) depending upon "C"(Cohesion) and ϕ (Angle of internal friction) of soil mass and depth of canal.
2. For partial filling, masonry retaining wall section can be provided up to 2.0 m height.
3. No plastering or pointing may be done on exposed surface of masonry. However joints should be properly filled, flushed and there should be no leakage. This ensures good quality of masonry.
4. Back fill may consist of excavated muck. No foreign material like clay etc. are required for this.
5. The maximum level of back filling shall be at par with F.S.L. or 0. 2 m high. Excess material to be utilized in widening of service road.
6. Service road etc. at ground level to be provided according to chapter 8.
7. Normally canal lining be avoided in such hard and compact strata.
8. If lining is essentail, normally excavation and lining should be taken in the same tender/estimate.
9. In the event of excavation & lining work in different tenders/agreement, excavation of canal lining shall be done at the time of canal lining. It helps in removal of roots, shrubs, trees and others deleterious or weathered deposits.
10. The canal bed lining may be as under (a) Bed width < 3.0 m.- 100 mm thick cement concrete in situ (b) Bed width > 3.0 m. – as at (a) or 200 mm thick R. R. Stone masonry in C.M. (1:6)/ pointed stone kharanja.
11. The lining of side slopes may be as under

    (a) Discharge < 200 cusecs- 75 m.m. thick C. C. (1:3:6) lining.

    (b) Discharge > 200 cusecs- 100 m.m. thick C. C. (1:3:6).
12. Panel joints in longitudinal direction shall be provided at 3.0 m interval.
13. Panel joint in transverse direction shall be decided by the engineer in charge, depending upon height.
14. At each panel joint a concrete sleeper 20 cm wide & 7.5 cm thick shall be placed/casted beneath canal lining.
15. In deep cuttings further flattening of slope may be required depending upon geological conditions.
16. Service road, drains etc. to be provided as per chapter 8.
17. For over all cross-section, see figures, in chapter 6 and chapter 8.
18. Sufficent weep holes shall be provided.

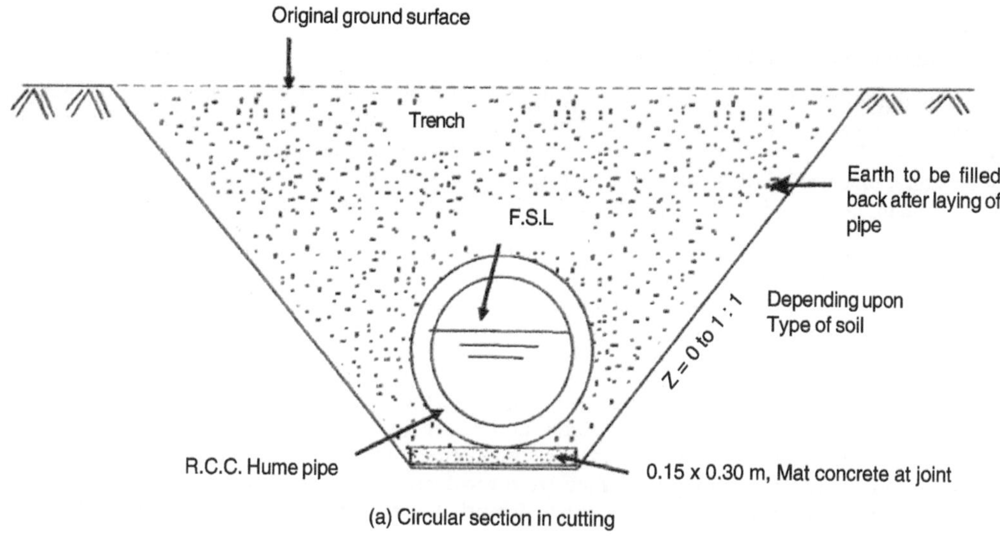

(a) Circular section in cutting

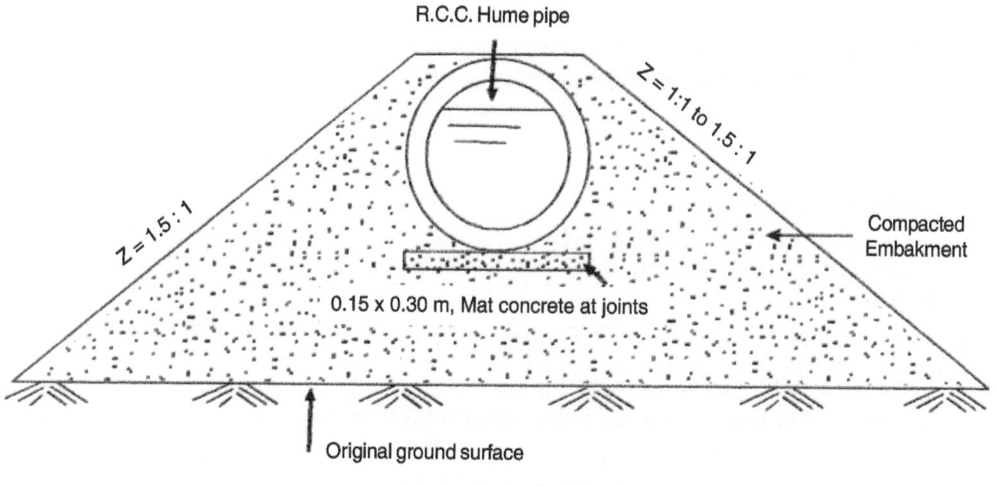

(b) Circular section in filling

Figure 9.12 (x) Circular section in heavy cutting/ filling for low to medium discharges.

Therefore with lot of discussions on various cross sections given in figure 9.4 and their adaptability, some economical and practical typical cross sections are proposed in figure 9.12. Also dowel, service (inspection) road, drains, suitable ramps etc. are shown in these figures. Enough notes are also given as a guideline for preparation of detailed estimates and project reports. These typical sections are based on experiences from many projects including remedial measures taken for effective operation and maintenance. These typical sections are not exhaustive to covers all situations. Readers should refer to many diagrams given in chapter 6, 8 and in various section of this chapter.

9.12 Practical Aspects of Canal Section and Alignment

A canal section and the alignment of canal are interacting i.e. influencing each other.

The alignment of canal is laid on the ground according to the various principles of irrigation, hydraulics, economics, with due consideration of topography, geology, soil characteristics, and ground water table. Principles of irrigation and hydraulics include due consideration to command the areas adequately, practical possible velocity, bed slopes, easy and efficient operation, and maintenance of canal system, farm water application, and minimum density of canal network.

Further principles of hydraulics include channel design i.e. internal channel geometry and flow characteristics. Types of soil, ground water level and topography influence them. Soil and sediment affect the rugosity coefficient, velocity, internal side slopes. Soil and ground water flow characteristics, along with service road influence the over all section.

Topography, geology and soils may not remain uniform on the ground as well as under the ground, all along the canal alignment in spite of a best possible alignment laid in a particular region. Topography may change from a rolling ground to a highly undulating ground. Similarly, geology and soils also change, from sandy blown sands to clayey, gypsepherious soils, very soft to hard rocks of various natures, and ground water may sometime cross the channel section. However, there are specific characteristics of a particular region. Too much detouring of canal may be very costly and some times may not be feasible.

In a long canal, bed slopes may change from mild to milder or vice verse, but steeper slopes are avoided except in exceptional cases, (see para 9.13). In the two slopes, full supply depth may remain the same or may decrease from head to tail, with decrease in discharge. However, the full supply levels (FSL), full supply depth (FSD) can not be changed considerably except at falls or structures. But some adjustments are definitely feasible and are generally accommodated as best as possible.

Thus, canal alignment and its sections (i.e. design) are complimentary to each other and work in hand to hand.

9.13 Longitudinal Bed Slope

Longitudinal slope is primarily fixed from the consideration of country slope consistent with economy and command. It generally, follows natural ground slope unless otherwise needed due to other considerations.

Steeper the slope, more is velocity and smaller is the cross-section of canal. But this is limited to critical slope, see para 2.3.2. There is also a limit to flattest possible slope. Beyond it, for flow to take place, either depth or bed width has to be increased considerably, see para 4.3, and expample in it, otherwise there will be heading of water in the upstream to create extra head or energy for the flow to take place.

Kennedy's critical velocity ratio phenomenon which may range between 1.1 at head to 0.8 near tail, also restricts the bed slope.

Many engineers advise use of Lacey's formula for slope adoption, but do not over rule the authority of experience.

A long open channel may contain two or more reaches of different bed slopes, due to changes in topography, see para 9.14. The bed slope can be changed at structures.

The bed slope is usually forced to conform within certain limits to the requirements of the general topography, the dimensions of the various structures concerned, and to keep within reasonable command. It is commonly not feasible for the channel to retain the same bed slope throughout its long length. This, of course, is very common in natural alluvial channels. In rocky rivers bed slope very considerably.

Based on a data of many canals given in chapter 6, characteristics of lined canals can be summerised as in table 9.3. Feasible/suitable section that can be adopted are proposed in the last column of the table.

Table 9.3 General practices of purpose, capacity, length, bed slopes and Velocities of lined canals.

Sl. No.	Capacity and Length	Purpose	Bed Slope	Velocity (m / s)	Feasible / Suitable Section
1.0	**High to Very high capacity (Q > 100 cumecs 300 to say 1000 cumecs)**				
1.1	Long Canals > 50 km	Generally Irrigation, feeder and P.H only at falls or tail	Restricted to very flat 1/12000 to 1/8,000	Generally 1 to 1.5	Depth is restricted, so only option is wide channels (compromise between D,B and Z)
1.2	Medium Length < 50 km > 10 km	-do-	Restricted to 1 in 8000 to 1/ 6000	Generally between 1.50 to 2.0	Little deep, still restricted so wide channel (compromise between S,D, B and Z).
1.3	Short length < 10 km	Generally feeder and power canals.	Generally 1/6000 to 1/ 3000	Between 1.0 to 2.0 or maximum upto 2.5	-do-
2.0	**Medium Capacity (Q ≤100 cumecs & > 50 cumecs)**				
2.1	Long Canals > 50 km	Generally Irrigation, feeder and P.H only at falls or tail	Restricted from flat to moderate, 1/8000 to 1/6,000	Generally near 1.5	-do- (As above)
2.2	Medium Length < 50 km > 10 km	- do-	- do-	Generally between 1.5 to 2.0	Depth need not be restricted, so best sections are most suitable
2.3	Short length < 10 km	Generally feeder and power canals	Generally 1/6000 1/3000	Between 1.5 to 2.5 or maximum upto 2.5	-do-
3.0	**Low Capacity (Q < 40 to 50 cumecs)**				
3.1	Long Canals > 50 km	Generally Irrigation, feeder and P.H only at falls.	Restricted 1 / 8000 to 1/5,000	Generally between 1.5 to 2.0	Depth need not be restricted, so best section are most suitable
3.2	Medium Length < 50 km > 10 km	-do-	Restricted to 1 in 6000 to 1/4000	-do-	Optimum best or little deep section
3.3	Short length < 10 km	Generally feeder and power canals	Generally 1/6000 1/3000	Between 1.5 to 2.0 or maximum upto 2.5	-do-
4.0	**Very low capacity (Q < 0.5 cumecs)**				
4.1	Short lengths < 10 km	Generally minors and water courses	1 in 3000 to 1000 or even 1/500 generally around 1/1000	0.20 to 0.5	Optimum best section

PH = Power House

Velocity can be increased by adopting steeper side slopes up to optimum slope of 1: 0.58 (V: H) and depth also up to D = 2R (since maximum R = D/2). Depth can also be further increased beyond 2 R to increase velocity, but perimeter will also increase see figure 1.3.

For power channels ; Masony (1967) proposed the practical velocity value according to the following principles.

If the mean velocity is increased by increasing the bed slope of the canal the effect on cost will be bilateral and contrasting. Any increase in canal slope reduces the available head at the turbine and consequently the output of the plant. On the other hand due to smaller cross-sections the aggregate cost of the development will be lower. The effect of decreased mean velocity in canals are contrary to those mentioned above, i.e. extra cost of canal but more power generation. Therefore a judicious judgement is needed based on factors given in para 9.1 and the table 9.3.

Because of mild slopes, velocity is always small therefore every effort is made to have bed slope as much as possible but less than critical slope.

Though above table is prepared for lined canals, still it is applicable to unlined canals, as allmost simmilar or indentical situation occur in unlined canals.

9.14 Flow Profiles in a Channel of Varying Bed Slopes

The type of flow profile depends in part upon bed slope and control gates, falls and other structures. It is important to recognize the different types of profiles that may occur simultaneously in different reaches of the same channel. An initial qualitative study of the complete profile will indicate what segments of the channel which may affect the system and may have to be changed. Finally it will be necessary to compute detailed profiles.

Several basic requirements should be kept in mind, as discussed in following cases:

1. A change from sub critical flow on a mild slope to supercritical flow on a steep slope is such that critical flow is attained essentially at the break in grade.
2. A change from supercritical to sub critical flow is accomplished through a hydraulic jump, located at that point where the flows of momentum in the sub critical and supercritical flows are equal, i.e. when ever d_1 corresponds to d_2 in the channel.
3. A change from one mild slope to another mild slope causes a flow profile transition curve that is completed entirely on the upstream slope.
4. A change from one steep slope to another steep slope causes a flow profile transition that is accomplished entirely on the downstream slope.
5. Flow profiles approaching, or diverging from, uniform flow for the given slope, do so asymptotically.
6. Flow profiles approaching, or diverging from critical depth tend to do so at a vertical slope, although not actually becoming vertical.
7. Flow profiles should be computed (or sketched) upstream if the depth is greater than critical and downstream if the depth is less than critical.

There are two basic types of profile analysis that may have to be made:

1. For a given discharge, determine the flow profile throughout.
2. For a given water elevation at some points in the channel system (the most common being a given surface elevation in the reservoir supplying the channel), compute the discharge and the balance of the flow profile.

The general problem of channel design involves trial analysis of several possible channel designs, with comparative economic studies based thereon.

By considering first the case of known discharge, the normal depth for each slope segment, as well as the critical depth, can be computed. A sketch should be made to scale, with vertical scale highly exaggerated, showing bed line as a solid line, critical depth line as a dotted line, and the various normal depth lines as dashed lines. Possible control points for segments of the flow profile should be noted, these being of two basic types, critical depth control, where the flow goes through critical and normal depth control, where the flow has become essentially uniform. Since both these depths are known numerically, if it is known where they may occur in the channel, they can be used as basic on the flow profiles.

Outlining of Surface Profiles

The first step is to draw the lines of critical and normal depth. These are parallel to the bed and when the channels are of uniform section the critical depth is the same throughout. Next, the control points are inserted at the appropriate entry and exit sections, at the change of sections and slopes, see figure 9.13.

Downstream of the sluice the position of the hydraulic jump between the M3 and M2 profiles is determined in a similar manner.

The general character of possible flow profiles for the channel is shown in the figure.

The critical depth and normal depth lines are shown by dotted and dashed lines, respectively. Possible control points on these lines are indicated by circled numerals. Sketching of profiles proceeds upstream for sub critical flow and downstream for super critical flow.

An M-2 curve is sketched upstream and an S-2 curve downstream from the critical depth control at point 1. The latter is followed by an S-3 curve on the second steep slope. This will begin from control point 2 if the first steep slope is long enough to attain normal depth; otherwise it will begin at the actual water elevation at the breakpoint. Similarly, point 3 may serve as a normal depth control for the H-3 curve that develops on the horizontal slope.

The latter may intersect the critical depth line at point 9, or it may continue throughout the horizontal reach and be replaced by an M-3 curve on the next mild slope, depending upon the particular conditions. Similarly the M-3 curve may intersect the critical depth line at point 10, or it may continue and be succeeded by an A-3 curve on the adverse slope, and so on. Wherever the profiles intersect the critical depth line, it may be stopped, since a hydraulic jump must occur somewhere before that point is reached.

To determine the sub critical profile, to which the supercritical flow must jump, a downstream control point must be used as a starting point. Possible controls are located at points 4, 5, 6 and 9. Point 4, the farthest downstream, would probably be used unless it was reasonably certain that uniform flow could be assumed at point 5, 6, or 9. Point 4 is a critical depth control, and causes an M-2 profile upstream from it, which may or may not; attain uniform flow at point 5 or 6 or both.

On the adverse slope, an A-2 profile will follow, then an M-1 curve on the mild slope upstream from it. If the A-2 curve should have come out below the normal depth line on the latter slope (very unlikely, however), then an M-2 curve would have followed, approaching point 9. The horizontal slope causes an H-2 profile; then S-1 profiles must develop on the steep slopes. The latter will end at point 8 on the critical depth line, which may turn out to be on either of the two steep slopes.

Detailed profile calculations would be necessary to elucidate the exact forms and locations of the various profiles. The qualitative analysis does indicate, however, that point 9 will not be a control and that there will be an M-1 rather than an M-2 profile on the mild slope succeeding it. It is quite possible, also, depending mainly on the lengths of slopes that computations may begin at point 2 or 3 for the supercritical flow and at point 6 for the sub critical flow, with the antecedent portions neglected, since this reach will contain both the maximum flow depths and the hydraulic jump.

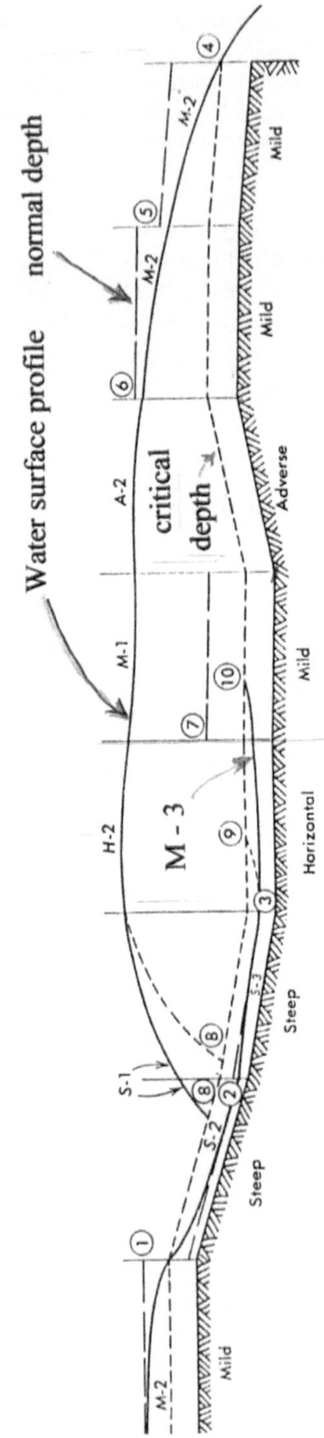

Note : There can be different slopes in a long canal of any capacity from very small to very high

To ensure flow water surface profile and energy line should be plotted

Figure 9.13 Possible flow profiles in channel

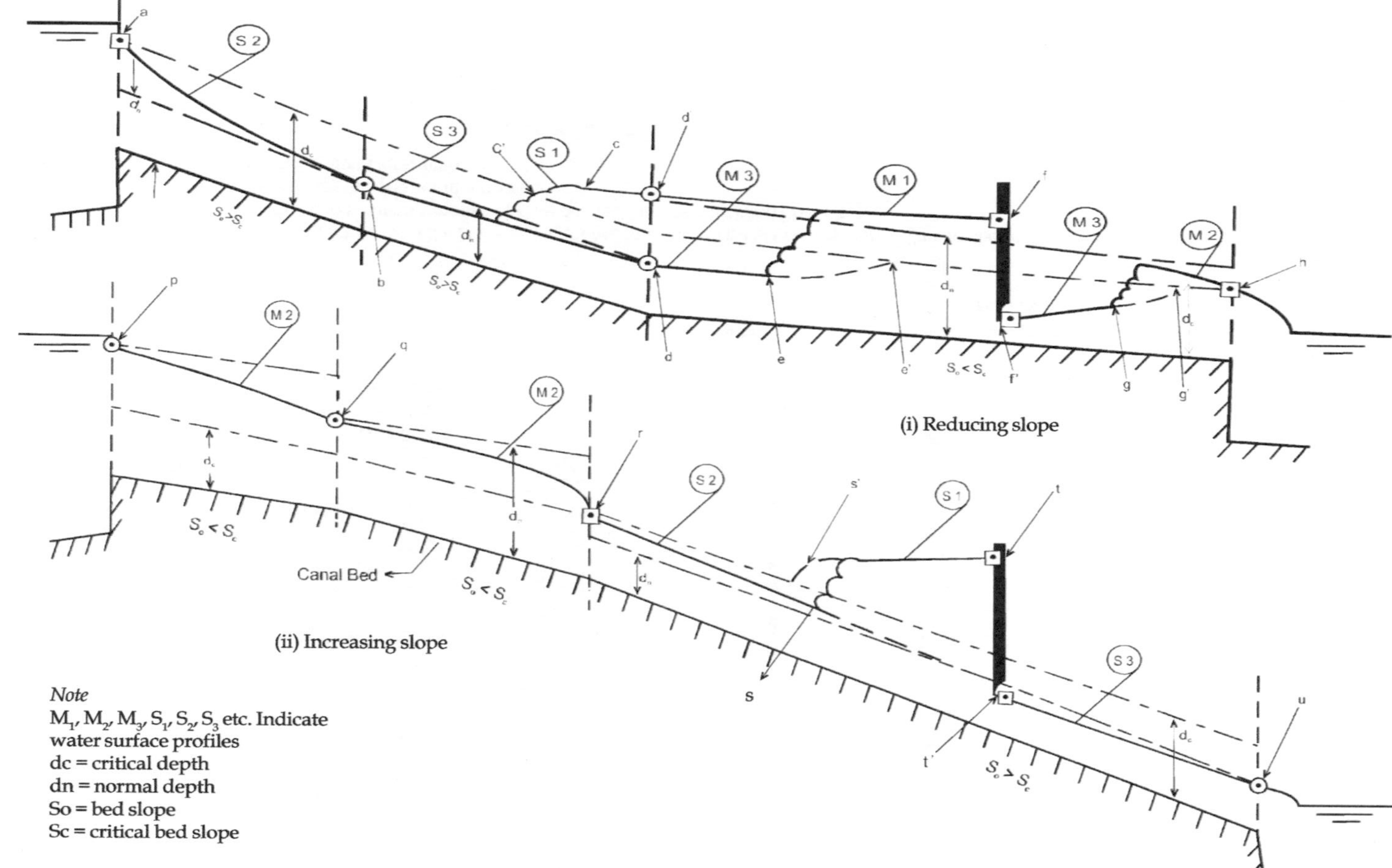

Figure 9.14 : Typical flow profiles in canals showing hydraulic curve identification and control points

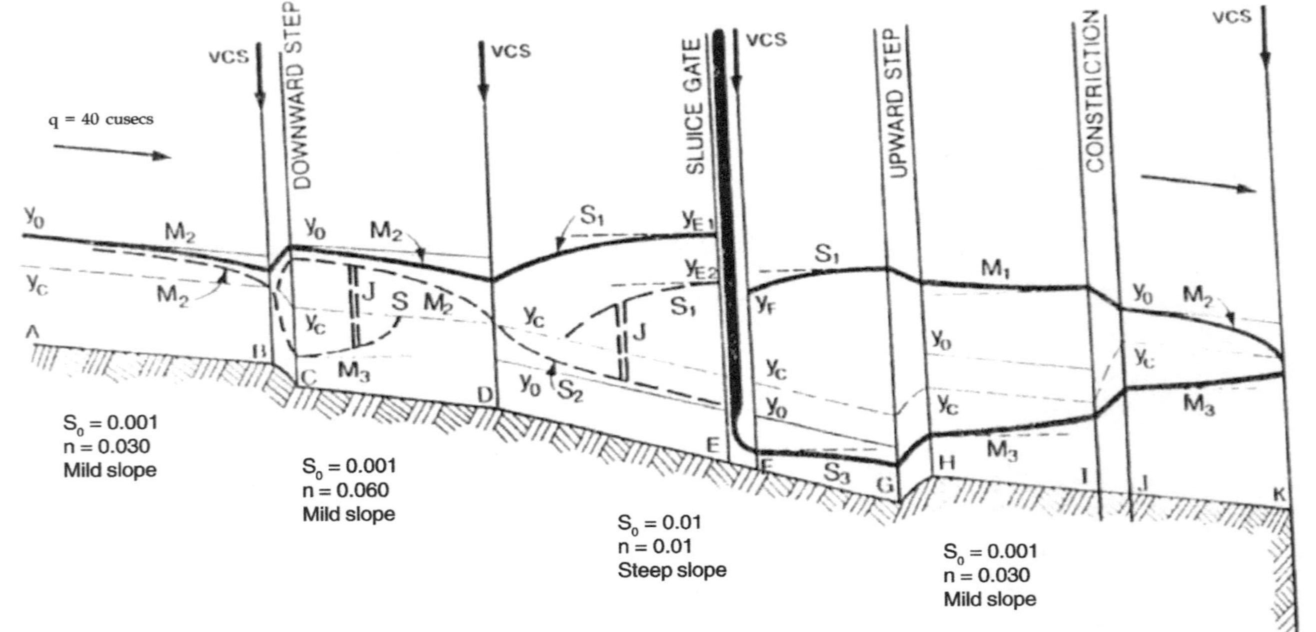

Notes : The gradually varied flow reaches AB, CD, DE, FG, HI, JK are interrupted by the rapidly varied flow sections BC a downward step, EF a sluice gate GH, an upward step, IJ a contraction. The solution process can be divided into the following phases.

Phase I : Determination of critical depths and normal depths

Phase II : Virtual/Actual Control Section Determination

Figure 9.15 Typical open channel flow profiles.

The jump, of course, must be located by determining a point where the depths on the sub critical and supercritical flow profiles satisfy the requirement for conjugate depths in the jump's equation. It will most likely occur on either the downstream steep slope or the horizontal slope, but it may possibly occur anywhere between the points (8 and 10) where the two profiles respectively reach the critical depth lines.

In the event that point 8 on the S-1 curve turns out to be above point 1 on the S-2 curve, it may be that the jump is "submerged" and cannot occur. The flow would follow the upper profile throughout, with either a truncated M-2 curve or an M-1 curve on the first mild slope, depending upon whether point 8 were below or above the normal depth line for that slope.

In the above example, it was assumed that the discharge was known. However, it may be that some particular surface elevation is fixed and the discharge will be limited thereby. The flow profile cannot be determined until this discharge is known.

Two more typical examples of flow profiles in canals are shown in figure 9.14. In figure 9.14 (i) the bed slope of the canal is steeper than the critical slope and is decreasing from steep to mild. In the second case figure 9.14 (ii) the bed slope is mild and increasing till becomes greater than critical slope. Though bed slopes greater than critical are normally not adopted, yet can be permitted in hilly area in short reaches. Even though a canal may have mild slope, the bed slope changes both in lined and unlined canals due to silt entry in the canal and its deposition/erosion. Typical example is upper Ganga canal, (U.P.) see Para 7.19.

In the two figures shown, there is a cross regulator (also some times called a sluice). It acts as a control point in both directions, since the upstream and downstream depths are governed by the sluice gate rating curve.

Referring to channel in figure 9.14 (i), proceeding in a downstream direction from the control point is because the flow is super critical, the S2, S3, and M3 profiles may be traced as far point e', the intersection with the critical depth line. Similarly, proceeding upstream from the control point *f* because the flow is sub critical, the M1 and S1 profiles extend to the critical depth line at e'. At some point between e and e' the upper profile has a depth sequent to the lower, and the hydraulic jump transition occurs. The actual position is determined later, when the detailed computation is made, by drawing the locus of sequent depth to the lower profile, making due allowance for the length of the jump.

Referring to Channel 2 in figure (i), providing that the depth at the sluice is not greatly in excess of the critical, a control point is to be found at point r, the transition from sub critical to supercritical flow. A hydraulic jump thus links the S1 and S2 profiles at a point s' where the upper profile is at sequent depth.

Proceeding in an upstream direction from r, because the flow is sub critical, there are two distinct M2 profiles linking with the reservoir at point p, the surface level of which must be appropriate to the discharge, due allowance being made for a small drop in level at entry to the channel.

Downstream of the sluice there is an S3 profile terminating in a free over fall to the lower reservoir.

The above treatment is typical, but of course the range of possible channel conditions is so large that there is considerable scope for detailed variation. After identifying the profiles in this manner the actual surface levels may be evaluated by one of the methods described by chow (1973).

9.14.1 Zero Bed Slope Canals

Water in canal flows due to the energy slope and not by the bed slope. Bed slopes is provided to provide sufficient energy slope. But energy slope can also be provided otherwise i.e. through the cross regulators/structures and by successively decreasing the depth of flow, i.e. from higher depth to low depth. Instead of decreasing the width, depth is decreased towards the tail ends. Such flows can be seen in the nature's way of flowing water in river near hill toes, where some times zero slopes and adverse slopes also occur. Typical how profiles for varying bed slopes, such as horizontal to adverse slopes are shown in figure 9.13 to 9.15. These canals are designed by the same Manning formulae using 'S' as the slope of energy gradient. Such canals are common in America. Such canals have following advantage:

1. Ensures easier and better regulation. The control being on the down stream side, see figure 7B. 6 and 7.B.10 for water surface profile at the cross regulator. The flow is regulated by automatic float gates. These gates are not costly.
2. Better safety of canals - flow reduces with the decrease in the energy slope and when the energy slope becomes horizontal no flow can take place and stops.
3. More safe as there are less chances of over topping.
4. Ensures more constant or uniform water level over a long times.
5. Tails are more easily fed, as the energy gradient increases; discharge increases towards the down stream side.

Since such canals are not so far prevalent in India, many engineers raise many doubts. Some doubts and their answers are as under:

The canal may be costly due to high banks and costly gates. Well the over all cost of gates is not high. It is practically same. The heavy cost of hoist, their plate form decreases in comparison to vertical lift gates. More over the cost of operation and maintenance decreases.

As regards the canal section the cost of earthwork over a long-section remains practically same due to adjustment or double use of earth work in cutting and filling. The bank level of such canals is practically horizontal. The energy gradient is provided by providing a drop at the control structure.

9.14.2 Top Level Bank Canals

In these canals, the bed slope is same as in normal canals, but the free board is increased gradually towards the tail ends and the top level is made horizontal. It is not parallel to FSL in long canals. It can also be done in suitable reaches of reasonable length so as to avoid excessive earthwork, and drop in top bank is provided, through a ramp.

Importance of higher free board or top Bank Canal as has been discussed in Para 7B.4.1. It also has the same advantage as the zero bed slope canal. The difference is that in these canals, normal vertical gates can be used with the normal/general present operation practices. Even the existing canals can be improved by raising the bank from silt clearance, without much cost.

Some engineers raise doubts about increase of earthwork in such canals. This doubt is easily removed, when the total earthwork of a long reach is analysed, in comparison to the conventional (top bank parallel to FSL). On the contrary the earthwork may reduce, as the same is balanced by reduction of earthwork in heavy cutting reaches, due to higher Bank level, and little increase in heavy filling reaches.

To reduce earthwork, an up and down grade of top Bank was provided in the Sundar Nagar Hydel Channel, see Para 6.9.

Therefore, this being a new concept in some countries, most of the doubts can only be removed by drawing the L-section, cross-sections making computation of earthwork, and analysis of flow characteristics at full supply discharge and less than full supply discharge.

9.15 Economics of Lining

9.15.1 Techno-economical Studies

Lining of canals at the time of original construction itself may be more advantageous than at a later stage. Major economics in construction is possible when lining is planned and placed at the time of the original canal construction. The size of associated structures such as cross-drainage works, bridges, pumping plants and distributaries etc., necessary with the canal can be reduced to some extent.

Unlined, earthen canals are generally provided much larger cross-sectional areas than lined canals of equal capacity, so as prevent damagingly high velocities. Unlined canals are usually wider, and shallower, but generally have naturally adjusted steep side slopes, due to silt deposition. Flatter side slopes are usually adopted in lined canals. Silt deposition on sides of lined canals and formation of steep natural stable slopes of silt can be seen.

Subsequent lining in the existing canal requires backfilling and reshaping the canal to accommodate the more efficient section for lining and many other problems, see para. 6.3 and 9.2. These alternatives are costlier than a properly designed lined canal.

It is almost impossible to stop the water in the canal for the time required to line the same. There may be large expenditure for construction of temporary bye passes to carry irrigation water during reconstruction, see para 6.4 and 9.2. These added costs may exceed the cost of lining itself. The problems depends upon the size of the canal and extent of irrigation facilities. It is extremely important that the lining of a canal be carefully considered at the time of its planning and construction. All these factors emphasize the importance of weighing the need of lining and including them in the original construction.

The economics of canal lining can be worked out by a careful assessment of losses or study. The calculations needed for the economic study of the canal lining are given in table 9.4 and 9.5.

Table 9.4 Estimation of additional annual cost of a lined canal over an unlined canal (original construction or modification of existing canal) and net annual saving.

S. No	Details of works	Cost in Rs. (a) Lined *	(b) Unlined *
1.	Excavation, cutting/filling, compaction and trimming etc (see Para 9.2.1)		
2.	Lining and other items, (see Para 9.2.1)		
3.	Cross drainage works, bridges and other necessary structures		
4.	Incidental Structures (see para 9.2.1)		
5.	Total construction cost = (1) + (2) + (3) + (4)		
6.	Difference in construction cost column (a) - (b)		
7.	Life expectancy, and depreciation	Years	Year
8.	Salvage value at life expectancy		
9.	Total depreciation during life = (6) - (8)		
10.	Annual depreciation charges = (9)/(7)		
11.	Annual interest charges $=[\frac{(6)+(8)}{2} X \text{ interest } rate]$		
12.	Total annual cost = (10) + (11)		

* The actual costs as calculated shall be filled in the table

For comparing annual costs of two different types of linings above table can be used.

Annual cost of a lining on existing canals and their net annual saving, can also be worked out with the above table with suitable modifications in relevant items and difference can be worked out. Also see Para 6.3. The above table should also be used for working out the cost of lining of watercourses.

Table 9.5 Estimation of annual benefits from lined canals

S. No. 1	Details of works 2	Cost in Rs. 3
I	**Land Saved**	
1.	Right of way, unlined canals in hectares	-
2.	Right of way, lined canals in hectares	-
3.	Right of way, saved (1) - (2) in hectares	-
4.	Reclaimed waterlogged land in hectares	-
5.	Total land saved in hectares (3) + (4)	-
6.	Annual value of total land saved = (5) net crop value per hectare per year.	-
II	**Water saved**	
7.	Flow when in use, in m3/s, weekly, monthly (see figure 9.16 and 9.18, generally may vary)	-
8.	Volume of water in whole year = cumecs x 24 x 3600 number of days of flow	-
9.	Average flow per year over 100 years = (8) x dependability factor (if dependability is 80% multiply by 0.80)	-
10 a	* Estimated annual loss from unlined canal in M m^3	-
10b	* Estimated annual saving in water by lining in M m^3	
11.	Annual value of water saved = (10b) cost per M m^3	-

* Assessment of losses in a canal should not be direct losses measured/estimated or % of head discharge. Rather it should be worked meticulously on the depth flow-duration characteristics of the channel, other wise these losses may be off by more than 100%. A typical discharge duration curve of upper Ganga Canal and one of its distributory is shown in figure9.16. Losses in a long length vary considerably on many factors, see Para 9.3. Therefore losses should also be worked out in reaches of suitable lengths. The canals of minor and medium irrigation projects in Rajasthan run for a short duration of one month to 4 or 6 months in a year. Therefore losses are during that period only.

Annual saving will not be 100%. Generally may be 60 to 80% of losses, and may be taken only up to half to ¾ of the life of lining. Losses increase with deterioration in lining.

III	**Savings in maintenance (include maintenance of necessary drainage facilities)**	
12.	+Annual maintenance cost of unlined canal	
13.	+Annual maintenance cost of lined canal	
14.	Annual savings in maintenance (12) - (13) + See Para 9.15.2 it can be + ve or ; - ve when cost of maintenance is less and -ve if more than (12)	
IV	**Labour saving in operation of canal networks** There may be very little saving if any. Generally this cost is proportional to Q, and to a small extent B/D ratio and length of channels. The major cost is the cost of operation/regulation of gates and monitoring of Gauges. Also it is very difficult to assess this cost and therefore normally not considered. However items are as under and attempt should be made to assess the cost.	
15.	Labour required in operation during irrigation from unlined canal in man days per year (if any)	
16.	Labour required in operation during irrigation from lined canal in man days per year (if any)	
17.	Annual savings in labour in man days per year = (15) - (16)	
18.	Total net annual savings / tangible benefits = (6) + (11) + (14) + (17)	
19.	Estimated annual intangible benefits (Socio-economic uplifts etc. including pride of the nation)	
20.	Total annual benefits = (18) + (19)	
21.	Benefit cost ratio = (20) (12 of table 9.6)	
22.	Net annual savings = (20) - (12 of table 9.6)	

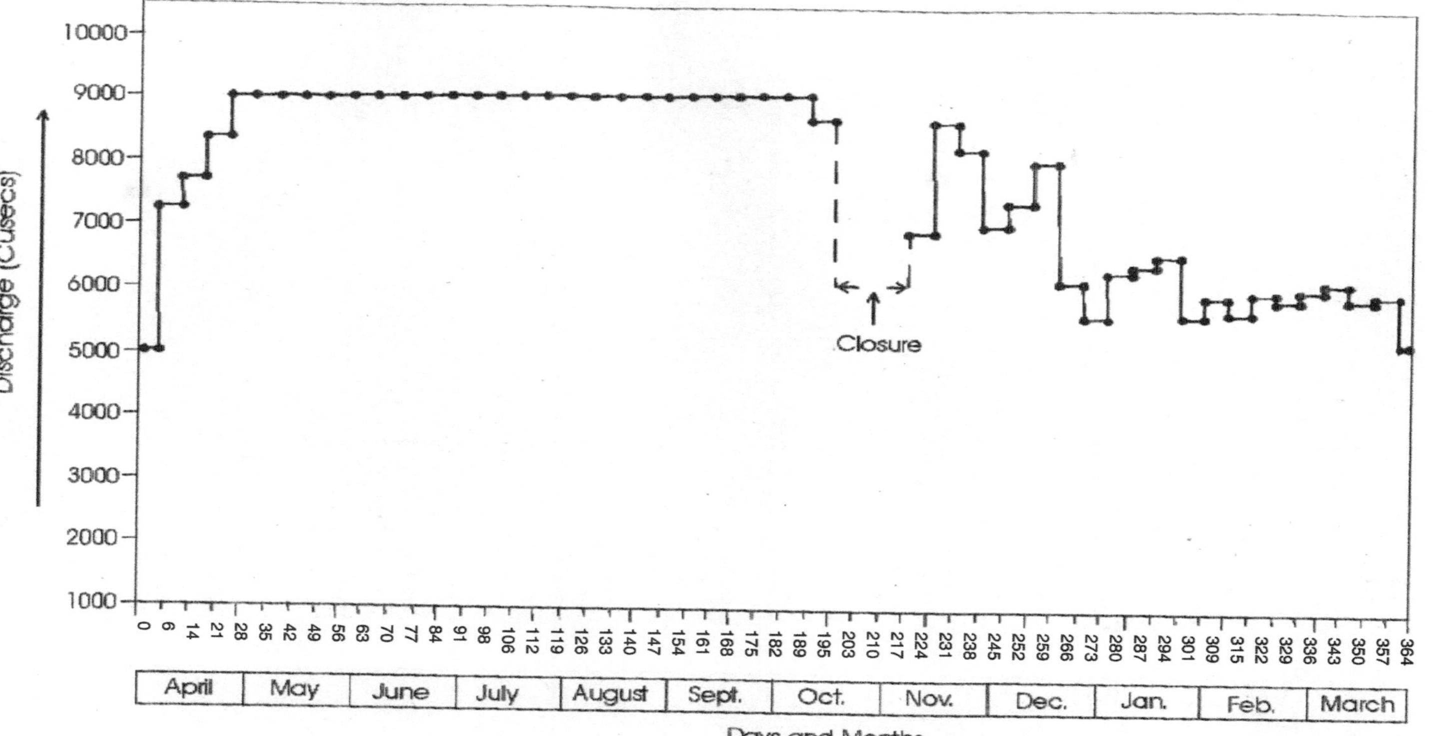

Notes

1 Discharge gradually reduces at tail. A flow duration curve of tail section can be prepared and used separately or superimposed on it for averaging. Discharge variation can also be represented in a L. Section for different time periods as Y coordinates.

2 Realistic losses should be worked out on the basis of depth flow duration curve, (preferable a realistic water surface level along the canal, it may not corresponed to discharge duration curve), and variation of water table along the canal during a year. Depth duration curve may not be similar to it in shape, as the FSL may have to be maintained. However closure periods may be identical.

Figure 9.16 (i) Flow (discharge) duration curves at head of upper Ganga canal, for the year 2000-01. (Locally known as roster in Uttar Pradesh, India).

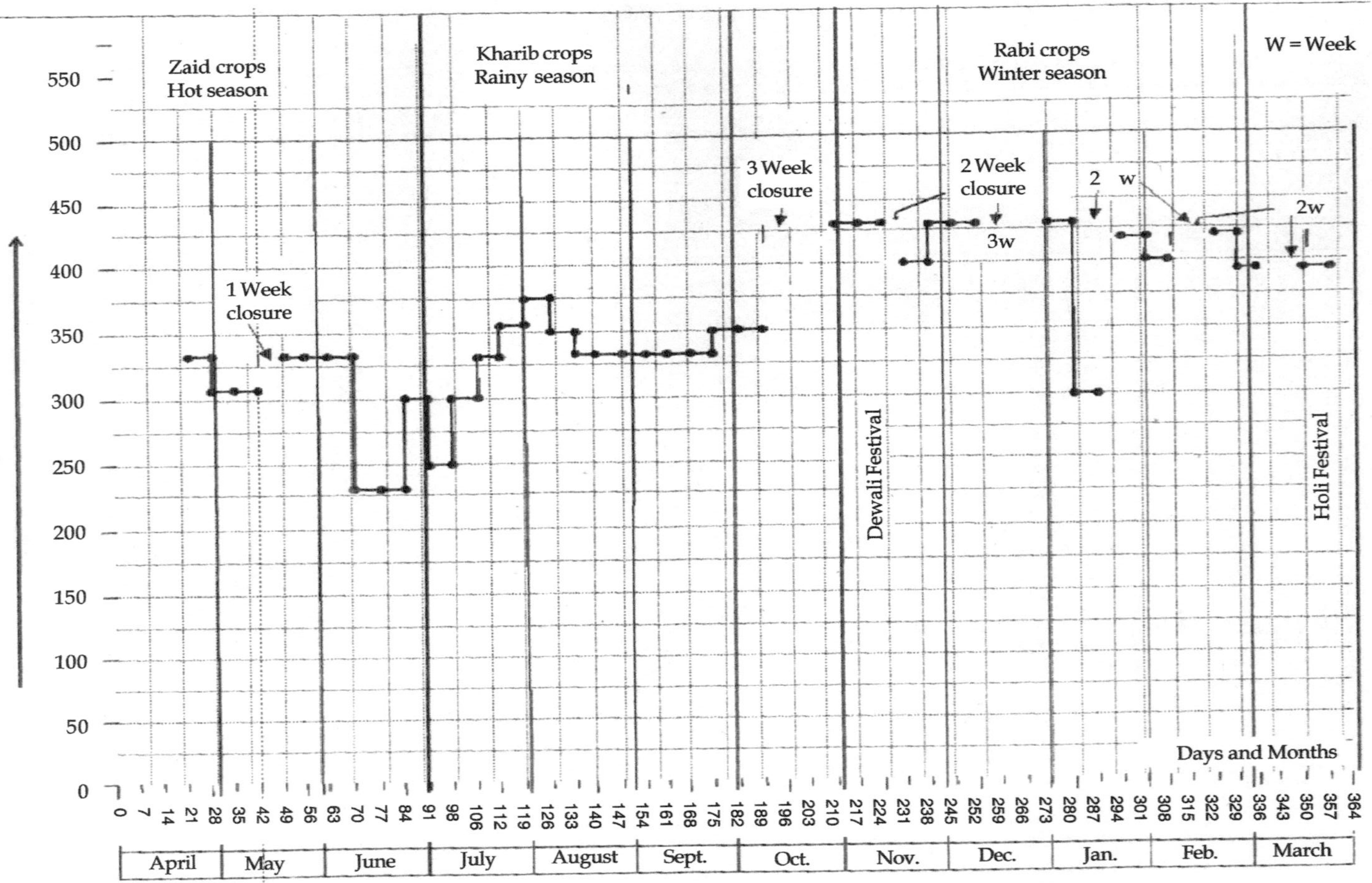

Figure 9.16 (ii) Flow (discharge) duration curve (Roster) at head of upper mat branch of UGC for the year 2000-01.

9.15.2 Cost of maintenance and repairs (M&R) of lined and unlined canals

The various problems posed by irrigation canals, lined or unlined may be listed as below.

(i) Silting of canals
(ii) Weeds and plants growth
(iii) Failure of banks - inner and outer slopes
(iv) Canal breaches
(v) Damages/Failure of structures like canal falls, cross drainage work etc.
(vi) Damage to lining and its filters, rock toe etc.

Cost of maintenance of lined canal needs very careful assessment. Some times it may be little less or same as unlined canals or may be even more. Initially, just on completion i.e. on new canals, the cost of maintenance of lined canal is low and looks attractive. But just after one or two years it may generally be more. All lined canals are also subject to silt deposition and weed growth. Silt deposition and weed growth may be more due to flatter side slopes adopted in lined canal. Initially silt gets deposited on flatter side slopes and than weed starts growing. Weed growth damages lining. Also lining is damaged by the roots of shrubs and trees on the earthen slopes, dowel etc. above lining and particularly too much in cutting reaches and more in good rainfall areas. Roots of trees penetrate behind lining and cannot be easily extracted or removed. When cut upto the root level, the shrubs and trees start re growing, with roots getting thicker. Typical examples are the canals of Mahi Bajaj Sagar Project, Banswara, particularly left main canal (LMC) where shrubs, trees (particularly Pepal and Babul trees, Indian local names) have created serious problems. These cause upheaval of lining. Also back water pressures (due to choking of drains) damages lining.

However masonry parapet walls on the top of lining and some earth backing behind dowel has helped in reducing growth of trees and penetration of roots. Typical examples are 10,000 Kilometres of lined main and distribution network of IGNP in Thar Desert.

Still silt clearance and weed removal is a common feature in most of lined and unlined canal system.

Leaves and seeds of trees in autumn fall in canal and help in germination of seeds and weed growth. Autumn season is generally March-April in India, but for some species it may be different for example for Shesam tree it is December and January

The problems of rock falls, slope slips are same in lined and unlined canals. In fact the damage and repairs to such lined canal is much more serious and costly. Typical examples of serious damage to lining are the canals of Som Kamla Amba projects and many other small projects in Dungarpur district, Rajasthan, India. Rock slopes slipped into the canal blocking the entire water way and damaging vertical walls of lining. Also see figure 8.6. It is advisable to avoid lining in such areas or better is to prefer closed canal (conduit) section.

Besides silt clearance, clearance of weed, grass, shrubs, trees on inner slopes, repairs of lining and drainage arrangements have to be carried out.

The maintenance, of surface banks, service road, surface drains and outer slopes including outer drains remains same in lined and unlined canal.

Thus a realistic assessment of maintenance of lined canal may usually be more than unlined canals. Routine maintenance is normally not less than in unlined canals.

Factors effecting cost of maintenance

The various factors affecting the cost of maintenance with their effect on lined and unlined canals are listed in table 9.6.

Table 9.6 Factors Affecting Cost of O & M

Factor	Remarks
i Size factors of the canal-B/D ratio; small, medium, high and very high capacity canals	i. Cost of maintenance may be same for lined and unlined canals for identical B/D ratio and capacity. Also see Table 9.5
ii Topography–partial cutting, filling, just cuttings /filling Heavy cutting Heavy filling, sloping ground Undulating ground	ii. Identical site conditions are likely to have equal cost of maintenance. Rather damage to lining is more in few cases.
iii Type of soil-Coarse/fine sands/gravel etc (GP, GM, GW, GC, SC, SP, SW ML, MH, etc). Clay (CL,CH), Organic silts and silt clays with low/ high plasticity, Hillocks/rock cutting	iii. -do-
iv (a) Stability of internal and outer slopes, in filling/ cutting reaches, (b) Silt deposition on slopes, weed growth on slopes in free board portion and above in cutting reaches	iv. (a) Internal side slope and bed adjusts in unlined canal (b) More silt deposition on flatter slopes in lined canals. Cost of maintenance of outer slopes is identical on both lined and unlined canal. In unlined canal weed growth above FSL does not damage banks In lined canal weeds grow on lined surface and roots penetrate there by damage lining.
v Silt entry from source and from area of canal passing, its carrying capacity and exclusion mechanism:	v. Practically identical cost. More in case of lined canals. In unlined canals silt is used to fill borrow pits/depressions in canal bed, strengthening slopes etc.
vi Climate and site specific conditions for weed germination	vi. -Do- More on flatter slopes of lined canal.
vii Drainage of rain water, rain cuts etc and repair of earth section	vii. -Do-
viii Repairs of lining	viii Extra in lined canals
ix Repairs of drainage behind lining	ix. -Do-
x Repairs of filter toe in heavy filling reaches :	x. Identical
xi Repairs to damage by insects, rats, crabs and borrowing animals.	xi. Identical

Maintenance costs of distribution systems is largely dependent upon their location, the climate, and soil conditions. The cost of operation i.e. petrolling, observation of ganges etc. in same in both lined and unlined canal.

The evaluation of maintenance costs should, in general, include above consideration and in the schdule given in table 9.7.

Table 9.7 Schedule for Cost of O & M

Item	Remarks
(a) For open canal systems.	
i. Operation of gauges and patrolling during running of canals	i. As per site requirement depending upon Q and rotational schedule.
ii. Removal of weeds, moss, etc. from canal section.	ii. Essential and to be done as and when required, normally once or upto 3 times in a year. Climate and region (area) specific.
iii. Removal of silt (wind blown sand etc.)	iii. May be once annually or at regular intervals according to site specific.
iv. Weed control on banks/grass cutting on banks and slopes	iv. Regular routine maintenance/auction of grass after first rains.
v. Maintenance of service/operating roads.	v. -do-
vi. Repairs to banks and linings, (including inner slopes, outer slopes, rain cuts, drains, dowel, etc.	vi. Once in a year or ocasional
vii. Cleaning/maintenance and repairs of drainage arrangement behind lining	vii. -do-
viii.Maintenance of structures	viii -do-
ix Maintenance of gates and their operation	ix -d o-
x Opening and closer of canals.	x As per regulation schedule. Strict control on rate of opening and closures required. Occassional Watch of heavy filling reaches may be required for high capcity canals.
xi Monitoring gauges (also see table 9.5 (iv)	
(b) For pipe systems i. Repairs and replacements of pipe lines and appurtenances	i Once every year or on busting of pipes. Regular watch and immediate repairs on notice is required
ii. Maintenance of structures	ii Once every year
iii. Removal of debris from structures.	iii Regularly

9.15.2.1 Maintenance Programme

To formulate a maintenance programme, the following steps are necessary.

i. Make an inventory of all the works with details of reaches Km wise that require maintenance.

ii. Determine the quantity of maintenance activities to be taken up and assess their cost i.e. prepare estimate.

iii. Fix the optimum cycle of maintenance for each work.

iv. Determine the machinery and Manpower requirements for each type of works.

v. Fix the maintenance priorities and periodicities.

vi. Ensure the provisions for funds, materials etc.

There should be inspection of all the masonry and concrete works at least once a year during the closure period of canal or whenever there is a sign of distress. Remedial measures should be designed where deep scours are noticed, so that they do not occur in future, such as foundations can be protected by dumping stone or brickbats or provide a launching apron in a slope of 2:1

vii. Log book of important works/structures should be maintained which will show the damages and repairs done each year. The record in the log book will serve as a guide for further inspections, effectiveness of the methods used for repairs in the past and remedial measures further needed.

It also helps in verification of old theories, development of new theories and concepts.

Cost of Maintenance Expenditures

Cost of maintenance can be divided in two broad heads (parts)

(i) Hydraulic Survey

This is essential to know the performance of canal.

Hydraulic survey is to know the overall cross section of a canal- lined/unlined and its water surface profile during the running. The overall cross sections show the position of silt deposition/erosion. Water surface profile indicate slope and with cross sections, flow characteristics (discharge) are determined. The overall cross sections i.e. of inner slope, outer slope, berms and banks etc are taken at suitable interval of 100 ft. to 500 ft. apart depending upon the size of the canal.

For better reliability it should be carried out by independent agency other than the normal operating agency and at a very regular time interval of 2 or 3 years, irrespective of any maintenance being carried out.

Hydraulic survey is carried out partly during running of canal and partly after the canal closure. For low depth canals, it may be carried out fully even when running. For more than this discharge it may not be practical. With digital computerised total survey equipment it has become easy.

It is also required to allot the works on contract. Though some routine maintenance may be carried out regularly round the year but that may not be sufficient.

Hydraulic survey helps in preparation of realistic cost estimates, and reporting the performance behaviour of the canal. Therefore it should be always preferred. Budget requirement must also be the kept separately for this purposes and for O & M.

The cost of hydraulic survey may be practically same in unlined and lined canal, or may increase by 10 to 15% for very wide or deep channels. Since hydraulic survey helps in knowing the flow characteristics of channel and may be carried out in accordance to the condition of channel once in a year.

Also a walk through survey of service road, inner slope and outer slopes should be carried after a rainy season and a detailed survey be carried out once in three or 5 years.

(ii) Actual cleaning operation and maintenance

This is allotment of works on contract, and its execution to be carried out once or twice a year that may be sufficient. Some works may be carried out once in three or five years.

Grass on outer slope and inner slope should be auctioned every year after first rains, so as to make as many grass cuts as possible during rains itself. It should not be allowed to grow 1′ to 2′ ft high and auctioned after rains, which is a general practice. Grazing of animals should not be allowed.

All the precautions taken in an earthen dam is also required in canal banks of lined or unlined canal.

The budget grant and expenditure should also be further sub classified as per table 9.7.

9.15.2.2 Present Practices of Allotment of Maintenance Funds

In India, there are various practices followed by different state governments in allocating funds for operation and maintenance of irrigation system. The practice also varies from project to project in the same state. In general following are the methods by which O & M charges are fixed.

(a) On the basis of gross annual area actually irrigated, a rate per hectare is fixed. This is being adopted in many states in India.

(b) In case of Bihar, the rate per ha is fixed on the basis of culturable command area. It was Rs. 110/Ha in the year 1995. It includes establishment charges also.

(c) In Uttar Pradesh a rate per Km. length of the canal, on the basis of the length of canal, is fixed.

(d) In Maharastra, the basis is of the irrigation potential created i.e. the area proposed to be irrigated in the command, a rate per ha is fixed.

It is a normal experience in most of the states in India, that the allocation of funds in the M & R head are given at the end or very end of the financial year (i.e. March). This results in the improper utilisation of the available funds, or it becomes difficult to utilise them in a planned and economic way.

They are also reappropriated for other works due to the fear of being lapsed.

The present practice of allotment of Budget is not sound or technically based. It should be based as per para 9.15.2 and table 9.7.

9.15.3 Systems approach to economics of canal lining

(Or Wholistic Approach or Conjunctive use of surface and ground water)

In general, it has been estimated broadly by Kennedy (1892), Irrigation commission (1992) that nearly 50% (1/2 nd) of water entering a unlined system canal is lost by way of seepage to join ground water and only 33% (1/3rd) is utilized by plants. Further, out of total loss, nearly 33% (1/3 rd) water is lost in actual water application in field and remaining 66% (2/3 rd) in the canal and water course conveyance system. Thus rearly 40 to 50% water is lost by seepage in canal network system and 21% to 15% in the fields. Actual figures of losses vary from project to project and may be from 40 to even 70%. This immediately raises the following issues.

i. Heavy wastage of costly water (huge money invested in canal net work)

ii. Water logging (rise in water table) and salinity, i.e degradation of land.

iii. Groundwater contamination.

All above problems are site specific and vary from project to project, area to area in a project, and time (with in a year and year after year) in the same area and are not common every where.

Most people and even, engineers at large, generally blame unlined canals as the root cause of it. The above problems can still be seen in a lined canal net work. Then they blame the type of lining and its quality.

Unlined canal net work is not the only cause to above problems and there are serveral other causes. This fact is more exiplicit when following more problems are seen in the same command (along with above 3 problems).

iv. Declining ground water levels

v. Land Subsidence due to Ground water Pumping

vi. Upconing of Saline Water.

For easyness of good understanding first, later three problems i.e (iv) to (vi) and subsequently the problems (i) to (iii) are discussed as under :

(i) Declining Water Levels

Over exploitation of ground water either for agricultural, domestic, industrial, and other uses is resulting in declining i.e. lowering of water table and creation of dark zones in certain pockets of a area, through out the world (Gambolati *et al.*, 1994; Luckey, 1982; Gorgensen, 1982; Navarro and Pazos, 1982; Prihar *et al.*, 1993).

A lowered water table increases the cost of lifting the ground water, thereby affecting both investment and operating costs. The deeper the water table, the more expensive the cost of drilling, pipes, pumps, and prime movers. Also, it results in an increase in the amount of energy required to pump the water. The risk of sea water intrusion in coastal areas is significant.The salinity of water in such areas may increase. The lowered water table may reverse the direction of groudnwater flow. When the flow is reversed, the water that normally percolates towards a progressively saline aquifer is drawn backwards. Therefore, the quality of freshwater aquifer may deteriorate with continued pumping.

(ii) Land Subsidence due to Groudwater Pumping

When water is pumped from a confined aquifer, the decrease in artesian pressure results in compression of the aquifer material. Most of the water released from storage is derived from compaction of the fine-grained sediments. Excessive pumping causes extensive lowering of the piezometric surface, resulting in land subsidence, i.e, the downward movement of land surface.

In the case of dolomite and limestone aquifers, the soluble rocks are slowly dissolved locally by ground water. Ground water pumping from such aquifers results in land subsidence, leading to the formation of cup-shaped depressions of the land surface (Foose, 1967).

Land subsidence can severely damage the wells resulting in their failure. One of the possible remedial measures as a safeguard against land subsidence is to reduce pumpage to safe yields (Gambolati et al., 1974), Poland (1984).

(iii) Upconing of Saline Water

In the canal-irrigated areas, good quality -canal seepage due to seepage from canal networks, water percolates to the groundwater aquifer. In the case of saline aquifer, it results in the formation of a thin layer of freshwater overlying the saline water. Pumping the well in the freshwater zone causes the fresh water/seawater interface to rise below the well. This is referred to as upconing. If the well penetrates close to the freshwater - salt water interface, the poor quality water may enter the well, causing the well discharge to be of poor quality. Th well needs to be designed such that saline water does not mix with freshwater. The upconing of saline water is described by Bear and Dagan (1968), Schmorak and Mercado (1999), McWhorter (1992), Bouwer (1998), and Goel et al. (1991).

(iv) Heavy wastage or useful recharge to groundwater

Still lining of canals is generally considered as one of the measures for water conservation and reduction of above problems, every where. This approach along with the economics of lining given in Para 9.15.1 is in isolation. It is without due regard to some positive aspects of seepage such as ground water recharge/redistribution/transfer to other areas, just like inter basin water transfer or canal water transport or conveyance.

All water which is lost by seepage in an unlined canal system is not total loss and a significant part of it joins groundwater by way of deep percolations, which can be recovered under favourable conditions.

Lining of canal thus affects groundwater availability and therefore, the advantage of lining needs proper evaluation in context of total surface and groundwater system performance. Further more extent of conveyance losses and their contribution to groundwater from different components of canal distribution network (main canal, distributaries & field channels) are different both in time and space. Therefore, the cost effectiveness of lining of various components individually and in combination needs careful assessment for deciding the extent of lining.

Main advantage with groundwater is that it can be tapped almost at any place as and when needed, a big flexibility to the farmers, depending upon the pumping capacity.

In fact, farmers, good agricultural production policy, and irrigation water management requires an irrigation system on demand, according to crops and weather, which is very costly.

All unlined canals are a great source to the recharge of ground water and also its conveyance over larger area like surface gravity canals, and economics should be examined in that light.

Only disadvantage with this is that, it requires energy for pumping. Advantage of lining therefore be viewed along with the associated loss in groundwater recharge due to reduced canal seepage, and their overall impact on ground and surface water resource availability of the area. An attempt may be made to study the economics of canal lining by using a suitable programming model associated with conjunctive use of surface and groundwater.

Conjunctive use of ground and surface water

Advantages and disadvantages of sub-surface and surface reservoir are given below in table 9.8.

Table 9.8 Advantages and Disadvantages of Subsurface and Surface Reservoirs

Sub-surface reservoir **Advantages**	**Surface reservoir** **Disadvantages**
1. Many large capacity sites available.	Few new sites available.
2. Slight to no evaporation loss.	High evaporation loss, even in humid climate. Generally between 10 to 30%.
3. Require no little land area (for recharge).	Require large land area.
4. Slight to no danger of catastrophic structural failure.	Ever present danger of catastrophic structural failure.
5. Uniform water temperature.	Fluctuating water temperature.
6. High ecological purity.	Easily contaminated.
7. Safe from immediate radio active fallout.	Easily contaminated by radio active material.
8. Serve as natural conveyance systems (canals or pipeline) across lands of other basins).	Water must be convened, through costly network of canals.
9. Can be pumped out as and when required.	High evaporation and seepage losses during storage.
10. Initially less costly, structures.	May cause water logging and salinity
11. Reduce water logging and salinity.	Initial heavy investment required.

- All the above advantages and disadvantages for groundwater are site specific; depending upon the nature of ground. Where as for surface water it varies from project to project. Typical examples of conjunctive use of surface and ground water are shown in figure 9.17 and 9.18. There are three options in figure 9.17 and figure 9.18 explains the canal capacity/regulation pattern.

Thus, lining must be done selectively where it is most beneficial, especially under saline groundwater conditions. The optimum length of canal network should be lined for maximizing the benefits (Malhotra, 1994; Khepar and Chaturvedi, 1999; Sur and Singh, 1993).

(v) Waterlogging and Salinity

Large parts of several command area are found to be waterlogged within ten years of commissioning in various countries. A few examples include the Bhakra, Chambal, Tawa, Mahi Right Bank Canal, Sarda Sahayak, and Indra Gandhi Canal Command areas in India; Bari Daob, Chenab, & Jhelum Canal Command areas in Pakistan (Yaseen and Chang, 1990); Andyuma Valley, Arizona, and San Luis Valley, Colorado, in the United States. It does not mean that the entire canal network should be lined. See above Para and 9.18.

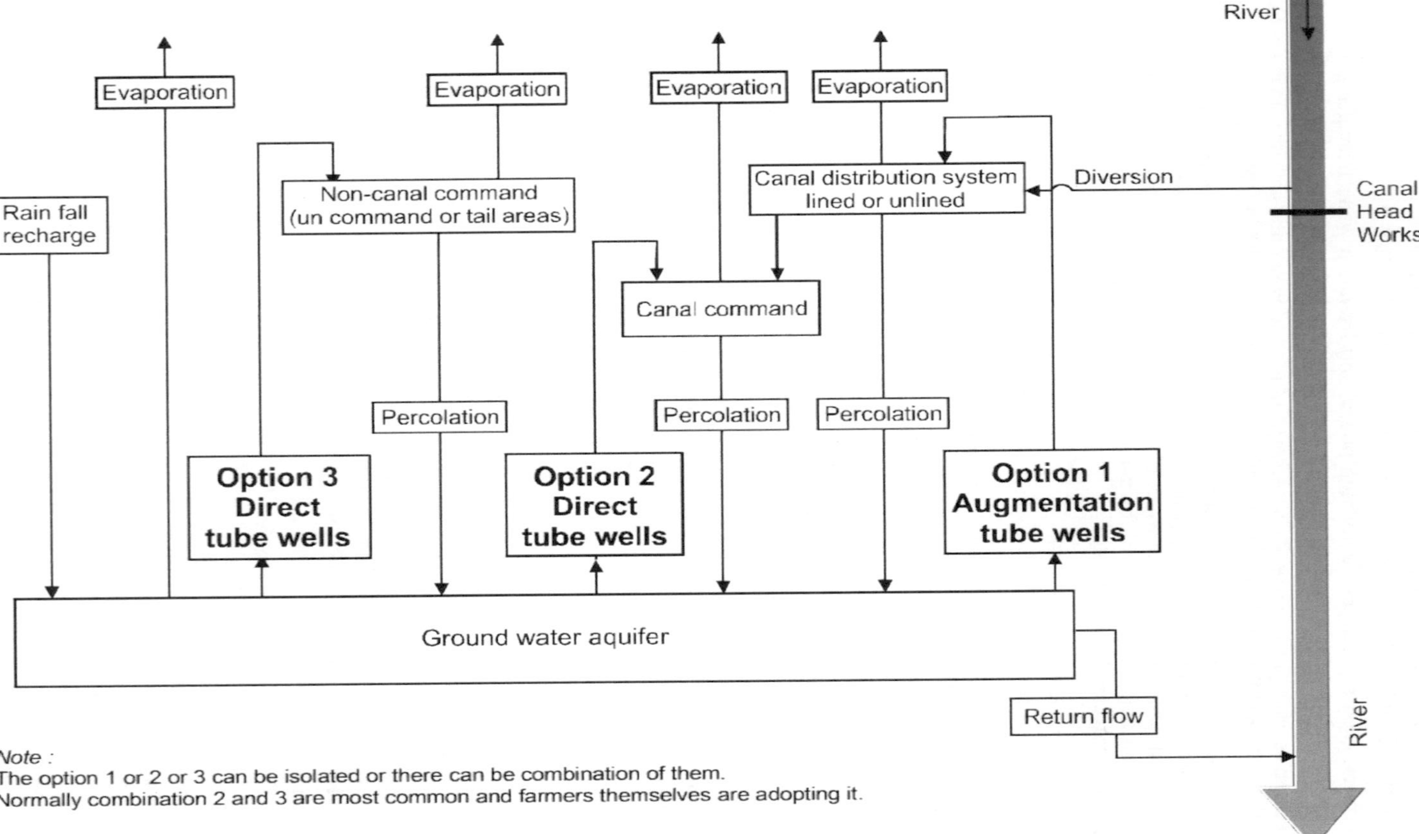

Note :
The option 1 or 2 or 3 can be isolated or there can be combination of them.
Normally combination 2 and 3 are most common and farmers themselves are adopting it.

Figure 9.17 Schematic Representation of Conjunctive Use or systems Model Approach to Economics of Lining.

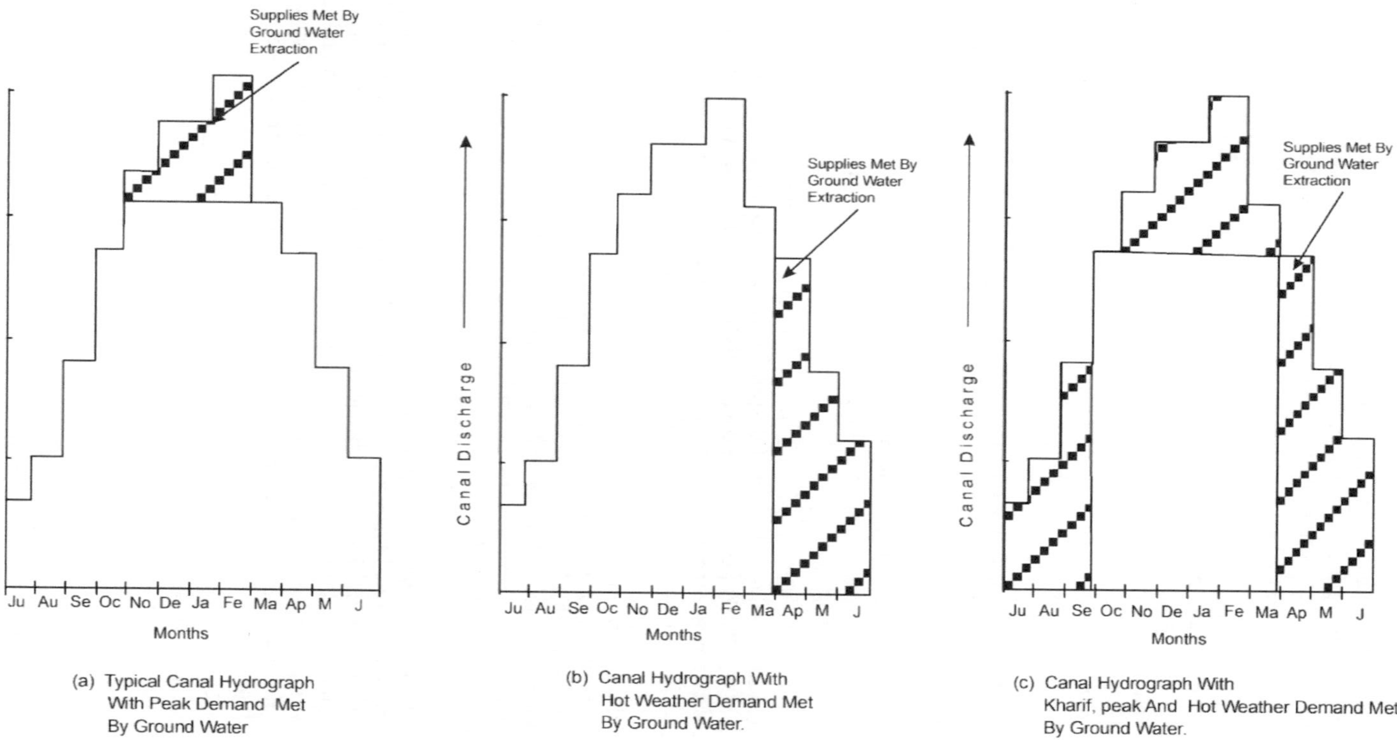

(a) Typical Canal Hydrograph With Peak Demand Met By Ground Water

(b) Canal Hydrograph With Hot Weather Demand Met By Ground Water.

(c) Canal Hydrograph With Kharif, peak And Hot Weather Demand Met By Ground Water.

Note : Canal hydrographs will change from region to region, depending upon ground water table, climate and crops

Figure 9.18 Typical examples of conjunctive use of ground and surface water in irrigation canal hydro graph.

(vi) Ground water Contamination

Intensive use of water resources and the unavoidable disposal of domestic as well as industrial efluents have contaminated the groundwater in many parts of the world.The groundwater is also being contaminated by deep percolation from intensively fertilized fields and due to excessive use of weedicides and pesticides (Bijay-Singh et al., 1995; Crampon, 1989). These contaminants have deteriorated the ground water quality to such an extent that it has become unfit for drinking purposes and other uses at several locations. It is a potential human health hazard in the long run for grown ups and can be fatal for infants. Water having large nitrate contents when consumed by infants during he first six months of life may produce a form of cyanosis due to methaemoglobinemia. Similarly, bacteriological pollution of groundwater results in water-borne and water-associated diseases such as cholera, gastroenteritis, typhoid fever, and round worm. The clean up of a contaminated aquifer is a difficult task, if it is not impossible at all. Efforts should be made to prevent, reduce, and eliminate groundwater pollution, but not by stopping the ground water recharge.

9.16 Recommendations of CWC (India), Regarding Lining

CWC (1989), the 8thplan working group on Major and Medium Irrigation Programme for India, recommended as under:

Reduction of seepage Losses (Para. 10.4.4 – (iii) of CWC)

The percentage reduction in the seepage by lining is much less than what is normally assumed in the formulation of the Project Reports.

Cost of Lining, (Para 10.4.5 of CWC): It is well known that cost of providing lining is very high and accounts for as much as 40% to 60% of the cost of the project. In some projects, cost of lining is working out as high as Rs. 20,000 - 25,000 per ha (around year 1980). Therefore, with the uncertainty of the saving in water, its cost effectiveness needs to be convincingly proved before large scale lining is resorted to especially in view of the limited resources available for the irrigation sector.

Conveyance system (Para 10.6.1 of CWC-1989)

i) *Keeping in view the limited financial resources and questionable efficacy and cost effectiveness of lining, normally the canal systems should not be lined. Lining should be restricted to the minimum, where this is proved to be absolutely necessary. Even in such cases, comparative studies should be made for lined as well as unlined canal system in different reaches and the economic analysis conducted. The decision for lining should be taken only after it is proved to be technically and economically more beneficial than unlined canal system.*

ii) A large scale national program of scientific observations of seepage losses over a long period in existing lined and unlined canal systems in various projects in each State should be undertaken in the beginning of the Eighth Plan under the head of Research and Development. The studies should be coordinated at the State level by the respective Design, Planning Organisations in collaboration with State Research Stations and at the National level by the Central Water Commission so that the results can be properly analysed and correlated for arriving at uniform National guidelines.

iii) Under the same program, exhaustive Action Research studies should be under taken to evaluate cost effectiveness and efficacy of various methods of lining using different materials including plastics.

iv) Based on these results and experience in India and other countries Central Water Commission should frame detail guidelines on lining for uniform adoption in the country.

v) There is already some evidence of achieving effective seepage insulation through selection of appropriate material for embankments and effectively compacting. The method should be advantageously used and more studies should be taken up.

vi) Where lining is considered essential in a system, main canal and branches should be lined in the initial phase. Lining of minors and distribution should be deferred until the command is fully developed.

vii) The work of lining should be carried out under strict quality control and the designs for the lining should confirm to the Indian standards.

viii) For effective seepage control of lined canal, use of LDP film of appropriate thickness is considered advisable. The lining with LDP should however be carried out under rigid quality control as per the standard specifications.

CWC (2001) – 10th Plan working group has recommended as under :

Conjunctive use of water to be given higher priority particularly in alluvial areas, as a very effective means of improving water use & basin efficiency, and of saving possible water logging & salination.

The policy of favouring only selective lining of canals be continued.

9.17 Claims on Advantages of Lining and Its Realistic Status

(Comparative analysis of lined canals versus unlined canals)

Economies of lining against unlined canals are discussed in para 9.15, cost of maintenance in Para. 9.15.2 and systems approach or wholistic approach along with conjunctive use are discussed in para 9.15.3. Here an effort in made to examine the advantage and disadvantages of lined canals versus unlined canals.

In spite of heavy cost, a large number of advantages of lining as listed below are claimed in its support.

I. Economy in cost of earthen section and structures

A reduction in cross-sectional area of flow due to

(i) Increase in permissible velocity-

It is higher in lined canals then in unlined canals.

(ii) Reduction in rugosity coefficient.

(iii) Availing steeper slopes wherever possible.

Consequently economy or saving in cost of

(a) land

(b) Earth work

(c) Smaller and lesser canal structures.

e.g. falls can be avoided, due to steep slope.

II. Reduction in seepage losses in conveyance (due to lining of canal)

(i) Saving of costly water (Huge money invested in canal network and head works)

(ii) Increase in command or intensity of irrigation

(iii) Prevention of water logging. (rise in water table) and Lower drainage requirement

(iv) Prevention of salinity and degradation of land.

(v) Reduction of contamination of ground and surface water

(vi) More structural safety, less chances of breaches, low interruption of regular supply.

III. Economy in Operation and maintenance cost due to

(i) Prevents erosion of bed and berms

(ii) Little or no silting due to high velocity

(iii) Prevention of weed growth due to hard lined surface

(iv) Little damage on lining or long life so less maintenance cost. (Depending upon type of lining)

(v) Improvement of operational efficiency due to higher velocity

(vi) Protection from bank failures, due to hard surface lining and low seepage.

(vii) Smaller and lesser structures.

All above advantages practically more than 15 are argued as strong (great) disadvantages of unlined canals. So there is a huge voice in favour of lined canals. Even the above arguments are placed as strong reasons for lining of existing canals under the Modernization / Rehabilitations, or renovations etc. Huge lengths of a large number of canals have been lined. See Para 6.4. Generally their true or realistic economic analysis as discussed in para 9.15 to 9.15.3 is missing/not available. Also a long-term performance evaluation is not available.

In actual field conditions/practice all above advantages may not be available everywhere and vary considerably from project to project or little may be achieved depending upon the design carried out. Also there are some more disadvantages besides heavy cost.

Their actual or realistic status is discussed, with respect to high, medium and low capacity (discharge) canals given in table 9.3. This table is self explanatory and gives a good idea of velocity, bed slopes etc.

1. Economy-Rugosity Coefficient, Velocity etc.

Although, theoretically, there is reduction in rugosity coefficient due to lining. But it is initially only and over a period of years it increases due to silt deposition and growth of vegetation particularly on flatter slopes generally adopted in lined canals etc. See 9.5 and 9.15.2. Thus offsetting to some extent initial advantage see para 2.02. More over with a higher value of N there is little saving in cost by adopting different R, i.e. changing A and P, see para 4.5 and a practical example 4.1. Also flow characteristics indicate a very little change in depth and total cost of canal; see Para 4.6.

2. Reduction in seepage losses, Water Logging and other problems

There are many major unlined canal systems in the country, like Upper and Lower Ganga Canal, Sone Canal etc. which have not experienced any major problems of water-logging inspite of their being in operation over a hundred years. These projects favour unlined canals for recharge and conjunctive use of ground water, see para 9.15.3, 9.16 and 9.18 (iv)

The seepage loss through canal system is not a total loss of water available for irrigation. A major portion forms a part of rechargeable ground water, which is used for irrigation at the convenience of farmer (on demand by crops), drinking purpose and mitigates draught. Also see para 9.15.3 (iv) for conjunctive use.

Though there is a reduction in seepage loss in the beginning, but reduces with development of cracks, damages to lining. In fact embankments are to be designed safe as without lining. That means seepage must be considerably reduced just like in earthen dams. Cutting sections may or may not be that much porous. More over seepage loss depends upon water table, see figure 9.19.

Thus a very judicious analysis is required reach wise to justify the same.

4. Experience in other countries

In most countries of the World, including U.S.A. large-scale lining of canals is not done and there are many major unlined canal systems or are lined with clay or any other low cost lining.

5. Maintenance

It is found that when hard surface lining is done without proper quality control, such as inadequate compaction of the embankments and insufficient /blockage of drains behind lining, serious problems of cracks, lining failures, vegetative root penetration etc. develop in the lining. Its maintenance is difficult and costly.

a

(a) WATER TABLE ABOVE F.S.L.

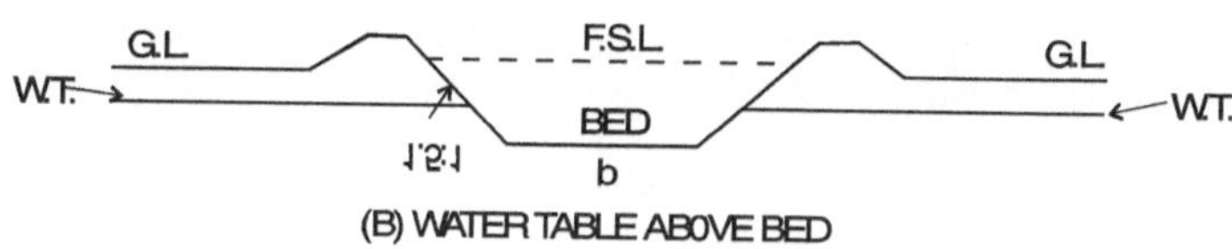

(B) WATER TABLE ABOVE BED

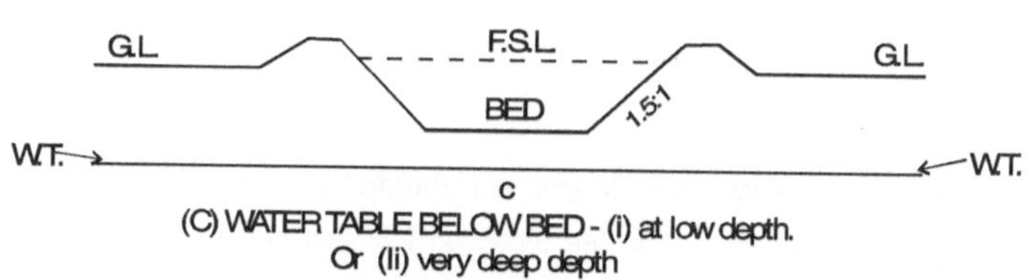

(C) WATER TABLE BELOW BED - (i) at low depth.
Or (Ii) very deep depth

Figure 9.19 Different Water- Table conditions in a canal.

Though adequate measures are or may be taken to rigidly monitor the quality of compaction and lining. Yet some times may be lacking or difficult due to scattered nature of work over long lengths/ distances and some shortfalls/omissions may occur at some/isolated locations. It causes serious damage to lining.

Replacement of damaged lining is again costly and is often not done through available maintenance grants. Consequently, patches of dislodged lining more or less defeat the original purpose.

In some of the lined canal systems, the following observations have been made in the Manual of Canal Lining by CBI&P in September 1995.

a) The failure and collapse have been caused due to unequal settlement of the area due to creation of cavities under the lining arising from the movement of fines through joints.

b) The bulging or slabs has been caused due to excessive hydrostatic pressure as a result of inadequate drainage against high water table outside the canal.

c) Due to continuous running of canal, the joints could not be attended repaired for several years.

d) Some leakage has been observed through the canal embankment and at the cross drainages during the past few years although the sources of leakage can only be located on closure of canals. It appears to be a result of gradual saturation of banks and the sub-soil due to continuous seepage through joints and cracks.

Therefore a realistic assessment of advantages/economics of linings, based on system approach (see Para 9.15.3) is needed before making any claims on it.

9.18 Need of Lining

With lot of discussions on cost of lining (Para9.2), economics of water saving etc. (Para 9.15) ground water recharge and its conjunctive use, water logging , soil saliniy and contamination of ground and surface water (Para 9.15.3) etc. and a realistic approach to the benefits of lining, see para 9.17 the need of lining can be based on following four important criteria.

(i) Based on type of sub-grade material to reduce seepage

Generally clean sands and gravels have the heaviest losses. Uniform gravels have the highest permeability followed by well-graded gravel and than uniform and well-graded sand. Therefore losses may be huge. Soil types with their soil classification symbols and a relative rating as to permeability and probable need for lining is given below in table 9.8. Porosity, Permeability, and Hydraulic Conductivity of naturally occuring soil and rock deposits are given in table 9.9.

Where K= Coefficient of hydraulic conductivity in Darcy's law Q = KAI

And $I = \Delta h / \Delta S$, unit hydraulic gradient

k= permeability and is related to

$K = k\rho g/\mu$ (ρ = fluid density and μ = fluid viscosity)

Clay has small seepage losses when continuously wet, but developes large shrinkage cracks on drying. Clay also poses problems of cracks in lining and stability of slopes. Linning can be avoided in clay soils.

(ii) Based on general geophysical and chemical condition (properties)of command area.

(a) In the areas of soils containing salts, carbonates, chrorides sulphates etc.

Seepage water gets contaminated and further contaminates ground water. This water, even when pupmed out can not be used either for drinking or irrigation without processing. Therefore lining is an essential requirement in such areas.

Before lining priority lies in improving the methods of water application i.e. farm designs because this water contains, fertilizers, pesticides, insecticides and is practically 1/3 rd of total loses. It also increases the degradation of land, reduces crop yield and conteminates ground water, more than canal seepage.

(b) In the areas of impermeable layers at low depth, say at 2 to 5 m below surface.

In such areas a perched water table is formed soon and causes water logging. There fore lining may be required. But before lining priority lies in improving the methods of water application, as discussed above in (a).

If lining is to be avoided the seepage water should be used back in irrigation by open wells or tube wells and perched water table should not be allowed to form. The impermeable layers can also be punchered by a suitable grid of inverted wells and seepage water can recharge ground according to geophysical/chmeical condition and ground water for exploitaion at other suitable time/locaion depending upon ground water flow.

Table 9.9 Soil classification symbols and a relative rating as to permeability and probable need for lining

S.No and relative rating to permeability	Soil Classification as per IS		Permeability Cm/s	Probable need of Lining
	Notations	Soil description	K	
1	2	3	4	5
1	GW	Well graded gravels, gravel sand mixtures; little or no fines	$>10^{-2}$	Extremely permeable, needs lining
2	GP	Poorly graded gravels or gravel sand mixtures; little or no fines.	$>10^{-2}$	Extremely permeable, needs lining.
3	GM	Silt gravels, poorly graded gravel sand-silt mixtures.	$>10^{-3}$ to 10^{-6}	Usually fairly impermeable but hard to hold on bank. Lining requirement is minimal.
4	GC	Clayey gravels or poorly graded gravel-sand –clay mixtures	$>10^{-6}$ to 10^{-8}	May range from moderate to very low permeability.
5	SW	Well-graded sands, or gravely sands, little or no fines.	$>10^{-3}$	Moderately permeable, usually requires lining
6	SP	Poorly graded sands or gravely sands; little or no fines.	$> 10^{-3}$	Moderate to highly permeable, usually requires lining
7	SM	Silt-sands, Poorly graded sand silt mixtures	$>10^{-3}$ to 10^{-6}	Usually impermeable good stability
8	SC	Clayey sands, poorly graded sand-clay mixtures.	$>10^{-6}$ to 10^{-8}	Usually impermeable, good stability.
9	ML	Inorganic slits and very fine sands, rocks flour, silt or clayey fine sands or clayey silts with none to low plasticity.	$>10^{-3}$ to 10^{-6}	Fairly impervious, but bank section is difficult to hold in place. Needs special stability consideration.
10	MH	Inorganic silts of high compressibility micaceous or diatomaceous fine sandy or silt soils, elastic silts.	$>10^{-4}$ to 10^{-6}	Fairly impervious, but bank section is difficult to hold in place. Needs special stability consideration
11	CH	Inorganic clay of high plasticity, fat clays.	$>10^{-6}$ to 10^{-8}	Very impermeable when wet or extremely permeable after drying, needs special stability consideration.
12	CL	Inorganic clays, gravely/sandy/silt clays, lean clays of low plasticity.	$>10^{-6}$ to 10^{-8}	Usually very impermeable. Lining requirement is minimal.
13	OL	Organic silts & silt clay of low plasticity.	$>10^{-4}$ to 10^{-8}	Permeability fairly low but stability is questionable.
14	OH	Organic clays of medium to high plasticity.	$>10^{-6}$ to 10^{-8}	Low permeability if soil is kept wet but stability is questionable and shrinkage cracks are probable.

Tabel 9.10 Values of Porosity, Permeability, and Hydraulic Conductivity of soils and rocks

Material	Porosity, n (%)	Permeability, k (cm^2)	Hydraulic Conductivity K (cm/s)
1	2	3	4
Unconsolidated deposit			
Gravel	25 - 40	10^{-3} - 10^{-6}	10^{2} - 10^{-1}
Sand	25 - 50	10^{-5} - 10^{-9}	1 - 10^{-4}
Silt	35 - 50	10^{-8} - 10^{-12}	10^{-3} - 10^{-9}
Clay	40 - 90	10^{12} - 10^{-15}	10^{-9} - 10^{-10}
Rock			
Fractured basalt	5 - 50	10^{-9} - 10^{-11}	10^{-2} - 10^{-6}
Karst limestone	5 - 50	10^{-5} - 10^{-9}	1 - 10^{-4}
Sandstone	5 - 30	10^{-9} - 10^{-13}	10^{-4} - 10^{-8}
Limestone, dolomite	0 - 20	10^{-9} - 10^{-12}	10^{-4} - 10^{-9}
Shale	0 - 10	10^{-12} - 10^{-16}	10^{-9} - 10^{-11}
Fractured crystalline rock	0 - 10	10^{-9} - 10^{-11}	10^{-2} - 10^{-6}
Dense crystalline rock	0 - 5	10^{-13} - 10^{-19}	10^{-8} - 10^{-12}

(iii) As a structural and or operational requirement of some topograhical situation, e.g. in sloping mountain of soft rocky strata

A closed section is more desirable, as far as possible see table 9.2. However for high discharges a lined canal is the only option.

(iv) Ground water table and extraction i.e. conjunctive use of ground and surface water

(a) When the ground water table is deep, and there are no impermeable layers at low depth, and soils do not contain salts, carbontes, chlorides, sulphates, etc,

Lining can be avoided. Seepage water can be allowed to recharge the ground waer, and used at other appropriate times. Ground has a good storage capacity and many advantage, see Para. 9.15.3 and table 9.8.

Typical such command is Upper Ganga canal, Uttatar Pardesh, 150 year old, and no water logging developed inspite of intesive irrigation to high water demand crops like sugar cane. There are number of tube wells pumping out the ground water in the period of low/no supply from the canal. Thus water balance is maintained. In such situation there appears no justification of lining.

(b) When ground table is at a shallow depth or may even cross canal section.

There is little saving in water by lining. Heavy drainage arrangement behined lining is also required. So over all cost in very heavy. Therefore lining can be avoided. A proper water balance can be achieved by conjunctive use.

(v) Selective lining of vulnerable reaches or part lining of sides or bottom

Considerable cost savings, effectiveness of seepage control (only where ever necessary as per Para 9.19), elimination of water logging, salinity and ground water contamination can be easily achieved by adopting selective liming of only highly vulnerable reaches in porous media or part lining of section i.e. either of sides (or even one side) or bottom. This approach besides resolving all the problems of canal network as indicated in Para 9.15.3 is also helpful in better irrigation water management, (minimum cost of supply per unit of water) low cost of operation and maintenance, flexibility of use of tube well water (conjunctive use) at the convenience of the farmer. It also ensures availability of good quality drinking water and a good assurance of surety of water availability of in periods of draught.

Some engineers feel problems of hydraulics in frequent changes of canal section, and preparation of long sections, This doubt get easily removed when water surface profile and grades are plotted in a long section of canal. Changes in cross-sections are inevitable with changes in topography and soils. These are easily adjusted at structures or by suitable transitions.

(vi) Based on running days of canal

On many minor and and medium projects, canals run only for few days, say 30 to 100 days in a year. For rest of the period the canal is dry and lining may not be justified.

All the above aspects are interacting and needs a judicious examination of geophysical conditions of ground variation and likely future fluctuation of water table before deciding open the lining. Considerable seepage losses can alos be saved by suitable design of canal embankments, which is an essential requirement of lined and unlined canal.

References

Alam, A.M.Z., and Kennedy, J.F., Friction Factors for Flow in Sand Bed Channels, Journal of The Hydraulics Division, ASCE, Vol. 95, No. HY6, Proc. Paper 6400, Nov., P. 1973 - 1992., 1969.

Alam, A.M.Z., Cheyer, T.F., and Kennedy, J.F., Friction Factors for in Sand Bed Channels, Hydrodynamics Laboratory Report No. 78, Massachusetts Institute of Technology, Cambridge, Mass., June, 1966.

Bairathi V.K., Hand Book of Applied Water Courses, Alok Technical Publishers, 1982

BIS (Bureau of Indian Standards), Code of Practice for Maintenance of Canals Unlined Canals, IS 4839-1968, Part I.

BIS (Bureau of Indian Standards), Guide for Planning and Layout of Canal System for Irrigation IS 5968 - 1970

CBIP, (Central Board of Irrigation & Power), Manual on Current Practices in Canal Design in India, Malcha Marg, Chanakyapuri New Delhi, June 1968.

Central Irrigation Hydrodynamic Research Station, Poona, India, Annual Report (Technical) of Work During the Year, 1942 – 43, P.4.

Chow, V.T., Handbook of Applied Hydrology, Mc. Graw - Hill Book Company Ltd., 1964.

CWPC (Central Water and Power Commission, Water Wing), Draft Manual on Irrigation Channels, New Delhi, 1960.

Dave, K.M. Outlets Through Embankments and Very Fills; M.E. Thesis, University of Roorkee, 1971.

Dhillon, G. S. Measurement of Seepage Losses from Irrigation Canals Vol. XVII No. 1, P. 23-28 & 37, Jan. 1967.

Dhillon, G.S. Estimate of Seepage Losses from Unlined Channels Vol. XVIII, No.8, P. 317-324, Aug., 1968.

Dhillon, G.S. Estimation of Seepage Losses from Lined Channels Vol. XVIII, No.12, P. 16-20, Dec., 1967.

Flynn, P.J. Irrigation Canals and Other Irrigation Works, George Spaulding & Co., San Francisco, Cal., 1892.

Fortier, S. and Scobey, F.C. "Permissible Canal Velocities", Trans. ASCE, Pp. 940-984, 1926.

Fortier, S. Conveyance of Water in Irrigation Canals, Flumes and Pipes Water Supply Paper No. 43, U.S. Geological Survey, Washington, D.C., 1901.

Gibson, A.H., Hydraulics and its Applications, Fifth Edition,.P. 342, 1952.

Hall W., Principles of Conjunctive Use in Water Resources Planning and Management

Holl, I.S, Sanford I.S Lining for Irrigation Canals, Vol. X, No. 12, P. 145-148, December 1960.

ICID, Canal Construction - Open Channels Construction Machinery and Techniques, Report of International Commission on Irrigation and Drainage, , Govt. of India, New Delhi, 48 Nyaya Marg Chanakyapuri, New Delhi - 110021, 1979.

Inglis, C.C., and Gokhale V.K., The Eradication of Water Weeds from Decon Canals, Technical Paper, P.W.D. Bombay, 1937-38.

Jain, R.K., Control of Peaking Station on Canals Vol. XVI, No. 6, P. 15-26, June 1966.

Kala, M.K. Study for Economic Section of Lined Canals; M.E. Thesis University of Roorkee, Roorkee, 1971.

Kalha, Ajit Singh, Report on, Lining of Bhakra Main Branch.

Lacey, J.M., Minutes of Proceedings Inst. C.E.P. 333, Vol. 229

Luthra, S.D.L. Conveyance Losses in Lined Irrigation Canal Systems. A Brief Review Journal of Central Board of Irrigation and Power, July 1980.

Modi, P.N., Irrigation Water Resource and Power Engineering, Forth Edition, Publised by Rajinder Kumar Jain, Prop. Standard Book House 1705 - A, Nai Sarak, Delhi - 110006, 2000.

Sharma, H. D., Varshney, D.V. and Agarwal, V.K., Study of Sediment Transport Capacity of Ganga Canal, IAHR., XV Congress Istambul, September, 1973.

Thomas, H.E., The Conservation of Ground Water, 1951.

U.S. Dept. of Agri. Tech. Bulletin No. 67 P. 44

Uppal, H.L. & Singh, Gajinder, Study of Seepage from Lined Canals Vol. XV, No. 12, P. 36-52, Dec. 1965.

USRS, 1915.

Virendra, Prasad, M.E. Dissertation, WRDTC, University of Roorkee, A Review of Canal Linings in India, 1975.

Woods, F.W., Engineer P. 648. Vol 40.

❑❑❑